Lecture Notes in Electrical Engineering

Volume 972

The book series *Lecture Notes in Electrical Engineering* (LNEE) publishes the latest developments in Electrical Engineering—quickly, informally and in high quality. While original research reported in proceedings and monographs has traditionally formed the core of LNEE, we also encourage authors to submit books devoted to supporting student education and professional training in the various fields and applications areas of electrical engineering. The series cover classical and emerging topics concerning:

- Communication Engineering, Information Theory and Networks
- Electronics Engineering and Microelectronics
- Signal, Image and Speech Processing
- Wireless and Mobile Communication
- Circuits and Systems
- Energy Systems, Power Electronics and Electrical Machines
- Electro-optical Engineering
- Instrumentation Engineering
- Avionics Engineering
- Control Systems
- Internet-of-Things and Cybersecurity
- Biomedical Devices, MEMS and NEMS

For general information about this book series, comments or suggestions, please contact leontina.dicecco@springer.com.

To submit a proposal or request further information, please contact the Publishing Editor in your country:

China

Jasmine Dou, Editor (jasmine.dou@springer.com)

India, Japan, Rest of Asia

Swati Meherishi, Editorial Director (Swati.Meherishi@springer.com)

Southeast Asia, Australia, New Zealand

Ramesh Nath Premnath, Editor (ramesh.premnath@springernature.com)

USA, Canada

Michael Luby, Senior Editor (michael.luby@springer.com)

All other Countries

Leontina Di Cecco, Senior Editor (leontina.dicecco@springer.com)

**** This series is indexed by EI Compendex and Scopus databases. ****

Chinese Society of Aeronautics and Astronautics
Editor

Proceedings of the 10th Chinese Society of Aeronautics and Astronautics Youth Forum

Springer

Editor
Chinese Society of Aeronautics and Astronautics
Beijing, China

ISSN 1876-1100 ISSN 1876-1119 (electronic)
Lecture Notes in Electrical Engineering
ISBN 978-981-19-7651-3 ISBN 978-981-19-7652-0 (eBook)
https://doi.org/10.1007/978-981-19-7652-0

This Springer imprint is published by the registered company Springer Nature Singapore Pte Ltd.
The registered company address is: 152 Beach Road, #21-01/04 Gateway East, Singapore 189721, Singapore

Contents

Experimental and Modeling Study of Open-Hole Composite Laminates Under Tension

Chen Fangyu(✉) and Jin Xiaowei

China Helicopter Research and Development Institute, Jingdezhen 333001, China
{chenfy,jinxw}@avic.com

Abstract. To study the failure behavior of open-hole carbon/epoxy composite laminate under tension, six pieces of open-hole composite laminates are tested, and the tension failure strength of every composite laminate is evaluated. Failure strength and damage phenomena of all composite laminates are examined. For all composite laminates, the failure phenomena is that the crack crosses the center hole transversely. The damage diffuses at the edge of the laminate composites. The main failure mode is fiber fracture and matrix cracking, with only a few delaminations. According to the experimental phenomenon, Hashin failure criteria and a stiffness degradation method are proposed to analyse the damage initiation and evolution of composite laminate. Finite element model is also established to simulate the failure propagation process of composite laminates. Compared with experimental data, a good agreement is achieved. Therefore, the finite element model established and the progressive failure method proposed in this paper can be applied to simulate the failure propagation process of composite laminates effectively.

Keywords: Open-hole · Composite laminate · Tension

1 Introduction

Though composites are chosen widely in aircraft, aerospace and other fields because of specific strength and light weight [1, 2], the fiber breakage caused by open-hole will lead to the serious decline of mechanical properties, which limits the application of composites.

In the past, several researches have been focused on the strength and failure behavior on open-hole composite laminates. Lekhnitskii et al. [3] projected a theoretical calculation formula of stress concentration near the hole of orthotropic plate. Whitney and Nuismer [4, 5] proposed the criterion of point stress and average stress, which combined with the test method can predict the strength of composite laminates. Sun et al. [6] proposed a non iterative failure element method based on micromechanics to study the progressive damage of composites. In the calculation of failure load of composite laminates, Petit and Waddoups [7] proposed the progressive failure analysis method. The classical laminate theory was used to calculate the stress distribution of laminates, and the final failure was judged by the singularity of in-plane matrix or the negative diagonal term of matrix. Chang et al. [8, 9] built a progressive damage model of notched

Chinese Society of Aeronautics and Astronautics: CASTYSF 2022, LNEE 972, pp. 1–7, 2023.
https://doi.org/10.1007/978-981-19-7652-0_1

laminates under tension to predict tensile load. When studying the strength of open-hole composite laminates, Tan [10] expressed the decline of material properties caused by different damage modes by parameter reduction. Mccarthy et al. [11] established the numerical analysis of open-hole tension based on Hashin [12] failure criterion and nonlinear material continuous degradation model, which was verified by experiments.

In the present study, tensile experiment of six open-hole composite laminates are completed. Failure strength and damage phenomena of all composite laminates are examined. According to the experimental phenomenon, Hashin failure criteria and a method of material property degradation proposed are applied to study the damage mode of open-hole composite laminates under tension. The numerical prediction of failure behavior and loading capacity are compared with the experimental data.

2 Experiment

The material we are studying is a composite fabricated with woven carbon fiber and EH301 epoxy resin. The composite panel has a layup of [0/90/45/0/90]S with an in-plane dimension of 300 mm by 36 mm. A hole with a diameter of 6mm is made in the center of the composite laminate (see Fig. 1).

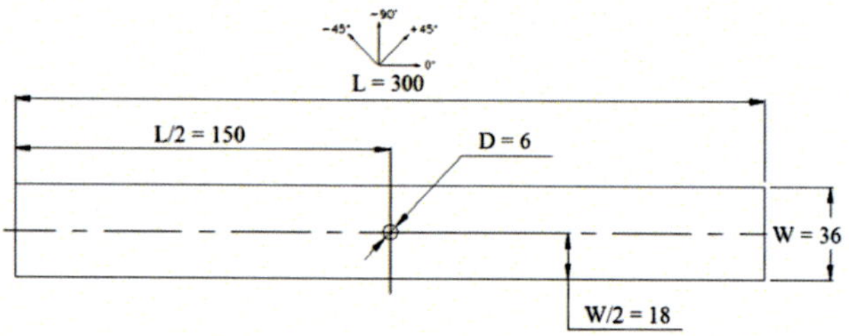

Fig. 1. Composite laminate

The method of tension experiment is set up by standard of ASTM D 5766-07, as from Fig. 2 shows. First, the composite laminate is installed in MTS-TestStar II machine. Then, displacement load with 2mm is brought to the composite laminate until fracture.

Fig. 2. Tension experiment

The calculation formula of tensile strength is shown as follow:

$$\sigma * = P_{\max}/[(W - D)T] \tag{1}$$

Here σ* is tensile strength, P_{max} is failure load, and T is the thickness of the composite laminate.

As shown in Table 1, six pieces of composite laminates are tested, and the tension failure strength of every composite laminate is evaluated. The Standard deviation is 10, and the dispersion coefficient is 0.022. Obviously, it is observed that the experiment has good repeatability.

Table 1. Experimental results of tension

Number	Failure strength (MPa)	Average strength (MPa)	Standard deviation	Dispersion coefficient
1	458	456	10	0.022
2	442			
3	464			
4	446			
5	464			
6	465			

As Fig. 3 shows, the damage phenomena of all composite laminates are examined. It's observed that for all composite laminates, the failure phenomena is that the crack crosses the center hole transversely. The damage diffuses at the edge of the laminate composites. The main failure mode is fiber fracture and matrix cracking, with only a few delaminations.

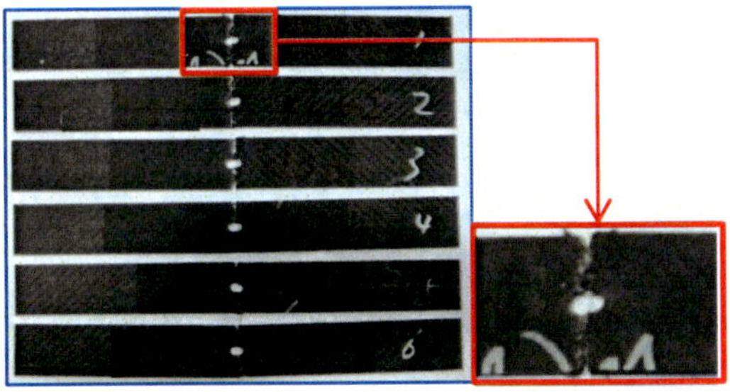

Fig. 3. Damage phenomena

3 Analytical Model

3.1 Damage and Failure Criteria

Consider the damage phenomena of experimental results, the influence between layers is ignored. So, Hashin failure criteria, which described four failure modes is applied, and proposed that the failure modes as follows:

Fiber tensile failure ($\sigma_{11} \geq 0$)

$$\frac{\sigma_{11}^2}{X_T^2} + \alpha\left(\frac{\sigma_{12}^2}{S_{12}^2}\right) \geq 1 \tag{2}$$

Fiber compressive failure ($\sigma_{11} < 0$)

$$\frac{\sigma_{11}^2}{X_C^2} \geq 1 \tag{3}$$

Matrix tensile and shear failure ($\sigma_{22} \geq 0$)

$$\frac{\sigma_{22}^2}{Y_T^2} + \frac{\sigma_{12}^2}{S_{12}^2} \geq 1 \tag{4}$$

Matrix compressive failure ($\sigma_{22} < 0$)

$$\left[\left(\frac{Y_C}{2S_{23}}\right)^2 - 1\right]\frac{\sigma_{22}}{Y_C} + \left(\frac{\sigma_{22}}{2S_{23}}\right)^2 + \frac{\sigma_{12}^2}{2S_{12}^2} \geq 1 \tag{5}$$

Here σ_{ij} is the stress component with respect to the material principal axis of the ply. X_T (X_C) is the axial tensile (compressive) strength along the x_1-axis. Y_T (Y_C) is the transverse tensile (compressive) strengths along the x_2-axis. S_{12} and S_{23}, respectively, are the shear strengths in the x_1x_2-plane and which in the x_2x_3-plane. α is the material nonlinear factor.

3.2 Degradation of Material Property

The initiation and growth of fiber damage and matrix damage will lead to degradation of material property. Degradation schemes are used to express combined failure modes:

$$E'_{11} = (1 - d_f)E_{11} \tag{6}$$

$$E'_{22} = (1 - d_f)(1 - d_m)E_{22} \tag{7}$$

$$G'_{12} = (1 - d_f)(1 - d_m)G_{12} \tag{8}$$

$$G'_{23} = (1 - d_f)(1 - d_m)G_{23} \tag{9}$$

The fiber damage variable d_f is given by:

$$d_f = \max\left[0, \min[\frac{\varepsilon_f(\varepsilon - \varepsilon_0)}{\varepsilon(\varepsilon_f - \varepsilon_0)}]\right] \tag{10}$$

where ε, ε_0 and ε_f are the strain at the computed step, at damage initiation and final strain respectively.

The matrix damage variable dm is given by:

$$d_m = \max\left[0, \min[\frac{\varepsilon_f(\varepsilon_m - \varepsilon_0)}{\varepsilon_m(\varepsilon_f - \varepsilon_0)}]\right] \tag{11}$$

$$\varepsilon_m = \sqrt{\varepsilon_{22}^2 + \gamma_{12}^2} \tag{12}$$

where ε_{22}, γ_{12} and ε_m are the normal strain in the transverse step, shear strain and effective strain respectively.

3.3 Finite Element Modeling

A finite element model based on the damage model described above was built in Abaqus. The edge left is pinned. Axial displacement is applied on the right side. The composite laminate is meshed with S4R elements available in Abaqus element library (see Fig. 4). Material properties assigned for the laminate have been given in Table 2.

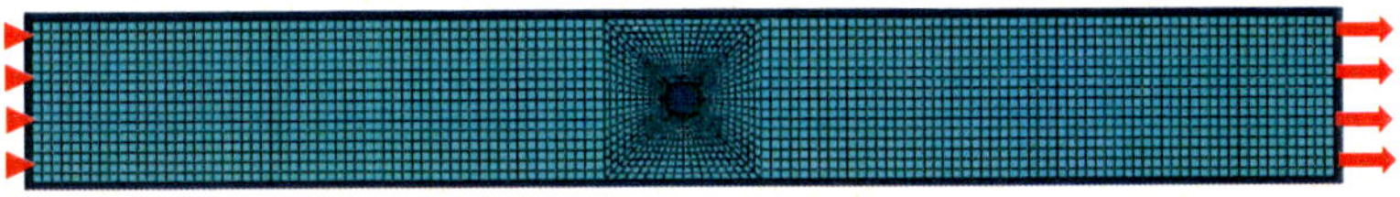

Fig. 4. FE model

Table 2. Properties of composite laminates

X_T/MPa	X_C/MPa	Y_T/MPa	Y_C/MPa	S_{12}/MPa	S_{23}/MPa	G_{12}/GPa	G_{23}/GPa
600	550	600	550	60	60	4	4
E_{11}/GPa	E_{22}/GPa	μ_{12}	μ_{13}	μ_{23}			
60	60	0.3	0.16	0.16			

4 Result and Analysis

4.1 Damage Mechanism During Tension

The finite element model established and the progressive failure method proposed in this paper are used to simulate the failure propagation process of composite laminates, some plies of which can be shown in Fig. 5. Due to the use of upper and lower symmetrical plies, this analysis only shows the upper plies, not the symmetrical part. Obvious, all the failures originate from the stress concentration near the hole. It can be seen from the figure that ply-3 (±45) contains matrix cracking and fiber fracture units. It also shows that ply-3 is the first ply to fail, and the failure mode is matrix tensile cracking, which induces the fiber tensile fracture later in this ply. This is because the ply bears both tensile load and shear load. With the increase of load, other plies begin to fail gradually. The failure mode of ply-1 and ply-4 is fiber tensile fracture, while the failure mode of ply-2 and ply-5 is matrix tensile cracking. For all plies, the damage diffuses at the edge of the laminate composites. The situation is similar to the experimental phenomenon in Fig. 3.

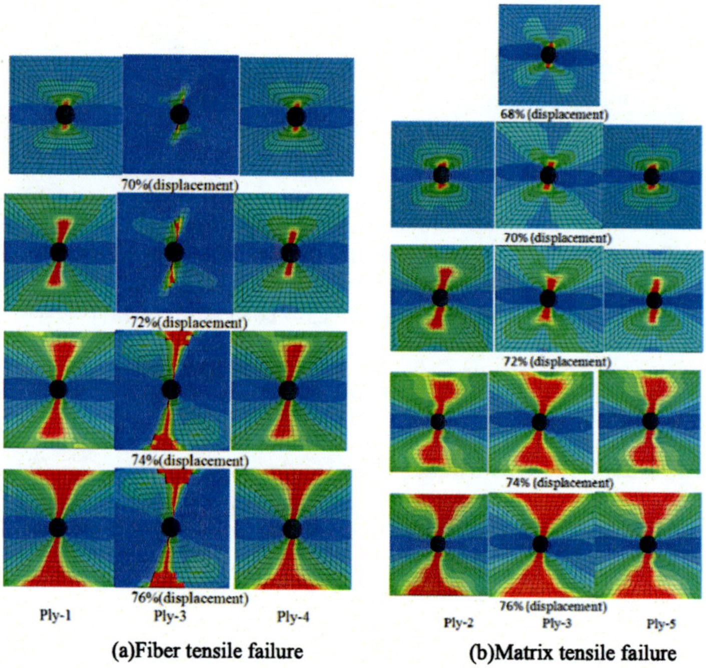

Fig. 5. Damage Mechanism

4.2 Failure Strength Analysis

The ultimate failure strength criterion adopted in this paper is that when the ply damage in the model extends to the whole plate width, the composite laminate loses its bearing capacity and fails. at this time, the load is the final tensile load. When the Axial displacement reaches 74% (1.48 mm), ply-3 fails completely. when the axial displacement reaches 76% (1.52 mm), all the plies are destroyed. The comparison result of experiment and calculated value are shown in Table 3. The error of 5.3% shows that the effectiveness of the modeling and analysis method.

Table 3. Comparison results of experiment and calculated values

Calculated value (MPa)	Test value (MPa)	Error (%)
480	456	5.3

5 Conclusion

For all composite laminates, the failure phenomena is that the crack crosses the center hole transversely under tension. The damage diffuses at the edge of the laminate composites. the main failure mode is fiber fracture and matrix cracking, with only a few delaminations. Under axial load, all the failures of composite laminates originate from the stress concentration near the hole. ply(± 45) contains matrix cracking and fiber fracture units. The first failure mode is matrix tensile cracking, which induces the fiber tensile fracture later in this ply. This is because the ply bears both tensile load and shear load. The comparison results of experiment and calculated values in this paper show the effectiveness of the modeling and analysis method.

References

1. Du, S.Y.: Advanced composite materials and aerospace engineering. Acta Mater. Compos. Sin. **24**(1), 1–12 (2007)
2. Du, S.Y., Guan, Z.D.: Strategic considerations for development of advanced composite technology for large commercial aircraft in China. Acta Mater. Compos. Sin. **25**(1), 1–10 (2008)
3. Lekhnitskii, S.G.: Theory of Elasticity of an Anisotropic Body. Mir Publisher, San Francisco (1981)
4. Whitney, J.M., Nuismer, R.J.: Stress fracture criteria for laminated composites containing stress concentration. Comp. Mater. **8**, 253–275 (1974)
5. Nuismer, R.J., Whitney, J.M.: Uniaxial failure of composite laminates containing stress concentrations. Fract. Mech. Compos. **49**(3), 117–142 (1975)
6. Sun, X.S., Tan, V.B.C., Tay, T.E.: Micromechanics-based progressive failure analysis of fibre-reinforced composites with non-iterative element-failure method. Comput. Struct. **89**(11–12), 1103–1116 (2011)
7. Petit, P.H., Waddoups, M.E.: A method of predicting the nonlinear behavior of laminated composites. J. Compos. Mater. **3**(1), 2–19 (1969)
8. Chang, F.K., Chang, K.Y.: A progressive damage model for laminated composites containing stress concentrations. J. Compos. Mater. **21**(9), 834–855 (1987)
9. Chang, K.Y., Liu, S.: Damage tolerance of laminated composites containing an open hole and subjected to tensile load. J. Compos. Mater. **25**(3), 2–19 (1991)
10. Tan, S.C.: A progressive damage model for composite laminates containing openings. J. Compos. Mater. **25**(5), 556–557 (1991)
11. Mccarthy, C.T., Ohiggins, R.M., Frizzell, R.M.: A pubic spline implementation of non-linear shear behavior in three-dimensional progressive damage models for composite laminates. Compos. Struct. **92**(1), 173–181 (2010)
12. Hashin, Z.: Failure criteria for unidirectional fiber composites. J. Appl. Mech. **47**(2), 329–334 (1980)

Research on Magnetic Tracking Algorithm Based on Double Transmitting Coils

Ranzhong Yao(✉), Hong Liu, and Jipeng Liu

AVIC Aerospace Life-Support Industries, Ltd., Xiangyang 441003, China
Azhong_1@126.com

Abstract. IT is necessary to track the head position and attitude of pilots during the operation of aviation simulation aircraft. There are large errors in the distance between the transmitting coil and the receiving coil of the magnetic tracking pose tracking system. The six dimensional magnetic tracking position and attitude tracking system based on weighted dual transmitting coils can select the appropriate transmitting coils by weighting, so as to reduce the positioning and attitude errors of the system. The electromagnetic positioning method used in this paper is to measure the sine wave phase difference between the square of the synthetic magnetic induction intensity of the sensor and the square of the excitation signal of the magnetic field source, so as to accurately determine the six degree of freedom position of the target. The algorithm has the characteristics of stable performance and strong anti-interference ability.

Keyword: Magnetic positioning · Double transmitting coil · Phase difference pose

1 Introduction

Positioning system is a device for spatial position measurement, which is widely used in virtual reality, artificial intelligence and other fields. At present, the more common positioning methods are ultrasonic positioning, photoelectric positioning, inertial positioning and electromagnetic positioning. Motion tracking technology is a practical research topic with wide application prospects. Its application fields include animation, games, virtual reality, medicine, biomechanics and so on. Positioning technology and navigation system are the methods to obtain the spatial position and attitude of the target. They play a very important role in many fields, such as mobile communication, medical instruments, transportation, wearable portable devices and military [1, 2]. It is necessary to track the head position and attitude parameters of pilots in aviation simulation aircraft. According to the implementation means of positioning, the system can be divided into:

(1) Electromagnetic navigation system
Use electric field, magnetic field or electromagnetic field to locate and navigate the target. The advantage of this navigation method is that it can penetrate some

Chinese Society of Aeronautics and Astronautics: CASTYSF 2022, LNEE 972, pp. 8–18, 2023.
https://doi.org/10.1007/978-981-19-7652-0_2

obstacles for information transmission and realize six degrees of freedom positioning and tracking; The disadvantage is that it is easy to be disturbed by metals or ferromagnetic substances in the environment. Electromagnetic positioning technology is widely used in radar positioning, GPS positioning system and navigation system of medical instruments. Electromagnetic positioning and navigation technology can be applied not only in long-distance positioning system, but also in short-distance precision positioning system. Because human physiological activities do not affect the distribution of magnetic field in the body, electromagnetic positioning technology has unique advantages in the field of medicine.

(2) Optical navigation system
Although light wave is also a kind of electromagnetic wave, the light wave navigation system is very huge, which is introduced and studied separately as a navigation system. Optical navigation system is a navigation system that uses visible light, infrared light or X-ray to locate. It is widely used in military and medical fields. The utility model has the advantages of simple structure, portability, easy carrying and high positioning accuracy; The disadvantage is that it is limited to the line of sight and cannot pass through or around obstacles. Although X-ray penetration is strong, it will cause radiation damage to human body when used in medical field.

(3) Ultrasonic navigation system
Ultrasound is widely used in the medical field, which can be used for medical imaging, medical diagnosis and other applications. For example, in gastroscope, ultrasonic probe is usually included for navigation and imaging. The advantages of ultrasonic positioning and navigation technology are simple, convenient and cheap; The disadvantage is that the transmission distance of ultrasonic wave is relatively short, and the attenuation speed is different in different substances, so the range of positioning and navigation is limited.

(4) Inertial navigation system
Inertial navigation system applies the gravity of the earth and the inertial navigation of moving objects. It is mainly used to judge the motion state and attitude of moving objects. Before, the inertial navigation system was relatively large and heavy, so it was inconvenient to apply. With the progress of microelectronic technology, there are micro acceleration sensors and micro gyroscopes, and even inertial sensors (make the accelerometer and gyroscope in one chip), so the inertial navigation system has become simple and lightweight. However, the application of inertial navigation in the measurement of head pose has some disadvantages, such as large volume, long-time use of gyroscope drift, alignment and poor accuracy of position parameters.

To sum up, electromagnetic tracking is a suitable means of head posture. The receiving coil is small and can be placed in the helmet, and the transmitting coil is placed on the engine room opening frame [3].

At present, in order to improve the accuracy, speed and anti-interference of electromagnetic positioning, researchers have put forward more and more electromagnetic positioning methods. According to different classification methods, electromagnetic positioning technology is divided into different categories. According to the number of magnetic field sources in the positioning system, the electromagnetic positioning system can be divided into two categories [4].

(1) Single source system
Single source system refers to the electromagnetic positioning system that contains only one electromagnetic field source. Early electromagnetic positioning systems used this design method. This kind of positioning system is relatively simple, with fast positioning speed and small amount of calculation, but the positioning accuracy is not high, so it can be used in occasions with low accuracy requirements.
(2) Multi source system
The multi-source system contains two or two electromagnetic field sources, which are combined into a certain form of array, which can greatly improve the positioning accuracy.
In terms of anti noise ability, the traditional model-based magnetic positioning system is vulnerable to the electromagnetic interference of the surrounding environment, so the service conditions are relatively harsh. If it is used in occasions with serious electromagnetic noise, the accuracy will be greatly reduced [5, 6].

Aiming at the main problems existing in the current electromagnetic tracking system, from the perspective of improving the positioning accuracy and tracking speed of the system, this paper proposes a six-dimensional magnetic tracking algorithm based on weighted double transmitting coils and a magnetic positioning algorithm based on phase discrimination technology.

2 Method

2.1 Positioning Principle

The standard attitude positioning system is composed of two three-axis magnetic field sources and three-axis magnetic sensors. As shown in Fig. 1, the coordinate systems S1 and S2 are established with the center point of the two magnetic field sources as the coordinate origin respectively. The three axes of the two coordinate systems coincide with the three axes of the magnetic field source respectively, and the coordinate system S3 is established with the midpoint of the magnetic sensor as the coordinate origin, and its three axes coincide with the three axes of the magnetic sensor. Assuming that the distance between the center points of two magnetic field sources is D, the spatial position (x, y, z) and attitude of the magnetic sensor can be calculated according to formulas (1) and (2).

$$
\begin{aligned}
x &= \frac{d\tan\gamma_2}{\tan\gamma_1 + \tan\gamma_2} \\
y &= \frac{d\tan\gamma_2\tan\gamma_2}{\tan\gamma_1 + \tan\gamma_2} \\
z &= \frac{x}{\tan\delta_1}
\end{aligned}
\tag{1}
$$

$$R_{13} = R_{43}R_{14} = R_{34}{}^{-1}R_{14} \quad (2)$$

As shown in Fig. 1, γ_j, j = 1, 2 is the included angle between the projection of the sensor on the xoy plane of the coordinate system SJ and the positive half axis of the X axis, and j represents the number of the coordinate system; δ 1 is the included angle between the projection of the sensor on the xoz plane of the coordinate system S2 and the positive half axis of the Z axis. In order to express the attitude of the sensor conveniently, that is, the relationship between the sensor coordinate system S3 and the system coordinate system S1, this paper defines two additional coordinate systems S4 and S5. S4 takes the center point of magnetic field source 1 as the coordinate origin, and its X-axis direction is connected along the coordinate origin of S1 and S3, that is, R1 and S5 take the center point of magnetic field source 2 as the coordinate origin, and its X-axis direction is connected along the coordinate origin of S2 and S3, R2. At the same time, Rab represents the rotation matrix from Sa to sb.

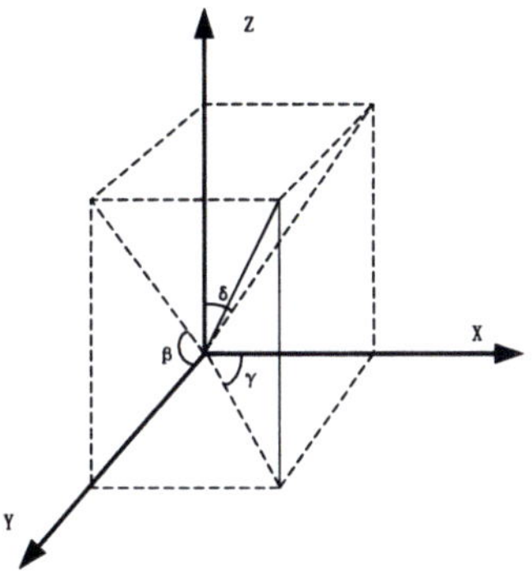

(a) Schematic diagram of projection angle

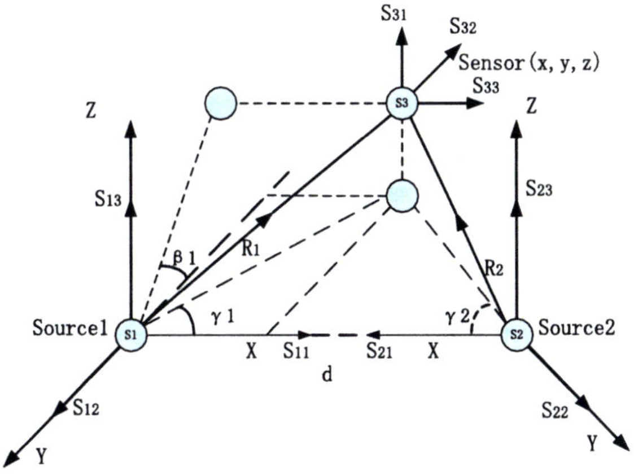

(b) Schematic diagram of coordinate axis

Fig. 1. Schematic diagram of electromagnetic positioning system

If the coil of magnetic field source 1 in the y-axis direction is excited with a sinusoidal signal of a certain frequency and amplitude, and the coil of magnetic field source 1 in the z-axis direction is excited with a cosine signal of the same frequency and amplitude, as shown in Fig. 2. Obviously, there is a phase difference between the square of the synthetic magnetic induction detected by the magnetic sensor and the square of the cosine excitation signal. We found that the magnitude of the phase difference will change with the relative position between the magnetic sensor and the magnetic field source. Therefore, it is assumed that there is some relationship between the two. After repeated theoretical derivation and experimental verification, it shows that there is not only a relationship between them, but also a linear relationship—the phase difference is twice the projection angle [7]. If the coil is excited in some way, the projection angle can be obtained by measuring the phase difference between the square of the synthetic magnetic induction intensity at the position of the sensor and the square of the excitation signal of the magnetic field source, and then the six degree of freedom position of the target can be calculated [8].

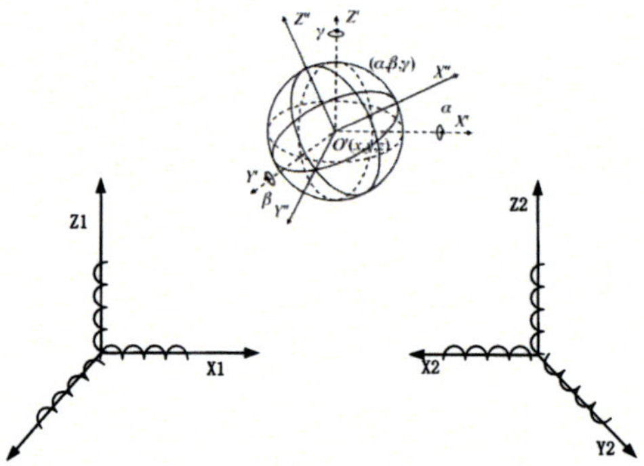

Fig. 2. Schematic diagram of transmitting coil and receiving coil

2.2 Location Algorithm

The projection angle is derived below $\gamma 1$. The same method can also be used to calculate $\gamma 2$ and $\delta 1$. Figure 1 shows a schematic of the process. If the coil 1 and coil 2 in the magnetic source 1 represented by S11 and S12 flow through the phase orthogonal currents I1 and I2, the phase difference between the square of the excitation of the magnetic source and the square of the total magnetic flux density at the sensor position will change when the target moves. Through theoretical derivation, it can be seen that this relationship is not only fixed, but also linear. The phase difference between the square of the cosine excitation of the magnetic source and the square of the total magnetic flux density at the sensor position is the projection angle γ Twice as much as 1. The derivation is as follows [9, 10].

We use B = (BX, BY, BZ) to represent the total magnetic flux density, which can be expressed as:

$$\begin{cases} B_x = \frac{\mu_0 SI}{4\pi r^3}\left(\frac{3x(mx+ny+pz)}{r^2} - m\right) \\ B_y = \frac{\mu_0 SI}{4\pi r^3}\left(\frac{3y(mx+ny+pz)}{r^2} - n\right) \\ B_z = \frac{\mu_0 SI}{4\pi r^3}\left(\frac{3z(mx+ny+pz)}{r^2} - p\right) \end{cases} \tag{3}$$

where (m, N, P) represents the magnetic moment of the magnetic dipole, μ 0 represents the vacuum permeability, s represents the area surrounded by circular current, I represents the current intensity, and R represents the distance between the center of the sensor and the center of the magnetic source. Now, suppose that S11 is fed with current I1 = I cos(ωt + ϕ 0), S12 is fed with current I2 = I sin(ωt + ϕ 0), and S13 is fed without current. Therefore, we can see that (m, N, P) = (COS(ωt + ϕ0), sin(ωt + ϕ 0), 0). For simplicity, we define k = (u_0SI/4 π R3) as a constant related to the target position and assume φ 0 equals zero. The magnetic field strength (BX,BY, BZ) can be expressed as:

$$\begin{cases} B_x = K\left(\frac{3x(x\cos\omega t + y\sin\omega t)}{r^2} - \cos\omega t\right) \\ B_Y = K\left(\frac{3y(x\cos\omega t + y\sin\omega t)}{r^2} - \sin\omega t\right) \\ B_z = K\left(\frac{3z(x\cos\omega t + y\sin\omega t)}{r^2}\right) \end{cases} \tag{4}$$

B generated by such excitation coils is the total flux density at the sensor position, which can be expressed as Bx, By and Bz as follows:

$$\begin{aligned} B^2 &= B_x^2 + B_y^2 + B_z^2 \\ &= \frac{k^2}{r^2}(3(x\cos\omega t + y\sin\omega t)^2 + r^2) \\ &= \frac{k^2}{r^2}(3\left(\sqrt{x^2+y^2}\sin(\omega t + \gamma 1)\right)^2 + r^2) \end{aligned} \tag{5}$$

where, tanγ1 = y/x. According to the double angle formula, B2 can be expressed in the following form:

$$\begin{aligned} B^2 &= -\frac{3k^2(x^2+y^2)}{2r^2}\cos(2\omega t + 2\gamma 1) + \frac{3k^2(x^2+y^2)}{2r^2} + k^2 \\ &= \cos(2\omega t + \Delta\varphi) + N \end{aligned}$$

where $\Delta\varphi = 2\gamma 1$, $M = -(3k^2(x^2 + y^2)/2r^2)$, $N = -M + k^2$.

M and N are constants related to the target position. S11 square can be expressed as (Fig. 3):

$$i_1^2 = I^2\cos^2\omega t = U\cos 2\omega t + V \tag{6}$$

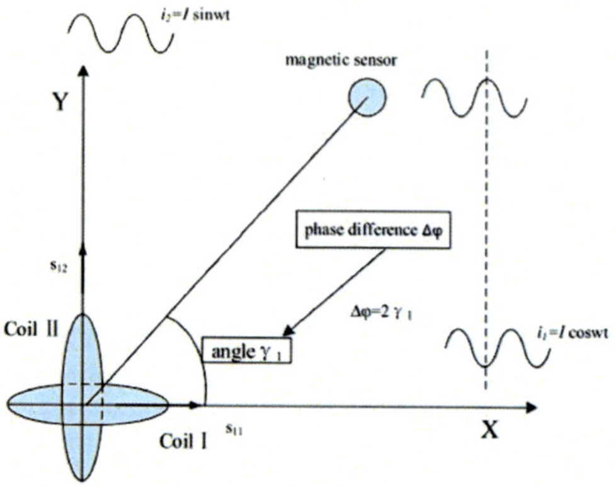

Fig. 3. Schematic diagram of emitter current

Where U and V are constants. It is obvious from formulas (5) and (6) that the phase difference between the square of the total flux density at the sensor position and the square of the excitation current i1 is 2γ1. γ1 is the included angle between the projection of the sensor on the xoy plane and the positive x-axis, as shown in Fig. 2. Therefore, we can obtain the projection angle by measuring the phase difference between B^2 and $i_1{}^2$.The position of the target can be determined from three angles γ1, γ2 and δ1. Confirm. The projection angle is obtained according to formula (5) and formula (6) γ1. When S11 is fed with current i1, S12 is fed with current i2. Because in our method, the two magnetic sources are constructed in the same way. Therefore, using the same method, when S21 feeds in the current i1 and S22 feeds in the current i2, it can be obtained γ2, and when S13 feeds in current i1 and S11 feeds in current i2, it can be obtained δ1.Then, the position of the target can be calculated according to formula (1).

Figure 4 shows the relationship between positioning error and signal-to-noise ratio of phase difference algorithm and amplitude algorithm.

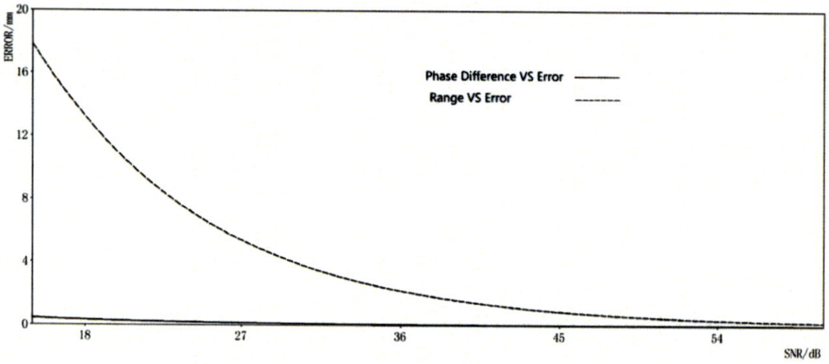

Fig. 4. Location error and signal-to-noise ratio of different algorithms

2.3 Orientation Algorithm

Define rotation matrix R = rot (y, β) Rot(x, α), among

$$\mathrm{Rot}(Z,\gamma)=\begin{pmatrix}\cos\gamma & -\sin\gamma & 0\\ \sin\gamma & \cos\gamma & 0\\ 0 & 0 & 1\end{pmatrix}$$

$$\mathrm{Rot}(y,\beta)=\begin{pmatrix}\cos\beta & 0 & \sin\beta\\ 0 & 1 & 0\\ -\sin\beta & 0 & \cos\beta\end{pmatrix}$$

$$\mathrm{Rot}(x,\alpha)=\begin{pmatrix}1 & 0 & 0\\ 0 & \cos\alpha & -\sin\alpha\\ 0 & \sin\alpha & \cos\alpha\end{pmatrix} \tag{7}$$

Then the component of the magnetic field in each direction in the positioning coordinate system is the product of the magnetic field component of the rotation matrix and the emission coordinate system, and the component of the magnetic field in each direction in the positioning coordinate system is the product of the magnetic field component of the rotation matrix and the emission coordinate system.

$$\begin{pmatrix}B_x'\\ B_y'\\ B_z'\end{pmatrix}=R\begin{pmatrix}B_x\\ B_y\\ B_z\end{pmatrix}$$
$$=\begin{pmatrix}\cos\alpha\cos\beta\times B_x+(-\sin\alpha\cos\gamma+\cos\alpha\sin\beta\sin\gamma)\times B_y+(\sin\alpha\sin\beta+\cos\alpha\sin\beta\cos\gamma)\times B_z\\ \sin\alpha\cos\beta\times B_x+(\cos\alpha\cos\gamma+\sin\alpha\sin\beta\sin\gamma)\times B_y+(-\cos\alpha\sin\gamma+\sin\alpha\sin\beta\cos\gamma)\times B_z\\ -\sin\beta\times B_x+\cos\beta\sin\gamma\times B_y+\cos\beta\cos\gamma\times B_z\end{pmatrix} \tag{8}$$

3 Different Coil Opening Options

The distance between the magnetic induction device and the transmitting coil is too close and too far, which will bring large errors. Figure 5 shows the curve between the distance and position error of one receiving device and two left and right transmitting devices.

It can be seen that taking the left transmitting coil as the coordinate zero point, when the detected distance is within 680 mm, the receiving coil receives the data of the left transmitting coil. When it exceeds 680 mm, the right transmitting coil is opened and the receiving coil receives the magnetic data of the right transmitting coil.

4 Build the Actual System

Build a magnetic sensing device in the laboratory as shown in Fig. 6. In the Fig. 1 is the magnetic induction system, 2 and 3 are two transmitting coils, and 4 is the signal acquisition equipment. Figure 7 is a comparison diagram of the position parameters actually calculated and tested by the laser positioning system. The attitude parameters are determined by the rotation angle of the manipulator. Table 1 shows the positioning and attitude errors of the magnetic position and attitude sensing system.

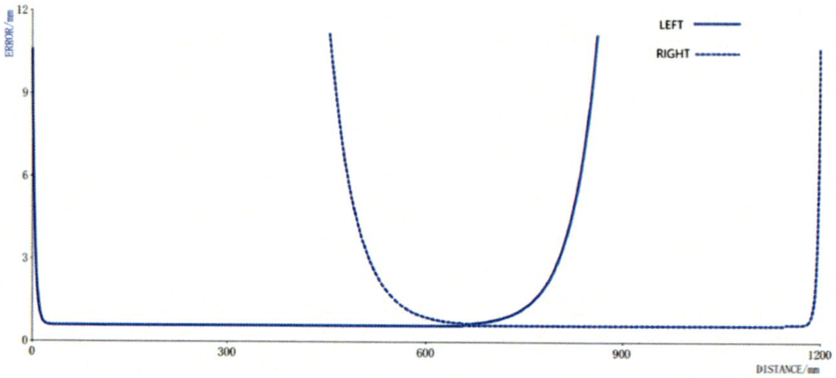

Fig. 5. Distance and position error curve of receiving device and transmitting device

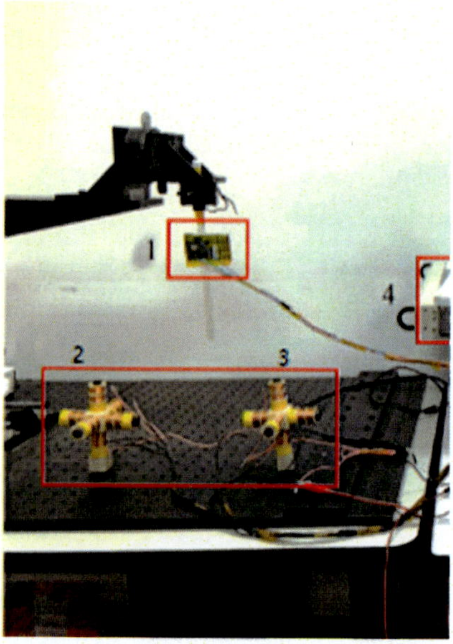

Fig. 6. Laboratory magnetic positioning system

5 Conclusion

This paper describes an algorithm based on phase difference to realize the pilot's head position and attitude test system of flight simulator. Compared with the amplitude calculation of position and attitude, the accuracy and anti-interference ability have been greatly improved. At the same time, in order to improve the accuracy, two transmitting coils and one receiving coil are used. The coils have different distances and the errors are inconsistent. Different weighting processing is used to reduce the system error.

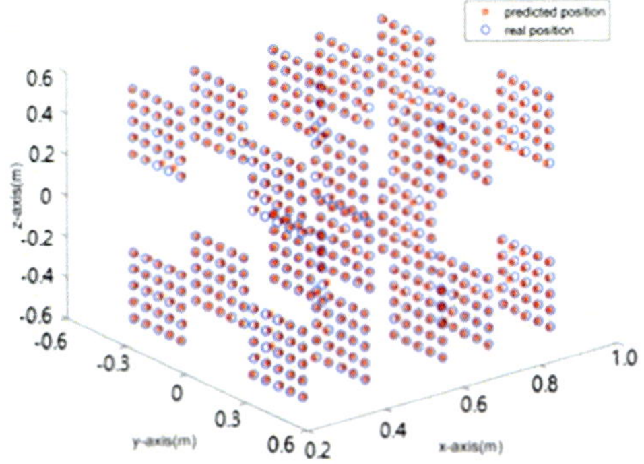

Fig. 7. Position error diagram of magnetic positioning system

Table 1. Positioning and attitude error table

Type	Mini error	Max error	Average error	Variance
Δx (cm)	0.0000	0.4545	0.0868	0.0068
Δy (cm)	0.0002	0.4350	0.1185	0.0070
Δz (cm)	0.0011	0.3952	0.1077	0.0069
Δr (cm)	0.0245	0.5635	0.2080	0.0089
Δγ1 (°)	0.0027	0.8689	0.2217	0.0335
Δγ2 (°)	0.0002	0.8878	0.2226	0.0318
Δδ1 (°)	0.0000	0.8156	0.2018	0.0287

References

1. Kim, W., Song, J., Park, F.: Closed-form position and orientation estimation for a three-axis electromagnetic tracking system. IEEE Trans. Ind. Electron. **65**(5), 4331–4337 (2018)
2. Huang, X., Wu, X.: An electromagnetic tracking method based on phase difference detection. IEEE Trans. Magnet. **55**(9) (2019)
3. Yu, Z., Xiao, C., Zhou, G.: Multi-objectivization-based localization of underwater sensors using magnetometers. IEEE Sensors J. **14**(4), 1099–1106 (2014)
4. Hu, C., Song, S., Wang, X., Meng, M.Q.-H., Li, B.: A novel positioning and orientation system based on three-axis magnetic coils. IEEE Trans. Magn. **48**(7), 2211–2219 (2012)
5. Schultze, V., Andrä, W., Peiselt, K., Gleichmann, N., Meyer, H.G.: Unambiguous position and orientation tracking using a rotating magnet. J. Appl. Phys. **114**(11) Art. no. 114502 (2013)
6. Tanase, D., et al.: Radiation dose reduction in minimally-invasive intravascular procedures using a magnetic guidance system. In: Proceedings of 2nd Annual International IEEE-EMB Special Topic Conference Microtechnologies in Medicine and Biology, pp. 305–308 (2002)
7. Tanase, D., Goosen, J.F.L., Trimp, P.J., Reekers, J.A., French, P.J.: Catheter navigation system for intravascular use. In: Proceedings of 2nd Annual International IEEE-EMB Special Topic Conference Microtechnologies in Medicine and Biology, pp. 239–242 (2000)
8. Schneider, M.: Measuring position and orientation using magnetic fields. U.S. Patent 6 073 043 (2000)

9. Dai, H., Song, S., Zeng, X., Su, S., Lin, M., Meng, Q.-H.: 6-D electromagnetic tracking approach using uniaxial transmitting coil and tri-axial magneto-resistive sensor. IEEE Sensors J. **18**(3), 1178–1186 (2018)
10. Wu, F., Robert, M., Frey, D.D., Foong, S.: Enhanced magnetic localization with artificial neural network field models. In: Proceedings of IEEE International Conference on Robotics and Automation, pp. 1560–1565 (2013)

Research on Hardware Drive Circuit of Electro-Hhydraulic Servo Controller for Aircraft Steering Gear

Fu Pinqing[1(✉)], Cui Wenjun[1], and Yang Feng[2]

[1] Electronic Engineering Department, AVIC NEIAS, Nanjing 211100, China
1690232355@qq.com

[2] Hydraulic and Actuation Systems Department, AVIC NEIAS, Nanjing 211100, China

Abstract. The research on the hardware drive circuit of electro-hydraulic servo controller for aircraft steering gear is carried out by means of functional analysis, principle design, circuit simulation and test verification. Firstly, the function analysis and principle design of the drive circuit are carried out from the level of electro-hydraulic servo control system and the control architecture of DSP(Digital Signal Processing)+DAC (Digital To Analog Converter) is defined in the design. Using discrete devices to build analog drive circuit, drive output and feedback, and finally form a closed-loop control current-type drive output mode. The mathematical model of the analog drive circuit is established, and the relationship between the drive output current and the control volage is obtained. The circuit simulation model is set up to verify the correctness of the circuit mathematical model. Finally, the controller hardware drive circuit is verified by setting up the test environment.

Keywords: Aircraft steering gear · Electro-hydraulic servo controller · Hardware drive circuit · Mathematical model · Simulation model

1 Introduction

With the increasing maturity of aviation technology and the gradual complexity of aviation equipment, the requirements for aircraft control system are becoming higher and higher [1]. Steering gear is one of the key components of the flight control system [2, 3], and its performance will directly affect the flight control accuracy and flight safety of the aircraft [4–6]. At present, aircraft steering gear mainly uses electro-hydraulic servo control system [7], which has the advantages of mature technology, high power density, fast response speed, wide speed regulation range and good environmental tolerance [8, 9]. Therefore, it is of great significance to reasonably design the electro-hydraulic servo system of steering gear adapted to the changes of aircraft flight environment, which can not only effectively improve the working capacity of relevant equipment in the aircraft, but also increase the safety factor of the aircraft [10, 11].

The electro-hydraulic servo control system of steering gear takes the servo controller as the control core and the electro-hydraulic servo valve as the direct control object [12].

Chinese Society of Aeronautics and Astronautics: CASTYSF 2022, LNEE 972, pp. 19–27, 2023.
https://doi.org/10.1007/978-981-19-7652-0_3

The controller hardware drive circuit directly controls the size and direction of the servo valve spool through the output drive current. According to the impedance characteristics of electro-hydraulic servo valve, the equivalent treatment is carried out, focusing on the design and research of the hardware drive circuit of electro-hydraulic servo controller.

2 Theoretical Research

2.1 Functional Description

According to the function analysis of the controller hardware drive circuit, the functional block diagram is designed as shown in Fig. 1.

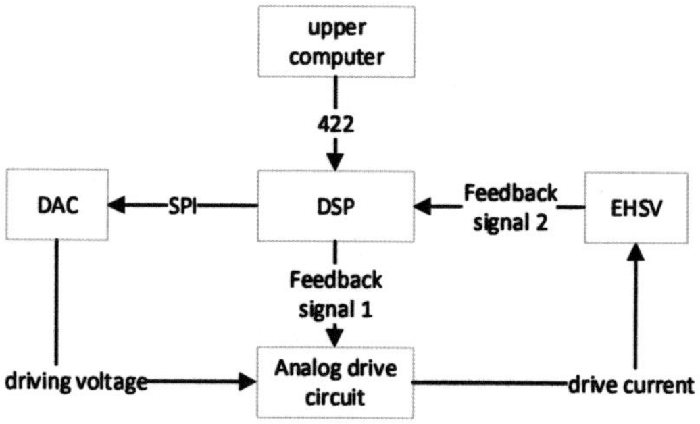

Fig. 1. Controller hardware drive circuit function diagram

In Fig. 1, EHSV is represented as electro-hydraulic servo valve. To describe the circuit function:

(1) Operators input operation instructions to DSP through upper computer;
(2) DSP collects analog drive circuit feedback signal 1 and electro-hydraulic servo valve feedback signal 2;
(3) DSP makes closed-loop calculation by docking with the received upper computer instruction and feedback signal 1;
(4) DSP transmits the closed-loop operation result to DAC through SPI, so as to output voltage signal;
(5) The DAC output voltage signal is converted by the analog driving circuit, and the corresponding driving current is output;
(6) EHSV (electro-hydraulic servo valve) makes corresponding actions under the action of driving current;

2.2 DAC Circuit Design

As A voltage-output DAC with SPI serial input interface, the chip has two output channels A and B, four control bits and 12 data bits, and its differential nonlinearity is less than 0.5LSB Typ. The data functions are shown in Table 1.

The DAC data function table is explained as follows:

Table 1. DAC data function sheet

Register selection	Speed control	Power control	Register selection	Data
P1	SPD	PWR	P0	DATA
D15	D14	D13	D12	D11-D0

$P1 = 0, P0 = 0$ Write data to DAC B and buffer;
$P1 = 0, P0 = 1$ Write data to buffer;
$P1 = 1, P0 = 0$ Write data to DAC A and update DAC B with buffer data;
$P1 = 1, P0 = 1$ Write data to register;
$SPD = 1$ Fast mode;
$SPD = 0$ Slow Mode;
$PWR = 1$ Power down;
$PWR = 0$ normal operation;

Output voltage calculation formula:

$$D_A = 2V_{REF}\frac{C_{ODE}}{0X1000}V \tag{1}$$

In formula (1), V_{REF} is 1.5 V voltage reference source and external input is adopted. C_{ODE} is the C_{ODE} value input by DSP to DAC through SPI. 0X1000 is a 12-bit full-scale hexadecimal code value. D_A is the DAC output voltage, and the code value needs to be converted to decimal value during calculation (Fig. 2).

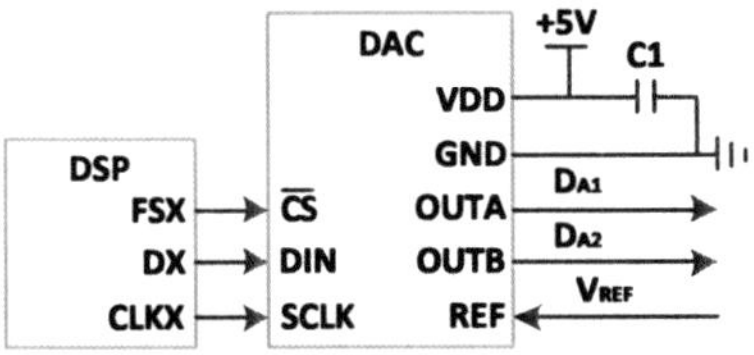

Fig. 2. DAC schematic diagram

2.3 Analog Drive Circuit Design

2.3.1 Functional Analysis

The analog drive circuit is divided into three parts by function:

(1) N1, N2 operational amplifiers and resistance of the amplifier circuit, its function is to amplify the D_A input value according to a certain proportion.

(2) Push pull circuit composed of Q1, Q2 triode and resistor, its function is based on positive and negative potential values V_F relative to AGND, providing corresponding drive current for external loads (note: the direction of current flowing through resistor R from point N to point M is positive).
(3) The amplifier circuit composed of N3 operational amplifier and resistance, its function is to collect V_D point voltage value according to a certain proportion of amplification, and the output voltage V_K as a feedback signal 1 transmission to DSP.

2.3.2 Schematic Diagram Designing

The DAC input voltage signal is converted to drive current by the analog drive circuit. The external load of the analog drive circuit is the electro-hydraulic servo valve. The analysis of the electrical characteristics of the electro-hydraulic servo valve can be equivalent to the load characteristic of 103Ω power resistance. According to the design requirements, the driving current flowing through the electro-hydraulic servo valve is in the range of (−40 ~+ 40) mA, and the circuit can collect the output current, and transmit it to DSP as a feedback signal for closed-loop control. Based on the above analysis, the schematic diagram of the simulated driving circuit is designed as shown in Fig. 3.

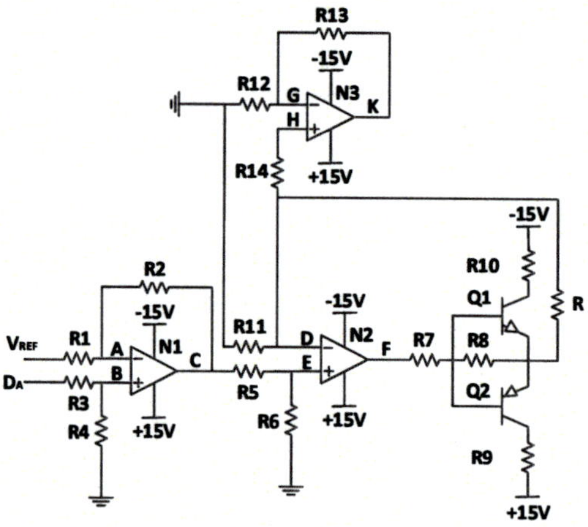

Fig. 3. Analog drive circuit schematic diagram

Analysis of Fig. 3 shows that:

$$V_C = \frac{R_4(R_1 + R_3)D_A}{R_1(R_3 + R_4)} - \frac{R_2 V_{REF}}{R_1} \tag{2}$$

$$V_D = V_C \frac{R_6}{R_6 + R_5} \tag{3}$$

$$I_{RL} = \frac{V_D}{R_{11}} \tag{4}$$

where: V_A, V_B, V_C, V_D, V_E, V_K denote voltage values of point A, B, C, D, E, K relative to AGND. D_A represents the input voltage of analog drive circuit. IRL represents the drive current flowing through the external load.

2.3.3 Mathematical Modeling

According to the driving current required by the external load, set the DAC output voltage value and the resistance value in the circuit reasonably. The resistance value of the analog driving circuit is set as shown in Table 2.

Table 2. Analog drive circuit organization parameters

Symbol	Value
R1	10 KΩ
R2	10 KΩ
R3	10 KΩ
R4	10 KΩ
R5	15 KΩ
R6	10 KΩ
R7	4.7 KΩ
R8	20 KΩ
R9	100 Ω
R10	100 Ω
R11	15 Ω
R12	2 KΩ
R13	10 KΩ
R14	91 KΩ
R	103 Ω

The mathematical model of I_R and D_A can be obtained according to the resistance value of the resistor set and the basic operation rule of the operational amplifier circuit, as shown in formula (5). The mathematical model of VK and IRL is shown in Eq. (6).

$$I_{RL} = \frac{2}{75} D_A - 0.04 \tag{5}$$

$$V_K = 90 I_{RL} \tag{6}$$

3 Simulation Analysis

3.1 Simulation Parameter Setting

According to Eq. (5), the mathematical relationship between the output driving current IRL and the input voltage D_A of the driving circuit to the external load is obtained. According to Eq. (6), the relation between the feedback signal V_K of the drive circuit and the drive current I_R is obtained. Based on this relationship, the forward maximum driving current of 40 mA, reverse maximum driving current of −40 mA and zero current of 0 mA are set respectively, and their corresponding input voltage value D_A and feedback voltage value V_K are shown in Table 3.

Table 3. Analog drive circuit mathematical model data sheet

Number	D_A (V)	I_R (mA)	V_K (V)
1	3	40	3.6
2	1.5	0	0
3	0	−40	−3.6

3.2 Simulation Model Construction

In order to verify the correctness of the mathematical model of the analog drive circuit and the validity of the data in Table 3, a circuit model is built in the simulation software Multisim for simulation, as shown in Fig. 4. According to the group 1, 2 and 3 D_A values in Table 3 as the input voltage, the simulation data of output current value I_R and feedback voltage value V_K obtained are shown in Table 4.

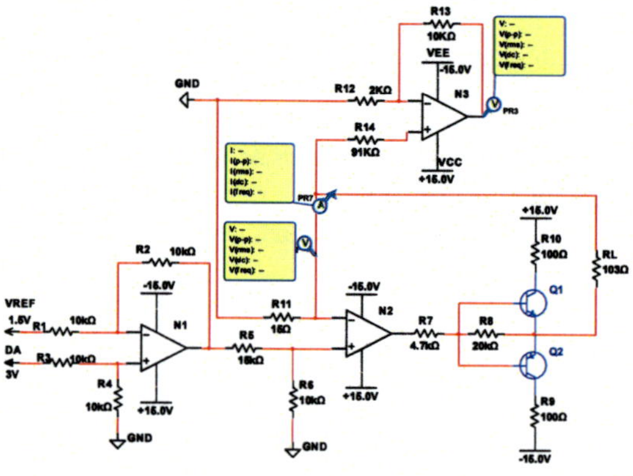

Fig. 4. Analog drive circuit simulation diagram

Table 4. Analog drive circuit simulation model table

Number	D_A (V)	I_R	V_K
1	3	40.0 mA	3.62 V
2	1.5	2.97 μA	371 μV
3	0	−40.0 mA	−3.60 V

3.3 Analysis of Simulation Result

By comparing the data in Tables 3 and 4, it can be seen that the I_R deviations of the three data groups are 0 mA, 0.00297 mA and 0 mA respectively. The V_K deviations were 0.02 V, 0.371 mV and 0 V, respectively. The relative values of deviations were all less than 0.6%, which could be ignored. Therefore, the correctness of the mathematical model of the analog drive circuit can be proved, and the rationality of the design of the analog drive circuit can be proved.

4 Experimental Verification

4.1 Setting up Test Environment

The test environment is built according to the simulation results, as shown in Fig. 5. The test instruments and tools corresponding to the digital label in Fig. 5 are shown in Table 5.

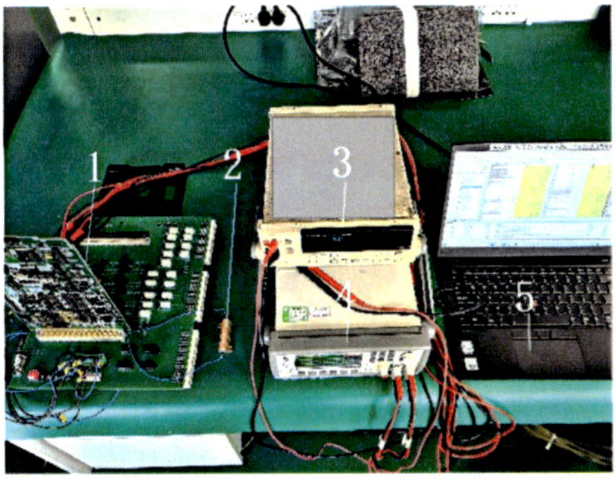

Fig. 5. Hardware drive circuit test environment

4.2 Test Procedure

(1) Complete the circuit connection without power output, adjust the power output voltage to the required value, confirm the connection is correct, and then power on.

Table 5. Hardware drive circuit test tool list

Number	Instruments	Amount
1	Hardware drive circuit board	1
2	103 Ω power resistor	1
3	Electronic multimeter	1
4	Power supply	1
5	PC	1

(2) The corresponding codes of D_A value are configured in the software code, and burned into THE DSP through the upper computer and the emulator.
(3) Use an electronic multimeter to measure the voltage V_R and corresponding V_K at both ends of the power resistor R under different input voltages of the drive circuit (feedback signal 1), and record the data;
(4) Calculate the current flowing through the power resistor R according to Ohm's law and record the data;

The data recorded and calculated during the above tests are shown in Table 6.

Table 6. Analog drive circuit test data sheet

Number	D_A (V)	I_R (mA)	V_K (V)
1	3	39.82	3.59
2	1.5	0.01	0.001
3	0	−39.80	−3.59

4.3 Analysis of Test Results

(1) There is a certain deviation between the test data in Table 6 and the preset data in Table 3. The IRL deviation of the three groups is 0.18 mA, 0.01 mA and 0.2 mA respectively, and the relative error is no more than 1%, and the one that meets the design requirements is no more than 1%. The V_K deviations were 0.01 V, 0.001 V and 0.01 V, respectively, and the relative errors were less than 1%.No more than 1% meet the design requirements.
(2) Through the analysis of simulation data and test data, it can be seen that the test data have a large deviation. The reason for this phenomenon is that there are certain unfavourable factors such as leakage current and certain tolerance of resistance in operational amplifier and triode in actual circuit. These parameters have an impact on the output accuracy, and the impact will continue to increase as the temperature

of the component rises. Therefore, heat dissipation treatment should be done during PCB design.

5 Conclusion

(1) Hardware drive circuit design of electro-hydraulic servo controller of aircraft steering gear adopts the closed-loop control current-type drive output architecture of DSP+DAC+analog drive circuit to meet the functional design requirements.
(2) Although there is some deviation between the test data and the preset value in Table 3, the relative error of all data is less than 1%, meeting the output accuracy requirements.
(3) Within the linear output range of the operational amplifier, the driving current provided by the analog driving circuit to the external load is not affected by the external load.
(4) Within the linear output range of the operational amplifier, the driving current and its feedback value provided by the analog driving circuit to the external load are positively linear correlated with the input voltage value.

References

1. Tan, C.W.: Research on the plane steering controller. Hunan University, Changsha (2015)
2. Li, J., Tang, G.C., Cao, J.F., et al.: Thermodynamic analysis and calculation on hydraulic steering units of aircraft hydraulic system. China Mech. Eng. **26**(15), 1–2 (2015). (in Chinese)
3. Liu, Z.W.: Research on electro-hydraulic servo loading system of aircraft steering gear based on BP-RBF neural network. Sci. Technol. Innovat. **9**, 1–2 (2017)
4. Chen, X.C.: Airborne hydraulic steering gear servo system optimization control research and simulation. Comput. Simul. **29**(6), 1–2 (2012). (in Chinese)
5. Hu, L.M., Cao, K.Q., Su, X.B., et al.: Research on fault simulation of hydraulic actuator for horizontal tail of an aircraft. Mach. Tool Hydraul. **44**(13), 1–2 (2016). (in Chinese)
6. Hu, L.M., Cao, K.Q., Xu, H.J.: Typical fault analysis for four redundant steering engines. Chin. Hydraul. Pneumat. **6**, 1–2 (2006). (in Chinese)
7. Wei, K.P., Hu, J., Yao, J.Y., et al.: Fast terminal sliding mode control of neural network for aeromechanical actuator. Acta Aeronaut. Astronaut. Sin. **42**(6), 1–2 (2021). (in Chinese)
8. Li, Y., Hao, S.Q., Wang, F.Q., et al.: Design of on-board model monitoring for electro-hydraulic servo system of aeroengine. Aeroengine **47**(2), 1–2 (2021). (in Chinese)
9. Wu, J.L., Zhang, J., Jin, L.S., et al.: Modeling and analysis of servo control of aeroengine actuator cylinder. Aeroengine **45**(2), 1–2 (2019). (in Chinese)
10. Huang, X.X., Long, Y., Zhang, Z.L., et al.: Summarization of distributed visual simulation technology. J. Syst. Simul. **22**(11), 1–2 (2010). (in Chinese)
11. Liu, X.L., Su, Y.: Application of artificial bee colony algorithms in aircraft steering electro-hydraulic servo system. Sci. Technol. Eng. **20**(30), 1–2 (2020). (in Chinese)
12. Huang, J.L.: Design of dual-core CPU hydraulic servo controller based on DSP. Xi an Polytechnic University, Xian pp. 1–2 (2017) (in Chinese)

Influence on the Flutter Characteristics of Hollow Fan Blade with Different Rib Number

Guo Xiao[1](✉), Wu Meng-Di[2], Shi Yong-Qiang[3], and Wen Zhen-Hua[1]

[1] School of Aeronautical Engineering, Zhengzhou University of Aeronautics, Zhengzhou 450046, China
guoxiao@zua.edu.cn

[2] School of Materials Science and Engineering, Zhengzhou University of Aeronautics, Zhengzhou 450046, China

[3] School of Power and Energy, Northwestern Polytechnical University, Xi'an 710072, China

Abstract. The wide-chord hollow fan blade was one of the new technology which had been applied on the aeroengine. It could effectively improve the performance of aeroengine. The energy method had been used to analysis the influence of the different rib number on the flutter characteristic of the hollow fan blade. The results show that the hollow structure does not change the distribution location of the unsteady aeronautical power. The rib can improve the anti-vibration characteristics of hollow blade. When the rib number is four, the aerodynamic damping ratio is higher than the basic solid blade under the same phase angle which can extend the adaptive capacity for the poor working contexts.

Keyword: Wide chord · Hollow-blade · Energy method · Flutter · Rib number

The fan blade is one of the key components of large bypass ratio turbofan engine. Compared with the original blade, the wide-chord blade can effectively increase the compressor surge margin, increase the redundancy degree of external damage, and improve the engine thrust. The use of wide-chord blade can also reduce the number of compressor blades. However, the height and the centrifugal load of fan/compressor blade is continuous increasing with the development of large bypass ratio turbofan engine towards high power, high load and high performance. In order to meet its reliability, it is necessary to further increase its disk mass. The wide-chord hollow fan blades were investigated to solve this problem. The weight loss of hollow blade leads to the reduction of blade rigidity which makes the possibility of blade flutter increasing [1, 2]. Therefore, the research on the blade flutter prediction occupies an important position in the design of hollow fan.

The flutter of the hollow fan blade is a new problem in turbomachinery flutter. Scholars at home and abroad have carried out relevant research on this problem. Vahdati M. and other researchers have [3] studied the flutter characteristics of wide-chord fan blades by means of fluid-solid coupling method. They think that the flow separation is the key factor affecting the stall flutter of blades, and the acoustic flutter is mainly related to the reflection of acoustic wave. The research results has a certain promotion effect on understanding the flutter mechanism of wide-chord blades. However the blade model is solid

Chinese Society of Aeronautics and Astronautics: CASTYSF 2022, LNEE 972, pp. 28–36, 2023.
https://doi.org/10.1007/978-981-19-7652-0_4

blade which cannot truly reflect the flutter characteristics of wide-chord hollow blades of turbofan engines. Chew J.W. and others have [4] studied the flutter characteristics of two different kind of the transonic fan blades at different speed conditions. The two kind of fan blades represent respectively typical fan blades used in military turbofan engines with small bypass ratio and civil turbofan engines with large bypass ratio. The advanced linear method and unsteady nonlinear method were used to calculate the unsteady flow field. The research direction of domestic scholars on wide-chord hollow fan blade mainly focuses on the structure of wide-chord hollow fan blade. Ma lei [5] discussed the influence of the number and thickness of rib on the structural strength, maximum deformation and vibration of hollow blades. Ji Fusen and others [6] explored the structural design and rapid strength analysis method of coreless hollow fan blade aiming at the application research of wide-chord hollow fan blade technology in the design of large bypass ratio aeroengine. The geometric characteristic parameters of the coreless hollow blade were defined. Also the influence of the number and width of ribs on the strength and stiffness of hollow blade were analyzed emphatically in the research. Wu Xinchen [7] carried out the research on the key manufacturing technology of two-layer TC4 titanium alloy hollow fan blades based on SPF /DB which captured some key manufacturing technologies. Yu Yang and others [8] analyzed the design parameters of hollow blade model cavity and stiffener in the structural design parameters of wide-chord hollow fan. The research results show that the increase of the number of ribs or the increase of the thickness ratio of ribs to skin has little influence on the maximum stress value of skin, but has a great influence on the maximum stress value of rib. Wang Ying and others [9] studied the transient response of wide-chord hollow fan blade under aerodynamic effect by using numerical simulation method. The deformation response under the unsteady boundary also was obtained. Yang Wen and others [10] used numerical simulation method to study the modal characteristics and response characteristics of wide-chord hollow fan blades under impulse excitation. The results show that the deformation of fan blades used in the study is mainly bending deformation during vibration.

To sum up, there is little research on the flutter characteristics of wide-chord hollow blades. Based on a solid blade, a set of wide-chord hollow fan blades with different numbers of ribs are designed. The flutter characteristics of wide-chord hollow blades with different numbers of ribs are calculated by using the energy method. The influence of the number of ribs on the flutter characteristics of hollow blades under the same compressor condition is emphatically studied, which provides some technical support for the flutter prediction of hollow blades in the design stage.

1 Numerical Simulation Method

1.1 Energy Method

The structural dynamics method was used to get the vibration mode, vibration frequency and amplitude distribution of the blade. These rules are incorporated into the numerical simulation of the three-dimensional unsteady viscous flow field. The moving rules of the grid near the blade wall are obtained by assuming the blade simple harmonic vibration. The harmonic balance method is used to calculate the three-dimensional unsteady flow field. It is a frequency domain method for unsteady flow field calculation in compressor internal flow channel. The method not only has good nonlinear simulation ability, but also has some advantages in computational efficiency.

The energy method is a commonly used flutter boundary prediction method. The energy method adopts the viewpoint of energy to establish the criterion of aerodynamic stability. In the case of neglecting mechanical damping, whether the blade flutter occurs or not is judged by the positive or negative energy exchange between the blade and the surrounding gas in a vibration period. The unsteady average accumulate power in a vibration period is used as the judgment quantity in the analysis. The unsteady mean accumulated power is defined as follows [11–13]:

$$P(\vec{x}) = \frac{1}{T}\int_{t0}^{t_0+T} p(\vec{x},t)\vec{V}(\vec{x},t)\cdot\vec{n}(\vec{x},t)\mathrm{dt}\ \forall\vec{x}\in\partial\mathrm{B} \tag{1}$$

In formula (1): T is the vibration period of the blade. t_0 refers to any time after the unsteady flow field is completely established. $p(\vec{\mathbf{x}}, t)$ refers to the static pressure. $\vec{V}(\vec{x}, t)$ refers to the velocity vector. ∂B refers to the surface of the blade. $\vec{n}(\vec{x}, t)$ refers to the unit vector in the direction of the inner normal of the blade surface. The total power of the unsteady force of the fluid on the blade L can be obtained by integrating the unsteady accumulated power along the surface of the blade:

$$L = \iint_{\partial B} P(\vec{\mathbf{x}})dS \tag{2}$$

When the unsteady total power is greater than 0, it is considered that flutter will occur according to the criterion of energy method., the unsteady time-average accumulated power per unit blade surface area Ls is used in order to analyze the blade flutter more conveniently. Ls is defined as below:

$$Ls = L/s \tag{3}$$

The aerodynamic damping coefficient Ξ is used to judge the aeroelastic stability of blades to compare the possibility of flutter of different blades under different conditions. The aerodynamic damping coefficient is defined as follows:

$$\Xi = \frac{-W}{\pi X_0 c(\overline{P_t} - \overline{P})} \tag{4}$$

In formula (4): W refers to the work done by the aerodynamic force in one period. X_0 refers to the amplitude of the blade. c is the chord length of the blade. $\overline{P_t}$ refers to the total pressure of the incoming flow. $\overline{P}$ refers to the static pressure of the incoming flow. If the mechanical damping and other factors can be neglected,, the blade is aeroelastic stable when the aerodynamic damping coefficient is greater than 0.

2 Calculation Model

According to different design requirements, the cavity inside the blade can be designed into different structural forms. With the development of wide chord hollow blades, the first generation of honeycomb hollow blades have been replaced by hollow blades with truss core and centerless structure. Figure 1 shows the hollow blades with different structures. Therefore, it is carried out that the research on aeroelastic stability of wide-chord hollow blade with coreless structure in this paper [14].

The design parameters of wide-chord hollow [15] blades with coreless structure mainly includes: the rib number N, the rib width W, the vane basin thickness of the cavity H1 the blade back thickness H2. A set of wide-chord hollow blade with different rib number are designed accord to the design parameters of wide-chord hollow blades with coreless structure based on that first stage blade of a fan (Fig. 2).

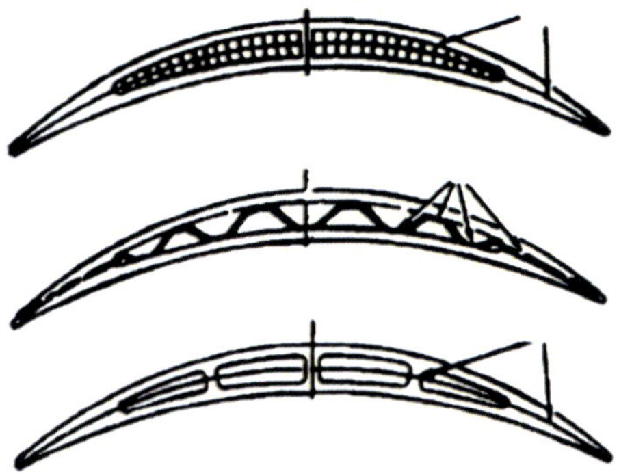

Fig. 1. Schematics diagram of hollow blade with different internal structure

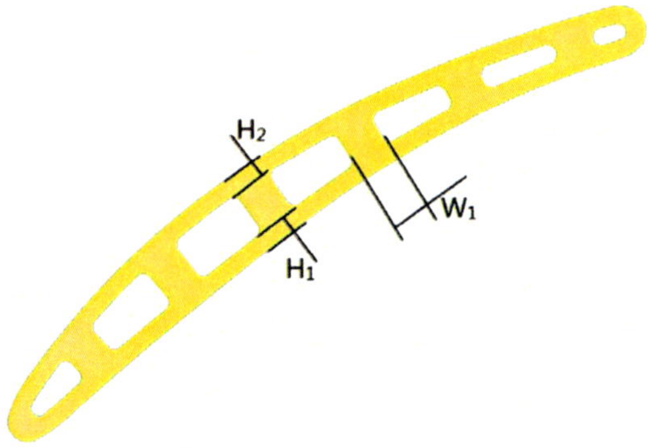

Fig. 2 The parameters of hollow blade with ribs

3 Calculation Results and Discussion

The main conditions of the design point are shown as below. The rotor design speed is 1122.596 rad/s. The inlet total pressure of the rotor is 101325 Pa. The total temperature is 288 K, and the outlet back pressure is 94700 Pa. The blade material is titanium alloy TC4 whose density 4440 kg/m^3, elastic modulus 109 GPA and Poisson's ratio 0.2386.

The first-order natural frequencies of different blades are obtained by structural dynamics calculation. Table 1 shows the first-order vibration frequency of different blades. As can be seen from Table 1, the first-order vibration frequency of hollow blades will increase with the increasing rib number which is mainly because the ribs effectively improve the rigidity of the blades themselves. When the rib number is greater than 2, the first-order vibration frequency of hollow blades is greater than that of solid blades.

Table 1. First frequency of solid blade and wide-chord hollow blade with different number of rib

Blade type	First-order vibration frequency/(Hz)
Solid blade	205.77
N = 0	183.14
N = 1	195.48
N = 2	201.65
N = 3	205.98
N = 4	207.83

Figure 3 shows the distribution of unsteady aerodynamic work on the blade surface of hollow blade and solid blade with different ribs. From Fig. 3, the change of the internal structure of the hollow blade does not change the main distribution area of the unsteady aerodynamic force on the blade surface. Blade, the area of the unsteady dynamic negative power region of the hollow blade surface decreases with the increasing rib number inside the cavity. It is mainly due to the increasing rib number which increases the displacement restriction on the vibration of the hollow blade surface skin, and then reduces the deformation of the hollow blade surface skin.

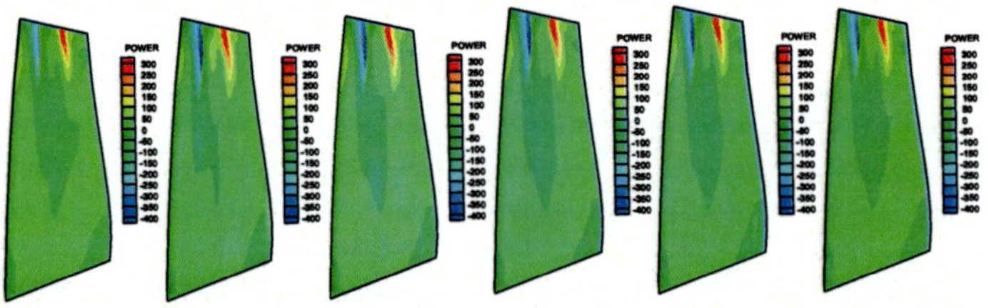

(A) solid blades (b)N=0 (c) N=1 (D) n=2 (e) n=3 (F) n=4

Fig. 3. Unsafe flow power distribution of the different blade

Table 2 shows the unsteady time-averaged accumulated power of per unit blade surface area for different blades at the design point. As can be seen from the table, the work exchange between the air flow and the blade is negative at the design point which means the blade does work on the air flow. All blades will not flutter at this time from the perspective of energy method. When the rib number is more than 1, the amplitude of the work done by the hollow blade to the air flow is larger than that of the solid blade to the air flow. Also the velocity of vibration amplitude attenuation of the hollow blade is faster than that of the solid blade.

Table 2. LS of different kinds of blades at design Point

Blade type	Unsteady time-averaged cumulative power per blade surface area LS (w/m^2)
Solid blade	−0.046341
N = 0	−0.004639
N = 1	−0.040398
N = 2	−0.056988
N = 3	−0.06737
N = 4	−0.068607

Figure 4 shows the characteristic line and Ls variation curve of different kind of blades at design rotation speed. The x-axis is the unsteady average accumulated power, and the y-axis is the compressor pressure ratio in the Ls figure. As can be seen from Fig. 4, when the pressure ratio reaches the maximum, the unsteady average accumulated power on all blade surfaces is greater than 0 which means the air flow does work on the blade. The vibration of the blade will not gradually weaken with the accumulation of time. Also the possibility of flutter occurring is the greatest. With the decrease of compressor pressure ratio, the unsteady average accumulated power Ls on the blade surface changes from positive to negative. The blade begins to do work on the flow during these pressure ratios. According to the criterion of energy method, it is considered that the blade is at the critical point of flutter and non-flutter when the unsteady average accumulated power equals to 0. It can be seen from Fig. 4a–f that the compressor pressure ratio corresponding to the flutter critical point of different blade is different. The air flow does positive work on the blade in a large range of compressor ratio in the Ls diagram when then rib number is zero as shown from Fig. 4b. The blade is in flutter state, which is mainly due to the lack of supporting effect of ribs on the blade skin, and the overall rigidity of the blade is greatly reduced. It can be seen from Fig. 4c–f that when the ribs in the cavity are supporting the hollow blade, the compressor pressure ratio corresponding to the flutter critical point of the hollow blade is effectively improved. The hollow blade can ensure the compressor in normal operation under most working state. The compressor pressure ratio corresponding to the flutter critical point of the hollow blade becomes larger with the increasing rib number of the hollow blade which means that the compressor can stably work under a wider working condition range. It can be seen that when the rib number is 1 and 2, the compressor pressure ratio corresponding to the flutter critical point of the hollow blades is smaller than that of the solid blades which does not show the corresponding advantages of hollow blades compared with solid blades. When the rib number is larger than 2, the pressure ratio corresponding to the flutter critical point of the hollow blades is close to or greater than that of the solid blade. The hollow blades have stronger adaptability to severe working conditions than solid blades.

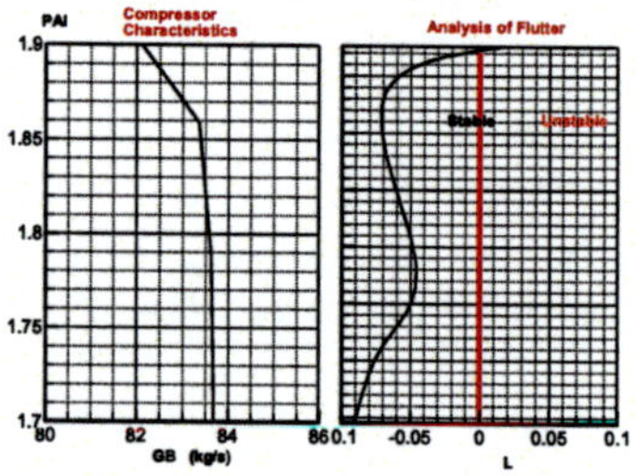

(a) solid blades

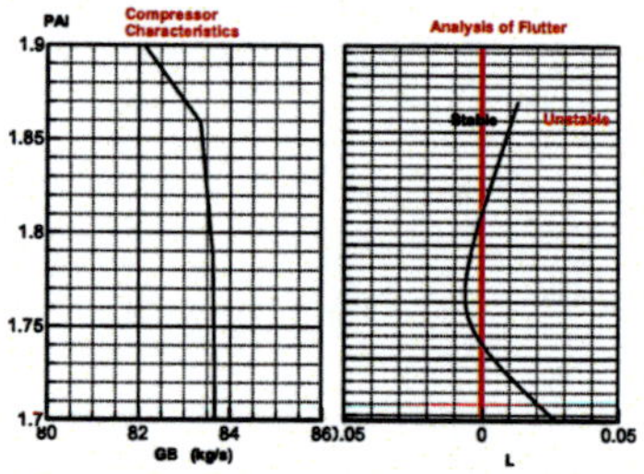

(b)N=0

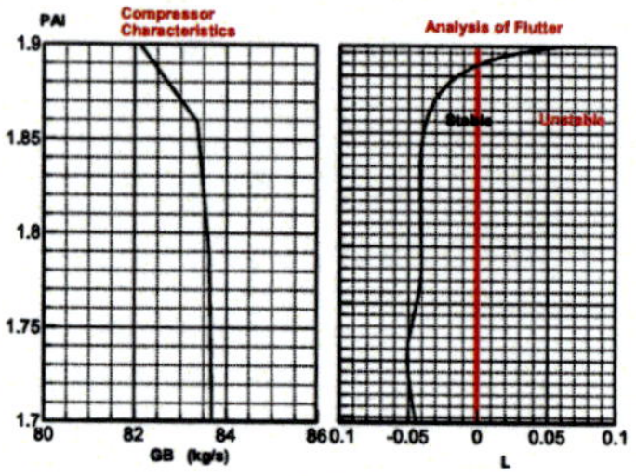

(c)N=1

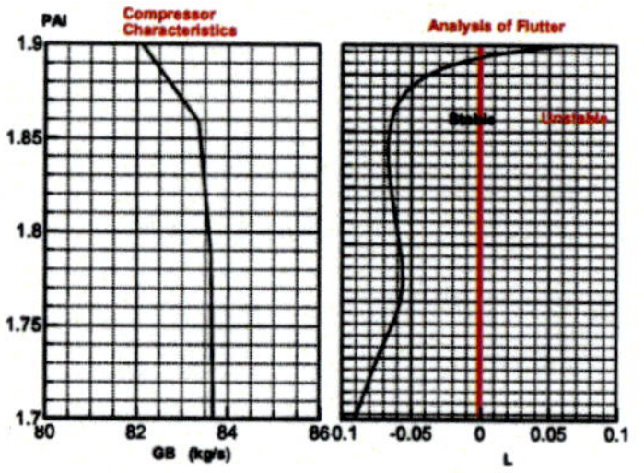

(d)N=2

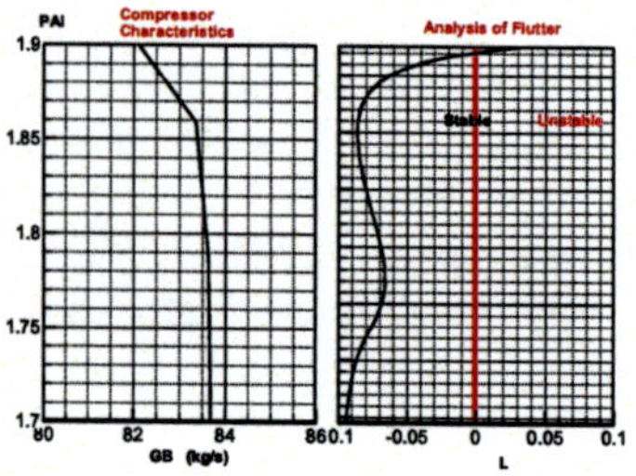

(e)N=3

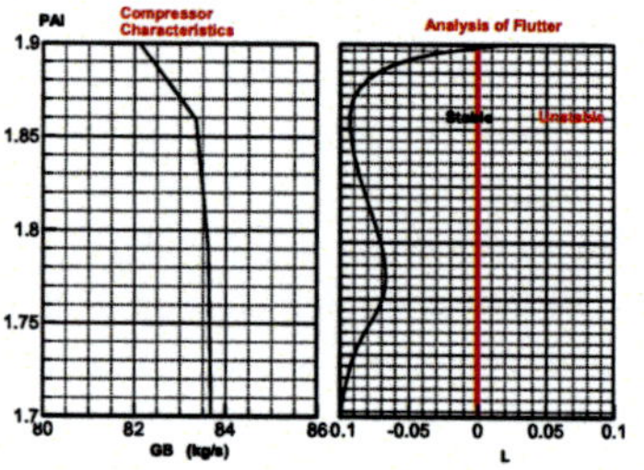

(f)N=4

Fig. 4. The Ls Curve And Characteristic Curve of Different Blade at 100% rev

Figure 5 shows the aerodynamic damping coefficient curve of different blades at different phase angles. It has not been considered that the effect of the interblade phase angle when analyzing the influence of rib number on the aeroelastic stability. The interblade phase angle reflects the interaction between oscillating blades which has a significant effect on the aeroelastic stability of turbomachinery. As can be seen from Fig. 5, the aerodynamic damping coefficients of different blades vary in a simple and harmonic manner with the change of phase angle. The aerodynamic damping coefficients reach a minimum value when the phase angle is 90°. The solid blades and hollow blades without ribs are in aeroelastic instability stage in a large angular phase angle range. The aerodynamic damping coefficient of the hollow blade with N = 4 is larger than that of the solid blade and the hollow blade with N = 0 at all phase angles which is mainly due to the fact that the rib effectively improves the aerodynamic stability of the hollow blade and reduces the phase angle range of the blade in the aeroelastic instability state. The aerodynamic stability of hollow blades with N = 4 is better than that of solid blades.

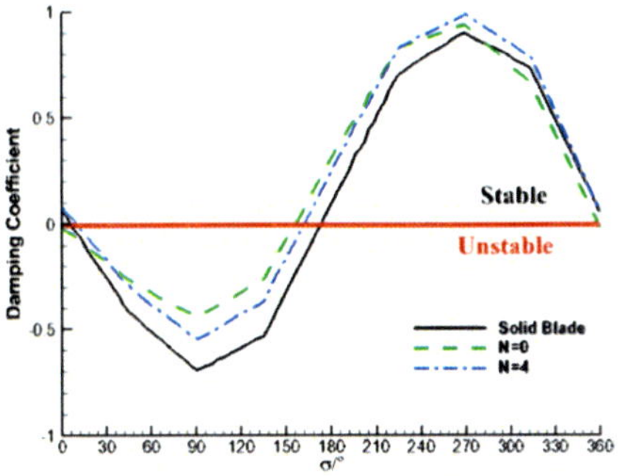

Fig. 5. The aerodynamic damping coefficient curve of different blade

4 Conclusion

In this paper, the number of stiffeners in the structural parameters of hollow wide chord blades is taken as the starting point, the unsteady flow field of hollow blades with different numbers of stiffeners is simulated numerically, and the flutter is predicted.

1. Although the weight of the blade is effectively reduced for the hollow blade without rib, the pressure ratio corresponding to the flutter critical point is far less than that corresponding to the solid blade. The hollow blade has no practicality.
2. The hollow blade with ribs improves the compressor pressure ratio corresponding to the flutter critical point, and effectively expands the normal working range of the blade compared with the hollow blade without ribs. Also the compressor pressure ratio corresponding to the flutter critical point of hollow blade increases with the increasing rib number.
3. The compressor pressure ratio corresponding to the flutter critical point of the hollow blade is greater than that corresponding to the solid blade at the design point when rib number is 4. The hollow blade has better anti-flutter performance than the solid blade.
4. The hollow blade with N = 4 is in the aerodynamic stability range in a larger phase angle range than the solid blade and the hollow blade without ribs at the design point.

Acknowledgements. This study was co-supported by the National Natural Science Foundation of China (72001192, 51975539), the Natural Science Foundation of Henan Province (202300410490), the Key Scientific and Technological Program of Henan Province (202102210237, 222102210050), the Henan Youth Talent Support Project (2021HYTP017) and the Aeronautical Science Foundation of China (2018ZD55008). The authors also wish to thank them for their financial support.

References

1. Guanqun, H., Bosheng, S.: Development of wide chord fan blade. Int. Aviat. (12), 45–47 (2002)
2. Chunhua, L.: Overview of advanced fan and compressor blade/vane in high performance aeroengine. Aeroengine **32**(3), 48–52 (2006)
3. Vahdati, M., Simpson, G., Imregun, M.: Mechanisms for wide-chord fan blade flutter. In: ASME Turbo Expo 2009: Power for Land, Sea, and Air, pp. 585–593.(2009)
4. Chew, J.W. Hamby, et al.: Part speed flutter of transonic fans. RTO Meeting Proceedings (1999)
5. Lei, M.: Xi'an: research on optimization design method of hollow blade structure. Northwestern Polytechnical University (2011)
6. Fusen, J., Huilian, D.Q.L.: Design and analysis of hollow fan blade structure. Aeroengine **39**(4), 42–44+66 (2013)
7. Xinchen, W.: Research on superplastic forming and diffusion bonding technology of hollow wide-chord fan blade. Nanjing University of Aeronautics and Astronautics, Nanjing (2015)
8. Yang, Y., Li, Z., Peng, Z., et al.: Analysis of design parameters of wide chord hollow fan blade structure. Aeroengine **36**(1), 24–26 (2010)
9. Ying, W., Zhi, T., Farong, D., et al.: Response analysis of fluid and solid coupling characteristics for a wide-chord hollow fan blade. J. Aerosp. Power **23**(12), 2177–2183 (2018)
10. Wen, Y., Farong, D., Yong, H., et al.: Investigation of dynamic response property of wide-chord hollow fan blade. J. Aerosp. Power **22**(3), 444–449 (2007)
11. Yongqiang, S., Qingzhen, Y., Xinhai, Z.: Influence of blade thickness on flutter characteristics in fan/compressor. Acta Aeronaut. Astronaut. Sin. **30**(6), 979–984 (2009)
12. Xiuquan, H., Xinhai, Z.: Unsteady flow analysis in oscillating cascade using Fourier-transform based method. J. Northwest. Polytech. Univ. **26**(3), 357–361 (2008)
13. Yongqiang, S.: Research on the Correlation between 3D Blade Flutter and Blade Design. Northwestern Polytechnical University, Xi'an (2006)
14. Fitzgerald, G.A., Broughton, T.: The Rolls-Royce wide chord fan blade. In: The 1st International Conference of the Titanium Development Association, San Francisco, CA, October (1986)
15. Fusen, J., Quan, D., Huilian, L.: Design and analysis of hollow wide-chord fan blade with two-layers and three-layers plate. Aeroengine **39**(2), 49–51 (2013)

Gas Ingestion Under Unsteady Rotor-Stator Interaction of Turbine Blade

Tao Bai[1,2], Qing Zhen Yang[1](✉), Jian Liu[3], and Miaoer Wang[2]

[1] School of Power and Energy, Northwestern Polytechnical University, Xi'an 710072, China
921036859@qq.com
[2] School of Aircraft, Xi'an Aeronautical, Xi'an, China
[3] China Aerodynamics Research and Development Center, Mianyang 621000, China

Abstract. Unsteady simulation of a 1.5 stage turbine with an axial cavity was conducted to illustrate the ingress mechanism under the blade-rotor unsteady interaction. The numerical approach was fist be validated by the experiment data. The pressure potential caused by the rotor blade plays a major role in externally induced ingress. The ingress area in the circumstance direction is rotated with the rotor blade, and the most serious ingestion would happen as the rotor leading edge and stator wake are close to each other. It is worth noting that a vortex would form in the situation that the ingress gas and sealing flow were both exit in the rim seal clearance. The rim seal vortex rotates from rotor wheel space to the stator side, as the result makes the ingress deeper near the rotor wheel space in rim clearance, which can be proved from the lower sealing efficiency near the rotor wheel space in the rim gap. Moreover, the rim seal vortex is mainly caused by shear in the gap. In addition to the tangential velocity gradient, the radial gradient also promotes the formation of rim seal vortex.

Keywords: Gas ingestion · Rotor-stator interaction · Seal efficiency · Rim seal vortex

1 Introduction

The operating life of a turbine can be shortened owing to hot gas ingestion into cavities. High-pressure air, known as purge flow (sealing flow), from a compressor, is introduced to cool the metal configuration as well as to suppress the ingress of cavities. In previous studies, the research by Bayley and Owen exhibited that gas ingress is mainly affected by the seal structure and the rotating Reynolds number, without considering the annulus flow effect [1, 2]. For a considerable time, researchers have considered that gas ingress is caused by the effects of rotationally-induced ingestion (RI) and externally-induced ingestion (EI) and the latter plays a major role in gas ingesting [3, 4]. Due to the interaction between rotor and stator, the non-axisymmetric pressure field is mainly generated. Ingress occurs where the pressure in the annulus is higher than that in the cavity; in the opposite situation egress would be happen.

Many studies have further focused on the contribution of the rotor and stator to the circumferential non-uniform pressure of the annulus channel. Green and Turner et al. [5]

Chinese Society of Aeronautics and Astronautics: CASTYSF 2022, LNEE 972, pp. 37–46, 2023.
https://doi.org/10.1007/978-981-19-7652-0_5

believe that the existence of rotor blade weakens the non-uniform pressure field caused by the wake of vane blades; therefore, the sealing efficiency is improved. However, hills et al. [6] pointed out that the rotating high-pressure area caused by the leading edge of the rotor blade would increase the degree of gas intrusion. Chew et al. [7] studied the relationship between gas ingestion and axial position of vane blade separately, without rotor blade. The ingress was weakened with the increase of the distance between the guide and the wheel space. Bohn et al. [8, 9] reached a similar conclusion on the influence of vane blade position on gas ingestion; Further research showed that both the rotor blade and guide vane have a very important impact on gas ingestion, although the degree of impact is related to the sealing structure. Hualca and Horwood et al. [10] studied the effect of guide vane position and rotor blade on gas ingestion. The result showed that compared with the rotating blade, the axial distance had a significant effect on the range between maximum and minimum pressure which would be decreased with the increasing distance from the trailing edge of the blade. However, in the absence of rotor blades, the guide vane position has little effect on gas ingestion.

As the rotating blade, the interactions between the stator wake and the pressure potential field at the leading edge of the rotor are unsteady. It is still of practical significance to explore the gas ingestion mechanism and sealing degree, in this actual unsteady environment.

2 Computational Models and Numerical Approach

2.1 Geometries and Models

The section numerically invested include a 1.5 stage turbine with a vane-blade-vane configuration was show in Fig. 1. For more detailed instructions referred to [11]. The number of blades in the numerical simulation was molded as 2:3:2 since the number of blades in the whole annulus was 36, 54, 36 respectively. Therefore, 1/18 sector was adopted in the unsteady calculation.

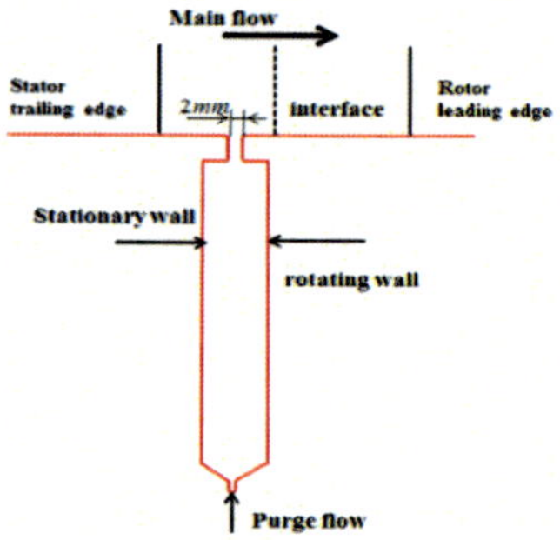

Fig. 1. Cross section of calculated model

2.2 Numerical Approach and Boundary Condition

Numerical simulations were performed in ANSYS CFX software. Reynolds-averaged Navier–Stokes equations for fully three-dimensional (3D), unsteady flow were solved. Structured grid generation was undertaken using NUMECA AutoGrid5 and ICEM as displayed in Fig. 2. For minimizing the interpolation error, a matching grid interface was defined between the stator and cavity domains. The total number of nodes was 8.0 million.The circumferential grid resolution was set as 0.34°/cell, Grids near the wall were characterized by a boundary layer grid with the first cell height of approximately $y^+ < 1$.

A total pressure (140 kPa) and total temperature (328 K) boundary condition was used at the main path inlet, whereas an average static pressure condition was set at the main path outlet. The cooling air mass flow (purge flow) and total temperature were applied at the cavity inlet. The cooling air mass flow rate (IR = 0.5%) and the rotational periodicity in the circumferential direction were assumed at appropriate boundary. The unsteady solution was obtained at 42 time steps per blade passing event. The CO2 tracer gas was employed, which was be set to o in the mainstream inlet, while 1 in the cavity inlet.

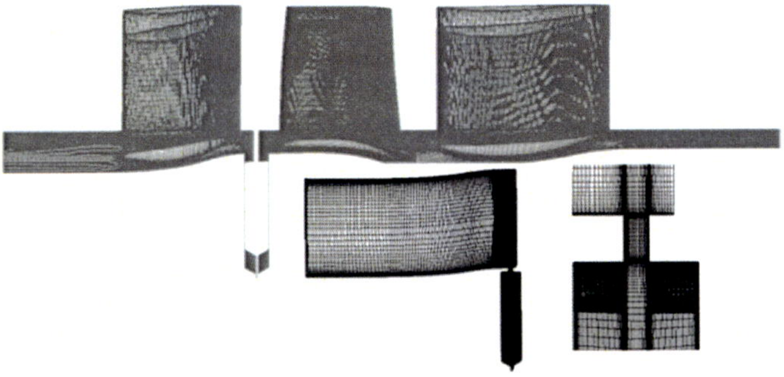

Fig. 2. Computation mesh of 1.5-stage turbine with seal cavities

2.3 Validation of Numerical Model

To verify the reliability of the numerical approach, the relative total pressure coefficient and flow angle at rotor exit were choose to compare with the experimental data, as show in Figs. 3 and 4. The relative total pressure coefficient was defined as follows.

$$c_{pt} = (p_t - p_{s,exit})/(p_{t,inlet} - p_{s,exit}) \quad (1)$$

The figure from flow angle indicates that the experimental and numerical predictions agree well in a wide range of span.Furthermore compared with the experimental results, the numerical simulation accurately captures the high loss area of the rotor exit. Therefore, it can be considered that the numerical simulation method in this paper satisfy the calculation requirements.

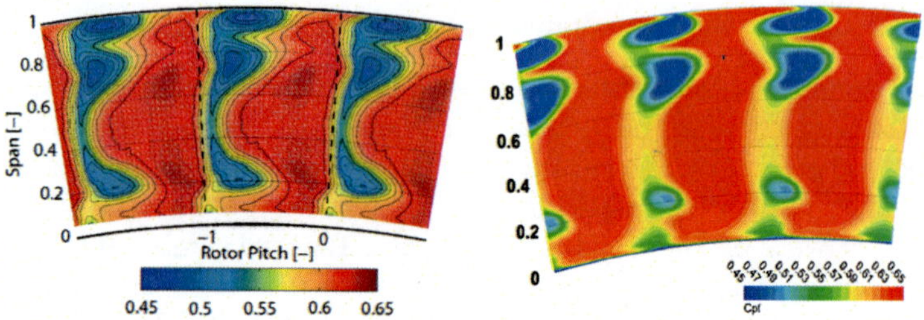

Fig. 3. Relative total pressure coefficient at rotor exit

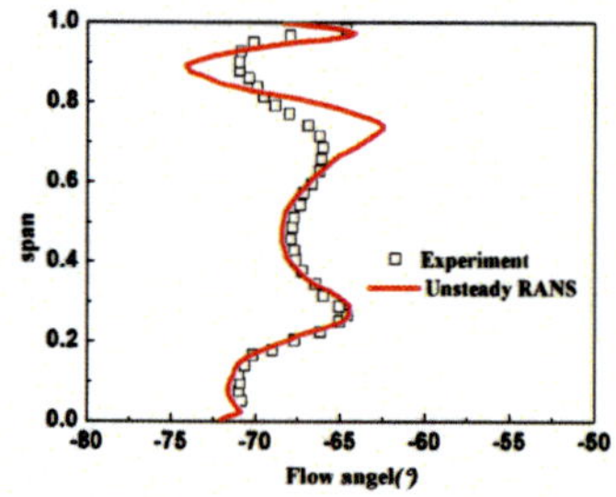

Fig. 4. Flow angle at rotor exit

3 Research and Discuss

TO study the influence of the relative position of rotor and stator on gas ingestion, the time of rotor seeped per stator sector was divided into seven equal parts and the unsteady analysis was carried out at four times: 0T, 2 / 7T, 4 / 7T, and 7T. Furthermore, the two circumferential surfaces close to the stationary wheel space and the rotor disc in the cavity seal are defined as S_Cavity, R_Cavity. The CO2 fraction contour on above surfaces and cavity exit was shown in Fig. 5.

At time t = 0T, the circumferential position of gas ingestion is 1, 2 and 3 areas as shown in the black wireframe in Fig. 5a, of which area 1 and 3 are located near the leading edge of the rotor, and area 2 is located at the intersection of the leading edge of the rotor and the trailing edge of the stator. At 2/7T, the gas ingestion area rotates along the circumferential position. Areas 1 and 2 are near the leading edge of the rotor, and area 3 is located at the junction area of rotor leading edge and stator trailing edge, while, the circumferential ingestion degree is deepened compared to 0T moment. The same conclusion can be obtained from 4/7T and 6/7T. It can be summarized as that the circumferential position of gas ingestion is mainly affected by the pressure potential field of the leading edge of the rotor, while the stator wake will increase the depth of gas ingestion when it is close to the leading edge of the rotor. It also can be see that the intrusion area rotates with the rotor blade, Meanwhile, as the rotating ingress area matched the rotating blade, therefore the pressure potential field at the leading edge of the rotor is the dominant factor of EI (externally-induced ingestion). The Circumferential location of gas intrusion near the stator and rotor is almost consistent however the ingested area neighboring rotor wheel space is larger than the stator side, which can be seen from Fig. 5b.

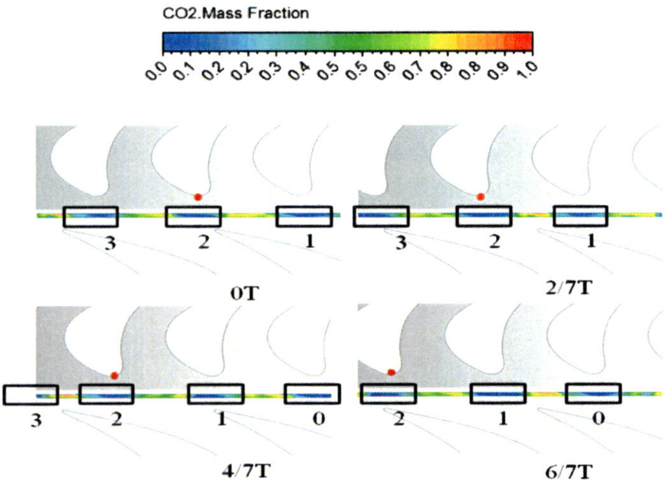

(a) CO2 concentration contour at the outlet of cavity

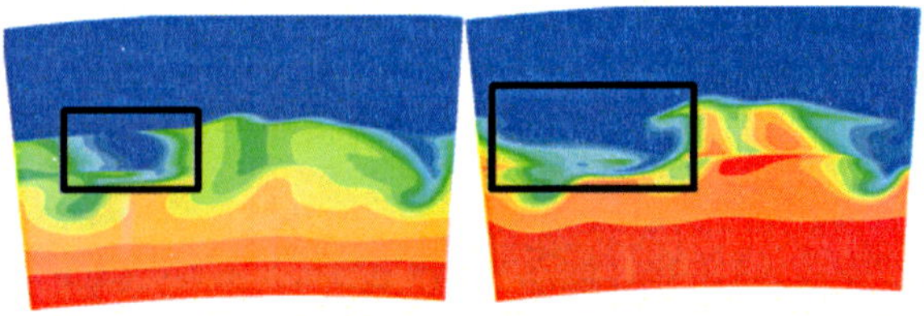

(b) CO2 concentration contour(left: S_Cavity;right:R_Cavity)

Fig. 5. **a** CO_2 concentration contour at the outlet of cavity. **b** CO_2 concentration contour (left: S_Cavity; right: R_Cavity)

The stator channel was divided into 5 equal sections along the circumferential direction, and 5 dimensionless circumferential angles: $\theta = 0.1, 0.3, 0.5, 0.7$, and 0.9 sections were taken into analysis. The CO2 fraction contour at each location was displayed in Fig. 6. The circumferential position and radial depth of gas ingestion change with the relative position change of rotor and stator. The gas ingest into the whole rim clearance at the location of $\theta = 0.3$(0T), $\theta = 0.5$(2/7T), $\theta = 0.7$ (4/7T), $\theta = 0.1$ (4/7T). Combined with Fig. 5a, the above areas mentioned are located near or at the intersection of the leading edge of the rotor and the trailing edge of the stator. The high pressure induced by the rotor leading edge and wake make the pressure in the main annulus is higher than that in the cavity, as the result the gas-filled the rim clearance. While the Circumferential position sited only adjacent to the leading edge, the ingress was mainly happened at rotor wheel space, such as $\theta = 0.9$(0T), $\theta = 0.1, 0.9$(2/7T), $\theta = 0.3$(4/7T), $\theta = 0.1, 0.3$(6/7T). In other positions, affected by the effect of rotor pumping, the purge flow was flowed out through rotor wheel space. In Conclusion: when the leading edge of the rotor approaches, the circumferential pressure of the main flow dominates ingress, and the gas ingested from the rotor side. When the leading edge of the rotor is far away, the pump effect is dominating, the sealing flow egress form rotor side. Rotor blade has significantly affected both gas ingress and sealing flow egress.

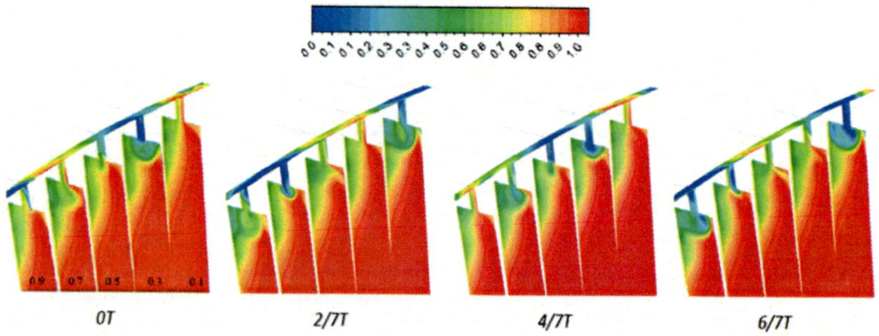

Fig. 6. CO_2 fraction contour at different time

At different times, there may be gas, sealing flow, or their mixture in the rim gap. To analyze the flow details in the cavity gap and the influence of its flow characteristics on gas ingestion, the streamline and axial vorticity at three circumferential positions (θ = 0.1, 0.5, 0.9) on four moments were taken into consideration, displayed in Fig. 7. The gap vortex structure is not likely formed when the gap is dominated by gas ingress or purge flow egress such as the positon $\theta = 0.1$ (egress) on 0T or 6/ 7T (ingress). The same conclusion is reached in another similar case. However, when the annulus gas and purged flow are both exist in the gap, a vortex structure occurred. It can be considered that the high-temperature gas ingested through rotor wheel space meets the sealing flow purged out through the stator side shape of the vortex structure which can be found through the positive axial vorticity value.

The above changes can be revealed as follows: when the gas ingestion occurs, the corresponding mainstream pressure on the rotating side is greater which could suppresses the rotor pump effect and makes the gas-driven the gas ingested from the rotor side. In another world, the pressure potential field generated by the rotation of the leading edge of the rotor blade is the dominant factor causing the gas ingestion.

Moreover, the gap(rim seal clearance) vortex should be caused by radial or tangential velocity shear in the gap, which is considered be K-H instability by much research. However, its unsteady characteristics are worth studying, because the causes include radial velocity gradient, while the K-H instability was thought to have been induced by tangential velocity gradient. It is certain the vortex structure certainly made the ingress deeper along the rotor side.

Figures 8 and 9 shows the sealing efficiency, radial velocity coefficient, and tangential velocity coefficient in the turbine cavity, the axial location is defined as: cavity Stator, Seal middle and cavity Rotor. The velocity coefficient is dimensionless with rotor hub rotation speed. In the middle section of the cavity, the radial velocity is small and remains

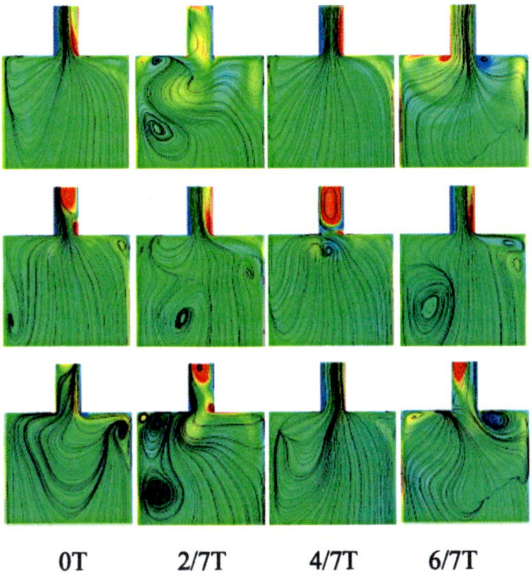

Fig. 7. **a** Streamline and axial vorticity ($\theta = 0.1, 0.5, 0.9$ from top to bottom)

unchanged when r/R is less than 0.8, which is coincident with the inviscid flow theory in the vortex core region.

At the high radius close to the rim, the radial velocity coefficient is greatly difference between the rotor wall and the stator wall due to the influence of gas ingestion and vortex structure in the gap. It can be concluded from the above analysis that the gas ingestion is affected by the unsteady pressure wave of the rotating blade, and most of the high-temperature gas enters the cavity along the rotor side. Therefore, the time average radial velocity coefficient on the rotor side decreases with the increase of radius, when r/R is greater than 0.8. Due to the above ingress mechanism, the sealing efficiency near the rotor wall is lower than that near the stator wall in the gap, although the sealing efficiency is higher near the stator wall for the most regions in the inner cavity.

As for tangential velocity: in most areas of the disk cavity, the tangential velocity of the stationary wall and the rotating wall is mainly affected by the wall friction, that is, the tangential velocity close to the stator wall is almost 0, for the same reason the tangential velocity near the rotor wall increase with the radius. In the high radius and gap of the cavity, the dominant factors affecting the tangential distribution include gas ingestion, sealing flow egress, and their interaction. The tangential velocity is increasing rapidly as a consequence of the ingested gas with higher circumferential velocity caused by the guide vane upstream, What's more, exaggerated is the tangential speed at the rotating wall will exceed the rotation speed of rotor hub. Unlike the rotor, the tangential velocity is still relatively small even at the cavity outlet. What needs more attention is that fluid near the stator wall flows out of the cavity into the mainstream, which could cause the K-H inability for the tangential velocity gradient of mainstream and purged flow. The vortex structure is located at the rim, as shown in the continuous blue surface vortex mechanism in Fig. 10.

In addition to the rim, the shear vortex structure induced by gas intrusion in the gap can also be seen from the figure, as shown in the elliptical wireframe in the figure. The vortex is not only related to the tangential velocity in the gap, but also related to the radial velocity gradient.

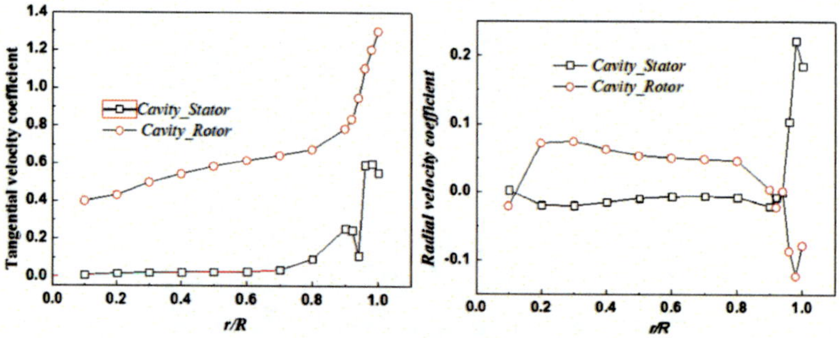

Fig. 8. Tangential velocity coefficient and Radial velocity coefficient in cavity

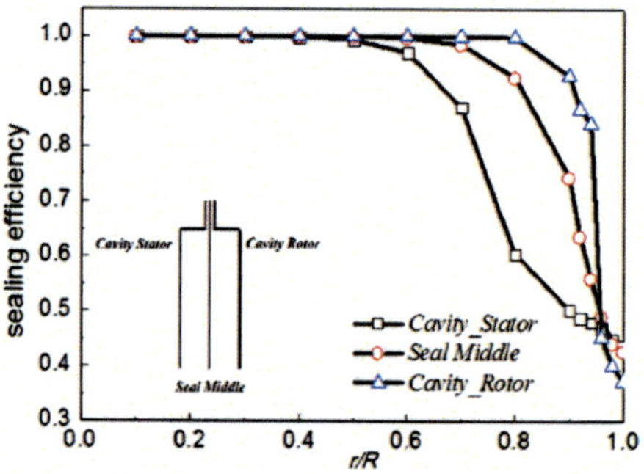

Fig. 9. Sealing efficiency in cavity

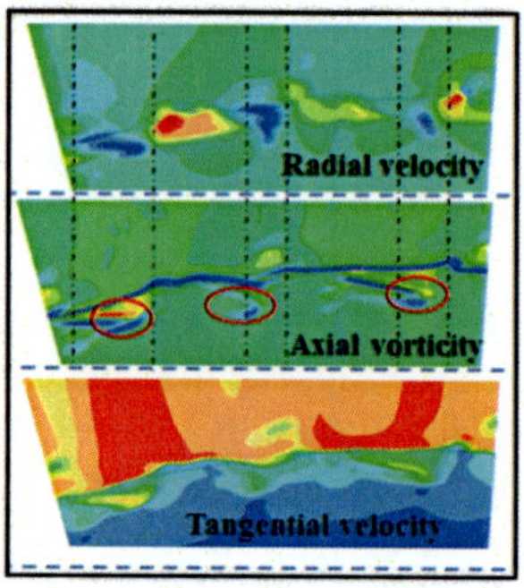

Fig. 10. Vortex in cavity wheel space

4 Conclusions

In order to carefully analyze the gas ingestion mechanism and the generation mechanism of rim seal vortex under rotor-stator interaction, an unsteady numerical calculation of a 1.5-stage turbine is carried out, and the following conclusions are obtained.

(1) The pressure potential field at the leading edge of the rotor dominates the gas ingestion and makes the ingested area rotates in the circumferential direction with the rotation of the rotor blade. The degree of gas intrusion near the leading edge of the rotor is aggravated when the leading edge of the rotor is close to the stator wake. Affected by the pump effect of the rotor, the purged flow egress still occurs near the rotor side when it is far from the leading edge of the rotor. That is to say, rotor is the key factor for both ingress and egress.
(2) Vortex in rim seal will form as the interaction of the radially inward ingestion flow and radially outward sealing flow. Which make the gas intrusion depth near the rotor wheel space, and the sealing efficiency in the cavity gap is reduced and lower than that of the stationary wall.
(3) Because the large radial velocity gradient in the gap promotes the formation of the gap vortex, it is worth discussing whether the gap vortex is the same as the K-H vortex proposed in the previous literature.

References

1. Phadke, U., Owen, J.: An investigation of ingress for an "air-cooled" shrouded rotating disk system with radial-clearance seals. J. Eng. Power **105**, 178–183 (1983)
2. Phadke, U., Owen, J.: Aerodynamic aspects of the sealing of gas-turbine rotor-stator systems: Part 2: The performance of simple seals in a quasi-axisymmetric external flow. Int. J. Heat Fluid Flow **9**(2), 106–112 (1988)
3. Bohn, D.E., Decker, A., Ohlendorf, N., Jakoby, R.: Influence of an axial and radial rim seal geometry on hot gas ingestion into the upstream cavity of a 1.5-stage turbine. ASME Paper No. GT2006-90453 (2006)
4. Patinios, M., Scobie, J.A., Sangan, C.M., et al.: Measurement and modeling of ingress in a new 1.5-stage turbine research facility. ASME J. Eng. Gas Turb. Power **139**(1), paper No. 012603 (2017)
5. Green, T., Turner, A.B.: Ingestion into the upstream wheelspace of an axial turbine stage. J. Turbomach. **116**(2), 327–332 (1994). https://doi.org/10.1115/1.2928368
6. Hills, N.J., Chew, J.W., Turner, A.B.: Computational and mathematical Modeling of turbine rim seal ingestion. ASME J. Turbomach. **124**(2), 306–315 (2002)
7. Chew, J.W., Green, T., Turner, A.B.: Rim sealing of rotor-stator wheel spaces in the presence of external flow. ASME Paper No. 94-GT-126 (1994)
8. Bohn, D.E., Rudzinski, B., S€urken, N., G€artner, W.: Experimental and numerical investigation of the influence of rotor blades on hot gas ingestion into the upstream cavity of an axial turbine stage. ASME Paper No. 000-GT-0284 (2000)
9. Bohn, D.E., Decker, A., Ma, H., Wolff, M.: Influence of sealing air mass flow on the velocity distribution in and inside the rim seal of the upstream cavity of a 1.5-stage turbine. ASME Paper No. GT2003-38459 (2003)

10. Hualca, F.P., Horwood, J.T.M., Sangan, C.M., et al.: The effect of vanes and blades on ingress in gas turbines. J. Eng. Gas Turb. Power **142**(021020-1), 12 (2020)
11. Behr, T.: Control of rotor tip leakage and secondary flow by casing air injection in unshrouded axial turbines

Hypersonic Vehicle Control Based on Deep Reinforcement Learning

Xianlong Ma, Weijun Hu(✉), and Zhiqiang Gao

Northwestern Polytechnical University, Xi'an 710072, China
maxianlong@nwpu.edu.cn, huweijun2021@126.com, 1256113589@qq.com

Abstract. The hypersonic vehicle is an important combat force in the future battlefield. At present, traditional guidance and control need to be designed prior to achieving the operational mission, lacking the ability of independent decision-making and rapid response, and not able to adapt to the development needs of complex battlefield situations in the future. With the rapid development of artificial intelligence, the decision-making ability of deep reinforcement learning has been applied in many aspects. In this paper, for the decision-making problem of autonomous flight maneuvering control of hypersonic aircraft, the deep deterministic policy gradient algorithm is used to design an autonomous flight maneuvering control decision-making algorithm to achieve the trajectory planning for the vertical climb and cruise tasks of the aircraft. Through the simulation test, the autonomous flight from the random initial position to the target position is realized, which proves that the training results have certain generalization. In the end stage, the longitudinal climbing section is extended to three-dimensional space, and the training simulation is carried out, showing the feasibility of the algorithm in the actual situation.

Keywords: Hypersonic vehicle · DDPG · Longitudinal climb · Cruise control

1 Introduction

Hypersonic flight Vehicle (HFV) has fast flight speed, high maneuverability and strong defense penetration ability. In the military field, it is an important weapon and equipment for accomplishing future space operations and global rapid strikes. With the gradual maturity of high-thrust rockets, high-temperature-resistant special materials and scrambling engines, the development of hypersonic vehicles has entered a new stage and has higher requirements for guidance and control systems [1].

There are currently a number of modern control methods for hypersonic vehicle control. Literature [2] used an inversion control structure to solve the trajectory optimal tracking control problem. Literature [3] proposed the use of nonlinear dynamic inverse control method to accomplish the decoupling of velocity and altitude channels, and to achieve precise tracking of altitude and velocity commands. Literature [4] used the idea of trajectory linearization to complete hypersonic nonlinear attitude tracking control. Literature [5] used a robust control method based on signal compensation to complete

Chinese Society of Aeronautics and Astronautics: CASTYSF 2022, LNEE 972, pp. 47–59, 2023.
https://doi.org/10.1007/978-981-19-7652-0_6

the longitudinal control of the aircraft. Literature [6] proposed the backstep sliding mode control of hypersonic vehicle based on cerebellar neural network to achieve longitudinal height and velocity control. Literature [7] proposed a new type of iterative learning control based on sliding mode control to achieve aircraft attitude control. Literature [8] used a middle-aged high-end sliding mode to design a limited-time controller to complete the aircraft attitude control problem. The current modern control methods cannot fully meet the high control accuracy and strong robustness requirements of the control system, so the intelligence and autonomy of hypersonic vehicles are the necessary trends in the development of guidance and control technology [9].

In this paper, for the autonomous flight of hypersonic vehicle, deep reinforcement learning algorithm is used to study its control decision-making problem. In the first place, the framework of the DDPG algorithm and its algorithm training process are established, and the design of the state space that meets the task requirements is selected. On this basis, different reward functions are established for different task learning problems to solve the sparse reward problem of deep reinforcement learning. Verification of the feasibility of deep reinforcement learning in the control problem of aircraft through the simulation test of the agent.

2 Problem Description

2.1 Hypersonic Vehicle Modeling

The atmospheric model for hypersonic vehicle modeling adopts the USSA76 standard atmospheric model developed by the United States in 1976 [10].

The motion equations of the aircraft are divided into the center of mass motion equations and the motion equations rotating around the center of mass. In this paper, the instantaneous equilibrium assumption is adopted, the aircraft is regarded as a mass point, and ignoring the attitude changes of the aircraft. Assuming that the aircraft maneuvers at a fixed speed in the flat, the simplified motion model is shown as below:

$$\begin{cases} \dfrac{d\theta}{dt} = \dfrac{g}{V}(n_y - \cos\theta) \\ \dfrac{dx}{dt} = V\cos\theta \\ \dfrac{dy}{dt} = V\sin\theta \end{cases} \tag{1}$$

Considering that the aircraft maneuvers in three-dimensional space in practice, the training of the longitudinal plane climb segment model is extended to three-dimensional space, and the trajectory declination, lateral motion coordinates and normal overload will be considered in the aircraft motion model. It is still assumed that the aircraft speed is fixed, the simplified motion model of the aircraft displays as following:

$$\begin{cases} \dfrac{d\theta}{dt} = \dfrac{g}{V}\left(n_y - \cos\theta\right) \\ \dfrac{d\psi_v}{dt} = -\dfrac{g}{V\cos\theta}n_z \\ \dfrac{dx}{dt} = V\cos\theta\cos\psi_v \\ \dfrac{dy}{dt} = V\sin\theta \\ \dfrac{dz}{dt} = -V\cos\theta\sin\psi_v \end{cases} \tag{2}$$

2.2 Reinforcement Learning Method

Reinforcement learning is a branch of machine learning [11]. Figure 1 explains the basic principle of reinforcement learning. Firstly, the agent interacts with the surrounding environment through an action to obtain a reward and a new state, and repeat until the end state. The interaction continues a lot of data is generated, and the reinforcement learning algorithm uses this data to improve the action strategy. Then it uses the new strategy to interact with the environment, generates new data, and uses the new data to optimize the action strategy. After several iterations, the agent can learn to get the maximum optimal strategy for return.

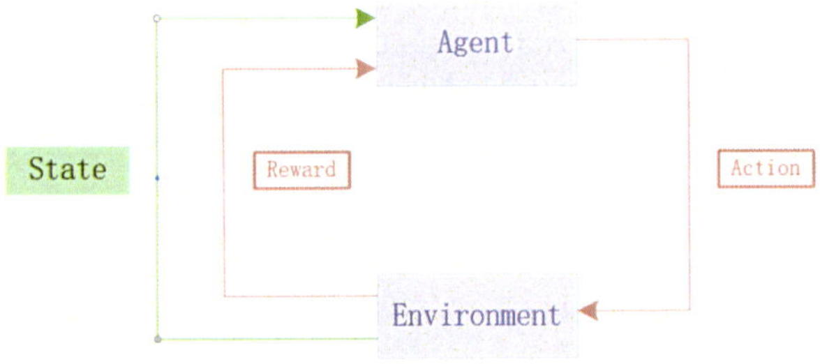

Fig. 1. Reinforcement learning basic framework

Deep Deterministic Policy Gradient (DDPG) [12] is a model-free policy, and Actor-Critic-based policy search method that can be used to solve continuous action space problems. The avoidance strategy means that the action strategy for generating data is not the same strategy as the evaluation and improvement strategy. The action strategy is a random strategy, and noise is added to the output of the strategy network to ensure sufficient exploration; the evaluation strategy is a deterministic strategy. DDPG integrates the successful experience of Deep Q-learning (DQN), that is, experience replay and setting up a separate network, which solves the correlation between data and the problem that the A-C algorithm is difficult to converge. Figure 2 shows the process framework of the DDPG algorithm.

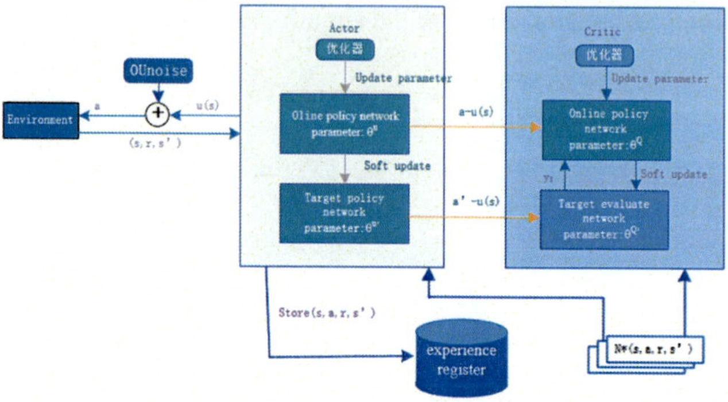

Fig. 2. DDPG algorithm framework

DDPG consists of Online policy network $\mu(s; \theta^{\mu})$, target policy network $\mu'(s; \theta^{\mu'})$, online evaluation network $Q(s, a; \theta^{Q})$, target evaluation network $Q'(s, a; \theta^{Q'})$ and the experience register $R = \{s, a, s', r\}$ consists of five parts. Where s is the current system state, a is the action, s' is the state of the system at the next moment, r is the current return, and θ^{μ}, $\theta^{\mu'}$, θ^{Q}, $\theta^{Q'}$ are the network trainable parameters, respectively. The algorithm process is as follows: firstly, the online decision network outputs action according to the current state, and then acts on the environment after adding noise. The environment feedback gets the reward and the next state, and then the current state, action, reward and next state are stored in the experience cache. Then, the data is randomly sampled from the experience buffer, the online policy network and the online evaluation network are updated by the optimization algorithm, and finally the target evaluation network and the target evaluation network are soft-updated.

The online evaluation network update adopts the method of TD target.

$$y_t = r + \gamma Q'\left(s', \mu'\left(s; \theta^{\mu'}\right); \theta^{Q'}\right) \tag{3}$$

where γ is the reward discount factor.

The loss function of the online evaluation network shows as following:

$$loss = \left(y_t - Q\left(s, a; \theta^{Q}\right)\right)^2 \tag{4}$$

The online evaluation network uses the back-propagation method to update the parameters θ^{Q} according to Eq. (6). a'-u(s).

The online decision network gradient update formula is as follows:

$$\nabla_{\theta^{\mu}} J = \frac{1}{M} \sum_{i=1}^{M} \nabla_a Q\left(s, a \middle| \theta^{Q}\right)|_{s=s_i, a=a_i} \nabla_{\theta^{\mu}} \mu\left(s \middle| \theta^{\mu}\right)|_{s_i} \tag{5}$$

However, it is difficult to implement this theoretical function in engineering. The goal of the decision network is to maximize the value function $Q(s, a)$ of the output action, so the loss function $loss = -Q(s, \mu(s; \theta^{\mu}))$ can be used to update parameter θ^{μ}.

The target network uses a soft update method to update the network parameters $\theta^{\mu'}$ and $\theta^{Q'}$.

$$\begin{cases} \theta^{\mu'} = (1-\tau)\theta^{\mu'} + \tau\theta^{\mu} \\ \theta^{Q'} = (1-\tau)\theta^{Q'} + \tau\theta^{Q} \end{cases} \tag{6}$$

where τ is the update parameter, which is generally taken relatively small, that is, the parameter is updated a little each time, the training is more stable, and the convergence is better.

3 Problem Description

3.1 Parameter Space Settings

Aiming at the problem of trajectory planning of hypersonic aircraft, this paper regards the aircraft as a mass point, and establishes a system of motion equations for the center of mass. The basic state parameters selected in this paper are as follows:

$$s = \left[V_x, V_y, ex, ey\right] \tag{7}$$

where V_x and V_y are the speed components of the aircraft, including the magnitude and direction of the speed; *ex* and *ey* are the distance error informations between the aircraft and the target point, which is beneficial to improve the generalization of the neural network.

According to the definition of the control quantity in the established motion model, the normal overload n_y is selected as the action parameter. Limited by the lift, drag, engine thrust and structural strength of the aircraft, there is a maximum overload limit $n_y \in [-N_{max}, N_{max}]$. N_{max} is the maximum overload.

The three-dimensional climbing state space is defined as $s = [\theta,\ \psi_v,\ x,\ y,\ z]$, the action space is defined as $a = [n_y, n_z]$, and the maximum overload constraint is satisfied $n_y, n_z \in [-N_{max}, N_{max}]$.

3.2 Reward Function Settings

For the aircraft climbing maneuver task in the longitudinal plane, this paper designs the termination reward function and the flight evaluation reward function, and uses the positive reward to promote the learning of the agent.

Termination reward function is following. The goal of the aircraft maneuvering decision is to guide to the target point, and the conditions that define the completion of the guidance task are as follows:

$$|X_{end} - X_T| \le D_{\min} \tag{8}$$

where X_{end} is the position of the aircraft at the end time, X_T is target x location, D_{min} is the maximum allowable off-target amount.

Set the task completion flag to done, and the description of the termination reward is shown as follows:

$$r_{termin} = \begin{cases} 10 & \text{if done} \\ 0 & \text{else} \end{cases} \tag{9}$$

Flight evaluation reward function is listed as following. The autonomous guidance of the aircraft is a difficult exploration problem. Only setting the sparse return such as the termination reward is difficult to learn, and the algorithm is even more difficult to converge. Therefore, setting the flight evaluation reward function can better guide the aircraft to the target position. The specific form is as follows.

$$r_{position} = \begin{cases} \dfrac{D_t - D_{t+1}}{\Delta T v_{\max}} & D_t > D_{t+1} \\ \dfrac{2(D_t - D_{t+1})}{\Delta T v_{\max}} & D_t \le D_{t+1} \end{cases} \tag{10}$$

where D_t and D_{t+1} are the distance between the aircraft and the target point at t and $t + 1$ time, respectively; ΔT is the decision-making cycle; v_{max} is the maximum flight speed of the aircraft.

Overall, the total reward function R used for climbing and diving training is shown below:

$$R = r_{termin} + r_{position} \tag{11}$$

For the fixed altitude cruise task, that is, the stable flight height and the ballistic inclination, so this paper designs a reward function for the altitude deviation and the ballistic inclination deviation. y_T is target y location.

$$\begin{cases} r_{\Delta y} = -\dfrac{|y - y_T|}{\Delta y_{\max}} \\ r_{\Delta\theta} = -\sin|\theta| \end{cases} \tag{12}$$

The total reward function used in the cruise task at fixed altitude is shown below:

$$R = r_{\Delta y} + r_{\Delta\theta} \tag{13}$$

The three-dimensional space climbing reward function adds a reward function $r_{\Delta z}$ to the original reward function in the longitudinal plane to eliminate the z-direction error, and the specific expression is as follows. z_T is target z location.

$$r_{\Delta z} = -\frac{|z - z_T|}{\max|z - z_T|} \tag{14}$$

Table 1. Network Hyperparameter Settings

Hyperparameter	Font size and style
Policy network learning rate	0.00001
Evaluate network learning rate	0.0001
reward decay factor	0.99
Soft update parameters	0.001
batch sample size	128
Experience buffer size	100000

4 Emulation Proof

4.1 Analysis of Training Results

Network Hyperparameter Settings is following (Table 1).

In order to reduce the size of the exploration space and accelerate the learning of the neural network, the size of the aircraft flight airspace is set to 70 km × 15 km, and the normalization method is used to integrate the input state data and action data into a dimensionless quantity to prevent the gradient explosion of the neural network. In the DDPG algorithm, in order to enable the agent to find a better strategy, exploration noise is added to the actions output by the actor network. This paper considers adding a simpler and easier to implement Gaussian noise.

After about eight hours of training, the algorithm is turning stable, the reward function gradually rises to a stable value, the loss function of the policy network gradually decreases and becomes stable, and the neural network tends to converge, which indicates that the maneuvering strategy of the aircraft is gradually optimized and tends to stabilize. The following figures are the reward functions and loss graphs for different tasks (Figs. 3, 4 and 5).

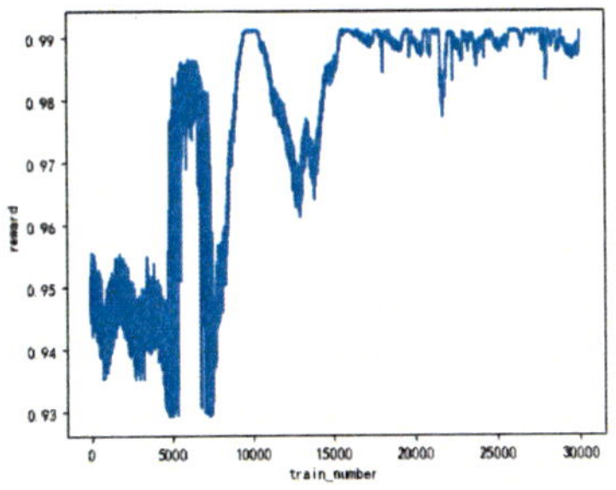

reward function curve

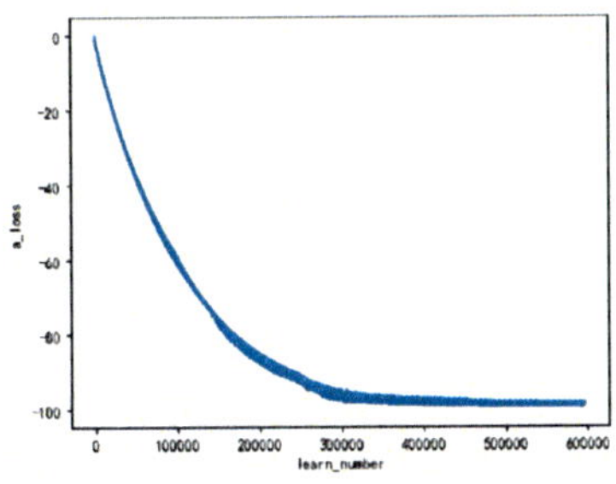

Policy Network loss curve

Fig. 3. Climbing section training convergence process curve

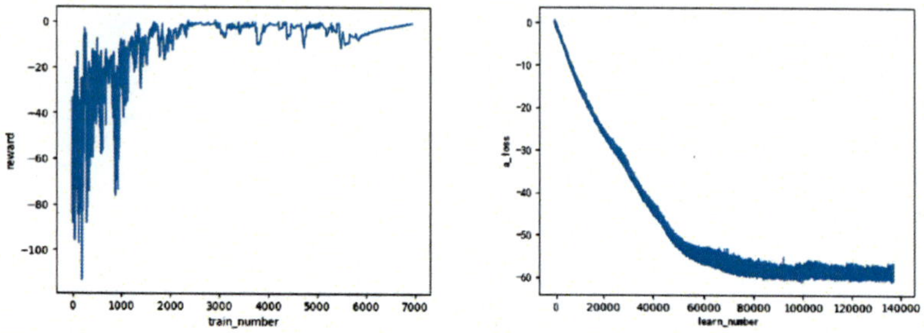

reward function curve Policy Network loss curve

Fig. 4. The curve of the training convergence process in the cruise segment

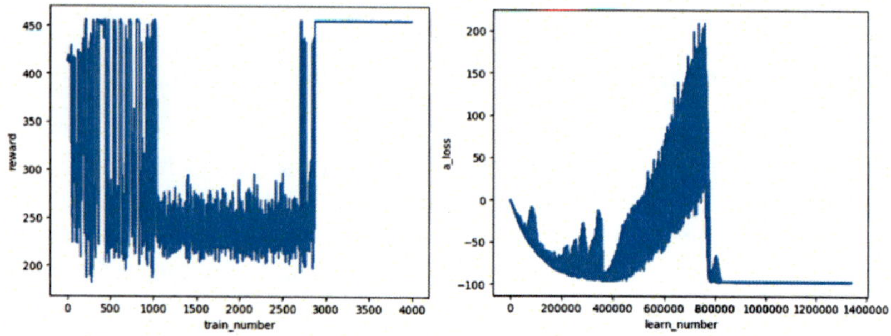

reward function curve Policy Network loss curve

Fig. 5. Convergence process curve of 3D climbing training

4.2 Longitudinal Plane Simulation Test

Climb Simulation Test

Taking the trained climbing strategy network, and carrying out a large number of simulation tests, the initial position of the aircraft is x = 0, and the height y is randomly generated within 5000–10000 m. A total of 1000 tests are carried out. The end point distribution of the aircraft is shown in Fig. 6.

The aircraft successfully flew 803 times within the target range required by the mission, with a success rate of 80.3%, including 165 times within 100 m and 174 times within 50 m, effectively completing the preset mission requirements. The simulation test results are as follows (Table 2 and Fig. 7).

From the simulation results of the four representative samples, it can be seen that the aircraft can fly from the starting point to the given target point. In the untrained airspace, the strategy network also has good performance and good generalization ability.

Constant Altitude Cruise Simulation Test

If taking the trained aircraft's constant-altitude cruise strategy model for simulation testing, and randomly generate the initial ballistic inclination in the range of −2° to 2° to reflect the performance of the trained strategy network. The ten random test flight

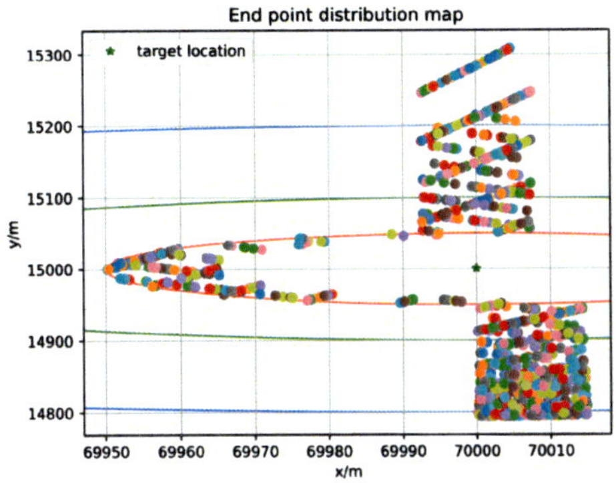

Fig. 6. Climbing simulation test shooting chart

Table 2. Four representative sample simulation test results.

Initial position (x,y)/km	Distance from target point/m	Flight duration t/s
(0,5)	137.54	47.26
(0,7)	202.81	47.03
(0,9)	47.32	46.85
(0,10)	308.40	46.84

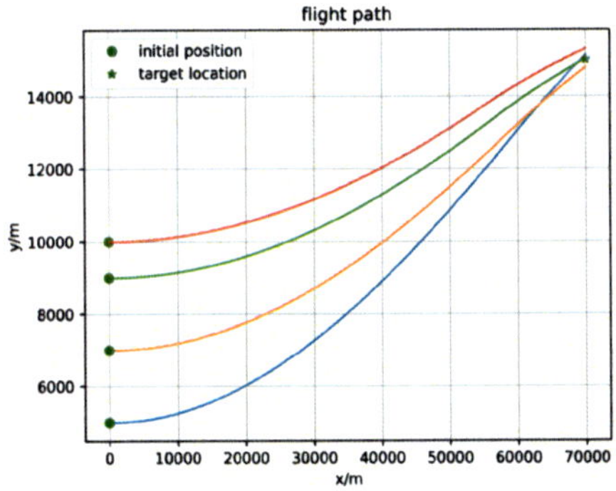

Fig. 7. Four representative sample flight trajectories

trajectories are shown in Fig. 8. It can be seen from the figure that starting from a random ballistic inclination, it can return to the cruising altitude within 20 s and continue to maintain a high cruise, and the maximum altitude difference does not exceed 150 m.

Two groups of representative samples are selected, and the initial ballistic inclination is $-2°$ and $2°$, the change of ballistic inclination and overload during flight are shown in Fig. 9.

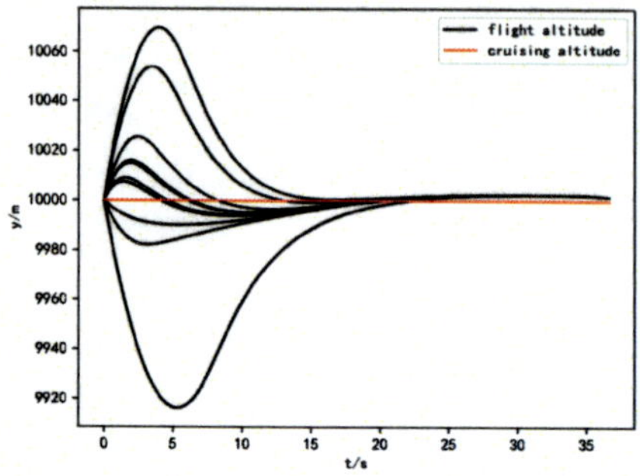

Fig. 8. Random ballistic inclination test flight height change curve

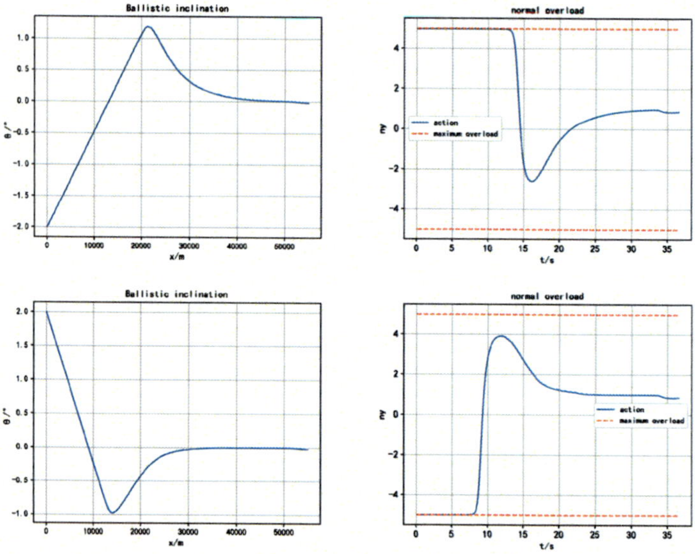

Fig. 9. Change curve of ballistic inclination and overload during initial angle flight

It can be seen from the figure that the aircraft immediately makes a decision based on the current ballistic inclination, executes the maximum overload to quickly correct the current ballistic inclination, then gradually reduces the maneuvering overload to smoothly return the aircraft to the predetermined cruise altitude, and finally maintains a constant overload and constant altitude cruise.

In order to verify the generalization performance of the policy network, set the flight task: keep cruising from the current cruising altitude of 11 km to the target altitude of 12 km, and the results are shown in Fig. 10.

By the given data, it is obvious that the aircraft maintains a cruise at an altitude of 12 km for about 50 s, which proves that the network has a certain generalization ability.

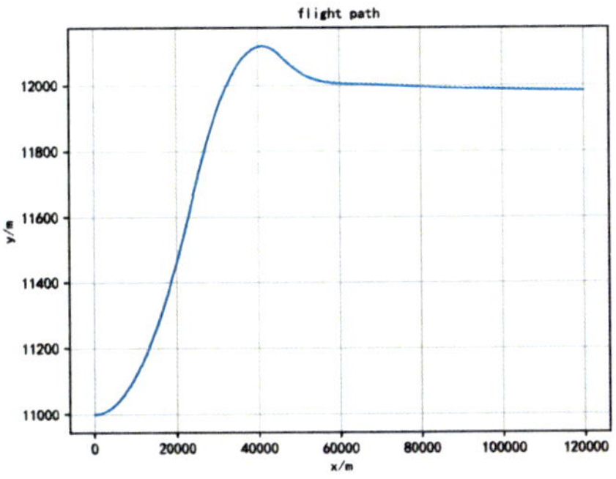

Fig. 10. The generalization test results of the cruise policy network

3D Space Climb Training Test

The trajectory planning is shown in Fig. 11. By the given figure, we can conclude that due to the fast flight speed and the large turning radius, the flight with the maximum normal overload is required for the whole journey. However, the decision-making ability of the policy network also needs to be further improvements.

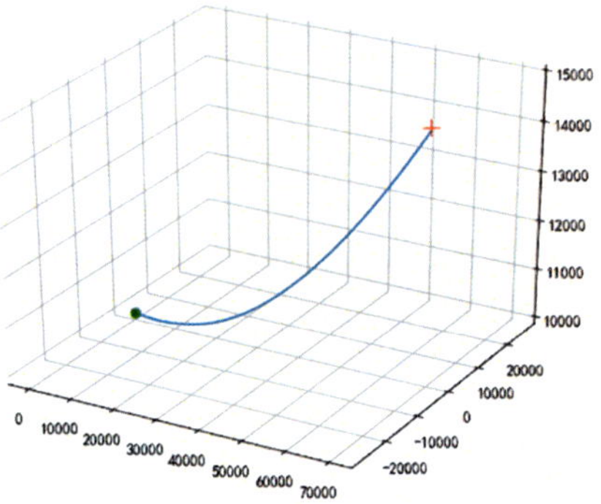

Fig. 11. Schematic diagram of ballistic planning

Loading the converged neural network into the algorithm model and perform 1000 random tests, the initial position of the aircraft is randomly generated within the range of 2 km × 5 km on the plane, and the position of the target point is fixed as (70 km, 15 km, 0). The horizontal coordinate is the number of tests and the vertical coordinate is the distance between the end position of the aircraft and the target point.

Figure 12 shows that 933 times the flight of the 1000 tests reached the target within 100 m, which means a success rate of 93.3% and it effectively completed the preset task. The result proves that the trained policy network enables the aircraft to have a certain autonomous flight ability and a certain generalization ability.

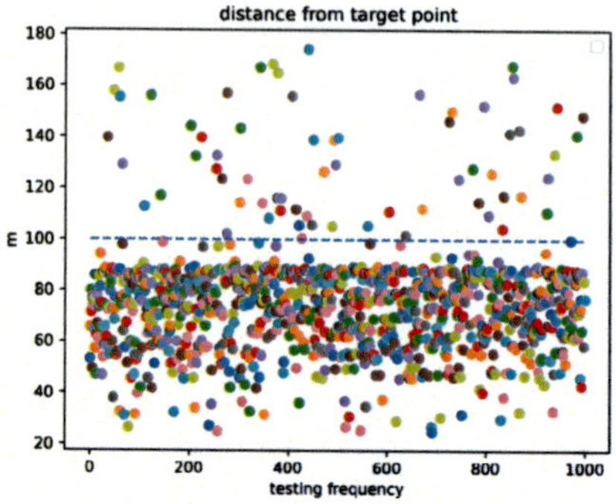

Fig. 12. Random test results

5 Conclusion

This paper conducts simulation training on three maneuvering situations of climbing, cruising. Finally, in order to further study the decision-making model in the three-dimensional space, the vertical plane climbing motion is extended to the three-dimensional space, and the simulation training is continued. The conclusions of this paper are as follows:

(1) State space, action space and different reward functions are designed for climbing, cruising The reward function avoids the sparse reward problem in deep reinforcement learning and enables the aircraft to learn an optimal strategy.
(2) For different tasks, the use of the DDPG algorithm proposed in this paper to make autonomous flight maneuver decisions for hypersonic aircraft, which can provide the ability that the aircraft learn a set of strategies to guide the aircraft to fly to the target point without any prior knowledge. Additionally, since the terminal distance error could meet the mission requirements, the autonomy of the maneuvering control of the aircraft is effectively improved, and the simulation test proves that the algorithm has a certain generalization.

References

1. Na, L., Mengmeng, F.: A review of research advances in hypersonic vehicle control technology. Aerodyn. Missile J. (12), 16–21+62 (2019)
2. Qiang, Q., Xiangwei, B., Baoxu, J.: Adaptive dynamic programming for hypersonic flight vehicle based on backstepping control. Tact. Missile Technol. (05), 102–112 (2021)
3. Wenjun, Y., Ke, Z., Minghuan, Z., et al.: ESO based nonlinear dynamic inversion control for hypersonic flight vehicle. J. Northwest. Polytech. Univ. **34**(05), 805–811 (2016)
4. Xingling, S., Jun, L., Dongguang, L.: Nonlinear attitude control for hypersonic vehicles based on trajectory linearization. Unmanned Syst. Technol. **3**(03), 56–66 (2020)
5. Jiansong, Z., Qinghua, M., Haiqing, L., et al.: Robust longitudinal control method for hypersonic vehicle. J. Project. Rockets Missiles Guid. **40**(02), 19–22 (2020)

6. Daqing, H., Dingguo, J.: Backstepping sliding mode neural network control system for hypersonic vehicle. Opt. Precis. Eng. **27**(11), 2392–2401 (2019)
7. Xiaodong, L., Hao, Z., Jianguo, G., et al.: Iterative learning control combination with adaptive sliding mode technique for a hypersonic vehicle. J. Northwest. Polytech. Univ. **37**(06), 1120–1128 (2019)
8. Guo, J.G., Lu, N.B., Zhou, J., et al.: Fuzzy control of finite time attitude coupling in hypersonic vehicles. Acta Aeronaut. Astronaut. Sin. **41**(11), 41–50 (2020)
9. Yangwang, F.A.N.G., Dong, C.H.A.I., Donghui, M.A.O., et al.: Status and development trend of the guidance and contol for air-breathing hypersonic vehicle. Acta Aeronaut. Astronaut. Sin. **35**(07), 1776–1786 (2014)
10. Bingwei, Y.: Formulization of standard atmospheric parameters. J. Astronaut. (01), 83–86 (1983)
11. Li, Y.: Deep reinforcement learning: an overview (2017). arXiv:1701.07274
12. Lillicrap, T.P., Hunt, J.J., Pritzel, A., et al.: Continuous control with deep reinforcement learning (2015). arXiv:1509.02971

3D Design Engine Developed for Aircraft Conceptual Design Phase Based on Visual Basic and OpenGL

Li Yueli[1,2](✉), Wang Qi[2], and Gao Yining[2]

[1] School of Aeronautics, Northwestern Polytechnical University, Xi'an 710072, China
rockandtcl@163.com
[2] The First Aircraft Institute, AVIC, Xi'an 710089, China

Abstract. This article develops a 3D design engine for the software requirements of the aircraft conceptual design phase, introducing its development environment, capabilities, frameworks, coordinate axis definition, data structure, component drawing, etc. The applications of the 3D design engine are introduced at last.

Keywords: Aircraft conceptual design phase · 3D design engine · Software

1 Introduction

1.1 Aircraft Design Process

In general, the aircraft design process involves several distinct phases. They are referred to as: preliminary design/project design/detail design. Preliminary design phase forms the initial draft of a document that will be subsequently revised to contain a thorough description of the aircraft [1]. Some literature divides the aircraft design process into the following processes: requirement, conceptual design, preliminary design, detail design [2]. Typically, the following characteristics are defined during conceptual design phase: Type of aircraft\Aesthetics\Special aerodynamic features etc.

1.2 State of the Art

The common aircraft design software can be categorized as follow:

The first type is windows software based on GUI, such as advanced aircraft analysis [3].

The second type is windows software based on OpenGL, such as OpenVSP and XFLR5 [4, 5].

The third type is Matlab Tookie's, such as Tornado [6].

The last one is windows software based on console application, uch as AVL and DATCOM [7, 8].

In the author's opinion, the Open VSP is the best software for conceptual design phase. However, there are great differences on usage habit internationally, such as the design ideas, human-machine interaction, display interaction, data format etc (Fig. 1).

Chinese Society of Aeronautics and Astronautics: CASTYSF 2022, LNEE 972, pp. 60–67, 2023.
https://doi.org/10.1007/978-981-19-7652-0_7

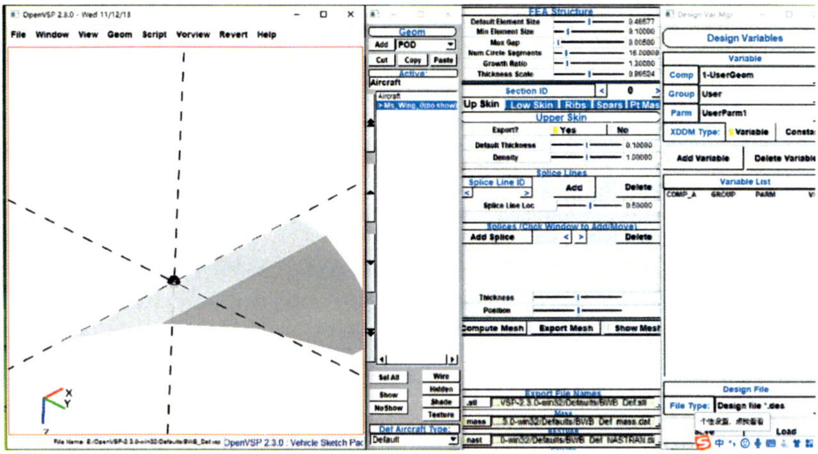

Fig. 1. Open VSP runtime interface.

The Conceptual Design software has been developed for decades in the domestic. Unfortunately, they are still at the primary level [9]. Therefore, it is necessary to develop software that combines the advantage of Open VSP as well as user habits (Fig. 2).

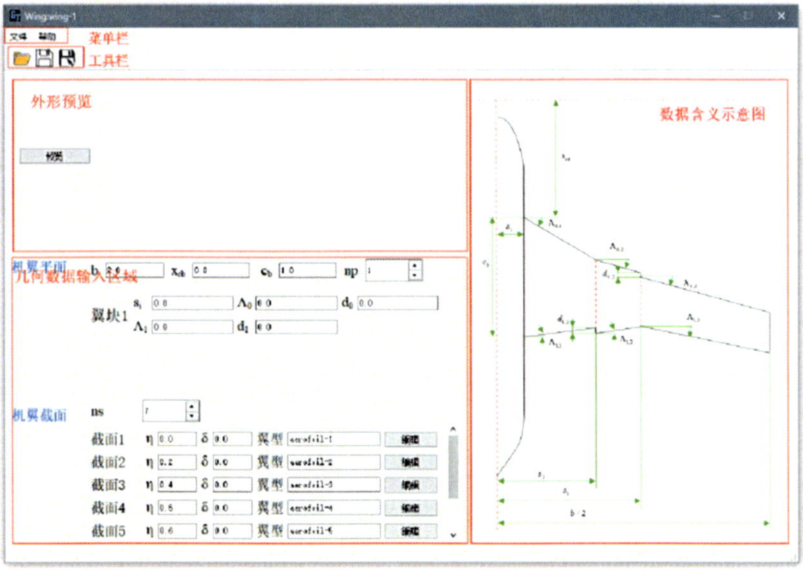

Fig. 2. Design runtime interface of Ref. [9].

In this paper, an aircraft design software with 3D display and interaction for the conceptual design phase is introduced, with a detailed description of the software's 3D design engine. Its runtime interface is shown in Fig. 3.

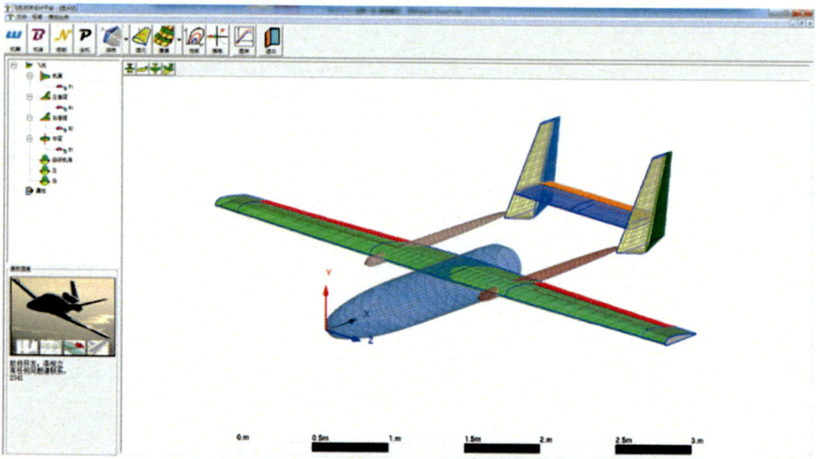

Fig. 3. Conceptual design phase aircraft design software runtime interface.

1.3 Development Environment

The common sense is all of the 3D engines are based on OpenGL framework. But, how does it works? The first step is to download a library of OpenGL, although different programming languages have different types of libraries, their standards are uniform. The second step is to reference this library during software development. The third step is to initialize OpenGL. The fourth step is to select the appropriate color RGB and alpha value, draw the points, lines and faces based on the data, and form the graph we need. In the fifth part, the graphic is displayed in an orthographic projection on the form. At present, in aircraft design and related software development, OpenGL is mainly used in several aspects—test data display, flight simulation, and virtual reality-based maintenance support and so on [10, 11].

Although most of the current 3D engines are based on C language, the author still chooses Visual Basic 6.0 as the development language, the reasons are: 1, Visual Basic language syntax is simple, 2, the data structure is relatively clear, there is no complex pointer, 3, the development process requires less effort; 4, more focused can be put on the description of the problem.

2 Methodology

2.1 Requirements and Capabilities

In the aircraft concept design phase, the main work includes the following aspects: the choice of airfoil, the layout of the wing, the layout of the engine, the style of the fuselage, the correspondence of the components, the definition of related parameters, and so on.

The requirements for 3D display include: each component's lines and surfaces that can be displayed; lines, and surfaces can be selectively displayed; component's color can be changed; support component selection prompts; support mouse-based scaling; support mouse-based rotation; support real-time display refresh after part position parameters are updated; and support screen graphic save as bmp picture.

2.2 Frameworks

The engine mainly contains two parts, one is how to describe the aircraft, and the other is how to draw the aircraft.

The aircraft in this 3D design engine contains three components, wing surface, body, and the engine. These physical parts need to be abstracted into mathematical descriptions. These components have a similar description structure, including index parameters, positional parameters, geometric parameters, and display parameters. The structure of specific parameters differs according to the types of components. For example, the data structure of the geometric parameters of the wing surface includes the individual wing surface data structure, the airfoil library data structure, the section information data structure, and the rudder data structure.

The step in drawing an aircraft is to load the component data, construct its 3D shape in a basically similar way based on the characteristics of the component, then the surface grid is generated from the 3D shape, finally surface's grid and cell is drawn and displayed on the screen (Fig. 4).

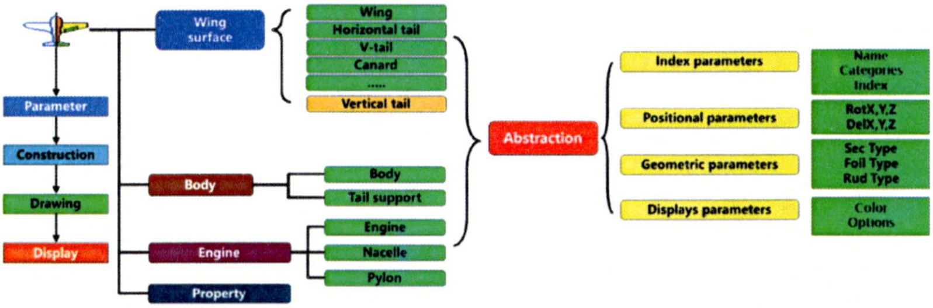

Fig. 4. Frameworks.

2.3 Coordinate Axis Definition and Projection Transform

In order to be compatible with 2D airfoil section and 3D wing surface, and to make the code more versatile, the coordinate axis is defined as follows: the X coordinate is the axial coordinate of the airframe, and the backward direction along the flow direction is positive. The Y coordinate is perpendicular to the X coordinate, positive upwards, and the XY plane forms the symmetrical plane of the aircraft. The Z coordinate is perpendicular to XY plane, with positive on the left.

There are two ways to project in OpenGL, perspective projection and orthographic projection. Perspective projection is similar to see objects in our daily lives, objects of the same size, larger at close range and smaller at a distance. Orthographic projection does not have such an effect, and the display is the same near and far. In computer-aided design software development, we do not want to deviate from the display of objects because of the distance and proximity of the vision, so we often use orthographic projection (Fig. 5).

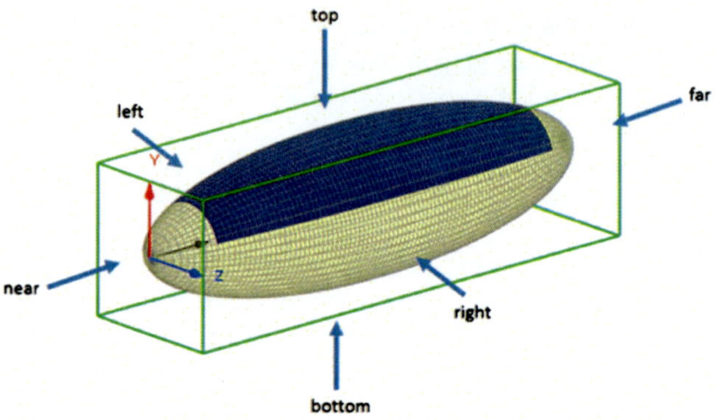

Fig. 5. Coordinate axis and projection transform.

2.4 Data Structure

Mathematically describing an airplane requires complex data structures and a variety of variable types. Defining variables or arrays one by one increases the complexity of data manipulation, so this article uses structure data structures while also using structure nesting.

Taking the wing structure as an example, the index parameters, position parameters, display parameters, wing feature parameters (projection area, span length, mean aerodynamic chord length, X coordinate of mean aerodynamic chord, Y coordinate of mean aerodynamic chord, infiltrated area), surface grid parameters, nested airfoil library structure, structural parameter structure, rudder surface structure are defined in the wing structure.

2.5 Component Drawing

Component drawing is introduced based on the example of the wing surface.

In order to display the features of the component independently, the drawing function of the grid, the drawing function of the section, the drawing function of the edge, and the drawing function of the cell are defined. Continuous unclosed polylines are used for drawing lines, and independent fill quads are used for drawing cells (Fig. 6).

2.6 Mouse Operation

Mouse operations are defined as follows:

When the left button is pressed, it means dragging, the mouse pointer changes to a four-way dimension line, and the mouse button value is 1.

When the scroll wheel is pressed, it means zooming, the mouse pointer changes to a vertical dimension line, and the mouse button value is 4.

When the right button is pressed, it means rotation, the mouse pointer is the default value, and the mouse button value is 2.

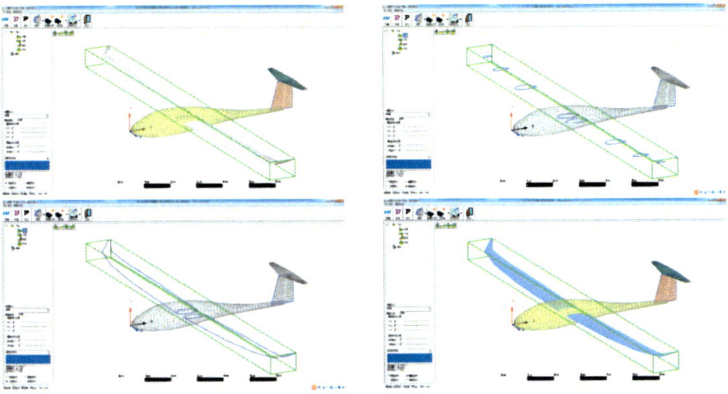

Fig. 6. Variable display modes.

When you release the mouse button, the mouse pointer reverts to its default value.

During the mouse movement, different mouse value, realizes translation, rotation, zooming. Mainly according to the distance of mouse movement, equivalent to the variation parameters of translation, rotation, and scaling, these parameters are passed to the projection function, and translation, rotation and scaling are realized. In this process, the distance equivalent parameter needs to be adjusted according to the user's habits and the developer's experience, so that the response of the mouse during translation, rotation and zoom is more in line with the operation intention.

3 Application Examples

3.1 Conceptual Design Phase Aircraft Design

AS mentioned above, the software can support the design and aerodynamic evaluation of any layout type of aircraft, and can directly and quickly generate CATIA models, OBJ models, and STL models for 3D printing of the whole aircraft and components (Fig. 7).

3.2 Solar Aircraft Design

Based on the 3D design engine, solar aircraft design software was developed. In addition to the above functions, it also can calculate and analyze solar radiation energy at any location, any date, any time, any height, any aircraft surface, and any attitude. In addition to solar aircraft design, it is also applicable to carry out computational evaluation on solar airships, solar vehicles, photovoltaic power generation, etc (Fig. 8).

3.3 Rotor and Propeller Design

Based on this 3D design, engine, rotor & propeller design based on leaf element momentum theory was developed. The software supports power-diameter relationship calculation, rapid analysis of airfoil and surface, rotor & propeller detailed design and analysis, and calculation of aerodynamic, power and torque in axial and non-axial states. It can output data, cloud map, screen graphic, CATIA module, etc (Fig. 9).

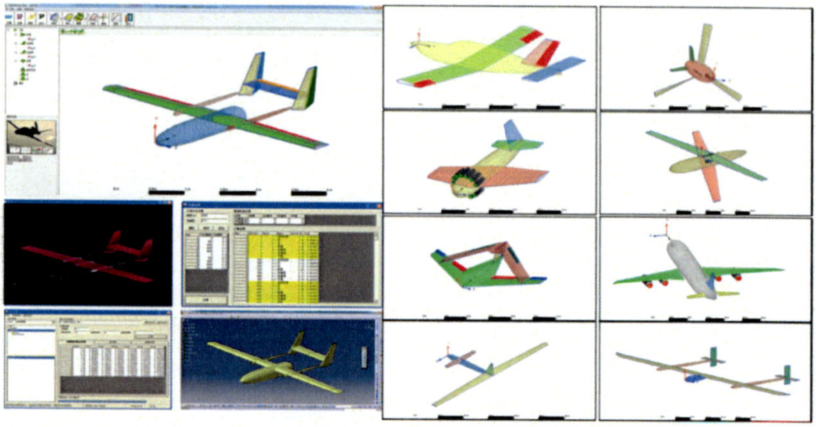

Fig. 7. Conceptual design phase aircraft design software.

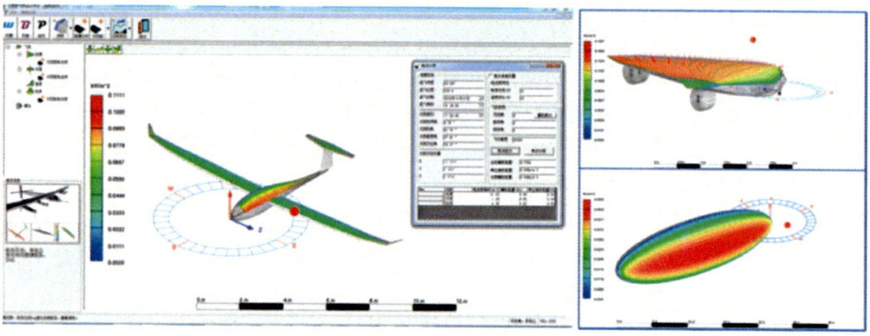

Fig. 8. Solar aircraft design software.

4 Conclusion

This paper develops a 3D design engine capable of being used in aircraft conceptual design phase that can be adapted to the design, description, and display of different types of aircraft. Based on this engine, the authors developed three different types of aircraft design software, improving the efficiency of conceptual design.

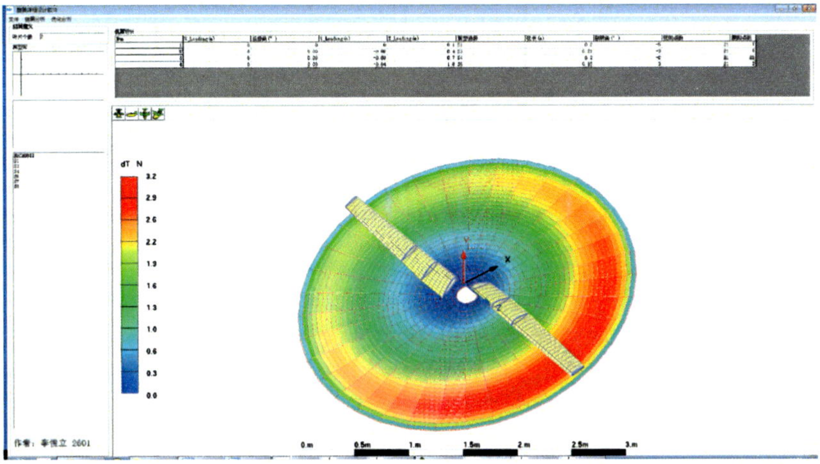

Fig. 9. Rotor and propeller design software.

References

1. Jenkinson, L.R., Marchman III, J.F..: Aircraft design projects for engineering students. Elsevier Science Ltd, UK (2003)
2. Gudmundsson, S.: General aviation aircraft design applied methods and procedures, 1st edn. Butterworth-Heinemann (2013)
3. DAR corporation Homepage. https://www.darcorp.com/. Accessed 30 Mar 2022
4. OpenVSP Homepage. http://openvsp.org/. Accessed 01 Apr 2022
5. Xflr5 Homepage. http://www.xflr5.tech/. Accessed 14 Jan 2021
6. Tornado Homepage. http://tornado.redhammer.se/. Accessed 19 Sept 2021
7. AVL page of MIT. https://web.mit.edu/drela/Public/web/avl/. Accessed 23 Feb 2017
8. PDAS Digital Datcom page. https://www.pdas.com/datcom.html. Accessed 19 Apr 2021
9. Jiawang, C.: Development of a rapid aerodynamic estimation tool for large civil transports (in Chinese). Master thesis. Nanjing University of Aeronautics and Astronautics, Nanjing (2020)
10. Xiaodan, Z.: The research and implement of 3D space mission simulation system based on OSG (in Chinese). Master thesis. Chongqing University, Chongqing (2015)
11. Yang, S.: Research and realization of virtual aviation instrument based on OpenGL (in Chinese). Master thesis. Huazhong University of Science and Technology, Wuhan (2014)

Research on Testing Technology for Noise Certification of Civil Aircraft

Chunzhuang Zhao(✉), Jiaxin Gao, Lei Wang, and Tengyuan Liu

Harbin Aircraft Industry Group Co., Ltd., Harbin 150066, China
zhaochunzhuang@126.com

Abstract. Aiming at the actual demand for noise certification of civil aircraft, the noise certification testing technology of propeller small aircraft and helicopters is mainly studied. Civil aircraft noise certification is mainly carried out in accordance with the provisions of CCAR-36, including two aspects of test system architecture and test flight method design. The qualification criteria in the noise certification test are mainly two points, one is that the noise data meets the restriction requirements of CCAR-36, and the other is that the measured track meets the method requirements of CCAR-36. An equivalent procedure for noise certification of small propeller aircraft is provided, and two specific implementation methods for determining the measured track are provided, which provide some effective references for noise certification of other types of aircraft.

Keywords: Small propeller plane · Helicopter · Noise · Airworthiness

1 Introduction

Due to the requirements of environmental protection, the outside noise generated by civil aircraft during flight needs to be limited within a certain range. The International Civil Aviation Organization has detailed regulations on civil aircraft noise in Annex 16, including four aspects: test flight methods, measurement system requirements, data processing methods and noise data limits. With reference to ICAO Annex 16 and FAR-36, the Civil Aviation Administration of my country has formulated the "Noise Regulations for Aircraft Type and Airworthiness Certification" (CCAR-36) [1, 2]. All civil aircraft must complete the noise by the test flight method specified in CCAR-36. Therefore, there are two main assessment points for the validity of the noise certification test, one is that the flight trajectory of the aircraft in the test must meet the requirements, and the other is that the measurement system that obtains noise data in the above flight trajectory must meet the requirements. This paper mainly studies the noise certification test technology of propeller small aircraft and helicopter.

2 Noise Measurement System Architecture

2.1 The Architecture of Noise Measurement System for Small Propeller Aircraft

The noise measurement system of propeller small aircraft consists of metal disc, microphone (amplifier), calibrator, data acquisition recorder, GPS system and cable assembly.

Chinese Society of Aeronautics and Astronautics: CASTYSF 2022, LNEE 972, pp. 68–74, 2023.
https://doi.org/10.1007/978-981-19-7652-0_8

Among them, the microphone must be 12.7 mm in diameter, with a protective fence; the surface of the metal disc is painted with white paint [3], 40 cm in diameter, and not less than 2.5 mm in thickness. The noise measurement system is located on the ground projection line of the flight trajectory of the small propeller aircraft, and the distance from the brake release point of the aircraft is 2500 m.

The microphone is installed upside down on the metal disc, the film is parallel to the surface of the disc, the axis is perpendicular to the surface of the disc, the projection point of the axis on the surface of the disc is located at three quarters of the radius and the line connecting with the center of the circle is perpendicular to the flight trajectory on the disc. Projection of the surface. The microphone is connected to the data acquisition recorder through a cable assembly, and the calibrator is used for on-site acoustic calibration before and after each experiment. The GPS system provides GPS time to the data acquisition logger.

2.2 Helicopter Noise Measurement System Architecture

The noise measurement system of the helicopter consists of a microphone (amplifier), a calibrator, a data acquisition recorder, a windshield, a tripod, a time synchronization device, a GPS system and a cable assembly. The noise measurement system includes three ground noise measurement stations, which are one flight trajectory noise measurement station and two sideline noise measurement stations. The flight trajectory noise measurement stations are located on the ground projection line of the helicopter's flight trajectory, and the distance from the helicopter is not less than 500 m; Two sideline noise measurement stations are located 150 m on both sides of the flight trajectory noise measurement station, and the line connecting the two is perpendicular to the ground projection line of the helicopter flight trajectory.

Each noise measurement station includes a microphone (amplifier), a windshield, a tripod, a data acquisition recorder and several cable assemblies. The microphone is mounted on a tripod, and the distance between the tripod and the ground is 1.2 m and the axis is adjusted by adjusting the tripod. It is perpendicular to the incident direction of helicopter noise. The GPS system is installed at the flight trajectory noise measurement station, and provides GPS time to the data acquisition recorder. The time synchronization device is connected to three ground noise measurement stations at the same time, and realizes the time synchronization of the three ground noise measurement stations by means of synchronizing marks.

3 Method of Flight Test

3.1 Method of Propeller Small Aircraft Flight Test

General Procedure. The propeller plane is adjusted to the maximum take-off weight (the weight needs to be re-adjusted after the flight time exceeds one hour), take off from the brake release point (point A) with the take-off power, and pass the height of 15 m above the runway (point C) after leaving the ground (point B). After adjusting to the optimum rate of climb speed (point D), continue to climb, and record the air position and noise data of the aircraft when it passes right above the noise measurement point (point O) (point E). Figure 1 shows a schematic diagram of the general procedure for the noise certification test for small propeller aircraft.

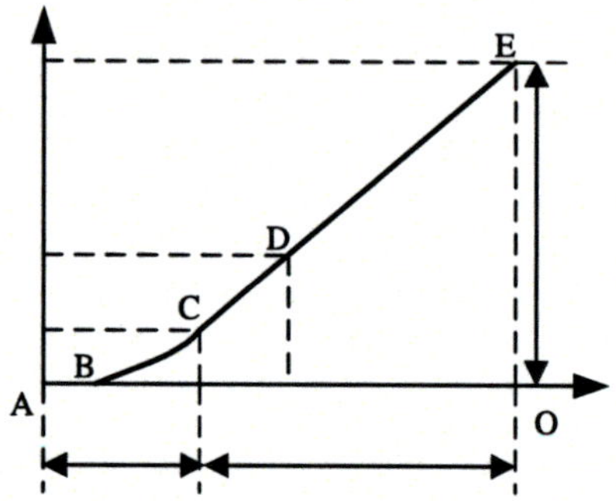

Fig. 1. Schematic diagram of the general procedure for noise certification test of propeller small aircraft.

Equivalent Procedure CCAR-36 stipulates that the effective number of flights in the noise certification test is not less than 6 times, so the number of cycles of the general procedure is not less than 6 times noise. In order to save the cost of airworthiness certification and reduce the impact of noise on the surrounding environment of the airport, the equivalent procedure can be used instead of the general procedure.

The equivalent procedure is realized by the method of track cut-in [4, 5]. After the propeller aircraft completes the first noise test according to the general procedure, it performs level flight at the level at which the best rate of climb speed is achieved in the first flight, and continues level flight to the height position (point A1) just above the original braking point., after reaching the position (point D) where the best rate of climb speed is reached in the first flight, turn to the climbing state, keep the power and speed unchanged, and record the air position and noisy data. The sequential method repeats subsequent noise tests. Figure 2 shows a schematic diagram of the equivalent procedure for noise certification of propeller small aircraft.

3.2 Method of Helicopter Flight Test

The helicopter noise certification test includes three sub-tests, namely take-off, fly-over, and approach noise tests, and the noise measurement results of the three sub-tests are in the 10 dB drop range [6, 7].

In the take-off noise test, the helicopter climbs from the starting point (point A) at a height of 20 m from the ground and 500 m from the flight path noise measurement station (point O) with the maximum take-off power, the best rate of climb, and the best rate of climb speed. Just above the flight path noise measurement station (point B), continue to climb until the end point (point C) of the take-off noise test is passed. Figure 3 shows the schematic diagram of the flight path of the take-off noise for helicopter noise certification.

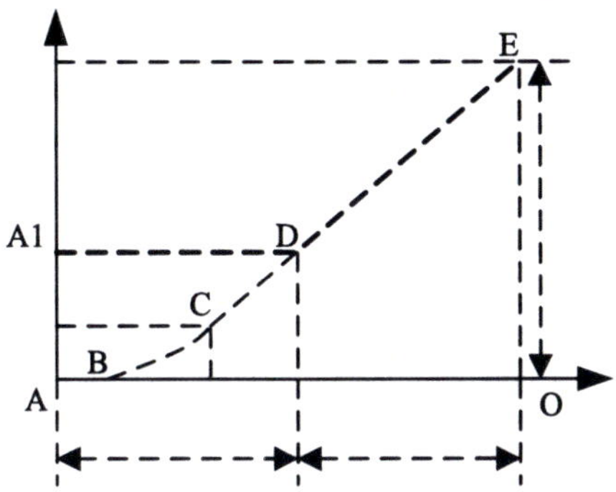

Fig. 2. Schematic diagram of the equivalent procedure for noise certification of propeller small aircraft.

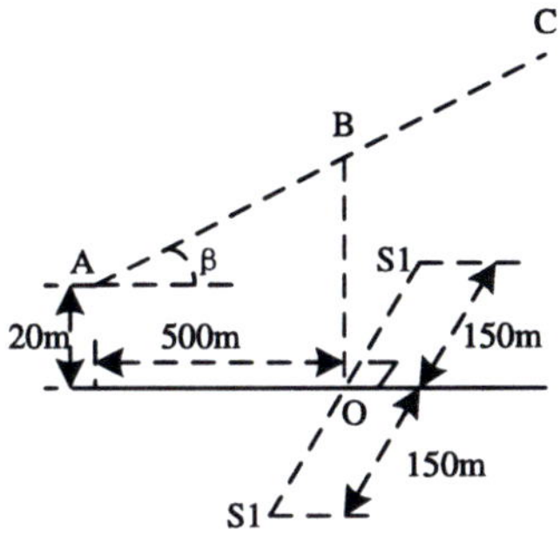

Fig. 3. Schematic diagram of the flight path of take-off noise for helicopter noise certification

In the flyover noise test, the helicopter flies level from the starting point (point D) at a height of 150 m from the ground and passes directly above the flight path noise measurement station (point O) (point E) and continues level flight until it passes the end point of the flyover noise test (point F). The level flight speed selects the minimum value of 0.9VH, 0.9VNE, 0.45VH+120 km/h, 0.45VNE+120 km/h, VH is generally referred to as the maximum level flight speed, and VNE is generally referred to as the unsurpassable speed. Obtain the flight manual. Figure 4 shows a schematic diagram of the flight path of the helicopter noise certification flyover noise.

In the approach noise test, the helicopter continues the approach from the starting point (point G) of the approach noise test, at an approach angle of 6°, after passing 120 m (point H) directly above the flight path noise measurement station (point O), until Pass the endpoint of the approach noise test (point I). Figure 5 shows a schematic diagram of the approach noise flight trajectory for helicopter noise certification.

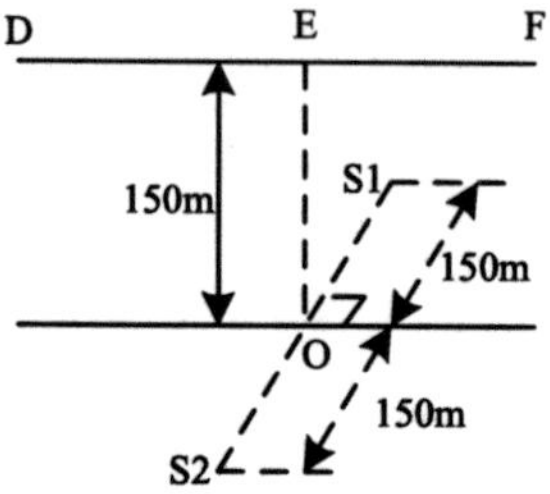

Fig. 4. Schematic diagram of flight path of helicopter noise certification flyover noise

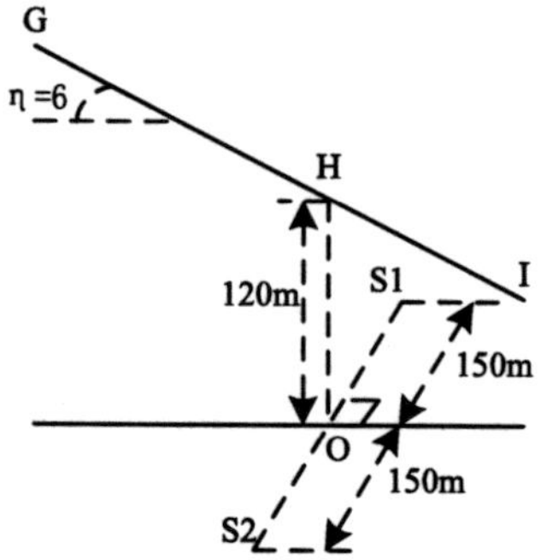

Fig. 5. Schematic diagram of flight path of helicopter noise certification approach noise

4 Method of Track Measurement

4.1 Method of Photographic Ratio

The photographic ratio method is a method of determining the position of the aircraft in the air by using the trigonometric function relationship and the imaging principle. When using it, it is necessary to select the significant structure on the aircraft as the reference benchmark. The propeller usually selects the wing of the small aircraft, and the helicopter generally selects the rotor. The photographic scale method is implemented based on the photographic scale system. The schematic diagram of the photographic scale system is shown in Fig. 6. The number is adjusted according to the different test objects. The propeller small aircraft only needs to use one photographic scale system, and the helicopter needs to use three sets of photographic scale systems. The starting point of each test, the point just above the flight path and the end point, the connection of the three is regarded as the flight path of the helicopter.

A photo-scale system consists of a rectangular frame (square frame), a stand, and a camera. The rectangular frame is divided into 10 equal parts by inelastic strings, and the height of the rectangular frame is supported by adjustable brackets. When the aircraft passes over the camera scale system, take a picture, apply the imaging principle and trigonometric function relationship to obtain the measured track height to judge the altitude deviation, and use the aircraft position on the photo to judge the heading

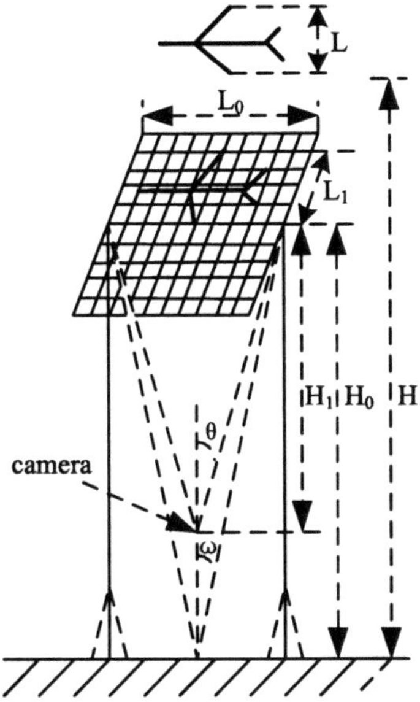

Fig. 6. Photographic scale system

deviation. The calculation formula of the measured track height is as follows:

$$H = \left(\frac{10 * L * L3}{L0 * L2} - 1\right) * \frac{L0}{2 * \tan\theta} + H0$$

Among them, ω is the angle between the edge of the rectangular frame and the vertical direction of the ground, θ is the angle between the edge of the rectangular frame and the vertical direction of the camera lens (the larger value of the corresponding angle of yaw 10° and 20 m), H is the measured track height, H0 is the vertical distance from the grid plane to the ground, H1 is the vertical distance from the camera lens to the grid plane, L0 is the side length of the rectangle, L is the length of the (wing) rotor diameter, and L1 is the wing (rotor) diameter at The length on the rectangular frame, L2 is the length of the diameter of the wing (rotor) in the photo, and L3 is the grid width of the rectangular frame in the photo.

4.2 Method of DGPS

A GPS system is placed at a point with a known precise location, an aircraft and a noise measurement station. Among them, one point with a known precise position is used as the base station, and the two points of the aircraft and the noise measurement station are used as the mobile station. The DGPS coordinates of any point on the measured track and the DGPS coordinates of noise measurement station A are obtained from the GPS system. Take point A as the calculation origin, take the aircraft heading as the X-axis direction, take the direction perpendicular to the ground as the Z-axis, and determine the Y-axis from the right-hand coordinate system. The allowable range of deviation under the GPS coordinate system is obtained from the allowable range of deviation between the measured track and the reference track [8]. Software such as matlab can be used to draw all the measured track points and the area in the same three-dimensional map. If a point is located inside the area, then It indicates that the measured track conforms to the reference track deviation, and vice versa.

5 Conclusion

Through the research on civil aircraft noise certification test technology, the following conclusions are drawn:

(1) All civil aircraft must meet the noise requirements of CCAR-36, and must use the prescribed test flight methods and measurement methods to obtain noise data.
(2) For propeller small aircraft and helicopter, a specific construction method of noise measurement system is provided respectively.
(3) The method of Photographic Ratio and the method of DGPS are designed to determine the measured track, and a specific implementation method is provided.
(4) For the propeller small aircraft, an equivalent procedure of the test flight method is introduced, which can reduce the number of take-off, landing and adjustment of the aircraft and improve the test flight efficiency.

References

1. CCAR-36-R2: Aircraft Model and Flight Qualification Verification Noise Regulations (2017)
2. AC-36-AA-2008-04: Consultation notice of Aircraft Model and Flight Qualification Verfication Noise Regulations (2008)
3. Haojun, X.: Introduction to the Airworthiness. Northwestern Practice University Press, Xi'an (2012)
4. Chen Dejun, W., Zhigang, W.K.: Design and application of helicopter noise airworthy test system. Comput. Measur. Control. **06**, 368–284 (2018)
5. Fang, W., Ning, Y., Long, Z.: Research on the development of civil aircraft noise. Air Stand. Qual. (01), 45 (2019)
6. Guohua, Y., Jie, L.: Application of the equivalent flight paths in airworthiness certification of civil aircraft noise. Acoust. Technol. (02), 163 (2018)
7. Guochi, G., Jingze, Q., Li, D. et al.: Airworthiness certification technology for the kakeoff noise concurrent flight testing of the Y12F aircraft. Civ. Aircraft Des. Res. (02), 57 (2019)
8. Enhui, Z., Jinying, Y., Shuren, X.: Noise and Vibration Control, vol. 47. Metallurgical Industry Press, Beijing (2012)

Research on Relative Navigation Technology of Air-Borne Missile in Satellite Denial Environment

Haibin Cheng(✉) and Hao Lu

China Airborne Missile Academy, Luoyang 471009, China
chbsian@163.com

Abstract. Aiming at the problem of relative information acquisition in air-borne missile cooperative navigation, a cooperative navigation method in satellite denial environment is proposed. Taking the radar seeker and ranging data link for relative measurement and data transmission, the relative navigation state equation between the leader and the slave is designed employing master-slave structure. The measurement equations derivate based on ranging data link and radar seeker ranging. The relative deviation is estimated by Kalman filter. Finally, the simulation is conducted to verify the algorithm. The simulation results show that the proposed method could estimate the relative deviation between missiles where the relative position estimation deviation is less than 20 m, the relative velocity estimation deviation is less than 5 üm/s, and the relative attitude estimation deviation is less than $8'$. The proposed method solves problems of obtaining the relative position, relative velocity and relative attitude error information of multiple missiles, and lays a foundation for realizing the cooperative detection of multiple missiles.

Keywords: Collaborative navigation · Satellite-jamming · Radar seeker · Datalink · Simulation and analysis

1 Overview

With the continuous acceleration of the integration process of information warfare, the confrontation between attack and defense is becoming intensely. The trend is expanding the concept of air-borne missile cooperative operation. The future air combat will change from the confrontation between single combat aircraft to multi-platform cooperative combat. The cooperative air-borne missile will become the main role of future air combat. The cooperative operation of air-borne missile can improve its search ability for moving targets, especially the ability of anti-stealth targets, and effectively improve the comprehensive combat effectiveness.

Current research on air-borne missile cooperative operation mainly focuses on two aspects: The concept and method of air-borne missile cooperative operation; Cooperative guidance model, strategy and control method. Reference [1] proposed a cooperative guidance operation method of air-borne missile; Ref. [2] introduced the technologies of

Chinese Society of Aeronautics and Astronautics: CASTYSF 2022, LNEE 972, pp. 75–86, 2023.
https://doi.org/10.1007/978-981-19-7652-0_9

multi missile cooperative collectivity, formation flight, multi missile data link communication and so on; Refs. [3, 4] studied the multi missile cooperative guidance strategy; Ref. [5] researched on target tracking based on cooperative positioning of two missiles; Ref. [6] studied the multi missile cooperative guidance law. The above research studied the multi missile cooperative guidance law, control model and control method in the hypotheses that the relative information between missiles has been obtained. There has no relevant research on the acquisition method of relative information between missiles. When the relative information between missiles cannot be obtained, the research of multi missile cooperative detection is difficult to achieve.

The sensors that can be used for measurement mainly include lidar [7], vision measurement [8–10], ultra-wideband (UWB) [11], satellite navigation [12–14]. Due to the strictly limited volume of air-borne missiles and the characteristics of high mobility in combat tasks, the optional auxiliary means are very limited. Satellite navigation, radar and infrared seeker and ranging data link can be selected as auxiliary measurement means, among which satellite navigation is the most practical relative navigation means. This paper focus on the relative navigation by using radar seeker and ranging data link as auxiliary measurement in the environment of satellite navigation rejection.

2 Cooperative Guidance Strategy of Air-Borne Missile

2.1 Master-Slave Collaboration Strategy

In this paper, a master-slave collaborative guidance strategy and a coordinated guidance formation are shown in Fig. 1, in which one is the leader and the other is the follower. The pilot in the middle guidance section broadcasts his own inertial navigation data through the high-speed data link. After receiving the data broadcast by the pilot, the follower forms the relative navigation between them, and estimates the relative attitude, relative speed and relative position deviation between missiles through filtering, so as to provide formation basis for multi-missile flight in the middle guidance section. After turning to terminal guidance, the follower uses the obtained relative deviation data to correct the target information measured by the pilot radar that can be used by the follower, finally cooperative detection and cooperative attack can be realized.

Fig. 1. Multi-missile flying formation

In order to facilitate the derivation of the algorithm, the navigation of the pilot and the follower are unified in the solidified geographic frame(n). The coordinate origin and axis are fixed relative to the earth, and X, Y and Z correspond to the north, up and east.

3 Relative Navigation Error Model

3.1 Inertial Navigation Error Equation

Owning to the short flight time and unpredictable trajectory, the estimation of position error, velocity error and misalignment angle can be applied to reduce the complexity of the algorithm without regard to device error. The pilot error equation can be seen as follows:

$$\begin{aligned} \dot{\varphi}_l &= -\omega_{ie}^n \times \varphi_l - \delta\omega_{ib}^n \\ \delta\dot{v}_l^n &= C_b^n f^b \times \varphi_l - 2\omega_{ie}^n \times \delta v_l^n + \delta f^n \\ \delta\dot{R}_l^n &= \delta V_l^n \end{aligned} \tag{1}$$

3.2 Relative Navigation State Equation

The state equation of the pilot and follower are:

$$\dot{x}_l = A_l x_l + G_l w_l \tag{2}$$

$$\dot{x}_f = A_f x_f + G_f w_f \tag{3}$$

We can get the Eq. (4) by take the pilot as the benchmark when subtract Eq. (2) from Eq. (3) to obtain:

$$\dot{x}_l - \dot{x}_f = A_l x_l - A_f x_f + G_l w_l - G_f w_f \tag{4}$$

where:

$$x_{lf} = x_l - x_f \quad w_{lf} = w_l - w_f \quad G_{lf} = G_l - G_f$$

Equation (4) contains both pilot state and follower state by transforming to eliminate follower state variables.

$$\begin{aligned} \dot{x}_{lf} &= A_l x_l - A_f x_l + A_f x_l - A_f x_f + G_l w_l - G_f w_l + G_f w_l - G_f w_f \\ &= (A_l - A_f)x_l + A_f(x_l - x_f) + (G_l - G_f)w_l + G_f(w_l - w_f) \\ &= (A_l - A_f)x_l + A_f x_{lf} + G_{lf} w_l + G_f w_{lf} \end{aligned} \tag{5}$$

It can be seen from Eq. (5) that in order to obtain the estimation of the state parameters of cooperative navigation, it is necessary to obtain the estimation of the follower's state first. Therefore, the final cooperative navigation state equation is as follows, in which the follower's state equation is combined:

$$\begin{aligned} &\dot{X} = AX + GW \\ &\begin{bmatrix} \dot{x}_l \\ \dot{x}_{lf} \end{bmatrix} = \begin{bmatrix} A_l & 0 \\ A_{lf} & A_f \end{bmatrix} \begin{bmatrix} x_l \\ x_{lf} \end{bmatrix} + \begin{bmatrix} G_l & 0 \\ G_{lf} & G_f \end{bmatrix} \begin{bmatrix} w_l \\ w_{lf} \end{bmatrix} \end{aligned} \tag{6}$$

4 Measurement Model

4.1 Data Link Distance Measurement Model

The data link uses pseudo code to measure the distance, which can obtain the distance information between missiles. The true position vector of the pilot $\boldsymbol{R}_l^n$ and follower $\boldsymbol{R}_f^n$ can be expressed as follows.

$$\boldsymbol{R}_l^n = \left[R_{lx}\ R_{ly}\ R_{lz}\right]^T\ \boldsymbol{R}_f^n = \left[R_{fx}\ R_{lf}\ R_{fz}\right]^T$$

The position vector of the navigator and follower inertial navigation output is:

$$\hat{\boldsymbol{R}}_l^n = \boldsymbol{R}_l^n + \delta R_l \tag{7}$$

$$\hat{\boldsymbol{R}}_f^n = \boldsymbol{R}_f^n + \delta R_f \tag{8}$$

The position Eq. (7) of missile inertial navigation output minus Eq. (8) is:

$$\hat{\boldsymbol{R}}_{lf}^n = \hat{\boldsymbol{R}}_l^n - \hat{\boldsymbol{R}}_f^n = \boldsymbol{R}_l^n - \boldsymbol{R}_f^n + \delta\boldsymbol{R}_l - \delta\boldsymbol{R}_f = \begin{bmatrix} R_{lx} - R_{fx} \\ R_{ly} - R_{fy} \\ R_{lz} - R_{fz} \end{bmatrix} + \begin{bmatrix} \delta R_{lx} - \delta R_{fx} \\ \delta R_{ly} - \delta R_{fy} \\ \delta R_{lz} - \delta R_{fz} \end{bmatrix}$$

where:

$$\boldsymbol{R}_{lf}^n = \boldsymbol{R}_l^n - \boldsymbol{R}_f^n \quad \delta\boldsymbol{R}_{lf}^n = \delta\boldsymbol{R}_l - \delta\boldsymbol{R}_f$$

Then we can get:

$$\hat{\boldsymbol{R}}_{lf}^n = \begin{bmatrix} R_{lfx} \\ R_{lfy} \\ R_{lfz} \end{bmatrix} = \boldsymbol{R}_{lf}^n + \delta\boldsymbol{R}_{lf}^n = \begin{bmatrix} \hat{R}_{lfx} \\ \hat{R}_{lfy} \\ \hat{R}_{lfz} \end{bmatrix} + \begin{bmatrix} \delta R_{lfx} \\ \delta R_{lfy} \\ \delta R_{lfz} \end{bmatrix} \tag{9}$$

The distance between bullets can be expressed as:

$$\hat{L} = \left|\hat{R}_{lf}^n\right| = \sqrt{\hat{R}_{lfx}^2 + \hat{R}_{lfy}^2 + \hat{R}_{lfz}^2}$$

The formula can be obtained by expanding the Taylor series at $\left(R_{lfx}\ R_{lfy}\ R_{lfz}\right)$ and removing the error of the first term:

$$\hat{L} = L + \frac{R_{lfx}}{L}\delta R_{lfx} + \frac{R_{lfy}}{L}\delta R_{lfy} + \frac{R_{lfz}}{L}\delta R_{lfz} \tag{10}$$

The distance measured by the data link ranging system can be discribed as:

$$\tilde{L} = L + \delta L \tag{11}$$

Subtract Eq. (11) from Eq. (10) to obtain:

$$Z_L = \hat{L} - \tilde{L} = \frac{R_{lfx}}{L}\delta R_{lfx} + \frac{R_{lfy}}{L}\delta R_{lfy} + \frac{R_{lfz}}{L}\delta R_{lfz} - \delta L \tag{12}$$

where: $\boldsymbol{M}_L = \left[\frac{R_{lfx}}{L}\ \frac{R_{lfy}}{L}\ \frac{R_{lfz}}{L}\right]$ $\boldsymbol{H}_L = \left[\boldsymbol{0}_{1\times3}\ \boldsymbol{0}_{1\times3}\ \boldsymbol{0}_{1\times3}\ \boldsymbol{M}_L\boldsymbol{0}_{1\times3}\ \boldsymbol{0}_{1\times3}\right]$

And: $Z_L = \boldsymbol{H}_L\boldsymbol{X} + \upsilon_L\ \upsilon_L = -\delta L$.

4.2 Radar Relative Position Measurement

The position vector between the leader and follower shells has.

The position vector $\hat{R}_{fl}^{n}$ between the leader and follower can be discribed:

$$\hat{R}_{lf}^{n} = R_{l}^{n} - R_{f}^{n} = R_{lf}^{n} + \delta R_{lf}^{n} \tag{13}$$

The vector b_f is formed when projected into the projectile body coordinate system of the slave projectile:

$$\hat{R}_{lf}^{b_f} = \hat{C}_{b_l}^{b_f}\hat{C}_{n}^{b_l}\hat{R}_{lf}^{n} \tag{14}$$

The relationship between the attitude misalignment angle of the leading projectile and the subordinate projectile and the attitude matrix is:

$$\hat{C}_{b_l}^{n} = (I - \Phi_l)C_{b_l}^{n} \tag{15}$$

$$\hat{C}_{b_f}^{n} = (I - \Phi_f)C_{b_f}^{n} \tag{16}$$

where Φ is the antisymmetric matrix composed of misalignment angle φ.

$$\Phi = \begin{bmatrix} 0 & -\varphi_z & \varphi_y \\ \varphi_z & 0 & -\varphi_x \\ -\varphi_y & \varphi_x & 0 \end{bmatrix}$$

The conversion matrix from the lead bomb system to the slave bomb system can be written as:

$$\hat{C}_{b_l}^{b_f} = \hat{C}_{n}^{b_f}\hat{C}_{b_l}^{n} = C_{n}^{b_f}(I + \Phi_f)(I - \Phi_l)C_{b_l}^{n} = C_{b_l}^{b_f} - C_{n}^{b_f}(\Phi_l - \Phi_f)C_{b_l}^{n} - C_{n}^{b_f}\Phi_f\Phi_l C_{b_l}^{n}$$

Ignore small product terms $C_{n}^{b_f}\Phi_f\Phi_l C_{b_l}^{n}$, Let $\Phi_{lf} = \Phi_l - \Phi_f$, and:

$$\hat{C}_{b_l}^{b_f} \approx C_{b_l}^{b_f} - C_{n}^{b_f}\Phi_{lf}C_{b_l}^{n} \tag{17}$$

After introducing Eqs. (13), (15) and (17) into Eq. (14), we get:

$$\begin{aligned}\hat{R}_{lf}^{b_f} &= \hat{C}_{b_l}^{b_f}\hat{C}_{n}^{b_l}\hat{R}_{lf}^{n} = \left[C_{b_l}^{b_f} - C_{n}^{b_f}\Phi_{lf}C_{b_l}^{n}\right]\left[C_{n_l}^{b} + C_{b_l}^{n}\Phi_l\right]\left(R_{lf}^{n} + \delta R_{lf}^{n}\right) \\ &= R_{lf}^{b_f} + C_{n}^{b_f}\Phi_l R_{lf}^{n} - C_{n}^{b_f}\Phi_{lf}R_{lf}^{n} + C_{n}^{b_f}\delta R_{lf}^{n} - C_{n}^{b_f}\Phi_{lf}\Phi_l R_{lf}^{n} \\ &\quad + \left(C_{n}^{b_f}\Phi_l + C_{n}^{b_f}\Phi_{lf} + C_{n}^{b_f}\Phi_{lf}\Phi_l\right)\delta R_{lf}^{n}\end{aligned}$$

After ignoring the small product term, we can get:

$$\hat{R}_{lf}^{b_f} = R_{lf}^{b_f} - C_{n}^{b_f}\left(R_{lf}^{n}\times\right)\varphi_l + C_{n}^{b_f}\left(R_{lf}^{n}\times\right)\varphi_{lf} + C_{n}^{b_f}\delta R_{lf}^{n} \tag{18}$$

The range $\tilde{d}$, azimuth $\tilde{\alpha}$ and elevation angle $\tilde{\beta}$ of the lead-in missile measured by the radar seeker of the slave missile are:

$$\tilde{d} = d + \varepsilon_d \quad \tilde{\alpha} = \alpha + \varepsilon_\alpha \quad \tilde{\beta} = \beta + \varepsilon_\beta$$

where: d, α, β are the true values of distance and azimuth; $\varepsilon_d, \varepsilon_\alpha, \varepsilon_\beta$ refers to ranging error and angle measurement error respectively.

The projection of the relative position vector of the lead-in missile measured by the radar seeker of the slave missile under the missile body system a of the slave missile is:

$$\tilde{r}_{lf}^{b_f} = \begin{bmatrix} \tilde{r}_x^{b_f} \\ \tilde{r}_y^{b_f} \\ \tilde{r}_z^{b_f} \end{bmatrix} = \begin{bmatrix} \tilde{d}\cos\tilde{\alpha}\sin\tilde{\beta} \\ \tilde{d}\cos\tilde{\alpha}\cos\tilde{\beta} \\ \tilde{d}\sin\tilde{\alpha} \end{bmatrix} \tag{19}$$

Owning to the ranging error and angle measurement error are small, the small product term is ignored and the sine and cosine of small angle are simplified as:

$$\begin{aligned} \tilde{r}_x^{b_f} &= (d + \varepsilon_d)\cos(\alpha + \varepsilon_\alpha)\sin\left(\beta + \varepsilon_\beta\right) \\ &\approx d\cos\alpha\sin\beta + d\varepsilon_\beta\cos\alpha\cos\beta - \varepsilon_\alpha\sin\alpha\sin\beta + \varepsilon_d\cos\alpha\sin\beta \end{aligned}$$

$$\begin{aligned} \tilde{r}_y^{b_f} &= (d + \varepsilon_d)\cos(\alpha + \varepsilon_\alpha)\cos\left(\beta + \varepsilon_\beta\right) \\ &\approx d\cos\alpha\cos\beta - d\varepsilon_\beta\cos\alpha\sin\beta - d\varepsilon_\alpha\sin\alpha\cos\beta + \varepsilon_d\cos\alpha\cos\beta \end{aligned}$$

$$\tilde{r}_z^{b_f} = (d + \varepsilon_d)\sin(\alpha + \varepsilon_\alpha) \approx d\sin\alpha + d\varepsilon_\alpha\cos\alpha + \varepsilon_d\sin\alpha$$

where:

$$\varepsilon_r = \begin{bmatrix} d\varepsilon_\beta\cos\alpha\cos\beta - \varepsilon_\alpha\sin\alpha\sin\beta + \varepsilon_d\cos\alpha\sin\beta \\ -d\varepsilon_\beta\cos\alpha\sin\beta - d\varepsilon_\alpha\sin\alpha\cos\beta + \varepsilon_d\cos\alpha\cos\beta \\ d\varepsilon_\alpha\cos\alpha + \varepsilon_d\sin\alpha \end{bmatrix}$$

$$\boldsymbol{R}_{lf}^{b_f} = \begin{bmatrix} r_x^{b_f} \\ r_y^{b_f} \\ r_z^{b_f} \end{bmatrix} = \begin{bmatrix} d\cos\alpha\sin\beta \\ d\cos\alpha\cos\beta \\ d\sin\alpha \end{bmatrix}$$

Then, Eq. (19) can be expressed as:

$$\tilde{\boldsymbol{r}}_{lf}^{b_f} = \boldsymbol{R}_{lf}^{b_f} + \varepsilon_r \tag{20}$$

Subtracting Eq. (20) from Eq. (18), the position measurement equation can be expressed as:

$$\boldsymbol{Z}_r = \hat{\boldsymbol{R}}_{lf}^{b_f} - \tilde{\boldsymbol{r}}_{lf}^{b_f} = \boldsymbol{H}_r\boldsymbol{X} + \boldsymbol{v}_r = -\boldsymbol{C}_n^{b_f}\left(\boldsymbol{R}_{lf}^n\times\right)\boldsymbol{\varphi}_l + \boldsymbol{C}_n^{b_f}\delta\boldsymbol{R}_{lf}^n + \boldsymbol{C}_n^{b_f}\left(\boldsymbol{R}_{lf}^n\times\right)\boldsymbol{\varphi}_{lf} - \boldsymbol{\varepsilon}_r \tag{21}$$

Then:

$$\boldsymbol{H}_r = \left[\boldsymbol{0}_{3\times3}\ \boldsymbol{0}_{3\times3}\ -\boldsymbol{C}_n^{b_f}\left(\boldsymbol{R}_{lf}^n\times\right)\ \boldsymbol{C}_n^{b_f}\ \boldsymbol{0}_{3\times3}\ \boldsymbol{C}_n^{b_f}\left(\boldsymbol{R}_{lf}^n\times\right)\right]$$

$$\boldsymbol{v}_r = -\boldsymbol{\varepsilon}_r$$

4.3 Radar Relative Velocity Measurement Model

The velocity measured by radar is the velocity in the direction of line of sight, and the tangential velocity of line of sight has missed, so the angular projection method cannot be used to decompose the velocity into other coordinate systems, see Fig. 2.

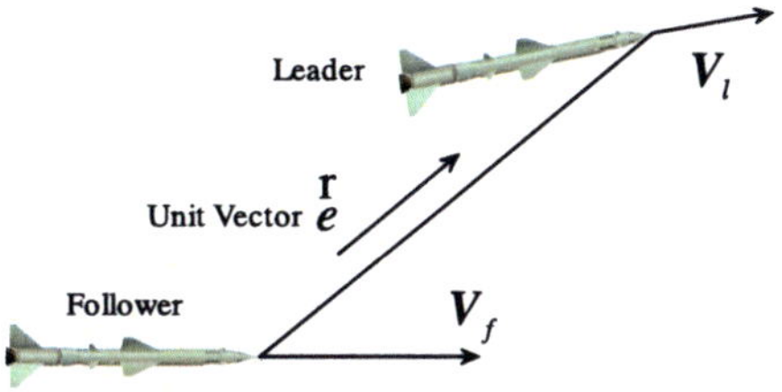

Fig. 2. Relative velocity

The output speeds of the pilot and the following inertial navigation are $\hat{V}_l^n$ and $\hat{V}_f^n$ respectively, which can be expressed as:

$$\hat{V}_l^n = V_l^n + \delta V_l \tag{22}$$

$$\hat{V}_f^n = V_f^n + \delta V_f \tag{23}$$

where V_l^n, V_f^n are the true velocity, δV_l, V_f are the errors of velocity.

$\vec{e}$ is the unit vector of the line of sight direction, which can be expressed as

$$\vec{e} = \frac{\vec{R}_{lf}^n}{\left|\vec{R}_{lf}^n\right|} = \left(\frac{R_{lfx}}{L}\ \frac{R_{lfy}}{L}\ \frac{R_{lfz}}{L} \right)$$

Then the relative velocity between the leader and follower shells can be expressed as:

$$\hat{V}_{lf}^e = \hat{V}_{lf}^n \cdot \vec{e} = \left(\hat{V}_l^n - \hat{V}_f^n\right)\vec{e} = \frac{V_{lfx}R_{lfx} + V_{lfy}R_{lfy} + V_{lfz}R_{lfz}}{\hat{L}} \tag{24}$$

Carry out Taylor series expansion of Eq. (23) at $\left(R_{lfx}\ R_{lfy}\ R_{lfz}\right)$ and $\left(V_{lfx}\ V_{lfy}\ V_{lfz}\right)$ respectively, and take the primary term error to obtain:

$$\begin{aligned} \hat{V}_{lf}^e = V_{lf}^e &+ \frac{\partial \hat{V}_{lf}^e}{\partial R_{lfx}} \cdot \delta R_{lfx} + \frac{\partial \hat{V}_{lf}^e}{\partial R_{lfy}} \cdot \delta R_{lfy} + \frac{\partial \hat{V}_{lf}^e}{\partial R_{lfz}} \cdot \delta R_{lfz} + \frac{\partial \hat{V}_{lf}^e}{\partial V_{lfx}} \cdot \delta V_{lfx} \\ &+ \frac{\partial \hat{V}_{lf}^e}{\partial V_{lfy}} \cdot \delta V_{lfy} + \frac{\partial \hat{V}_{lf}^e}{\partial V_{lfz}} \cdot \delta V_{lfz} \end{aligned} \tag{25}$$

where:

$$\frac{\partial \hat{V}_{lf}^e}{\partial R_{lfx}} = \frac{\left(R_{lfy}^2 + R_{lfz}^2\right)V_{lfx}}{\left(R_{lfx}^2 + R_{lfy}^2 + R_{lfz}^2\right)^{\frac{3}{2}}} - \frac{\left(V_{flty}R_{flty} + V_{fltz}R_{fltz}\right)R_{lfx}}{\left(R_{lfx}^2 + R_{lfy}^2 + R_{lfz}^2\right)^{\frac{3}{2}}}$$

$$\frac{\partial \hat{V}_{lf}^{e}}{\partial R_{lfy}} = \frac{\left(R_{lfx}^{2} + R_{lfz}^{2}\right)V_{lfy}}{\left(R_{lfx}^{2} + R_{lfy}^{2} + R_{lfz}^{2}\right)^{\frac{3}{2}}} - \frac{\left(V_{lfx}R_{lfx} + V_{lfz}R_{lfz}\right)\hat{R}_{lfy}}{\left(R_{lfx}^{2} + R_{lfy}^{2} + R_{lfz}^{2}\right)^{\frac{3}{2}}}$$

$$\frac{\partial \hat{V}_{lf}^{e}}{\partial R_{lfz}} = \frac{\left(R_{lfx}^{2} + R_{lfy}^{2}\right)V_{lfz}}{\left(R_{lfx}^{2} + R_{lfy}^{2} + R_{lfz}^{2}\right)^{\frac{3}{2}}} - \frac{\left(V_{lfx}R_{lfx} + V_{lfy}R_{lfy}\right)R_{lfz}}{\left(R_{lfx}^{2} + R_{lfy}^{2} + R_{lfz}^{2}\right)^{\frac{3}{2}}}$$

$$\frac{\partial \hat{V}_{lf}^{e}}{\partial V_{lfx}} = \frac{R_{lfx}}{\sqrt{R_{lfx}^{2} + R_{lfy}^{2} + R_{lfz}^{2}}}$$

$$\frac{\partial \hat{V}_{lf}^{e}}{\partial V_{lfy}} = \frac{R_{lfy}}{\sqrt{R_{lfx}^{2} + R_{lfy}^{2} + R_{lfz}^{2}}}$$

$$\frac{\partial \hat{V}_{lf}^{e}}{\partial V_{lfz}} = \frac{R_{lfz}}{\sqrt{R_{lfx}^{2} + R_{lfy}^{2} + R_{lfz}^{2}}}$$

So, we can immediately write

$$M_r = \left[\frac{\partial \hat{V}_{lf}^{e}}{\partial R_{lfx}} \; \frac{\partial \hat{V}_{lf}^{e}}{\partial R_{lfy}} \; \frac{\partial \hat{V}_{lf}^{e}}{\partial R_{lfz}} \right] \quad M_v = \left[\frac{\partial \hat{V}_{lf}^{e}}{\partial V_{lfx}} \; \frac{\partial \hat{V}_{lf}^{e}}{\partial V_{lfy}} \; \frac{\partial \hat{V}_{lf}^{e}}{\partial V_{lfz}} \right]$$

The radar seeker line can measure the velocity of the leading projectile relative to the subordinate projectile in the line of sight, it can be seen as follow

$$\tilde{V}_{lf}^{e} = V_{lf}^{e} + \varepsilon_{v} \tag{26}$$

Then subtract Eq. (26) from Eq. (25) to obtain:

$$\begin{aligned} Z_v &= \hat{V}_{lf}^{e} - \tilde{V}_{lf}^{e} \\ &= \frac{\partial \hat{V}_{lf}^{e}}{\partial R_{lfx}} \cdot \delta R_{lfx} + \frac{\partial \hat{V}_{lf}^{e}}{\partial R_{lfy}} \cdot \delta R_{lfy} + \frac{\partial \hat{V}_{lf}^{e}}{\partial R_{lfz}} \cdot \delta R_{lfz} + \frac{\partial \hat{V}_{lf}^{e}}{\partial V_{lfx}} \cdot \delta V_{lfx} \\ &+ \frac{\partial \hat{V}_{lf}^{e}}{\partial V_{lfy}} \cdot \delta V_{lfy} + \frac{\partial \hat{V}_{lf}^{e}}{\partial V_{lfz}} \cdot \delta V_{lfz} - \varepsilon_{v} \end{aligned}$$

$$H_v = \left[\mathbf{0}_{1\times3} \; \mathbf{0}_{1\times3} \; \mathbf{0}_{1\times3} \; M_r \; M_v \; \mathbf{0}_{1\times3} \right]$$

5 Simulation Analysis

The state variables of cooperative navigation are 18 dimensions in total, and the measurement equation has 5 dimensions, which includes data link ranging measurement dimension, radar position measurement dimension and radar speed measurement dimension. Kalman filter is used for processing in a sequential manner.

5.1 Construction of Simulation Environment

The cooperative navigation algorithm is verified by software simulation. The simulation environment structure is shown in Fig. 3. First the simulation trajectory of the pilot and follower is setted, and IMU inertial device errors are generated. The navigator carries out inertial navigation calculation independently, the pilot and follower transmit data through a high-speed data link capable of ranging. The follower uses its own seeker to measure the distance, angle and line of sight speed of the pilot. Data from follower participate in the collaborative navigation filtering to obtain the estimation of follower error and relative error, so as to realize the collaborative navigation filtering calculation.

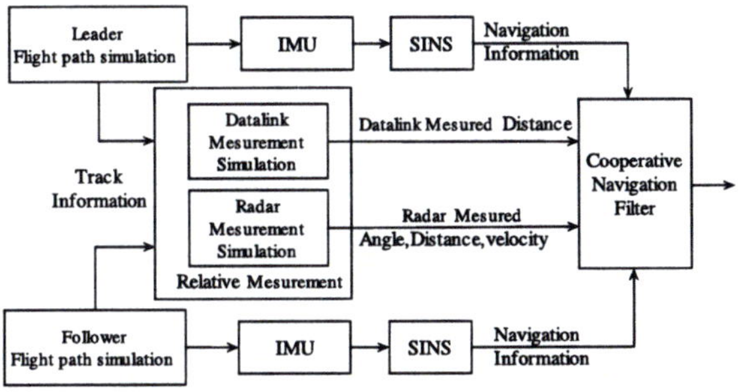

Fig. 3. Simulation platform structure diagram

5.2 Simulation Conditions

In the simulation, the gyro zero drift is set to 1°/h, the acceleration zero offset is 0.5 mg, and the speed measurement error is set to <5 m/s(1σ), rang measurement error <10 m(1σ). The relative attitude deviation between the pilot and the follower, and the relative attitude deviation angle of roll, heading and pitch are (20, 20, 20)′. The flight path of the pilot and follower in the solidified geographical system is shown in Fig. 4. The initial transverse spacing is about 500 m, the total simulation time is 200 s, and the flight distance is about 250 km.

The filtering estimation of cooperative navigation is carried out under the flight trajectory of the follower to obtain the estimation of the relative position, relative speed and relative attitude deviation of the follower relative to the pilot, as shown in Figs. 5, 6 and 7.

It can be seen from the figure that the relative position estimation error of the follower relative to the pilot is less than 20 m. The relative velocity estimation error is less than 5 m/s, with the cooperation of missile maneuver, the relative attitude estimation error is less than 8′.

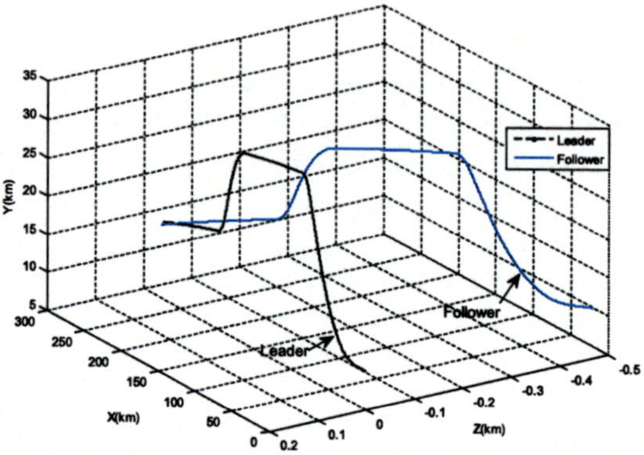

Fig. 4. Trajectory of Missile

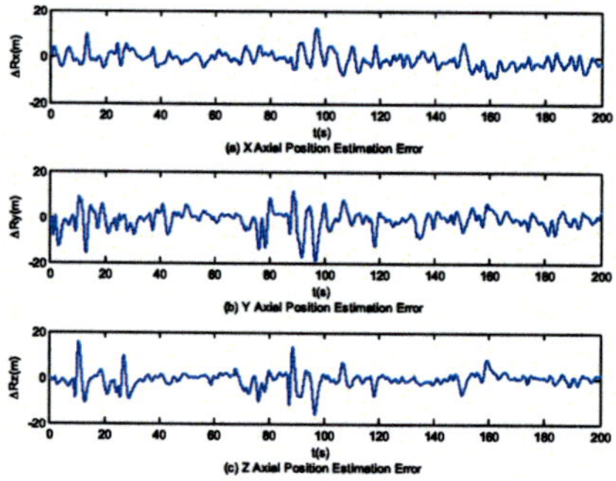

Fig. 5. Relative position estimation error

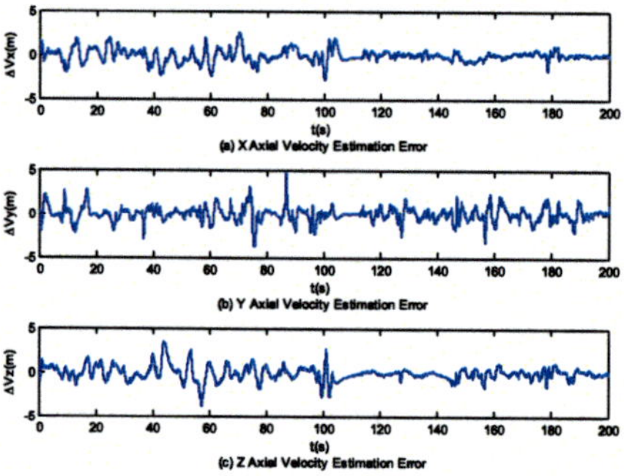

Fig. 6. Relative velocity estimation error

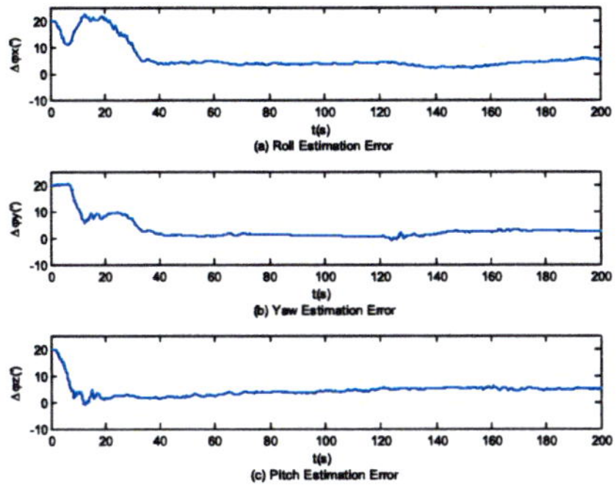

Fig. 7. Relative attitude estimation error

6 Summary

Cooperative detection, cooperative guidance and cooperative attack will be the necessary capabilities of future air-borne missiles to adapt to the new situation of future multi platform cooperative air combat. The basis of realizing multi missile cooperation is the real-time acquisition of relative information between multiple missiles. This paper presents a new method to obtain the relative information of Multiple Missiles Based on satellite navigation. This method can solve the problem of obtaining the relative velocity, relative position, especially the relative attitude deviation between air-borne missiles; it lays a foundation for the realization of multi missile cooperative operation of air-borne missiles.

References

1. Jun, W., Tonglin, R., Jin, L., et al.: The networking cooperative guidance combat pattern for air-borne missile. Aero Weaponry **5**, 32–35 (2015)
2. Xushengli, C.F., et al.: Application of multi missile cooperative technology in the development of air-defense missile. Electron. Opt. Control **2017**(2), 55–59
3. Hong, W., Jun, W., Wen bo, L., et al.: Multi-missile cooperative attacking strategy of new type of air-borne missile. Aero Weaponry **2011**(2), 7–11
4. Jianbo, Z., Shuxing, Y.: Review of multi-missile cooperative guidance[J]. Acta Aeronautica ET Astronautica Sinica **38**(1), 17–29 (2017)
5. Jipeng, D., Dong, T.: An Algorithm of Target Tracking Based on Two Missiles Cooperative Location Information[J]. Aero Weaponry **3**, 3–7 (2014)
6. Junhong, S., Song Shenmin, X., Shengli.: A Cooperative Guidance Law for Multiple Missiles to Intercept Maneuvering Target[J]. Journal of Astronautics **37**(12), 1432–1440 (2016)
7. Woods, J.O., Christian, J.A.: Lidar-based relative navigation with respect to non-cooperative objects[J]. Acta Astro- nautica **126**, 298–311 (2016)
8. Scaramuzza, D., Achtelik, M.C., Doitsidis, L., et al.: Vision-controlled micro flying robots: from system design to autonomous navigation and mapping in GPS-denied environments[J]. IEEE Robot. Autom. Mag. **21**(3), 26–40 (2014)
9. Fosbury, A.M.: Relative navigation of air vehicles. J. Guid. Control Dyn. **31**(4), 824–833 (2008)

10. Wu, T.-H., Flewelling, B., Leve, F., Lee, T.: Spacecraft relative attitude formation tracking on SO(3) based on line-of-sight measurements. In: American Control Conference (ACC)Washington, DC, USA, June 17–19, 2013
11. Jun, X., Xiong Zhi, Y., Yongjun, et al.: Close relative navigation algorithm for unmanned aerial vehicle aided by UWB relative measurement. J. Chin. Inert. Technol. **26**(3), 346–351 (2018)
12. Yandong, W., Yang Hongyu, F., Haifeng, et al.: Pseudorange differential GPS relative navigation without base-station for formation flight. J. Chin. Inert. Technol. **6**, 763–768 (2014)
13. Gross, J.N., Gu, Y., Rhudy, M.B.: Robust UAV relative navigation with DGPS, INS, and peer-to-peer radio ranging. IEEE Trans. Autom. Sci. Eng. **12**(3), 935–944 (2015)
14. Xiusen, W., Hongjin, Z., Shangyue, Z.: Relative navigation between vessels based on GPS single difference. J. Chin. Inert. Technol. **20**(4), 464–467 (2012)

Numerical Simulation of Air Flow Field in High-Speed Aviation Tank

Sheng-Zhe Shi[1,2](✉), Fan Wang[1], Zai-Bin Zuo[1], and Jun Jiao[1]

[1] AVIC Special Vehicle Research Institute, HuBei JingMen 448035, China
Shishengzhe05011232@126.com

[2] Key Aviation Scientific and Technological Laboratory of High Speed Hydrodynamic, HuBei JingMen 448035, China

Abstract. In order to quantify the increment of wind speed for high-speed trailer, this paper proposes to set sixteen measurement points of wind speed increment on it. Then, the experimental results are compared with those obtained by using the CFD method. It is shown in this paper that calculation results are 2% smaller than the ideal values obtained at different wind speed experimental points under the trailer, which leads to a wind speed increment steady region. Besides, the experimental and calculation results of wind speed increment are 8–10% smaller than the ideal values in this region, respectively. CFD shows feasibility and practicability in the analysis of air flow field for high-speed aviation tank. Surface crafts tend to be significantly affected by wind speed, such as seaplane, wing-in-ground effect ship, planning hull. Therefore, trailer front equipment can be used to mitigate the impact of wind speed on the test result.

Keywords: High speed aviation trailer · Air flow field · Experiment · Computational fluid dynamics (CFD)

1 Introduction

AS the basic component of hydrodynamic experiment conducted on seaplane, high-speed aviation tank enables the movement of seaplane model on the water. Trailer is powered by a motor fitted with high-precision speed control system, which ensures that the velocity error falls below 1% for the trailer. Trailer operation has an immediate effect on the precision of model movement speed and the accuracy of experimental results.

The high-speed hydrodynamic laboratory at Special vehicle research institute (605) is the only laboratory engaged in the study of hydrodynamics and experimental aviation science and technology. It is experimental speed highest (25 m/s) and the longest (510 m) towing tank in China. It focuses on the research of seaplane and amphibious aircraft, aircraft ditching experiment, the other surface ships and underwater device experiment. High-speed aviation tank is used to conduct hydrodynamic experiment and pneumatic experiment in close proximity to water[1]. All wind tunnel experiments require a satisfactory flow field of wind tunnel, which means that the flow parameters in time and space must be consistent[4, 5]. The ship travelling on the water surface at a low speed

Chinese Society of Aeronautics and Astronautics: CASTYSF 2022, LNEE 972, pp. 87–101, 2023.
https://doi.org/10.1007/978-981-19-7652-0_10

is usually affected insignificantly by wind speed [6, 7]. The quality of air flow field is essential for the seaplane, wing-in-ground effect ship, planning hull and so on. In order to reduce the impact of wind speed increment under the trailer, some high-speed aviation tank put model in the lead of trailer in seaplane experiment in overseas (Fig. 1).

Fig. 1. Towing tank at the Russian hydrodynamic research institute

2 Air Flow Field Experiment in High Speed Aviation Tank

The air flow field experiments on high-speed aviation tank were conducted on 3–5 November 2009 at the high-speed hydrodynamic laboratory of Special Vehicle Research Institute (605). The measurement equipment included five hole pitot tubes and a micro pressure sensor complete with amplifier and temperature compensation. Five hole pitot tubes were demarcated in Wuhan university of technology fluid mechanics lab wind tunnel. The calibration of computer data acquisition software could be displayed in real time to record wind speed and its direction.

Two experimental sections are at before and after in the region of model (Fig. 2). There were 8 points in each experimental section, denoted as A1, B1, C1, D1, E1, A2, B2 and C2, respectively (Fig. 3). The stability value of trailer speeds is 3, 6, 9, 12, 14, 15, 16 and 18m/s, respectively. A coordinate system is established at the middle profile bottom of the trailer. The trailer travels in the positive direction of X axis, the left direction is in the positive direction of Y axis, and the up direction is in the positive direction of Z axis. The direction of air flow on the horizontal and vertical planes of each measurement point oscillates within the plus or minus 2 degrees. It is acceptable in the error range of air flow and installation. It is considered that there is no deflection to the measurement region for air flow (Table 1).

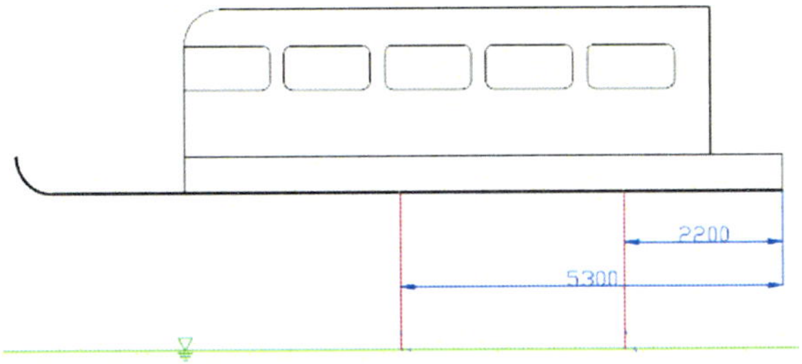

Fig. 2. Longitudinal position schematic diagram of air flow field measurement point

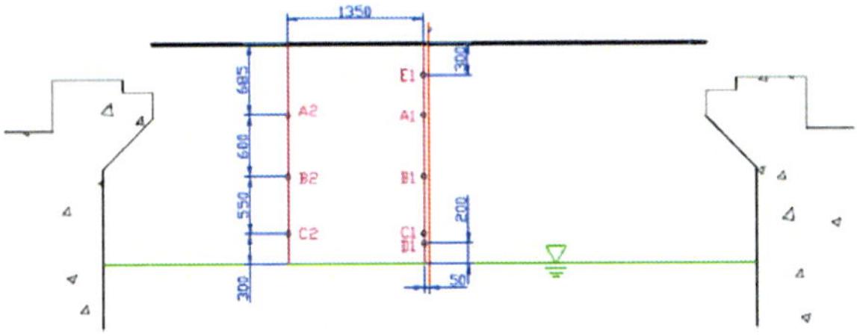

Fig. 3. Transverse position schematic diagram of air flow field measurement point

Table 1. Three-coordinate of air flow measurement points (unit: m)

Measurement point	X	Y	Z	Measurement point	X	Y	Z
1-A1	5.300	0.050	−0.685	2-A1	2.200	0.050	−0.685
1-A2	5.300	1.400	−0.685	2-A2	2.200	1.400	−0.685
1-B1	5.300	0.050	−1.285	2-B1	2.200	0.050	−1.285
1-B2	5.300	1.400	−1.285	2-B2	2.200	1.400	−1.285
1-C1	5.300	0.050	−1.835	2-C1	2.200	0.050	−1.835
1-C2	5.300	1.400	−1.835	2-C2	2.200	1.400	−1.835
1-D1	5.300	0.050	−1.935	2-D1	2.200	0.050	−1.935
1-E1	5.300	0.050	−0.300	2-E1	2.200	0.050	−0.300

3 Numerical Simulation of Air Flow Field in High Speed Aviation Tank

3.1 Basic Principles of Numerical Simulation

The core idea of numerical simulation is as follows: there are a series of variable value set of discrete points, instead of continuous physical field in the original time domain and space domain. The algebra equations of variable are established through certain principle and mode. Then, the algebra equations are solved to obtain the approximate solution to variables. Numerical simulation mainly includes pre-treatment, solution and post-processing.

3.2 Mesh Model and Computational Domain of High Speed Aviation Tank

Compared with the actual situation, the computational domain of high-speed aviation tank is simplified in order to create mesh while reducing the calculation time. The breadth of computational domain is the breadth of workshop. The actual tank breath under the trailer is 6.5 m. In order to reduce the difficulty in dividing mesh and computing time, however, the tank breath under the trailer is set to the same level as workshop breadth. There is little impact made on the result (Table 2).

Table 2. Principal dimensions of mesh model and computational domain of high speed aviation tank (unit: m)

Project	Parameter	Project	Parameter
Length of trailer pedestal	8.7	Distance between trailer stern and outlet	10.0
Total length	9.2	Distance between trailer bottom and water surface	2.135
Breath of control room	2.2	Length of computational domain	30.0
Breath of trailer pedestal	7.0	Breath of computational domain	10.2
Height of trailer(include pedestal)	2.4	Height of computational domain	8.365
Height of trailer pedestal	0.4		

3.3 Divide Mesh

The computational domain range is an extremely wide one. The computing region is the air flow field under the trailer. By increasing model mesh under the trailer, the flow field accuracy of air flow under the trailer can be improved (Fig. 4).

4 Comparison Between Calculation Result and Experimental Result

Time model is stability, material model is air, and flow model is separation flow. The equation of state is ideal air, viscous model is turbulence, and Renault average turbulence model is K-Epsilon turbulence.

There are 16 measurement points calculated. The calculation results of wind speed are compared with the experimental results of wind speed at all points.

The comparisons performed between the ideal value, the calculation result and the experimental result of wind speed at all 16 points are showed from Figs. 5, 6, 7, 8, 9, 10, 11, 12, 13, 14, 15, 16, 17, 18, 19 and 20. It can be observed from these figures that:

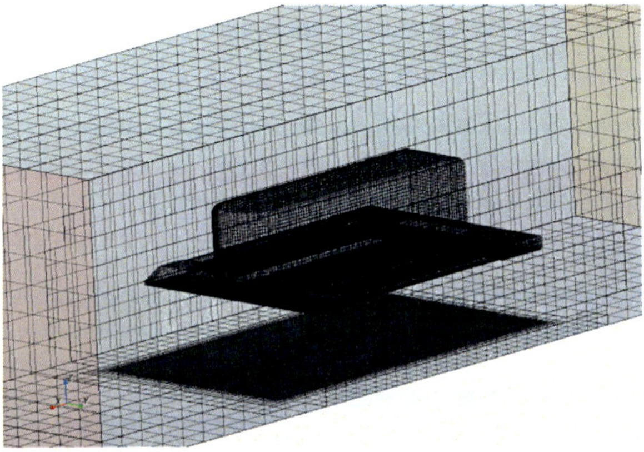

Fig. 4. Mesh model and computational domain of high-speed aviation tank

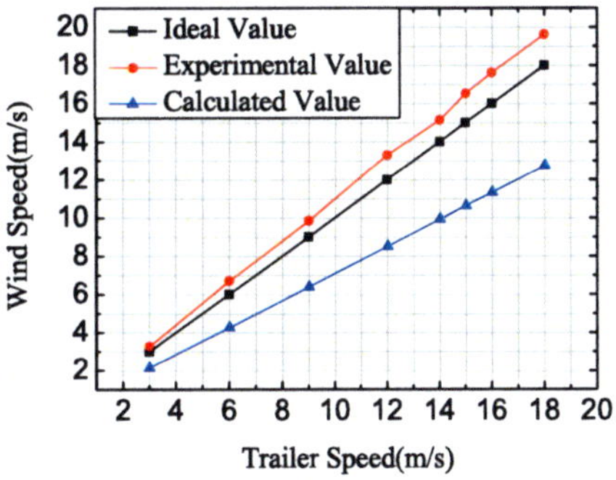

Fig. 5. Comparison of ideal value, calculation result and experimental result of wind speed at 1-A1

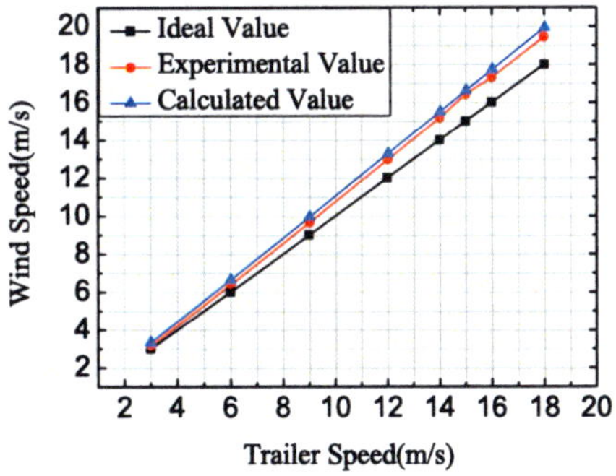

Fig. 6. Comparison of ideal value, calculation result and experimental result of wind speed at 1-A2

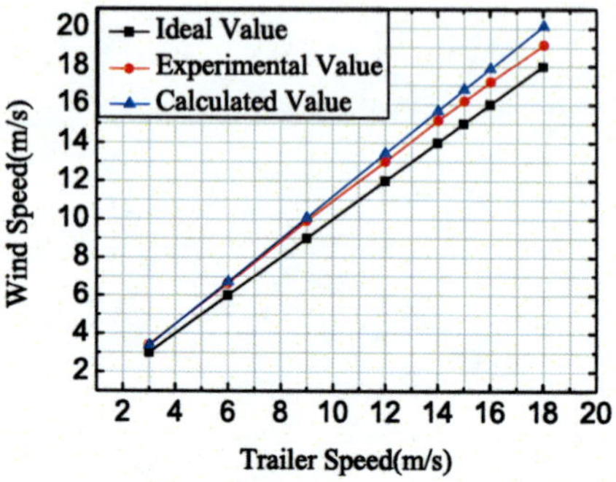

Fig. 7. Comparison of ideal value, calculation result and experimental result of wind speed at 1-B1

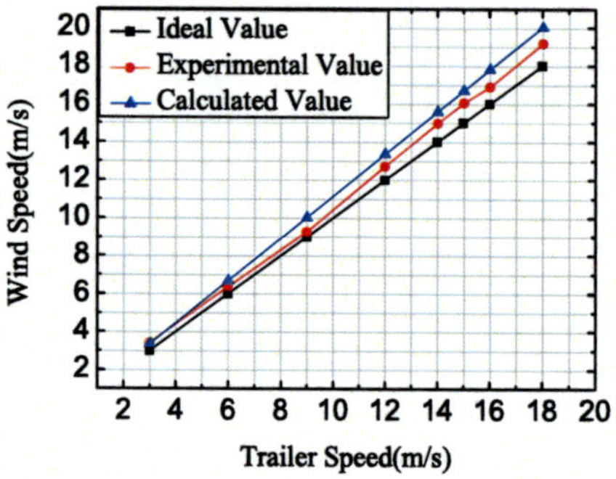

Fig. 8. Comparison of ideal value, calculation result and experimental result of wind speed at 1-B2

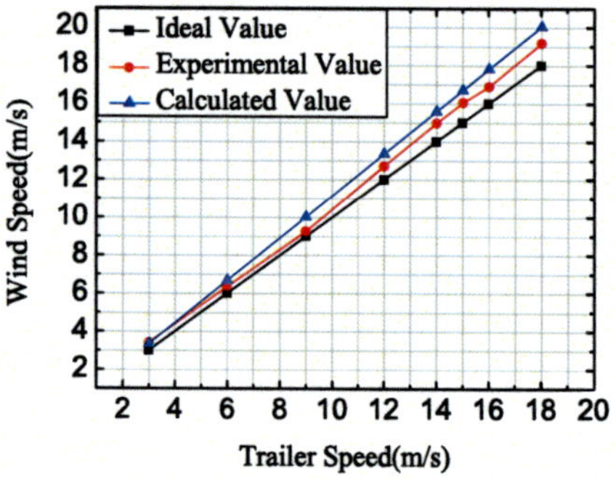

Fig. 9. Comparison of ideal value, calculation result and experimental result of wind speed at 1-C1

(a) Except the measurement points 1-A1, 1-E1, 2-A1, 2-E1, the calculation results are 2% less than the ideal value at the other measurement points. Since the measurement points are closer to water surface, the errors are less significant. It is demonstrated

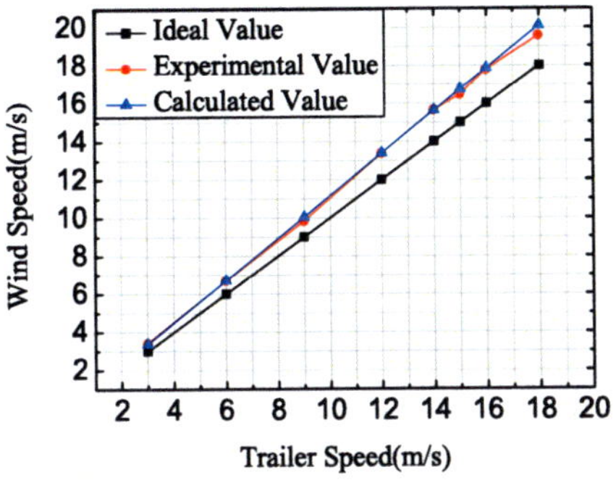

Fig. 10. Comparison of ideal value, calculation result and experimental result of wind speed at 1-C2

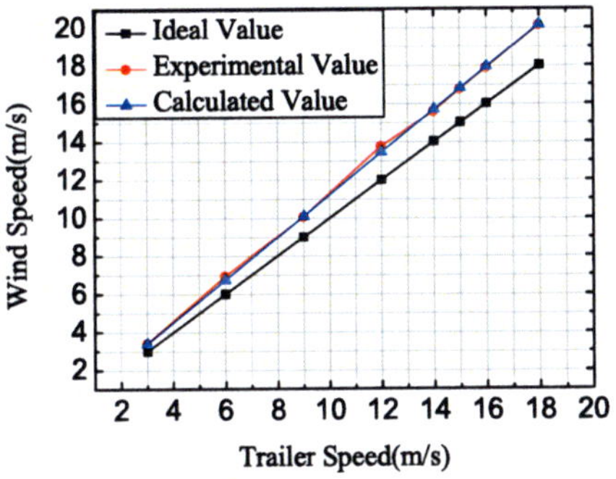

Fig. 11. Comparison of ideal value, calculation result and experimental result of wind speed at 1-D1

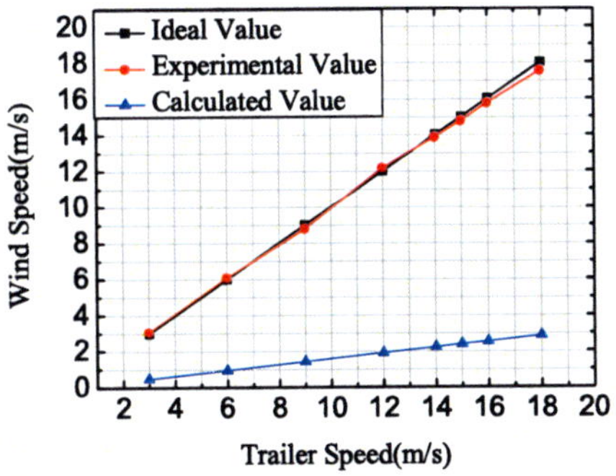

Fig. 12. Comparison of ideal value, calculation result and experimental result of wind speed at 1-E1

that CFD numerical simulation is suitable for analyzing the air flow field in high-speed aviation tank. The error of ideal value, calculation result and experimental result of wind speed at all measurement points makes no significant difference to

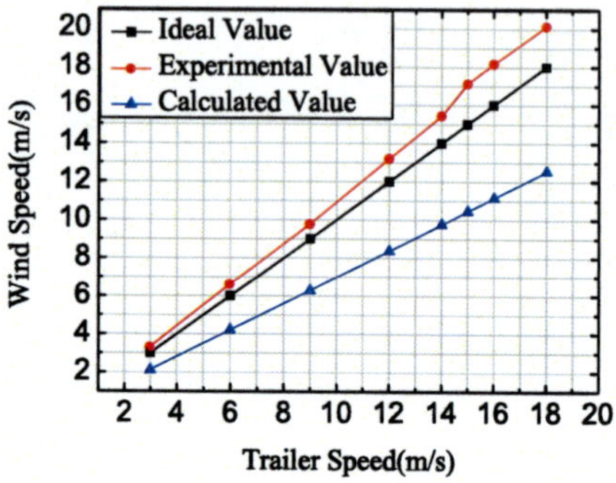

Fig. 13. Comparison of ideal value, calculation result and experimental result of wind speed at 2-A1

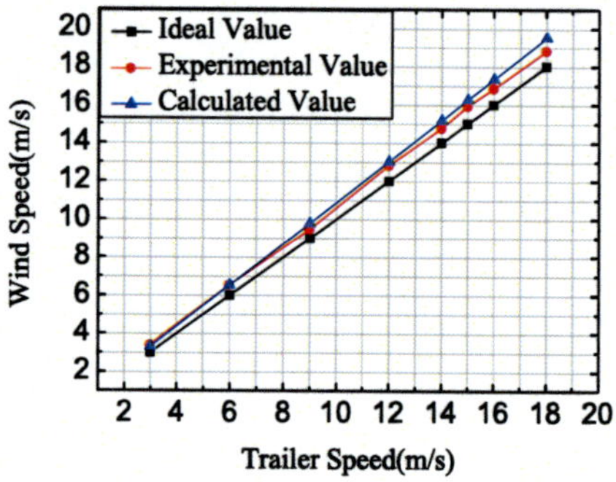

Fig. 14. Comparison of ideal value, calculation result and experimental result of wind speed at 2-A2

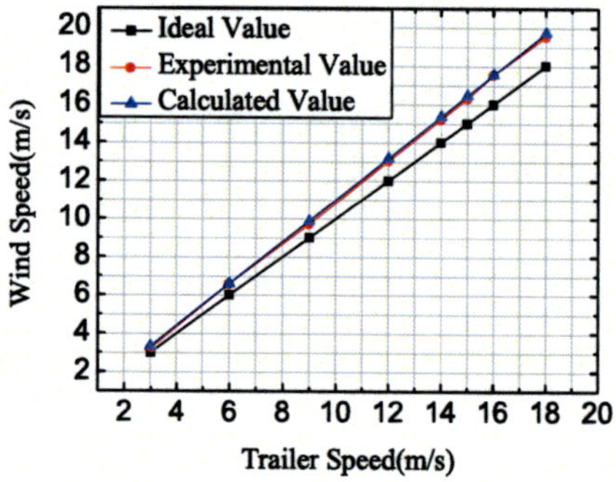

Fig. 15. Comparison of ideal value, calculation result and experimental result of wind speed at 2-B1

travel speed of the trailer. It shows that the increment of wind speed under the trailer is a natural result from the blocking effect.

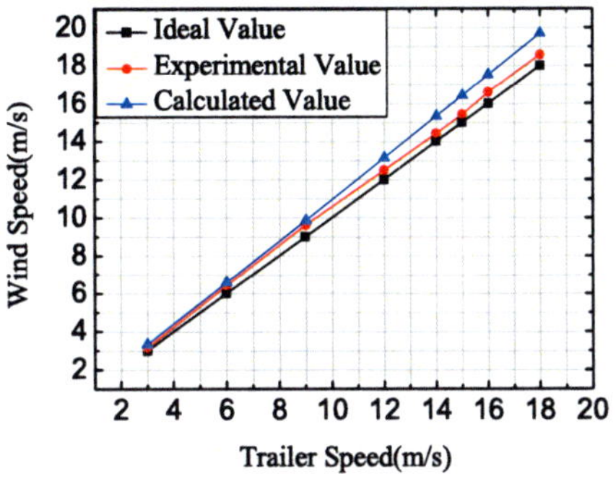

Fig. 16. Comparison of ideal value, calculation result and experimental result of wind speed at 2-B2

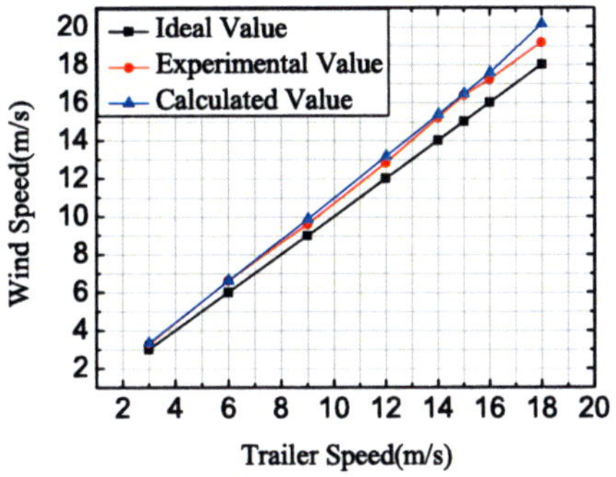

Fig. 17. Comparison of ideal value, calculation result and experimental result of wind speed at 2-C1

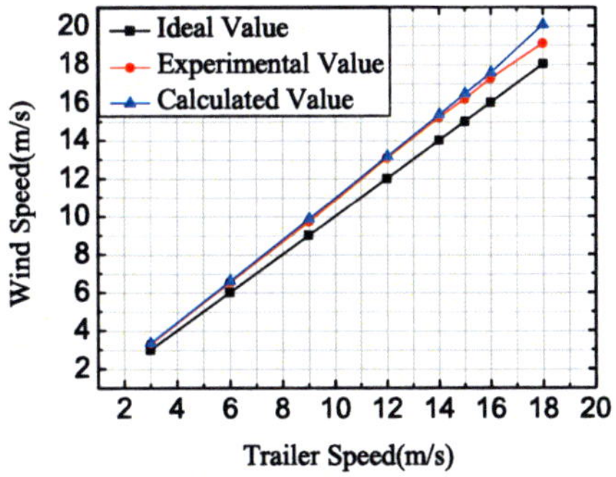

Fig. 18. Comparison of ideal value, calculation result and experimental result of wind speed at 2-C2

(b) As shown in Figs. 5, 7 and 9, the measurement point 1-A1 is closer to the trailer than the measurement points 1-B1 and 1-C1. Due to the effect of trailer boundary layer viscous, the calculation results of wind speed at the measurement points close

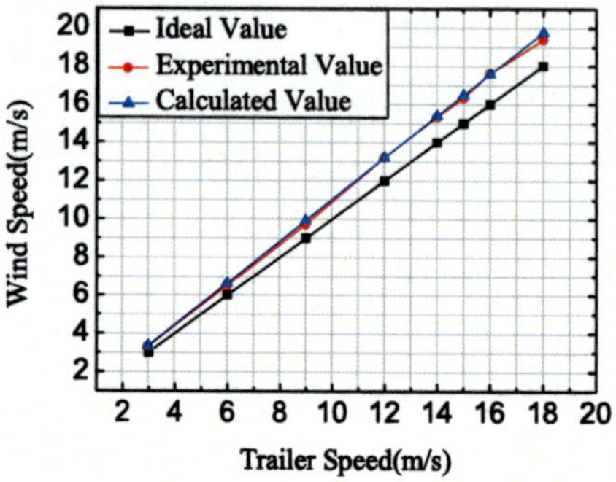

Fig. 19. Comparison of ideal value, calculation result and experimental result of wind speed at 2-D1

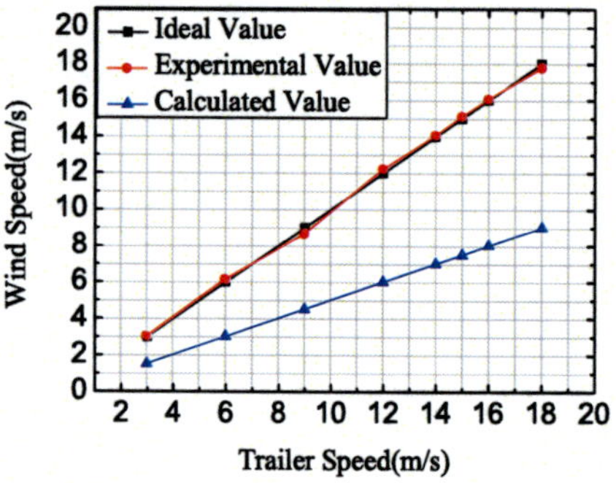

Fig. 20. Comparison of ideal value, calculation result and experimental result of wind speed at 2-E1

to the trailer are smaller compared to those far from the trailer. In Fig. 12, it is more obvious that the calculation result of wind speed is smaller at the measurement point 1-E1. However, the difference isn't significant in the experiment, because the bottom of trailer isn't a plane in practice. In order to facilitate installation and observation, the bottom of the trailer is made open at about 0.8m breadth. When air flows through the opening, it will enter into the trailer, and wind speed wouldn't be sharply reduced. The bottom of the trailer is close flat at modelling, with no pressure inlet at the bottom of the trailer for air to flow inwards. It can cause the error of calculation and experimental results. In Fig. 12, the error of wind speed at the measurement point 1-E1 approaches 80%. As shown in Fig. 20, the error of wind speed at the measurement point 2-E1 is close to 50%. This is mainly because the mesh model of the trailer is different from the actual situation.

(c) In Figs. 7, 8 and 11, the measurement points 1-B1, 1-C1 and 1-D1 are far from the bottom of the trailer but close to the water surface. It is less significant that the measurement points are affected by the trailer boundary layer, but the blocking effect caused by the trailer persists. The calculation and experimental results are about 8–10% larger than the ideal value at the three measurement points, respectively. The calculation result is almost identical to the experimental result. The error of three experimental results with each other is close to 2%, which shows that this height range is far from the bottom of trailer but close to the water surface. Differently, wind speed is stable at different positions of the range, which makes it usually the most important area for seaplane experiment.

(d) In Figs. 5, 7 and 9, the measurement points 1-A1, 1-B1, 1-C1 are in the front experimental section under the trailer. In Figs. 13, 15 and 17, the measurement points 2-A1, 2-B1, 2-C1 are in the after experimental section under the trailer. The corresponding calculation results and experimental results are comparable at the measurement points 1-A1, 1-B1, 1-C1 and measurement points 2-A1, 2-B1, 2-C1. The experimental results at the front point are about 2% larger than experimental results at the after point. It shows that wind speed increment is highly stable along the direction in which the trailer travels.

(e) In Figs. 5, 7 and 9, the measurement points 1-A1, 1-B1, 1-C1 are close to the middle longitudinal profile under the trailer. In Figs. 6, 8 and 10, the measurement point 1-A2, 1-B2, 1-C2 are far from the middle longitudinal profile under the trailer. The experimental error at the measurement point 1-A2 is less significant than the error at the measurement point 1-A1. This is possibly because the measurement point 1-A2 falls outside the region of breath of the control room and the impact of trailer boundary layer viscous is insignificant. This phenomenon persists at the measurement points 1-B1, 1-C1, 1-B2 and 1-C2. These measurement points are within the stable range of wind speed, and wind speed increment is stabilized at 8–10%.

The wind speed isolines and wind speed streamlines obtained in different sections when the trailer speed is 3 m/s, 9 m/s, or 16 m/s are showed from Figs. 21, 22, 23, 24, 25, 26, 27, 28 and 29, respectively. Through observation, it can be found out that:

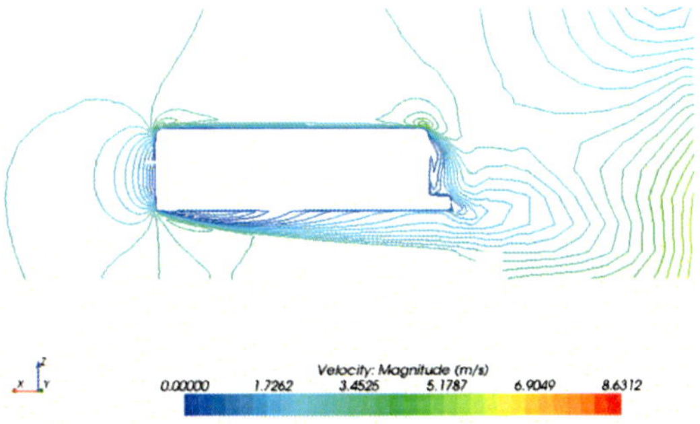

Fig. 21. Wind speed isoline in section Y = 0.05 m, when the trailer reaches a speed of 3 m/s

(f) In Figs. 21, 22, 24, 25, 27 and 28, the wind speed increment region shows no significant change with the speed at which the trailer travels. Wind speed isolines

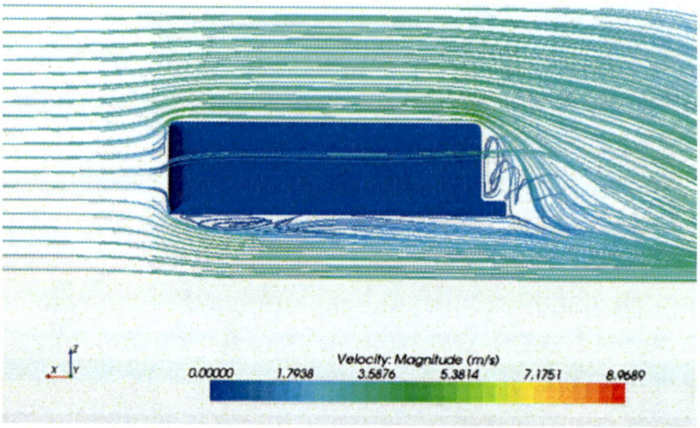

Fig. 22. Wind speed streamline in section Y = 0.05 m, when the trailer reaches a speed of 3 m/s

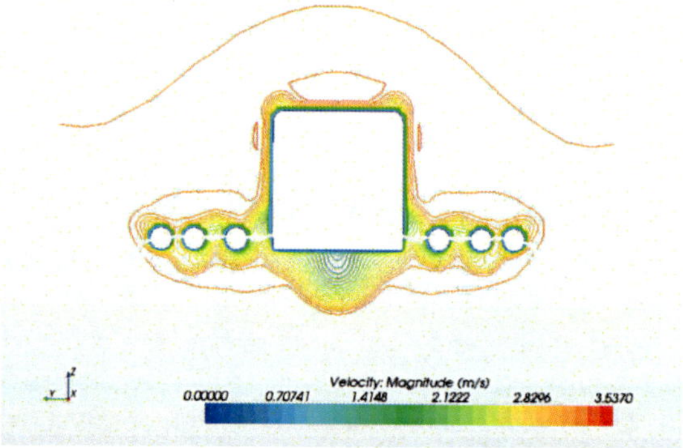

Fig. 23. Wind speed isoline in section X = 2.20 m, when the trailer reaches a speed of 3 m/s

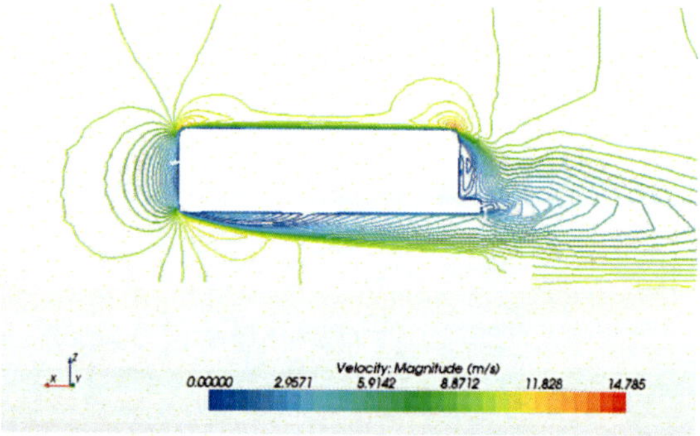

Fig. 24. Wind speed isoline in section Y = 0.05 m, when the trailer reaches a speed of 9 m/s

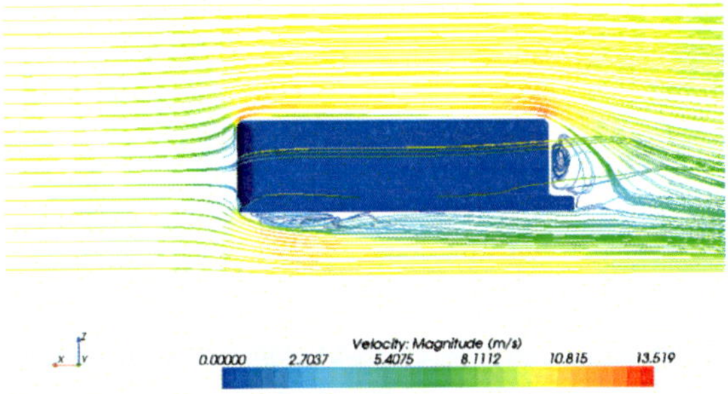

Fig. 25. Wind speed streamline in section Y = 0.05 m, when the trailer reaches a speed of 9 m/s

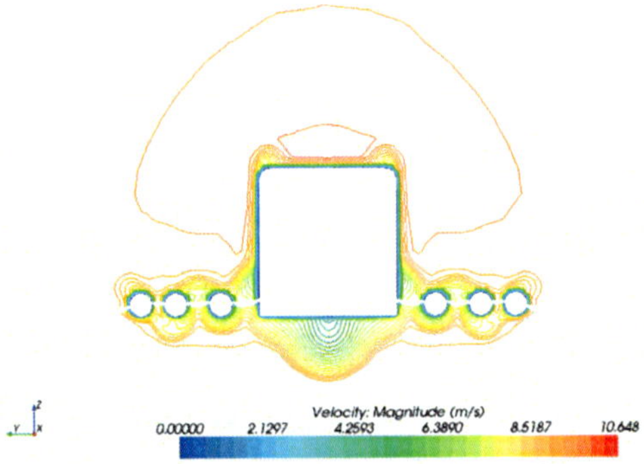

Fig. 26. Wind speed isoline in section X = 2.20 m, when the trailer reaches a speed of 9 m/s

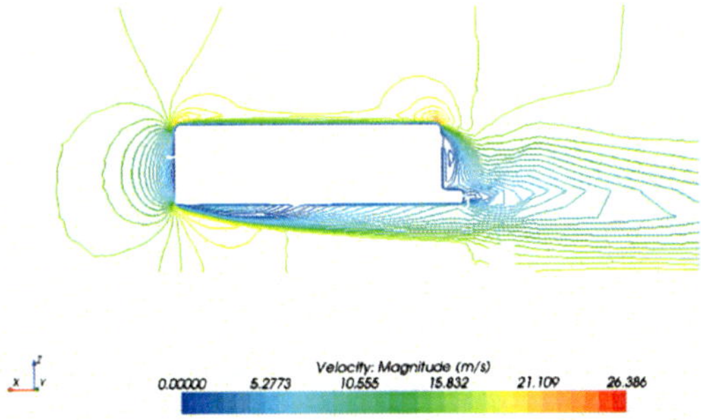

Fig. 27. Wind speed isoline in section Y = 0.05 m, when the trailer reaches a speed of 16 m/s

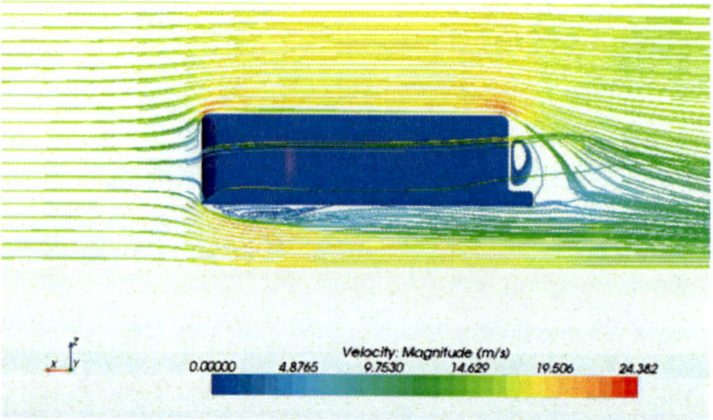

Fig. 28. Wind speed streamline in section Y = 0.05 m, when the trailer reaches a speed of 16 m/s

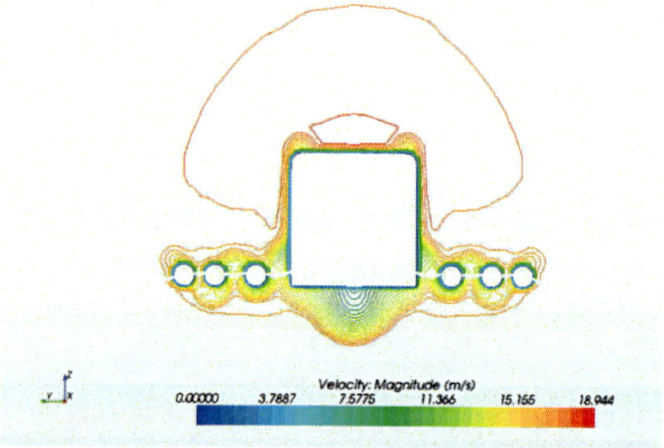

Fig. 29. Wind speed isoline in section X = 2.20 m, when the trailer reaches a speed of 16 m/s

and wind speed streamlines figure the result of comparison between the ideal value, the calculation result and experimental result of wind speed in a).

(g) Figures 21 and 22 show the effects of trailer boundary layer viscous. The region close to the bottom of the trailer is where the calculation result of wind speed is smaller. The range close to water surface but far from the bottom of the trailer is where wind speed is calculated to stabilize. Wind speed isolines and wind speed streamlines confirm the result of comparison between the ideal value, calculation result and experimental result of wind speed in (b) and (c).

(h) Figures 23, 26 and 27 show the effect of trailer boundary layer viscous in the region near the bottom of the trailer. There is a low wind speed range middle region formed under the trailer.

5 Conclusions

The numerical simulation of air flow field in high-speed aviation tank indicates that:

(1) Except 4 measurement points close to the bottom of the trailer, the calculation results are less than 2% of the ideal value at the other measurement points. The closer to water surface, the less significant the error. It is suggested that CFD numerical simulation is suitable for the analysis of air flow field in high-speed aviation tank.
(2) The errors between the ideal value, the calculation result and the experimental result of wind speed at all measurement points don't change significantly with the speed at which the trailer travels. It shows that the wind speed increment under the trailer is related to the blocking effect.
(3) Due to the effect of trailer boundary layer viscous, the calculation results of wind speed are smaller at a point close to the bottom of the trailer than at the point far from the bottom of the trailer. Besides, the calculation results of wind speed at a transverse position close to the middle longitudinal profile are smaller than at a transverse position far from the middle longitudinal profile.
(4) The experimental results are 2% smaller than the ideal value at different measurement points distant from the bottom of the trailer but close to water surface. This leads to a stable wind speed increment region. The experimental results and calculation results are 8–10% smaller than the ideal value in the region.
(5) Wind speed has a significant impact on surface crafts, such as seaplane, wing-in-ground effect ship and planning hull. Therefore, trailer front equipment can be built to reduce the impact of wind speed on the test result.

References

1. Chu, L.-T., Gu, B.: Model aerodynamic characteristics test in high speed aviation tank. In: Seaplane Symposium. Aviation Industrial Publishing, Beijing, pp. 349–357 (2011) (in Chinese)
2. Li, S.O.N.G., Rui, T.I.A.N., Zhan-Feng, L.I.U., Rui-Xing, W.A.N.G.: The research is based on the CFD for the speed uniformity on the exports section of low-speed wind tunnel. J. Inner Mongolia Agricult. Univ. **32**(2), 199–201 (2011). (in Chinese)
3. Xi-Fang, L.Y.U., Fan, L.I.U.: The design and establishment of air dynamic wind tunnel. J. Environ. Hygiene **3**(4), 356–360 (2013). (in Chinese)
4. Xun-Nian, W.A.N.G., Ming-Hong, Z.H.U., Xiao-Dong, G.O.N.G.: Current state and shortage of low-speed wind tunnel testing on large civil aircraft. Aeronaut. Manuf. Technol. **2**, 95–98 (2009). (in Chinese)
5. Kun, H.U., Zhi-Jun, Z.H.A.O., Ting-Wei, Q.I.A.N.: Experiments investigation of the flow field in a simple wind tunnel. Mech. Res. Appl. **118**(2), 74–75 (2012). (in Chinese)
6. Lu, C., Jiang, Z., Wang, T.: Influences of different airflow situations for ship airwake. Ship Sci. Technol. **31**(9), 38–42 (2009) (in Chinese)
7. Hong, W.H., Jiang, Z.-F., Wang, T.: Infulence on air-wake with different layout of ship superstructure. Chin. J. Ship Res. **4**(2), 54–68 (2009) (in Chinese)

Experimental Study on Low-Speed Wing Rock Characteristics of Low Aspect Ratio Flying Wing

Yanjie Shen[1,2](✉), Chen Bu[1,2], Yanling Wang[1,2], Shuai Feng[1,2], and Hao Chen[1,2]

[1] AVIC Aerodynamics Research Institute, Harbin 150001, China
shenyanjie0123@163.com
[2] Aviation Key Laboratory of Aerodynamics for Low Speed and High Reynolds Number, Harbin 150001, China

Abstract. Because of its high stealth and high aerodynamic efficiency, the low-aspect-ratio flying wing layout has become an advanced fighter development platform. The particularity of the layout reduces the lateral stability, which makes it face the instability of wing rock, and affects the maneuverability and flying quality of the aircraft. Therefore, it is necessary to study the wing rock characteristics of the low-aspect-ratio flying wing layout. Through various wind tunnel tests, the wing rock characteristics of the standard wing of the flying wing with low aspect ratio are comprehensively analyzed. The results show that there are sudden stall problems such as wing rock and wing drop in the low aspect ratio flying wing configuration, which are mainly due to the stable equilibrium point of zero-crossing of the rolling moment coefficient near the wing rock equilibrium position, and the zero-crossing of the rolling damping derivative. At the equilibrium point, the slope of the rolling moment coefficient is negative and stable.

Keywords: Low aspect ratio flying wing · Wing rock · Limit cycle oscillation · Unsteady aerodynamic · High angle of attack

Nomenclature

C_l	Roll moments coefficient
C_{l0}	Roll moments coefficient at zero angle of attack
$C_{l\phi}$	Roll moments coefficient slope, rad^{-1}
$C_{l\dot{\phi}}$	Roll-damping coefficient, rad^{-1}
θ	Pitch angle, deg
ϕ	Roll angle, deg
f	Frequency of oscillation, Hz
t	Time, sec
V	Free stream velocity, m/s
α	Angle of attack, deg
β	Angle of sideslip, deg

Chinese Society of Aeronautics and Astronautics: CASTYSF 2022, LNEE 972, pp. 102–114, 2023.
https://doi.org/10.1007/978-981-19-7652-0_11

1 First Section

Future fighter platforms are generally required to have the capabilities of long voyage, high speed, and Omni-directional broadband stealth [1]. The low aspect ratio flying wing layout inherits the advantages of high stealth and high aerodynamic efficiency of the flying wing layout, which can effectively improve the range and flight time, and is one of the main development directions of future fighter platforms.

High maneuverability has always been the goal of advanced fighters, and the application of thrust vector makes fighters have the ability of post-stall maneuvering at high angle of attack [2]. However, with the increase of flight angle of attack, due to the influence of flow separation and vortex breakdown, the aircraft is faced with complex and varied unsteady aerodynamic problems, among which the wing rock problem has become one of the focuses of the research on the instability characteristics of advanced layout fighter at high angle of attack because of its serious influence on flight safety and flying quality.

Early in Europe and America, static and dynamic wind tunnel tests and mathematical modeling and simulation methods were used to study the mechanism of wing rock problems of Harrier, Gnat, F-4, F-5, F-14, X-29A and flying wings with vertical tail. The results showed that the wing rock problems were related to the flow separation on the wing surface induced by the auxiliary fuel tank, leading edge slats, leading edge strakes and nose tips. Through dynamic stability tests, it was found that Supported by the U.S. Department of Defense and led by NASA and the U.S. Navy, the wing sudden stall (AWS) project developed the free rock test (FTR) method and the unsteady solution code, determined the criterion of wing rock problem, developed a comprehensive analysis method based on wind tunnel test, CFD and modeling simulation, and passed the tests of four layouts of F/A-18E, F/A-18C, AV-8B and F-16C [4–18]. Owens et al. evaluated the wing rock quality factor of F-35 under various wing configurations by using the above comprehensive analysis method, and extracted the dynamic data by using the parameter identification method, and studied and verified the control effect of leading edge slot-wing clearance on wing rock movement [19]. Chung et al. used free roll test and CFD to study the rocking characteristics and mechanism of T-50 coach wing, and analyzed the influence of head strake and dorsal fin on the rocking characteristics of wing [20]. Bing Wang et al. put forward a mathematical model of sine wave superimposed wing rock movement based on the spectrum analysis of free rolling time course [21]. Xueying Deng, Baofeng Ma, etc. conducted an experimental study on wing rock characteristics of wing-body combination at high angle of attack, focusing on the analysis of the influence of tip disturbance, wing position and Reynolds number on wing rock movement [22–25]. Zhiwei Shi, Dechen Wei, etc. conducted a wind tunnel test study on the rock-and-roll characteristics of canard aircraft, and analyzed the influence mechanism of pitch angle on the rock-and-roll characteristics of canard aircraft [26–28].

The layout of low-aspect-ratio flying wing is faced with the problem of verified rolling stability due to the cancellation of horizontal tail and vertical tail. In this paper, the variation law of static aerodynamic force (moment) and dynamic derivative at typical wing pitching angle is studied by experimental method, and the characteristics of aerodynamic force (moment) and dynamic derivative of low-aspect-ratio flying wing standard model under different free rolling motion states are analyzed.

2 Test Equipment and Model

2.1 Test Model

Fig. 1. FDM model

Based on the clean configuration designed by China Aerodynamics Research and Development Center, flaps, ailerons, all-moving wing tips and leading-edge flaps are added to the model, and the center of gravity is adjusted from 25% to 35% of the average aerodynamic chord length. The redesigned model is named Flying Wing Dynamic Standard Model (FDM), with an aspect ratio of 1.54, a leading-edge sweep angle of 65° and a trailing-edge sweep angle of 47°. Figure 1 shows the test model, with a scale of 1:11, a total length of 1.3923 m, an extension length of 1.0391 m, an average aerodynamic chord length of c = 0.8691 m, a reference area of 0.6991 m^2, a measured mass of 11.14 kg.

2.2 Wind Tunnel

The test was carried out in FL-51 wind tunnel of AVIC Aerodynamics Research Institute. FL-51 wind tunnel is a low-speed reflux wind tunnel characterized by dynamic test. It has interchangeable open/closed double test sections (Fig. 2). The cross section of the test section is 4.5 m × 3.5 m, and the maximum test wind speed is 100 m/s (closed test section) and 85 m/s (open test section). Static force measurement and free rolling test are carried out in the open test section, and dynamic derivative test is carried out in the closed test section.

2.3 Test Equipment

The static force test is carried out in FL-51 opening test section by using high angle of attack test support system (Fig. 3). The range of angle of attack is ±110°, and the range of sideslip angle is ±40° respectively. The aerodynamic force of the model is measured by a high-precision six-component balance. The repeatability of comprehensive loading of the balance is less than 0.05%F.S, and the comprehensive loading error of the balance is less than 0.07%F.S.

(a) Closed test section of FL-51 wind tunnel

(b) FL-51 wind tunnel opening test section

Fig. 2. FL-51wind tunnel closed and open test sections

Fig. 3. Test of high angle of attack

The dynamic derivative test is carried out in FL-51 closed test section by using a two-degree-of-freedom large amplitude oscillation test system (Fig. 4). The range of angle of attack is $\pm 130°$, sideslip angle is $\pm 40°$ and maximum oscillation frequency of the system is 2.5 Hz. Dynamic derivative test and static test use the same balance.

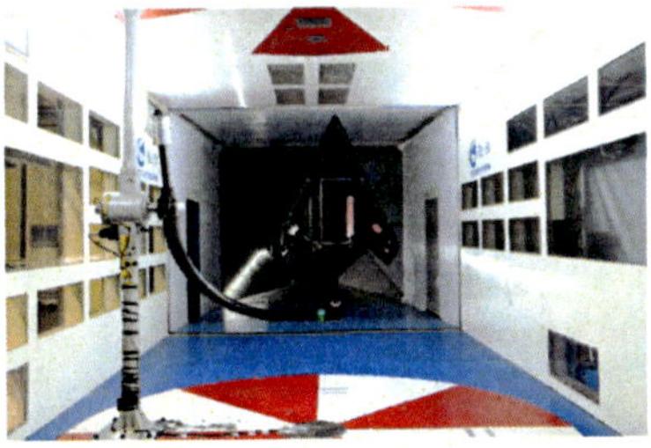

Fig. 4. Two-degree-of-freedom large amplitude oscillation test

The free rolling test adopts a high angle of attack test support system. The size and shape of the free rolling device are the same as those of the test balance, and the front and rear ends are connected with the model and the strut through cone fitting, as shown in Fig. 5. The front end of the free rolling device is the free end, which is connected with other parts through bearings. The rolling damping and inertia of the device are ignored in the test. In the free rolling wind tunnel test device, the center of mass of the model is located on the axis of the tail support mechanism, and the rolling freedom along the longitudinal axis of the model shaft system is unconstrained. When measuring the time history, the time history of rolling angle is recorded by rotary encoder, and after frequency doubling, it is collected by Advantech PCI1784 card and processed by the processing system developed based on LABVIEW platform.

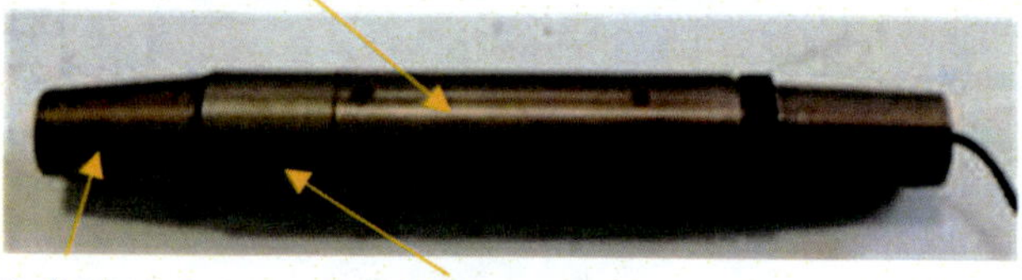

Fig. 5. Free rolling mechanism

3 Test Results and Discussion

3.1 Static and Dynamic Derivative Test

Figures 6 and 7 show the rolling moment curve and cloud image respectively. From Fig. 7, it can be seen that the rolling moment changes little with the roll angle when the angle of attack is 15° to 22°. As the roll angle needs to reach 12° to 19° when the angle of attack is 5° at low angle of attack, the influence data of the roll angle obtained at 5° intervals can't accurately reflect the change of static stability characteristics with the roll angle at small and medium pitch angles. Therefore, Figs. 6 and 7 can only be used to judge the change trend of rolling moment at large pitch angles. When the angle of attack is between 20° and 45°, the rolling moment increases and then decreases with the sideslip angle. Figure 7 shows there is a peak rolling moment near the rolling angle of 20° between the angles of attack of 28° and 45°. In this range, the rolling moment is in a static unstable state, and it returns to a static stable state after the angles of attack 45°. The model enters a deep stall state, and the wing flow is completely separated. At this time, the flow change on the windward side plays a decisive role in the rolling moment.

Because the results of conventional static force measurement test at low angle of attack are sparse, it can't be used to analyze the variation of rolling moment coefficient with rolling angle. Therefore, a force measurement method with fixed pitch angle and variable rolling angle is proposed in this paper, and the results are shown in Fig. 8. It

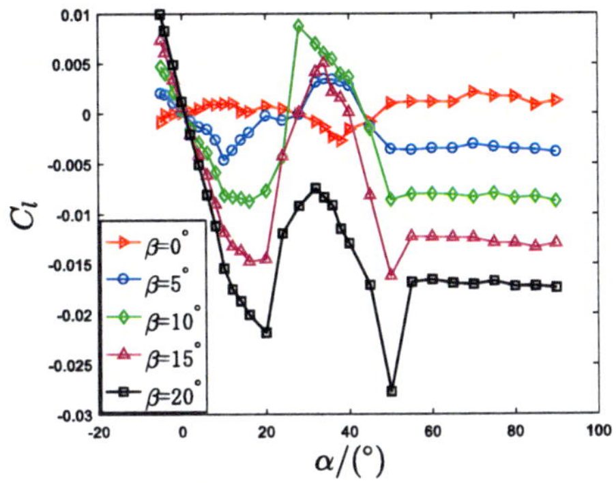

Fig. 6. Rolling moment coefficient of static force measurement

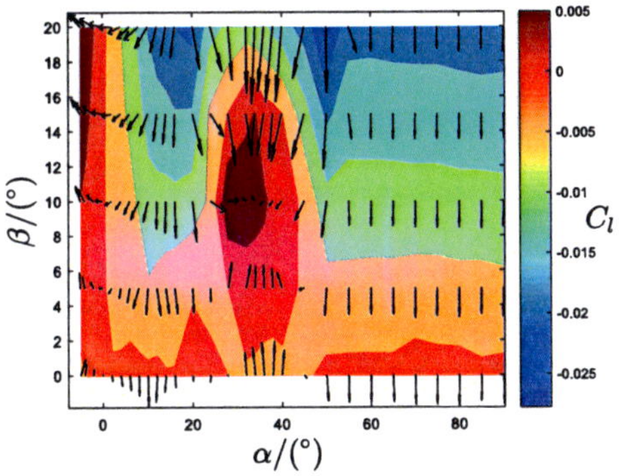

Fig. 7. Cloud picture of rolling moment coefficient in static force measurement

can be seen from the figure that when $\theta = 15°$, the rolling moment is in a static unstable state between $\phi = 5°$, and becomes a static stable state with the increase of rolling angle, and there are two equilibrium states at positive and negative rolling angles. When $\theta =$ 20° and 25°, $\phi = 0°$ is a static stable equilibrium state, and there is a symmetrical static stable equilibrium point near $\phi = 15°$. When $\theta = 30°$ and 40, $\phi = 0°$ is in a rolling static unstable state, and there are two static stable equilibrium points around $\phi = 30°$ and ϕ = 25° respectively. When $\theta = 50°$, $\phi = 0°$, there is a static equilibrium point of rolling moment.

Figures 9 and 10 show the rolling damping derivative curves and nephograms respectively. It can be seen from the figure that the rolling moment damping derivative is basically unchanged within β = 10°. When α ≥ 15° and β = 20°, the rolling damping derivative decreases significantly, and it becomes divergent at most angles of attack. When $\theta \geq 50°$, the rolling damping derivative changes little with the rolling angle. According to Fig. 9, it can be found that at different angles of attack, there is damping in the range of small roll angle, and there is a critical roll angle. Before the critical roll angle, the roll damping changes little, but when it is larger than the critical roll angle,

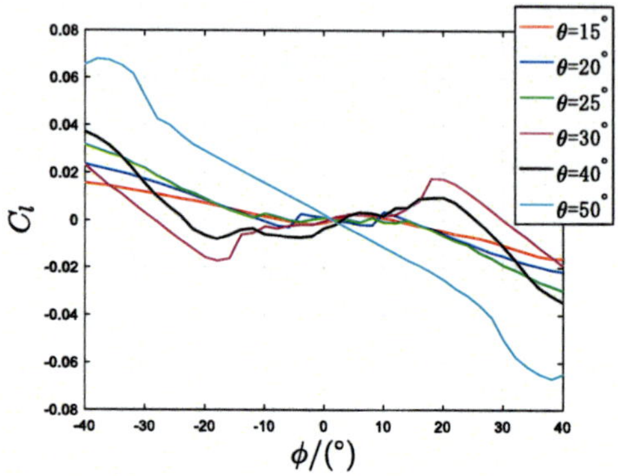

Fig. 8. Rolling moment coefficient varying with rolling angle

the roll damping derivative decreases sharply, and the dynamic derivative changes from damping to divergence.

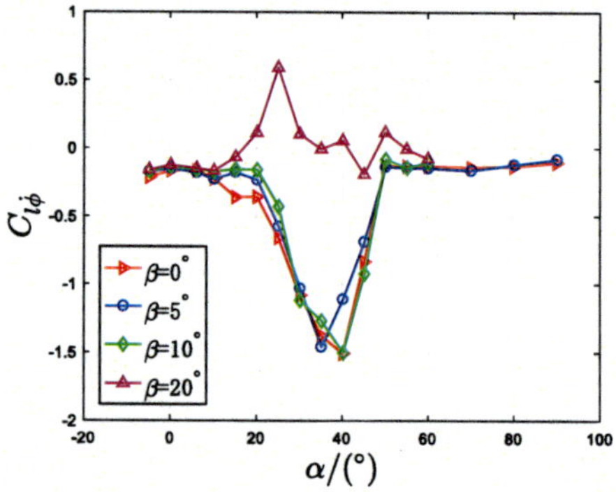

Fig. 9. Roll damping derivative of rolling moment

Figure 11 shows the test results of the rolling damping derivative with fixed pitch angle and variable rolling angle. The rolling damping derivative is close to 0 near the rolling angle of 30°, that is, the sign-changing area of the dynamic derivative is near the rolling angle of 30°. Figure 12 shows the curve of rolling damping derivative with dimensionless reduction frequency. It can be seen that with the increase of reduction frequency, the trend of rolling damping derivative with pitch angle tends to be gentle, and the reduction frequency has a great influence on the rolling damping derivative in the region ($15° \leq \alpha \leq 45°$) where the rolling static stability derivative changes sharply, because the vortex breakdown and flow separation in this region are affected by the oscillation frequency, and the hysteresis characteristics change.

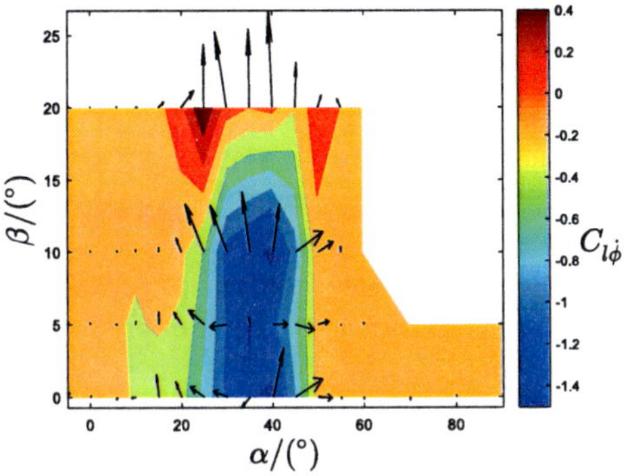

Fig. 10. Cloud picture of roll damping derivative of rolling moment

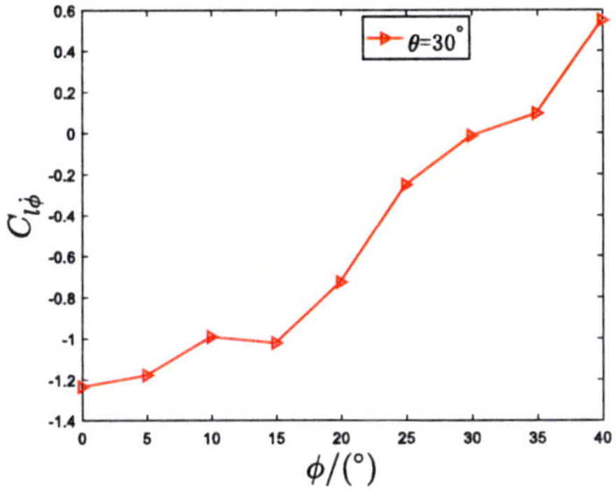

Fig. 11. Roll damping derivative of roll moment varying with roll angle

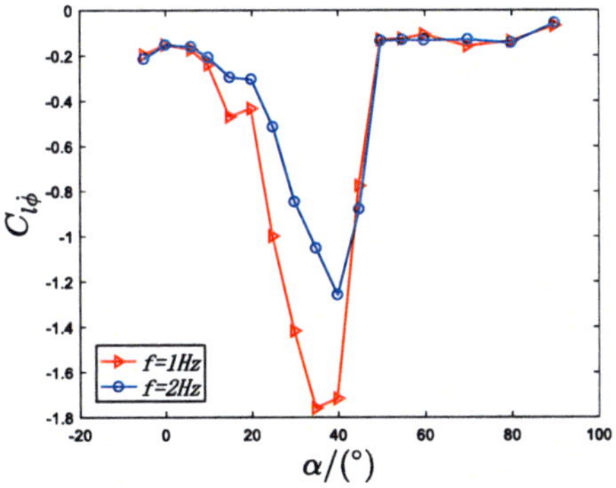

Fig. 12. Rolling damping derivative of rolling moment varying with oscillation frequency

3.2 Free-to-Roll Test

During the free rolling test, the wind speed is kept constant, the rolling freedom of the model is always released, and the pitch angle of the model increases stepwise from 0°.

At a given pitch angle, the time course of the rolling angle of the model is stabilized and recorded, with a recording time of 20s and a recording frequency of 100 Hz. The pitch angle first increases from 0° to 90° and then decreases to 0°, and the sequence of pitch angle is the same in the ascending and descending process. During the whole test, the deflection of the model rudder surface remains unchanged. Figure 13 shows the curve of roll balance angle and motion amplitude in the state of all zero rudder, in which θ is the pitch angle, ϕ is the roll angle, the model rolls to the right and turns positive, the fishbone line represents the range of model roll oscillation motion, the diamond represents the center position of model oscillation motion, and the test wind speed V = 20 m/s.

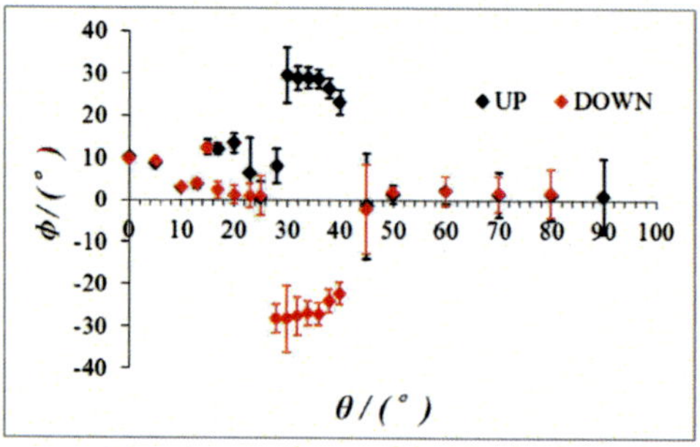

Fig. 13. Rolling balance angle and motion amplitude

From Fig. 13, we can find some rules of the free rolling motion of the standard model of the flying wing with low aspect ratio: (1) When the pitch angle is within 15°, the equilibrium points of the model are basically the same when going up and down, and no oscillation occurs; (2) From the pitch angle of 15° to 40°, the wing rock phenomenon occurs, where the balance position deviates from the roll angle of 0°, and the oscillation balance positions in the ascending and descending processes are different, and some angle oscillation balance positions show a certain symmetrical relationship with the roll angle of 0 as the symmetry plane; (3) From the pitch angle of 45° to 90°, the balance position of wing rock recovered to the vicinity of the roll angle of 0, and the oscillation balance position in the ascending and descending process was basically the same; (4) From the pitch angle of 28° to the pitch angle of 30°, the roll equilibrium position of the wing suddenly changes. The main reason is that the leading edge vortex breakdown position develops to the critical state of the wing area, resulting in a sudden and large increase of the rolling static instability. At the same time, due to the instability of the vortex breakdown position, the symmetry of the balance point between the positive and negative roll angles of the pitch angle of 28° is poor, which is the sensitive area for aircraft deviation.

According to the static test results, it can be seen in Fig. 13 that each roll balance angle corresponds to a roll static stability trim point (roll moment is zero), and the up-down roll balance angle is determined by the roll balance position of the pitch angle at the previous moment, that is, the roll balance angle is related to the initial release position. If the roll moment coefficient of the release position is less than zero, it will move towards the negative roll angle; if the roll moment coefficient is greater than zero, it will move towards the positive roll angle until it reaches the static stability balance point. Therefore, the equilibrium angle of wing rock oscillation is determined by the static equilibrium point of rolling moment and the release position.

Figures 14, 15 and 16 show the time history, phase plane and spectrum characteristic curves of wing rock at typical pitch angles, respectively.

It can be seen that when the pitch angle is 15°, there is weak oscillation and intermittent amplitude overshoot. This phenomenon is called wing tip drop [7], which is mainly caused by the asymmetric and unsteady change of wing separation area or leading edge vortex position, and the frequency of oscillation process is unstable.

When the pitch angle is 30°, it is a typical limit cycle oscillation state. Although the amplitude changes slightly with time, the frequency is always stable at 1.40 Hz. The concentric circles with varying diameters can be clearly seen on the phase plane, which shows that the oscillation equilibrium position is always around $\phi = 30°$. The main reason for the occurrence of limit cycle oscillation is the asymmetry of leading edge vortex rupture in the model area. When $\phi = 0°$, the $C_{l\phi}$ is unstable, while $\phi = 30°$, the $C_{l\phi}$ is stable and the $C_{l\dot{\phi}}$ is near zero.

When the pitch angle is 45°, the wing tip of the model drops, the frequency of wide oscillation state is low, and the oscillation equilibrium angle changes frequently. This phenomenon is due to the fact that the model has just entered the state of complete stall, and the flow on the wing surface of the model begins to show a state of complete separation intermittently. The $C_{l\phi}$ changes unsteadily between stability and instability, and the C_{l_0} (rolling moment at zero rolling angle) also changes frequently. At this time, the $C_{l\dot{\phi}}$ is in a damping state.

When the pitch angle is 50°, the model is still in the intermittent state of wing rock and wing tip drop, but there is a significant dominant oscillation frequency, which indicates that wing rock is dominant and the oscillation equilibrium position has an unsteady change. The main reason is that the model has entered a deep stall state, and the wing surface is completely separated, so that $C_{l\phi}$ is in a strong stable state, and the $C_{l\dot{\phi}}$ is close to zero. However, the instability of separated flow makes C_{l_0} change unsteadily.

4 Conclusion

The wing tip of the flying wing standard model with low aspect ratio drops when the pitch angle is 15°, and the wing rocks when the pitch angle is greater than 28°, and a serious roll deviation occurs when the pitch angle is 28° to 45°.

The change of rolling static stability can't be captured by the 5° sideslip angle interval at medium and small angles of attack, so it is suggested to carry out a force test with fixed pitch angle and variable roll angle to obtain a more accurate change rule of rolling moment.

The balance angle of wing rock of the flying wing standard model is determined by the static stable balance point of rolling moment and the release position, and the oscillation characteristics of wing rock at each balance angle have nothing to do with the release position.

When the pitch angle is between 20° and 45°, the roll damping derivative is significantly affected by the reduction frequency.

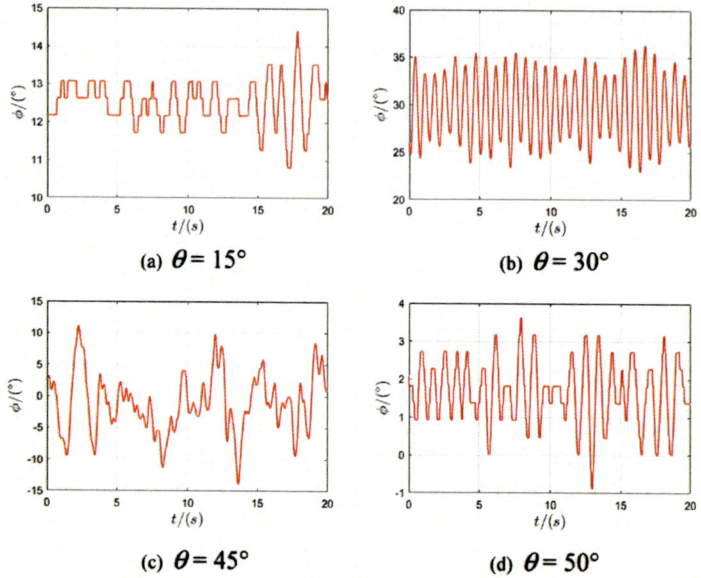

Fig. 14. Time history of wing rock under typical pitch angle

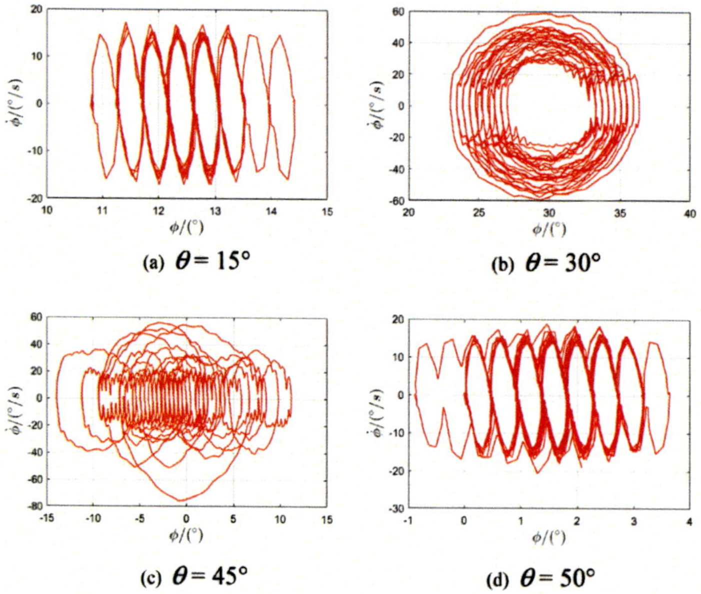

Fig. 15. Wing rock phase plane under typical pitch angle

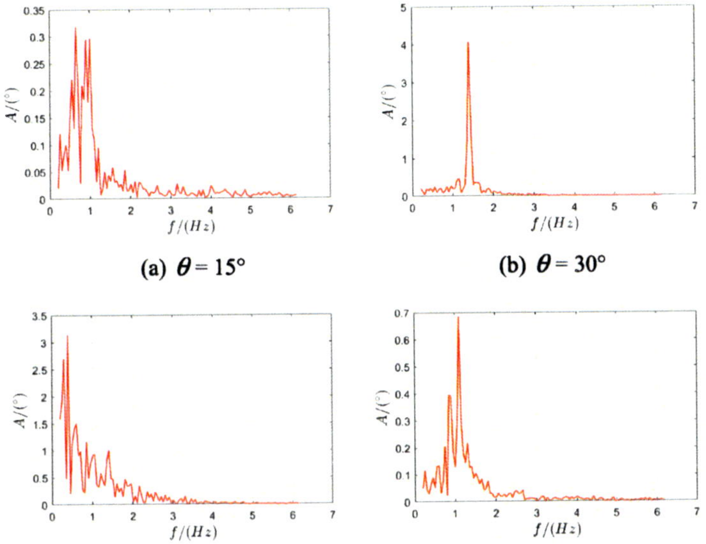

(a) $\theta = 15°$ (b) $\theta = 30°$

(c) $\theta = 45°$ (d) $\theta = 50°$

Fig. 16. Spectrum characteristic of wing rock under typical pitch angle

References

1. Yang, W.: Development of future fighters. Acta Aeronaut. Astronaut. Sin. **41**(6), 524377 (2020)
2. Wang, H.: Key technology and application prospect of fighter thrust vector. Acta Aeronaut. Astronaut. Sin. **41**(6), 524057 (2020)
3. Robert, C.N., Alain, P.: The unsteady aerodynamics of slender wings and aircraft undergoing large amplitude maneuvers. Progr. Aerosp. Sci. **39**(2003), 185–248 (2003)
4. Mcmillin, N., Hall, R., Lamar, J.: Transonic experimental observations of abrupt wing stall on an F/A-18E model. AIAA-2003-0591. Reston (2003)
5. Schuster, D., Byrd, J.: Transonic unsteady aerodynamics of the F/A-18E at conditions promoting abrupt wing stall. AIAA-2003-0593. Reston (2003)
6. Lamar, J., Capone, F., Hall, R.: AWS figure of merit (FOM) developed parameters from static transonic model tests. AIAA-2003-0745. Reston (2003)
7. Owens, B., Capone, F., Hall, R.: Free-to-roll analysis of abrupt wing stall on military aircraft at transonic speeds. AIAA-2003-0750. Reston (2003)
8. Capone, F., Owens, B., Hall, R.: Development of a free- to- roll transonic test capability. AIAA-2003-0749. Reston (2003)
9. Capone, F., Hall, B., Owens, B.: Recommended experimental procedures for evaluation of abrupt wing stall characteristics. AIAA-2003-0922. Reston (2003)
10. Woodson, S., Green, B., Chung, J.: Understanding abrupt wing stall (AWS) with CFD. AIAA-2003-0592. Reston (2003)
11. Forsythe, J., Woodson, S.: Unsteady CFD calculations of abrupt wing stall using detached-eddy simulation. AIAA-2003-0594. Reston (2003)
12. Green, B., Ott, J.: F/A-18C to E wing morphing study for the abrupt wing stall program. AIAA-2003-0925. Reston (2003)

13. Parikh, P., Chung, J.: A computational study of the AWS characteristics for various fighter jets: Part I, F/A-18E & F-16C. AIAA-2003-0746. Reston (2003)
14. Chung, J., Parikh, P.: A computational study of the abrupt wing stall (AWS) characteristics for various fighter jets: Part II, AV-8B and F/A-18C. AIAA-2003-0747. Reston (2003)
15. Woodson, S., Green, B., Chung, J.: Recommendations for CFD procedures for predicting abrupt wing stall (AWS): AIAA- 2003-0923. Reston (2003)
16. Roesch, M., Randall, B.: Flight test assessment of lateral activity. AIAA-2003-0748. Reston (2003)
17. Kokolios, A., Cook, S.: Use of piloted simulation for evaluation of abrupt wing stall characteristics. AIAA-2003-0924. Reston (2003)
18. Cook, S., Chambers, J.: An integrated approach to assessment of abrupt wing stall for advanced aircraft. AIAA-2003-0926. Reston (2003)
19. Owens, D.B., Mcconnell, J., Brandon, J.M.: Transonic free-to-roll analysis of the F-35(joint strike fighter) aircraft. J. Aircr. **43**(3), 608–615 (2006)
20. Chung, H.S., Cho, D., Kim, J.: Study on wing rock phenomenon of a fighter aircraft using free-to-roll wind tunnel test and dynamic CFD method. In: American Institute of Aeronautics and Astronautics AIAA Aviation 2019 Forum, Dallas (2019)
21. Wang, B., Deng, X.Y., Ma, B.F.: Modeling of wing rock based on frequency spectrum analysis to free-to-roll time history. J. Aircr. **48**(3), 805–811 (2011)
22. Ma, B.F., Deng, X.Y.: Tip-disturbance effects on asymmetric vortex breakdown around a chined forebody. J. Aircr. **45**(4), 1098–1104 (2008)
23. Ma, B.F., Deng, X.Y., Wang, B.: Effects of wing locations on wing rock induced by forebody vortices. Chin. J. Aeronaut. **29**(5), 1226–1236 (2016)
24. Wang, B., Deng, X.Y., Ma, B.F.: Effects of tip perturbation and wing locations on rolling oscillation induced by forebody vortice. Acta Mech. Sin. **26**(5), 787–791 (2010)
25. Ma, B.F., Wang, B., Deng, X.Y.: Effects of Reynolds numbers on wing rock induced by forebody vortices. AIAA J. 1–12 (2017)
26. Wei, D., Shi, Z., Ang, H.: Analysis of the influence and mechanism of pitch angle on the rock movement of canard layout aircraft. J. Aerodyn. **35**(2), 198–204 (2017)
27. Wei, D.C., Shi, Z.W., Ang, H.S.: Effects of pitch angle on wing-body rock of canard-configuration aircraft. Acta Aerodyn. Sin. **35**(2), 198–204 (2017)
28. Wei, D., Shi, Z., Geng, X.: Wind tunnel test on the rock characteristics of the wing body of a canard-type layout aircraft. Acta Aeronaut. **37**(10), 3003–3010 (2016)

Study on Swaging Parameters and Vibration Characteristics of 0.25inch Civil Aircraft Hydraulic Pipe

Yu Lingjie[1](✉), Chen Shikang[1], and Wang Yuan[2]

[1] Civil Aviation University of China, Tianjin 300300, China
15510928918@163.com
[2] AAA Dornier Technology (Tianjin) Co.LTD, Tianjin 300300, China

Abstract. For the dynamic response of hydraulic pipe, experts have a certain research foundation. However, for the research on the swaging repair of hydraulic pipe, the specific swaging parameters are not defined and analyzed in the manual. The research on the vibration characteristics of swaged pipe is almost blank. In this paper, the sine sweep experiment is carried out on a section of civil aircraft hydraulic pipe under typical swaging parameters, and the first six-order natural frequencies are obtained. Then, based on the basic principle of fluid-structure interaction, the modal analysis of the hydraulic pipe is carried out in ANSYS Workbench, the rationality of the Finite Element (FE) model is verified by comparing the simulation data with the experimental data. Taking the Swaging Length (SL) and Swaging Gap (SG) as the research object, 64 groups of hydraulic pipe models are established. Through modal analysis, it is determined that the maximum stress and strain of the hydraulic pipe is the pressure pipe joint, and the reasonable swaging parameters are obtained based on the actual swaging process. Finally, the harmonic response of the pipe model is analyzed based on the modal analysis results. It is found that the resonance is easier to occur at the first-order natural frequency, which is helpful to provide some references for the actual maintenance of aircraft hydraulic pipe.

Keywords: Hydraulic pipe · Swaging parameters · Sine sweep experiment · Modal analysis · Harmonic response

1 Introduction

1.1 Swaging Definition

Swaging is the controlled uniform reduction of the fitting end over the tube whereby forces are applied equally, 360 degrees around the fitting. Plastic deformation is the result of forming a permanent connection between the tubing and the pipe fitting during assembly. The swaging process also controls the elastic rebound of the tube within the fitting bore to create a sound mechanical connection and permanent fluid containment seal [1].

Chinese Society of Aeronautics and Astronautics: CASTYSF 2022, LNEE 972, pp. 115–135, 2023.
https://doi.org/10.1007/978-981-19-7652-0_12

1.2 Research Status of Pipe Swaging Technology

The metal hydraulic pipes in different areas of the fuselage of a fleet have been worn for many times and even caused leakage. For this kind of failure, using swaging to repair the pipe in pipe maintenance can save the time and manpower consumed by replacing the whole hydraulic pipe section.

The installation forms of aircraft hydraulic pipe include permanent installation and non permanent installation. For the pipe that does not need to be disassembled, the permanent installation form is usually adopted. During the installation of hydraulic pipe, there are often scratches, pits, cracks, scratches and other defects on the pipe surface. These defects will lead to the rupture and leakage of the pipe and have different effects on the service life of the pipe. Since the permanently installed pipe cannot be replaced directly, when the pipe is defective, it is usually maintained or restored to its reliability in use through maintenance [2].

At present, the repair of damaged hydraulic pipe mainly adopts two repair processes: welding and swaging. Swaging repair is to cut off the damaged part of the hydraulic pipe on the aircraft. The standard joint with the same material as its pipe shall be adopted, and the two sections of pipe after cutting shall be connected together by swaging. In this way, the process of disassembling and assembling the hydraulic pipe is avoided and the maintenance man hours are reduced. At the same time, the swaging equipment is convenient, flexible, easy to carry and easy to repair in other places. It is an economic, effective and convenient maintenance method [3].

In terms of the current international application degree, permanent connection has been widely used, not only in aviation manufacturing enterprises, but also in major airlines. Foreign advanced large and medium-sized aircraft manufacturing or maintenance enterprises mostly use permanent connection to manufacture and maintain hydraulic pipes. The leakage problem of pipe joints has been well solved and the reliability of the loop has been increased [1].

1.3 Research Status of Pipe Vibration

In the service process of civil aircraft, mechanical vibration and hydraulic oil pressure pulsation constantly excite the hydraulic pipe system to cause vibration. When the excitation frequency overlaps with the natural frequency of the pipe system, resonance will occur. According to the data, the vibration damage caused by resonance is the main cause of pipe damage [4].

Professor Ashley et al. [5]established the beam model and obtained the differential equation of transverse vibration of pipe based on the basic principle of fluid-structure interaction. Professor Harold combined theory with experiment in order to further analyze the pipe vibration phenomenon based on fluid-structure interaction [6]. Professor Paidousis et al. [7]studied the vibration of liquid filled pipe based on the added mass method, ignoring the interaction between oil and pipe. Bezborodov et al. [8]established a mathematical model based on the complexity of pipe shape, multi-axis linear and torsional stiffness and the influence of joint support damping, which can solve the pipe vibration and its stress amplitude. S. Riedelmeier et al. [9]designed and built a test device to analyze the influence of fluid-structure interaction on the pipe during water hammer impact. The results show that in the pipe with strong influence of bending geometry, it is of great significance to study the fluid-structure interaction.

Some domestic scholars analyze the vibration response of the pipe to solve the natural frequency of the pipe, so as to prevent the resonance caused by its coincidence with the excitation frequency of the system, and finally cause the pipe to be broken due to excessive stress. Professor Quan et al. [10] established a mathematical model based on fluid-structure interaction to obtain the natural frequency of hydraulic pipe. Ding [11] explored the mechanical vibration transmission law of civil aircraft hydraulic pipe system based on the vibration transmission path analysis method, and solved the natural frequency of typical hydraulic pipe based on the transmission matrix method. Song et al. [12] established an aircraft pipe design analysis model to quantitatively analyze the pipe dynamic characteristics and pressure loss, which provides an idea for the analysis of mechanical properties of civil aircraft pipe. Sang et al. [13] established the one-way fluid-structure interaction mathematical model of hydraulic pipe, and studied the variation law of pipe natural frequency with its structural parameters. Yuan et al. [14–20] studied the vibration of hydraulic pipe under different boundary conditions, solved the natural frequency of the pipe, and laid a theoretical foundation for the characteristic analysis, design and installation of hydraulic pipe. Li [21] selected a section of aviation hydraulic elbow for fluid-structure interaction analysis, and summarized the influence laws of pipe structure, material and fluid on pipe vibration characteristics. Zhang [22] used finite element analysis software to conduct modal simulation analysis of aircraft hydraulic pipe system, and compared the analysis results with the test data to prove the necessity of fluid-structure interaction simulation.

Taking the hydraulic pipe system of an aircraft as the object, Zhong [23] analyzed the vibration response of the pipe under pressure, frequency and fluid-structure interaction, and conducted experimental research. Through the comparison of the two data, the rationality of pipe system modeling and the accuracy of simulation were verified.

1.4 Research Content of This Paper

This pipe pressing technology is not generally mastered by route maintenance personnel, and the specific swaging parameters are not defined and analyzed in the Aircraft maintenance manual. Taking the swaging equipment of Deutsh company in Germany as an example, this paper discusses two typical permanent swaging repair parameters of hydraulic pipe, namely swaging length (SL) and swaging gap (SG).

This paper is arranged as follows: Firstly, in Sect. 2, six repeated sweep frequency vibration experiments are carried out on the hydraulic pipe of civil aircraft under typical swaging parameters, and the average value of the first six-order natural frequencies of the pipe is obtained. Next, in Sect. 3, the FE modeling of the civil aircraft hydraulic pipe in Sect. 2 is carried out in ANSYS Workbench, and the modal analysis of the pipe model is carried out based on the basic principle of fluid-structure interaction to obtain the first six-order natural frequencies. The rationality of the FE model is verified by comparing the experimental results with the numerical analysis results. Then, in Sect. 4, for the pipe models under different swaging parameters, modal analysis and harmonic response analysis are carried out respectively. Through the comprehensive analysis of the deformation corresponding to the first six-order main vibration modes, the maximum value of stress and strain and the first six-order natural frequencies of the model, the reasonable swaging parameters are obtained. And the frequency section prone to resonance is found out through the displacement-frequency response curve. Finally, the Sect. 5 summarizes the previous conclusions.

2 Experiment

2.1 Preparation and Process of Experiment

The pressure pipe joint used in this paper is MS21902 standard joint, so the maximum allowable gap of the pressing area is 7.6 mm [24]. The research object is a section of 0.25 inch aircraft hydraulic pipe, which is made of stainless steel, with an inner diameter of 6.35 mm and a thickness of 0.41 mm. According to the swaging repair operation manual of a company, the maximum swaging outer diameter of the pressure pipe joint used for the 0.25 inch hydraulic pipe is 8.014 mm, which is taken as 8 mm, and the thickness of the pressure pipe joint is the same as that of the hydraulic pipe. Two sections of 400 mm hydraulic pipes are pressed together with a 35 mm long pressure pipe joint, the SG is 2 mm, the pipe is filled with hydraulic oil, and both ends of the pipe are sealed with plugs.

The experimental system used in this paper is a three-dimensional comprehensive shaking table system, with a frequency range of 5–3000Hz and an acceleration amplitude of 1G. The z-axis is selected as the vibration direction.

The tooling used for vibration test shall be fixed on the upper surface of the three comprehensive vibration table, and the hydraulic pipe shall be fixed on the experimental tooling in the way of two-point constraint, which shall follow the principle of stress-free installation.

Stick one end of the IEPE acceleration sensor to the pressure pipe joint of the hydraulic pipe, connect the other end of the acceleration sensor to the signal acquisition instrument, and debug the signal acquisition instrument. The acceleration sensor here measures the radial acceleration of the hydraulic pipe.

Another acceleration sensor is glued to the experimental tooling, and the other end is connected with the vibration controller for control. Set the basic parameters of vibration test on the vibration control software, set the sweep frequency range to 5–3000Hz, set the type of acceleration sensor to charge type, and the maximum self-test voltage is 0.5 V (here, it can be adjusted appropriately according to the situation during the experiment, and the maximum self-test voltage shall not exceed 10 V), the frequency sweep is generally from low frequency to high frequency.

Start the vibration controller, conduct frequency sweep test on the hydraulic conduit, collect the acceleration response of the pipe along the Z-axis direction with the signal acquisition instrument, convert the time domain signal into the frequency domain signal, extract the natural frequency of the pipe and record it.

Repeat the test for 5 times according to the above steps and record the experimental data.

2.2 Results and Analysis

The results of the first and sixth experiments are imported into Origin, and the Acceleration-Frequency response is obtained through data processing, as shown in Fig. 1.

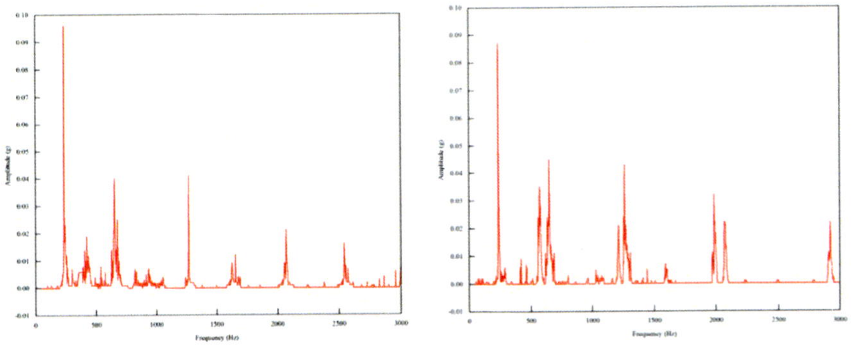

Fig. 1. Acceleration-frequency response curve

When the external excitation frequency is close to the natural frequency of the pipe, the acceleration amplitude will fluctuate greatly, and resonance may occur at this time. Therefore, the natural frequency of the pipe can be extracted according to the acceleration response curve. Here, the first six-order natural frequencies are taken (generally, the maximum frequency of the exciting force of civil aircraft in normal flight can reach 2000 Hz). The six experimental data of the first six-order natural frequencies of the pipe are shown in Table 1.

Take the average of the second to sixth test data as the test results of the pipe natural frequency. Due to the large difference between the first experimental data and the experimental data of other groups, it is ignored. The possible reasons are inaccurate equipment debugging and human factors during the first experiment.

Table 1. The first six-order natural frequencies of the pipe

Order	Natural frequency (Hz)						Average value (Hz)
	Test 1	Test 2	Test 3	Test 4	Test 5	Test 6	
1	236.08	236.09	236.09	236.14	236.04	236.08	236.088
2	423.17	530.84	576.02	616.86	532.29	569.24	565.05
3	647.61	647.62	647.62	647.69	647.56	647.6	647.618
4	955.98	1144.6	1226.6	1262.2	1147.2	1214.1	1198.94
5	1262.1	1262.1	1262.1	1301.6	1262	1262.1	1269.98
6	1656.3	1896.9	2004.9	2071.4	1900.3	1988.3	1972.36

3 Simulation and Discussion

3.1 Modeling and Modal Analysis

Firstly, the hydraulic pipe model is preliminarily established in CATIA 3D modeling software and saved as a suffix STP file. Then import the model into the Geometry module of ANSYS Workbench and fill the fluid in the pipe with the "fill" command. In order to facilitate the calculation, the ZX plane is set as the symmetrical plane, and each plane of the model is named. Figure 2 shows the hydraulic pipe model.

Then define the material parameters of pipe and fluid, as shown in Table 2. Mesh the fluid in Mechanical, as shown in Fig. 3.

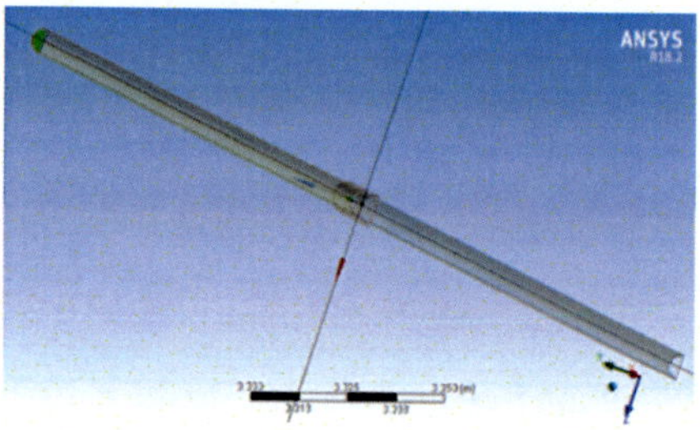

Fig. 2. Diagram of pipe model

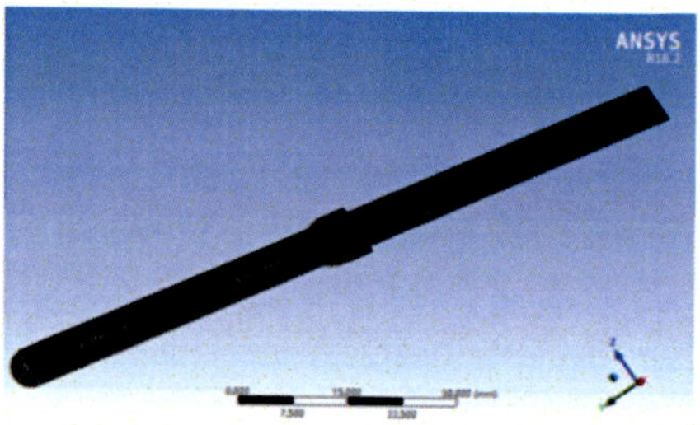

Fig. 3. Meshing of fluid

Table 2. The material parameters of pipe and fluid

Material	Density (kg/m^3)	Elastic/bulk modulus (Pa)	Poisson's ratio	Dynamic viscosity (Pa·s)
0Cr18Ni9	7750	1.93e11	0.31	–
Hydraulic liquid	960	1.69e11	–	0.048

Next, solve the fluid part in fluent. Set the inlet velocity of the fluid as 0.5m/s and the outlet pressure as 101325 Pa. Select aviation hydraulic oil as the fluid and set the fluid-solid interface as the wall. After the solution is completed, the velocity cloud diagram of the fluid symmetry plane is shown in Fig. 4.

Finally, carry out modal analysis of the pipe model in Mechanical. Bonding contact is set between the pressure pipe joint and the pipe and between the two pipes to be swaged. Fixed constraints are applied at both ends of the pipe and at both ends of the pressure pipe joint. Apply 70MPa pressure on the surface of the pressure pipe joint, and guide the

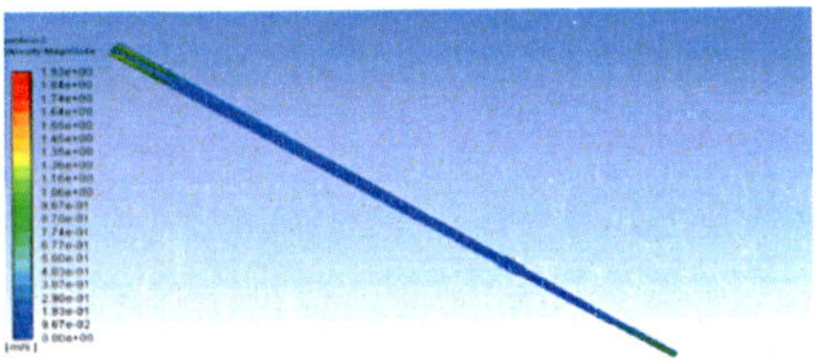

Fig. 4. Velocity cloud diagram

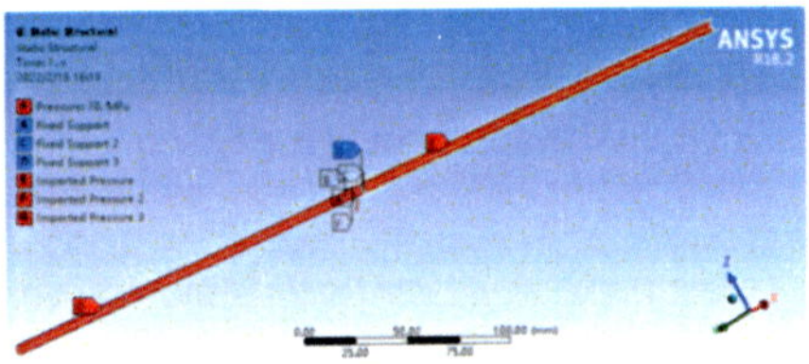

Fig. 5. Diagram of constraints of pipe

pressure of the fluid-solid interface obtained after fluent solution to the inner surface of the pipe. All constraints on the pipe are shown in Fig. 5. After the solution is completed, the first six-order natural frequencies of the pipe model are obtained. Figure 6 shows the first six-order main vibration modes of the pipe model.

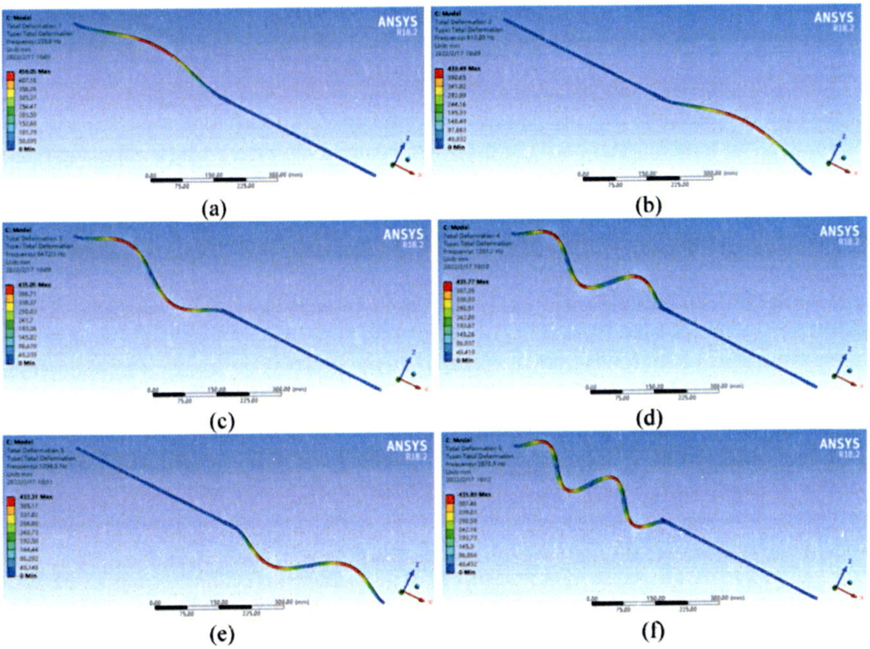

Fig. 6. First six-order main vibration modes of the pipe model

3.2 Analysis and Consistency Discussion of Simulation Results

Table 3. Comparison between simulation results and experimental results

Order	Natural frequency (Hz)		Comparison (%)
	Average value	Simulation value	
1	236.088	235.8	−0.12
2	565.05	613.08	8.5
3	647.618	647.23	−0.06
4	1198.94	1261.7	5.23
5	1269.98	1294.5	1.93
6	1972.36	2070.9	4.99

Table 3 shows the comparison of the first six-order natural frequencies between the simulation results and the average value of the experimental data of the hydraulic pipe. It is found that the error is within 10%. Therefore, the rationality of the swaged hydraulic pipe modeling and the accuracy of simulation are verified.

4 Simulation and Discussion

4.1 Modeling and Modal Analysis

According to the requirements of insertion depth and insertion position of pipe in the connection of compression type casing joint specified in "aircraft assembly process pipe installation" in national standard [25]. For 0.25 inch civil aircraft hydraulic pipe, the gap between the two sections of pipe after swaging shall not exceed 7.6 mm, and the insertion depth range is 17.5 ± 5 mm. Figure 7 is the two typical swaging parameters schematic diagram of hydraulic pipe. The SL is 33–47 mm and the SG is 0–7 mm, 64 sets of hydraulic pipe models are established.

In order to facilitate the research, 64 hydraulic pipe models are numbered, L33G0, L33G1, L33G2, L33G3, L33G4, L33G5, L33G6, L33G7, L35G0, L35G1, L35G2, L35G3, L35G4, L35G5, L35G6, L35G7, L37G0, L37G1, L37G2, L37G3, L37G4, L37G5, L37G6, L37G7, L39G0, L39G1, L39G2, L39G3, L39G4, L39G5, L39G6, L39G7, L41G0, L41G1, L41G2, L41G3, L41G4, L41G5, L41G6, L41G7, L43G0, L43G1, L43G2, L43G3, L43G4, L43G5, L43G6, L43G7, L45G0, L45G1, L45G2, L45G3, L45G4, L45G5, L45G6, L45G7, L47G0, L47G1, L47G2, L47G3, L47G4, L47G5, L47G6 and L47G7 respectively.

4.2 Model Analysis

This paper studies the vibration characteristics of civil aircraft hydraulic pipe under flight conditions, regulations on pressure pipe: when the system working pressure is

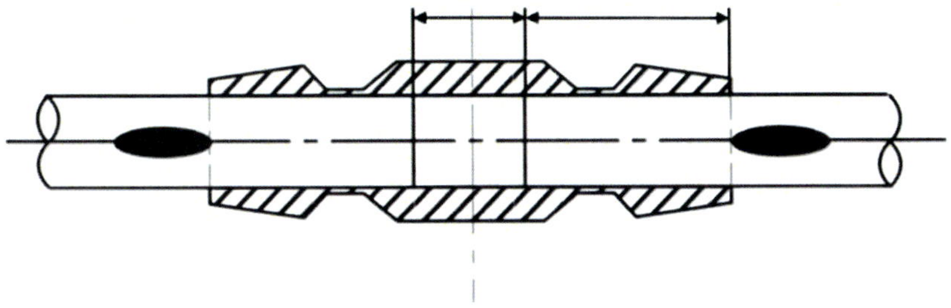

Fig. 7. The SL and SG diagram of hydraulic pipe

$P < 2.5MPa$, take $v = 2m/s$; when the system working pressure is $2.5MPa < P < 14MPa$, take $v = 3\ 4m/s$; when the system working pressure is $P > 14MPa$, take $v \leq 5m/s$ [26]. Therefore, when the pipe is solved by fluid-structure interaction in fluent, the inlet speed of fluid is set to $4m/s$ and the outlet pressure is set to $21MPa$.

Theoretical Basis of Modal Analysis Modal analysis is to solve the vibration characteristics of the structure under free vibration. The main parameters are natural frequency and modal shape [27].

The dynamic equation is shown as follows

$$M\ddot{x}(t) + C\dot{x}(t) + Kx(t) = F(t)\sin(\theta t) \tag{1}$$

where M,C,K represent the mass matrix, damping matrix and stiffness matrix of the pipe respectively; x, $\dot{x}$, $\ddot{x}$ represent the displacement vector, velocity vector and acceleration vector respectively;F is the external force vector. In the modal analysis, it is assumed that the external force vector F is 0 and the damping matrix C is 0. At this time, the dynamic equation can be written as follows

$$M\ddot{x} + Kx = 0 \tag{2}$$

The equation of simple harmonic motion is assumed as follows

$$x(t) = \phi_i \sin(\omega_i t) \tag{3}$$

It can be seen that the undamped equation of motion can be written as follows

$$\left(K - \omega_i^2\right)M\phi_i = 0 \tag{4}$$

Since the amplitude of each node is not all 0, the following expression can be obtained from Eq. (4) [28]

$$\left|K - \omega_i^2 M\right| = 0 \tag{5}$$

where ω_i and ϕ_i represent the ith natural frequency and the ith modal shape respectively.

Results and Discussion of Modal Analysis Through the modal analysis of 64 pipe models with pre-stress, the vibration characteristics of hydraulic pipe under different swaging parameters are obtained. Generally, the maximum frequency of excitation force received by civil aircraft during normal flight can reach 2000 Hz, the first four-order vibration modes of L35G2 model are shown in Fig. 8. It can be seen from the figure:

(1) The first-order natural frequency is 285.92 Hz, the main vibration mode is bending along the Z-axis, and the maximum deformation occurs in the left half of the pipe, and about 1/2 near the inlet;
(2) The second-order natural frequency is 718.55 Hz, and S deformation occurs along the Z-axis. The maximum deformation is located in the left half of the pipe, and about 1/4 near the inlet;
(3) The third-order natural frequency is 1340.9 Hz, and S deformation occurs along the Z-axis. The maximum deformation is located in the left half of the pipe, and about 3/4 near the inlet;
(4) The fourth-order natural frequency is 1790.6 Hz, the main vibration mode is bending along the Z-axis, and the maximum deformation occurs in the right half of the pipe, and about 1/2 near the outlet.

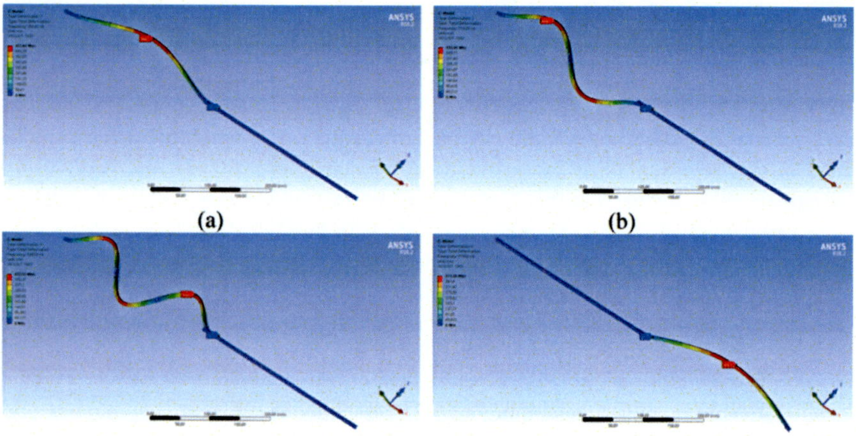

(a) (b) (c) (d)

Fig. 8. The first four-order main vibration modes of L35G2 model

The first four-order deformation results of 64 pipe models are shown in Table 4. It can be seen from the table:

(1) When the SG is 0 mm, the maximum deformation of the models under the first two-order natural frequencies are almost the same, and the deformation under the third-order and fourth-order natural frequencies are also similar.
(2) When the SG is 1–7 mm, the maximum deformation of the models under the first-order natural frequency are slightly larger than those of the other orders.
(3) When the SL is 35 mm, the variation trends of the maximum deformation of pipe models at each-order natural frequency are relatively stable.

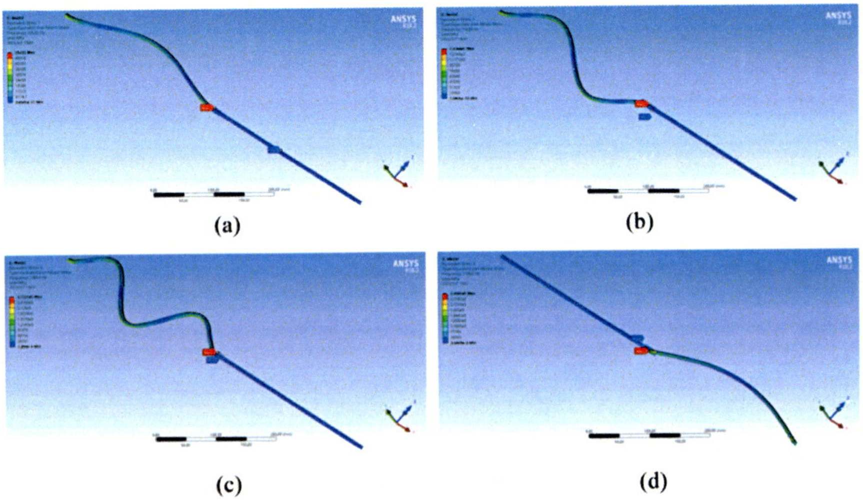

Fig. 9. The first four-order equivalent stress of L35G2 model

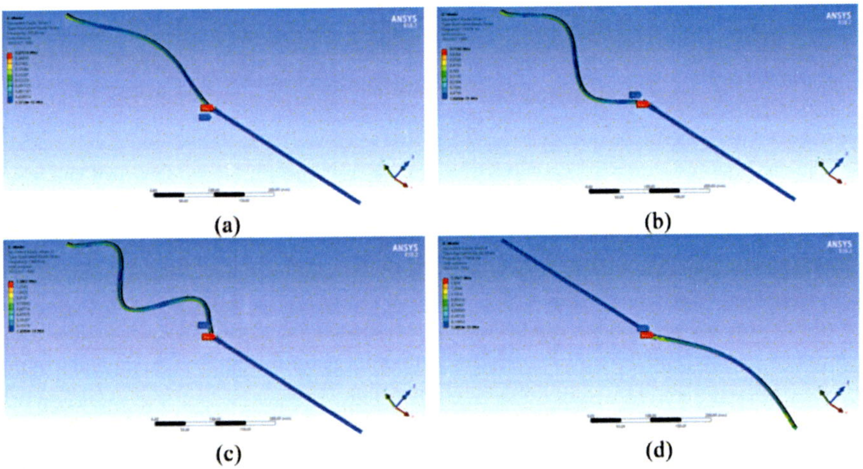

Fig. 10. The first four-order equivalent elastic strain of L35G2 model

(4) When the SL is 45 mm, the maximum deformation at the third-order and fourth-order natural frequencies fluctuate greatly.

Similarly, Figs. 9 and 10 show equivalent elastic strain, equivalent stress and the position of the maximum stress-strain of L35G2 model when reaching the first four-order natural frequencies. It can be seen from Figs. 9 and 10 that the place where the maximum stress and strain are generated by the pipe is located at the pressure pipe joint. At the same time, the first four-order maximum stress-strain values of 64 pipe models are shown in Tables 5 and 6, the following conclusions can be drawn from the tables:

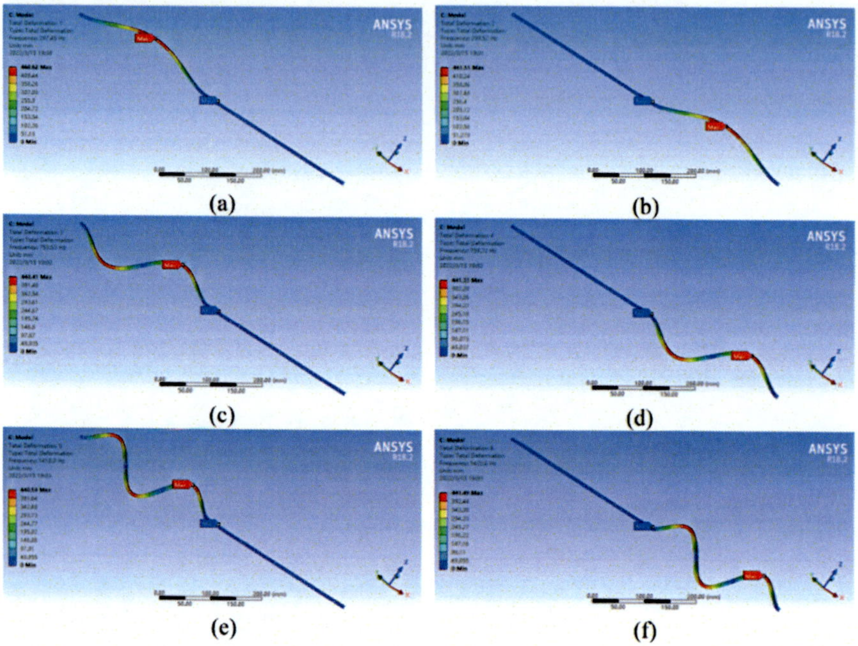

Fig. 11. The first six-order main vibration modes of L35G0 model

(1) At the first two-order natural frequencies, the maximum stress-strain of the pipe are relatively small. At the third and fourth natural frequencies, the total variation trend of the maximum stress-strain of the pipe model is to increase.
(2) When the SG is 0 mm, under the first four-order natural frequencies, the maximum stress-strain values are relatively small, that is, the strain at the pressure pipe joint is small when there is no gap between the two swaged pipe.
(3) When the SL is 35 mm, under the first four-order natural frequencies, the maximum stress-strain values are relatively small, in other words, a small strain occurs at the pressure pipe joint.

Finally, the first six-order natural frequencies of 64 pipe models are shown in Table 7, and the first six-order main vibration modes of L35G0 model is shown in Fig. 11. It can be seen from the table and the figure:

(1) When the SG is 0 mm, the first six-order natural frequencies of the pipe are all less than 2000 Hz. This moment, the first two-order natural frequencies are close, the left and right main vibration modes are symmetrical and the deformations are basically the same. The third and fourth order natural frequencies and their main vibration modes also have the same law. Similarly, the fifth and sixth orders also accord with the above law.

Table 4. Max deformation of 64 pipe models

SL	Order	Max deformation (mm)							
		G0	G1	G2	G3	G4	G5	G6	G7
L33	1	453.76	454.1	459.18	454	453.9	453.74	453.6	453.9
	2	453.75	434.29	435.31	434.02	434.11	434.07	434.11	434.11
	3	434.01	433.76	436.34	433.72	433.63	433.54	433.49	433.63
	4	434.05	434.4	418.33	401.1	434.49	411.16	411.61	434.49
L35	1	460.62	454.69	453.69	454.19	454.08	453.12	453.85	454.63
	2	461.51	434.25	433.92	434.09	434.05	433.85	433.97	434.29
	3	440.41	434.06	433.54	433.8	433.74	433.35	433.64	434.02
	4	441.33	410.89	413.32	411.95	407.1	410.39	413.46	408.77
L37	1	460.75	454.29	454.52	459.08	454.38	453.51	453.36	453.92
	2	461.19	434.17	434.2	439.24	434.19	433.89	433.95	434
	3	440.59	433.9	433.92	438.92	433.89	433.48	433.46	433.63
	4	441.05	434.41	412.91	405.8	434.44	434.07	425.9	434.28
L39	1	462.99	453.66	460.9	460.99	461.12	453.82	454.06	453.46
	2	463.17	434.1	441.19	440.79	440.86	434.14	434.1	434.17
	3	442.72	433.56	440.96	440.88	440.96	433.6	433.82	433.6
	4	442.77	434.34	419.41	425.63	407.56	421.8	415.85	434.29
L41	1	462.66	455.72	463.01	461.58	460.4	453.85	453.69	454.37
	2	463.04	435.16	442.52	441.1	440.4	434.1	434.03	434.24
	3	442.28	434.41	442.67	441.32	440.43	433.59	433.63	433.97
	4	442.66	434.9	729.75	1357.6	440.46	434.46	402.8	434.6
L43	1	462.66	462.98	453.36	454.37	462.8	453.9	454.23	460.44
	2	463.04	442.05	434.05	434.18	442.22	433.99	434.12	440.12
	3	442.28	441.98	433.81	433.9	442.13	433.63	433.82	440.25
	4	442.66	408.81	408.82	407.68	409	434.24	434.36	440.48
L45	1	463.66	453.78	464.2	463.26	454.08	452.59	453.88	453.7
	2	463.87	434.28	443.64	442.94	434.11	433.82	434.05	434.17
	3	443.27	433.63	443.71	544.77	434.12	433.1	433.73	1030.2
	4	443.45	434.34	407.26	442.93	434.42	553.48	399.17	433.71

(continued)

(2) When the SG is 1–7 mm, the fourth-order natural frequency of most pipe models can reach 2000 Hz, and the fifth and sixth natural frequencies are much greater than 2000 Hz.

Table 4. (*continued*)

SL	Order	Max deformation (mm)							
		G0	G1	G2	G3	G4	G5	G6	G7
L47	1	465.51	453.88	464.46	463.88	463.77	454.17	461.46	461.79
	2	465.76	434.22	443.62	443.49	443.67	434.13	441.9	441.75
	3	445.02	433.67	443.74	443.51	443.58	433.77	441.58	1538.8
	4	445.19	434.38	400.98	409.44	415.52	434.58	441.93	441.65

4.3 Harmonic Response Analysis

According to the results of modal analysis, eight pipe models with the SL of 35 mm are taken as the research object for harmonic response analysis. According to the first six-order natural frequencies, set the loading frequency to 0–4000 Hz, the number solving steps to 500 steps, and apply 1G acceleration along the Z-axis. The output format of displacement solution along the Z-axis of harmonic response analysis is selected as amplitude and phase angle output, and the results are imported into origin Pro software to obtain the Displacement-Frequency curve, as shown in Fig. 12.

It can be seen from the figure that the parts with large vibration response of the pipes are at the first-order natural frequency, which are 297.45 Hz, 275.51 Hz, 285.92 Hz, 279.9 Hz, 281.52 Hz, 290.79 Hz, 284.07 Hz and 275.99 Hz respectively. However, the vibration response at other-order natural frequencies is more than one order of magnitude smaller than that at the first-order natural frequency, so it can be inferred that the pipe is more prone to resonance at the first-order natural frequency.

5 Conclusion

(1) According to the model analysis, when the swaging length (SL) is 35mm, the maximum stress-strain value of the pipe models are very small compared with those under other SL. When the swaging gap (SG) is 0 mm, although the first six-order natural frequencies are all less than 2000 Hz, the maximum stress-strain values of the pipe models are much more smaller than those of the pipe models under other SG. Considering that the SL is easy to control in the actual swaging process, when the SL is 35 mm, the SG can be less than 7.6 mm. When the SL takes other values, in order to make the pressure pipe joint bear less stress and strain, it must ensure that there is no gap between the two sections of hydraulic pipes to be swaged, that is, the SG is 0 mm.

(2) According to the harmonic response analysis, the vibration response of the pipe at the first-order natural frequency is the largest. In other words, the low-order natural vibration mode has a greater impact on the vibration response of the pipe than the high-order natural vibration mode, and resonance is easier to occur at the first-order natural frequency, resulting in damage to the pressure pipe joint of the pipe.

Table 5. Max equivalent elastic strain of 64 pipe models

SL	Order	Max equivalent elastic strain (mm/mm)							
		G0	G1	G2	G3	G4	G5	G6	G7
L33	1	0.27357	0.273	0.24455	0.2728	0.27126	0.27461	0.27561	0.27615
	2	0.27508	0.71518	0.67253	0.7146	0.71223	0.7205	0.72192	0.72292
	3	0.71597	1.3622	1.3216	1.3617	1.3589	1.3763	1.3778	1.3789
	4	0.71973	2.2076	44.55	71.505	30.04	2.2326	1.8488	5.4448
L35	1	0.43795	0.26927	0.27516	0.27144	0.2722	0.27831	0.27384	0.2692
	2	0.38756	0.70784	0.7182	0.71227	0.71374	0.7233	0.71619	0.70785
	3	1.1185	1.3566	1.3661	1.3587	1.3605	1.3725	1.3637	1.3554
	4	0.96697	2.7239	1.7877	12.605	10.758	2.3229	7.7632	9.2632
L37	1	0.37001	0.27062	0.27047	0.90222	0.27054	0.27623	0.27654	0.27394
	2	0.35717	0.71131	0.70993	2.312	0.71062	0.71992	0.72065	0.71613
	3	0.91801	1.3579	1.3573	4.3127	1.3569	1.3682	1.3691	1.3634
	4	0.88457	2.2024	11.526	50.691	2.2039	2.2125	765.3	2.2077
L39	1	0.40671	0.27485	0.35394	0.41104	0.38255	0.27482	0.27173	0.27406
	2	0.32275	0.71836	0.87431	1.0278	0.95292	0.72092	0.71207	0.71674
	3	1.0728	1.371	1.5648	1.8618	1.7183	1.3767	1.3613	1.3649
	4	0.82306	2.2233	200.16	77.444	772.67	320.58	397.64	2.2116
L41	1	0.39763	0.26709	0.39348	0.54357	0.75184	0.27508	0.27387	0.26972
	2	0.35158	0.70616	1.0172	1.4181	1.9372	0.7212	0.71577	0.70982
	3	1.0528	1.3597	1.9112	2.6953	3.6311	1.3771	1.3636	1.3594
	4	0.93117	2.2142	1045.3	2755.9	5.7679	2.2334	121.61	2.2075
L43	1	0.39763	0.47559	0.27508	0.27055	0.42233	0.27365	0.27139	0.43229
	2	0.35158	1.2936	0.71892	0.71065	1.1028	0.71589	0.71234	1.1595
	3	1.0528	2.5647	1.371	1.357	2.0934	1.3631	1.359	2.2641
	4	0.93117	3.1959	40.731	28.759	26.643	2.2068	2.2041	3.7541
L45	1	0.36811	0.27436	0.38432	0.35141	0.27388	0.2818	0.27283	0.27261
	2	0.36835	0.71732	0.95737	0.87502	0.71979	0.7307	0.71392	0.7145
	3	0.91214	1.3698	1.7243	541.47	1.3757	1.3876	1.3626	1770.6
	4	0.91312	2.2224	415.06	1.5784	2.235	326.41	46.786	1.3624
L47	1	0.33005	0.2736	0.52726	0.39374	0.41163	0.27355	0.44444	0.5498
	2	0.33054	0.71648	1.2548	0.97883	1.0253	0.71918	1.1387	2.2666
	3	0.84238	1.369	2.1625	1.7621	1.8522	1.375	2.1434	2485
	4	0.84463	2.2218	146.05	36.596	112.98	2.2319	3.4365	3.2228

In a word, the swaging length and gap obtained in this paper can provide a technical reference for the maintenance process of civil aircraft hydraulic pipeline. The parameters such as the width and depth of the pressure groove in the actual swaging process are not discussed here, which can be further studied in the future.

Table 6. Max equivalent stress of 64 pipe models

SL	Order	Max equivalent elastic stress (MPa)							
		G0	G1	G2	G3	G4	G5	G6	G7
L33	1	54713	54538	48671	54561	54252	54921	55121	55230
	2	55016	1.43e+5	1.35e+5	1.43e+5	1.42e+5	1.44e+5	1.44e+5	1.45e+5
	3	1.43e+5	2.72e+5	2.64e+5	2.72e+5	2.71e+5	2.75e+5	2.76e+5	2.76e+5
	4	1.44e+5	4.42e+5	8.91e+6	1.30e+7	3.71e+6	4.47e+5	3.66e+5	8.92e+5
L35	1	87397	53663	55033	54253	54441	55662	54767	53711
	2	77237	1.42e+5	1.44e+5	1.42e+5	1.43e+5	1.45e+5	1.43e+5	1.42e+5
	3	2.23e+5	2.71e+5	2.73e+5	2.71e+5	2.72e+5	2.74e+5	2.72e+5	2.71e+5
	4	1.93e+5	4.86e+5	3.50+5	2.52e+6	2.15e+6	4.11e+5	1.50e+6	1.08e+6
L37	1	73905	54124	53994	1.19e+5	54062	55246	55308	54788
	2	71210	1.42e+5	1.42e+5	3.06e+5	1.42e+5	1.44e+5	1.44e+5	1.43e+5
	3	1.83e+5	2.72e+5	2.71e+5	5.73e+5	2.71e+5	2.74e+5	2.74e+5	2.73e+5
	4	1.76e+5	4.40e+5	2.02e+6	1.01e+7	4.41e+5	4.42e+5	8.13e+7	4.42e+5
L39	1	80429	54971	70508	81743	76409	54964	54262	54812
	2	63519	1.44e+5	1.74e+5	2.04e+5	1.90e+5	1.44e+5	1.42e+5	1.43e+5
	3	2.12e+5	2.74e+5	3.12e+5	3.70e+5	3.43e+5	2.75e+5	2.72e+5	2.73e+5
	4	1.62e+5	4.45e+5	3.96e+7	1.54e+7	1.54e+8	6.40e+7	7.92e+7	4.42e+5
L41	1	73771	53418	78081	1.05e+5	1.27e+5	55016	54774	53780
	2	67213	1.41e+5	2.02e+5	2.69e+5	3.20e+5	1.44e+5	1.43e+5	1.42e+5
	3	1.96e+5	2.72e+5	3.79e+5	5.08e+5	6.01e+5	2.75e+5	2.73e+5	2.72e+5
	4	1.69e+5	4.43e+5	2.09e+8	4.71e+8	9.67e+5	4.47e+5	2.42e+7	4.41e+5
L43	1	73771	88439	55016	5.41e+4	8.30e+4	54731	54278	86183
	2	67213	2.27e+5	1.44e+5	1.42e+5	2.17e+5	1.43e+5	1.42e+5	2.24e+5
	3	1.96e+5	4.32e+5	2.74e+5	2.71e+5	4.12e+5	2.73e+5	2.72e+5	4.37e+5
	4	1.69e+5	6.39e+5	8.08e+6	3.88e+6	4.58e+6	4.41e+5	4.41e+5	7.23e+5
L45	1	73440	54872	76425	70148	54775	56359	54545	54522
	2	73312	1.43e+5	1.90e+5	1.75e+5	1.44e+5	1.46e+5	1.43e+5	1.43e+5
	3	1.82e+5	2.74e+5	3.43e+5	1.01e+8	2.75e+5	2.78e+5	2.73e+5	3.44e+8
	4	1.82e+5	4.44e+5	7.89e+7	3.15e+5	4.47e+5	5.14e+7	9.34e+6	2.72e+5
L47	1	64618	54719	72192	78424	81964	54709	88563	1.08e+5
	2	65373	1.43e+5	1.79e+5	1.95e+5	2.04e+5	1.44e+5	2.27e+5	4.50e+5
	3	1.61e+5	2.7e+5	3.21e+5	3.51e+5	3.69e+5	2.75e+5	4.27e+5	4.93e+8
	4	1.62e+5	4.44e+5	2.05e+7	7.29e+6	2.26e+7	4.46e+5	6.85e+5	6.40e+5

Table 7. The first six-order natural frequencies of 64 pipe models

Model code	Order	Natural frequency (Hz)							
		G0	G1	G2	G3	G4	G5	G6	G7
L33	1	282.81	280.37	222.98	282.58	280.25	281.89	283.61	284.29
	2	283.16	709.68	630.1	713.58	710.23	712.35	714.9	715.97
	3	714.45	1331.1	1243	1335.2	1331.4	1334.6	1337.5	1338.7
	4	714.92	2143.9	1985.8	1847.1	1908.1	2147.2	2047.2	1942.5
	5	1336.3	2150.1	2051.4	2148.2	2144.1	3145.6	2150.3	2151.6
	6	1336.8	3142.3	3047.8	3147	3142.8	4322.1	3148.7	3150.1
L35	1	297.45	275.51	285.92	279.9	281.52	290.79	284.07	275.99
	2	299.52	702.92	718.55	709.77	712.05	725.94	715.82	703.93
	3	753.53	1323.3	1340.9	1330.9	1333.5	1349.3	1337.8	1324.3
	4	759.22	1971.5	1790.6	1873.3	1906.8	1904.3	1900.2	1927.1
	5	1410.9	2135.6	2154.1	2143.6	2146.3	2163.1	2150.9	2136.6
	6	1422.6	3134	3153.1	3142.2	3145.1	3162.4	3149.7	3135.1
L37	1	297.68	279.53	278	302.95	278.68	287.63	288.07	283.67
	2	298.88	709.25	706.51	758.91	707.88	721.08	721.95	715.14
	3	754.37	1330.2	1327.3	1411.9	1328.8	1343.8	1344.7	1337
	4	757.43	2142.9	2121	2055.3	2141.3	2157.2	2114.5	2150
	5	1413	2167.4	2139.8	2263	2262.8	2159.2	2158.2	2252.6
	6	1419.1	3141.6	3138.4	3305.2	3139.9	3156.3	3157.3	3148.9
L39	1	302.18	283.84	304.97	298.95	299.29	282.21	280.57	284.96
	2	302.95	715.93	766.65	756.34	756.52	712.91	710.58	717.47
	3	766.96	1338.1	1430.4	1416.6	1417.9	1335.2	1332.3	1339.8
	4	768.55	2150.9	2218.4	2172.8	2084.2	1817.4	2039.6	2152.9
	5	1438.3	3049.6	2297.3	2278.1	2280.6	2147.9	2144.5	2306.9
	6	1440.8	3149.3	3360.7	3334.9	3338.9	3146.3	3143.1	3151.7
L41	1	301.58	269.31	300.46	298.31	299.47	282.44	284.43	277.19
	2	302.7	692.51	763.38	756.27	755.52	713.09	716.29	705.67
	3	765	1311.5	1433.2	1417.7	1413.6	1335.4	1338.8	1326.4
	4	767.84	2123.8	1886	2074.1	2270.7	2148.1	2098.6	2138.8
	5	1433.8	2210.1	2309.3	2280.2	2796.9	2574.2	2151.3	2346.1
	6	1439.4	3121.9	3384.7	3337.3	3320.6	3146.5	3150.1	3137.2
L43	1	301.58	297.74	286.18	278.71	300.3	283.52	280.51	298.42

(continued)

Table 7. (*continued*)

Model code	Order	Natural frequency (Hz)							
		G0	G1	G2	G3	G4	G5	G6	G7
	2	302.7	758.17	720.2	707.87	762.34	714.96	710.48	753.85
	3	765	1424.2	1342.6	1328.8	1430.2	1336.8	1331.7	1409.9
	4	767.84	2068.4	1969.5	1928.7	2034.8	2149.8	2144.5	2265.7
	5	1433.8	2293.7	2155.7	2141.3	2303	3148.7	2261.5	2311
	6	1439.4	3358.9	3154.5	3140	3373.6	3225.3	3143.2	3314.3
L45	1	303.69	283.34	304.84	305.51	280.54	294.37	282.7	283.18
	2	304.53	715.24	773.31	772.16	710.85	731.11	713.67	714.82
	3	771.01	1337.2	1450.6	1213.9	1332.3	1356.2	1335.8	762.32
	4	772.85	2150.1	2101.4	1446.2	2144.8	1469.3	1892.4	1336.8
	5	1446.1	3148.4	2336.3	2326.2	2146.5	2046.1	2148.2	2149.7
	6	1449.2	3187.8	3423.4	2845.6	3143.4	2170.1	3146.9	2503.5
L47	1	307.9	282.1	298.87	307.12	309.89	279.67	314.01	301.85
	2	308.88	713.18	763.68	776.4	780.49	708.98	781.39	763.29
	3	782.4	1335	1438.3	1454.1	1460.1	1330.7	1451.3	1094.8
	4	784.56	2147.7	2061.3	1963.2	2065.6	2143.2	2323.8	1428
	5	1468.3	2317.8	2321.1	2338.9	2346.7	2603.7	2345	2294.7
	6	1471.8	3146.1	3405	3425.3	3435.3	3141.5	3392.5	2929.7

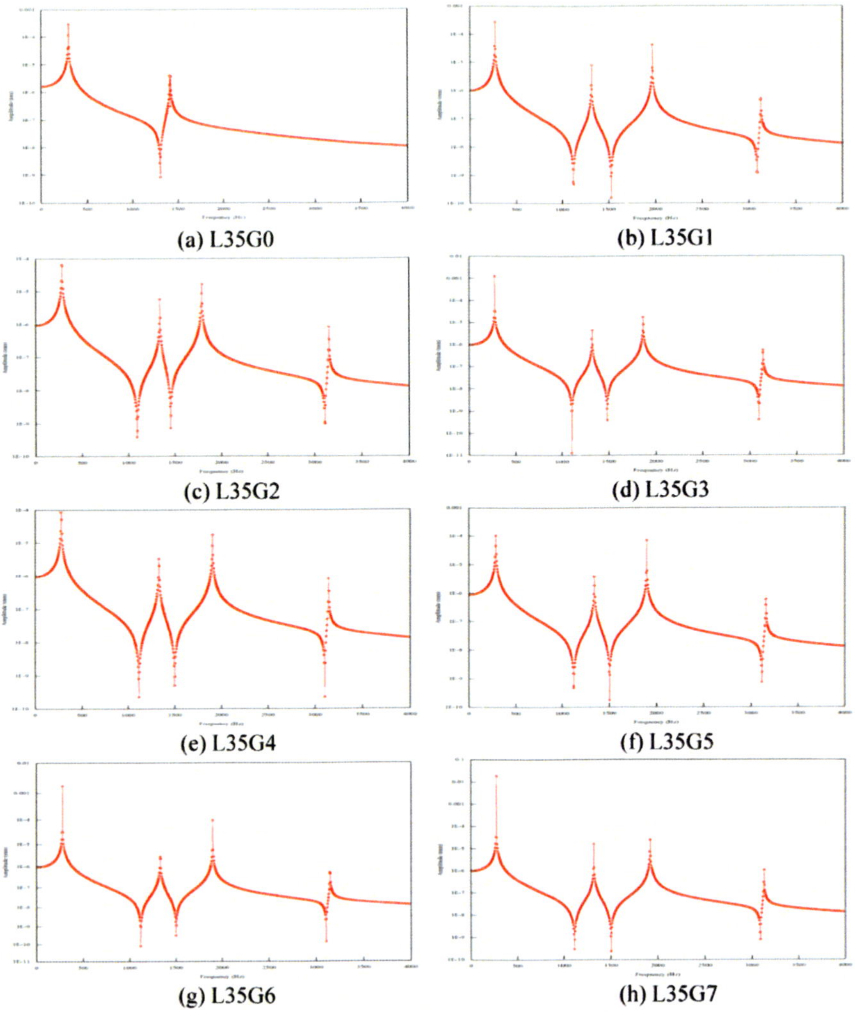

Fig. 12. Displacement-Frequency response of pipe models along Z-axis

References

1. Yue, Y.J.: Application of seamless swaging technology in aircraft assembly process hydraulic system. Sci. Technol. Innov. Herald **19**, 85–86 (2011)
2. Li, G.C.: Study on repair method of permanent installation hydraulic pipe defects. Sci. Technol. Vis. (03), 101 (2017)
3. Zheng, Y.: Airplane hydraulic tube maintenance. Aviation Maintenance **06**, 28–31 (2003)
4. Shi, K.J.: Study on stress analysis method of 3000 PSI hydraulic pipe of civil aircraft based on typical airworthiness load. Yanshan University **5**, 7 (2020)
5. Asheley, H., Haviland, G.: Bending vibrations of a pipe containing flowing fluid. J. Appl. Mech. **17**(3), 229–232 (1950)
6. Harold, L.D., Harry, J., Runyan, L.: Effect of high-velocity fluid flow on the bending vibrations and static divergence of a simply supported pipe. NASA-Langley, pp. 435–446 (1965)
7. Paidoussis, M.P., Laithier, B.E.: Dynamic of Timoshenko beams conveying fluid. J. Mech. Eng. Sci. **18**(4), 210–220 (1999)
8. Bezborodov, S.A., Ulanov, A.M.: Calculation of vibration of pipe bundle with damping support made of MR material. Proc. Eng. **176**, 169–174 (2017)
9. Riedelmeier, S., Becker, S., Schlucker, E.: Measurements of junction coupling during water hammer in piping systems. J. Fluids Struct. **48**, 156–168 (2014)
10. Quan, L.X., Sun, B.J., Zhao, J.S., et al.: Frequency response analysis of fluid-structure interaction vibration in aircraft bending hydraulic pipe. J. Northwest. Polytech. Univ. **3**, 487–495 (2018)
11. Ding X. Study on Vibration Transfer Path and Regularity of Aircraft Hydraulic pipe System. Yanshan University, 5–10 (2017)
12. Song, Y.S., Ren, X.D., Ren, F.: Analysis and test of dynamic characteristics of hydraulic pipe. China Metalform. Equip. Manuf. Technol. **4**, 79–82 (2018)
13. Sang, Y., Shao, L.L., Wang, X.D.: Vibration analysis of fluid-structure interaction for large flow pipe with ANSYS. Hydraul. Pneumat. Seals **7**, 1–5 (2018)
14. Yuan, H., Zhang, J.Y.: Pipe vibration analysis of cylinder cover hydraulic system. China Rubber/Plast. Technol. Equip. (Plastics) (20), 61–63 (2017)
15. Zhao, Q.L., Sun, Z.L., Chai, X.D.: Vibration analysis of fluid conveying pipe elastic support. J. Vib. Measur. Diagnosis **6**, 1222–1226 (2017)
16. Xia, Y.S., Zhang, C.L.: Hydraulic pipe fluid-structure interaction vibration calculation based on ANSYS workbench. Fluid Power Transm. Control **3**, 38–41 (2017)
17. Zhao, M.W., Fan, Z.M.: Analysis of pipe Stress under the Influence of Vibration and Hydraulic P ulse. J. Shandong Ind. Technol. **6**, 284–285 (2017)
18. Han, T., Liu, W., Zhang, Z.J.: Natural frequency analysis of complex hydraulic pipes based on straight-curved pipe assembly algorithm. J. Vib. Shock **7**, 13–22 (2018)
19. Gao, Y.K., Jing, H.S., Zhang, S.L.: Experimental analysis on vibration of ships steam pipe system. Ship Ocean Eng. **04**, 72–75 (2017)
20. Yu, S.M., Lu, T., Wei, G., et al.: Research on pipes vibration based on strain analysis. J. Dyn. Control **15**(6), 512–517 (2017)
21. Li, D.: Frequency domain characteristics analysis of fluid structure coupling vibration of civil aircraft hydraulic bending pipe. Yanshan University (2016)
22. Zhang, D.Q., Zhong, L.L.: Modal simulation analysis of aircraft hydraulic pipe system. Sci. Technol. Style **26**, 134–135 (2017)
23. Zhong, L.L.: Vibration simulation analysis of aircraft hydraulic pipe system. Shenyang Aerospace University (2018)
24. China Eastern Airlines. A320 aircraft maintenance manual of Chinese Eastern Airlines. 162–188 (2012)

25. Aircraft Assembling Technology Conduct Installation: Aviation industry standard of the People's Republic of China **223**(20), 9–11 (2002)
26. Du, L.L., Song, X.J.: Aircraft Accessory Maintenance. Aviation industry press, Beijing (2006)
27. Zheng, B., E J Y. Finite element modal analysis and harmonic response analysis of engine connecting rod. J. Mach. Des. (37), 98–101 (2020)
28. Zhao, Y.C., Huang, S.F.: Vibration model and harmonic response analysis on vertical pillar of vertical hill. Mining Res. Dev. **32**(06), 94–97 (2012)

Dynamic Modeling and Longitudinal Stability Analysis of a Stratospheric Dual-Hull Airship

Li Hang[1](✉), Cui Yining[2], and Hong Guanxin[3]

[1] AVIC The First Aircraft Institute, Xi'an 710089, China
453126878@qq.com
[2] AVIC XAC Commercial Aircraft CO., LTD, Xi'an 710089, China
[3] Beijing University of Aeronautics and Astronautics, Beijing 100191, China

Abstract. This paper provided a design of a dual-hull airship concept. It contained two traditional airships hulls that were connected in the middle by a wing, and had more control surface area than the traditional airship of the similar weight. In addition, the dual-hull airship's yawing and pitching moments can be provided by changing directions and creating different thrust by four propulsion units combined with fins. Then a 6-Degree of Freedom (DOF) dynamic model for the dual-hull airship was described in this paper. It updated the estimation method of the added mass, force and torque (aerodynamic and propulsion unit's thrust) of the dual-hull airship. The open-loop disturbance feature of the longitudinal mode simulated by Simulink and compared the the dual-hull airship's added mass with the traditional airship of the similar weight. The results show that the dual-hull airship has less resistance and better flexibility than the traditional airship when they are flying forward.

Keywords: Dual-hull airship · 6-DOF dynamic model · Simulation · Dynamic characteristic · Longitudinal stability

1 Introduction

Many researchers have shown an interest in the stratospheric airship in recent years. Dynamic models of the airship are numerous because of its many different possible shapes and structures. The airship most researched in China is the waterdrop-like single-hull airship. In Italy, Battipede has done some research modeling a dual-hull airship called Nautilus [1]. In this paper, we describe a new concept for a dual-hull airship that is formed by combining two waterdrop-like single-hull airships. And its unique features are presented after comparing with the similar weight single-hull airship.

Chinese Society of Aeronautics and Astronautics: CASTYSF 2022, LNEE 972, pp. 136–146, 2023.
https://doi.org/10.1007/978-981-19-7652-0_13

2 The Features of the Dual-Hull Airship Structure Design

There are three types of airships in terms of the frame: the non-rigid airship, the semi-rigid airship, and the rigid airship. The airship in this paper is based on the waterdrop-like single-hull airship and is designed in the semi-rigid style in order to ensure frame stability while simultaneously decreasing the weight.

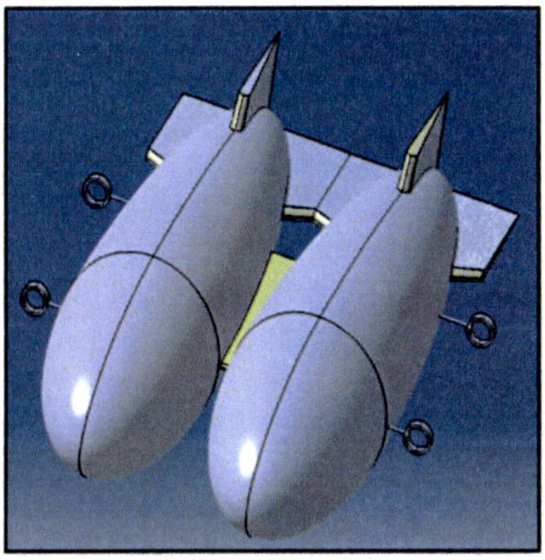

Fig. 1. The dual-hull airship's profile

As Fig. 1 shows, the dual-hull airship in this paper has two same single-hull airships, and they are connected by a wing in the middle of them. The hull is designed like a waterdrop. There are two groups of wings that form a "+" in the tail of each hull, which are equipped with elevators and rudders to control the airship's pitching and yaw-ing. The horizontal fins are fixed, whereas the elevators in each hull can deflect in different directions. This feature enhances the horizontal fins' structural rigidness and the available control area. The airship includes a secondary air cell for the purposes of controlling the flying altitude, maintaining the inner pressure of the airship by regulating the amount of air in the secondary air cell and avoiding the loss of helium in the main cell.

3 The Dynamic Model of the Dual-Hull Airship

In body coordinates, we define the center of volume as the origin. It lies at the center of the line connecting the two volume centers of the hulls. The X body axis is parallel to each hull axis and points to the front of the airship, the Z body axis is perpendicular to the X body axis and points downward from the airship. The Y body axis is perpendicular to both the X and Z body axes and points to the right of the airship.

u, v, w are components of translation velocity along body axes X, Y, Z; p, q, r are roll, pitch, and yaw angular rate, respectively; θ, ψ, φ are Euler angles; α is the angle of attack; β is the sideslip angle; I_X, I_Y, I_Z are main inertia moments; I_{xy}, I_{yz}, I_{xz} are products of inertia.

Three assumptions were used in the dynamic modeling [3]:

1. We see the dual-hull airship in this paper as a rigid body and neglect the influence of flexibility.
2. The volume center and the buoyancy center of the airship are located at the same point.
3. The XZ is a symmetrical plane, and the gravity center lies in this plane; $I_{xy} = I_{yz} = 0$.

3.1 The Analysis of Forces and Moments

In the condition of cruising along a horizontal line, there are 5 forces and 5 moments on the airship. 5 forces include added inertia force $\boldsymbol{F_I}$, propulsion unit thrust $\boldsymbol{F_T}$, gravity $\boldsymbol{F_G}$, buoyancy $\boldsymbol{F_B}$, and aerodynamic force $\boldsymbol{F_A}$. 5 moments caused by these forces include $\boldsymbol{M_I}, \boldsymbol{M_T}, \boldsymbol{M_G}, \boldsymbol{M_B}$ and $\boldsymbol{M_A}$.

3.1.1 $\boldsymbol{F_G}, \boldsymbol{F_B}, \boldsymbol{M_G}$ and $\boldsymbol{M_B}$

It can deduce that $\boldsymbol{MB} = \boldsymbol{0}$ from the assumptions. Other models including ***FG***, ***FB*** and ***MG*** are the same with the single-hull airship [3].

3.1.2 $\boldsymbol{F_A}$ and $\boldsymbol{M_A}$

The aerodynamic forces on the hulls, the gondolas, the fins and the middle wing that connects the two hulls can be defined as follows:

$$\boldsymbol{F_A} = \boldsymbol{F_{A-hulls}} + \boldsymbol{F_{A-gondolas}} + \boldsymbol{F_{A-fins}} + \boldsymbol{F_{A-midwing}} \tag{1}$$

$$\boldsymbol{M_A} = \boldsymbol{M_{A-hulls}} + \boldsymbol{M_{A-gondolas}} + \boldsymbol{M_{A-fins}} + \boldsymbol{M_{A-midwing}} \tag{2}$$

In this paper, the definitions of $C_{X(i)}$ $(i = 1, 2)$, $C_{Y(i)}$ $(i = 1, 2, 3, 4)$, $C_{Z(i)}$ $(i = 1, 2, 3, 4)$, $C_{L(i)}$ $(i = 1, 2)$, $C_{M(i)}$ $(i = 1, 2, 3, 4)$, and $C_{N(i)}$ $(i = 1, 2, 3, 4)$ are the same as Ref. [4]. $C_{X(i),Y(i),Z(i)}$ are force coefficients, and $C_{L(i),M(i),N(i)}$ are moment coefficients.

$$C_{X1} = -(C_{Dh0}S_h + C_{Df0}S_f + C_{Dg0}S_g + C_{Dpy0}S_{py}) \tag{3}$$

$$C_{Z2} = -\frac{1}{2}\left(\frac{\partial C_L}{\partial \alpha}\right)_f S_f \eta_f - \frac{1}{2}\left(\frac{\partial C_L}{\partial \alpha}\right)_{py} S_{py}\eta_{py} \tag{4}$$

$$C_{Z3} = -(C_{Dch}J_1 S_h + C_{Dcf} S_f + C_{Dcpy} S_{py}) \tag{5}$$

$$C_{M2} = -\frac{1}{2}\left(\frac{\partial C_L}{\partial \alpha}\right)_f S_f \eta_f l_{f1} + \frac{1}{2}\left(\frac{\partial C_L}{\partial \alpha}\right)_{py} S_{py}\eta_{py} l_{py} \tag{6}$$

The definitions of these functions are same as in Ref. [4], except for those functions relating to the mid-wing. $S_{fg,}$, S_{fh}, and S_{fpy} are fin, gondola, hull, and mid-wing reference area, respectively. C_{Dch} is the hull cross-flow drag coefficient; C_{Dcf} is the fins cross-flow drag coefficient; C_{Dcpy} is the mid-wing drag coefficient; C_{Dh0} is the hull zero-incidence drag coefficient; C_{Df0} is the fin zero-incidence drag coefficient; C_{Dg0} is the gondola zero-incidence drag coefficient; C_{Dpy0} is the mid-wing zero-incidence drag coefficient; η_{py} is the mid-wing efficiency factor accounting for the effect of the hull on the fins; and η_f is the fin efficiency factor accounting for the effect of the hull on the fins. l_{f1} is X-distance from origin to aerodynamic center of fins, and l_{py} is X-distance from origin to aerodynamic center of the mid-wing.

Each group of fins on each hull can only deflect in the same direction in order to enhance their efficiency. The deflection of the elevators can be defined as $\delta_{E(i)} = \delta_{EL(i)} + \delta_{ER(i)}$, while the deflection of the rudders can be defined as $\delta_{R(i)} = \delta_{RL(i)} + \delta_{RR(i)}$. The variable $i = 1, 2$ refers to the left and right hull, respectively.

Other aerodynamic coefficients' definitions are same with Ref. [4].

The thrust and the moment provided by the propulsion units

There are many strategies for controlling the propulsion units. This paper presents one of these strategies for controlling the airship.

As seen in Fig. 2, the propulsion of the airship is provided by four propellers powered by four direct-current propulsion units. These propulsion units are on the XY plane. Two propulsion units at the front of the airship provide thrust and the yawing moment by rotating at different speeds.

The thrust and moment are given as follows:

$$\boldsymbol{F_{Tq}} = \begin{bmatrix} \kappa_{q1}T_{max} + \kappa_{q2}T_{max} \\ 0 \\ 0 \end{bmatrix} \tag{7}$$

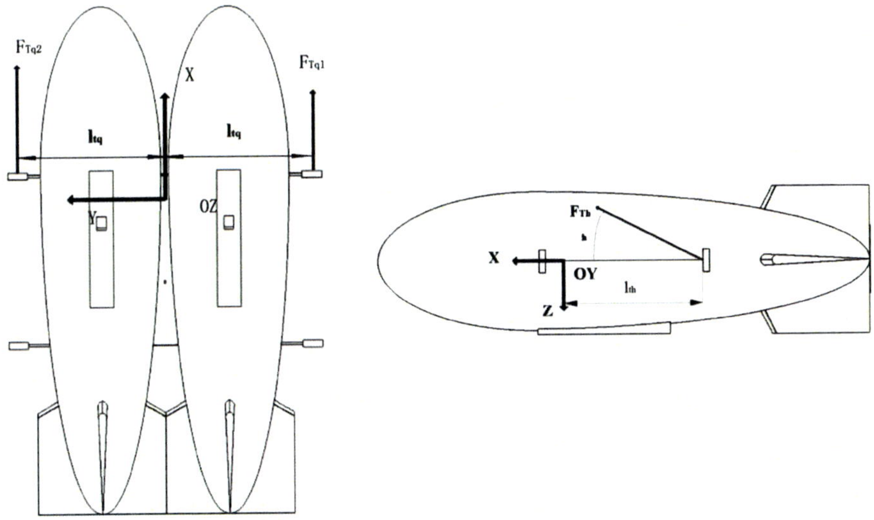

Fig. 2. The propulsion system of the dual-hull airship

$$\boldsymbol{M_{Tq}} = \begin{bmatrix} 0 \\ 0 \\ \kappa_{q1} T_{max} l_{tq} - \kappa_{q2} T_{max} l_{tq} \end{bmatrix} \tag{8}$$

Two propulsion units at the back of the airship can rotate in the XZ plane to provide thrust and the pitching moment for the airship, which can be defined as follows:

$$\boldsymbol{F_{Th}} = \begin{bmatrix} 2\kappa_h T_{max} cos\mu_h \\ 0 \\ 2\kappa_h T_{max} sin\mu_h \end{bmatrix} \tag{9}$$

$$\boldsymbol{M_{Th}} = \begin{bmatrix} 0 \\ 2\kappa_h T_{max} sin\mu_h l_{th} \\ 0 \end{bmatrix} \tag{10}$$

$\boldsymbol{F_{Tq}}$, and $\boldsymbol{F_{Th}}$ represent the front and back propulsion units' thrust vector, respectively. T_{max} represents the maximum thrust of the propulsion unit. $\boldsymbol{M_{Tq}}$ and $\boldsymbol{M_{Th}}$ are Moment of front and back propulsion units' thrust, respectively. L_{tq} is a distance. It is from front propulsion units to the XZ plane. L_{th} is a distance. It is from back propulsion units to the YZ plane. μ_h is an angle. It is from the back propulsion unit thrust vector to the XY plane. We define the angle is plus when it is below the XY plane.

$\kappa_{q1,q2}$ are the left and right front propulsion units' thrust coefficient, respectively, $\kappa_{q1,q2}\in[0,1]$. κ_h is the back propulsion units'thrust coefficient, $\kappa_h\in[0,1]$.

In summary, the force and the moment provided by the propulsion units are defined as follows:

$$\begin{cases} \boldsymbol{F_T} = \boldsymbol{F_{Tq}} + \boldsymbol{F_{Th}} \\ \boldsymbol{M_T} = \boldsymbol{M_{Tq}} + \boldsymbol{M_{Th}} \end{cases} \tag{11}$$

3.1.3 $\boldsymbol{F_I}$, $\boldsymbol{M_I}$, and $\boldsymbol{m_{add}}$

Because of the airship's large volume and its density, which is similar to the surrounding air's density, the inertial feature is noticeable. Thus, the fluid inertial force and the moment should be considered.

The added mass m_{ij} [5] can be understood as the added mass, the added static moment and the added rotary inertia in the *j* direction when the object moves toward the *i* direction at a unit acceleration rate. The numbers 1, 2, and 3 represent movements along the X, Y, and Z directions, while 4, 5, and 6 represent rotation around the X,Y, and Z axes.

Every object has 36 added mass terms when it is moving. Because the dual-hull airship in this paper has two symmetrical planes, including XY and XZ planes, as well as m_{26} and m_{35}, which are relatively small terms, the dual-hull airship's added mass can be simplified as follows:

$$\boldsymbol{m_{add}} = diag[m_{11}\ m_{22}\ m_{33}\ m_{44}\ m_{55}\ m_{66}] \tag{12}$$

These terms, except m_{44} (the added rotary inertial coefficient), can be estimated according to the semi-empirical methodology of estimating the added mass of the single-hull airship [5]. They can be expressed as follows:

$$\begin{cases} m_{11} = k_{11}\rho U \times 2 \\ m_{22} = k_{22}\rho U \times 2 \\ m_{33} = k_{33}\rho U \times 2 \\ m_{55} = k_{55}\rho U^{5/3} \times 2 \\ m_{66} = k_{66}\rho U^{5/3} \times 2 \end{cases} \tag{13}$$

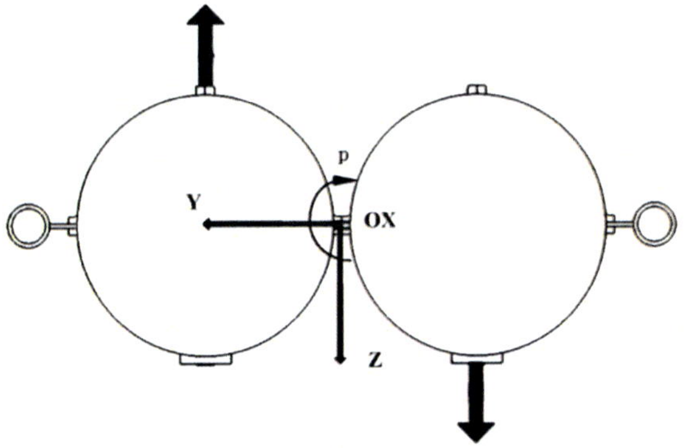

Fig. 3. The moving direction of each hullwhen the airship is rotating

ρ is the density of air. U is the volume of each hull. K_{ii} (i = 1, 2, 3, 5, 6) is the coefficient of added inertial for the single airship according to Ref. [5]. Actually, it is the multiple of the airship's added density divide by the apparent density, and it has a close relation with the airship's shape [6]. As Fig. 3 shows, because the X axis is in the middle of the two hulls, the added rotary inertia around the X axis has to be considered. When the airship is rotating around the X axis at a constant angular velocity, the moving direction of any hull is parallel with the Z axis. Therefore, in a short period, the effect on the air for each hull of the dual-hull airship under that condition can be considered the same as a single-hull moving along the Z axis. Thus, it can be estimated as follows:

$$k_{44-dual} = k_{33} \tag{14}$$

The dual-hull airship's added rotary inertia around the X axis is as follows:

$$m_{44} = k_{44-dual}\rho U^{5/3} \tag{15}$$

The kinetic energy of the fluid disturbed is as follows:

$$T = \frac{1}{2}\sum_{i=1}^{6}\sum_{j=1}^{6} m_{ij}v_i v_j \tag{16}$$

The momentum (or the moment of momentum) B_i and the kinetic energy T have the following relationship:

$$B_i = \frac{\partial T}{\partial v_i}\ (i = 1, 2 \ldots 6) \tag{17}$$

$\boldsymbol{B_*} = [B_1\ B_2\ B_3]^{\mathrm{T}}$ is the momentum. $\boldsymbol{K} = [B_4\ B_5\ B_6]^{\mathrm{T}}$ is the moment of momentum.

According to the principles of mechanics, $\boldsymbol{F_I}$ and $\boldsymbol{M_I}$ can be expressed in body coordinates as follows:

$$\boldsymbol{F_I} = -\frac{d\boldsymbol{B_*}}{dt} \tag{18}$$

$$\boldsymbol{M_I} = -\frac{d\boldsymbol{K}}{dt} \tag{19}$$

The added inertial force and the moment can be derived from the added mass.

$$\begin{cases} \boldsymbol{F_I} = [X_I\ Y_I\ Z_I]^T \\ \boldsymbol{M_I} = [L_I\ M_I\ N_I]^T \end{cases} \tag{20}$$

3.2 The Dual-Hull Airship Dynamic Modeling and Longitude Stability Analysis

In this paper, the 6 Degree of Freedom (DOF) dynamic model can be expressed as follows:

$$\begin{cases} m\boldsymbol{V'_G} = \boldsymbol{F_A} + \boldsymbol{F_I} + \boldsymbol{F_T} + \boldsymbol{F_B} + \boldsymbol{F_G} \\ \boldsymbol{H'} = \boldsymbol{M_A} + \boldsymbol{M_I} + \boldsymbol{M_T} + \boldsymbol{M_B} + \boldsymbol{M_G} \end{cases} \tag{21}$$

$\boldsymbol{V'_G}$ is the acceleration of the dual-hull airship's gravity center. $\boldsymbol{H'}$ is the derivative of the moment of the dual-hull airship's momentum.

After combining the model with the dual-hull airship's kinematics functions, the following is obtained:

$$\begin{cases} \dot{\phi} = p \\ \dot{\theta} = q \\ \dot{\psi} = r \end{cases} \tag{22}$$

The dual-hull airship's state function can be determined:

$$\dot{\boldsymbol{x}} = \boldsymbol{Ax} + \boldsymbol{Bu} \tag{23}$$

After the airship's longitudinal movement and latitudinal movement have been divided, the state function can be defined as follows:

$$\begin{cases} \dot{\boldsymbol{x}}_{\boldsymbol{lon}} = \boldsymbol{A}_{\boldsymbol{lon}} \boldsymbol{x}_{\boldsymbol{lon}} + \boldsymbol{B}_{\boldsymbol{lon}} \boldsymbol{u}_{\boldsymbol{lon}} \\ \dot{\boldsymbol{x}}_{\boldsymbol{lat}} = \boldsymbol{A}_{\boldsymbol{lat}} \boldsymbol{x}_{\boldsymbol{lat}} + \boldsymbol{B}_{\boldsymbol{lat}} \boldsymbol{u}_{\boldsymbol{lat}} \end{cases} \tag{24}$$

In these functions, state variables are the following:

$$x_{lon} = [u\ w\ q\ \theta\ ]^T \tag{25}$$

$$x_{lat} = [v\ p\ r\ \phi\ \psi\ ]^T \tag{26}$$

The control elements are the following:

$$u_{lon} = [\kappa_q\ \kappa_h\ \mu_q\ \mu_h\ \delta_{E1}\ \delta_{E2}\ ]^T \tag{27}$$

$$u_{lat} = [\kappa_q\ \kappa_h\ \mu_q\ \mu_h\ \delta_{R1}\ \delta_{R2}\ ]^T \tag{28}$$

In the condition of altitude at 20000 m and velocity range from 5 m/s to 15 m/s, Fig. 4 shows the calculation results of the latent root distribution of $\boldsymbol{A}_{lon}$.

The results point out the stability of the longitudinal mode for a velocity varying from 5 m/s to 15 m/s at an altitude of 20000 m.

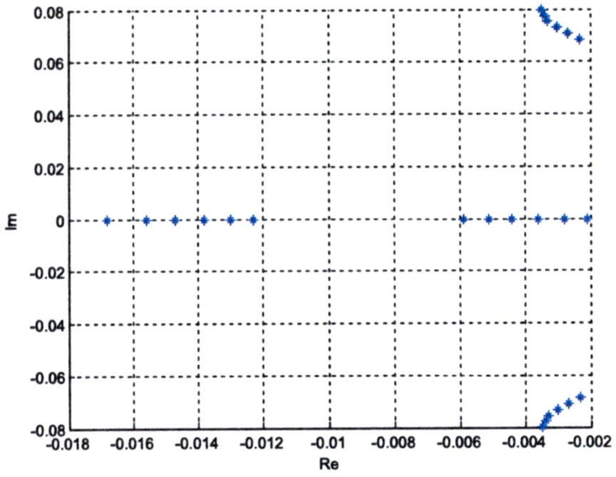

Fig. 4. The latent roots distribution of $\boldsymbol{A}_{lon}$

4 The Dual-Hull AIRShip's Features

The dual-hull airship has 222 m length, 60.5 m max cross section's diameter, and 425,463 m^3 single hull volume. The total volume of the dual hull airship can be reach to 851,026 m^3. The added mass of the dual-hull airship and the sample of the single-hull airship described in reference 5 are expressed in Table 1.

Although the volume is smaller, the dual-hull airship's m_{44} is remarkable. From Table 1, we can deduce that the airship designed in this paper is more flexible in yawing and pitching. However, the dual-hull airship is harder to rotate as its m_{44} is dramatically large compared with the sample.

Table 1. The two airships' added mass

Item	The single hull airship [4]	Airship designed in this paper
Volume/m^3	888490	851026
Length/m	287.1	222
m_{11}/kg	6726	7261
m_{22}/kg	81951	63296
m_{33}/kg	81951	63296
m_{44}/(kg.m^2)	7×10^5	17.89×10^7
m_{55}/(kg.m^2)	38×10^7	24×10^7
m_{66}/(kg.m^2)	38×10^7	24×10^7

Using the data in Ref. [4], another 6-DOF dynamic model for a single-hull airship with a similar weight, a 250 m length and a 75 m maximum cross-sectional diameter was created. When these two airships are moving along a line in the condition of 10 m/s velocity and 20 km altitude, their longitudinal open-loop responses to a 10-degree deflection of the elevators are shown below (Fig. 5).

The line with the larger amplitude is the dual-hull airship's open-loop amplitude, and the other is the single-hull airship's open-loop amplitude. The velocity range in the X axis of body coordinate is larger than the sample's because its pitching angle amplitude is larger than the sample's. Because the dual-hull airship's reference area in the body coordinate's Z axis is bigger than the single-hull airships', the velocity along the Z axis of the dual-hull airship is slower than the single-hull airship's.

The dual-hull airship's longitudinal response to the 10-degree deflection of the elevators is more noticeable than the sample. Although its longitudinal surge lasts longer than the single-hull airship's, the amplitude is not too large to control. On the contrary, the elevators' controlling efficiency of the dual-hull airship is higher than that of the single-hull airship.

5 Conclusions

A kind of dual-hull airship is designed. Its unique features are compared with the same weight class traditional airship.

This dual hull airship has more control surface area than the traditional airship does.

Compared with the traditional airship, which has the same propulsion unit distribution, the dual-hull airship's front propulsion units have the ability to provide a bigger yawing moment through differential rotation because it has a longer distance between the two propulsion units on the front of the airship.

Combined with fins, the four propulsion units of the dual-hull airship can provide more controlling methods.

A simulation test revealed that the dual-hull airship in this paper is stable in longitudinal mode, which is similar to the single-hull airship, but it is more flexible and has a higher control plane efficiency.

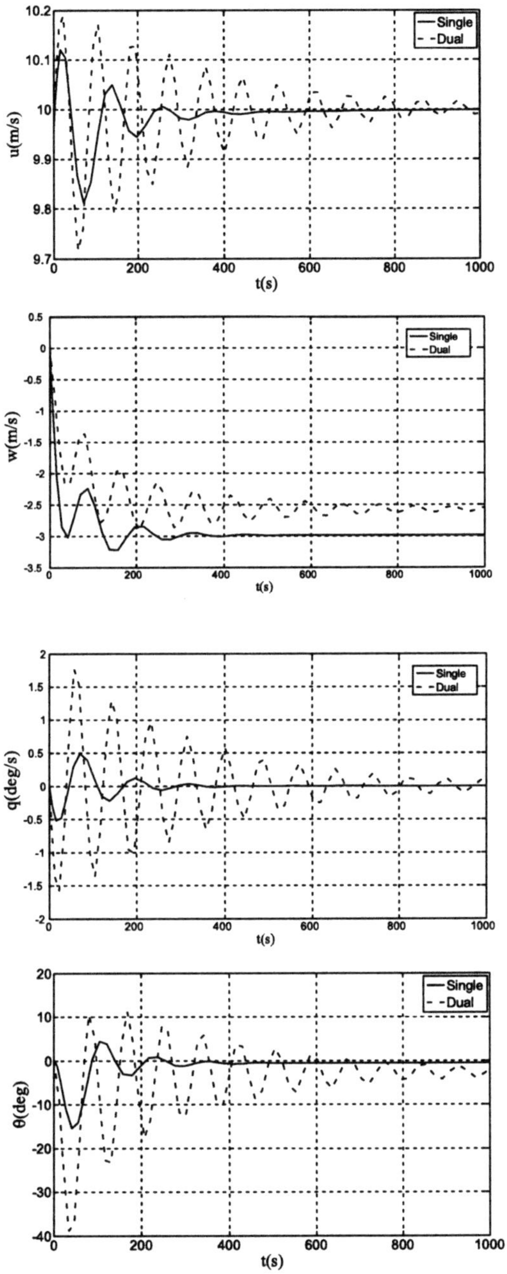

Fig. 5. The single- and dual-hull airship's response to a 10-degree deflection of the elevators.

In general, when they are moving forward, the dual-hull airship has less resistance, faster velocity and it has better flexibility than the single-hull airship in the same power situation. As its unique frame characteristic, the load could distribute more equally, the airship could have more control plane area, and the mid-wing could provide additional lift when the airship is flying forward. But its self stability is not better than the single-hull airship. Thus the dual-hull airship needs more control manipulation to keep stable, and it is not efficient for the long term station keeping mission because of large power consumption. Therefore the single-hull airship is suitable to fulfill missions such as high altitude station keeping detection while the dual-hull airship is more suitable to fulfill missions that require excellent flexibility such as transportation.

References

1. Battipede, M., Gili, P.A., Massotti, L., et al.: Dynamic modelling of a non conventional thrust-vectored airship. In: AIAA Modeling and Simulation Technologies Conference and Exhibit. Austin, Texas (2003)
2. Ouyang Jin, Q., Weidong, X.Y.: Stratospheric verifying airship modeling and analysis. J. Shanghai Jiaotong Univ. (Chin. Ed.) **6**, 956–960 (2003)
3. Mueller J.B., Paluszek, M.A., Yiyuan, Z.: Development of an aerodynamic model and control law design for a high altitude airship. In: AIAA 3rd "Unmanned Unlimited" Technical Conference. Workshop and Exhibit, Chicago, Illinois (2004)
4. Zhuang, L., Hengfa, Z.: Computational methods study on added mass of stratospheric platform airship. In: Proceedings of 2007 China Lighter-Than-Air Vehicles Conference, pp. 238–243 (2007)
5. Ziniu, W.: Aerodynamics, vol. 2. Qing Hua Press, Beijing, pp. 547–548 (2008)

Design of Pitch Angle Rate Control Law Based on L1 Adaptive Control Theory

Liangliang Wang[1(✉)], Yuanxun Wang[1], Shuang Zhang[1,2], and Wanling Gao[1]

[1] Xi'an Aeronautics Computing Technique Research Institute, Xi'an 710100, China
15991750675@163.com

[2] School of Computer Science, Northwestern Polytechnical University, Xi'an 710100, China

Abstract. Aiming at the problems of high-frequency oscillation easily caused by the traditional model-referenced adaptive control and the poor uncertainty and robustness of the traditional PID control model. This paper takes L1 adaptive control theory as the background and a scaled-down UAV in the laboratory as the object, first linearizes the UAV leveling to obtain the UAV longitudinal equations, then designs the pitch angle rate control law based on L1 adaptive theory, and designs a low-pass filter based on Liapunov's positive definite principle. Finally, perturbations and uncertainties are added to the model to verify the robustness and resistance to model uncertainties of the L1 adaptive control law, and compared with the traditional model-referenced adaptive and PID control methods, respectively. The results show that the designed L1 adaptive control system can still control the pitch angle rate well with the addition of disturbance and model uncertainty. And it ensures fast adaption with certain robustness and avoids the high-frequency oscillation problem brought by the traditional model-referenced adaption.

Keywords: Flight control · Pitch angle rate · L1 adaptive control · Longitudinal

Flight control system as the "brain" of the aircraft, its performance has an important impact on the stability and safety of the UAV. The current control methods for UAVs are mainly linear and nonlinear methods. The classical PID, LQR and other linear control methods are generally weak against model uncertainty. Nonlinear control methods such as dynamic inverse and backstepping generally require high model accuracy, and it is difficult to exclude the problem of insufficient system robustness caused by errors and external disturbances.

To solve the above problems, L1 adaptive control theory was proposed by Cao Chengyu and Naria Hovakimyan in 2006 [1]. It decouples fast adaptability and robustness, defines adaptive law with error input, introduces feedback loop into low-pass filter, and improves the transient performance of the system by increasing high adaptive gain. Therefore, it has been widely used in wing flutter suppression [2], path tracking [3], longitudinal control design [4], super maneuver control law [5], etc. At present, the research on lateral control of UAV mainly focuses on linear UAV control system [6, 7].

This paper focuses on the linear longitudinal control problem of UAV. In the paper, a laboratory type of scaled-down UAV is used as the research object, and the UAV pitch angle rate controller is designed based on the L1 adaptive theory, and compared

Chinese Society of Aeronautics and Astronautics: CASTYSF 2022, LNEE 972, pp. 147–155, 2023.
https://doi.org/10.1007/978-981-19-7652-0_14

with the MRAC method and PID control method under disturbance respectively, and the effectiveness and robustness of the L1 adaptive control algorithm are verified by simulation.

1 Construction of the Controlled Object Model

First of all, at the state point of velocity V = 30 m/s and altitude H = 2000 m, a certain type of UAV in the laboratory is leveled and linearized, and the mathematical model of the longitudinal motion of the aircraft is obtained as follows:

$$\dot{x} = Ax + bu \tag{1}$$

Selecting one set of state vectors, i.e., $x = \begin{bmatrix} \alpha \ q \end{bmatrix}^T$ and $u = [\delta_e]$, the collocation linearization results in a longitudinal linear model of the aircraft calculated as follows:

$$A = \begin{bmatrix} -6.1426 & 0.9599 \\ -74.0624 & -6.4824 \end{bmatrix}, b = \begin{bmatrix} -0.4454 \\ -123.5795 \end{bmatrix} \tag{2}$$

where, α denotes the angle of approach, q denotes the pitch rate, and δ_e denotes the elevator deflection. The simulation analysis of the longitudinal L1 adaptive longitudinal pitch angle rate flight control law design can be done based on the above linear model.

2 Design of Pitch Angle Rate Control Law Based on L1 Adaptive Control Theory

Pitch rate maintenance means that the aircraft system is given a pitch reference input q_{REF}, the system output can track well on the system input, and the control system works to maintain this attitude according to the pilot's demand.

2.1 Design of the State Predictor

The uncertain mathematical model of the aircraft system is constructed as follows.

$$\begin{cases} \dot{x}(t) = Ax(t) + b(wu(t) + \theta^T(t)x(t) + \delta(t)) \\ y(t) = c^T x(t) \quad x(0) = x_0 \end{cases} \tag{3}$$

where, $x(t) \in R^{2\times1}$ is the observable system state vector; $b, c \in R^{2\times1}$ is the known constant vector; $A \in R^{2\times2}$ is the known system matrix; $w \in R$ is the unknown input gain; $\theta_1(t) \in R^{2\times1}$ is the unknown time-varying parameter vector; $\delta(t) \in R$ is the time-varying disturbance; $y(t) \in R$ is the system output; $u(t) \in R$ is the control signal.

Because the system matrix does not satisfy the Hurwitz matrix condition, the control law $u(t) \in R$ of the controller can be divided into two parts, the linear state feedback control law $u_1(t)$ and the adaptive law $u_2(t)$ as follows.

$$u(t) = u_1(t) + u_2(t) \tag{4}$$

where, $u_1(t) = -K^T x(t)$, $u_2(s) = kD(s)\bar{r}(s)$.

2.2 Design of the Adaptation Lows

From Eqs. (1) and (2), the model of the controlled object at the state point of velocity V = 30 m/s, height H = 2000 m can be described as:

$$\begin{bmatrix} \dot{\alpha} \\ \dot{q} \end{bmatrix} = \begin{bmatrix} -6.1426 & 0.9599 \\ -74.0624 & -6.4824 \end{bmatrix} \begin{bmatrix} \alpha \\ q \end{bmatrix} + \begin{bmatrix} -0.4454 \\ -123.5795 \end{bmatrix} [\delta_e] \tag{5}$$

First check the controllability of the system and whether A satisfies the Hurwitz condition, and then configure the poles of the closed-loop system according to the desired system stability and dynamic quality so that $A_m = A - bK^T$ satisfies the Hurwitz condition.

Since $rank([b, Ab]) = 2$, the system is controllable by L1 adaptive theory, and thus the poles of the matrix $A_m = A - bK^T$ can be arbitrarily configured. The LQR method is used to determine the feedback gain and to configure the poles of $A_m = A - bK^T$ in a reasonable way. Letting Q = diag(50,300) and R = 90, the calculation is able to obtain K as follows.

$$K^T = [0.6250{-}1.7382] \tag{6}$$

At this time, the matrix A_{m} satisfies the Hurwitz matrix.

Since the goal of system control is to enable output $y(t)$ to track a given bounded segmented continuous input signal $r(t) \in R$ by selecting the control signal $u(t)$ of the state feedback controller, the state observer model is as follows.

$$\begin{cases} \dot{x}_m(t) = A_m x_m(t) + bk_g r(t) \\ y_m(t) = c^T x_m(t) \; x_m(0) = x_0 \end{cases} \tag{7}$$

where the role of k_g is to ensure that $y_m(t)$ can track the upper step reference control input with zero static error.

From Eqs. (5), (6) and (7), calculate the value of k_g as follows.

$$k_g = -\frac{1}{c^T A_m^{-1} b} = -1.7180 \tag{8}$$

2.3 Design of the Low-Pass-Filter

In the case where the system control gain is unknown, the low-pass filter cannot be applied directly to the control signal. To solve this problem, the control signal is transformed into a form containing a filter. Using the L1 adaptive controller to generate the control signal, the control input can be described as:

$$u_2(s) = kD(s)\overline{r}(s) \tag{9}$$

where $\overline{r}(s)$ is the Rasch transform of $\overline{r}(t)$, $u_2(s)$ is the Rasch transform of $u_2(t)$, $k > 0$ is the feedback gain and $D(s)$ is a strictly canonical transfer function.

In order to make the selected filter strictly canonical and stable, the filter can be described as:

$$C(s) = \frac{kwD(s)}{1 + kwD(s)} \quad \forall w \in \Omega \tag{10}$$

Here $D(s) = 1/s$ is chosen, then $C(s)$ is a first order filter and satisfies:

$$C(s) = \frac{kw}{s + kw} \quad \forall w \in \Omega \tag{11}$$

Further determination of the parameters in the low-pass filter, using the L1 gain stabilization requisite.

$$\lambda = \|G(s)\|_{L_1} L < 1 \tag{12}$$

$L = 20$ can be calculated. Let $\omega_c = k\omega$, Fig. 1 shows the relationship curve between λ and ω_c:

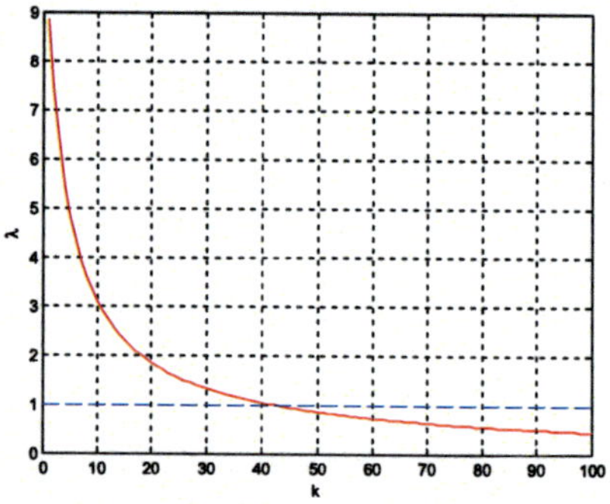

Fig. 1. The relationship curve between λ and ω_c.

As the bandwidth value ω_c of $C(s)$ gradually increases, the value of λ decreases accordingly, and $\omega_c = 90$ can be chosen in the simulation.

2.4 L1 Control Structure and Parameter Setting

The L1 adaptive pitch angle rate control structure consisting of Eqs. (1), (4), (7) and (11) is shown in Fig. 2.

Further set the range of unknown parameters:

$$w \in \Omega = [w_l, w_u], \ \theta(t) \in \Theta, \ |\delta(t)| \leq \Delta, \ \left\|\dot{\theta}\right\| \leq d_\theta, \ \left\|\dot{\delta}\right\| \leq d_\delta \tag{13}$$

where Θ is a given convex set when $t \geq 0$; $0 < w_l < w_u < \infty$ and d_θ, d_δ is a known boundary. So the convex set of set uncertainty parameters is chosen to be $\Omega = [0, 1, 2]$, $\Theta = [-10, 10]$, $\Delta = 10$.

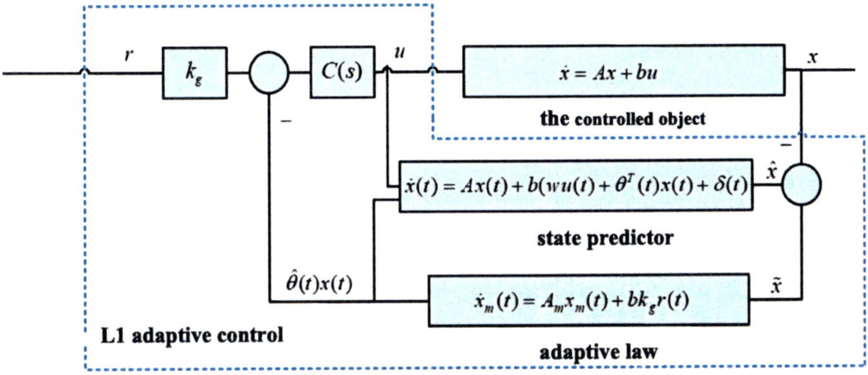

Fig. 2. L1 adaptive pitch angle rate control structure

As the L1 adaptive control effect is to be verified and compared with other control methods. Set the unknown input gain $w = 1$ for the aircraft uncertainty mathematical model of Eq. (2), and the corresponding uncertainty parameter $\theta(t)$ is:

$$\theta(t) = [0.1\sin(\pi t + {}^{\pi}/_{2})\ \ 0.3\sin(0.1\pi t)]^T \tag{14}$$

The time-varying perturbation $\delta(t)$ is:

$$\delta(t) = 0.1\sin(\pi/2t) \tag{15}$$

3 Simulation Comparison

Taking a UAV as the research object, the L1 adaptive control structure established above is used to control the UAV. The initial state of UAV is set as:

$$V = 30\text{ m/s},\ \alpha = 0.024\,rad,\ \beta = 0,\ p = q = r = 0$$
$$\phi = 0,\ \theta = 0.024\,rad,\ \psi = 0,\ x = 0,\ y = 0,\ z = 2000\text{ m}$$
$$\delta_e = 0.049\ rad, \delta_a = \delta_r = 0$$

Further collocation linearization yields the longitudinal state matrix of the aircraft as follows: $A = \begin{bmatrix} -6.1426 & 0.9599 \\ -74.0624 & -6.4824 \end{bmatrix}, b = \begin{bmatrix} -0.4454 \\ -123.5795 \end{bmatrix}$.

The L1 longitudinal control model is simulated using Matlab/Simulink, and the following scenarios are analyzed.

3.1 Comparison of the Effect of L1 Adaptive Control in the Case of System Without and with Disturbance Factors

The controlled object is the object defined by Eq. (1), the adaptive law takes $\Gamma = 100000$, the input pitch angle rate signal is selected as a square wave signal with 5/57.3. If the system has disturbance, the selection of the object uncertainty parameter $\theta(t)$ and external time-varying disturbance $\delta(t)$ are shown in Eqs. (14) and (15), respectively; otherwise, both $\theta(t)$ and $\delta(t)$ are made to be 0.

The effect of L1 adaptive control in the case of system without and with disturbance is compared as follows (The left figure shows the system without disturbance).

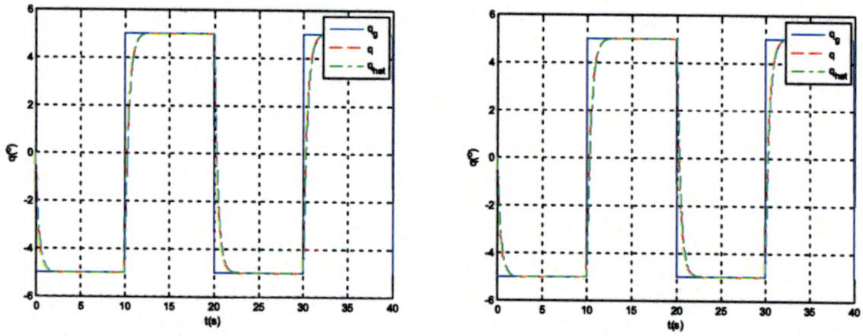

Fig. 3. Comparison of the pitch rate tracking effect of L1 adaptive control in the case of system without and with disturbance

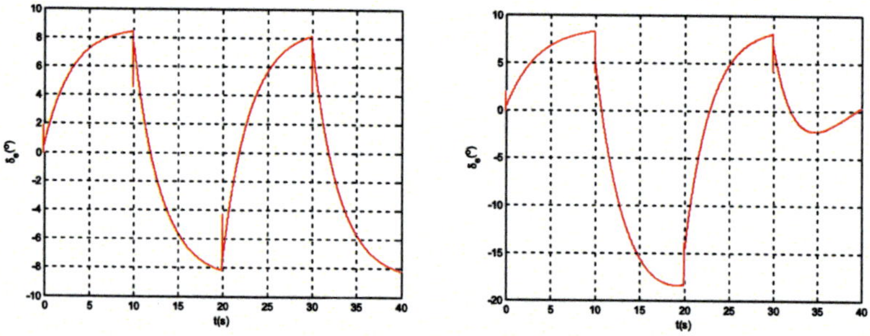

Fig. 4. Comparison of the rudder response effect of L1 adaptive control in the case of system without and with disturbance

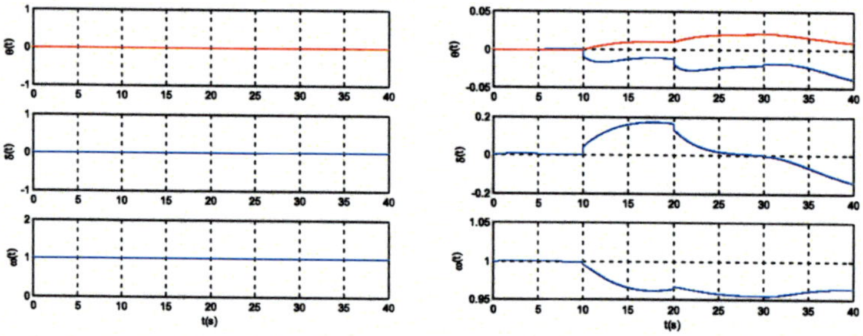

Fig. 5. Comparison of parameter estimation results of L1 adaptive control in the case of system without and with disturbance

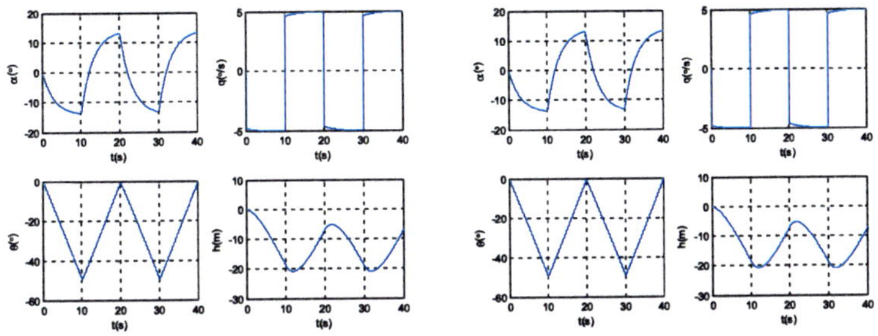

Fig. 6. Comparison of aircraft output response results of L1 adaptive control in the case of system without and with disturbance

From Fig. 3, it can be seen that the system with L1 adaptive control method can ensure good robustness, system output response and rudder adaptive deflection in both the system without and with disturbance, and the system output response can track the reference input well with zero static difference and can achieve fast adaption. Figure 4 shows that the rudder deflections of the elevator are in the controllable range and consistent and bounded when the system without and with disturbance. The rudder deflection is higher when the disturbance is added to the system than when it is not, to counteract the effect of the disturbance. Figure 5 shows that all parameter estimation results are consistently bounded and are guaranteed to vary within the set parameters. Figure 6 shows that all the aircraft output response results are stable and reliable, and all are within the acceptable range.

3.2 Comparison of the Effect of L1 Adaptive Control with Other Control Methods in the Case of System with Disturbance Factors

In order to fully illustrate the superiority of the L1 adaptive control method, the L1 adaptive algorithm is compared with the conventional model-referenced adaptive (MRAC) and PID in the case of the disturbance is added to the system.

The controlled object still selects the object defined in formula (1), the object uncertainty parameter $\theta(t)$ and external time-varying disturbance $\delta(t)$ are shown in Eqs. (14) and (15), the adaptive law is $\Gamma = 100000$ and the given input of the system is $q_g = 5/57.3$, The results of the comparison of L1 adaptive control method with MRAC control and PID control respectively are as follows.

Figure 6 shows that after adding the disturbance, the pitch angle rate output of the system with L1 adaptive control method and MRAC method can track the input command signal well without overshoot and the control effect is excellent, while the PID method has been unable to control the controlled object and the pitch angle rate output of the control system cannot track the upper reference input.

Figure 7 shows that the system with MRAC control method showed obvious high-frequency oscillation of the rudder surface, and the control effect was poor. Comparing MRAC method and PID method, the rudder output of the system with L1 adaptive control method has slow rudder response without spikes, smaller maximum peak of rudder deflection, and higher rudder efficiency (Fig. 8).

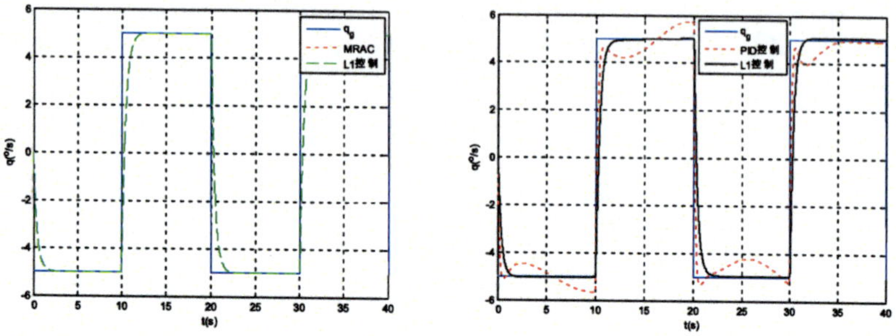

Fig. 7. Comparison of pitch rate control of the L1 adaptive control method with the traditional model-referenced adaptive (MRAC) method and PID method

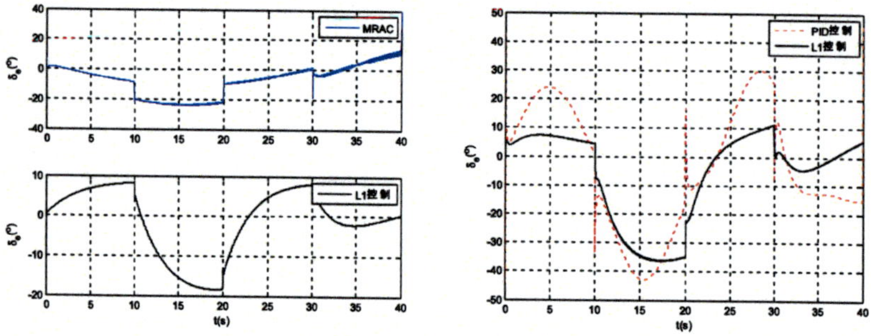

Fig. 8. Comparison of the rudder output effect of the L1 adaptive control method with the traditional model-referenced adaptive (MRAC) method and PID method

Thus, the L1 adaptive control method is superior to the MRAC method and the PID method on balance.

4 Conclusion

In the paper, a pitch angle rate controller based on L1 adaptive theory is designed with a laboratory type of scaled-down UAV as the research object, and the comparison of different disturbance conditions and different methods is carried out. From the simulation results, we can see that the L1 adaptive control method can still control the pitch angle rate well and the rudder adaptive deflection is stable and effective, which can reduce the influence caused by the disturbance well under the condition of adding model uncertainty and disturbance again. In contrast, after the disturbance is added, the traditional MRAC control method shows high frequency oscillation of the rudder surface, and the PID control can no longer control the pitch rate effectively under the disturbance addition. The effectiveness and superiority of the designed L1 adaptive theory algorithm are verified.

Acknowledgments. This study was supported in part by Aeronautical Science Foundation of China under Grant Project 2020Z069031001 and a key research project MJ-2018-34 in China.

References

1. Hovakimyan, N., Cao, C.: L1 Adaptive Control Theory: Guaranteed Robustness with Fast Adaptation. Siam (2010)
2. Cao, C., Hovakimyan, N.: L1 Adaptive controller for multi–input multi–output systems in the presence of unmatched disturbances. In: 2008 American Control Conference, Westin Seattle Hotel, Seattle, Washington, USA, June 11–13 (2008)
3. Cao, C., Hovakimyan, N.: L1 Adaptive controller for a class of systems with unknown nonlinearities: Part I. In: 2008 American Control Conference. Westin Seattle Hotel, Seattle, Washington, USA, June 11–13 (2008)
4. Cao, C., Hovakimyan, N.: Design and analysis of a novel L1 adaptive controller, part 1:control signal and asymptotic stability. In: American Control Conference, Institution of Electrical and Electronics Engineers, Piscataway, NJ, pp. 3397–3402 (2006)
5. Cao, C., Hovakimyan, N.: L1 adaptive controller for a class of systems with unknown nonlinearities: part II. In: 2008 American Control Conference, Westin Seattle Hotel, Seattle, Washington, USA, June 11–13 (2008)
6. Banerjee, S., Vanyai, T.: L1 augmentation configuration for a lateral/directional manoeuvre of a hypersonic glider in the presence of uncertainties. In: 19th AIAA International Space Planes and Hypersonic Systems and Technologies Conference, Atlanta, GA, June 16–20 (2014)
7. Patel, V., Cao, C., Hovakimyan, N.: L1 adaptive controller for tailless unstable aircraft in the presence of unknown actuator failures. In: AIAA Guidance, Navigation, and Control (GNC) Conference, August 19–22 (2007)
8. Kharisov, E., Dobrokhodov, V., Pascoal, A.: Limiting behavior of L1 adaptive controllers. In: AIAA Guidance, Navigation, and Control Conference, Portland, Oregon, 08–11 August (2011)
9. Kitsios, I., Dobrokhodov, V.: Experimental validation of a metrics driven L1 adaptive control in the presence of general unmodeled dynamics. In: AIAA Guidance, Navigation, and Control Conference, Chicago, Illinois, 10–13 August (2009)
10. Mallikarjunan, S., Nesbitt, B., Kharisov, E.: L1 adaptive controller for attitude control of multirotors. In: AIAA Guidance, Navigation, and Control Conference, Minneapolis, Minnesota, 13–16 August (2012)
11. Zhanqi, F., Lin, L. etc.: Design of super maneuvering flight control law based on L1 adaptive method. Comput. Meas. Contr. **21**(12), 3281–3283 (2013). (in Chinese)
12. Xuesong, L., Yinghui, L., Shigang, Z., etc.: L1 adaptive longitudinal control design of small UAV. Flight Mech. **29**(2), 58–62 (2011)
13. Weiguo, Z., Aijun, L., Guangwen, L., Xiaoxiong, L.: Modern Flight Control System Design. Northwest University of Technology Press, Xi'an (2009).(in Chinese)
14. Liang-liang, W., Xiao-xiong, L., Qing-yuan, M., et al.: Longitudinal damper design of UAV based on L1 adaptive method. Flight Mech. **33**(6): 131–135 (2015). (in Chinese)

A Review of Electric Propulsion for Air Vehicle Research

Chen Xiao-li[1,2](✉), Zhou Chao[2], and Bai Cheng-an[2]

[1] Chongqing Aerospace Polytechnic, Chongqing 400021, China
592829123@qq.com
[2] Peking University, Beijing 100871, China

Abstract. Electric propulsion technology possesses the potential of high efficiency, low noise and low emissions. Distributed electric propulsion technology can not only provide the thrust required by the aircraft, but also the flexibility of power transmission, which makes the design of the aircraft from a larger dimension. Therefore, electric propulsion aircraft is one of the main development directions of future aircraft. Based on the electric propulsion method, the aerodynamic layout of the aircraft can be changed, and the flight efficiency can be improved by improving the lift-to-drag ratio, making the aircraft safer, more efficient, and low-carbon. This article reviews the development status of aviation electric propulsion vehicles, focusing on the characteristics of distributed electric propulsion technology and aerodynamic layout changes of electric aircraft.

Keywords: Electric propulsion technology · Distributed electric propulsion technology · Air vehicle

1 Introduction

With the development of society and the advancement of science and technology, the development of *green aviation* has become the consensus of today's society. The specific implementation methods include the development of advanced aircraft design technology, power system technology using clean energy, material manufacturing technology and flight landing technology. Noise technology, etc. [1]. In recent years, the goal of low carbon has become increasingly clear. Adopting carbon offset schemes, switching to efficient aircraft, and using renewable energy are all ways to reduce carbon dioxide emissions. A series of research projects have been carried out internationally, such as Carbon Offset and Reduction Scheme of International Aviation (CORSIA) [2], clean skies programme of the European Union, environmentally responsible aviation of NASA (ERA) and *2016 Green Aviation White Paper* and so forth.

Electric propulsion power plant is one of the feasible ways to realize low-carbon green aviation. Compared with the traditional fuel power system, the electric propulsion system has potential advantages such as high efficiency, environmental protection, low noise, low vibration, and easy maintenance [3]. It is one of the research hotspots in the field of aviation technology in the world. E-Fan aircraft of Airbus uses lightweight,

Chinese Society of Aeronautics and Astronautics: CASTYSF 2022, LNEE 972, pp. 156–169, 2023.
https://doi.org/10.1007/978-981-19-7652-0_15

high-efficiency generator and electric motor technologies, including the Cri-Cri, which has four electric motors to drive the propellers successfully tested flight in September 2010 and the E-Fan has four electric motors to drive propellers, which successfully test flight at the end of 2013 [4]. At the same time, Germany's Siemens, Germany's Siemens and Airbus jointly developed the E-Airbus 100-seats regional airliner concept based on a distributed hybrid electric propulsion system. Airbus and Rolls-Royce jointly developed the hybrid concept aircraft E-Thrust, which makes Track 2050 is a big step forward. Electric propulsion vehicles are also being considered as a possible way to achieve the ambitious goals of the European Union's *Flightpath 2050* [5].

Electric power for electric propulsion can be provided by the aircraft's on-board batteries or by a hybrid system using an engine. When the propulsion system is driven by an electric motor, aircraft designs that are difficult to achieve with conventional gas turbine power can be used, resulting in increased efficiency and reduced emissions. For example, Single-aisle Turboelectric Aircraft with Aft Boundary Layer (STARC-ABL) developed by NASA in the Fig. 1. Engines under the wings provide most of the thrust during takeoff, some during cruise, and the remainder is provided by electric propulsion in the tail. The aircraft tail duct of STARC-ABL is also known as Boundry Layer Ingesting (BLI), It can not only generate forward thrust when the fan rotates at high speed, but also extract and accelerate the boundary layer flowing through the fuselage and engine and then discharge it backward, improving the tail flow field, reducing the wake resistance, and increasing the fuel efficiency by 10%., significantly reducing environmental pollution [6, 7]. A similar tail propulsion system would be difficult to implement if powered by a gas turbine.

Fig. 1. The single-aisle turboelectric aircraft with aft boundary layer of NASA

The blended-wing-body N3-X aircraft concept, which studied by NASA, adopts gas turbine electric distributed propulsion system (TeDP) and full composite material, laminar flow, and wing body fusion technology, and its fuel consumption rate is reduced by more than 70% compared with Boeing 777-200LR aircraft [7].

Zunum believes that 40% of current aviation pollution comes from short-haul flights. In response to the demand for short-haul flights, Boeing and Zunum Aero have developed small hybrid electric passenger aircraft to significantly reduce flight expenses and reduce aviation carbon emissions. Compared with conventional propulsion aircraft, hybrid electric aircraft is expected to reduce carbon emissions by up to 80%, and by reducing fuel consumption, it can reduce operating costs by 40–80%. If the pure electric design is adopted, it is expected to achieve 75% noise reduction and zero carbon emissions [8] (Fig. 2).

Fig. 2. The concept picture of Zunum Aero

Compared with the mechanical transmission method, the convenience of transmitting energy through electricity is much higher, so a more flexible power arrangement can be adopted on the aircraft, and the distributed electric propulsion (DEP) system aircraft came into being. Distributed electric propulsion systems are powered by several motor-driven fans, propellers, and other devices on the wings or fuselage [9, 10] The distributed electric propulsion system can control the flow field of the aircraft more finely, so as to achieve higher efficiency, which is a very promising solution to improve the efficiency and performance of the aircraft. Compared with traditional powered aircraft, the electric propulsion system is compact, reliable and safe, and it is also conducive to the beneficial coupling between the two traditionally independent disciplines of the airframe and the propulsion system [11]. As an emerging technology, distributed electric propulsion technology involves a series of issues such as power system and propulsion layout [12]. This paper reviews the composition of electric propulsion system and the research status of distributed electric propulsion aircraft, focusing on the research progress of aerodynamic layout of distributed electric propulsion aircraft, in order to provide reference for the development of distributed electric propulsion technology.

2 The Common Distributed Propulsion System

Generally, the aircraft with distributed propulsion has more than four or even as many as dozens of propulsion units, which can better integrate the propulsion system into the airframe, to raise aerodynamic efficiency, improve flight performance and fuel consumption rate. The engines of distributed propulsion aircraft are propellers or ducted fans. Biser et al. [5] found that the DEP-6 and DEP-12 project aircraft both have a potential of reducing the specific block fuel consumption by approx. 4.2% and 5.8% respectively in case of direct driven propellers, while for a geared-drive scenario a total reduction of up to 7.6% in case of DEP-6 and 8.5% in case of DEP-12 is possible, all compared to the direct-driven baseline (BSL) aircraft SE-CS23 turboprop commuter aircraft (as shown in Fig. 3) adopted a distributed electric propulsion design concept. In a simplest way the typical twin-engine configuration could be modified by having in total four or six propellers facilitated by electrical power transfer. And as the number of propellers increases, the new aerodynamic layout has potential advantages [13]. And a reasonably spanwise-invariant loading can be produced by the installation of the overwing ducted fan DEP system [14]. The interaction of the propeller slipstream with wing aerodynamics for improvement of high-lift capabilities or the reduction of induced drag by the installation of wingtip propellers, alleviate the wing root bending moment caused by high aerodynamic forces acting on the wing and decrease the structural weight of the wing. This kind of flight control system has the potential to reduce the size of the vertical tail plane, thus, reducing its structural mass and drag contribution. The studied shows a turbo-electric aircraft configuration can only perform better than a comparable turobprop if performance and efficiency of the electrical components are improved [15]. Erhard et al. [16] evaluated the aero-propulsive efficiency of a general aviation aircraft under various propeller distributions and found propeller-wing interactions can bring about beneficial aerodynamic effects. The power unit can be located at the leading edge, trailing edge and other places in a distributed propulsion system, which will have different effects.

Fig. 3. The turboprop aircraft of CS23

2.1 The Distributed Electric Propulsion is Located on the Leading Edge of the Wing

Multiple distributed propellers are installed on the leading edge of the wing is a common distributed electric propulsion aircraft structure [17], which is characterized by a plurality of propellers distributed along the span of the wing, and these propellers are located at the leading edge of the wing. The rear wing of the propeller is in the slipstream region, and the high-speed airflow can increase the lift of the wing. The distributed propeller system increases the wing area affected by the slipstream to improve the effect of slipstream lift. There are multiple trim schemes for distributed power units, which is easily installed [16].

This concept consists in distributing propulsive power and thrust through several relatively small propellers that are installed ahead of the leading edge of the wing with the purpose of blowing their slipstream on it during takeoff and landing, increasing the generated lift.

Stoll et al. [27] tested on a four-seat general aviation aircraft to study the effect of different wing and propeller layout. It shows the aircrafts with distributed power layout are more efficient than complex traditional designs, and the lower propeller tip speed can also reduce noise. Murilo et al. [18] studies the performance of distributed electric propulsion on general aviation aircraft. It shows that when the aircraft is flying at low speed, the effect of slipstream enhancement is improved by installing multiple small propellers on the leading edge of the wing, which can reduce the wing area by 62% and increase the wing load by 131%. As shown in Fig. 4, the EcoPulse distributed hybrid electric propulsion demonstrator has three electric propellers installed on both sides of the wings to replace the original standard engine and propeller power unit. The slipstream generated by the distributed propeller has a larger area of action on the surface of the wing to increase the lift coefficient [19]. Meanwhile, the four propellers which near the fuselage are adopt a folding design, and can be folded according to the requirements of flight conditions to reduce the cruise resistance.

Fig. 4. EcoPulse demonstrator

NASA X-57 demonstrator (as shown in Fig. 5) is a typical configuration of a distributed electric propulsion system with engines located on the leading edge of the wing, which adopts a large aspect ratio wing layout to reduce flight resistance and improve efficiency. There are 12 small high-lift motors with propellers are distributed along the leading edge of the wing, which can provide lift during take-off and landing and the propeller blades are retracted into the nacelle to reduce resistance during cruising. The propellers on the wingtips on both sides can reduce the adverse effect of the wingtip vortex on the aerodynamic performance. X-57 aircraft not only takes advantage of the advantages of high battery energy utilization, low or even zero emissions, and low motor noise, but also uses generated by the addition of motor propellers through the slipstream effect during takeoff and landing to significantly improve the maximum lift coefficient of the wing. Under the same take-off and landing performance conditions, the length of the wing chord can be greatly shortened, the wing area can be reduced to 40% of the original, and the weight of the body structure can be reduced [20]. Distributed electric propulsion systems increase the degree of freedom associated with flight controls as well. NASA Scalable Convergent Electric Propulsion Technology Operations Research (SCEPTOR) [21, 22] converted the internal combustion engine-powered propulsion of an existing light aircraft into a structure with 12 small electric propellers and two large propellers at the tip. The 12 small high-lift electric-driven propellers are distributed along the leading edge of the left and right wings to increase the dynamic pressure on the wing surface and improve the wing performance when the aircraft is flying at low speed. The 2 electric-driven cruise motors and propellers are located on the wingtips to provide main propulsion power and reduce the adverse effects of wingtip vortices to improvie the take-off and cruising performance of the aircraft. NASA builds a SCEPTOR 4.0 aircraft by creating SCEPTOR demonstration geometry and determining the modeling design parameters for different layout forms [21]. Flight tests have shown a significant increase in efficiency over the prototype at a cruise point of 150 knots and an altitude of 8,000 feet. The performance of distributed electric propulsion aircraft with different structures will be different. Compared with the aircraft powered by 2 propellers, the take-off distance of the aircraft powered by 16 propellers is shortened by at least 50%; the aircraft with 8 propellers has the best noise reduction effect, and the take-off distance is 43% shorter than that of the aircraft with 2 propellers [23].

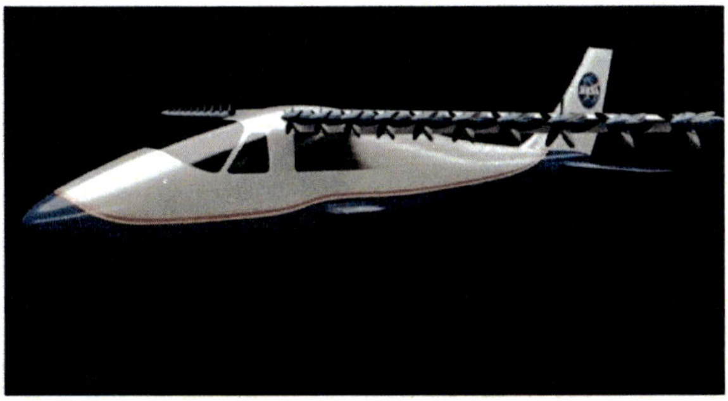

Fig. 5. NASA X-57 Maxwell aircraft demonstrator

2.2 The Distributed Electric Propulsion is Located on the Trailing Edge of the Wing

The distributed power enables a high degree of coupling between the propulsion system and the fuselage when the propulsion system is along the fuselage or trailing edge of the wing. It adopts a ducted engine configuration, and the distributed power is usually located at the tail of the wing or fuselage. The boundary layer at the fuselage or tail of the wing are suck into ducted engines to increase the lift-to-drag ratio of the wing. Hill et al. [24] studied the propulsion system of an advanced civil aircraft and found that the boundary layer inhalation effect can reduce the energy input by 9%, and the distributed arrangement of the power system can also delay the stall of the aircraft. However, the positions between the spanwise engines will still be separated when the angle of attack reaches the stall point, and the three-dimensional flow may affect the consistency of engine operating conditions [25].

SAX40 (as shown in Fig. 6) and N3-X (as shown in Fig. 7) are the typical representatives which the propulsion system of the aircraft is located on the fuselage or the trailing edge of the wing. The SAX40 is powered by nine fans. The N3-X aircraft concept utilizes a turbine electric distributed propulsion system (TeDP), with two wingtip-mounted turbine generators driving 14 electric fans. The devices that generate thrust and power are separated by TeDP. Two turboshaft engines mounted on the wingtip drive superconducting generators to generate electricity and 15 superconducting electric thrusters embedded in the fuselage to generate thrust. This new layout can achieve the goal of reducing the fuel consumption rate of Boeing 777-200LR aircraft by more than 70% [26].

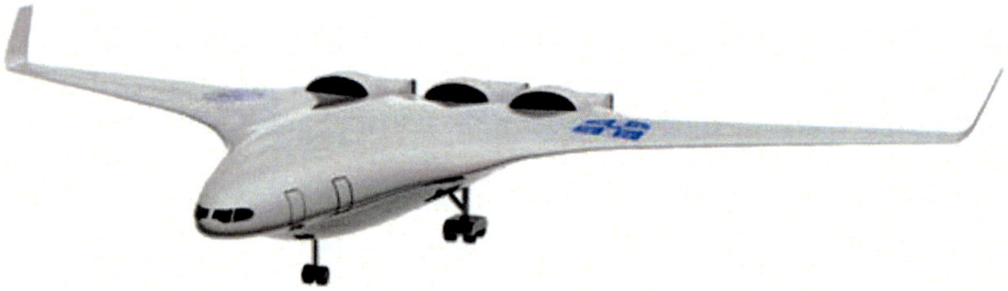

Fig. 6. SAX-40

In a word, when the engine is located on the leading edge of the wing, the distributed propulsion power system of the propeller is usually used, which can effectively obtain a better effect of increasing slip flow, control the tip vortex system, and facilitate structural design and engineering implementation. When the engine is located on the trailing edge of the wing or the rear of the fuselage, the structure of the ducted fan is often used, which can effectively offset the resistance of the fuselage mainly by controlling the boundary layer.

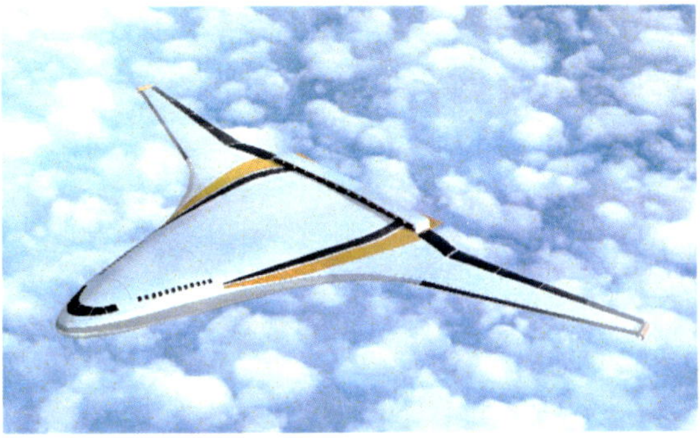

Fig. 7. N3-X

3 Application and Effect of Electric Propulsion Aerodynamic Configuration

Modern conventional civil airliners typically have an approximately cylindrical fuselage, swept-down monoplane, single vertical tail, wing crane or tail crane engine shape (as shown in Fig. 8). From piston propellers to turbojet aircraft engines, the design of aircraft aerodynamic configuration and aerodynamic efficiency are directly influenced by different types of engines. Modern large-scale civil airliners reduce drag by installing winglets, attaching deflectors to the wing and engine nacelle, and adopting a fusion of engine wing suspension configuration and blended wing body [27, 28]. The application of electric propulsion increases a new dimension to aircraft operation and aerodynamic configuration.

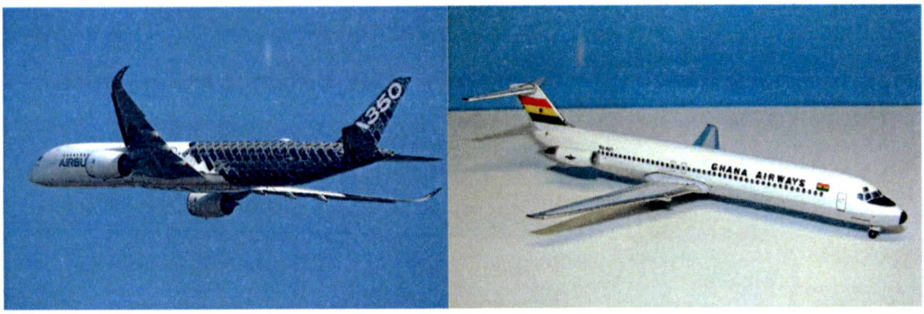

Fig. 8. The appearance of modern conventional civil airliners

3.1 Normal Aerodynamic Configuration

In the traditional aerodynamic configuration, in order to minimize the influence of airflow near the engine on the overall aerodynamics of the aircraft, the propulsion system is usually installed in front of the wing or near the fuselage. Airbus and engine manufacturer

Rolls-Royce's E-Fan X (as shown in Fig. 9) program, based on the BAe 146 aircraft, attempts to reduce carbon emissions with a hybrid electric propulsion system. One or two of the four engines of conventional aircraft are changed to electrically driven bypass fans, which can better meet the aircraft power requirements of the aircraft during take-off, landing and cruise phases, by taking advantage of the convenient and coordination feature of the motor to reduce fuel consumption.

Fig. 9. The concept aircraft of Airbus' E-Fan X

The hybrid aircraft concept E-Thrust (as shown in Fig. 10) was proposed based on the ducted fan installed at the tail of the aircraft to improve the performance of the aircraft by increasing energy to the boundary layer. A total of 6 electric ducted fans are installed above the wing roots on both sides to provide power for the aircraft. This electric distributed propulsion system can achieve low fuel consumption, low emissions and low noise, the overall bypass ratio of the engine can reach 20, when the aircraft needs to slow down, it can even generate electricity by the wind [13].

Fig. 10. The concept of E-Thrust

3.2 Blended Wing Body Configuration

The Blended Wing Body (BWB) blends the fuselage and wing structure into a flying wing shape. The original intention of the BWB concept is designed to maximize the overall efficiency of the aircraft by integrating the engine, wing and airframe into a single lift surface. The BWB configuration aircraft has the potential advantages of low aerodynamic resistance, high flight efficiency, light structure weight, large loading space, energy saving, environmental protection and noise reduction. It is one of the revolutionary technologies of civil aircraft that are expected to achieve the *economy, environmental protection, comfort and safety* requirements of green aviation in the future, and it is also a field that the international aviation community has been studying for many years. NASA studies have shown that the BWB can save fuel by 10%, and the noise of the BWB aircraft is 37 decibels lower than the *round* + *wing* aircraft [29].

The blended wing-body configuration created new possibilities for propulsion system innovation and integration. As shown in Fig. 11, the MAVERIC (an aircraft model used to verify and test robust innovation control) *BWB* demonstrator displayed by Airbus at the Singapore Airshow. The configuration design with a smooth transition from the wing to the fuselage shape is extremely revolutionary. This enables the entire fuselage to generate lift and reduce drag. Meanwhile, the engines are located above the fuselage, reducing noise. Compared with current single-aisle aircraft, the aircraft is expected to reduce fuel consumption by more than 20%, and its multifunctional cabin provides a new onboard passenger experience. Another example is the Airbus ZEROOe 3 zero-emission commercial airliner concept (as shown in Fig. 12), which adopts a *blended wing body* configuration and is equipped with a distributed turbofan engine power system. Its ultra-wide fuselage provides multiple options for storing and distributing hydrogen fuel.

Fig. 11. The model of MAVERIC

3.3 The Truss-Braced Wing (TBW) Configuration

The appearance of the Truss-Braced Wing (TBW) configuration is similar to the traditional *tubular fuselage* + *wing* configuration, but a truss structure is used to support the

Fig. 12. The blended wing body passenger aircraft concept scheme of ZEROe

wings, as shown in Fig. 13. Compared with the traditional cantilever beam wing, TBW bears part of the load on the truss, which can reduce the bending moment of the wing root, and can increase the wing aspect ratio with the same weight. The larger the aspect ratio, the larger the lift-to-drag ratio and the smaller the induced drag, the TBW configuration can reduce fuel consumption by 10% or even higher than the traditional cantilever wing configuration. In addition, the TBW configuration can reduce the thickness of the airfoil, which will significantly reduce the wave resistance during transonic flight, and the thin airfoil is more conducive to the realization of natural laminar flow [5].

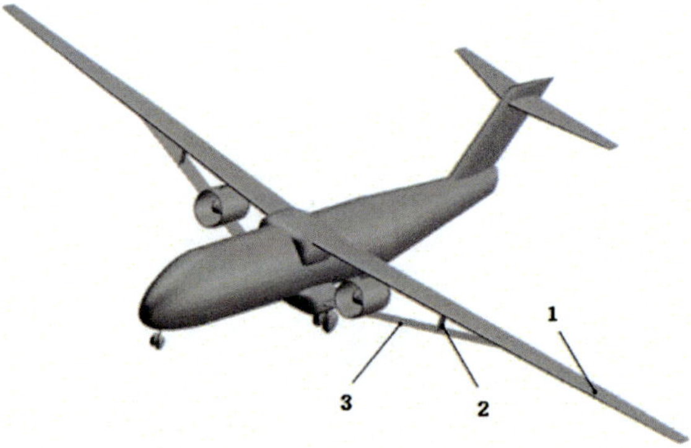

Fig. 13. The configuration of Truss-Braced Wing

The Boeing and NASA studied the Subsonic Ultra-Green Aircraft (SUGAR) (in Fig. 13) [30], which own Truss-Braced Wing (TBW), T-tail layout, and high aspect ratio. One end of the wing is supported by a strut. Compared with the traditional wing, the TBW can reduce the thickness and sweep angle of the wing, increase the wingspan, and improve the lift. The support beam can effectively reduce the bending deformation of the wing, relieve the load on the root of the wing, and significantly reduce the weight of the wing structure. Meanwhile, the wing can be folded and tucked when the aircraft land to fit into the airport space and reduce the footprint [31]. SUGAR Volt adopts parallel gasoline-electric hybrid technology, which is mainly powered by electricity for short-haul flights, while jet fuel will replace electricity for long-term flights. The fuel engine and battery will jointly provide power during take-off. The extra thrust added by the battery can shorten the take-off time and achieve the purpose of energy conservation and environmental protection. When the aircraft reaches cruising altitude, it can be completely switched to the battery power supply mode to reduce fuel consumption and nitrogen oxide emissions [32]. Compared with traditional aircraft, the efficiency of SUGAR Volt is improved by 55%, releasing 60% less CO_2 emissions and 80% less NOx emissions during flight [33] (Fig. 14).

Fig. 14. Subsonic Ultra-Green Aircraft (SUGAR) volt

4 Conclusions

Electric propulsion technology is a potential way to improve high-efficiency energy utilization, low life-cycle carbon emissions and high noise of aircraft. It make the travel through air transportation more environmental an more efficient. This paper reviews the distributed propulsion and conventional aerodynamic layout of electric propulsion aircraft. The main conclusions are as follows:

1. The aircraft aerodynamic efficiency can be improved via the synergistic integration of thrusters and aerodynamic control surfaces. It is expected to save fuel or energy, reduce emissions, lower noise, and improve take-off performance.
2. The distributed propulsion system aircrafts are more flexible than traditional aircraft. Enabling a high degree of aircraft and engine coupling design. The distributed propulsion layout involves the research on the integration of aircraft overall, power, electromechanical and other technologies, and is likely to bring revolutionary changes to the aviation industry. Many researchers who are from different countries are actively exploring related technologies and applications, but there are still many technical bottlenecks that need to be solved urgently. After the continuous exploration of scientists, people can take an electric propulsion system aircraft which is *low-emission*

or even *zero-emission* to fly around the world, and take an *air-taxis* to achieve commuting and short-distance travel. It can contribute to the harmony between man and nature.

3. Due to the limitation of battery technology, it is difficult to meet the requirements of long-distance aviation operations for the all-electric propulsion aircraft. However, the hybrid electric propulsion is a compromise solution that comprehensively considers environmental protection, range, flight time, and economy. If the lithium battery technology can be increased by two to three times on the existing basis in the future, the all-electric aircraft can realize air transportation with higher payload and range.

Acknowledgment. This work was financially supported by Chongqing Municipal Education Commission (KJQN201903004).

References

1. Liu, Z.M.: A greener road for aviation. Manufacture **7**, 50–55 (2020)
2. Ploetner, K.O., Urban, M., Roth, A., et al.: Fulfilling long-term emission reduction goals in aviation by alternative fuel options: an evolutionary approach. In: 2018 Aviation Technology, Integration, and Operations Conference, 3990 (2018)
3. Li, Y., Duan, C.L., Chen, L.: NASA launches X-57 aircraft to demonstrate and test distributed electric propulsion technology. China Aviation News, 2021-01-01(009)
4. http://www.cannews.com.cn/2014/12/02/99113022.html
5. Biser, S., Atanasov, G., Hepperle, M., et al.: Design space exploration study and optimization of a distributed turbo-electric propulsion system for a regional passenger aircraft. In: AIAA Propulsion and Energy 2020 Forum (2020)
6. Yildirim, A., Gray, J.S., Mader, C.A., et al.: Performance analysis of optimized STARC-ABL designs across the entire mission profile. In: AIAA Scitech 2021 Forum, 0891 (2021)
7. Schnulo, S.L., Chapman, J.W., Hanlon, P., et al.: Assessment of the impact of an advanced power system on a turboelectric single-aisle concept aircraft. In: AIAA Propulsion and Energy 2020 Forum (2020)
8. http://tech.chinadaily.com.cn/a/201903/28/WS5c9c6335a310e7f8b15733ba.html
9. Huang, J., Yang, F.T.: Development and challenges of electric aircraft with new energies. Acta Aeronaut. Astronaut. Sin. **37**(01), 57–68 (2016)
10. Huang, J.: A Survey of design technology on distributed electric propulsion aircraft. Acta Aeronaut. Astronaut. Sin. **42**(1), 624037 (2021). (in Chinese)
11. Kruger, M., Uranga, A.: The feasibility of electric propulsion for commuter aircraft. In: AIAA SciTech 2020 Forum, 1499 (2020)
12. Zhang, X.W.: Distributed electric propulsion technology oriented to 2030. In: Proceedings of the 2nd China Aviation Science and Technology Conference 2015 (2015)
13. Herzog, N., Reeh, A., Kümmel, A., et al.: Analysis of Distributed Electric Propulsion on Commuter Aircraft. In: AIAA Scitech 2021 Forum, 1199 (2021)
14. Yu, D., Ansell, P.J., Hristov, G.: Aero-propulsive integration effects of an overwing distributed electric propulsion system. In: AIAA Scitech 2021 Forum (2021)
15. Habermann, A.L.: Effects of distributed propulsion on wing mass in aircraft conceptual design. In: AIAA Aviation 2020 Forum, 2625 (2020)
16. Erhard, R.M., Clarke, M.A., Alonso, J.J.: A low-cost aero-propulsive analysis of distributed electric propulsion aircraft. In: AIAA Scitech 2021 Forum (2021)

17. Stoll, A.M., Bevirt, J.B., Moore, M.D., et al.: Drag reduction through distributed electric propulsion. In: 14th AIAA Aviation Technology, Integration, and Operations Conference (2014)
18. Gallani, M.A., Góes, L.C.S., Nerosky, L.A.R.: Effects of distributed electric propulsion on the performance of a general aviation aircraft. In: 2020 AIAA/IEEE Electric Aircraft Technologies Symposium (EATS). IEEE, pp. 1–16 (2020)
19. Liao, Z.Q.: The development of Safran electric propulsion technology. Electr. Propuls. **3**, 31–36 (2020)
20. http://www.cannews.com.cn/2021/01/04/99318140.html
21. Patterson, M.D., Derlaga, J.M., Borer, N.K.: High-lift propeller system configuration selection for NASA's SCEPTOR distributed electric propulsion flight demonstrator. In: 16th AIAA Aviation Technology, Integration, and Operations Conference, 3922 (2016)
22. Borer, N.K., Patterson, M.D., Gibson, A.R., et al.: Design and performance of the NASA SCEPTOR distributed electric propulsion flight demonstrator. In: 16th AIAA Aviation Technology, Integration, and Operations Conference (2016)
23. Moore, K.R., Ning, A.: Takeoff and performance trade-offs of retrofit distributed electric propulsion for urban transport. J. Aircr. **56**(5), 1880–1892 (2019)
24. Hall, D.K., Huang, A.C., Uranga, A., et al.: Boundary layer ingestion propulsion benefit for transport aircraft. J. Propul. Power **33**(5), 1118–1129 (2017)
25. Kerho, M.F.: Aero-propulsive coupling of an embedded, distributed propulsion system. In: 33rd AIAA Applied Aerodynamics Conference, 3162 (2015)
26. Felder, J., Tong, M., Chu, J.: Sensitivity of mission energy consumption to turboelectric distributed propulsion design assumptions on the N3-X hybrid wing body aircraft. In: 48th AIAA/ASME/SAE/ASEE Joint Propulsion Conference & Exhibit, 3701 (2012)
27. Zhang, S., Xia, M., Zhong, B.: Evolution and technical factors influencing civil aircraft aerodynamic configuration. Acta Aeronaut. Astronaut. Sin. **37**(1), 30–44 (2016)
28. Jiang, J., Zhong, B.W., Fu, S.: Influence of overall configuration parameters on the aerodynamic characteristics of a blended-wing-body aircraft. Acta Aeronaut. Astronaut. Sin. **37**(1), 278 (2016)
29. Wang, G., Zhang, B., Zhang, M.H., et al.: Research progress and prospect for conceptual and aerodynamic technology of blended-wing-body civil aircraft. Acta Aeronaut. Astronaut. Sin. **9** (2019)
30. Bradley, M.K., Droney, C.K.: Subsonic ultra green aircraft research: phase I final report. National Aeronautics and Space Administration, Langley Research Center (2011)
31. http://www.comac.cc/xwzx/gzdt/201703/17/t20170317_4978508.html
32. Bradley, M.K., Droney, C.K.: Subsonic ultra green aircraft research: phase II–volume II–hybrid electric design exploration. NASA CR-218704 (2015)
33. Jain, S., Crossley, W.A.: Predicting fleet-level carbon emission reductions from future single-aisle hybrid electric aircraft. In: 2020 AIAA/IEEE Electric Aircraft Technologies Symposium (EATS). IEEE, pp. 1–15 (2020)

Development of DNN Accelerator and Its Application in Avionics System

Zhao Yixuan(✉), Liu Feiyang, and Gao Han

Xi'an Aeronautics Computing Technique Research Institute, AVIC, Xi'an 710068, China
850510285@qq.com

Abstract. Future avionics systems must have high-performance intelligent data processing capability, especially for various intelligent application scenarios, such as image/radar target recognition, big data analysis, control and decision. Deep neural network (DNN) is a widely used AI algorithm due to its accurate fitness of nonlinear functions. It can introduce huge computation and memory costs as the increase of network depth. Recently, DNN hardware accelerator is proposed as a domain-specific accelerating platform. It can provide a flexible and reconfigurable operating environment for different algorithms, including convolutional neural network and recurrent neural network, with the advantages of high performance, low latency, and low power cost. The key technologies of DNN accelerators mainly include: parallel computing architecture, intelligent processing elements design, efficient memory structure, and dataflow scheduling optimization. This paper studies the design methodology of DNN accelerators, summarizes the key technologies, and discusses the prospective applications in avionics computing systems. This paper will provide some technique supports for the design of next-generation intelligent airborne computing systems.

Keywords: Avionics system · Intelligent computing · DNN accelerator

1 Introduction

The ever-increasing aircraft applications requirements and continuous advancement of computer system technologies pushes the fast development of avionics systems. Various intelligent application scenarios, such as image/radar target recognition, big data analysis, control and decision, and fault self-diagnosis, will play an important role in the new generation aircrafts. Future distributed modular airborne computing system must have high-performance intelligent data processing capability. The hardware components of future avionics system will be intelligent microsystem modules with highly integrated functions, strong computing capability, compact size, and low power and weight [1]. In recent research of intelligent computing, deep neural network (DNN) with the multi-layer network structure is becoming a widely used AI algorithm due to its accurate fitness of nonlinear functions. However, deep neural network model can introduce huge computation and memory costs as the increase of network depth. DNN hardware accelerator is proposed as a domain-specific accelerating platform. It can provide a flexible

Chinese Society of Aeronautics and Astronautics: CASTYSF 2022, LNEE 972, pp. 170–177, 2023.
https://doi.org/10.1007/978-981-19-7652-0_16

and reconfigurable operating environment for different algorithms, with the advantages of high performance, low latency, and low power cost [2]. Google needed 1,000 servers for the training of cat image recognition, but only 64 servers were required after using the TPU accelerator.

On the consideration that avionics systems have strict constraints on the volume of computing resources, power consumption, hardware size and weight, this paper studies the design methodology of DNN accelerators and key technologies, and analyzes the design requirements of avionics-specific DNN accelerator. This paper can support the design of next-generation intelligent airborne computing systems.

2 Typical DNN Hardware Accelerator

DNN hardware accelerator is a hardware acceleration platform dedicated for the AI applications. It can provide a high-performance operating environment for neural network algorithms, such as deep convolutional neural networks (CNNs) and recurrent neural networks (RNNs). Based on the operation characteristics of DNN accelerator, it is usually called IPU (intelligent processing unit), NPU (neural processing unit), TPU (tensor processing unit), DPU (data stream processing unit), and XPU (extended processing unit). Compared with the general-purpose computing hardware, e.g., CPU and GPU, the DNN accelerator has great advantages of outstanding performance, low latency, and low power consumption. Since CPU and GPU architecture separates the computation and storage, it leads to high computation/communication overheads. DNN hardware accelerator adopts parallel intelligent computing units, and achieves the near deployment of computing and storage resources, with the efficient scheduling of dataflow and data reuse. It can significantly improve the operation speed of matrix multiplication and addition, and reduce the delay and power cost of parameters transfer.

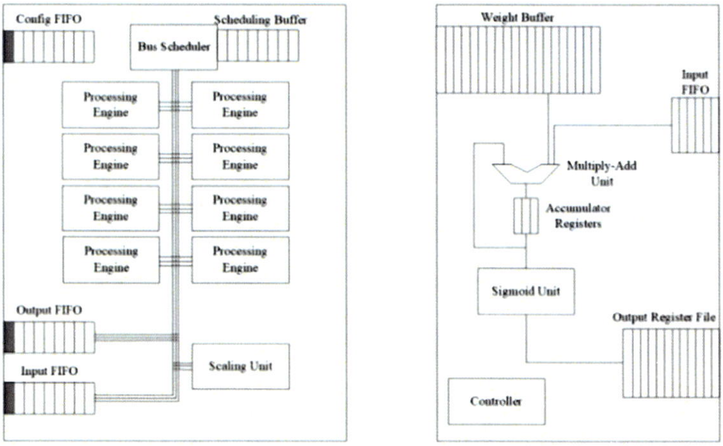

(a) 8-PE NPU (b) Single processing engine

Fig. 1. Architecture of a typical DNN hardware accelerator.

Figure 1 shows the architecture of a typical DNN accelerator. It contains multiple intelligent PEs (processing engines) and an extension unit, where the PE is used to perform most neuron operations, such as convolution, multiplication, accumulation and sigmoid activation functions, the extension unit is used to meet some special requirements of data input and output in different algorithm models. The topology of neural network model determines the time schedule of PEs computation, bus access and queue access. In the schedule of DNN accelerator, each neuron of the algorithm is assigned to a specific PE to complete computation. The DNN accelerator stores the bus scheduling information in its circular scheduling buffer, and each entry in the buffer sends data from the PE or input FIFO to the target PE or output FIFO via the scheduling bus. Each entry in the scheduling buffer contains a source address and a destination address. The source address can be an input FIFO or PE identifier, and the index of output register file. The destination address could be an output FIFO or a bit field representing the destination PE.

3 Parallel Computing Structure of DNN Accelerators

The parallel computing architecture of DNN accelerator improves the efficiency of data access between computing engines and memory. Several high performance DNN accelerators have been proposed recently, such as Google TPU [3, 4], and DianNao series of accelerators [5–8]. Tensor processing unit (TPU) is a high-performance AI accelerator for web searching, image classification, voice recognition, Go game, and decision making applications [3]. It can accelerate massive vector operations and matrix operations of deep neural network algorithms. The internal architecture of TPU accelerator and its dataflow are presented in Fig. 2. Its key modules include a 256 × 256 8bit MAC matrix multiplication unit, and a 28 MB software-controlled on-chip memory. Matrix multiplication unit is essentially a systolic array, which can efficiently carry out massive regular multiplication and addition operations. Experiments show that the processing speed of TPU can reach 92TOPS. Google has already developed four generations of TPUs, which can be used in mobile, edge, and cloud scenarios [4].

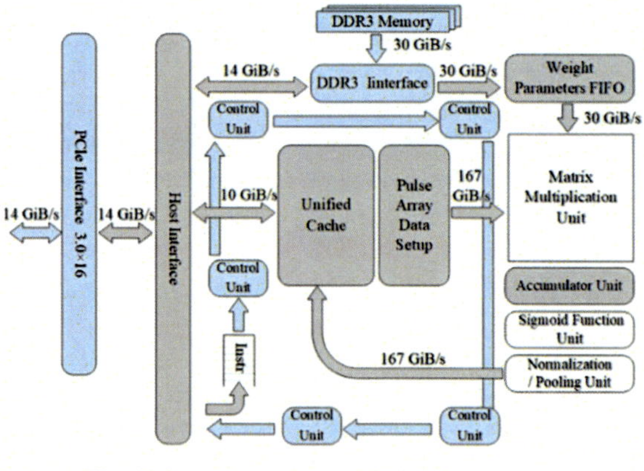

Fig. 2. Internal structure of TPU accelerator.

The research team of Institute of Computing Technology CAS has released four intelligent computing accelerators since 2014, i.e., DianNao [5] for edge computing, DaDianNao [6] for multi-core cloud computing, and ShiDianNao [7] for video signal, and PuDianNao [8] applicable to multiple AI algorithms. DianNao is an energy- efficient single-core DNN hardware accelerator, and its internal structure is as shown in Fig. 3. The computation acceleration is realized via a tight coupling of computing units and the cache. The internal cache consists of three areas: input buffer (NBin), output buffer (NBout), and synaptic weight buffer (SB), which can significantly reduce the data migration (by 10 ~ 30 times) in the computation process. In DianNao, the data flow processing is scheduled by the control logic (CP), and the multiplication, addition, and activation function operations are implemented in the neuron functional unit (NFU), in the form of a three-stage pipeline. The efficiency of matrix operations is improved by configuring multiple computing units within the NFU. Experiments show that DianNao can support various deep CNN algorithms, with a peak performance of up to 452 GOP/s and a power consumption of only 0.485 W.

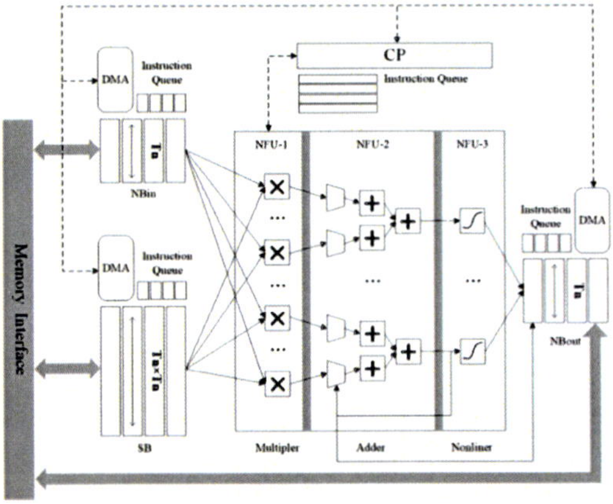

Fig. 3. Internal structure of DianNao accelerator.

DaDianNao is a multi-core DNN accelerator with 16 DianNao processing units [6]. In DaDianNao, a high-bandwidth fat tree network is used for data communication among the intelligent processing units and the large-capacity distributed storage units. It can support parallel computing of single instruction multiple data (SIMD) in the form of broadcasting. Experiments show that the peak performance of DaDianNao can reach 5.58 TOPS/s, and its power consumption is only 1/330 of mainstream GPUs.

ShiDianNao is an embedded DNN hardware accelerator for smart video sensors [7]. It contains 4 intelligent processing units and hardware components for video signal processing. Thus, it can directly obtain image information from the camera and fulfill DNN operations. Experimental results show that when ShiDianNao is used in image target recognition with a resolution of 1280 × 720, it can reach a speed of 56 frames/s.

PuDianNao is an accelerator for general machine learning algorithms [8]. It supports not only deep neural network, but also other mainstream AI algorithms, e.g., K-means,

KNN, SVM, linear regression, and classification tree. PuDianNao includes 16 basic functional units (FUs), and each FU integrates a machine learning unit (MLU) and an arithmetic logic unit (ALU). MLU matches with various algorithms and computational primitives, such as vector inner product, vector distance, counting, sorting, nonlinear functions. ALU fulfills other types of arithmetic operations with low-frequency.

From the architecture of TPU and DianNao series, it can be seen that the DNN accelerator design methodology should concern the following aspects: (i) optimize the MAC structure to improve the computing efficiency, including improving parallelism and increasing pipelines; (ii) increase the on-chip storage space and optimize its storage architecture and physical layout, to improve the writing and reading efficiencies of data and parameters; (iii) increase the communication network bandwidth to improve the data transmission speed and throughput.

4 DNN Accelerator with New Storage Structure

Deep neural network algorithms have the operation characteristics that the weight parameters are fixedly stored in the neurons, and the feature map/matrix are input to the neurons in the form of data stream to complete operations, such as, multiplication, addition, and activation function in parallel. It can be seen that the distributed storage of weight parameters and high-speed access of feature map/matrix have a significant impact on the computing performance and power consumption of DNN accelerators. Thus, the hardware acceleration can be achieved by optimizing the data storage. Many DNN accelerators based on new storage structures have been designed in recent years, such as resistive random access memory (ReRAM) and hybrid memory cube (HMC).

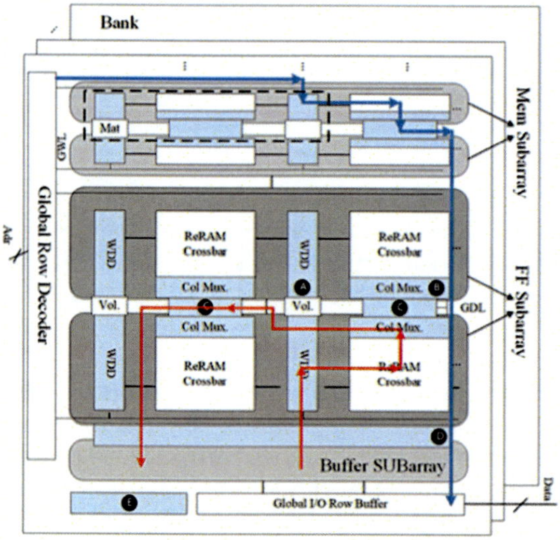

Fig. 4. Internal Structure of PRIME accelerator with ReRAM.

ReRAM is a non-volatile memory which can record the data information according to changes of the resistance of physical materials. Its key property is the memristive, i.e., the data storage is related to its resistance which can be programmed. This is similar to the operation process of neurons, in which strength of biological synapses is variable. Thus, ReRAM is one of the main technical approaches for DNN accelerator with the ability of in-memory computing. PRIME is a configurable DNN hardware accelerator based on ReRAM memory [9]. It modifies the design of ReRAM main memory to achieve both data storage and in-memory computing. The architecture of PRIME accelerator is as shown in Fig. 4. PRIME can configure part of the ReRAM cross matrix as a neural network acceleration unit or a large-capacity storage unit. In Fig. 4, **A** is a word line decoder and driver, with a multi-level voltage source, and the input feature map can be input into the memory array for calculation; **B** is a column multiplexer configured with analog subtraction and Sigmoid circuits, part of the results of two arrays can be combined through **B** and sent to the nonlinear activation unit; **C** is a reconfigurable amplifier with a multi-level output counter, and it can reconfigure the sensing resolution and perform the function of an analog-to-digital converter; **D** is the connection between the register and buffer sub-array; **E** is the overall controller. The word line (WL) decoder and driver in PRIME are configured with multi-level voltage sources, which can carry out computing of input feature in the memory array.

HMC is an ultra-high-speed memory bank based on the 3D stacking process, which utilizes the through-silicon-vias (TSV) technology to stack multi-layer DRAM chips. It can significantly improve the storage density and access speed. With the technology of HMC, the intelligent computing unit and the storage unit in a DNN accelerator can be integrated three-dimensionally at different chip layers, which can realize the near data processing and obtain high storage capacity, high storage bandwidth and low latency. Tetris is a DNN accelerator based on HMC technology [10]. It integrates 16 PEs for neuron computation, and uses a mesh network for the interconnection among PEs and eDRAM memory blocks. An out-of-order data access method is proposed in Tetris as well, which can simultaneously access the input feature map and output feature map blocks, realize data reuse, and reduce the frequency of remote data access.

5 Design Requirements of Avionics-Specific DNN Accelerator

The next-generation avionics system has urgent and broad requirements for high- performance intelligent computing applications. However, different from the other commercial products, an avionics system has strict limitations on the hardware components. According to the development trends of future avionics system, the unified intelligent micro-system (IMS) module will be the main form of hardware components. IMS is a highly integrated, high-performance, multiple AI algorithms supported, compact, and low- power hardware module. Thus, the avionics-specific DNN accelerator should not only provide outstanding intelligent computing power, but also have low hardware cost, low power consumption. The design requirements of avionics-specific DNN accelerator are summarized as following.

(1) Avionics-specific DNN accelerator should support the sparse operations.

Avionics-specific DNN accelerator should support sparse operations, or network pruning, since it can significantly reduce the computation complexity and hardware costs of DNN accelerator. Han proposed a compression algorithm for convolutional neural network model, and designed an efficient inference engine (EIE) for high-speed image recognition [11]. It can alleviate the information redundancy in the compressed network model, and reduce the frequency of DRAM accesses and system power consumption. The throughput of EIE is 2.9 times that of SRAM, and its power consumption is only one third of DaDianNao. Chen designed a sparse CNN hardware accelerator-Eyeriss [12]. It uses a new dataflow model to replace the traditional single-instruction-multiple-data architecture, and can effectively reduce the hardware overhead by reusing the massive data in convolution operations.

(2) Avionics-specific DNN accelerator should support low bit-width computing.

Avionics-specific DNN accelerator should use low bit-width data, since quantizing the data type from 64-bit double or 32-bit float to 8-bit, 4-bit, 2-bit integer can significantly reduce the memory storage and hardware costs. The concept of binary neural network (BNN) is proposed recently. It uses one-bit or two-bit weight quantization for network calculation, and can achieve fast and efficient classifications especially when using the FPGA platform. Xilinx research lab designed a binary CNN accelerator, FINN [13], with the idea of BNN model. It can efficiently map the binary neural network model into hardware logic circuits, and achieve the deployment of fully-connected, convolutional and pooling layers automatically. This accelerator can be integrated to an embedded FPGA platform with a total power of less than 25 W.

(3) Avionics-specific DNN accelerator should support the computation-storage integrated structure.

Avionics-specific DNN accelerator should optimize the storage structure, since the neural network models usually have a large number of layers and a large amount of data for storage. Most neural network accelerators separate the computing and storage units, and store weights and intermediate data on the off-chip memory, leading to high power cost and large memory space. In comparison, the computing and storage integrated architecture based on DRAM, SRAM, or non-volatile memory can couple the computing units with the storage units, greatly reducing the cost of memory access, and convert computing in a lower energy consumption form.

(4) Avionics-specific DNN accelerator should support the algorithm evolution.

Since the applications of artificial intelligence algorithms in aircrafts are still under development, the avionics-specific DNN accelerator should have the capability to be adapted to new algorithms. At present, almost all neural network accelerators only complete the forward inference process of neural network, while training the algorithm in the cloud environment in advance. Pre-trained algorithms cannot well adapt to the rapidly changing application scenarios. For example, a large amount of data in the real-time navigation systems involves the user privacy, and the statically trained algorithm cannot effectively handles changes in the surrounding environment. To address this problem, the hardware should use a self-evolving architecture, which can achieve mutual improvement of hardware and algorithm.

6 Conclusion

Next-generation aircrafts have huge requirements for high-performance intelligent computing applications. However, current avionics system has poor computing power to support the various intelligent applications. This paper studies the design methodology of DNN accelerator by analyzing some widely used DNN hardware accelerators, and summarizes the key technologies. According to the development trends of future avionics system hardware components, four aspects of requirements of avionics-specific DNN accelerator design are concluded. This paper will guide the design of next-generation intelligent airborne computing systems.

References

1. Feiyang, L., Xiaodong, Z., Yahui, L.: Study on the architecture of airborne intelligent network-based microsystem chip. Aeronaut. Comput. Techniq. **48**(5), 145–149 (2018)
2. Han, D., Zhou, S., Zhi, T., et al.: A survey of artificial intelligence chip. J. Comput. Res. Dev. **56**(1), 7–22 (2019)
3. Norman, J., Cliff, Y., Nishant, P., et al.: In-datacenter performance analysis of a tensor processing unit. In: IEEE International Symposium on Computer Architecture, pp. 1–12 (2017)
4. Norman, J., Doe, Y., Matthew, A., et al.: Ten lessons from three generations shaped Google's TPU v4i. In: IEEE International Symposium on Computer Architecture, pp. 1–14 (2021)
5. Tianshi, C., Zidong, D., Ninghui, S., et al.: DianNao: a small-footprint high-throughput accelerator for ubiquitous machine-learning. In: ACM International Conference on Architectural Support for Programming Languages and Operating Systems, pp. 269–284 (2014)
6. Yunji, C., Tao, L., Shaoli, L., et al.: DaDianNao: a machine learning supercomputer. In: International Symposium on Microarchitecture, pp. 609–622 (2014)
7. Zidong, D., Robert, F., Tianshi, C., et al.: ShiDianNao: shifting vision processing closer to the sensor. In: International Symposium on Computer Architecture, pp. 92–104 (2015)
8. Daofu, L., Tianshi, C., Shaoli, L., et al.: PuDianNao: a polyvalent machine learning accelerator. In: International Conference on Architectural Support for Programming Languages and Operating Systems, pp. 369–281 (2015)
9. Chi, P., Li, S., Xu, C., et al.: PRIME: a novel processing-in memory architecture for neural network computation in ReRAM-based main memory. In: IEEE International Symposium on Computer Architecture, pp. 27–39 (2016)
10. Lu, H., Wei, X., Lin, N., et al.: Tetris: re-architecting convolutional neural network computation for machine learning accelerators. In: IEEE International Conference on Computer-Aided Design, pp. 1–6 (2018)
11. Song, H., Xingyu, L., Huizi, M., et al.: EIE: efficient inference engine on compressed deep neural network. In: International Symposium on Computer Architecture, pp. 243–254 (2016)
12. Yu-Hsin, C., Tushar, K., Joel, E., et al.: Eyeriss: an energy-efficient reconfigurable accelerator for deep convolutional neural networks. IEEE J. Solid-State Circuits **52**(1), 127–138 (2017)
13. Umuroglu, Y., Fraser, N.J., Gambardella, G., et al.: FINN: a framework for fast, scalable binarized neural network inference. In: ACM International Symposium on Field Programmable Gate Arrays, pp. 65–74 (2017)

Research on Robust Adaptive Guidance Law Against Target Arbitrary Maneuvering

Zhang Peng(✉), Lu Hao, Fu Shu Tang, and Luo Xu Tao

China Airborne Missile Academy, Luoyang 471009, China
zhangpeng1060@163.com

Abstract. The augmented proportional navigation guidance law(APNGL), which is widely used at present, has poor robustness and large miss distance when it against escape target with large overload at the end of the trajectory. In order to solve this problem, firstly, the planar guidance model is established, a new guidance error is defined. Secondly, based on the principle of parallel approach method, a sliding mode adaptive guidance law is designed. In this process, there is no need to assume a target motion model or maneuver style, so the new guidance law is robust to target maneuvering and is suitable for engineering application. Thirdly, the miss-distance-index (MDI) is defined to evaluate the miss distance, and the guidance parameters are adjusted adaptively according to MDI. Finally, digital simulation was done in planar and 6-DOF, simulation results show that the designed guidance law has excellent guidance performance and can greatly reduce the average miss distance (the maximum can be more than 60%).

Keywords: Arbitrary maneuvering target · Guidance error · Miss-distance-index · Adaptive guidance law · Sliding mode control

1 Introduction

At present, our neighboring countries and regions have already equipped or are developing a large number of third and fourth-generation fighter with maneuverability more than 9 g. The escape strategy is also becoming more and more intelligent, which poses a serious challenge to the precise strike of the air-to-air missile on the target [1, 2].

The missile guidance law design problem can be transformed into a control problem that satisfies certain constraints. Generally, these constraints are control energy, terminal miss distance, attack time, attack angle and so on. In fact, these constraints are the guidance errors chosen when the guidance law is designed, which is very important. The purpose of guidance is achieved by controlling the guidance error to zero. The control methods used are all tools to reduce the guidance error, and different control methods require different guidance information.

For the guidance law of attacking maneuvering targets, the control energy and terminal miss distance (or its equivalent) are generally used as guidance errors. Specifically, the guidance error based on optimal control theory and game theory is generally zero effort miss (ZEM), and the guidance error based on nonlinear control theories, such as,

Chinese Society of Aeronautics and Astronautics: CASTYSF 2022, LNEE 972, pp. 178–193, 2023.
https://doi.org/10.1007/978-981-19-7652-0_17

sliding mode control, backstepping control, and dynamic surface control is the line-of-sight (LOS) angular rate. These guidance laws have been extensively studied in recent years.

Based on modern control theory, DV Stallard, P. Zarchan and others have studied the APNGL to compensate the target acceleration. It is not only necessary to assume the target maneuvering model beforehand, but also to accurately estimate the target maneuvering acceleration [3–6]. The guidance performance is poor when the actual target maneuvering style is mismatched with the model. Based on the differential game theory, the differential game guidance law is studied, but the two-point boundary value problem is encountered when solving the equation, and it is difficult to obtain an analytical solution [7–12]. Guidance law based on differential games need to know target maneuver information and accurate missile model, therefor, this method has certain limitations.

The study of guidance laws based on fuzzy control, sliding mode control, feedback linearization control, backstepping control are particularly attractive [13, 14]. Feedback linearization methods require an accurate known system control model, which is difficult for complex nonlinear missile systems [15–17]. The sliding mode and backstepping guidance law are robust to system uncertainty, and a large number of research results have been achieved [18–23]. In the design process, only the upper bound of the target acceleration is required, but the switching item needs to be handled well to avoid control command chatter. The advantage of fuzzy control is that it does not rely on an accurate system model, but the design of fuzzy rules relies on rich experience [24–26]. Using artificial intelligence method to design guidance law only trains and learns one of the parameters, the design cycle is long, and the interpretability is low [27–29]. Fuzzy control and artificial intelligence method are only auxiliary means to optimize certain parameters, but cannot fundamentally solve the problem of sensitivity to target maneuvering.

In the above studies, the reasons for the poor guidance performance are that the actual maneuvering of the target does not match the model, the target maneuvering acceleration information is difficult to accurately estimate, and the missile is not sensitive to the target maneuvering, etc.

In this paper, an adaptive guidance law that is robust to arbitrary maneuvering of the target based on the sliding mode control theory was studied. By defining a new guidance error, the missile can be made to respond quickly to the maneuver of the target, and the overload demand is reasonably distributed during the flight. In addition, MDI which can evaluate the magnitude of the guidance error is defined, and the guidance parameters are adaptively adjusted according to MDI, which not only ensures a small guidance error but also can well suppress the chatter caused by sliding mode control.

2 Guidance Process Description

The geometry of vertical planar interception is depicted in Fig. 1, respectively.

In the figure, R and q denote the relative distance and the LOS angle, respectively; V_m and V_t denote the velocities of the missile and the target, respectively; θ_m and θ_t denote the velocity inclination angles of the missile and the target, respectively.

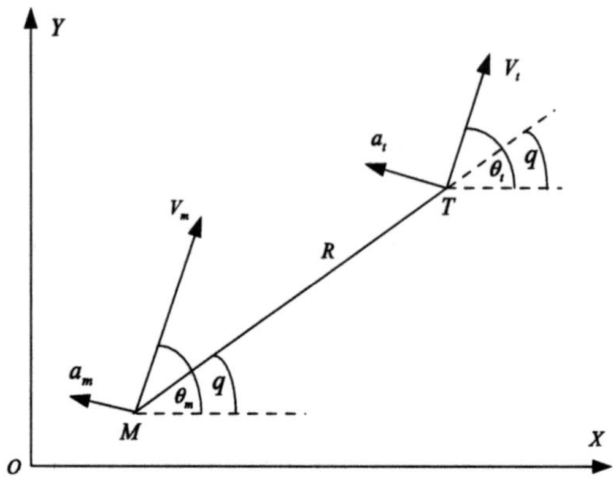

Fig. 1. Planar interception geometry

The planar interception equations satisfy the following relation

$$R\dot{q} = V_m \sin(q - \theta_m) - V_t \sin(q - \theta_t) \tag{1}$$

$$-\dot{R} = V_m \cos(q - \theta_m) - V_t \cos(q - \theta_t) \tag{2}$$

Define $\eta_m = q - \theta_m$, $\eta_t = q - \theta_t$, η_m and η_t denote the lead angles of the missile and the target, respectively. Differentiating Eqs. (1) and (2) with respect to time, and there is

$$R\ddot{q} + 2\dot{R}\dot{q} = -a_m \cos\eta_m + a_t \cos\eta_t + \dot{V}_m \sin\eta_m - \dot{V}_t \sin\eta_t \tag{3}$$

$$\ddot{R} = R\dot{q}^2 - a_m \sin\eta_m + a_t \sin\eta_t - \dot{V}_m \cos\eta_m + \dot{V}_t \cos\eta_t \tag{4}$$

where $a_m = V_m\dot{\theta}_m$ and $a_t = V_t\dot{\theta}_t$ denote the normal accelerations of the missile and the target, respectively.

According to the principle of parallel approach guidance method, the ultimate goal of guidance control is

$$\dot{q} = 0 \tag{5}$$

Substituting Eq. (5) into Eq. (1) yields

$$V_m \sin(q - \theta_m) = V_t \sin(q - \theta_t) \tag{6}$$

or equivalently

$$\theta_m = q - a\sin\left[\tau \sin(q - \theta_t)\right] \tag{7}$$

where $\tau = \frac{V_t}{V_m}$. If Eq. (7) holds, the LOS angular rate is zero, and the ideal velocity inclination angle θ_m^p is defined as follows

$$\theta_m^p = q - a\sin(\tau \sin\eta_t) \tag{8}$$

The optimal interception can be achieved as long as is gradually satisfied during the flight. Thus, the guidance angle error ε can be defined as

$$\varepsilon = \theta_m^p - \theta_m \tag{9}$$

The guidance law should ensure ε as small as possible.

3 Design of Robust Guidance Law

3.1 Guidance Law Design

It can be seen from Eqs. (1) and (2) that the planar interception equations are nonlinear. Due to the maneuvering of the target, the attack situation changes drastically at the terminal trajectory, and linearizing the nonlinear equations will bring large error, and can not achieve the expected purpose. Therefore, using the sliding mode control method to derive a guidance law with strong robustness to target maneuvering. The sliding mode control has strong robustness to a class of bounded disturbances, and is suitable for nonlinear interception systems. The design index is related to the selection of the sliding surface function, which can theoretically ensure the trajectory convergence.

Firstly, choosing a sliding surface function as given by

$$s = \dot{q} + \lambda\left(\theta_m^p - \theta_m\right) \tag{10}$$

where λ is a positive number. The purpose of this design is to meet the requirements of $\dot{q} = 0$ and $\theta_m = \theta_m^p$ on the sliding surface. It should be noted that, unlike the APNGL, which takes the ZEM as the guidance error, it will use $\dot{q} + \lambda\left(\theta_m^p - \theta_m\right)$ as a new guidance error here, and the goal of the control is to make the guidance error tends to zero.

Secondly, designing the control law to ensure the attraction characteristics of the sliding mode hyperplane $s = 0$, so that the system can enter the sliding mode as soon as possible. Constructing a Lyapunov function $V = \frac{s^2}{2}$, then the missile acceleration command should satisfy $V(t) \geq 0$ and $\dot{V}(t) < 0$ for any $t \in \left[t_0, t_f\right]$. Then the LOS angular rate reach to the sliding surface in finite time and converge to zero. That is, there exists $t_1 \in \left[t_0, t_f\right]$, $\dot{q} = 0$ and $\theta_m(t) = \theta_m^p$ is available when $t \geq t_1$.

For ease of derivation, let $\phi = a\sin(\tau \sin\eta_t)$, deffierentiating ϕ with respect to time yields

$$\dot{\phi} = \frac{d(\tau \sin\eta_t)}{\sqrt{1 - (\tau \sin\eta_t)^2}} = \frac{\dot{\tau}\sin\eta_t + \tau\cos\eta_t\dot{\eta}_t}{\cos\phi} = \frac{\dot{\tau}\sin\eta_t + \tau\dot{q}\cos\eta_t - \frac{a_t}{V_m}\cos\eta_t}{\cos\phi} \tag{11}$$

where $\dot{\tau} = \frac{\dot{V}_t V_m - \dot{V}_m V_t}{V_m^2}$, $\dot{\eta}_t = \dot{q} - \dot{\theta}_t$.

Due to $\dot{\theta}_m^p = \dot{q} - \dot{\phi}$, $\dot{\theta}_m = \frac{a_m}{V_m}$, there is

$$\begin{aligned} \dot{\theta}_m^p - \dot{\theta}_m &= \dot{q} - \frac{1}{\cos\phi}\begin{pmatrix} \dot{\tau}\sin\eta_t + \tau\dot{q}\cos\eta_t - \\ \frac{a_t}{V_m}\cos\eta_t \end{pmatrix} - \frac{a_m}{V_m} \\ &= \left(1 - \frac{\tau\cos\eta_t}{\cos\phi}\right)\dot{q} - \frac{1}{\cos\phi}\begin{pmatrix} \dot{\tau}\sin\eta_t - \\ \frac{a_t}{V_m}\cos\eta_t \end{pmatrix} - \frac{a_m}{V_m} \end{aligned} \tag{12}$$

According to Eq. (3), there is

$$\ddot{q} = \frac{1}{R}\left(-2\dot{R}\dot{q} - a_m \cos\eta_m + a_t \cos\eta_t + \dot{V}_m \sin L - \dot{V}_t \sin A\right) \tag{13}$$

$$\begin{aligned}\dot{s} = \ddot{q} + \lambda\left(\dot{\theta}_m^p - \dot{\theta}_m\right) = \frac{1}{R}\begin{pmatrix}-2\dot{R}\dot{q} - a_m \cos\eta_m + a_t \cos\eta_t + \\ \dot{V}_m \sin L - \dot{V}_t \sin A\end{pmatrix} - \frac{\lambda a_m}{V_m} + \\ \lambda\left(1 - \frac{\tau \cos\eta_t}{\cos\phi}\right)\dot{q} - \frac{\lambda}{\cos\phi}\left(\dot{\tau}\sin\eta_t - \frac{a_t}{V_m}\cos\eta_t\right)\end{aligned} \tag{14}$$

Adding the term $Ks + L\mathrm{sgn}(s) - Ks - L\mathrm{sgn}(s)$ to the right-hand side of Eq. (14), where K and L are both positive numbers, and $\mathrm{sgn}(s)$ is the sign function of s, and there is

$$\begin{aligned}\dot{s} = \frac{1}{R}\left\{\left[K - 2\dot{R} + \lambda R\left(1 - \frac{\tau\cos\eta_t}{\cos\phi}\right)\right]\dot{q} + K\lambda\left(\theta_m^p - \theta_m\right) - \left(\cos\eta_m + \frac{\lambda R}{V_m}\right)a_m\right\} + \\ \frac{1}{R}\left[\left(1 + \frac{\lambda R}{V_m \cos\phi}\right)a_t\cos\eta_t + \dot{V}_m \sin\eta_m - \dot{V}_t \sin\eta_t - \frac{\lambda R\dot{\tau}\sin\eta_t}{\cos\phi} + L\mathrm{sgn}(s)\right] + \\ \frac{1}{R}\left[-Ks - L\mathrm{sgn}(s)\right]\end{aligned} \tag{15}$$

If the missile acceleration command in Eq. (15) satisfies

$$\begin{aligned}\left(\cos L + \frac{\lambda R}{V_m}\right)a_m = \left[K - 2\dot{R} + \lambda R\left(1 - \frac{\tau\cos\eta_t}{\cos\phi}\right)\right]\dot{q} + K\lambda\left(\theta_m^p - \theta_m\right) + \\ \left(1 + \frac{\lambda R}{V_m\cos\phi}\right)a_t\cos\eta_t + \dot{V}_m\sin\eta_m - \dot{V}_t\sin\eta_t - \frac{\lambda R\dot{\tau}\sin\eta_t}{\cos\phi} + L\mathrm{sgn}(s)\end{aligned} \tag{16}$$

then

$$\dot{s} = \frac{1}{R}[Ks + L\mathrm{sgn}(s)] \tag{17}$$

$$\dot{V} = s\dot{s} = -\frac{1}{R}\left[Ks^2 + L\mathrm{sgn}(s)s\right] = -\frac{1}{R}\left[Ks^2 + L|s|\right] \tag{18}$$

Because K and L are positive numbers, $\dot{V} \le 0$ is true for all s except s=0.

That is, Eq. (16) can ensure that the derivative of the Lyapunov function is a value not greater than zero and not constant to zero, which meets the design requirements. Then the designed guidance law is given by

$$a_m = \frac{1}{Ka}\left\{\begin{aligned}\left[K - 2\dot{R} + \lambda R\left(1 - \frac{\tau\cos\eta_t}{\cos\phi}\right)\right]\dot{q} + K\lambda\left(\theta_m^p - \theta_m\right) + L\mathrm{sgn}(s) \\ \left(1 + \frac{\lambda R}{V_m\cos\phi}\right)a_t\cos\eta_t + \dot{V}_m\sin\eta_m - \dot{V}_t\sin\eta_t - \frac{\lambda R\dot{\tau}\sin\eta_t}{\cos\phi}\end{aligned}\right\} \tag{19}$$

where $Ka = \cos L + \frac{\lambda R}{V_m}$.

3.2 Guidance Law Stability Analysis

The stability of the guidance law is analyzed by using the Lyapunov stability theory. We obtain the results presented in the following Throerm 1.

Theorem 1. For the guidance equations described by Eqs. (3) and (4), when the parameter K and L in Eq. (19) is greater than zero, the guidance law ensures that the state of the guidance system converges to the sliding mode surface before the terminal time of the guidance process, and optimal interception can be achieved.

Proof. Firstly, it is proved that the sliding mode surface function can converge to zero.

According to the designed sliding mode guidance law, the derivative of the Lyapunov function is $\dot{V} = -\frac{K}{R}\left[s^2 + \rho|s|\right] \leq 0$, which is true for any $R > 0$, where $\rho = \frac{L}{K}$, that is $\frac{1}{2}\frac{ds^2}{dt} = -\frac{K}{R}\left(s^2 + \rho|s|\right) \leq 0$.

If $s(t) = 0$, the initial state of the system is on the sliding surface, and it holds for any $t \geq 0$.

If $s(t) \neq 0$, suppose $s(t) > 0$, then $\dot{s} = -\frac{K}{R}(s + \rho)$, let $\sigma = s + \rho$, there is

$$\frac{\dot{\sigma}}{\sigma} = -\frac{K}{R} \tag{20}$$

Let $K = -k\dot{R} > 0$, then $\frac{\dot{\sigma}}{\sigma} = k\frac{\dot{R}}{R}$. Integrate Eq. (20) yields.
$\ln(\sigma(t))\big|_0^t = k\ln(R(t))\big|_0^t$, then

$$\sigma(t) = \sigma(0)\left(\frac{R(t)}{R(0)}\right)^k \tag{21}$$

Due to $s = \sigma - \rho$, and consider the situation of $s(t) < 0$, we can get

$$|s(t)| = -\rho + [|s(0)| + \rho] \cdot \left|\frac{R(t)}{R(0)}\right|^k \tag{22}$$

It can be seen that the convergence of $s(t)$ is exponential, the convergence speed is fast when $k > 1$, and the convergence speed is slow when $k < 1$, but it will eventually converge to zero.

Secondly, it is proved that $s(t)$ will converge to zero before the interception point, not at the interception point.

For any non-zero relative distance $\overline{R}$ satisfy

$$\overline{R} \geq R(0)\left(\frac{\rho}{|s(0)| + \rho}\right)^{1/k} \tag{23}$$

Substituting Eq. (23) into Eq. (22) to have, it is shown that before the interception occurs(), the sliding mode guidance law has already controlled the system states to the sliding surface, and then will move on the hyperplane. That is, exists, there is when.

Finally, it is proved that when, there must be and, that is, the optimal interception can be achieved.

When $s(t) = 0$, there is $\dot{q} + \lambda(\theta_m^p - \theta_m) = 0$, obviously the special solution is and . Below we have to show that this particular solution is also a general solution according to the given sliding mode guidance law.

By using counterevidence, assuming that on the sliding mode $s(t) = 0$ there are $\dot{q} \neq 0$ and $\theta_m \neq \theta_m^p$, from the interception Eq. (1), $\frac{R\dot{q}}{V_m} = \sin\eta_m - \tau\sin\eta_t$, and substitute it into $s(t)$ gives $\sin\eta_m - \tau\sin\eta_t + \frac{\lambda R}{V_m}(\theta_m^p - \theta_m) = 0$, due to $\theta_m^p = q - a\sin(\tau\sin\eta_t)$, let $\eta_m^p = q - \theta_m^p$, then $\tau\sin\eta_t = \sin\eta_m^p$, there is $\sin\eta_m - \sin\eta_m^p + \frac{\lambda R}{V_m}(\theta_m^p - \theta_m) = 0$, or equivalently

$$\frac{\lambda R}{V_m} = \frac{\sin\eta_m^p - \sin\eta_m}{\eta_m^p - \eta_m} \tag{24}$$

The left side of Eq. (24) is a number that is always greater than zero, while the right side may be less than zero depending on the value of η_m, which contradict each other. So when $s(t) = 0$, there must be $\dot{q} = 0$ and $\theta_m = \theta_m^p$.

In addition, the acceleration command of the missile on the sliding surface is

$$a_m = \frac{1}{Ka}\left\{\left(1 + \frac{\lambda R}{V_m\cos\phi}\right)a_t\cos\eta_t + \dot{V}_m\sin\eta_m - \dot{V}_t\sin\eta_t - \frac{\lambda R\dot{\tau}\sin\eta_t}{\cos\phi}\right\} \tag{25}$$

Which is bounded obviously, means that the required control energy is limited.

In the above proof process, the assumptions used are $K = -k\dot{R} > 0$ and $L = \rho K$. Substituting $V_c = -\dot{R}$ into Eq. (19) gives

$$a_m = \frac{1}{Ka}\left\{\begin{array}{l}\left[(k+2) + \lambda R\left(1 - \frac{\tau\cos\eta_t}{\cos\phi}\right)\right]V_c\dot{q} + kV_c\lambda\left(\theta_m^p - \theta_m\right) + \rho kV_c\mathrm{sgn}(s) + \\ \left(1 + \frac{\lambda R}{V_m\cos\phi}\right)a_t\cos\eta_t + \dot{V}_m\sin\eta_m - \dot{V}_t\sin\eta_t - \frac{\lambda R\dot{\tau}\sin\eta_t}{\cos\phi}\end{array}\right\} \tag{26}$$

Let $a_{m1} = \left[(k+2) + \lambda R\left(1 - \frac{\tau\cos\eta_t}{\cos\phi}\right)\right]V_c\dot{q}$, $a_{m2} = kV_c\lambda\left(\theta_m^p - \theta_m\right)$, $a_{m3} = \rho kV_c\mathrm{sgn}(s)$, $a_{m4} = \left(1 + \frac{\lambda R}{V_m\cos\phi}\right)a_t\cos\eta_t$, $a_{m5} = \dot{V}_m\sin\eta_m - \dot{V}_t\sin\eta_t - \frac{\lambda R\dot{\tau}\sin\eta_t}{\cos\phi}$, then

$$a_m = \frac{a_{m1} + a_{m2} + a_{m3} + a_{m4} + a_{m5}}{K_a} \tag{27}$$

Based on the analysis above, the derived sliding mode guidance law with robustness to target maneuvering is mainly composed of five parts, namely proportional guidance term (a_{m1}), guidance angle error compensation term (a_{m2}), sliding mode control term (a_{m3}), target maneuver acceleration compensation item (a_{m4}) and velocity change rate compensation item (a_{m5}).

Remark 1. 1. Compared with APNGL, Eq. (26) contains a_{m2}, a_{m3} and a_{m5}, the guidance information is more abundant;

2. The guidance error is the LOS angular rate and the guidance angle error. The guidance error is more sensitive to the target maneuver, and only when both ε and $\dot{q}$ are relatively small, the miss distance is small;
3. The guidance law has a large overload command in the early stage to eliminate the guidance angle error as soon as possible, and a small overload command after that to save missile energy, and the overload demand distribution is more reasonable.

4 Implementation of Guidance Law

4.1 Chatter Adaptive Suppression

From Eq. (27), it can be seen that the sign function appears in a_{m3}, which is easy to cause chatter and affect guidance accuracy. In this paper, the sign function is first replaced by the hyperbolic tangent function, namely

$$a_{m3} = \rho k V_c \tanh(s/d) \tag{28}$$

where $\tanh(s/d) = \frac{e^{s/d} - e^{-s/d}}{e^{s/d} + e^{-s/d}}$, d is a positive number, the slope of the hyperbolic tangent curve at the origin can be changed by adjusting the size of d. Figure 2 shows a comparison of the hyperbolic tangent curves at different d.

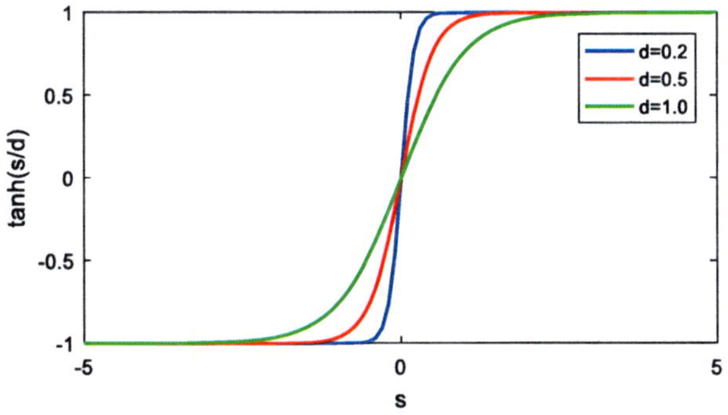

Fig. 2. Hyperbolic tangent curve with different d

Furthermore, in order to adaptively suppress system chatter, the MDI is given by

$$\upsilon = \left|\mu\dot{q} \cdot t_{go} + (1 - \mu)\varepsilon\right| \tag{29}$$

where μ is a positive number less than 1, $t_{go} = -\frac{R}{V_c}$ is the time-to-go, and υ can be used to evaluate the magnitude of the guidance error, d is adaptively adjusted according to υ

$$d = 1 - 10\upsilon \tag{30}$$

Limit the range of d between 0.1 and 0.8. In this way, when υ is large, reduce d to speed up the command response and quickly eliminate the guidance error; Otherwise, increase d to reduce the system chatter, so as to achieve the purpose of adaptively suppressing the chatter.

4.2 Simplification of Guidance Law

The guidance law described by Eq. (26) is the complete form, but it requires a lot of guidance information, some of which are not easy to obtain, which will bring many difficulties to practical application, and reasonable simplification must be made.

In the terminal guidance stage of intercepting, the velocity change rate and the lead angle of missile and target is not large, so the term a_{m5} is ignored. As an external disturbance with an upper bound, the term a_{m4} suppressed by the term a_{m3} does not need to know the specific value.

Both a_{m3} and a_{m2} need to calculate the guidance angle error $\varepsilon = \theta_m^p - \theta_m$, in which the velocity inclination angle of the target is used, and this information is also difficult to obtain accurately, so it's necessary to simplify it.

It can be seen from Eqs. (8) and (9):

$$\varepsilon = \theta_m^p - \theta_m = q - \theta_m - a\sin\left(\frac{V_t \sin(q - \theta_t)}{V_m}\right) \tag{31}$$

According to Eq. (1),

$$\frac{V_t \sin(q - \theta_t)}{V_m} = \sin(q - \theta_m) - \frac{R\dot{q}}{V_m} \tag{32}$$

Substituting Eq. (32) into Eq. (31) gives

$$\varepsilon = \theta_m^p - \theta_m = q - \theta_m - a\sin\left(\sin(q - \theta_m) - \frac{R\dot{q}}{V_m}\right) \tag{33}$$

It can be approximately obtained

$$a\sin\left(\sin(q - \theta_m) - \frac{R\dot{q}}{V_m}\right) \approx q - \theta_m + a\sin\left(-\frac{R\dot{q}}{V_m}\right) \tag{34}$$

Substituting Eq. (34) into Eq. (33) gives

$$\varepsilon \approx a\sin\left(\frac{R\dot{q}}{V_m}\right) \tag{35}$$

Then the sliding surface function is

$$s = \dot{q} + \lambda\varepsilon \approx \dot{q} + \lambda a\sin\left(\frac{R\dot{q}}{V_m}\right) \tag{36}$$

Let $K_N = \frac{(k+2)+\lambda R\left(1-\frac{\tau \cos \eta_t}{\cos \phi}\right)}{K_a}$ and $K_M = \frac{k}{K_a}$, the simplified form of the guidance law is given by

$$a_m = K_N V_c \dot{q} + K_M V_c \lambda\varepsilon + \rho K_M V_c \tanh(s/d) \tag{37}$$

5 Simulation Results

In order to verify the guidance performance of the robust guidance law derived in this paper, a lot of simulation work was done in planar and three-dimensional space under different simulation conditions.

5.1 Planar Simulation

Simulation Description. Assuming that both the missile and the target move in the vertical planar, the target maneuvering acceleration is 12 g, the maneuvering form is S-type, and the period of S-type maneuvering is 10 s. The maneuvering moment of target is 8s~0.5 s before encountering it, with an interval of 0.5 s. The time constant of missile is 0.15 s, the missile overload command limit is 36 g.

The initial parameters and the values of the relevant guidance parameters during the simulation are shown in Table 1.

Table 1. Initial simulation parameter

Variable name	Value	Variable name	Value
R	10 km	λ	0.5
q	10°	ρ	2.0
θ_m	0°	μ	0.8
θ_t	0°	N	4.0
V_m	800 m/s	K_N	4.0
V_t	280 m/s	K_M	5.0

The guidance laws used in the simulation are as follows.

1. APNGL, $a_m = NV_c\dot{q} + \frac{N}{2}a_t$
2. Robust adaptive guidance law(RAGL), Eq. (26)
3. Simplified robust adaptive guidance law(SRAGL), Eq. (39)

Analysis of Simulation Results. The miss distance and flight time corresponding to the three guidance laws are given, as shown in Fig. 3, the x-axis represents the target maneuver moment befor the interception point, and the y-axis represents the miss distance or flight time.

It can be seen from Fig. 3 that the robust adaptive guidance law can significantly reduce the amount of miss distance caused by the large overload of maneuvering. And more importantly, the performance of the SRAGL is comparable to RAGL, making it more valuable for engineering applications.

The reason for the large amount of miss distance when maneuver moment is 5.5 s before the interception point is that, the direction of the maneuvering acceleration of

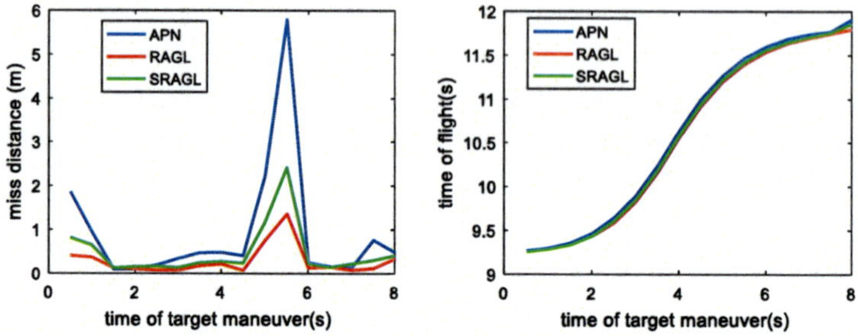

Fig. 3. Miss distance and time of flight under different guidance laws

the target suddenly changes about 0.5 s before encountering, the missile is too late to respond, which is the "optimal maneuver" it can do for the target.

The following takes target maneuver moment is 4 s before the interception point as an example to compare the simulation curves in detail. It can be seen from Fig. 4 that under the action of the RAGL or SRAGL, the missile flight trajectory is more flat.

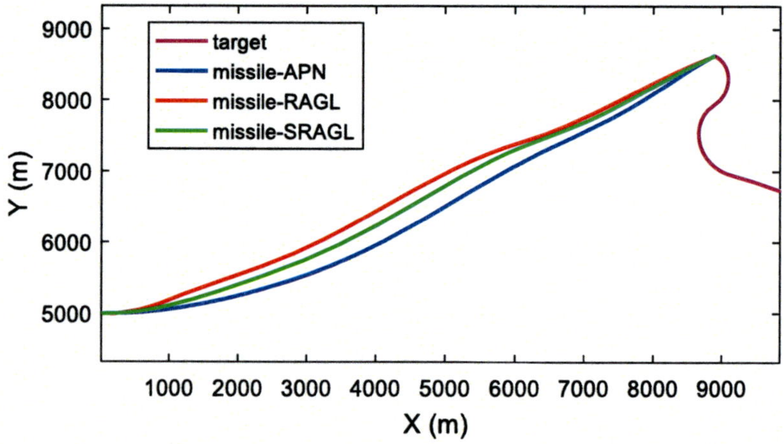

Fig. 4. Relative motion trajectory of missile and target

It can be seen from Fig. 5 that RAGL responds faster to the target maneuver, and the overload requirement at the end of the trajectory is small. Under the action of the RAGL, the LOS angular rate is basically maintained near 0°/s.

It can be seen from Fig. 6 that, the guidance angle error can converges to within 2° quickly under the action of RAGL1, the variation of the MDI of RAGL is relatively small, and it basically remains within 0.02 before encountering the target. The simulation shows that if the MDI is less than 0.05 when encountering the target, the miss distance will not be large.

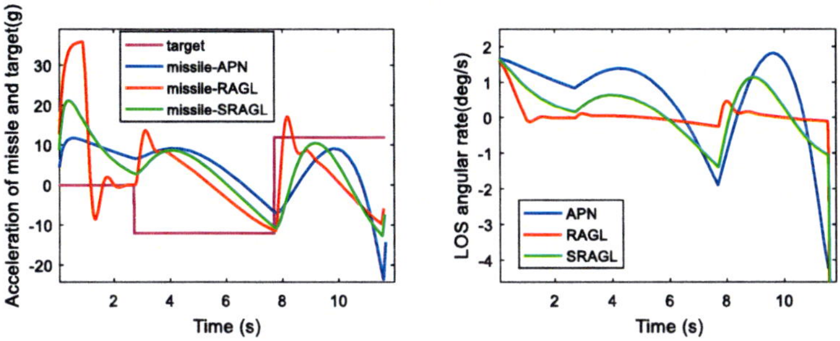

Fig. 5. Missile and target acceleration and LOS angular rate

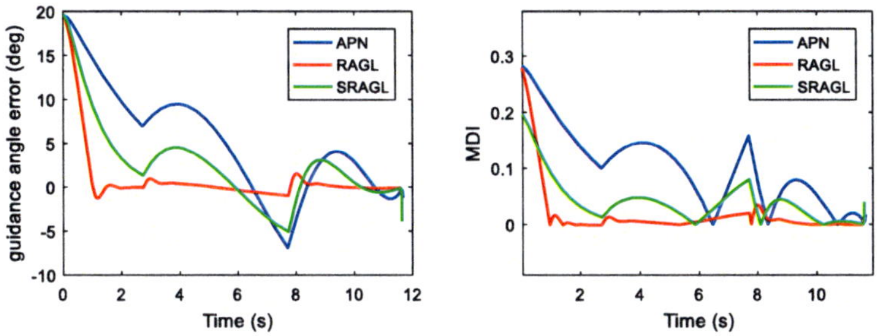

Fig. 6. Guidance angle error and MDI

5.2 Simulation in 6-DOF

The designed RAGL is verified in a 6-DOF simulation environment. The simulation conditions are shown in Table 2, where H_m denotes missile height, V_m denotes missile velocity, H_t denotes target height, V_t denotes target velocity, D denotes distance, A_t denotes target maneuvering acceleration.

The guidance laws used are APNGL and SRAGL. The target maneuvering style include U-type maneuvers and S-type maneuvers, and the maneuvering planar includes horizontal and vertical planars, which are the typical maneuvering styles of the target, and the rest are various combinations of these maneuvering styles. The S maneuver requires a 50° change in the direction of the target velocity. In the simulation, the maneuvering time is 15 s ~ 0.5s before encountering the target, and the interval is 0.5 s.

Table 2. Simulation condition

Condition No	H_m (km)	V_m (Ma)	H_t (km)	V_t (Ma)	D (km)	A_t (g)
1	15	1.5	14	1.5	100	3
2	10	1.2	9	0.8	45	6
3	5	0.8	6	0.8	35	9

Comparing the missile terminal velocity and miss distance obtained by simulation, the simulation results of condition 1 is shown in Fig. 7.

In the figure, H-U denotes U-style maneuver in the horizontal planar, V-U denotes U-style maneuver in the vertical planar, H-S denotes S-style maneuver in the horizontal planar, V-S denotes S-style maneuver in the vertical planar. The blue line is the result of APNGL, and the red line is the result of SRAGL.

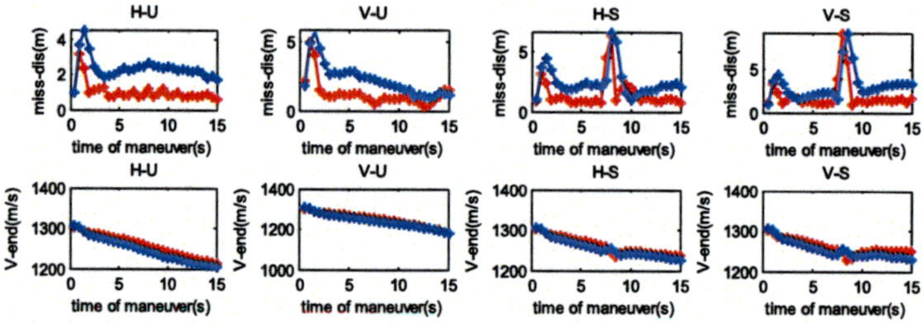

Fig. 7. Miss distance and terminal velocity comparison of condition1

The results of condition 2 is shown in Fig. 8.

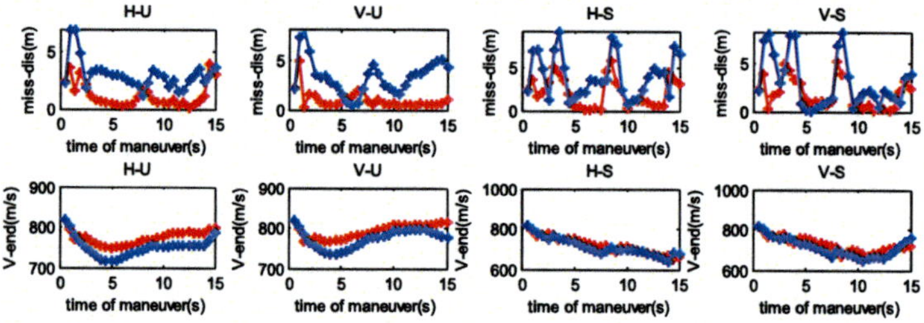

Fig. 8. Miss distance and terminal velocity comparison of condition2

The results of condition 3 is shown in Fig. 9.

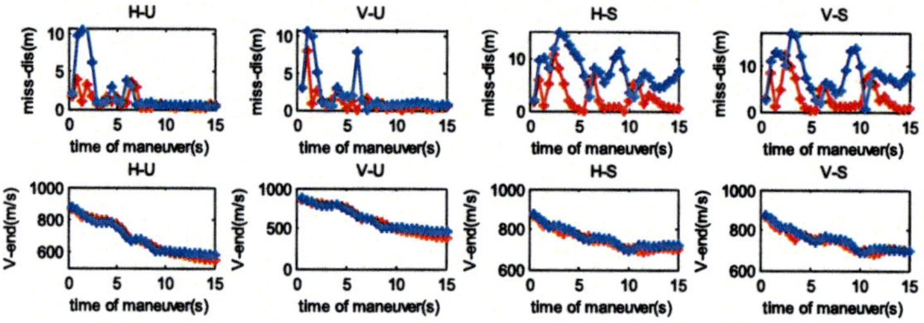

Fig. 9. Miss distance and terminal velocity comparison of condition3

It can be seen from the figures that the SRAGL can significantly reduce the miss distance caused by the target's large overload maneuver at the end of the trajectory, and is insensitive to target maneuvering style and maneuvering planar. The missile terminal velocity is also higher than APNGL, indicating that less control energy is required for the new guidance law.

In order to further illustrate the problem, a statistical analysis was carried out on the average miss distance in the above simulation results, as shown in Table 3.

It can be seen from Table 3 that the SRAGL can greatly reduce the average miss distance by up to 60% or more. It fully shows that in most cases, the SRAGL has higher guidance accuracy.

Table 3. Average miss distance(m) comparison

	H-U	V-U	H-S	V-S
1-APNGL	2.36	2.29	2.49	3.15
1-SRAGL	1.13	1.27	1.54	2.01
2-APNGL	3.03	3.35	4.11	3.02
2-SRAGL	1.72	1.16	2.22	1.89
3-APNGL	2.16	2.11	7.84	8.35
3-SRAGL	1.09	1.16	2.85	3.57

6 Conclusion

In this paper, an adaptive guidance law that is robust to arbitrary maneuver of the target is studied by using the sliding mode control theory and the principle of "parallel approach guidance". By defining a new guidance error, the robust adaptive guidance law is derived from the Lyapunov stability, and the convergence proof are carried out. The guidance parameters are adaptively adjusted according to MDI, and the guidance law is more suitable for engineering applications after being appropriately simplified.

The simulation results show that the RAGL or SRAGL has better performance than APNGL. However, in a complex battlefield environment, especially when the guidance information is inaccurate, the performance of the new guidance law is more critical for practical engineering applications, which is a topic that needs to be studied further in the next step.

References

1. Fan, H., Zhang, P.: The challenges for air-to-air missile. Aero Weaponry **2**, 3–7 (2017). (in Chinese)
2. Fan, H., Yan, J.: The important development direction of airborne missile: autonomization. Aero Weaponry **26**(1), 1–10 (2019). (in Chinese)
3. Zarchan, P.: Tactical and Strategic Missile Guidance. 6th edn. American Institute of Aeronautics and Astronautics, Inc., (2012)

4. Stallard, D.V.: Discrete optimal terminal guidance for maneuvering target. J. Spacecr. Rockets **1**, 381–382 (1977)
5. Asher, R.B., Matuzeusk, J.H.: Optimal guidance for maneuvering targets. J. Spacecr. Rockets **6**, 204–206 (1974)
6. Li, F., Jia, S., Bu, K., et al.: Maneuvering and guidance dynamics compensated for guidance law. J. Ordnance Equipment Eng. **39**(12), 20-24 (2018). (in Chinese)
7. Oshman, Y., Rad, D.A.: Differential-game based guidance law using target orientation observations. IEEE Trans. Aerosp. Electron. Syst. **42**(1), 316–326 (2006)
8. Zhang, Q.: Study of high-maneuvering target tracking and differential game guidance law design. Harbin Engineering University, pp. 30–44 (2017). (in Chinese)
9. Guo, Z., Zhou, S., Yu, Y.: Research of cooperative norm differential games guidance law for intercepting a maneuvering target. Comput. Simul. **37**(3), 23-26 (2020). (in Chinese)
10. Zhang, S., Liu, X., Li, B., et al.: Study on maneuver penetration guidance law of ballistic missile based on differential games in the terminal of interception. Missiles Space Vehic. 2, 81–84 (2015). (in Chinese)
11. Kang, S., Kim, H.J.: Differential game missile guidance with impact angle and time constraints, 3920–3925 (2011)
12. Anderson, G.M., Zong, Y.D.: Comparison of missile guidance law between optimal control interception and differential game interception. Syst. Eng. Electron. **5**, 16–24 (1982)
13. Yuan, Y.: Nonlinear guidance law for interception of maneuvering target. Harbin Institute of Technology, pp. 81–100 (2008). (in Chinese)
14. Li, J., Chen, J., Hu, H.: A nonlinear terminal guidance law for maneuver targets. J. Astronaut. **19**(2), 37–42 (1998). (in Chinese)
15. Li, X., Xu, B., Li, S.: Feedback linearization with active disturbance rejection for entry guidance. In: Control Conference (CCC), 2019 38th Chinese. Nanjing University of Aeronautics and Astronautics, pp. 4202–4207 (2019)
16. Cardona, S., Ospina, S., Giraldo, E.: Sliding mode control based on feedback linearization for a ball and beam system. In: 2019 IEEE 4th Colombian Conference on Automatic Control (CCAC), October 15–18 (2019)
17. Zhang, X., Liu, M., Li, Y., et al.: Impact angle control over composite guidance law based on feedback linearization and finite time control. J. Syst. Eng. Electron. **29**(5), 1036– 1045 (2018)
18. Shtessel, Y.B., Shkolnikov, I.A., Levant, A.: Guidance and control of missile interceptor using second-order sliding modes. IEEE Trans. Aerosp. Electron. Syst. **45**(1), 110–124 (2009)
19. Harl, N., Balakrishnan, S.N.: Impact time and angle guidance with sliding mode control. IEEE Trans. Control Syst. Technol. **20**(6), 1436–1449 (2012)
20. Zhang, W., Fu, S., Li, W., et al.: An impact angle constraint integral sliding mode guidance law for maneuvering targets interception. J. Syst. Eng. Electron. **31**(1), 168–184 (2020)
21. Hu, Q., Han, T.: Sliding-mode impact time guidance law design for various target motions. J. Guidance, Contr. Dyn. **42**(1), 136–148 (2019)
22. He, C., Shi, Z., Zheng, Y., et al.: Adaptive sliding mode cooperative guidance law against maneuvering target. In: Proceedings of the 37th Chinese Control Conference July 25–27, 2018, Wuhan, China
23. Wei, X., Yang, J.: Time-varying terminal sliding mode based on salvo attack cooperative guidance law for unknown uncertainty upper bounds. In: Proceedings of the 35th Chinese Control Conference July 27–29, 2016, Chengdu, China
24. Becan, M.R.: Fuzzy guidance in the aerodynamic homing missiles. In: Preceedings of International Conference on Computational Intelligence, pp. 266–269 (2004)
25. Rajasekhar, V.: Fuzzy logic implementation of proportional navigation guidance. Acta Astronacitica **46**(1), 17–24 (2000)

26. Suiyang, K., Zhihong, M., Shenkui, Z.: Adaptive fuzzy multi-surface sliding mode control for a class of nonlinear systems. In: 2007 Second IEEE Conference on Industrial Electronics and Applications, pp. 1095–1100
27. Xu, B., Li, X., Li, S., et al.: Intelligent guidance method based on nonlinear model predictive control for Mars atmospheric entry. Syst. Eng. Electron. **43**(7), 1943–1953 (2021). (in Chinese)
28. Fang, Y., Deng, T., Fu, W.: An overview on the intelligent guidance law. Unmanned Syst. Technol. **3**(6), 36-42 (2020). (in Chinese)
29. Li, S.,Yang, B., et al.: Space craft guidance and control based on artificial intelligence: review. Acta Aeronautica et Astronautica Sinica **42**(4), 524201 (2021). (in Chinese)

Design for Brake Device of Rudder Pedal Based on Torsion Bar

Sun Ming[1], Zhang Wannian[1](✉), Zuo Qiang[2], Liu Yuehua[1], and Liu Fan[1]

[1] AVIC Beijing KEEVEN Aviation Instrument Co., LTD, Beijing 101300, China
zhangwannian2000@163.com
[2] The Fifth Military Representative Office of the Air Force Equipment Department in Beijing, Beijing 101300, China

Abstract. As airborne equipment, the rudder pedal is used by pilots to control the rudder deflection by foot, and it has two functions of apply rudder and brake. As many disadvantages existing in traditional brake device, such as complex structure, large volume, heavy weight, a brake device of integrated rudder pedal based on torsion bar is designed. The device, using a torsion bar and an angular command sensor, realizes the brake device function of integrated rudder pedal. The design methods of spare functional parts of brake device are introduced in detail. The check calculation of torsion bar, design index of brake command sensor, kinematics simulation analysis of brake device and weight comparison after lightweight design are given. The practical application shows that the principle of the brake device is correct and it has the advantages of compact structure, high utilization, lightweight and high linearity.

Keywords: Integrated pedal · Torsion bar · Brake device · Lightweight design

1 Introduction

The rudder pedal belongs to the aircraft flight control and operation system. It is an air borne equipment for pilots to send heading deflection control and brake command signals. The integrated rudder pedal generally integrates two functions: rudder and brake [1, 2]. The brake function is realized by the action of the pilot turning the pedal. With the deflection of the pedal, the corresponding electrical signal is generated to control the brake of the left and right wheels. The existing brake device has large body, heavy mass and poor signal transmission linearity, which can't meet the requirements of aviation finished product design [3]. Therefore, in order to realize the lightweight, miniaturization and linearity requirements of integrated rudder pedal, it is urgent to develop a new integrated rudder pedal brake device.

Chinese Society of Aeronautics and Astronautics: CASTYSF 2022, LNEE 972, pp. 194–201, 2023.
https://doi.org/10.1007/978-981-19-7652-0_18

2 Mechanism Analysis of Brake Device

Figure 1 shows the mechanism of rudder pedal brake. As can be seen from Fig. 1, the pedal footboard is connected with the pedal footboard support through a dumpling chain, and the pedal footboard can burn the pedal support for rotation. A brake command sensor is installed between the pedal footboard and the pedal footboard support, measure the change of the rotation distance of the pedal footboard, generate a voltage signal and send it to the brake control box [4]. At the same time, the brake command sensor provides the pilot with brake force reaction, and ensures that the pedal footboard can return to the initial state when the pilot does not step on the pedal footboard.

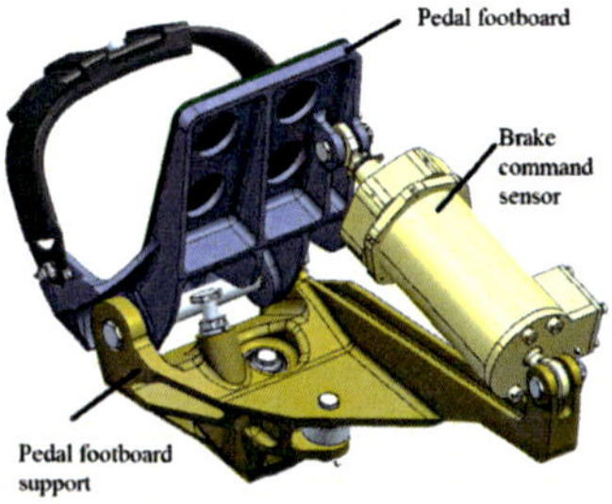

Fig. 1. Tradition structure of brake device

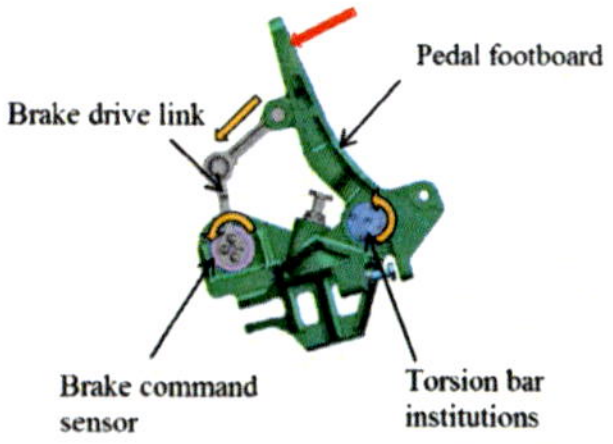

Fig. 2. Principle of brake device

At present, the brake command sensor in the form of linear spring is almost used in aircraft rudder pedal brake devices at home and abroad [5]. The installation and commissioning of this mechanism is simple, but the pedal deflection angle is directly transformed into the linear compression movement of the spring, resulting in poor linearity of brake signal transmission, complex structure of brake device, large space and heavy weight.

3 Principle of Brake Device Based on Torsion Bar

3.1 Principle of Brake Device

Figure 2 shows the principle of brake device based on torsion bar. As can be seen from Fig. 2, the weaving action of the pilot's toe pressing the pedal causes the pedal to deflect forward, and the rotation of the pedal is transmitted to the input shaft of the brake command sensor through the brake connecting rod transmission mechanism. The brake command sensor generates an output electrical signal after being sensitive to the rotation angle of the input shaft, At the same time, when the pedal rotates, it drives the torsion bar to twist and deform, resulting in a sense of brake force proportional to the weaving displacement, which is reversed to the pilot.

3.2 Torsion Bar and Supporting Mechanism

The torsion bar is an elastic straight bar, which generates torque through torsional deformation, so as to provide brake force for the pedal. The torsion bar is characterized by lightweight, simple structure and small space.

The core part of the torsion bar and support mechanism is the torsion bar, which produces torsional deformation under the action of external force, so as to provide torque. The support mechanism is designed as an empty tube (hollow shaft) to facilitate the passage of the torsion bar, and ensure that the torsion bar does not bear the bending load during the rudder process. Figure 3 shows the cross-linking relationship between torsion bar, support mechanism, pedal and pedal support.

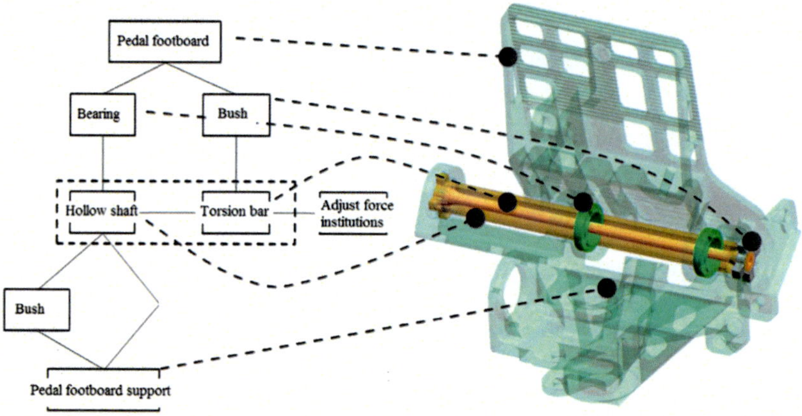

Fig. 3. Cross-linking relationship of brake device

3.3 Brake Transmission Mechanism

The brake transmission mechanism transmits the movement of the pedal to the brake command sensor, and the input shaft of the torso brake command sensor rotates. Because the torsion bar is at the rotation center of the pedal and pedal support, in order to make full use of the space of the brake mechanism, the brake command sensor is arranged parallel to the torsion bar, and the rotation of the pedal is transmitted to the input shaft of the brake command sensor through the four-bar mechanism between the pedal and pedal support. At the same time, according to the brake operation stage stroke and the effective working stroke of the brake command sensor, the rod system size of the brake connecting rod transmission mechanism is allocated, and the transmission relationship of the mechanism is obtained, as shown in Fig. 4.

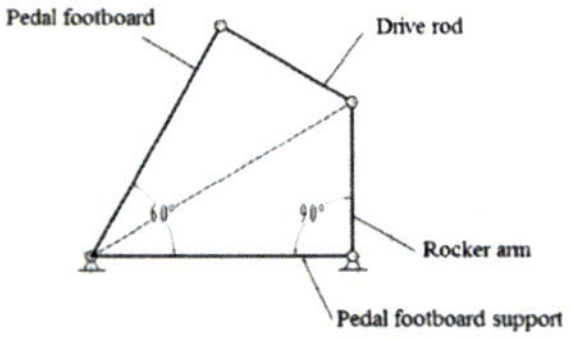

Fig. 4. Transmission diagram of brake linkage

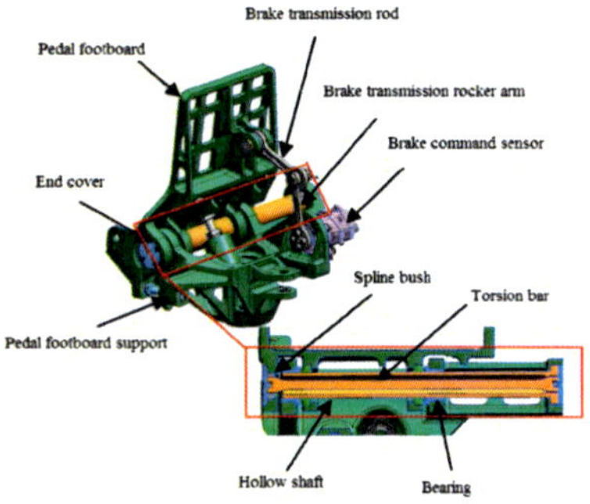

Fig. 5. Component of brake device

4 Detailed Design of Brake Device Based on Torsion Bar

4.1 Composition of Brake Device

The brake device based on torsion bar is mainly composed of pedal footboard, pedal footboard support, brake transmission rod, brake transmission rocker arm, brake command sensor, hollow shaft, torsion bar, spline bush, end cover, bearing, etc., as shown in Fig. 5.

4.2 Detailed Design of Torsion Bar and Supporting Mechanism

The torsion bar and supporting mechanism are shown in Fig. 6. The torsion bar is the core part to provide brake force for the rudder pedal brake. During the working process of the rudder pedal brake, the torsion bar is subjected to torsional load and torsional deformation occurs. At the same time, it should be avoided to apply bending load to the torsion bar during the operation of rudder and brake. The hollow shaft passes through the mounting hole on the rotation axis of the rudder pedal and supports the rudder pedal footboard. The spline end of the hollow shaft is embedded in the pedal support to fix it in the circumferential direction; The torsion bar passes through the hollow shaft, and its spline end is matched with the internal spline of the hollow shaft; The spline bush is embedded into the mounting hole of the rudder pedal footboard support, and its internal spline is matched with the spline at the end of the torsion rod; When the brake works, the pedal deflection drives the spline bush to rotate, and drives the torsion bar to rotate close to the end of the spline bush, so as to produce torsional deformation of the torsion bar and provide a sense of brake force for the pilot.

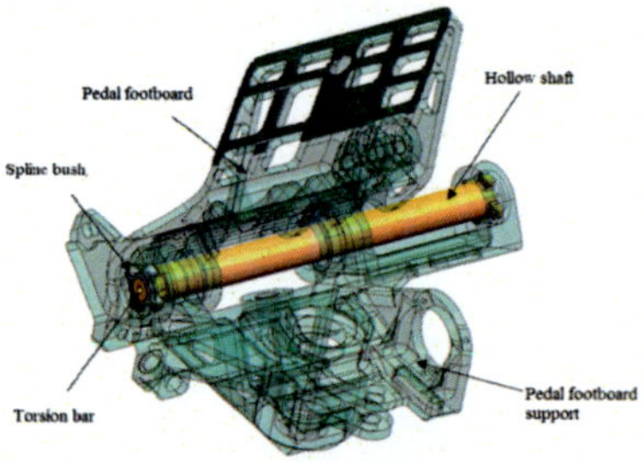

Fig. 6. Torsion bar and support structure

Fig. 7. Structure of torsion bar

Torsion bar is the core part to provide brake force for the rudder pedal, and its torsional stiffness and fatigue strength are very important. Considering the design, processing degree and pedal installation space, the section shape of the torsion bar structure is a solid circular straight bar, and its end (installation connection part) is an open line spline. Arc transition is selected between the torsion bar body and the end spline, as shown in Fig. 7.

The performance of torsion bar material has an important impact on the shear strength and fatigue life of torsion bar. The material parameters are obtained [6–8], as shown in Table 1.

Table 1. Material properties of elastic alloy

Material mark	Material standard	Tension strength Mpa	Yield strength Mpa	Fatigue limit Mpa	Young's modulus Gpa	Shear modulus Gpa
60Si2MnA	GB/T1222-2007	1570	1375	575	163 ~ 170	52 ~ 63

The stiffness K formula of torsion bar is:

$$K = \frac{1000\pi d^4 G}{32LT180} \tag{1}$$

where T is the torsional moment, L is the effective length of the torsion bar, d is the diameter of the torsion bar, and G is the shear modulus of the torsion bar material. Check the fatigue strength of the torsion bar, according to the fatigue limit of material mark 60Si2MnA, $\delta_{-1}=$ 575 Mpa (the number of stress cycles is 10^7), so the allowable shear fatigue limit $\tau_{-1}= 0.58 \times \delta_{-1}=$ 334 Mpa. It can be seen that the maximum shear stress of torsion bar is greater than the shear fatigue limit. According to the S-N curve, the calculation shows that under the maximum shear stress cyclic load, the fatigue life is far greater than the life requirements of the rudder pedal brake device.

4.3 Design of Brake Command Sensor

The brake command sensor of the brake device is a sensor sensitive to the brake weaving stroke. The brake command sensor is fixed on the rudder pedal footboard support through flange. The technical indicators of brake sensor are obtained, as shown in Table 2.

Table 2. Technical indicators of brake sensor

Serial number	Project	Index requirements
1	Excitation voltage	10V 2000 Hz
2	Electrical stroke	±40°
3	Weight	≤200 g
4	Gradient	(0.2 ± 0.01) Vrms/deg
5	Precision	≤1%F.S
6	Tracking accuracy	≤1%F.S
7	Asymmetry degree	≤0.5%F.S
8	Zero voltage	≤50 mV
9	Sum voltage	(10 ± 1) Vrms
10	Cross coupling	≤0.5%F.S
11	Phase shift	≤10°
12	Input current	≤30 mA

4.4 Kinematics Simulation Analysis of Brake Device

After the brake device is designed in detail, Simulation software is used to conduct kinematics simulation analysis of brake device [9, 10], and the results are shown in Fig. 8. According to the analysis results, the kinematic relationship between brake stroke and rotation angle of RVDT input shaft is consistent with the calculation results of theoretical design.

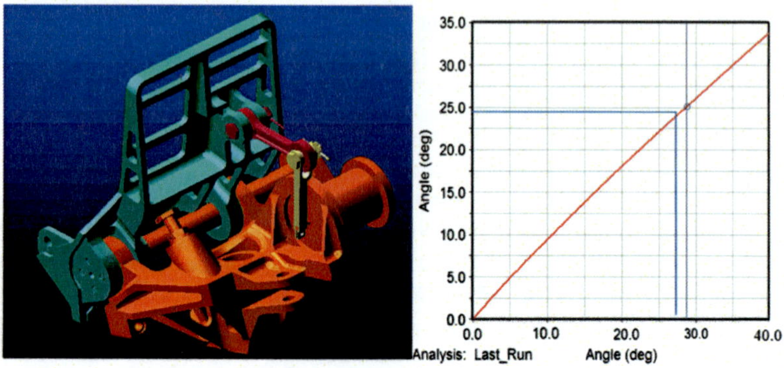

Fig. 8. Kinematics simulation analysis result of brake device

5 Weight Comparison of Brake Device

The lightweight effect of the brake device based on torsion bar is remarkable [11]. At the same time, on the premise of ensuring the strength of parts and components, the lightweight design of parts and components is carried out by optimizing the architecture design and removing redundant materials. The weight of the whole brake device based on torsion bar has been reduced considerably, and the lightweight design efficiency achieves 25%.

6 Conclusion

The integrated rudder pedal brake device based on torsion bar has been designed and verified, and applied to specific models of finished products. The practical application shows that the brake device based on torsion bar has correct principle, compact structure, high space utilization and lightweight. It solves the shortcomings of traditional brake device, such as complex structure, heavy weight and poor signal linearity. It has important engineering practical value and good economy.

References

1. Sun, M., Feng, Y., Zhao, Z.: The performance test and requirements analysis for plane rudder-pedals. Aviat. Precis. Manuf. Technol. **53**(6), 35–41 (2017)
2. Swan, W.M.: Airline Demand distributions: passenger revenue management and spill. Transport. Res. Part E **38**(3–4), 253–263 (2002)
3. Wu, F., Qiu, J., Xie, C.: Failure Analysis and Improvement on the Brake System of a Light Aircraft, 07-0001–03 (2014)
4. Weihao, C., Lianyue, J., Qiang, Z.: Flight test technology research on air minimum control speed limited by rudder deflection. Eng. Test **51**(3), 29–32 (2011)
5. Li, P., Zhang, K., Feng, H.: Transport aircraft rudder pedel design ergonomic research **123**(4) (2016)

6. Editorial Committee of China Aeronautical Materials Handbook, China Aeronautical Materials Handbook, Second Edition, First Volume, Structural steel, Stainless steel, Standards Press of China (2002)
7. Qinshan, F.: Mechanics of Materials. Higher Education Press (2005)
8. GB/T 17855-1999 Calculation method of spline load capacity
9. Wang, L., Yuan, X.: Emulation about the Controlling of the Course Pedal in the Cockpit, 06-0091-02 (2002)
10. Li, F., Huang, W., Liang, P.: Study on the kinematics simulation of six-DOF tandem robot based on ADAMS. Heavy Mach. **2**, 38–42 (2019)
11. Wang, F., Shao, H., Cui, H.: Topology optimization weight loss research of the floating raft isolation system. Ship Sci. Technol. **41**(4), 66–70 (2019)

Calibration of Optical Testing System for Dynamic Impact Experiment of Aviation Seat

Xiang Liu(✉), Hongyi Gao, Rubing Wang, and Shaoyin Wu

Aerospace Life-Support Industries, Ltd., Xiangyang 441003, China
liux.ali@hotmail.com

Abstract. In this paper, a complete set of optical testing system calibration method combining three-dimensional control field design and space resection calculation program development for dynamic impact experiment of aviation seat have been formed, which can obtain high-precision calibration parameters of high-speed camera: exterior orientation elements, interior orientation elements and optical distortion coefficients, and the effectiveness and accuracy of this method were verified by tests. This research content provides important data support for the subsequent use of intersection method to solve the high-precision three-dimensional trajectory and attitude of the occupant head in the dynamic impact experiment of aviation seat.

Keywords: Calibration · Optical testing system · Aviation seat · Dynamic impact experiment · Space resection

1 Introduction

Aviation seat is an important part of aircraft airborne system, which not only provides a comfortable experience for the occupants, but also plays an important role in safety protection and emergency evacuation of the occupants when the aircraft is in emergency state [1, 2]. In accordance with the requirements of crashworthiness provisions in the airworthiness regulations of CCAR-23-R3 [3], CCAR-25-R4 [4], CCAR-27-R2 [5], CCAR-29-R2 [6], etc., dynamic impact experiment is a part of airworthiness certification process of aviation seat [7, 8]. The necessary link mainly verifies the safety performance of aviation seat for occupant protection during simulated crash and impact process.

In the dynamic impact experiment of aviation seat, the optical testing system usually consists of lateral high-speed camera, top high-speed camera and vehicle-mounted high-speed camera. In addition to recording the experiment process, lateral high-speed camera is also used to calculate two-dimensional or three-dimensional occupant head trajectory and attitude according to the requirements of test task. Calibration of exterior orientation elements, interior orientation elements and optical distortion coefficients of lateral high-speed camera can restore correct shape of each image beam [9, 10], which is an effective guarantee for the solution of high-precision occupant head trajectory and attitude by close-range photogrammetry processing method.

Chinese Society of Aeronautics and Astronautics: CASTYSF 2022, LNEE 972, pp. 202–213, 2023.
https://doi.org/10.1007/978-981-19-7652-0_19

2 Principle of Calibration Model

2.1 Calibration Content of High-Speed Camera

Calibration parameters mainly include three parts: interior orientation elements, exterior orientation elements and optical distortion coefficients [11].

- **Interior orientation elements.** Interior orientation elements of high-speed camera are parameters that describe relative position between photography center and image, including principal distance f and image principal point coordinates (x_0, y_0).
- **Exterior orientation elements.** Exterior orientation elements are parameters of spatial position and attitude of image photography moment in space coordinate system, which contain six parameters, namely three line elements (X_S, Y_S, Z_S) and three angle elements $(\varphi, \omega, \kappa)$.
- **Optical distortion coefficients.** Due to manufacturing process of high-speed camera lens, actual image point deviates from ideal point, resulting in optical distortion difference [12]. Optical distortion mainly includes radial distortion and tangential distortion. In actual high-precision photogrammetry, nonlinear imaging model considering optical distortion should be used to describe imaging relationship.

2.2 High-Speed Camera Imaging Model

The imaging model of high-speed camera used for optical measurement is usually central projection imaging model, and its mathematical basis is collinearity condition equation [13]. The central projection imaging model considering optical distortion correction is expressed as:

$$\begin{cases} x - x_0 + \Delta x = -f\frac{a_1(X-X_s)+b_1(Y-Y_s)+c_1(Z-Z_s)}{a_3(X-X_s)+b_3(Y-Y_s)+c_3(Z-Z_s)} \\ y - y_0 + \Delta y = -f\frac{a_2(X-X_s)+b_2(Y-Y_s)+c_2(Z-Z_s}{a_3(X-X_s)+b_3(Y-Y_s)+c_3(Z-Z_s)} \end{cases} \tag{1}$$

In the formula: x, y are measured image coordinates; x_0, y_0, f are interior orientation elements; X_S, Y_S, Z_S are exterior orientation line elements; X, Y, Z are space coordinates; Δx, Δy are optical distortion correction values of image coordinates; a_i, b_i, c_i $(i = 1, 2, 3)$ are 9 direction cosines composed of exterior orientation angle elements.

2.3 Principle of Space Resection

Linearization of Collinearity Condition Equation. Since collinearity condition equation is a nonlinear function, it is necessary to expand collinearity condition equation into a linear form by Taylor formula [14]. Linearized error equation is obtained as follow:

$$\begin{bmatrix} v_x \\ v_y \end{bmatrix} = \begin{bmatrix} a_{11}\ a_{12}\ a_{13}\ a_{14}\ a_{15}\ a_{16} \\ a_{21}\ a_{22}\ a_{23}\ a_{24}\ a_{25}\ a_{26} \end{bmatrix} \begin{bmatrix} \Delta X_s \\ \Delta Y_s \\ \Delta Z_s \\ \Delta\varphi \\ \Delta\omega \\ \Delta\kappa \end{bmatrix} + \begin{bmatrix} a_{17}\ a_{18}\ a_{19} \\ a_{27}\ a_{28}\ a_{29} \end{bmatrix} \begin{bmatrix} \Delta f \\ \Delta x_0 \\ \Delta y_0 \end{bmatrix} + \begin{bmatrix} b_{11}\ b_{12}\ b_{13}\ b_{14} \\ b_{21}\ b_{22}\ b_{23}\ b_{24} \end{bmatrix} \begin{bmatrix} \Delta k_1 \\ \Delta k_2 \\ \Delta p_1 \\ \Delta p_2 \end{bmatrix} - \begin{bmatrix} x-(x) \\ y-(y) \end{bmatrix} \tag{2}$$

In the formula: v_x, v_y are correction values of image coordinates; ΔX_S, ΔY_S, ΔZ_S, $\Delta\varphi$, $\Delta\omega$, $\Delta\kappa$, Δf, Δx_0, Δy_0, Δk_1, Δk_2, Δp_1, Δp_2 are correction values of calibration parameters; x, y are image coordinate measurement values; (x), (y) are image coordinates calculated by substituting approximate value of unknown into collinearity equation.

Basic Method of Space Resection. In order to improve the solution accuracy, redundant equations are usually required, so the least squares adjustment method should be used for calculation [15]. Considering that observed values of all image coordinates are generally regarded as equal-precision measurements, the expression of normal equation solution is obtained as follow:

$$\boldsymbol{X} = (\boldsymbol{A}^T\boldsymbol{A})^{-1}\boldsymbol{A}^T\boldsymbol{L} \tag{3}$$

Thereby, correction values of approximate quantities of calibration parameters are obtained. In each iteration, the sum of unknown approximate quantity and correction value calculated in the previous iteration is used as a new approximation value, and calculation process is repeated until the correction value is less than a certain limit. Finally, solutions of 13 parameters are obtained.

The above is basic method of space resection, and the implementation of subsequent calibration scheme is carried out according to this method. Space resection solution based on collinearity condition equation is an analytical photogrammetry calibration method of test field for optical testing system. This method is easy to implement in test field and has high universality.

3 Technical Scheme of High-Speed Camera Calibration

3.1 Technical Route of High-Speed Camera Calibration

Technical route of high-speed camera calibration is mainly divided into two parts: on-site operation in control field and indoor data processing. Technical route diagram is shown in Fig. 1.

On-site operation in control field mainly focuses on the layout of three-dimensional control field and erection of high-speed camera. The operation process starts with the design and construction of three-dimensional control field and the reasonable arrangement of a certain number of mark points; then mark points space coordinates are measured by the total station; and finally, the images of mark points are recorded by high-speed camera.

Indoor data processing mainly relies on the interpretation software and self-written resection calculation program to calculate space coordinates and image coordinates of previous recorded mark points. The processing flow is as follows: firstly, the image coordinates of mark points are interpreted by interpretation software; then, the calibration parameters are obtained by using self-written space resection calculation program to process relevant data; and finally, the calibration parameters and image coordinates of verification points are substituted into intersection calculation program to calculate space coordinates of verification points.

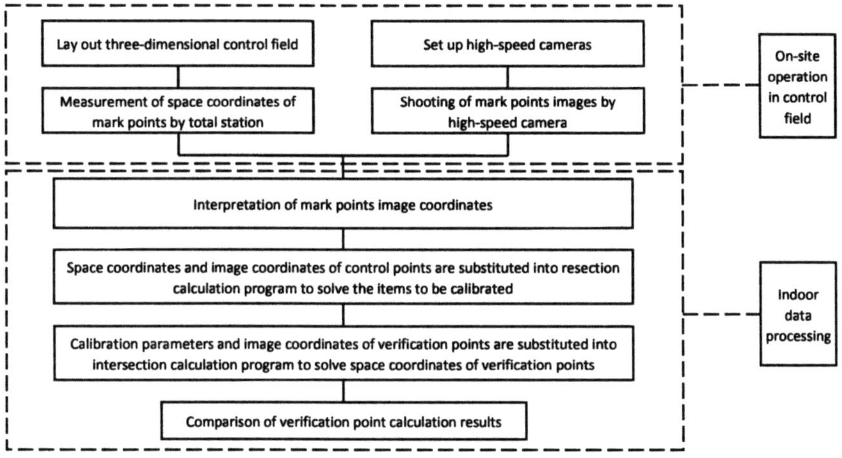

Fig. 1. Technical route of high-speed camera calibration

3.2 Three-Dimensional Control Field Design

In view of dynamic impact test scene of aviation seat, this calibration scheme arranges three-dimensional control field in the dynamic impact test platform, and the placement state of optical testing system is consistent with the specific test layout during the test. The design can ensure the accurate connection between the calibration of optical testing system and the measurement of dynamic targets.

Establishment of three-dimensional Control Field Coordinate System. In the dynamic impact experiment of aviation seat, absolute spatial orientation of target is not needed to be considered in measurement of three-dimensional position and attitude. Therefore, the space coordinate system of three-dimensional control field adopts a customized independent coordinate system according to a certain setting principle.

The setting diagram of the three-dimensional control field coordinate system is shown in Fig. 2. Three-dimensional control field adopts right-hand coordinate system, with the horizontal position of the vertical axis of total station as the plane origin, and the intersection of vertical and horizontal axes of total station as the elevation origin. Heading direction of dynamic impact test platform is selected as *X*-axis of the coordinate system, *Y*-axis is perpendicular to the *X*-axis and in the vertical plane, *Z*-axis is perpendicular to the *X*-axis and in the horizontal plane.

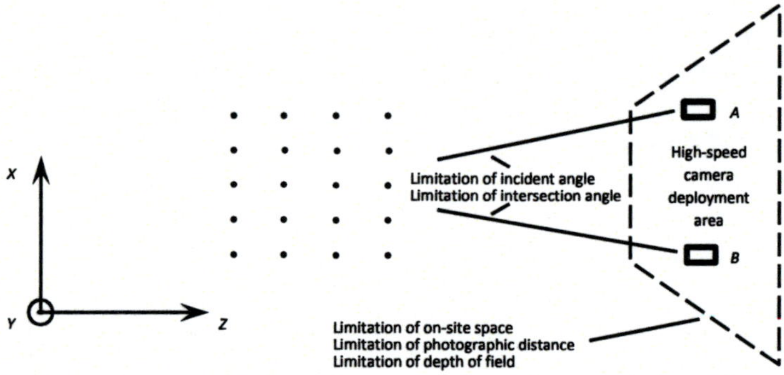

Fig. 2. Schematic diagram of layout position of three-dimensional control field

Selection of Mark Point. This calibration scheme adopts dual-color cross mark. In the scheme, TEMA motion image analysis software is used to interpret marks of control point and verification point in the high-speed images, and it sets up corresponding automatic image recognition and tracking mode for dual-color cross mark points. recognition accuracy of this mode is able to reach 1/3 pixel. meanwhile, in order to verify the accuracy of attitude calculation, a set of target plates is also designed in this calibration scheme (see Fig. 3). Centers of three mark points are distributed on target plate, which are ranked as 1, 2 and 3, and line 1–2 is perpendicular to line 2–3.

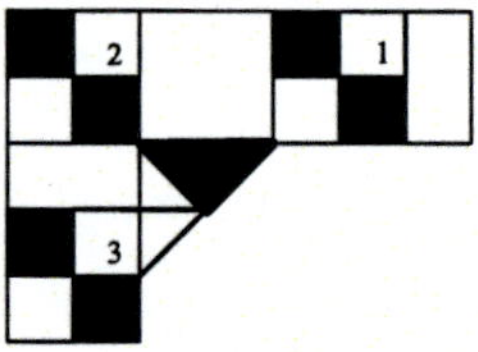

Fig. 3. Design of target plate

Technical Conditions for Establishment of Three-Dimensional Control Field. The quantity, quality and distribution of mark points directly affect the calibration accuracy [16]. Therefore, establishment of three-dimensional control field should meet the following technical conditions:

- A sufficient number of mark points should be laid out.
- Mark points are evenly distributed in the three coordinate directions and have sufficient extension.

- Coordinate measurement of mark point should be stable and reliable to improve measurement accuracy.
- Artificial light source is used for illumination in control field, which can improve light intensity to make image clear.
- Intersection angle of each mark point is preferably 60° ~ 120°.
- Photographic incident angle of each mark point is preferably less than 45°.
- Motion range of dynamic target should be roughly the same as photograph distance.

In conclusion, the calibration scheme should uniform distribution of mark points, as well as maximizing the measurement accuracy of control point coordinates, so as to ensure high precision and high reliability of the calibration. Based on the above constraints, the layout diagram of three-dimensional control field is shown in Fig. 2.

3.3 Space Resection Calculation Program Development

This calibration scheme uses Python language to write space resection calculation program, which can be used to input space coordinates of control points and coordinates of image points to calculate corresponding high-speed camera calibration parameters.

Algorithm of Space Resection Calculation Program. Combined with theoretical derivation of high-speed camera imaging model, the overall solution process of this algorithm is as follows:

1. Obtain the known data: Obtain image coordinates of the control point x_i, y_i from the image data; and obtain space coordinates of the control point X_i, Y_i, Z_i from three-dimensional control field observation results.
2. Determine the initial value of the unknowns: Since coordinates of principal point x_0, y_0 and optical distortion coefficients k_1, k_2, p_1 and p_2 are small values, these initial value are set to 0; the initial value of principal distance f is set to nominal focal length of lens; the initial values of exterior orientation line elements X_S, Y_S, and Z_S are measurement results of total station to estimated photographic center; the initial values of exterior orientation angle elements φ, ω, and κ are measurement values of protractor.
3. Calculate the rotation matrix $\boldsymbol{R}$: Calculate the direction cosine value by using the approximate value of exterior orientation elements and form matrix $\boldsymbol{R}$.
4. Calculate the approximate value of image coordinate: Calculate the approximate value of image coordinate of the control point (x), (y) according to collinearity condition equation.
5. Compose the error equation: Calculate the coefficients and constant terms of the error equation point by point.
6. Compose the normal equation: Calculate the coefficient matrix $\boldsymbol{A}^{\mathrm{T}}\boldsymbol{A}$ and constant terms of the normal equation $\boldsymbol{A}^{\mathrm{T}}\boldsymbol{L}$.
7. Solve the calibration parameters: According to the normal equation, solve the correction value of exterior orientation elements, interior orientation elements and optical distortion coefficients, and sum up with the corresponding approximate values to obtain new the approximate value of each calibration parameter.

8. Check the convergence of the calculation: Compare the correction value of each calibration parameter with the specified tolerance. If the correction value is less than the tolerance, the calculation is terminated; otherwise, repeat the calculation steps 3 to 7 with new approximation value until the requirements are met.

File for Space Resection Calculation Program. The input and output file format of space resection calculation program is CSV file format. During data processing, first convert image coordinates interpreted by Tema software into standard CSV format, and record space coordinates into CSV file. Space resection calculation program imports CSV file and reads image coordinates and space coordinates of each point in turn, and finally generates CSV file with calibration parameters. Calibration parameter file can be used as basic file for subsequent intersection calculation, and can also be directly called to participate in subsequent space coordinate calculation. Flow diagram of space resection calculation program file is shown in Fig. 4.

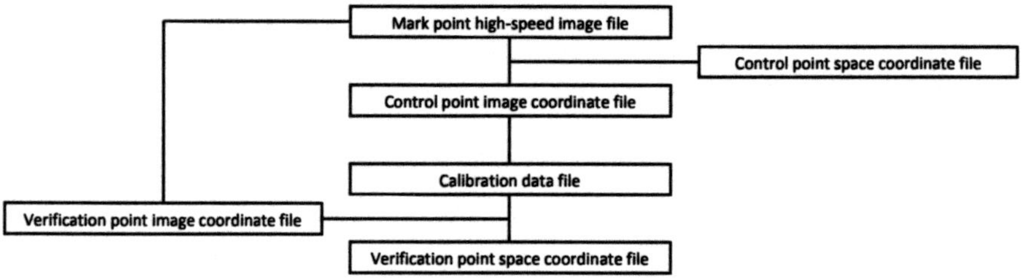

Fig. 4. Program file flow diagram

4 Test Validation and Analysis

4.1 Design of Verification Test Scheme

Considering requirements of dynamic impact experiment of aviation seat and the application scenario of optical testing system, the verification test was carried out on dynamic impact platform. According to layout principle and technical requirement of the three-dimensional control field, high-speed cameras on both sides were erected symmetrically with the center line of three-dimensional control field. The distance from center plane of three-dimensional control field was about 6.2 m, and the intersection angle was about 30°. In order to adapt to the automatic identification and tracking of image point coordinates, the imaging size of mark point should be 20 pixels, and the size of mark was finally determined to be 5 cm. Based on the requirement of redundant observation, a total of 13 control points and 14 verification points were arranged in this verification test. The high-speed camera used in the test was Photron BC2-HD, with resolution of 2048 × 1536 pixels and individual pixel size of 10 μm. The nominal focal length of lens was 35 mm.

In this verification test, a total of 5 sets of verification point images were taken, including 4 sets of images of different placement states of the target plate and 1 set of benchmark length verification point images. The target plate was set up in 4 placement states: no yaw-no pitch-roll, no yaw-pitch-roll, yaw-no pitch-roll, yaw-pitch-roll, which basically covered the actual situation.

After loading the images into TEMA software, the dual-color mark tracking mode was selected to automatically identify and measure mark points. The measured image point coordinates were exported in tabular form. The previous data files were substituted into the space resection calculation program and space intersection calculation program to obtain calibration parameters of high-speed cameras and space coordinates of verification points.

4.2 Verification and Analysis of Calibration Results

After calculating space coordinates and image coordinates of the 13 control points, the calibration parameters of high-speed cameras on both sides were obtained, as shown in Table 1. This verification scheme was mainly compared and verified from three aspects: space coordinates, structural angle, size and length. In the comparison and verification, since the measurement accuracy of total station was higher than that of photogrammetry method, the measurement result of total station was regarded as true value and used as reference value for comparison.

Table 1. Calibration parameters of high-speed camera (unit: mm)

Calibration parameters	Right-side high-speed camera	Left-side high-speed camera
X_s	5821.99365	2793.34057
Y_s	−351.63909	−363.83842
Z_s	−33.56439	−72.68824
φ	−0.30104	0.23281
ω	−0.07185	−0.06741
κ	0.01000	0.00742
f	34.81280	34.11824
x_0	0.05435	−0.25367
y_0	−0.38352	0.17955
k_1	8.8737823203e−05	1.7251928477e−04
k_2	−3.7038371663e−07	−9.9204570813e−07
p_1	7.7781131501e−06	1.2798750490e−04
p_2	−8.4795365993e−06	−7.0196636780e−05

Verification of Space Coordinates. Table 2 lists space coordinate values of the 12 verification points in X, Y, and Z directions of four groups of target plates in different placement states. By comparing with the measurement data of total station, the maximum difference is −1.78 mm and minimum difference is −0.10 mm in X direction, the maximum difference is 0.80 mm and minimum difference is 0.01 mm in Y direction, and the maximum difference is −1.92 mm and minimum difference is −0.13 mm in Z direction. The root mean square error in each direction RMS_X is 1.20 mm, RMS_Y is 0.51 mm, and RMS_Z is 1.02 mm. According to the comparison data, the space coordinates calculated by the calibration parameters of high-speed camera are able to reach an accuracy better than ±2 mm, and this accuracy range can meet the test accuracy requirements of current routine experiments.

Table 2. Comparison of space coordinates (unit: mm)

Number	$X_{measure}$	$Y_{measure}$	$Z_{measure}$	X_{true}	Y_{true}	Z_{true}	ΔX	ΔY	ΔZ
T1–1	2771.81	−804.59	−5630.37	2772.90	−804.60	−5629.70	−1.09	0.01	−0.67
T1–2	2674.50	−805.22	−5627.75	2674.60	−805.50	−5626.70	−0.10	0.28	−1.05
T1–3	2675.44	−868.09	−5591.54	2676.80	−868.60	−5592.80	−1.36	0.51	1.26
T2–1	2771.11	−1034.20	−5496.23	2772.10	−1034.50	−5496.10	−0.99	0.30	−0.13
T2–2	2703.80	−972.94	−5531.01	2704.70	−973.40	−5531.60	−0.90	0.46	0.59
T2–3	2651.32	−1016.26	−5504.96	2652.20	−1016.80	−5506.50	−0.88	0.54	1.54
T3–1	2805.53	−804.47	−5709.62	2806.20	−804.70	−5707.70	−0.67	0.23	−1.92
T3–2	2711.62	−805.38	−5682.58	2712.50	−805.90	−5682.30	−0.88	0.52	−0.28
T3–3	2721.82	−868.12	−5647.85	2723.60	−868.90	−5648.70	−1.78	0.78	0.85
T4–1	2839.12	−1034.10	−5579.95	2840.70	−1034.70	−5580.60	−1.58	0.60	0.65
T4–2	2764.76	−972.95	−5596.37	2766.10	−973.50	−5596.10	−1.34	0.55	−0.27
T4–3	2720.48	−1016.20	−5557.95	2722.20	−1017.00	−5559.20	−1.72	0.80	1.25
RMS							1.20	0.51	1.02

Verification of Structural Angle. The verification of structural angle requires the use of target plate, and the structural angle is three inner angles of the triangle connecting the center of target plate mark point. Table 3 lists three structural angles of four groups of target plates in different placement states. By comparing with theoretical calculated values, the maximum and minimum differences of ∠123 are 0.35° and 0.14°, the maximum and minimum differences of ∠213 are −0.22° and 0.02°, and the maximum and minimum differences of ∠132 are −0.26° and 0.01°; the root mean square error of each inner angle $RMS_{\angle 123}$ is 0.26°, $RMS_{\angle 213}$ is 0.16°, and $RMS_{\angle 132}$ is 0.16°; standard deviation of inner angle $\sigma_{\angle 123}$ is 0.25 °, $\sigma_{\angle 213}$ is 0.18°, $\sigma_{\angle 132}$ is 0.11°. The structural angle of target plate calculated by calibration parameters of high-speed camera can achieve an accuracy better than ±0.4°, and standard deviation of structural angle is within 0.3°. The above shows that the measurement data is consistent.

Verification of Size and Length. The comparison and verification of target plate size and length mainly examines the relative position accuracy, which to a certain extent avoids systematic error compared with true value of absolute space coordinate in object space coordinate system. Size of target plate is defined as the length of three sides of line connecting target plate mark point center. The size of target plate and distance between verification points are measured with a tape measure as true value of length. Table 4 lists four groups of target plate sizes in different placement states, and Table 5 lists

Table 3. Comparison of structural angle (unit:°)

Number	$\angle 123_{measure}$	$\angle 213_{measure}$	$\angle 132_{measure}$	$\angle 123_{true}$	$\angle 213_{true}$	$\angle 132_{true}$	$\Delta\angle 123$	$\Delta\angle 213$	$\Delta\angle 132$
T1	90.35	36.57	53.08	90.00	36.66	53.34	0.35	−0.09	−0.26
T2	89.79	36.87	53.35	90.00	36.66	53.34	−0.21	0.21	0.01
T3	90.30	36.44	53.26	90.00	36.66	53.34	0.30	−0.22	−0.08
T4	90.14	36.68	53.19	90.00	36.66	53.34	0.14	0.02	−0.15
σ	0.25	0.18	0.11			RMS	0.26	0.16	0.16

separation distance of the benchmark verification points. By comparing with the tape measured value, the maximum difference and minimum difference of L_{13} are 0.57 mm and 0.02 mm, the maximum difference and minimum difference of L_{12} are 0.33 mm and 0.03 mm, the maximum difference and minimum difference of L_{23} are 0.36 mm and 0.06 mm; the root mean square error of each side length RMS_{L13} is 0.43 mm, RMS_{L12} is 0.22 mm, and RMS_{L23} is 0.26 mm; The standard deviation of the same side length in different states σ_{L13} is 0.25 mm, σ_{L12} is 0.18 mm, and σ_{L23} is 0.22 mm; the difference of reference length $L_{S1\text{-}S2}$ is 1.86 mm, and the relative error is 0.09%. The length of line connecting center of target plate mark points calculated by high-speed camera calibration parameters can achieve an accuracy better than ±0.5 mm. The standard deviation of target plate size in different placement states is within 0.3 mm, and the relative accuracy of reference length is within 1/1000. The above reflects the high accuracy of relative positional relationship of object points in the digital photogrammetry results based on calibration parameters.

Table 4. Comparison of target plate size (unit: mm)

Number	$L_{13measure}$	$L_{12measure}$	$L_{23measure}$	L_{13true}	L_{12true}	L_{23true}	ΔL_{13}	ΔL_{12}	ΔL_{23}
T1	121.77	97.35	72.56	121.42	97.40	72.50	0.35	−0.05	0.06
T2	121.44	97.43	72.86	121.42	97.40	72.50	0.02	0.03	0.36
T3	121.95	97.73	72.43	121.42	97.40	72.50	0.53	0.33	−0.07
T4	121.99	97.67	72.86	121.42	97.40	72.50	0.57	0.27	0.36
σ	0.25	0.18	0.22			RMS	0.43	0.22	0.26

Table 5. Comparison of reference length (unit: mm)

Number	$X_{measure}$	$Y_{measure}$	$Z_{measure}$	$L_{S1-S2measure}$	$L_{S1-S2true}$	ΔL_{S1-S2}
S_1	3150.60	−192.05	−5518.37	2098.36	2096.50	1.86
S_2	5248.96	−191.43	−5519.17			

5 Conclusion

The main research content of this paper is the calibration of exterior orientation elements, interior orientation elements and optical distortion coefficients of high-speed camera in the dynamic impact experiment of aviation seat based on the space resection principle. Firstly, starting from the fact that collinearity condition is mathematical basis of central projection imaging, high-speed camera imaging model is constructed, and space resection relation is deduced. Next, according to the principle of space resection, the operation process of high-speed camera calibration is determined. Three-dimensional control field is laid out, and space resection calculation program is written to realize calibration of high-speed camera. Finally, intersection calculation results obtained based on calibration parameters are compared and analyzed with measurement data of total station to verify effectiveness and accuracy of this method. The research on calibration of optical testing system for dynamic impact experiment of aviation seat provides data basis for the subsequent three-dimensional trajectory and attitude calculation of the occupant head.

References

1. Li, M.X., Xiang, J.W., Ren, Y.R., et al.: Crashworthiness evaluation and analysis method of aircraft seat. J. Beijing Univ. Aeronaut. Astronaut. **42**(2), 383–390 (2016)
2. Feng, Z.Y., Yang, Y.P., He, Y.L., et al.: Research on vertical dynamic characteristics of aviation dummy under seat constraint. Eng. Mech. **37**(08), 246–256 (2020)
3. Civil Aviation Administration of China: CCAR-23-R3 Airworthiness standards of normal, utility, acrobatic, and commuter category airplanes. Civil Aviation Administration of China, Beijing (2005)
4. Civil Aviation Administration of China: CCAR-25-R4 Airworthiness standard of transport aircraft. Civil Aviation Administration of China, Beijing (2011)
5. Civil Aviation Administration of China: CCAR-27-R2 Airworthiness standards of normal category rotorcraft. Civil Aviation Administration of China, Beijing (2017)
6. Civil Aviation Administration of China: CCAR-29-R2 Airworthiness standards of transport category rotorcraft. Civil Aviation Administration of China, Beijing (2017)
7. Wang, Y.F., Miao, J., Zhang, J.T.: Research on dynamic airworthiness test technology of aviation seat. Equip. Manuf. Technol. **2020**(6), 236–239 (2020)
8. Wang, X.Z., Dong, D., Li, F., et al.: Relationship between the HIC and the head path of the dummy in civil aircraft seats during the 16g dynamic test. Modern Mach. **2020**(02), 47–51 (2020)
9. Luhmann, T., Fraser, C., Maas, H.G.: Sensor modelling and camera calibration for close-range photogrammetry. ISPRS J. Photogramm. Remote Sens. **115**(05), 37–46 (2016)
10. Liu, Z., Wu, Q., Wu, S., et al.: Flexible and accurate camera calibration using grid spherical images. Optics Express **25**(13), 15269 (2017)
11. Li, A.: Research on Calibration Technology of Non-metric Digital Camera. Shandong University of Science and Technology, Qingdao (2019)
12. Yang, B.W., Guo, X.S.: Overview of nonlinear distortion correction of camera lens. J. Image Graph. **10**(3), 269–274 (2012)
13. Abdel-Aziz, Y.I., Karara, H.M.: Direct linear transformation from comparator coordinates into object space coordinates in close-range photogrammetry. Photogramm. Eng. Remote Sens.: J. Am. Soc. Photogramm. **81**(2), 103–107 (2015)

14. Wang, L.H., Li, L.L., Guo, H.L.: A target-based calibration method. Beijing Surv. Map. **35**(3), 372–375 (2021)
15. Yao, N., Lin, Z.R., Ren, C.F., et al.: A distortion model suitable for nonlinear distortion correction of digital video camera. Laser Optoelectr. Progr. **51**(2), 171–178 (2014)
16. Wu, S.Y.: Research on Optical Sensor Geometric Calibration Technology. JiMei University, Xiamen (2013)

Flutter Analysis of Large Flexible Aircraft Based on Reduced Order Model

An Chao(✉), Zhang Duoyao, and Xie Changchuan

School of Aeronautic Science and Engineering, Beihang University, Beijing 100191, China
ac@buaa.edu.cn

Abstract. This paper developments approach for stable analysis in flight of large flexible aircraft. Considering geometrical nonlinearity, aircraft is divided into non-linear wing components and linear fuselage component. Structural ROM method is used for wing structure modeling and nonlinear substructure method is used for comprehensive assembling wing ROM and fuselage linear modes together to obtain integrated aircraft dynamic equations. Non-planar double lattice method (DLM) is used as aerodynamic model. Stability analysis is based on the linearization around the trim configuration. The numerical results for a flexible flying wing aircraft model indicate coupling effects between rigid-body motions and elastic modes are important for this type of aircraft.

Keywords: Flutter · Reduced order model · Substructure · Geometrical nonlinearity

1 Introduction

As the representative of the very flexible airplane, high-altitude long-endurance (HALE) aircrafts usually attract extensive attention. Because of its weight and large flexibility, geometric nonlinearities become very important and affects aeroelastic stability characteristics and dynamic responses. With the analysis requirements of HALE, Hodges, Cesnik and Patil proposed the concept of geometric nonlinear aeroelastic problem in 1999 [1, 2]. Lots of research considering geometric nonlinearities in aeroelastic analysis has been carried out [3–6]. For large flexible aircraft, especially flying wing aircraft, the frequency differences between elastic modes and rigid-body motions tend to be sufficiently small such that the coupled effect cannot be ignored [7]. Furthermore, the structure becomes nonlinear due to the large deformation. Nonlinear aeroelasticity and flight dynamics should be considered simultaneously [8]. Besides, aircrafts with flying-wing configuration which have relatively low fuselage pitch inertias and relative low elastic mode frequencies show difference in dynamic instability compared to traditional aircrafts, which is known as body-freedom flutter (BFF) [9].

Structural reduced order model (ROM) is an efficient method to analysis nonlinear response problem of large flexible structure. It shows us computational efficiency advantages of structure analysis especially in nonlinear aeroelastic issue and offers the

Chinese Society of Aeronautics and Astronautics: CASTYSF 2022, LNEE 972, pp. 214–223, 2023.
https://doi.org/10.1007/978-981-19-7652-0_20

potential for real-time domain analysis. Muravyov et al. developed an algebraic polynomial expression to characterize the structural nonlinear dynamic behavior [10]. Besides, Mignolet et al. showed that the nonlinear stiffness of large flexible structure can be described using an equation with the quadratic and cubic terms of the basis modes [11]. Based on that formulation, McEwan et al. presented the Modal/FE (MFE) approach with serial static test cases and regression approaches [12]. Cooper et al. developed the MFE approach in aeroelastic analysis [13]. An et al. modified MFE method to evaluate the accuracy of analysis and applied the method into static trim and nonlinear gust response problem [14, 15].

For more complex aircraft structures, such as the entire aircraft system, decomposing it into several simple substructure and using the boundary conditions between the substructures to assemble dynamic equations could be an efficient method for structure modeling. In practical applications, the component mode synthesis (CMS) method including the fixed interface CMS and the free interface CMS is widely used.

Applications of CMS in local nonlinear dynamic problem have attracted attention of many scholars. Clough et al. investigated substructure method to local nonlinear problem first [16]. Tan et al. divided the structure into linear substructure components and nonlinear substructure components [17]. The interactions between substructures were replaced by boundary forces to solve the transient response problem of local nonlinear structures. Fey et al. reduced the cantilever structure with nonlinear support and studied the nonlinear frequency domain response problem [18].

Most nonlinear substructures method focuses on the connection nonlinearities. As long as the nonlinear interface connection relationship is given, substructure itself is still a linear structure. The major source of geometric nonlinearities for large flexible aircraft is the wing component. The fuselage and other components of aircraft maintain linearity in analysis of geometric nonlinear problems. Characterizing the whole aircraft structures with nonlinearity will decrease the computational efficiency severally. The CMS method considering geometric nonlinearities in structure domain should be developed. Karpel et al. added virtual mass elements to the subcomponents to characterize the dynamic behaviors of other subcomponents [19]. The approach divides the wing component into segments processing rather than as a whole structure in order to interface with the fuselage, so it is more like a combination of nonlinear CMS and finite segment method. Kantor et al. developed this method and extended it into a simple model [20]. At present, substructure method considering geometric nonlinearities is still immature, and applications in aeroelastic analysis are still in the exploratory stage.

This paper is committed for aeroelastic problem associated with stability including geometric nonlinearities based on structural ROM and nonlinear substructure method. Aircraft is divided into nonlinear wing components and linear fuselage components. Structural ROM is used for wing structure modeling. Non-planar double lattice method (DLM) is used as aerodynamic model. To validate the method introduced, a very flexible flying wing model is taken as numerical model. Stability analysis results with linearization of dynamic equations are provided.

2 Theory

2.1 Structural Reduced Order Model

The structural reduced order model can be obtained with Galerkin method [10]. The structural dynamics equation can be expressed as:

$$\frac{\partial(F_{ij}S_{jk})}{\partial X_k} + \rho_0 b_i^0 = \rho_0 \ddot{u}_i \tag{1}$$

where tensor S is the second P-K stress tensor, tensor F representatives the deformation gradient tensor, b^0 representatives the force vector and ρ_0 representatives the density. X representatives the position vector and x representatives the deformed vector. A truncated basis of the linear modes is used and a third polynomial form describes the nonlinear stiffness and the structural dynamics equation can be given in modal form:

$$M_{ij}q_j + E_{ij}^{(1)}q_j + E_{ijl}^{(2)}q_jq_l + E_{ijlp}^{(3)}q_jq_lq_p = F_i \tag{2}$$

where M_{ij} are the reduced mass matrix, F_i is the modal force, and $E_{ij}^{(1)}$, $E_{ijl}^{(2)}$ and $E_{ijlp}^{(3)}$ are the reduced stiffness tensor. Einstein summation expression is introduced.

M_{ij} and $E_{ij}^{(1)}$ can be expressed in the formulation as:

$$\begin{cases} M_{ij}=M_i, i=j \\ M_{ij}=0, i\neq j \end{cases} \tag{3}$$

$$\begin{cases} E_{ij}^{(1)}=E_i, i=j \\ E_{ij}^{(1)}=0, i\neq j \end{cases} \tag{4}$$

The formulation of the nonlinear dynamic equations corresponding to the $i-th$ basis function can be written as:

$$M_iq_i + E_iq_i + E_{ijl}^{(2)}q_jq_l + E_{ijlp}^{(3)}q_jq_lq_p=F_i \tag{5}$$

The structure dynamic equations in modal space have been obtained and generalized coordinates q_i namely.

Two orthogonal spanwise modes are taken into the structural reduced order model to characterize the foreshortening effects of the large flexible structure. A combination of truncated linear modes and orthogonal spanwise modes is generated as a basis function in the nonlinear structural ROM.

Regression analysis is introduced to obtain the nonlinear stiffness coefficients $E_{ijl}^{(2)}$ and $E_{ijlp}^{(3)}$, the static formulation of Eq. (5) is:

$$E_{ijl}^{(2)}q_jq_l + E_{ijlp}^{(3)}q_jq_lq_p = F_i - E_iq_i \tag{6}$$

Evidently, if there are serials of test loads and structural deformations, nonlinear stiffness can be obtained by regression approach. The serials of test loads and structural deformations can be calculated by a commercial FEM software package.

The accuracy of the nonlinear stiffness coefficients directly depends on the selected test loads. The selection of test loads must emphasize that the aerodynamic force on the wing should be a follower force, which describes the actual characteristics of the aerodynamic force. In this paper, the aerodynamic force under the deformation combined bending and torsion modes is chosen as the test load case. The formulation of the wing deformation, which generates aerodynamic forces, should be:

$$f_{AIR}\{w\} = f_{AIR}\left(\sum a_i\{\phi_i\}_{bend} + \sum a_j\{\phi_j\}_{torsion}\right) \tag{7}$$

Here, $\{\phi_i\}_{bend}$ denote the bending modes and $\{\phi_j\}_{torsion}$ denote the torsion modes. $a_{i,j}$ denotes the weight factors.

2.2 Nonlinear Substructure Method

For large flexible aircraft, take wing components as nonlinear components and fuselage component as linear component. Considering substructure ($\alpha = 1, \cdots l_1, l_2, \cdots n$) system, each subcomponents dynamic equation in deformation coordinates can be given as:

$$\begin{pmatrix} \mathbf{m}_{ii} & \mathbf{m}_{ib} \\ \mathbf{m}_{bi} & \mathbf{m}_{bb} \end{pmatrix}^{(\alpha)} \begin{Bmatrix} \ddot{\mathbf{u}}_i \\ \ddot{\mathbf{u}}_b \end{Bmatrix}^{(\alpha)} + \begin{pmatrix} \mathbf{k}_{ii} & \mathbf{k}_{ib} \\ \mathbf{k}_{bi} & \mathbf{k}_{bb} \end{pmatrix}^{(\alpha)} \begin{Bmatrix} \mathbf{u}_i \\ \mathbf{u}_b \end{Bmatrix}^{(\alpha)} + \begin{Bmatrix} \mathbf{g}_i(u) \\ \mathbf{g}_b(u) \end{Bmatrix}^{(\alpha)} = \begin{Bmatrix} \mathbf{O} \\ \mathbf{G}_b \end{Bmatrix}^{(\alpha)} + \begin{Bmatrix} \mathbf{f}_i \\ \mathbf{f}_b \end{Bmatrix}^{(\alpha)} \tag{8}$$

Subscript i, b representative interior and boundary coordinates, $\mathbf{g}(u)$ is nonlinear section, $\mathbf{G}_b$ is the boundary load vector, and $\mathbf{f}$ is external force vector. Displacement and Force coordination conditions on interface can be expressed as:

$$\mathbf{u}_b^{(l_1)} = \mathbf{u}_b^{(l_2)} \tag{9}$$

$$\mathbf{G}_b^{(l_1)} + \mathbf{f}_b^{(l_1)} + \mathbf{G}_b^{(l_2)} + \mathbf{f}_b^{(l_1)} = 0 \tag{10}$$

Applied fixed interface CMS method to nonlinear wing components analysis, when give $\mathbf{u}_b^{(nl)} = 0$, structure dynamic equation is:

$$\mathbf{m}_{ii}^{(nl)}\ddot{\mathbf{u}}_i^{(nl)} + \mathbf{k}_{ii}^{(nl)}\mathbf{u}_i^{(nl)} + \mathbf{g}_i(\mathbf{u}_i, \mathbf{u}_b = 0)^{(nl)} = 0 \tag{11}$$

Use linear modes to reduce order with $\mathbf{u}_i^{(nl)} = \mathbf{\Phi}_i^{(nl)}\mathbf{q}_i^{(nl)}$, low-order equation is:

$$\ddot{\mathbf{q}}_i^{(nl)} + \mathbf{\Omega}_i^{(nl)}\mathbf{q}_i^{(nl)} + \tilde{\mathbf{g}}_i^{(nl)}(\mathbf{q}_i)^{(nl)} = 0 \tag{12}$$

Equation (12) has the similar formulation with reduced order model in Sect. 2.1. Introduce restrained modes $\mathbf{\Psi}_b$ to translate displacement:

$$\begin{Bmatrix} \mathbf{u}_i \\ \mathbf{u}_b \end{Bmatrix}^{(nl)} = [\mathbf{\Phi}_i \ \mathbf{\Psi}_b]^{(nl)} \begin{Bmatrix} \mathbf{q}_i \\ \mathbf{u}_b \end{Bmatrix}^{(nl)} = \begin{pmatrix} \mathbf{\Phi}_{ii} & \mathbf{\Psi}_{ib} \\ \mathbf{O}_{bi} & \mathbf{I}_{bb} \end{pmatrix}^{(nl)} \begin{Bmatrix} \mathbf{q}_i \\ \mathbf{u}_b \end{Bmatrix}^{(nl)} \tag{13}$$

The structure dynamic equation can be reformulated as:

$$\begin{aligned}&\begin{pmatrix} \mathbf{I} & \mathbf{M}_{ib} \\ \mathbf{M}_{bb} & \mathbf{M}_{bb}\end{pmatrix}^{(nl)} \begin{Bmatrix}\ddot{\mathbf{q}}_i \\ \ddot{\mathbf{u}}_b\end{Bmatrix}^{(nl)} + \begin{pmatrix}\boldsymbol{\Omega}_i & \mathbf{O} \\ \mathbf{O} & \mathbf{K}_{bb}\end{pmatrix}^{(nl)}\begin{Bmatrix} \mathbf{q}_i\\ \mathbf{u}_b\end{Bmatrix}^{(nl)} \\ &+ \begin{Bmatrix}\tilde{\mathbf{g}}_i(\mathbf{q}_i,\mathbf{u}_b)\\ \tilde{\mathbf{g}}_b(\mathbf{q}_i,\mathbf{u}_b)\end{Bmatrix}^{(nl)} = \begin{Bmatrix}\mathbf{O}\\ \mathbf{G}_b\end{Bmatrix}^{(nl)} + \begin{Bmatrix}\tilde{\mathbf{f}}_i\\ \tilde{\mathbf{f}}_b\end{Bmatrix}^{(nl)}\end{aligned} \tag{14}$$

where $\mathbf{M}_{ib} = \boldsymbol{\Phi}_{ii}^T(\mathbf{m}_{ii}\boldsymbol{\psi}_{ib} + \mathbf{m}_{jj})$ $\mathbf{M}_{bi} = \mathbf{M}_{ib}^T$, $\mathbf{M}_{bb} = \boldsymbol{\psi}_{ib}^T(\mathbf{m}_{ii}\boldsymbol{\psi}_{ib} + \mathbf{m}_{ij}) + \mathbf{m}_{ji}\boldsymbol{\psi}_{ib} + \mathbf{m}_{jj}$, $\tilde{\mathbf{f}}_i = \boldsymbol{\Phi}_{ii}\mathbf{f}_i$, $\tilde{\mathbf{f}}_b = \boldsymbol{\psi}_{ib}\mathbf{f}_i + \mathbf{f}_b$.

Apply linear mode reduction method to fuselage component analysis with main linear modes $\boldsymbol{\Phi}_k$ and residual modes $\boldsymbol{\Psi}_d$. The structural dynamics equation can be reformulated as:

$$\begin{pmatrix} \mathbf{I} & \mathbf{O} \\ \mathbf{O} & \mathbf{M}_{dd}\end{pmatrix}^{(l)} \begin{Bmatrix}\ddot{\mathbf{q}} \\ \ddot{\mathbf{u}}_b\end{Bmatrix}^{(l)} + \begin{pmatrix}\boldsymbol{\Omega} & \mathbf{O} \\ \mathbf{O} & \mathbf{K}_{dd}\end{pmatrix}^{(l)}\begin{Bmatrix} \mathbf{q}\\ \mathbf{G}_b\end{Bmatrix}^{(l)} = \begin{bmatrix}\tilde{\mathbf{f}}_k\\ \tilde{\mathbf{f}}_d\end{bmatrix}^{(l)} \tag{15}$$

where $\mathbf{M}_{dd} = \boldsymbol{\Psi}_d^T\mathbf{m}\Psi_d$, $\mathbf{K}_{dd} = \boldsymbol{\Psi}_d^T\mathbf{k}\Psi_d$, $\tilde{\mathbf{f}}_k = \boldsymbol{\Phi}_k^T\begin{Bmatrix}\mathbf{f}_i\\ \mathbf{G}_b + \mathbf{f}_b\end{Bmatrix}$, $\tilde{\mathbf{f}}_d = \boldsymbol{\Psi}_d^T\begin{Bmatrix}\mathbf{f}_i\\ \mathbf{G}_b + \mathbf{f}_b\end{Bmatrix}$. Considering the specified aircraft formulation, structure can be divided into left wing component, right wing component and fuselage component. Structure dynamic equations can be expressed as:

$$\begin{pmatrix} \mathbf{I}_{lw} & \mathbf{M}_{iblw} \\ \mathbf{M}_{bilw} & \mathbf{M}_{bblw}\end{pmatrix}\begin{Bmatrix}\ddot{\mathbf{q}}_{ilw} \\ \ddot{\mathbf{u}}_{blw}\end{Bmatrix} + \begin{pmatrix}\boldsymbol{\Omega}_{ilw} & \mathbf{O} \\ \mathbf{O} & \mathbf{K}_{bblw}\end{pmatrix}\begin{Bmatrix} \mathbf{q}_{ilw}\\ \mathbf{u}_{blw}\end{Bmatrix} + \begin{Bmatrix}\tilde{\mathbf{g}}_{ilw}(\mathbf{q}_{ilw})\\ -\mathbf{G}_{blw}\end{Bmatrix} = \begin{Bmatrix}\tilde{\mathbf{f}}_{ilw}\\ \tilde{\mathbf{f}}_{blw}\end{Bmatrix} \tag{16}$$

$$\begin{pmatrix} \mathbf{I}_{rw} & \mathbf{M}_{ibrw} \\ \mathbf{M}_{birw} & \mathbf{M}_{bbrw}\end{pmatrix}\begin{Bmatrix}\ddot{\mathbf{q}}_{irw} \\ \ddot{\mathbf{u}}_{brw}\end{Bmatrix} + \begin{pmatrix}\boldsymbol{\Omega}_{irw} & \mathbf{O} \\ \mathbf{O} & \mathbf{K}_{bbrw}\end{pmatrix}\begin{Bmatrix} \mathbf{q}_{irw}\\ \mathbf{u}_{brw}\end{Bmatrix} + \begin{Bmatrix}\tilde{\mathbf{g}}_{irw}(\mathbf{q}_{irw})\\ -\mathbf{G}_{brw}\end{Bmatrix} = \begin{Bmatrix}\tilde{\mathbf{f}}_{irw}\\ \tilde{\mathbf{f}}_{brw}\end{Bmatrix} \tag{17}$$

$$\begin{pmatrix} \mathbf{I}_{fu} & \mathbf{O} \\ \mathbf{O} & \mathbf{M}_{ddfu}\end{pmatrix}\begin{Bmatrix}\ddot{\mathbf{q}}_{fu} \\ \ddot{\mathbf{u}}_{bfu}\end{Bmatrix} + \begin{pmatrix}\boldsymbol{\Omega}_{fu} & \mathbf{O} \\ \mathbf{O} & \mathbf{K}_{ddfu}\end{pmatrix}\begin{Bmatrix} \mathbf{q}_{fu}\\ \mathbf{G}_{bfu}\end{Bmatrix} = \begin{bmatrix}\tilde{\mathbf{f}}_{kfu}\\ \tilde{\mathbf{f}}_{dfu}\end{bmatrix} \tag{18}$$

Subscript *lw*, *rw*, *fu* represent left wing, right wing and fuselage. Displacement and force coordination conditions are:

$$\begin{aligned}\mathbf{u}_{blw} &= \mathbf{u}_{bfulw}\\ \mathbf{u}_{brw} &= \mathbf{u}_{bfurw}\end{aligned} \tag{19}$$

where the $\mathbf{u}_{bfulw}$, $\mathbf{u}_{bfurw}$ is connection freedom between fuselage with left wing and right wing.

2.3 Aerodynamics Model

Non-planar DLM can be used to complete stable analysis. Unsteady aerodynamic force is given as:

$$\boldsymbol{\Phi}_S\mathbf{f}_A = q_\infty[\mathbf{A}_{SS0}\mathbf{q}_S + \frac{c_{ref}}{2V_\infty}\mathbf{A}_{SS1}\dot{\mathbf{q}}_S + \frac{c_{ref}^2}{4V_\infty^2}\mathbf{A}_{SS2}\ddot{\mathbf{q}}_S) + \mathbf{D}\mathbf{x}_a] \tag{20}$$

Structure model and aerodynamic model can be linearized around trim configuration [9] and comprehensive assembled to a state-space model:

$$\dot{\mathbf{x}}_{ae} = \mathbf{A}_{ae}\mathbf{x}_{ae} \tag{21}$$

where x_{ae} denotes the state vector and $\mathbf{A}_{ae}$ denotes the state matrix. The eigenvalues of state matrix represent the stability of system.

3 Numerical Results

3.1 Flying Wing Model

A numerical example of flying wing model is presented here to describe stable characteristics of rigid-elastic coupled problem. Figure 1 shows the aerodynamics model of aircraft with flying wing configuration. Main design parameters have been given in Table 1. This aircraft model has two large-aspect-ratio wings. Two control surfaces are set at the trailing edge of the wing. The FEM model of the flying wing is constructed with CBEAM elements and CONM elements in MSC.NASTRAN. The non-planar DLM are introduced to characterize aerodynamics behavior. For symmetric flight condition, degrees of freedom of x-axis, z-axis and pitch motions and elastic modes are considered. Linearized is complemented based on the nonlinear trim statement with large deformations.

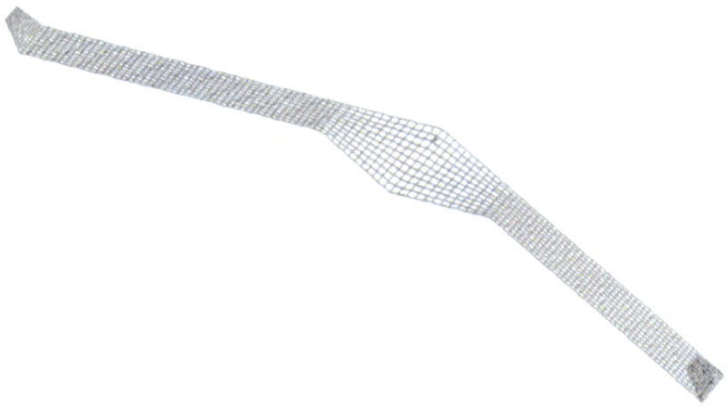

Fig. 1. Flying wing

Table 1. Design parameters of flying wing

Item	Parameters	Item	Parameters
Wing span/m	4.80	Mass/kg	20.0
Wing area/m^2	1.345	I_{xx}/kgm^2	3.572
Aspect ratio	17.1	I_{yy}/kgm^2	0.834
Airfoil of wing	EMX-07	I_{zz}/kgm^2	4.005

3.2 Numerical Results

Analysis results are obtained under the symmetric flight conditions. In elastic analysis, only elastic modes are involved in dynamic equations and rigid-body motions are not consideration. In the integrated analysis, both elastic modes and rigid-body motions are considered. In linear analysis, there are no nonlinear stiffness terms in structural ROM of wing components. Figure 2 shows the linear flutter analysis results with only elastic modes. The critical flutter speed is 49.0 m/s, and mode 5 participates in flutter. Figure 3 shows the nonlinear flutter analysis results with only elastic modes. The critical flutter speed is 36.5 m/s, which is different form linear analysis results. Nonlinearities affect stability obviously.

Considering rigid body motions, Fig. 4 shows the linear analysis results with rigid-body motion and elastic modes. Two branches of root locus cross the imaginary axis within the calculation range. When the airspeed velocity is 30.0 m/s, the short period mode locus crosses the imaginary axis, and the rigid-elastic coupling becomes unstable. Figure 5 shows the linear analysis results with rigid-body motion and elastic modes. Two branches of root locus cross the imaginary axis within the calculation range. When the airspeed velocity is 25.5 m/s, the short period mode locus crosses the imaginary. Nonlinearities affect stability obviously as well. Compared with elastic flutter analysis results, the critical stability flight speed of the coupled aeroelasticity and flight dynamics system is lower.

4 Conclusions

A practical approach for the stability analysis of the very flexible aircraft is established in this paper. The structural ROM is used for wing components modeling and nonlinear substructure method is used for aircraft system modeling. Non-planar DLM is used for aerodynamics modeling. The state-space equations are obtained, which can deal with rigid-elastic coupling problem.

A very flexible aircraft with flying wing configuration is selected to describe the stability problem. Because of large flexible characteristics and small pitching inertial of such aircraft, the critical stability flight speed is affected obviously by nonlinearities and rigid-motions. The flight dynamics and aeroelastic analysis should be performed simultaneously.

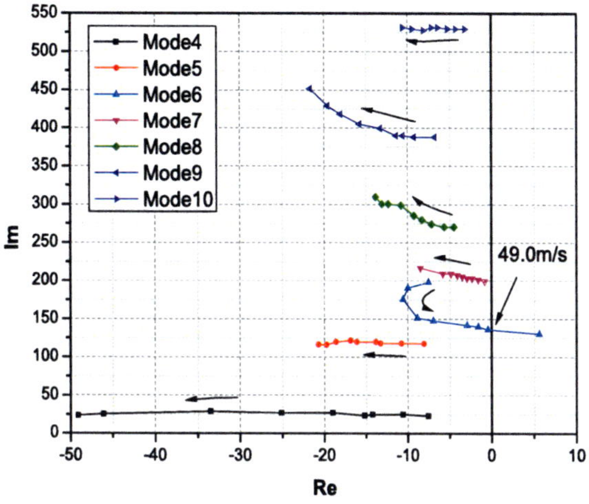

Fig. 2. Elastic analysis results

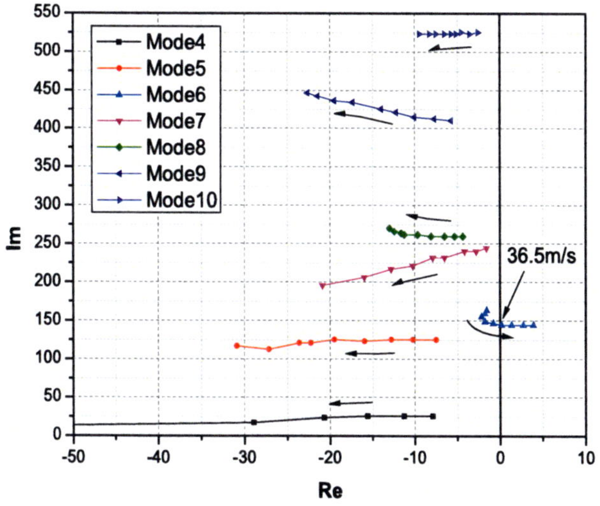

Fig. 3. Nonlinear elastic analysis results

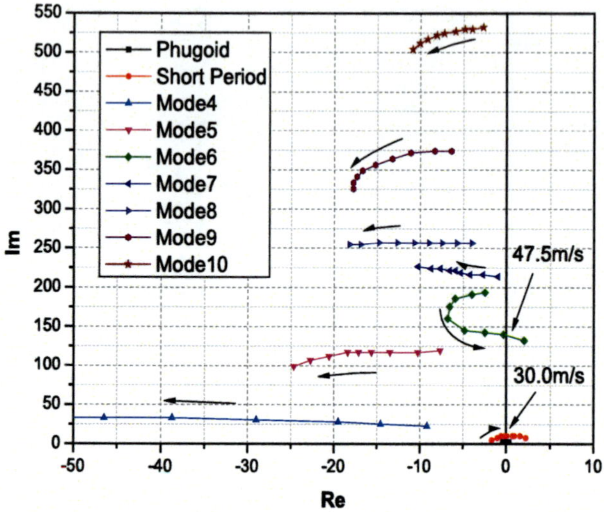

Fig. 4. Linear integrated analysis results

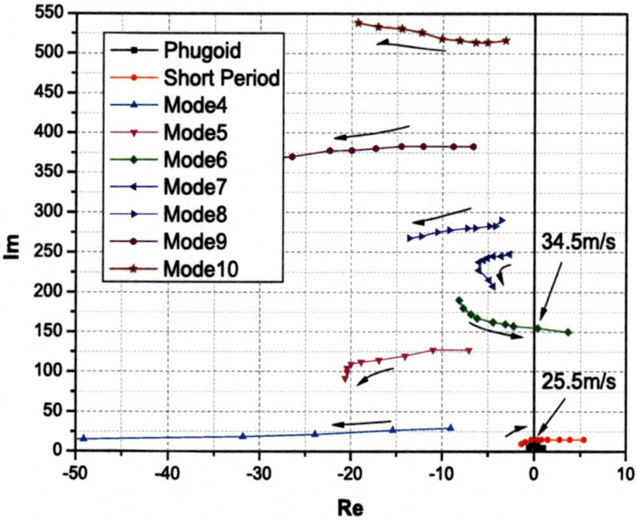

Fig. 5. Nonlinear integrated analysis results

References

1. Patil, M.J., Hodges, D.H.: On the importance of aerodynamics and structural geometrical nonlinearities in aeroelastic behavior of high-aspect-ratio wings. In: 41st AIAA/ASME/ASCE/AHS/ASC Structures, Structural Dynamics, and Materials Conference and Exhibit, AIAA, Reston (2000)

2. Patil, M.J.: Nonlinear Aeroelastic Analysis, Flight Dynamics, and Control of a Complete Aircraft. Georgia Institute of Technology, Atlanta (1999)
3. Tang, D.M., Dowell, E.H.: Experimental and theoretical study on aeroelastic response of high-aspect-ratio wings. AIAA J. **39**(8), 1430–1441 (2001)
4. Frulla, G., Cestino, E., Marzocca, P.: Critical behavior of slender wing configurations. Proc. I Mech. Eng. Part G: J. Aerosp. Eng. **224**(1), 587–600 (2009)
5. Abbas, L., Chen, Q., Marzocca, P., Milanese, A.: Nonlinear aeroelastic investigations of stores-induced limit cycle oscillations. Proc. I Mech. Eng. Part G: J. Aerosp. Eng. **222**(1), 63–80 (2008)
6. Kim, K., Strganac, T.: Aeroelastic Studies of a cantilever wing with structural and aerodynamic nonlinearities. In: 43rd AIAA/ASME/ASCE/AHS/ASC Structures, Structural Dynamics, and Materials Conference and Exhibit, AIAA, Reston (2002)
7. Yang, C., Huang, C., Wu, Z.G.: Progress and challenges for aeroservoelasticity research. ATCA Aeronaut. Astronaut. Sin. **36**(4), 1011–1033 (2015)
8. Patil, M.J., Hodges, D.H.: Flight dynamics of highly flexible flying wings. J. Aircr. **43**(6), 1790–1799 (2006)
9. Xie, C.C., Yang, L., Liu, Y., Yang, C.: Stability of very flexible aircraft with coupled nonlinear aeroelasticity and flight dynamics. J. Aircr. (2018)
10. Muravyov, A.A., Rizzi, S.A.: Determination of nonlinear stiffness with application to random vibration of geometrically nonlinear structures. Comput. Struct. **81**, 1513–1523 (2003)
11. Mignolet, M.P., Soize, C.: Stochastic reduced order models for uncertian geometrically nonlinear dynamical systems. Comput. Methods Appl. Mech. Eng. **197**, 3951–3963 (2008)
12. McEwan, M.I., Wright, J.R., Cooper, J.E., Leung, A.Y.T.: A combined modal/finite element analysis technique for the dynamic response of a nonlinear beam to harmonic excitation. J. Sound Vib. **243**(4), 601–624 (2001)
13. Harmin, M.Y., Cooper, J.E.: Aeroelastic behavior of a wing including geometric nonlinearities. Aeronaut J. **115**(1174), 767–777 (2011)
14. Xie, C.C., An, C., Liu, Y., Yang, C.: Static aeroelastic analysis including geometric nonlinearities based on reduced order model. Chin. J. Aeronaut. **30**(2), 638–650 (2017)
15. An, C., Yang, C., Xie, C.C., Yang, L.: Flutter and gust response analysis of a wing model including geometric nonlinearities based on a modified structural ROM. Chin. J. Aeronaut. **33**(1), 48–63 (2020)
16. Clough, R.W., Wilson, E.L.: Dynamic analysis of large structural systems with local nonlinearity. Comput. Method Appl. Mech. Eng. **17**(18), 107–129 (1979)
17. Tan, M.A., Cai, C.W.: Substructure method in local nonlinear system response analysis (in Chinese). Vib. Shock **23**, 25–34 (1987)
18. Fey, R.H.B., Van Campen, D.H., Kraker, D.: Long term structural dynamics of mechanical systems with local nonlinearities. J. Vib. Acoust. **118**, 147–153 (1996)
19. Bernhammer, L.O., Breuker, R.D., Karpel, M.: Geometrically nonlinear structural modal analysis using fictitious masses. AIAA J. **55**(10), 3584–3593 (2017)
20. Kantor, E., Raveh, D.E., Cavallaro, R.: Nonlinear structural, nonlinear aerodynamic model for static aeroelastic problem. AIAA J. (2018)

Bending Properties of Representative Composite Aircraft Structures Under Oil Pressure

Liu Han[1,2](✉), Hui Qi[2], Shifeng Li[1], Zhen Zhang[1], and Shaobo Gong[1]

[1] AVIC Harbin Aircraft Industry Group CO., LTD, Harbin 150066, China
348020580@qq.com

[2] College of Aerospace and Civil Engineering, Harbin Engineering University, Harbin 150001, China

Abstract. Composite structures adjacent to the fuel tank of a helicopter fuselage experience a substantial bending load created by the oil pressure. Thus, these structures may undergo considerable deformation. The present study investigated the bending characteristics of typical structures (spar cap) adjacent to the fuel tank. The spar cap specimens (under two conditions) were subjected to four-point bending, and the strength and failure mode were examined. Test results shown that all samples fractured at the middle of the curved part under Condition 1, which was consistent with the working conditions of an actual structure adjacent to the fuel tank. Progressive failure analysis was performed based two-dimensional Hashin failure for intralaminar failure model and bilinear traction separation constitutive model is adopted for interlaminar failure model. The results shown that the appearance and internal damage morphology obtained by the simulation are consistent with experimental results.

Keywords: Composite material · Four-point bending · Progressive failure · Intralaminar failure · Interlaminar failure · Bilinear traction separation constitutive model

1 Introduction

The percentage and configuration of composite parts used are important parameters indicating aircraft performance [1, 2]. Because of their high strength, design flexibility, and resistance to fatigue failure, as well as their ability to be manufactured in large integral (i.e., one-piece) parts, advanced composite materials are widely used in the integral tank structure to effectively reduce structure weight [3, 4]. Nevertheless, composites are associated with their own issues. For example, during use composite parts may occasionally experience impacts by foreign objects, such as birds, stones (on the runway), or tools (accidentally dropped during repairs). Such impacts can cause matrix cracking, delamination, fiber rupture, or other damage modes in composite structures. This damage severely affects the structural safety of an aircraft and, thus, has been extensively investigated. The residual mechanical properties (under tension, compression, and shear) of composite structures after a low-energy impact are a major current topic of research [5,

Chinese Society of Aeronautics and Astronautics: CASTYSF 2022, LNEE 972, pp. 224–234, 2023.
https://doi.org/10.1007/978-981-19-7652-0_21

6]. In addition, the structures adjacent to the fuel tank work under substantial bending loads due to fuel pressure and can deform considerably [7, 8]. Therefore, the out-of-plane bending properties of these structures and the residual bending performance after impact must be considered in the structural design.

The previous research were mainly focused on the bending of composite plate and sandwich structure. The typical structures adjacent to the fuel tank were seldom studied [9]. Luo Liang studied the compression after impact performance for T300/NY9200Z, T300/QY8911 and T700S/5228. Fracture mechanics behavior by means of mechanical numerical simulation The research results indicated that there are homologous knee points in these curves and these points which the damage mechanisms change significantly [10, 11]. Gong Xiaohui the studied low-velocity impact properties and three-point bending properties of the composite foamed-sandwich foam composites. The bending strength and stiffness of the foam sandwich composite board can be effectively improved, and the bending performance of the sandwich structure is effectively improved because of stiffeners [12]. Meng Xiangyao studied the three-point bending damage evolution and fracture mechanics behavior by means of mechanical numerical simulation [13]. The present study investigated the post-impact bending characteristics of typical structures (spar cap) adjacent to the fuel tank. The spar cap specimens (under two conditions) were subjected to four-point bending, the test results shown that it was consistent with the working conditions of an actual structure adjacent to the fuel tank under condition 1. Progressive failure analysis was performed based two-dimensional Hashin failure for intralaminar failure model and bilinear traction separation constitutive model is adopted for interlaminar failure model. The results shown that the appearance and internal damage morphology obtained by the simulation are consistent with experimental results.

2 Four-Point Bending

2.1 Structure of Test Specimens

Structures adjacent to the fuel tank experience high loads generated by the fuel pressure and, thus, can deform considerably. The out-of-plane bending properties of these structures are important considerations in the structural design of aircraft. Here, two typical structures adjacent to the tank, a spar cap and a spar web, were studied (Fig. 1).

Samples were prepared using prepregs fabricated from carbon fiber fabric and unidirectional tapes. Prepregs were plied and cured in an autoclave (SCHOLZ, Germany; diameter: 4 m, temperature departure: ±3 °C, ramp: 0–3 °C/min, cooling: 0–5 °C/min).

The spar cap layer:

- P1: $[\pm 45/0_2/\pm 45/0_2/\pm 45_2/0_2/\pm 45/0_2/\pm 45]$ (thickness: 2.427 mm);
- P2: $[\pm 45/0_2/\pm 45_2/0/\pm 45_2/0_2/\pm 45]$ (thickness: 2.545 mm).

Please note that the first paragraph of a section or subsection is not indented. The first paragraphs that follows a table, figure, equation etc. does not have an indent, either. Subsequent paragraphs, however, are indented.

The materials used for preparation of test specimens are domestic carbon fiber fabric 5224/CF3052 and unidirectional tape 5224/U3160; the mechanical properties of these materials are shown in Table 1.

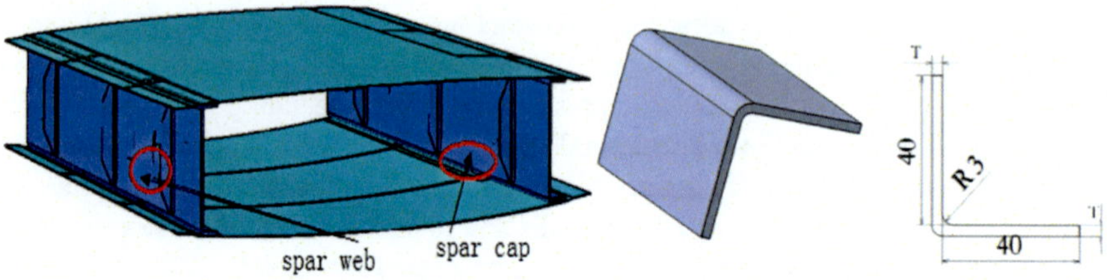

(a) wing oil tank (b) spar cap

Fig. 1. Schematic representation

Table 1. Mechanical properties of composite materials

		Unidirectional tape	Fabric
Density	(t/mm^3)	1.58e−9	1.58E−9
Longitudinal stiffness E_{11}	(GPa)	120	55
Transverse modulus E_{22}	(GPa)	9.0	55
Poisson ratio v		0.3	0.052
Shear stiffness G	(GPa)	4.1	510
Longitudinal tension strength X_T	MPa	1000	400
Transverse tension strength Y_T	(MPa)	40	500
Longitudinal compression strength X_C	MPa	700	400
Transverse compression strength Y_C	(MPa)	120	80

2.2 Spar Cap Bending Testing

Considering the spar cap deforms considerably under the bending load during operation, we tested the bending properties of this structure using a four-point bending test (ASTM D 7264) under two constraint conditions (Fig. 2). Under Condition 1 (Figs. 2a and 3a), two supporting pins (6 mm) were placed on a horizontal plane, and the distance between the pins (support span) was adjusted and fixed to 44 mm (±0.1 mm). The sample was allowed to straddle the pins, and a U-shaped loading head was inversely placed on the sample. The head contacted the sample via its round ends (6.0 ± 0.1 mm). Then a compressive load was applied at 7.5 m/min, and the sample was allowed to flatten and widen freely. Under Condition 2 (Figs. 2b and 3a), two protruding bars were engineered on the plane to prevent the sample from widening during compression; other parameters were identical to Condition 1.

Table 2 shown that all samples fractured at the middle of the curved part under Condition 1, which was consistent with the working conditions of an actual structure adjacent to the fuel tank. Under Condition 2, most samples failed at the regions contacting the loading head, exhibiting a local shearing failure mode. This was inconsistent with the out-of-plane bending condition experienced by a typical structure adjacent to the fuel tank.

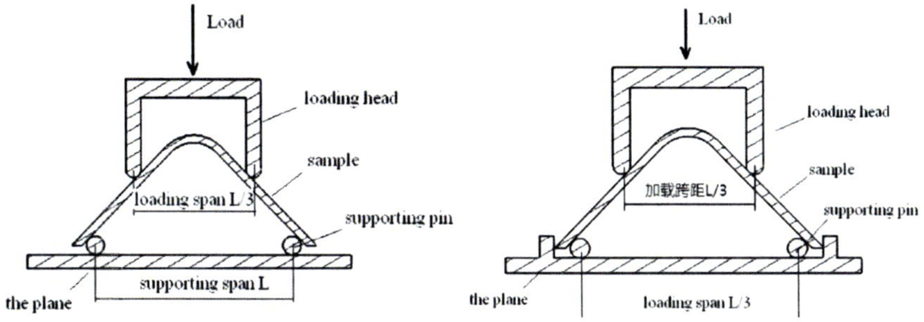

(a) Condition 1 (b) Condition 2

Fig. 2. Diagrams showing four-point bending tests

Figure 3 shown the load-displacement curve at condition 1, and Fig. 4 shown the load-displacement curve at condition 2. It can be seen that the samples exhibited only one failure mode under Condition 2: local shearing failure at the loading head-sample contact region, and the load at failure was 4–5 times higher than those tested under Condition 1 (out-of-plane bending). Additionally, comparison of results obtained from different ply arrangements (P1 versus P2) indicates that increasing the proportion of 0° plies (P1) also significantly increased the bending strength.

Table 2. Bending test results of spar cap samples

Structure		Sample code	Failure load (KN)	Displacement (mm)	Testing condition	Damage mode
Spar cap	P1	YT101	0.713	2.98	1	Fractured at the middle of the curved part
		YT102	0.784	3.21	1	
		YT103	0.798	3.04	1	
		YT104	3.270	3.61	2	Shearing failure at the regions contacting the loading head
		YT105	3.487	3.57	2	
		YT106	3.660	3.41	2	
	P2	YT201	0.732	3.82	1	Fractured at the middle of the curved part
		YT202	0.760	3.65	1	
		YT203	0.730	3.93	1	
		YT204	4.087	3.97	2	Shearing failure at the regions contacting the loading head
		YT205	4.209	3.58	2	
		YT206	3.662	4.11	2	

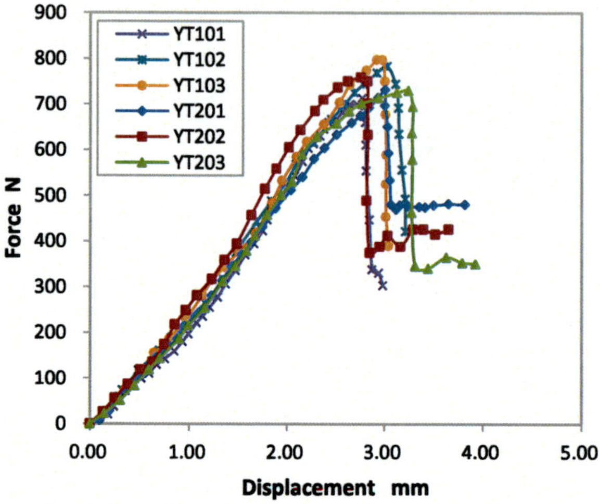

Fig. 3. Load-displacement curve at condition 1

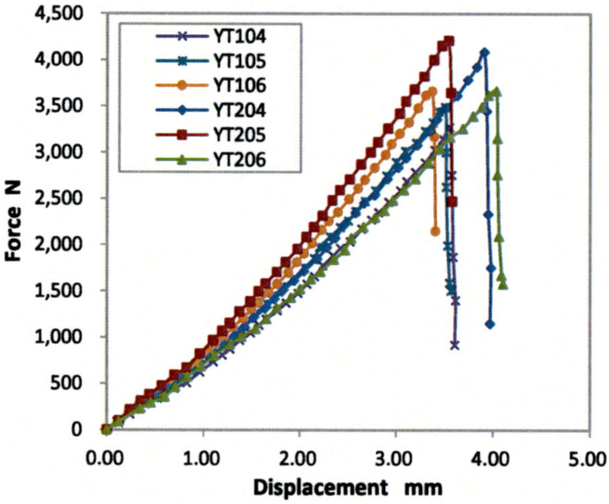

Fig. 4. Load-displacement curve at condition 2

3 Bending Failure Analysis

3.1 Interlaminar Failure Model

The damage constitutive relations of Interlaminar behavior include bilinear, trapezoid, Parabola and exponential models [14, 15]. Considering the damage of adhesive layer is elastic-brittle, the bilinear constitutive model is chosen. σ_n, σ_s and σ_t are used to describe the normal traction stress and shear stress respectively, and the stress-strain relationship of the adhesive interface behavior is [15–17]:

$$\boldsymbol{\sigma} = \mathbf{K} \cdot \boldsymbol{\delta} \tag{1}$$

where, $\mathbf{K} = [K_{nn}, K_{ss}, K_{tt}]$, $\cdot\boldsymbol{\sigma} = [\sigma_n, \sigma_s, \cdot\sigma_t]$, $\boldsymbol{\delta} = [\delta_n, \delta_s, \delta_t]$. **K** is the elastic stiffness coefficient of the adhesive interface in each stress direction, **δ** is the opening displacement in each stress direction [15].

When the adhesive interface is damaged, the material shows the softening stage of damage, and the constitutive relation is [15]:

$$\boldsymbol{\sigma} = (1 - \mathbf{D})\mathbf{K} \cdot \boldsymbol{\delta} \tag{2}$$

where, **D** is the damage state variable, **D** = 0, the adhesive layer is not damaged, **D** = 1, which means that the adhesive is completely failure [15].

According to the description of the properties of the interface adhesive, two criteria are needed to determine the final failure of the adhesive layer [15]: The second Nominal stress Criterion (Quadratic Nominal stress Criterion) is used to determine whether the plastic layer begins to fail; B-K criterion is used to determine whether a rubber layer unit is completely failed and destroyed as the damage propagation criteria [15, 18, 19].

$$(\sigma_n/N_{\max})^2 + (\sigma_s/S_{\max})^2 + (\sigma_t/T_{\max})^2 = 1 \tag{3}$$

where, $N_{\max}$, $S_{\max}$, $T_{\max}$ are the strength of the interface in each orthogonal direction [15, 18, 19].

$$G_{IC} + (G_{IIC} - G_{IC})(G_{\text{Shear}}/G_T)^{\eta} = G_{TC} \tag{4}$$

$$G_T = G_I + G_{\text{Shear}}\,;\; G_{\text{Shear}} = G_{II} + G_{III} \tag{5}$$

where, G_{TC} is the total release energy; G_I, G_{II} and G_{III} are the fracture toughness respectively of Type I, Type II and Type III; G_I, G_{II} and G_{III} are the critical energy release rates of Type I, Type II and Type III; G_{Shear} is the shear energy release rates; the η is a constant relative to the material (η = 1.45) [15, 18, 19].

3.2 Intralaminar Failure Model

Progressive failure analysis was performed based two-dimensional Hashin failure criteria for intralaminar failure model. Four different failure models are analyzed and processed for initiation mechanism: fiber stretching (*FT*), fiber compression (*FC*), matrix stretching (*MT*) and matrix pressure Shrink (*MC*).Two-dimensional Hashin failure criteria are as follows 15 [20, 21]:

$$\begin{aligned}
FT &= \left(\hat{\sigma}_{11}/X_1^+\right)^2 + \left(\hat{\tau}_{12}/S^L\right)^2 = 1,\ (\hat{\sigma}_{11} \geq 0)\\
FC &= \left(\hat{\sigma}_{11}/X_1^-\right)^2 = 1,\ (\hat{\sigma}_{11} < 0)\\
MT &= \left(\hat{\sigma}_{22}/X_2^+\right)^2 + \left(\hat{\tau}_{12}/S^L\right)^2 = 1,\ (\hat{\sigma}_{22} \geq 0)\\
MC &= \left(\hat{\sigma}_{22}/2S^T\right)^2 + \left[\left(Y^c/\left(2S^T\right)\right)^2 - 1\right]\left(\hat{\sigma}_{22}/X_2^-\right) + \left(\hat{\tau}_{12}/S^L\right)^2 = 1,\ (\hat{\sigma}_{22} < 0)
\end{aligned} \tag{6}$$

where, X_1^+ X_1^-, X_2^+, X_2^-, S^L and S^T are longitudinal tensile strength, longitudinal compressive strength, and transverse tensile strength, transverse compression strength, longitudinal shear strength and transverse shear strength. $\hat{\sigma}_{11}$, $\hat{\sigma}_{22}$ and $\hat{\tau}_{12}$ is the component of the effective stress tensor along the fiber direction, the transverse stress and in-plane shear stress., which can be calculated from the real stress $\boldsymbol{\sigma}$ and damage parameters M [15, 20, 22].

$$\hat{\boldsymbol{\sigma}} = \mathbf{M} \cdot \boldsymbol{\sigma} \tag{7}$$

$$\mathbf{M} = \begin{bmatrix} 1/(1-d_f) & 0 & 0 \\ 0 & 1/(1-d_m) & 0 \\ 0 & 0 & 1/(1-d_s) \end{bmatrix} \tag{8}$$

where, d_f, d_m and d_s are damage variables related to fiber, matrix and shear damage respectively. Typical curves of linear evolution were used for damage.

3.3 Numerical Simulation Results

The three-dimensional numerical simulation finite element model is established according to the size of the experimental specimen (shown in Fig. 1). The width of cap was 50 mm, and the two lengths of the angel clip were 40 mm. SC8R elements were used to simulation the fabric and unidirectional tape. The loading stick and supporting roller were simulated by the S4R shell elements and the loading stick and supporting roller were in contact with the upper surface and under surface of the specimen by Penalty function (Fig. 5). Cohesive interface contact behavior was defined to simulate the intralaminar failure model. The cohesive interface parameters of the adhesive bonding interface are shown in Table 3.

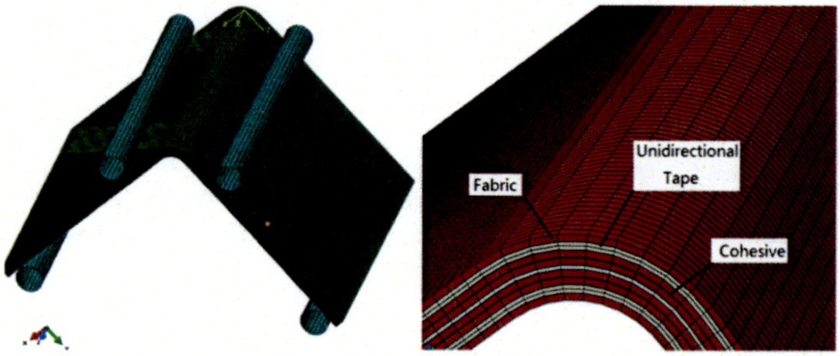

Fig. 5. Finite element model of bending failure analysis

Figure 6 shown the damage photographs under condition 1. Figure 7 shown the Different failure modes: matrix tensile failure, matrix compression failure, fiber tensile failure and fiber compression failure under condition1. It can be seen that all samples fractured at the middle of the curved part. In the simulation, the load was applied and

Table 3. Cohesive interface parameters [15]

σ_n/ MPa	σ_s/ MPa	σ_t/ MPa	K_{nn}/ N/mm^3	K_{ss}/ N/mm^3	K_{tt}/ N/mm^3	G_I/ mJ/mm^2	G_{II}/ mJ/mm^2	G_{III}/ mJ/mm^2
22	30	30	2.25e7	2.25e7	2.25e7	0.301	1.139	1.139

increased incrementally, up to an ultimate load of about 800 N. This was higher than the failure load measured experimentally, and the displacement is lower than load measured experimentally.

Figure 8 shown the load-displacement curves comparison between bending test and numerical simulation of four points bearing under condition 1.The results shown that the appearance and internal damage morphology obtained by the simulation were consistent with experimental results.

Fig. 6. The damage photographs under condition 1

4 Conclusions

Representative composite structures adjacent to the fuel tank were studied for failure mode and strength under out-of-plane bending. Based on the results, the following conclusions are drawn:

(1) The bending strengths of the spar cap samples were tested under two experimental conditions. Under Condition 2 (constrained lateral widening), the loading head caused local shearing failure, not satisfactorily simulating the loading pattern of this structure during actual use in an aircraft (out-of-plane bending by fuel pressure).
(2) Progressive failure analysis was performed based two-dimensional Hashin failure for intralaminar failure model and bilinear traction separation constitutive model was adopted for interlaminar failure model.

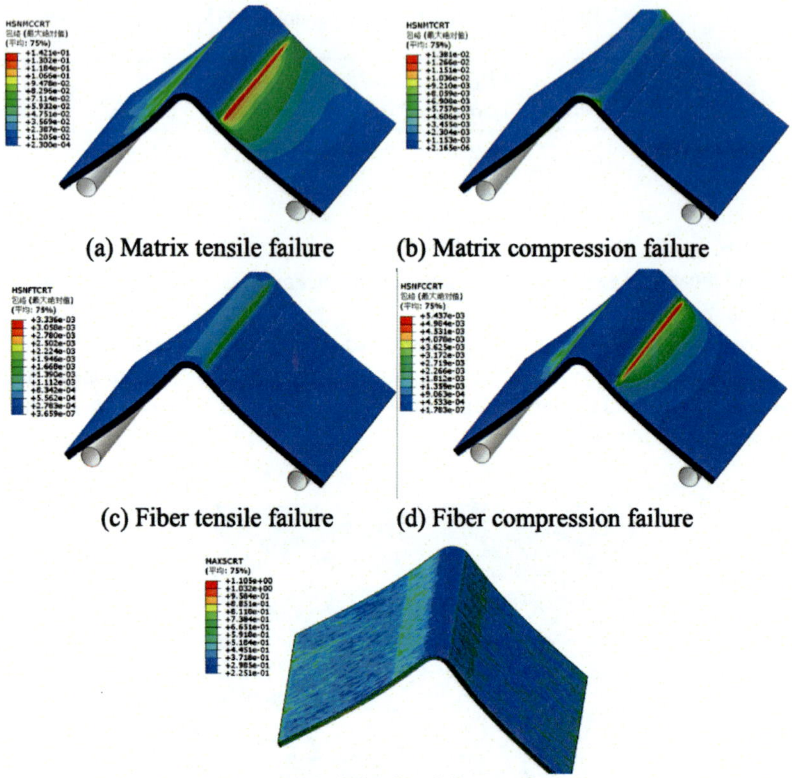

(a) Matrix tensile failure (b) Matrix compression failure

(c) Fiber tensile failure (d) Fiber compression failure

(e) Cohesive failure

Fig. 7. Load-displacement curve at condition 2

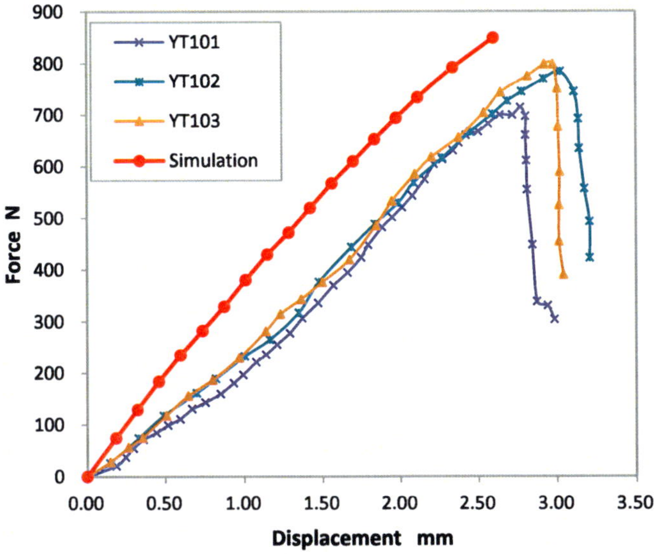

Fig. 8. Load-displacement curve at condition 2

References

1. Zhen, S., et al.: Handbook of Composite Structural Design. Aviation Industry Press, Beijing (2001)
2. Qusen, Z., et al.: Handbook of Advanced Composites. China Machine Press, Beijing (2001)
3. Lin, L.: Protection design of lightning and static of composite integral fuel tank for large civil aircraft. Civil Aircr. Design Res. **1**, 29–33 (2019)
4. Xinglong, H.: Analysis on Composite Wing Fuel Tank Seal Deformation. Nanjing University of Aeronautics and Astronautics, Nanjing (2013)
5. Serge, A.: Impact on laminate composite materials. Appl. Mech. Rev. **44**(4), 155–190 (1991)
6. Richardson, M., Wisheart M.J.: Review of low-velocity impact properties of composite material. Composites: Part A 27A, 1123–1131 (1996)
7. Xianggui, T.: Seal Failure Analysis and Test Verification on the Integral Fuel of Composite Wing. Naning University of Aeronautics and Astronautics, Nanjing (2018)
8. Jianbiao, G.: Research on composite fuel tank access door for civil aircraft. Civil Aircr. Design Res. **27**(2), 114–116 (2017)
9. Rui, X.: Sealing Performance Analysis and Structural Optimum Design of Composite Integral Fuel Tank. Naning University of Aeronautics and Astronautics, Nanjing (2017)
10. Liang, L., Zhen, S., Naichun, Y., et al.: Experimental study on low-velocity impact performance of carbon fiber reinforced composite laminates. Acta Mater. Compos. Sin. **25**(3), 20–24 (2008)
11. Shokrieh, M.M., Torabizadeh, M.A., Fereidoon, A.: Progressive failure analysis of glass/epoxy composites at low temperatures. Strength Mater. **44**(3), 134–324 (2012)
12. Xiaohui, G.: Study on Low-Velocity Impact and Bending Property of Foam Sandwich Structure. Nanchang University, Nanchang (2018)
13. Xiangyao, M., Wanpeng, D.: Damage behavior analysis of vehicle composite laminates under bending load. Agri. Equip. Veh. Eng. **59**(11), 123–126 (2021)

14. Kaixue, Y.: Investigation on Compression Strength of Honeycomb Sandwich Structure with Closing Frame. Nanchang University (2019)
15. Han, L., Qi, H., Li, R., Liu, P., Li, S., Zhan, Z., Wang, J.: Research on typical joint technology of composite stiffened skin-fuselage frame. In: Proceedings of the 5th China Aeronautical Science and Technology Conference: Lecture Notes in Electrical Engineering (2022)
16. Camanho, P., Dávila, C.: Mixed-mode decohesion finite elements for the simulation of delamination in composite materials. NASA Report, 9–12 (2002)
17. Camanho, P.P., Davila, C.G., de Moura, M.F.: Numerical simulation of mixed-mode progressive delamination in composite materials. J. Compos. Mater. **37**(16), 1415–1438 (2003)
18. Lan, J.: Research on Progressive Damage and Bonding Repair of Perforated Composite Laminates. Wuhan University of Technology, Wuhan, pp. 14–18 (2019)
19. Liang, Z., Yan, Y., Zhang, T., et al.: Experimental investigation and numerical simulation of composite laminate adhesively bonded single-lap joints. J. Beijing Univ. Aeronaut. Astronaut. **40**(12), 1786–1792 (2014)
20. Schwab, M., Todt, M., Wolfahrt, M., Pettermann, H.E.: Failure mechanism based modeling of impact on fabric reinforced composite laminates based on shell elements. Compos. Sci. Technol. **128**, 131–137 (2016)
21. Chang, F.K., Springer, G.S.: The strengths of fiber reinforced composite beads. J. Compos. Mater. **20**(1), 30–45 (1986)
22. Benzeggagh, M.L., Kenane, M.: Measurement of mixed-mode delamination fracture toughness of unidirectional glass/epoxy composites with mixed-mode bending apparatus. Compos. Sci. Technol. **56**(4), 439–449 (1996)

Impact Energy Absorption Features for Tensor-Skin Fuselage Box Structure

Liu Han[1,2](✉), Hui Qi[2], Shifeng Li[1], Kunfa Men[1], and Jinliang Wang[1]

[1] AVIC Harbin Aircraft Industry Group CO., LTD, Harbin 150066, China
348020580@qq.com

[2] College of Aerospace and Civil Engineering, Harbin Engineering University, Harbin 150001, China

Abstract. Finite-element software was used to conduct a simulated analysis of hard ground, soft soil, and water impacts for the composite tensor-skin fuselage box structure of helicopters; to study the energy absorption of the tensor skin in hard ground, soft soil, and water impacts; and to compare the accelerated velocity of the tensor-skin fuselage box structure under these impact conditions to establish whether the acceleration peak of the water impact is smaller than that of the hard ground and soft soil impacts. For the water impact, the well-designed tensor skin absorbs energy by stretching the fold area, so the tensor skin provides a greater advantage over a traditional skin in improving safety in helicopter impact. It is important to ensure that all wrinkles are spread successfully in the design of the tensor skin. A tensor ply spread is important for the tensor skin to absorb energy and to maintain structural integrity. The tensor skin offers greater safety advantages over traditional skins in helicopter crashes at the water surface.

Keywords: Helicopter · Tensor skin · Fuselage box section · Energy-absorption features

1 Introduction

Helicopters are highly mobile aircraft that are becoming increasingly popular, but, helicopter flight accidents threaten the lives of occupants. Anti-crash design is an important consideration for military or civilian helicopters. Only the impact of firm ground is taken into account in current anti-crash design despite accident statistics showing that more than 80% of military and civilian helicopter crash accidents occur in soft soil and at water surfaces [1, 2]. When a helicopter crashes into firm ground, the impact kinetic energy is often absorbed by devices such as the landing gear, fuselage and seats. The combined effect of these energy-absorbing devices reduces the impact energy that is delivered to occupants. However, when a helicopter collides with a water surface, a huge lateral pressure load is experienced after the landing gear enters the water and the fuselage impacts onto the water surface. In this case, the metal skin undergoes a large plastic deformation, and the composite laminate fails because of the large lateral load, which results in the failure of load transmission to other energy-absorbing devices beneath the floor. Most

Chinese Society of Aeronautics and Astronautics: CASTYSF 2022, LNEE 972, pp. 235–244, 2023.
https://doi.org/10.1007/978-981-19-7652-0_22

impact energy is absorbed by the seats, which reduces the chances of survival of the occupants [3]. According to statistics, 10% of civilian and 33% of military helicopter accidents are water-surface impacts with low occupant survival rates [4, 5].

In the early 1990s, Thuis et al. proposed a concept of a "tensor-skin panel", in which the skin can transfer water impact loads to nearby energy-absorbing structures without breaking. In 1995, Thuis performed a quasi-static test to verify that this structure can transfer a water load [6–10]. In 1998, Michielsen et al. conducted dynamic tests on a composite tensor-skin panel. The water-impact resistance of the tensor-skin panel was much better than that of a traditional honeycomb panel, and it could be used in the impact-sensitive areas of a helicopter sub-floor structure to provide sufficient film strength to transfer a pressure load to an energy-absorbing structure to absorb impact energy [11–13]. In 2000, the German Aerospace Center (DLR) carried out a sub-floor dynamic impact test to verify whether the helicopter sub-floor structure that is comprised of a tensor skin and a corrugated beam improves the helicopter crashworthiness [14, 15].

The behavior of hard-ground, soft-soil, and water-surface impacts for the composite tensor-skin fuselage box structure of helicopters were simulated; to study the loading conditions of tensor skin in water-surface impacts; and to analyze the velocity, accelerated velocity, and energy changes of the tensor-skin fuselage box structure in hard-ground, soft-soil, and water-surface impacts to provide a foundation to study subsequent applications of a tensor-skin structure on the composite fuselage structure of helicopters.

2 Structure Style

Composite material differs from metal. Its resin matrix is brittle and allows for almost no plastic deformation. Because reinforced fiber can only withstand a tensile load, parts made of composite materials resist tension well but resist compression poorly. A composite-material structure anti-impact design method is different from a conventional metal material structure. For example, structures that include sine wave beams and hollow column-reinforced beams are used under the floor to improve the structural stability. Foam or honeycomb is used to fill some structures. Some components, such as roll tubes and sandwich beams, which experience controllable compression deformation are used to enhance the structural capacity for impact-energy absorption and dissipation.

We designed a helicopter composite corrugated sub-floor structural component with anti-impact standards, and with energy-absorption characteristics of a composite energy-absorbing structure in mind. The component was contained in a 1240 mm × 940 mm × 300 mm box structure, and was composed of two longitudinal beams, two bulkheads, and upper and lower panels. A beam web plate in corrugated beam form and honeycomb sandwich structure was the main energy-absorbing component. A cross structure was designed at the intersection of the beam and bulkhead. Because the upper and lower panels were not the main energy-absorbing components, the upper panel was aluminum, and the lower panel was glass. The bottom skin structure of the fuselage was tensor skin. The CATIA model and structural components for the composite fuselage box structure are shown in Fig. 1 [10]. According to the research results of the composite corrugated beam structure elements, a ply detailed design of corrugated beams (web thickness, material composition, laying direction, initial triggering position, etc.) was applied to design the sub-floor components.

(a) CATIA digital model structure (b) Structural entity diagram

Fig. 1 Composite fuselage box structure of helicopter [10]

2.1 Tensor-Skin Structure

A tensor-skin plate consists of three parts: cover ply, tensor ply, carrying ply. The plate is a sandwich plate that comprises a corrugated cover ply, a hat-shaped inner, and with a structure as shown in Fig. 2a. The tensor ply is made of special high-tensile polyethylene fiber that is injected with epoxy resin. The tensor ply was a new bullet-proof material, S-FDC88, which is a fourth-generation fiber. The inner and outer surface materials were carbon fiber prepreg cloth. Experimental data were obtained from tensile, compressive, and shear tests on these materials [12, 16]. A 620 mm × 500 mm tensor-skin sandwich panel was the focus of this study, with its size and cross-section shown in Fig. 2c.

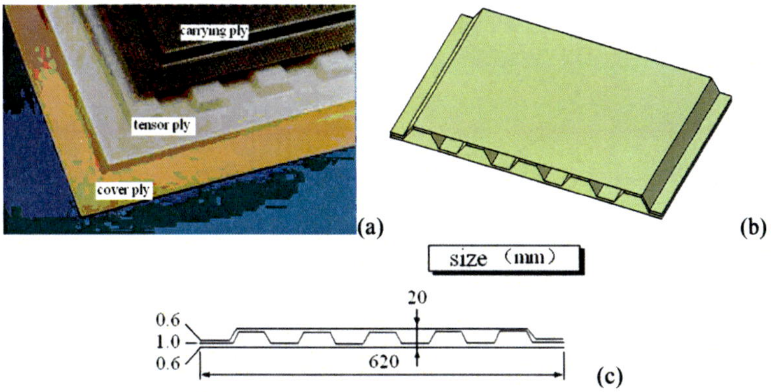

Fig. 2 Tensor-skin structure

2.2 Tensor Ply-Forming Process

For the tensor ply, an appropriate epoxy glue was selected from which to fabricate the new bullet-proof material, S-FDC88, into prepreg cloth for easy laying. A tensor ply was used as the core structure of the tensor skin. The material and structure of ply were unique. Each tensor ply consisted of four plies of new ballistic material, S-FDC88, and three filling sandwich material plies, and the structure forming process was as follows:

(1) The filling sandwich material consisted of three independent honeycombs. The honeycomb, as a buffer zone, could not be bonded with the S-FDC88 skin. According to prior techniques, the three honeycombs were cut according to the size on the drawings, and these were coated with a Teflon material. (2) The four plies prepreg cloth made from S-FDC88 was bound together according to the drawing requirements, and was cut according to the unfolded size of the tensor ply for use. (3) The S-FDC88 material that was cut based on the previous steps was preformed according to the shape of the filling sandwich material, and the sandwich material was coated (Fig. 3).

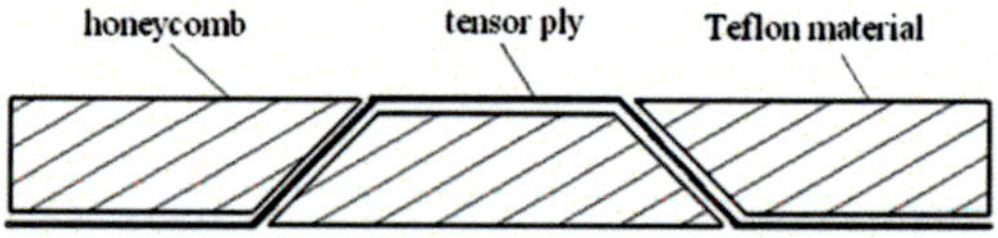

Fig. 3 Digital model for tensor skin forming

The outer edge of the 15-mm-thick pressure releasing honeycombs that exist on both sides of the tensor ply was a 45° chamfer. To increase the pressure resistance of the honeycomb edge, the length direction of the honeycomb should be perpendicular to the grid.

A tooling fabric was added at the honeycomb chamfering. To prevent honeycomb collapse during curing, LTM12 process glass cloth was used at the product honeycomb chamfering to paste the product edge profile (two to three plies under normal circumstances) on the demoulding film after completion of the product laying, with a staggered laying location as shown in Fig. 4.

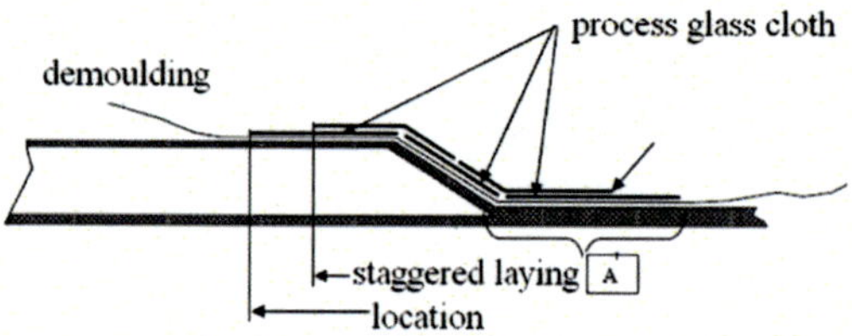

Fig. 4 Pressure-resistance process that is formed at the honeycomb edge

3 Analysis of Fuselage Box Section Impact

3.1 Finite-element Model

Isotropic shell elements were selected for the fuselage, because many theoretical models exist for such elements, such as the Belytschko–Tsai, Hughes–Liu, and Key–Hoff models. Large deformations, rotations, and elastic–plastic strains are allowed, and many failure criteria can be obtained. We used the Belytschko–Tsai shell-element theory by default. The Belytschko–Tsay, Hughes–Liu or Key–Hoff equations were used in the CQUAD4 element. Belytschko–Tsay is the most efficient in the shell equations and should be used in most cases. Key–Hoff is more time-consuming, but it performs better at large strains (>5%). When a portion of the structure is subjected to a large strain, the Key–Hoff shell should be considered for use in that area and the Belytschko–Tsay shell should be used elsewhere. Although more time-consuming than the two methods mentioned above, the Hughes–Liu shell displays advantages only when the element thickness is varied [17].

3.1.1 Fuselage Model

A finite-element model was established based on the composite corrugated beam sub-floor structure. The nonlinear transient dynamic program, MSC.Dytran, was used for the simulated analysis of a rigid surface impact, with the finite-element model as shown in Fig. 5. Four-node shell elements were used for the lower tensor skin, upper cover, and corrugated beam. Eight- and six-sided element simulations were used for the middle horizontal honeycomb, and a certain number of mass elements was set on the upper Cover's corrugated beam to simulate the mass of the occupant, seat, equipment, and structure above floor. The total element and sub-floor structure mass was 1031 kg. By considering the structure symmetry, half of the test article was used for finite-element analysis. The finite-element model consisted of 39872 nodes, 34940 four-node shell elements, and 4932 eight-node body elements. The contact relationship between a composite corrugated Beam's sub-floor structure and the rigid surface was simulated by a planar rigid wall. A DMATEP material model of the composite material corrugated beam was used, and the material property was of equivalent metal performance. The falling velocity complied with the anti-impact design and verification requirements of the *Military Helicopter Crash-Resistance Requirements* (GJB 2681-96) on the military helicopters. The design impact velocity variable quantity along the vertical direction was 7.9 m/s.

To prevent cover ply, tensor ply and carrying ply of the tensor skin from penetrating each other, a self-contact was set to the tensor-skin plate. To enable the transfer of a water-impact load between the support and the tensor ply of the tensor skin, a breakable contact setpoint was required. A self-contact was established between the corrugated beam and the lower skin to prevent the corrugated-beam penetration into the lower skin and failure to reach a predetermined effect.

3.1.2 Soil and Water Models

Because the soil mechanical properties are affected by parameters such as the Soil's composition, humidity, and depth, loose sandy soil was selected as an object of study in this paper. Because loose sandy soil has a smaller shear strength and no obvious rebound phenomenon in a drop test, a FOAM2 material model as provided by the simulation analysis program was selected. The model was used for isotropic compressible materials with a Poisson's ratio of almost zero. The unloading process followed a user-defined, nonlinear hysteresis-response stress–strain curve. The yield stress was defined to have a strain-rate dependence.

The Euler method was used to describe fluid flow and sputtering. The interaction between a liquid and a solid structure was simulated in combination with a general fluid–solid coupling method. A Eulerian element served as a fixed-reference system, and material could flow in a Eulerian element. During the analysis, the material mass, momentum, and energy could be transferred among Eulerian elements. Because the Euler method simulates material movement in Eulerian elements with a constant volume, the shape of the Eulerian element does not change with the material deformation at any time. Therefore, no elemental distortion occurs, and the model is suitable for simulating a fluid material that is prone to large deformation.

Water waves can rise above the water surface. To simulate the effect of splashing water, an air area was established above the water, with the length, width, and depth of the Euler air volume being 30, 30, and 3 fluid elements, respectively (total of 2700 elements). For the Euler water volume, the length, width, and depth were 30, 30, and 18 fluid elements, respectively (total of 16200 elements). The finite-element model is shown in Fig. 5c.

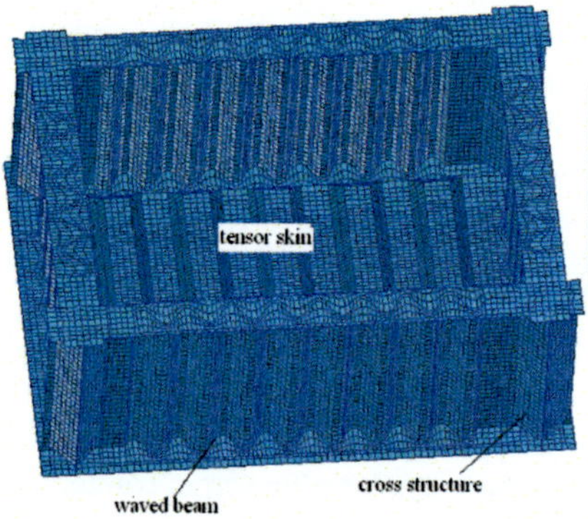

(a) Finite-element model of Fuselage box section

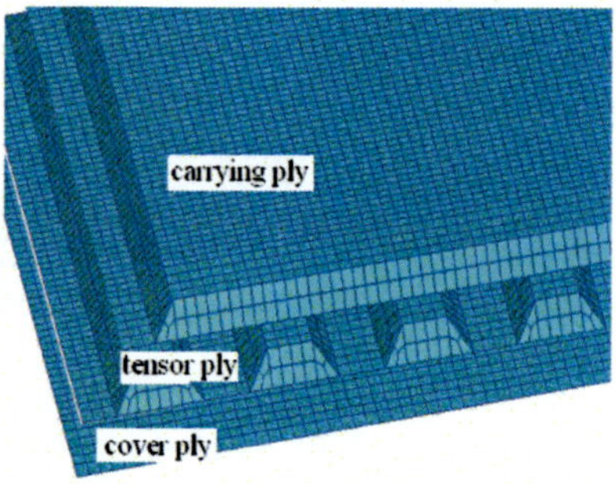

(b) Finite-element model of tensor skin

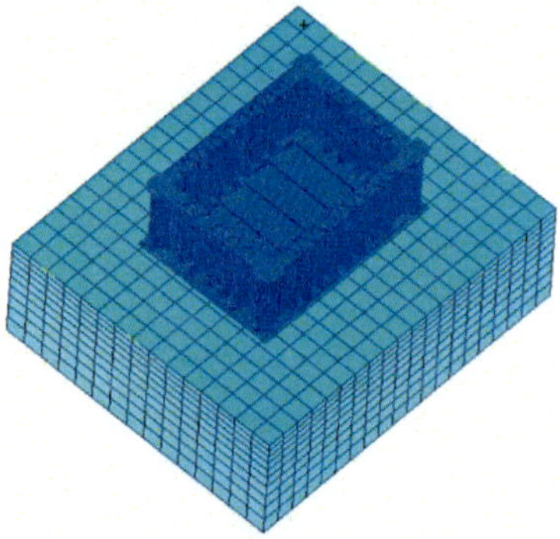

(c) Water-surface impact of fuselage box section

Fig. 5 Finite-element model

3.2 Material Properties

(1) Properties of corrugated beam material: density: 1.66×10^{-9} t/mm^3, elastic modulus: 14600 MPa, Poisson ratio: 0.229, yield strength: 180 MPa, maximum plastic strain: 0.1.
(2) Properties of honeycomb material: density: 4.8×10^{-11} t/mm^3, elastic modulus: 100 MPa, Poisson's ratio: 0.3, yield strength: 10 MPa, maximum plastic strain: 0.5.
(3) Properties of loaded material: density: 3.36×10^{-6} t/mm^3, elastic modulus: 71000 MPa, Poisson ratio: 0.3, yield strength: 325 MPa, and maximum plastic strain: 0.5.
(4) Material of upper cover: density: 2.77×10^{-9} t/mm^3, elastic modulus: 2×10^5 MPa, Poisson ratio: 0.3, yield strength: 600 MPa, and maximum plastic strain: 0.5.
(5) Materials of cover ply and carrying ply of tensor-skin plate: density: 1.37×10^{-9} t/mm^3, elastic modulus: 30300 MPa, Poisson ratio: 0.32, yield strength: 345 MPa, and maximum plastic strain: 0.02.
(6) Material of tensor-skin Plate's tensor ply: density: 1.4×10^{-10}t/mm^3, elastic modulus: 31000 MPa, Poisson ratio: 0.32, yield strength: 425 MPa, and maximum plastic strain: 0.3.
(7) Water properties: density: 1000 kg/m^3 = 1×10^{-9} t/mm^3, and bulk modulus: $K = \alpha 1 = 2.2 \times 10^9$ N/m^2 (Pa) = 2.2×10^3 MPa.
(8) Model parameters of soil material: density of 7.16947×10^{-11} t/mm^3, bulk modulus of 3.675 MPa, termination stress of -6.895×10^{-2} MPa, and energy dissipation factor of 0.99 [18, 19].

3.3 Simulation Results

Figure 6 shows the deformation of the box structure after hitting firm ground, soft soil, and the water surface. Figures 6a, b show that when the tensor-skin fuselage box structure impacts on firm ground and soft soil, the impact force acts a point instead of as a surface force. This behavior occurs because the firm ground and soft soil are not as pervasively as the water, and the tensor skin is less effective in energy absorption. The tensor skin does not function when it impacts on firm ground and soft soil. Figure 6c shows that the fuselage section and tensor skin suffer from significant deformation under the effect of water impact. The tensor skin is spread to a certain extent, but the tensor ply is not spread completely because of structural design factors, and therefore it only reflects some advantages of the tensor skin water-surface impact. Figure 7 shows the underwater deformation of the carrying ply, tensor ply and cover ply of the tensor skin. Under water impact, the cover ply is firstly damaged, and the fold parts of the tensor ply open to transfer load to the corrugated beam for energy absorption.

Figure 8a shows the contact force–displacement curve of the tensor-skin fuselage box structure in a soft-soil impact [2]. The trends are similar, which indicates that this analysis method is feasible. Figure 8b shows the accelerated velocity curve for the firm-ground, soft-soil, and water-surface impacts. The accelerated velocity peak for the water-surface impact is less than that for the soft-soil impact, and the accelerated velocity peak for the soft-soil impact is less than that for the firm-ground impact. Figure 8c shows the comparative changing curve of the deformation energy for parts of the fuselage box structure. The tensor skin absorbs a part of the energy, and most of the energy

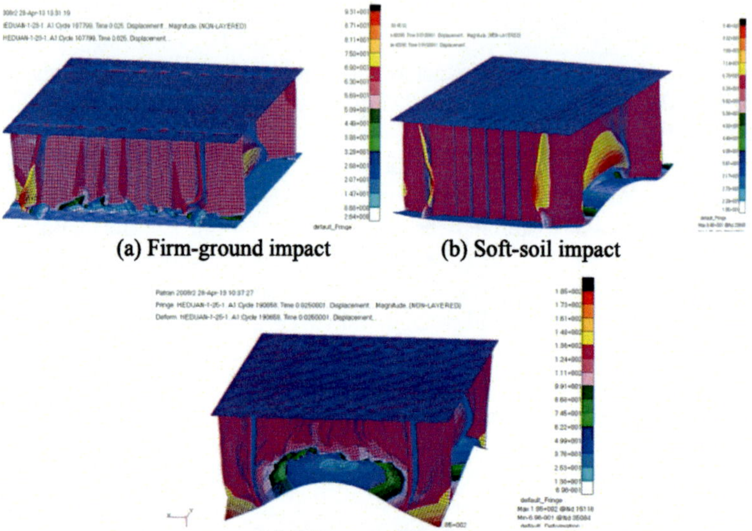

(a) Firm-ground impact (b) Soft-soil impact

(c) Water-surface impact

Fig. 6 Deformation of composite tensor-skin fuselage structure during the impact

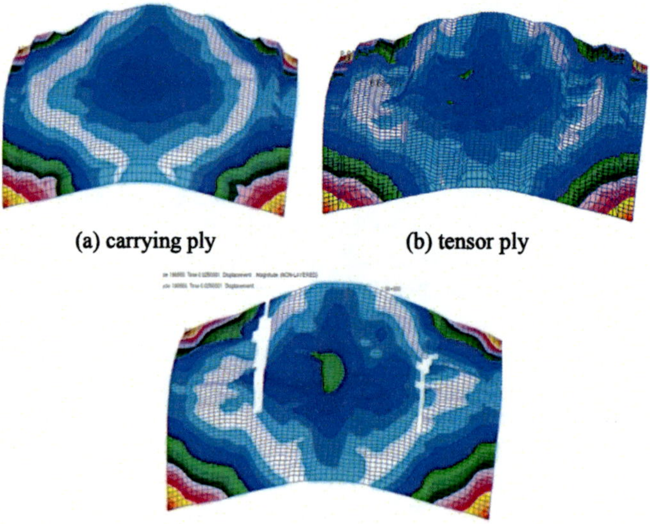

(a) carrying ply (b) tensor ply

(c) cover ply

Fig. 7 Deformation of tensor-skin plies

is transferred to the energy absorption structure of the surrounding corrugated beam through the deformation of tensor skin. Thus, the tensor skin exhibits greater safety advantages over traditional skin for a water-surface helicopter impact.

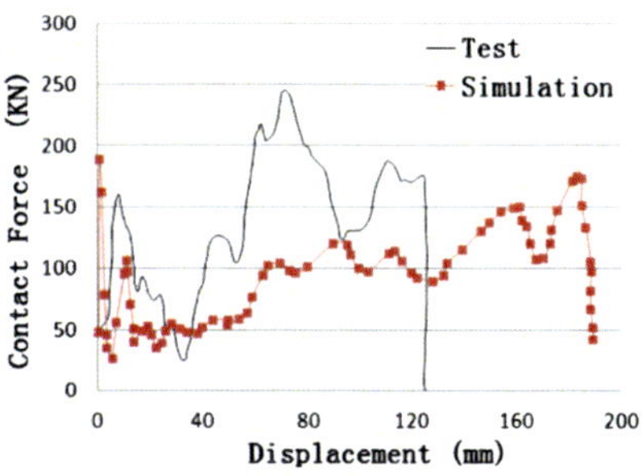

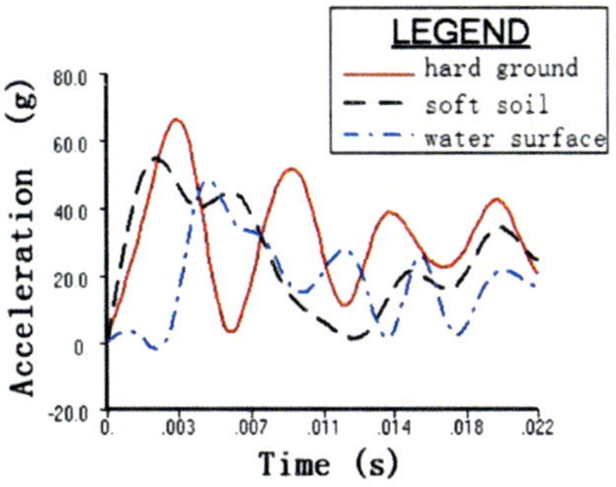

(a) Simulated test contrast of soil impact [2] (b) Curve of accelerated velocity over time

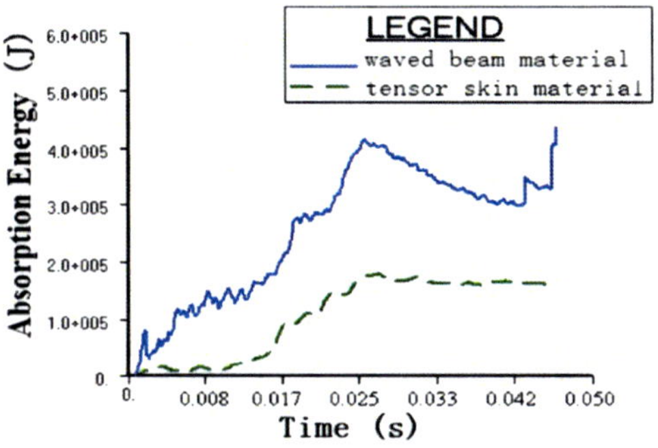

(c) Comparison of deformation energy for fuselage's parts

Fig. 8 Comparison of results from firm-ground, soft-soil, and water-surface impacts

4 Conclusions

The accelerated velocity peak for the water-surface impact is less than that for the firm-ground and soft-soil impacts.

A well-designed tensor skin can spread to absorb energy in a water-surface impact. This tensor skin cannot spread for soft-soil and firm-ground impacts, and so does not work, and the resultant effect is similar to that of pressure on a traditional sandwich structure.

According to the above analysis, the tensor skin is a special composite laminated plate structure, and its deformation mode can be flattened or fractured such as that of an ordinary laminated plate structure. If the structural design is reasonable, in a water crash, the tensor skin can achieve surface fractures and the tensor ply in the middle spreads under the action of hydraulic pressure. During the spread of a tensor ply, the large plastic deformation can absorb impact energy. In addition, because of the large deformation capacity of the tensor ply, the structural integrity can be maintained to prevent water from impacting other structures directly. However, whether the energy absorption of the tensor ply is effective is related to the structural design and stiffness, namely, we need to take into account issues that include the fuselage stiffness, floor stiffness, and tensor skin matching in the actual design process.

References

1. Bolukbasi, A.O.: Development of analysis methodology for crash impacts on soft soil. In: Presented at the 54th Annual Forum of the American Helicopter Society (1998)
2. Wiggenraad Ma Ma, J.F.M.: Numerical investigation of a crash test of a composite helicopter sub_floor structure. Compos. Struct. **51**(2001), 345–349 (2001)
3. Michielsen, A.L.P.J, W, J.F.M., Ubels, L.C., Frijns, R.H.W.M, Kohlgrüber, D., Labeas, G., Mccarthy, M.A.: Design, Test and Analysis of Tensor Skin Panels for Improved Crashworthiness in Case of Water Impact. National Aerospace Laboratory NLR (1998)
4. Wittlin, G., Smith, M., Sareen, A., Richard, M.: Airframe water impact analysis using a combined MSC/DYTRAN—DRI/KRASH approach. In: Proceedings of the AHS Forum 53, Virginia Beach (1997)
5. Brite-Euram project CAST: Crashworthiness of Helicopter on Water: Design of Structures Using Advanced Simulation Tools. G4RD-CT2000-00178 (2000)
6. Brite-Euram project CAST: Crashworthiness of Helicopter on Water: Design of Structures using Advanced Simulation Tools, G4RD-CT-2000-0178 (2003)
7. Brite-Euram project CRASURV: Commercial Aircraft—Design for Crash Survivability. CT96-0207 (1999)
8. Brite-Euram project CRAHVI: Crashworthiness of Aircraft for High Velocity Impact. G4RD-CT2000-00395 (2004)
9. Thuis, H.G.S.J., Wiggenraad, J.F.M.: A tensor-skin concept for crashworthiness of helicopters in case of water impact. NLR TP 94045 L (1994)
10. Labeas, G., Kermanidis, T.H.: Crushing behavior of the energy absorbing 'tensor skin' panels. Labor. Technol. Strength Mater. (2002)
11. Ubels, L.C., Wiggenraad, J.F.M.: Increasing the survivability of helicopter accidents over water'. LR-TP-2002-110, February (2002)
12. Ubels, L.C.: CRASURV – D.4.1.2, Behaviour of shear panels. NLR TR 97534 L (1997)
13. Kohlgrüber, D., Weissinger, H.: CRASURV–D.4.1.3, Results of dynamic tests of tensor skin panels. DLR-IB 435-97/31 (1997)
14. Lestari, W., Thuis, H.G.S.J., Wiggenraad, J.F.M.: Development of a trigger mechanism to reduce peak forces in crash loaded composite sine-wave spars'. NLR TP 94319 L (1994)
15. Wiggenraad, J.F.M., Michielsen, A.L.P.J., Santoro (Alenia), D., Lepage, F. (CEAT), Kindervater, C. (DLR), Beltran, F. (Principa).: Development of a crashworthy composite fuselage structure for a computer aircraft. NLR TP 99532 L (1999)
16. Kohlgrueber, D., Weissinger, H.D.: 4. 1. 3. Results of Dynamic Tests of 'Tensor Skin' Panels. DLR-IB 435±97/31 (1997)
17. Vigliott, A, Pentecote, N., Clifford, S.: Crashworthiness of Rotorcraft on Water, An Experimental Test Campaign: Bristol. UK: 28th European Rotorcraft Forum (2002)
18. Sareen, A.K., Fasanella, E., Sparks, C. etc.: Comparison of Hard Surface and Soft Soil Impact Performance of a Crashworthy Composite Fuselage Concept. UK: 28th European Rotorcraft Forum (2002)
19. Naghipour, P., Aktay, L., Johnson, A.: Numerical investigation of structural crash response of thin-walled structures on soft soil. Mater. Design **29**, 2052–2060 (2008)

ADS-B Anomaly Detection Algorithm Based on LSTM-ED and SVDD

Jiao Yi(✉), Lin Lin, Li Nisi, and Wang Jintao

CCAC Academy of Flight Technology and Safety, Civil Aviation Flight University of China, Guanghan 618300, Sichuan Province, China
jy18980203626@gmail.com

Abstract. With the rapid development of radio broadcasting technology and information technology, the possibility of ADS-B surveillance system suffering from radio frequency interference and network attack is increasing. Firstly, the types of attacks that ADS-B system is vulnerable to are analyzed, and the abnormal data of the corresponding state are simulated. Then, in order to detect ADS-B data attacks accurately, an anomaly detection algorithm based on Long Short-Term Memory Encoder–Decoder (LSTM-ED) and hypersphere classifier is proposed. Using the normal historical data to train the LSTM-ED model so that the reconstructed values can be obtained. Then, the difference values between the actual values and the reconstructed values are put into SVDD (support vector data description) for training, and a hypersphere classifier that can detect abnormal data is obtained. It is judged whether there is an abnormality according to the set threshold. Experiments show that the LSTM-ED-SVDD model can detect abnormal ADS-B data generated by various attacks and has high accuracy.

Keywords: ADS-B · Anomaly detection · LSTM-ED · SVDD classifier

1 Introduction

As one of the important components of air traffic monitoring, ADS-B system has been widely used around the world with the advantages of high monitoring accuracy, low deployment cost, wide coverage, and data sharing [1]. In order to reduce the system cost, the ADS-B system chooses to broadcast data information in plaintext, which leads to security communication problems such as information leakage and information tampering [2]. With the development of radio technology and the wide application of ADS-B system, the security drawbacks left in the early stage of system development are gradually exposed and amplified, becoming the biggest obstacle to the full application of ADS-B.

As of January 2020, the Federal Aviation Administration (FAA) has abandoned about 175 secondary surveillance radars. According to regulations, aircraft in specific types of airspace need to be forcibly equipped with ADS-B avionics to provide accurate information for air traffic management to replace primary and secondary radar systems. At the same time, China is also building a complete civil aviation ADS-B operation

Chinese Society of Aeronautics and Astronautics: CASTYSF 2022, LNEE 972, pp. 245–257, 2023.
https://doi.org/10.1007/978-981-19-7652-0_23

monitoring system and information service system. At present, most countries around the world have been equipped with ADS-B equipment. The comprehensive application of ADS-B system not only improves the performance of aviation surveillance, but also increases the risk of aviation safety.

McCallie et al. [3] analyzed ADS-B security vulnerabilities and their overall impact on transport aviation, classified ADS-B attack types according to the difficulty of implementation, and put forward suggestions for improving ADS-B security. Andrei [4] sent the modification information by a universal software radio peripheral (USRP) unit and completed the message reception by a commercial receiver. Magazu and Schäfer et al. [5, 6] demonstrated the generation and injection of fake ADS-B messages, and show information injection attacks in specific scenarios. The authors [7, 8] analyzed the flow of European air traffic on the 1090 MHz data link, and discussed the challenges and safety problems of ADS-B on the 1090es data link. The results show that in dense airspace, the information loss rate of more than 50% is also a serious security problem. Mohsen et al. [2] used Piccolo autopilot to run simulations on hardware to study the impact of ADS-B message injection attacks. The study showed that such attacks can significantly distract pilots and ground controllers, forcing them to execute emergency handling plans away from ghosts Aircraft, in the case of heavy traffic load, will endanger airspace safety.

All of the above studies have shown that ADS-B data can be easily corrupted by existing cheap hardware and software. An attacker can receive and modify ADS-B messages, delete legitimate ADS-B messages, and inject misleading ADS-B messages. Security flaws in ADS-B systems raise concerns about the direction and deployment of next-generation aviation. New security protocols take a long time and are expensive to roll out. Therefore, the research on ADS-B data anomaly detection algorithm has certain practical significance. Ensure the accuracy and validity of ADS-B data from the data analysis level, so as to assist controllers and flight personnel to make correct decisions and maintain the safety and order of air traffic.

2 Related Works

2.1 ADS-B Security Vulnerabilities

The rapid development of software radio provides potential attackers with a low-cost attack method. The attacked objects in the ADS-B system are mainly ground stations and aircraft. Attacks can be divided into passive and active. Passive attacks are mainly message eavesdropping; active attacks include tampering, injection, deletion and DOS attacks [9].

- Message eavesdropping: The message is not encrypted and broadcast on a fixed frequency. An attacker can regularly monitor the ADS-B message and obtain the location information of a flight without interrupting the system.
- Message tampering: The ADS-B system does not have an inquiry mechanism similar to the secondary radar, and neither the sender nor the receiver can authenticate. Attackers use stronger jamming signals to obscure the original signal, or modify aircraft position information and warning codes by bit flipping. Gradually steer the plane away from its original course without the controller and pilot noticing it.

- Message deletion: The message information is a radio frequency signal, and an attacker can attenuate and damage the message information by superimposing the reverse signal and the error signal.
- Message injection: Attackers use software to simulate and generate fake aircraft information in the same format, and use transceiver equipment to inject fake information into the air traffic control surveillance system and collision avoidance system, eventually creating ghost planes and disrupting the normal operation of air traffic control.
- DOS attacks: A denial of service attack refers to interfering with data link communications in an ADS-B system, preventing participants from receiving or sending messages. Attackers target ground stations or aircraft. For busy airspace or airports, ground radio devices are used to transmit messages to achieve message blocking, which will have a huge impact on the normal operation of air traffic control.

2.2 Anomalous Data Construction

The Opensky network is primarily an air traffic control data provider that collects, processes and stores flight data [10]. The data used for the experiments in this paper comes from the Openky dataset. The ADS-B data is basically normal data. The data that has suffered from abnormal attacks is rare and cannot be obtained. In order to verify the effectiveness of the anomaly detection model, the anomaly data is simulated and generated according to the ADS-B attack types. The test data set in this paper selects the data set of a certain flight for construction processing. For example, select the middle 100 pieces of data from the 1600th to 1900th pieces of a certain flight to simulate the following four types of anomalies, as illustrated in Fig. 1.

- Message injection attack: The route replacement method is used to simulate, and the flight data of another flight with a similar flight status is replaced with the original part of the track.
- Message tampering (location): Modify the original track data from the characteristics of longitude and latitude, and add Gaussian noise with a mean value of 0 and a variance of 0.1 to the flight longitude and latitude within a certain range, so that the attacked trajectory deviates from the actual trajectory.
- Message tampering (altitude): Gradually increase or decrease the flight height in a certain proportion within the flight range, so that the attacked flight trajectory deviates from the original flight trajectory in height.
- DOS attack: simulate message blocking and interfere with normal communication, choose to delete flight data within a certain period of time, and replace it with a value of 0, that is, the receiver cannot receive any messages.

2.3 Long Short-Term Memory Encoder–Decoder

Long Short Time Neural Network (LSTM) is a variant of RNN recurrent neural network [11]. It adds a memory component (forgetting gate, input gate, output gate) on the basis of RNN network to help the network transfer the information learned at T time to $T+n$

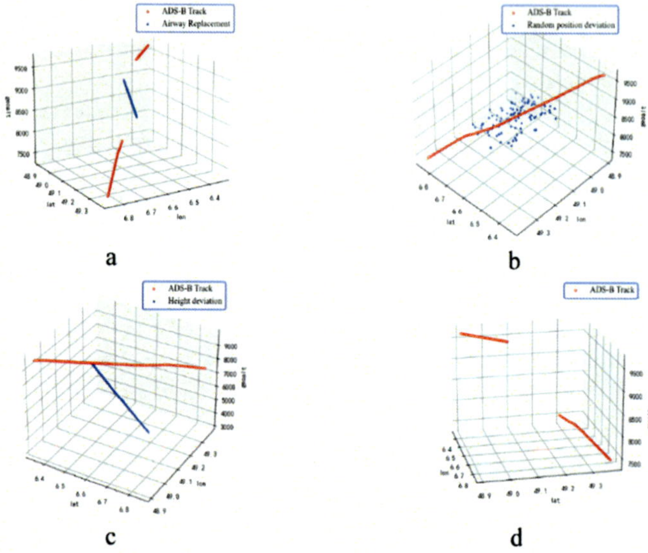

Fig. 1. Types of ADS-B data exceptions. **a** route replacement, **b** random position offset, **c** height offset, **d** DOS attack

time. In this way, the state can be selectively updated, and the important part of the state can be transmitted to the future stage, so that LSTM has the ability of long-term memory information, It works as follows:

$$f_t = \sigma_g\left(W_f x_t + U_f h_t + b_f\right) \tag{1}$$

$$i_t = \sigma_g(W_i x_t + U_i h_{t-1} + b_i) \tag{2}$$

$$o_t = \sigma_g(W_o x_t + U_o h_{t-1} + b_o) \tag{3}$$

$$c_t = f_t * c_{t-1} + i_t * \sigma_c(W_c x_t + U_c h_{t-1} + b_c) \tag{4}$$

$$h_t = o_t * \sigma_c(c_t) \tag{5}$$

x_t represents the LSTM cell input vector; f_t represents the activation vector of the forget gate; i_t represents the activation vector of the input gate; o_t represents the activation vector of the output gate; h_t represents the hidden state vector, which is also the output vector of the LSTM cell; c_t represents the cell state vector; W and U represents the weight matrix and Bias vector parameter, that is, what needs to be learned in the training process; σ_g represents the sigmoid function; σ_c represents the tanh function, the LSTM structure is shown in Fig. 2.

When using LSTM models for long sequence predictions, prediction errors accumulate over time. Order to solve this problem, this paper uses a model called Long

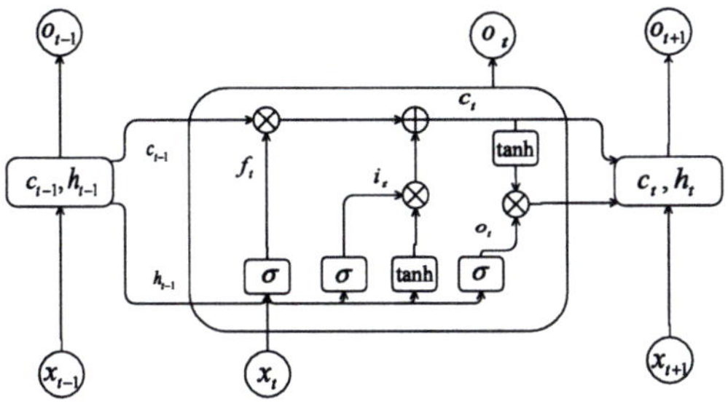

Fig. 2. LSTM network

Short-Term Memory Encoder-Decoder [12], whose structure is shown in Fig. 3. The encoder and decoder are two independent parts. Each part consists of one or more LSTM networks. In the model, the sequence is input to the encoder and outputs an internal state, which is used as the output of the decoder to predict the state at the current moment by the input of the previous.

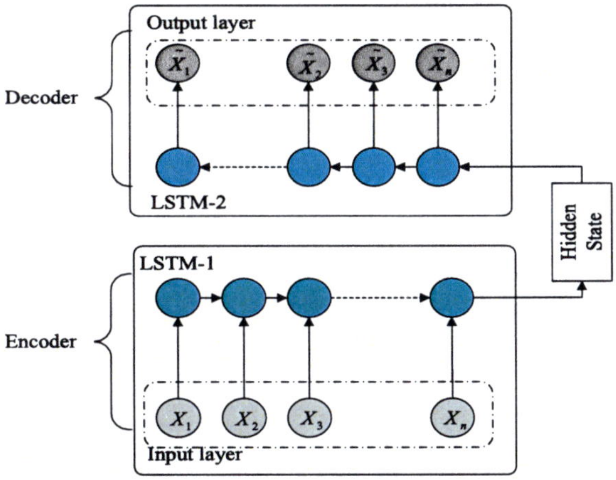

Fig. 3. Long short-term memory encoder–decoder

2.4 SVDD Classifier

The key to anomaly detection of ADS-B data lies in how to distinguish abnormal values from normal values. The most common method for classification is to judge according to the set threshold and the calculation of similarity, but the artificial threshold setting has certain limitations. The classifier obtains the classification boundary through the characteristics of the data itself, which has better adaptability as a threshold. The abnormal data detection only includes normal and abnormal data, which belongs to the binary classification problem.

SVDD (Support Vector Data Description) is an unsupervised machine learning method that can solve binary classification problem [13]. SVDD mainly maps the normal feature data $x \in R^{n \times d}$ (n represents the number of samples, d represents the feature dimension) into a high-dimensional feature space, and find a hypersphere that can contain the target samples. The hypersphere is minimized while including the target samples as much as possible, and the non-target samples are excluded from the hypersphere, as to achieve the purpose of distinguishing between normal and abnormal classes. As shown in Fig. 4, the hypersphere is used as the decision boundary, and the points on the boundary of the sphere are used as support vectors. When the distance between the test sample and the center of the sphere is less than the radius of the sphere, it is judged as a normal sample, and when the distance is greater than the radius of the sphere, it is judged as an abnormal sample.

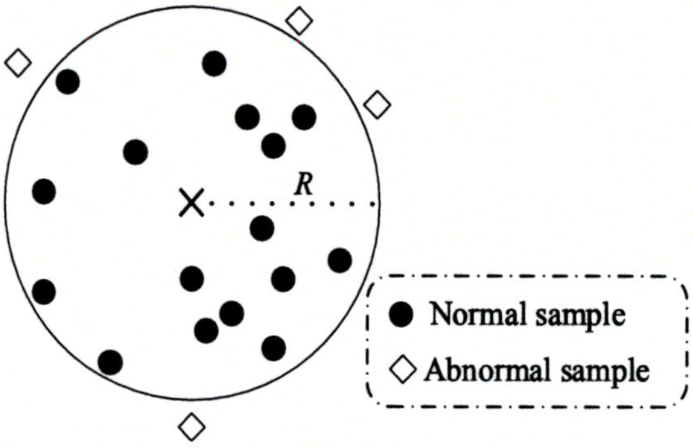

Fig. 4. SVDD hypersphere

Order to construct the smallest volume hypersphere, the optimization problem that SVDD needs to solve is shown in Eqs. 6 and 7:

$$\min_{a,R,\xi} R^2 + C\sum_{i=1}^{n} \xi_i \tag{6}$$

$$s.t.\|\phi(x_i) - a\|^2 \le R^2 + \xi_i, \xi_i \ge 0, \forall i = 1, 2, \ldots, n \tag{7}$$

R represents the radius of the hypersphere, *a* represents the center of the hypersphere, ξ Represents the relaxation factor, and *C* represents the penalty coefficient for weighing the hypersphere volume.

Combined with the Lagrange multiplier method, the original optimization problem is transformed into a Lagrange dual problem:

$$\min_{\alpha_i} \sum_{i=1}^{n} \sum_{j=1}^{n} \alpha_i \alpha_j K(x_i, x_j) - \sum_{i=1}^{n} \alpha_i K(x_i, x_i) \tag{8}$$

$$s.t. 0 \le \alpha_i \le C, \sum_{i=1}^{n} \alpha_i = 1 \tag{9}$$

α_i represents the Lagrangian coefficient corresponding to the sample, $K(x_i, x_j) = \langle \phi(x_i), \phi(x_j) \rangle$ represents the kernel function used to map samples into a high-dimensional space. Solve to obtain the Lagrangian coefficient α_i corresponding to all samples, and the samples that satisfy the conditions in Eq. 9 are the support vector x_v.

The center a and radius R of the hypersphere are derived from the sample set S composed of support vectors:

$$a = \sum_{i=1}^{n} \alpha_i \phi(x_i) \tag{10}$$

$$R = \sqrt{K(x_v, x_v) - 2\sum_{i=1}^{n} \alpha_i K(x_v, x_i) + \sum_{i=1}^{n} \sum_{i=1}^{n} \alpha_i \alpha_j K(x_i, x_j)} \tag{11}$$

3 ADS-B Abnormal Data Detection Algorithm Based on LSTM-ED-SVDD

3.1 Model Building

The ADS-B anomaly detection model based on LSTM-ED-SVDD is mainly divided into two parts: time series data reconstruction and reconstruction error classification. Using the sliding window mechanism, the data of the next moment is constructed from the data of the previous moments. The difference between the reconstructed data and the original data is used as the sample of the classifier, and finally it is determined whether the reconstructed data is an outlier according to the trained classifier, The general framework of LSTM-ED-SVDD model is shown in Fig. 5.

3.2 Model Training

The data used for the training of the anomaly detection model is the flight data of 10 flights belonging to the same route. The flight time of each flight is about one hour, the amount of data is 3000 ~ 4000, and the division ratio of training set and test set is 7:3. The training set is used to train the anomaly detection model, and the divided test set data is also normal flight data, so it cannot be used for testing directly, and anomaly data simulation is required.

According to the analysis of ADS-B message attack type and flight safety impact factor, five basic features are screened out. In this paper, ADS-B feature data at each time is used as a feature vector includes longitude, latitude, ground velocity, vertical velocity and heading:

$$M_{t_i} = [lon, lat, geo\_alt, vel, ver\_rate, true\_track]^T$$

All feature vectors form a feature sequence:

$$M = \{M_{t_1}, M_{t_2}, \cdots M_{t_i}\}$$

In this model, min-max normalization is used to limit each feature data to [0, 1]. The normalized eigenvector M'_{t_i} is shown in Eq. 12.

$$M'_{t_i} = \frac{M_{t_i} - \min(M_{t_i})}{\max(M_{t_i}) - \min(M_{t_i})} \tag{12}$$

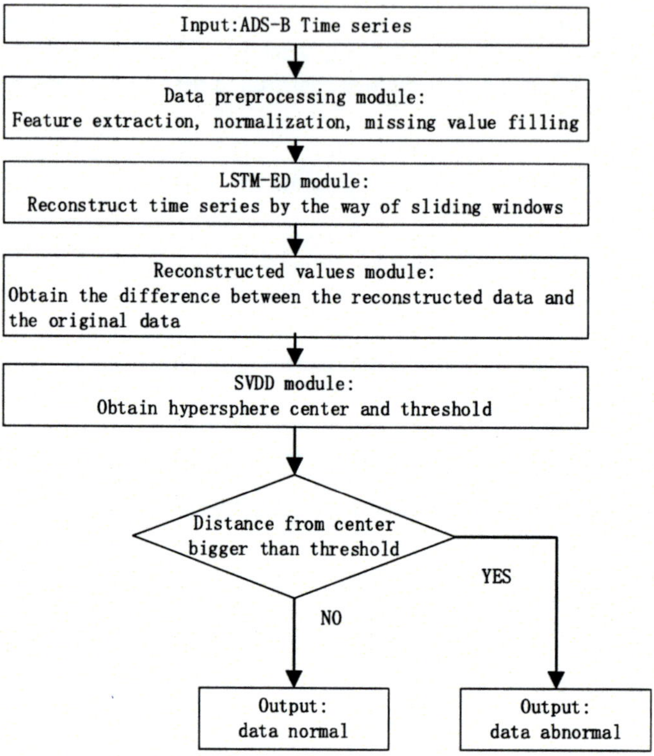

Fig. 5. Model framework

After the training data is preprocessed, the encoder in the LSTM-ED model learns the change trend in the normal historical data. Sliding window mechanism in the time series learning is the key to reconstructing and amplifying the abnormal data. This paper adopts the data sliding window mechanism. The window uses two sliding endpoints to identify the time series of specified length, as shown in Fig. 6. When new data is added in the window, the old data is deleted, that is, the data within a range is used to predict the next long data. According to the experimental calculation of the accuracy of reconstructed data with different window sizes, the window length is gradually increased from the window length of 2, and finally the length of the sliding window is determined to be 8, and the sliding step is set to 1.

The number of training epoches of the LSTM-ED network is 50, and the training batch size is 20. The change of the loss function value of the model during the training process is shown in Fig. 7. With the continuous increase of the number of training iterations, the value of the loss function gradually tends to be stable. The exact value of the training set can reach 0.928.

The training samples D used for SVDD classification are the difference sequence between the data reconstructed by LTSM-ED and the original data.

$$D = \{D_{t_1}, D_{t_2}, \cdots, D_{t_i}\},$$

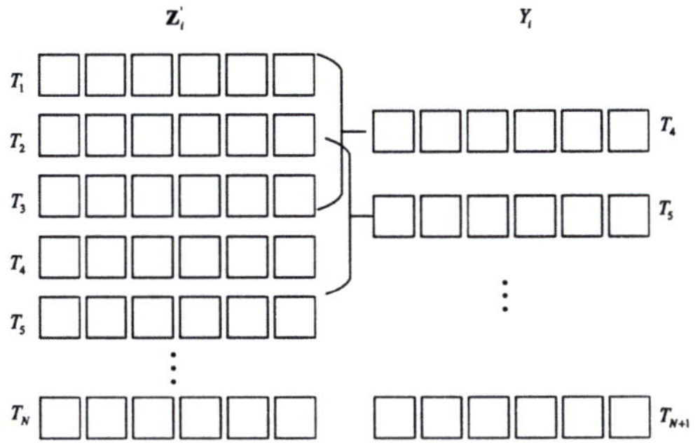

Fig. 6. Sliding window

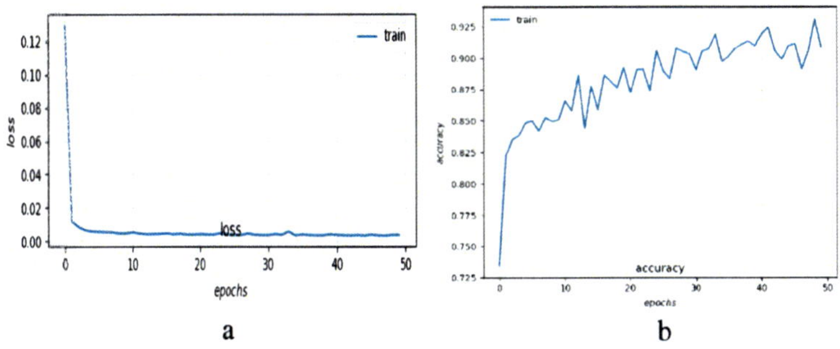

Fig. 7. Model training. **a** The change of loss value, **b** The change of accuracy

The sample data constructed by the difference sequence does not contain labels, and the samples need to be labeled before entering the classifier training. The samples used for training are all normal data, so the label needs to be added as 1. The classifier uses the Gaussian kernel function for model training. The model training mainly includes two parameters: The penalty coefficient C represents the tolerance to error. If its value is too large or too small, it will lead to the deterioration of the generalization ability of the model; The Gaussian kernel parameter gamma determines the distribution of the original data after mapping to the new feature space. The larger the gamma value, the fewer the support vectors, and the smaller the value, the more the support vectors. The number of support vectors will affect the speed of model training and prediction. According to the parameters of the training model, the grid search method is used to determine, that is, the punishment coefficient and Gaussian kernel parameters in a given range are arranged and combined, and all possible combinations generate a "grid".The accuracy of the model training corresponding to parameters shown as Table 1.

The Gaussian kernel function maps the difference samples to a high-dimensional space to find a hypersphere. The grid search method obtains the penalty coefficient of the optimal model $C = 0.15$, the Gaussian kernel parameter $g = 10$, and the radius of the hypersphere obtained by training is 0.221.

Table 1. Accuracy of model training with different parameters under Gaussian Radial Basis Kernel function

Penalty coefficient C	Gaussian kernel parameter g	Model training accuracy
0.15	1	78.4%(±0.525)
0.15	1	80.4%(±0.477)
0.15	0.5	95.4%(±0.142)
1	0.5	69%(±0.570)
1	0.1	88.6%(±0.513)
4	1	77.1%(±0.564)
4	0.2	87%(±0.449)
10	0.15	99.7%(±0.006)
10	1	91.7%(±0.214)
10	10	55.2%(±0.569)

3.3 Model Testing

In practice, it is difficult to obtain abnormal data after ADS-B is attacked. The test set used for model checking is the abnormal data constructed by simulation. Among the 300 pieces of flight data, the 100th to 200th pieces are abnormal data, and the rest are normal data. As a binary classification problem, the actual ADS-B data will be divided into normal classes or abnormal classes, and the data of the test set after passing through the anomaly detection model will be divided into four cases: The actual normal ADS-B data is accurately classified and predicted to be normal TP, the actual normal ADS-B data is misclassified and predicted to be abnormal FP, the actual abnormal ADS-B data is misclassified and predicted to be normal FN, and the actual abnormal ADS-B data is correct Classification predicted as abnormal TN.

- True positive rate TPR: It indicates the proportion of normal data that the classifier correctly attributes to all actual normal data.
- False positive rate FPR: indicates the proportion of abnormal data wrongly classified as normal class by the classifier to all abnormal data.
- Accuracy: Indicates the probability that the prediction is correct in the entire test data set.
- Precision: Indicates the ratio of actual normal data to predicted normal data.
- Recall: Represents the probability that the actual normal data is predicted correctly.

The anomaly detection judgment in this model is to calculate whether the distance from the feature data at each moment to the center of the hypersphere is greater than the trained sphere radius 0.221. The detection result of each type of attack consists of two parts, where the left graph represents the distance from each sample in the test set to the center of the sphere, where the red line represents the classification threshold (the radius of the hypersphere for training), and the points above the threshold line are abnormal samples, the normal samples at points below the threshold line. On the right is a receiver operating characteristic (ROC) curve formed by true positive rate TPR and false positive rate FPR. The area AUC under the curve represents the data probability of true normal class or true abnormal class predicted in the anomaly detection model. The closer the AUC of the model is to 1, the higher the accuracy of the model prediction.

1. Random position offset: The attack detection result is shown in Fig. 8. Among the 100 simulated abnormal data, 89 were correctly predicted as abnormal, and 195 of the 200 normal data were predicted as normal. The accuracy of the model test was 94.67%, the recall rate of the model was 97.5%, and the accuracy rate of the model was 94.67%.
2. High attack: The attack detection results are shown in Fig. 9. 96 of the 100 abnormal data in the middle are correctly predicted as abnormal, and 196 of the 200 normal data are predicted as normal. The accuracy of the model test is 97.33%, the recall rate of the model is 98%, and the accuracy rate of the model is 98%.
3. Route replacement: The attack detection results are shown in Fig. 10. 94 of the 100 abnormal data in the middle are correctly predicted to be abnormal, and 198 of the 200 normal data are predicted to be normal. The accuracy of the model test is 97.33%, and the recall rate of the model is Recall. is 99%, and the model accuracy is 97.0588%.
4. DOS attack: The attack detection results are shown in Fig. 11. 92 of the 100 abnormal data in the middle are correctly predicted as abnormal, and 198 of the 200 normal data are predicted as normal. The accuracy of the model test is 96.667%, the recall rate of the model is 99%, and the accuracy rate of the model is 96.165%.

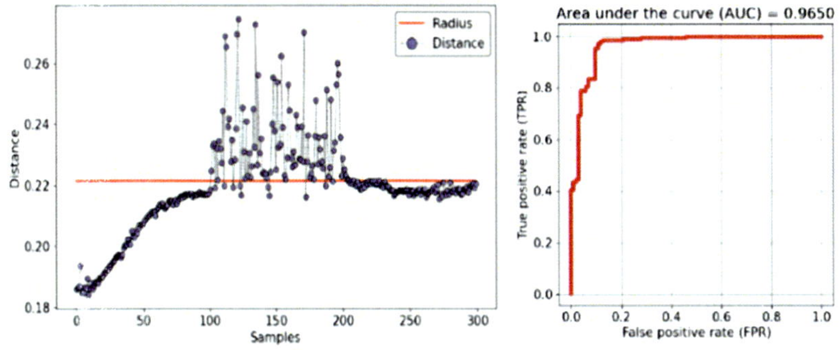

Fig. 8. Random offset detection result

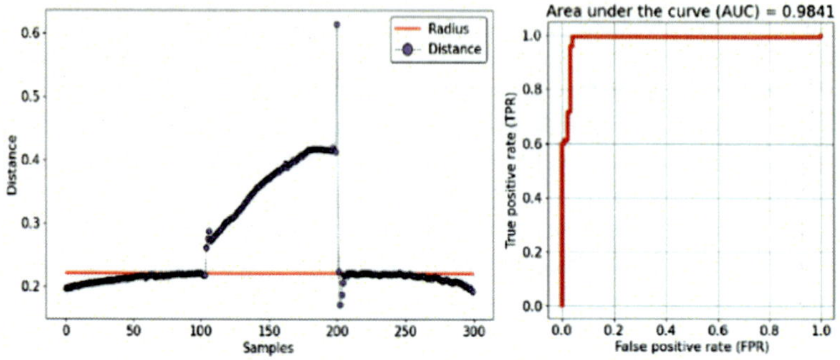

Fig. 9. Height attack detection result

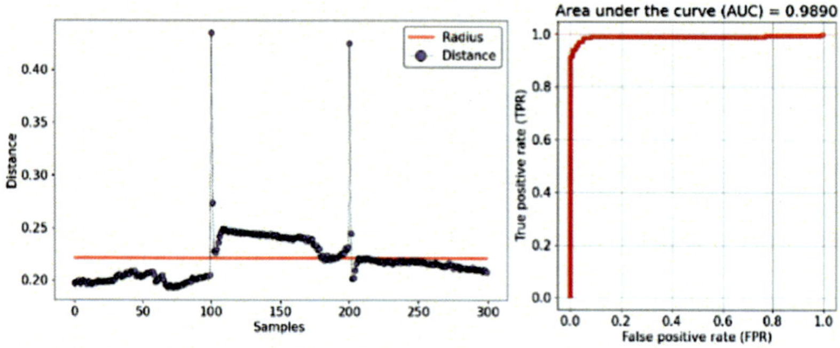

Fig. 10. Route replacement test result

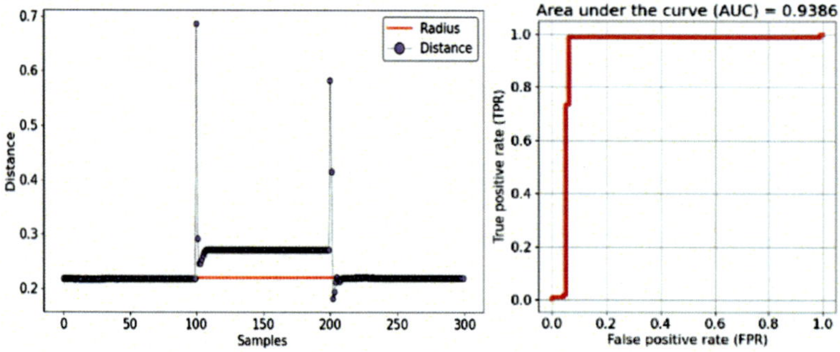

Fig. 11. Dos attack test result

4 Conclusions

This paper proposes an ADS-B anomaly data detection algorithm that combines the LSTM-ED model and the SVDD classifier. First, LSTM-ED is used to reconstruct the ADS-B data. Then, SVDD is used to solve the adaptive problem of threshold. The performance of the LSTM-ED-SVDD model is verified by simulating four common attacks.

In the future, we will continue our work in the following three aspects. First, we plan to use the radio software to simulate the abnormal data after the attack, which is more conducive to the verification of the accuracy of the model. Second, the model in this paper is anomaly detection for fixed routes, which lacks certain generality in practical engineering applications. Third, it is still in the analysis and research stage of offline data. With the development of deep learning and explainable artificial technology, the problems of optimal detection model and real-time detection remain to be solved.

Funding. Innovation and Entrepreneurship Training Program for Chinese College Students, Grant Number: S202110624266, Science and technology planning project of Sichuan province, Grant Number: 2021YFG0180.

References

1. Strohmeier, M., Lenders, V., Martinovic, I.: Security of ADS−B: State of the Art and Beyond. DCS (2016)
2. Mccallie, D., Butts, J., Mills. R.: Security analysis of the ADS-B implementation in the next generation air transportation system. Int. J. Critical Infrastruct. Protect. **4**(2), 78–87 (2011)
3. McCallie, D., Butts, J., Mills, R.: Security analysis of the ADS-B implementation in the next generation air transportation system. Int. J. Crit. Infrastruct. Prot. **4**(2), 78–87 (2011)
4. Costin, A., Francillon, A.: Ghost in the air (Traffic): On insecurity of ADS-B protocol and practical attacks on ADS-B devices. Black hat USA, pp. 1–12 (2012)
5. Magazu III, D.: Exploiting the automatic dependent surveillance-broadcast system via false target injection (2012)
6. Schäfer, M., Lenders, V., Martinovic, I.: Experimental analysis of attacks on next generation air traffic communication. In: International Conference on Applied Cryptography and Network Security. Springer, Berlin, Heidelberg, pp. 253–271 (2013)
7. Strohmeier, M., Schäfer, M., Lenders, V., et al.: Realities and challenges of next gen air traffic management: the case of ADS-B. IEEE Commun. Mag. **52**(5), 111–118 (2014)
8. Strohmeier, M., Lenders, V., Martinovic, I.: On the security of the automatic dependent surveillance-broadcast protocol. IEEE Commun. Surv. Tutor. **17**(2), 1066–1087 (2014)
9. Li, T., Wang, B.: Sequential collaborative detection strategy on ADS-B data attack. Int. J. Critical Infrastruct. Protect. (2018)
10. Strohmeier, M., Schafer, M., Fuchs, M., et al.: OpenSky: a Swiss army knife for air traffic security research. In: 2015 IEEE/AIAA 34th Digital Avionics Systems Conference (DASC). IEEE (2015)
11. Yu, Y., Si, X., Hu, C., Zhang, J.: A review of recurrent neural networks: LSTM cells and network architectures. Neural Comput. **31**(7), 1235–1270 (2019)
12. Habler, E., Shabta,i A.: Using LSTM encoder-decoder algorithm for detecting anomalous ADS-B messages. Comput. Secur. **78**(sep.), 155–173 (2017)
13. Tax, D.M.J., Duin, R.P.W.: Support vector data description. Mach. Learn. **54**(1), 45–66 (2004). https://doi.org/10.1023/B:Mach.0000008084.60811.49

Influence of the Nail Hole Adhesive Layer on Mechanical Properties of Bonded-Bolted Hybrid Joint

Dang Xiaogang[1](✉), Xu Jing[1], Wang Xi[1], Luo Bin[2], and Zhao Di[2]

[1] Chengdu Aircraft Industrial Co. Ltd, Chengdu, China
330344839@qq.com
[2] Northwestern Polytechnical University, Xi'an, China

Abstract. The bonded-bolted hybrid joint with the nail hole glue layer is widely used in the overall fuel tank structure of the UAV wing due to its both bearing and sealing capabilities. During the service of the aircraft, the bonded-bolted hybrid joint is prone to breakage and debonding of the adhesive layer, which causes problems such as load-bearing and leakage. Based on the design of the thickness of the nail hole adhesive layer, this paper analyzes the variation law of the tensile load of the ordinary bolt and the high-lock bolt hybrid joint with the thickness of the adhesive layer, and uses the scanning electron microscope to observe the failure mode of the nail hole adhesive layer. The results show that for ordinary bolts, the tensile load is optimal when the thickness of the nail hole adhesive layer is 0.03 mm, while for high-lock bolts, the tensile load increases with the increase of the adhesive layer thickness in the range of 0–0.05 mm adhesive layer thickness. During the tensile failure of the joint, the cohesive failure of the nail hole adhesive layer and the debonding failure of the composite material-adhesive layer interface are the main factors, and the bonding performance of the titanium alloy and the adhesive layer interface is better.

Keywords: Nail hole adhesive layer · Bonded-bolted hybrid connection · Tensile load · Adhesive layer failure

1 Introduction

In modern aircraft assembly, integral fuel tanks in the fuselage and wing structure are used to store fuel. Compared with the original rubber fuel tank, it has the characteristics of more fuel capacity and lighter weight, which can effectively improve the performance of the aircraft and reduce the weight of the aircraft. The sealing performance of the overall fuel tank is the key to ensuring the normal operation of the fuel system, which is related to flight safety and service reliability. The application of composite materials in aviation main load-bearing structures is becoming more and more mature. The new generation of integral fuel tanks are all composite materials or composite/metal hybrid structures. The structure is usually connected by bolts, and polysulfide sealant is filled at the joint surface of the fastener to form an adhesive layer, so as to ensure the sealing

Chinese Society of Aeronautics and Astronautics: CASTYSF 2022, LNEE 972, pp. 258–269, 2023.
https://doi.org/10.1007/978-981-19-7652-0_24

performance of the structure [1]. The research on the glue-screw hybrid connection structure in aircraft assembly originated from Hart-Smith [2] to improve the damage tolerance [3, 4]. Tan [5] and Steward [6] conducted further studies on this hybrid linking structure. The mixed connection form of glue and screw can improve the bearing capacity of the joint to a certain extent [7]. Chowdhury [8] studied the mechanical properties of the bolted connection in the glued joint, and the bolted connection in the glue-screw hybrid connection structure can reduce the peeling stress of the glue layer.

For the bonded-bolted hybrid joint, there are many related articles on the influence of the thickness of the adhesive layer between the laminates on the tensile strength of the structure [9]. Huang Wenjun [10] et al. by establishing a three-dimensional finite element model of the damage accumulation of the bonded-bolted hybrid connection, it is concluded that the thickness of the interlayer glue layer of the composite material has no effect on the structural strength and damage form. And Bodjona et al. [11–13] study showed that when the thickness of the interlayer adhesive is less than 0.4 mm, the maximum load transfer amount of the adhesive layer decreases with the increase of the thickness of the adhesive layer. Compared with the thickness of the interlayer adhesive layer, the impact of the nail hole clearance on the load sharing method is equally important, but the article does not specifically study it. Yuan Hui [14] also qualitatively described the relationship between the joint bearing capacity and the thickness of the adhesive layer in the article, and the analytical calculation results are in good agreement with the test results. To sum up, researchers pay more attention to the adhesive layer between the layers, but there are few studies on the bonded-bolted hybrid joint coated with glue layer between the nail holes. The influence of nail hole clearance (the thickness of the nail hole adhesive layer) on the bearing capacity of the structure needs to be further studied.

In this paper, the nail hole clearance is used as the test variable to design the tensile test of two kinds of bonded-bolted hybrid joints of ordinary bolts and high-lock bolts, and to analyze the influence of the thickness of the nail hole glue layer on the tensile strength of the structure. The distribution of the adhesive layer in the damaged area was measured by scanning electron microscopy to explore the damage mechanism of the adhesive layer.

2 Experimental Design and Preparation

2.1 Experimental Design

The bonded-bolted hybrid connection bonded-bolted method used in this test is inconsistent with the traditional bonded-bolted hybrid connection bonded-bolted method. In order to ensure the fuel sealing requirements, the test piece in this paper, in addition to the need for an adhesive layer between the composite laminates, the gap between the nail holes also needs to be filled with a sealant layer. In this paper, the purpose of controlling the thickness of the adhesive layer is achieved by controlling the size of the gap between the nail holes.

In this paper, the influence of the thickness of the adhesive layer of the nail hole on the tensile strength of the structure under the connection mode of the ordinary bolt and the pallet nut and the connection mode of the high-lock bolt is respectively studied.

The ordinary bolts and high-lock bolts selected in the test are standard bolts used in the production site of Chengfei Company.

The nominal diameter of ordinary bolts is 5.0 mm, and the diameter of the bolt hole is set to $d_b = 5.00$ mm, 5.06 mm and 5.10 mm. The nominal diameter of the high-lock bolt is 4.76 mm, and its bolt hole diameters are set to $d_h = 4.76$ mm, 4.82 mm and 4.86 mm.

Design experiments for each of three different connection apertures of two types of bolted connections are shown in Table 1.

Table 1. The specific planning of the test piece for the two connection methods.

No	Aperture size (mm)	Bolt size (mm)	Gap amount (mm)	Connection method
A1XG (1–5)	4.76	4.76	0	High lock bolt
A1X03G (1–5)	4.82	4.76	0.06	High lock bolt
A1X05G (1–5)	4.86	4.76	0.10	High lock bolt
A1XL (1–5)	5.00	5.00	0	Pallet nut
A1X03L (1–5)	5.06	5.00	0.06	Pallet nut
A1X05L (1–5)	5.10	5.00	0.10	Pallet nut

2.2 Preparation of Test Pieces

According to the tensile strength test standard of composite materials, a single-lap bonded-bolted hybrid connection bonded-bolted test piece of composite material is designed. The specific size of the test piece and the overlapping method are shown in Fig. 1. The layering method of the composite test piece is as follows: (0°, ±45°, 90°), and the fiber volume fraction ratio is 37%.

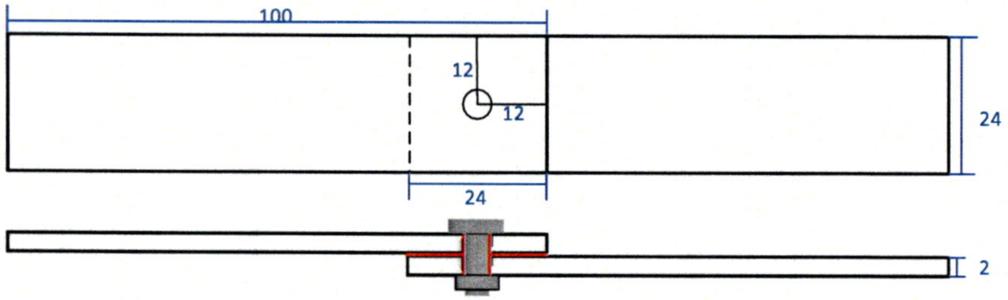

Fig. 1. Bonded-bolted hybrid joint test piece

The connecting holes of the test pieces were prepared by manual air drilling method. After roughing, reaming, reaming, deburring and other processes, the hole diameter that meets the tolerance requirements is obtained. The final nail hole clearance ranges are 0 mm–0.04 mm, 0.05 mm–0.09 mm, 0.10 mm–0.14 mm, which correspond to the aperture sizes set in the test respectively.

2.3 The Connection Process of the Test Piece

The glue coating process consists of two parts, the adhesive layer between the laminates and the adhesive layer between the nail holes. First, use alcohol to clean the overlapping area of the composite test piece, and wipe it with a clean cloth. Then apply a layer of NJD-3 tackifier on the lap area of the composite material and the wall of the bolt hole. The tackifier is pink and needs to wait for 30 min to improve the surface bonding performance of the composite material. Next, apply the well-stirred four-component XM-22 polysulfide sealant evenly on the overlap area with a tool, use a fixture to control the thickness of the adhesive layer to 0.1 mm, and then stick another composite material test piece in the overlap area to the adhesive layer, to ensure that the bolt holes are in the same position. Next, for the high-lock bolts, coat a layer of sealant with a thickness higher than the gap between the nail holes on the bolt rod. The length of the adhesive layer is greater than 4 mm, slowly insert the bolts into the bolt holes, and tighten the nuts until they break. For ordinary bolts with pallet nuts, it is necessary to first prepare positioning holes on the composite material test piece where the pallet nut is installed, and use small countersunk head bolts to fix the pallet nut on the composite material test piece. Next, apply sealant on the polished rod of the bolt, and slowly tighten the bolt into the pallet nut to complete the ordinary bolt installation process.

The connected specimen needs to be placed in a room temperature range with good openness and ventilation. A certain extrusion force is applied to the specimen with a clip, and it is placed at room temperature for 7 days until the sealant is completely cured. The cured sealant is in an elastic state, and the Shore hardness can reach 40. Use tools such as blades to remove the excess adhesive layer of the structure to make the surface of the connector smooth and flat.

3 Experimental Procedure

According to the ASTM6591 composite tensile test standard, the composite single lap tensile test was carried out on the SENS CTM500 tensile testing machine. The test process is shown in the figure. To counteract the secondary bending effect, a composite gasket of the same thickness was fixed at the clamping end of the specimen. The tensile speed of the testing machine is set to 2 mm/min. After the tensile load drops by 30% from the peak value, the loading stops and the load-displacement curve is output. Five replicates were set up for each experiment (Fig. 2).

4 Test Results and Analysis

In this section, the tensile load-displacement curves of ordinary bolts and high-lock bolts are analyzed respectively, and the tensile load and failure mechanism of the bonded-bolted hybrid joint with nail hole glue layer are revealed. Secondly, the failure mode of

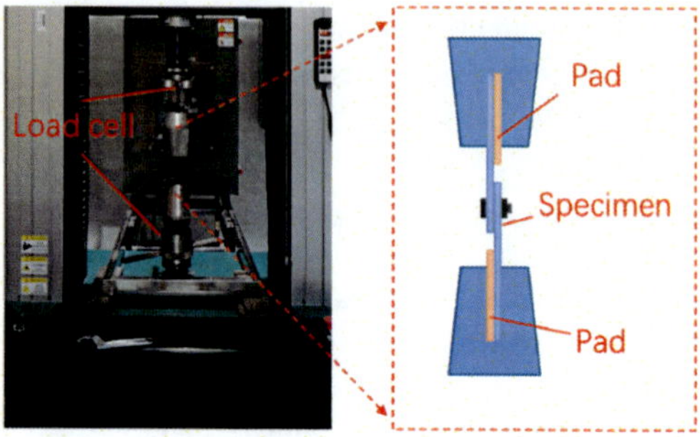

Fig. 2. Test loading process and schematic diagram.

the nail hole adhesive layer was analyzed through the electron microscope images of the two bolts after tensile failure, and the failure mechanism of the bonded-bolted hybrid connection bonded-bolted adhesive layer was explored.

4.1 Ordinary Bolts

4.1.1 Load-displacement Curve Analysis

The load-displacement curves of ordinary bolted hybrid joints with different diameters are shown in Fig. 3. It can be seen from the figure that the tensile process of the bonded-bolted hybrid connection hybrid joint can be divided into three stages, and the failure process of the joint with nail hole glue layer and the joint without nail hole glue layer is slightly different.

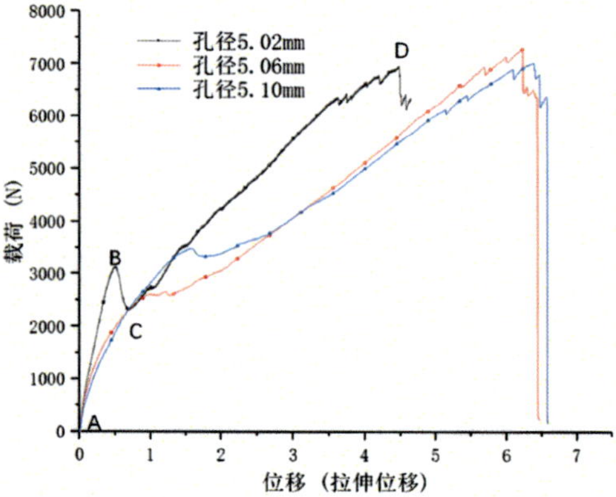

Fig. 3. Load-displacement curves of unloaded specimens under different apertures

A typical curve with a pore diameter of 5.02 mm was selected for analysis first.

(1) The co-bearing stage of the interlayer adhesive layer and the bolt. At the beginning of the initial stretching, the interlaminar adhesive layer is in direct contact with the composite material test piece and begins to bear the load first. At this time, the adhesive layer is the main bearing structure. After the tensile displacement reaches 0.01 mm, the bolts are in contact with the hole wall, and the bolts begin to bear the load. Under the combined action of the interlayer adhesive layer and the bolts, the structural stiffness is superimposed and the tensile load rises rapidly.

(2) Adhesive layer failure stage. After the tensile load reaches point B, the interlayer adhesive layer fails, and the bearing capacity of the joint decreases. At the same time, due to the small proportion of bolt bearing capacity, the tensile load begins to decrease rapidly until the bearing capacity of the bolt is consistent.

(3) Bolt bearing stage. After point C, the bolt load begins to dominate, and the tensile load increases with the bolt and composite specimen load capacity. With the further increase of the tensile displacement, the bearing capacity of the bolt gradually increases, as shown in the CD segment in the figure. When approaching the maximum load, there is a certain fluctuation in the tensile load due to fiber breakage of the composite specimen and a slight noise can be heard during the tensile process.

(4) The complete failure stage of the joint. After point D, due to the fracture and crushing of the composite specimen, the joint fails completely and the tensile load drops sharply.

The entire load-displacement curve has two peaks, The first time is when the bearing capacity of the adhesive layer structure reaches the maximum value, and the second time is when the bearing capacity of the bolt structure reaches the maximum value. The first peak is lower than the second peak mainly because the adhesive layer strength is lower than the bolt strength, The rapid decline of the first peak value may be due to the complete destruction of the cohesion of the interlayer adhesive layer, which cannot continue to maintain the bearing capacity.

Secondly, for the connection joint of the adhesive layer with nail holes, Analysis of the load-displacement curve shows that the failure process is similar to the above, but the strength decrease in the BC stage is not obvious. The reason is that the extrusion load of the nail hole rubber layer shares part of the tensile stress, so the joint bearing of the bolt and the nail hole rubber layer delays the downward trend of the tensile load.

Finally, this paper compares the maximum tensile strength of the ordinary bolt-and-screw hybrid structure under the influence of different nail-hole clearances. Take the average value of the ultimate load of each group of samples as the strength of the connector under the thickness of the nail-hole adhesive layer, As shown in Table 2, when ordinary bolts are used, the failure load with a clearance of 0.03 mm is significantly larger than that of a tight fit and a clearance of 0.05 mm. Therefore, proper clearance helps to improve the strength of ordinary bolted joints.

Table 2. The failure load of ordinary bolts after tensile fatigue loading with different diameters.

Nail hole clearance (mm)	0.01	0.03	0.05
Tensile strength (N)	6978.1	7141.1	6972.4

4.1.2 Analysis of Joint Failure Mechanism

Figure 4 shows the tensile damage behavior and microscopic damage morphology of the ordinary bolt single-lap bonded-bolted hybrid joint, revealing the failure mechanism of the bonded-bolted hybrid joint nail hole glue layer. It can be observed from the figure that when the composite test piece fails completely, the staggered displacement of the 0.01 mm gap joint is the smallest. The failure mode of the joint is mainly the cohesion failure of the adhesive layer and part of the adhesive layer debonding. The main failure site exists at the interface between the adhesive layer and the laminate. In addition, some test pieces have a certain warpage phenomenon.

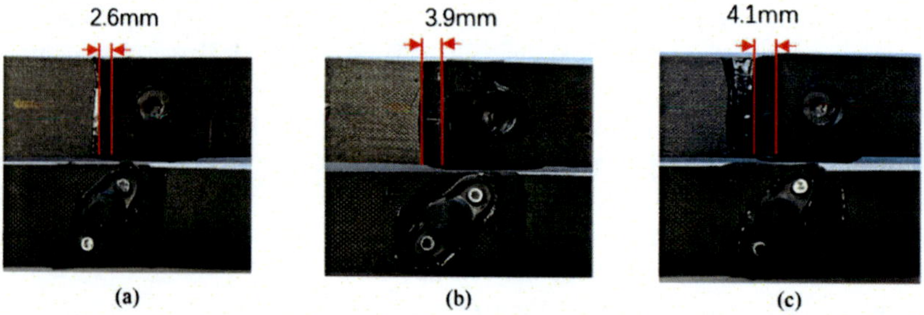

Fig. 4. Tensile fracture morphologies of ordinary bolt specimens: **a** 0.01 mm gap. **b** 0.03 mm gap. **c** 0.05 mm clearance

4.2 High Lock Bolts

4.2.1 The Influence Analysis of the Load-Displacement Curve of the Adhesive Layer Thickness Bonded-Bolted Hybrid Joint

The load-displacement curves of the high-lock bolt hybrid joints with different adhesive layer thicknesses are shown in Fig. 5. It can be seen that there is a significant difference between the typical connection test members of different rubber layer thickness. The overall loading process can be divided into the following stages:

(1) **The bearing stage of the adhesive layer between the laminates**. In the initial stage of loading, the three load-displacement curves almost overlap, as shown in the AB section. At this time, it mainly relies on the friction between the two-layer CFRP laminate and the adhesive layer and the cohesive force of the adhesive layer caused by the high viscosity of the adhesive layer to carry the load.
(2) **The joint bearing stage of the bolt glue layer**. After point B, the load-displacement curve with an adhesive layer thickness of 0 mm gradually deviates from the original trajectory, and the joint stiffness decreases significantly, while the joint curve with an adhesive layer between the nail holes still maintains the original trajectory, so point B is presumed to be a load mutation point. After point B, the joint carrying position is changed from the contact surface of the two layers to the contact position of the bolt and the hole. Since the contact position of the bolt and the hole is not filled,

the inner wall of the composite hole is directly squeezed, resulting in a different trend in its curve. With the gradual increase of the displacement, we found that the load-displacement curve of the adhesive layer test piece changed at point C. The 0.05 mm adhesive layer test piece remains unchanged along the initial trajectory, while the typical test piece with 0.03 mm adhesive layer has a phenomenon of reduced stiffness at point C. It is speculated that the main reason is that the bearing capacity of the 0.03 mm adhesive layer is lower than that of the 0.05 mm adhesive layer.

(3) **The adhesive layer fails, and the composite material is initially damaged**. When the point D_i is reached, the three load-displacement curves all show a sudden downward trend. This is mainly because with the gradual increase of tensile displacement, the composite specimen begins to be damaged. As a result, the curve suddenly oscillates, and the stiffness of the structural member decreases significantly as the structural member continues to stretch.

(4) **Structural parts fail completely**. The D_i-E_i segment is a progressive damage stage. When reaching the E_i point, the composite material fails completely due to the expansion of the damage, and the bearing capacity of the structural member decreases sharply.

The failure process of the hybrid connection of high-lock bolts is similar to that of ordinary bolts, but the stiffness of the high-lock bolt joint decreases significantly after the composite material begins to fail, indicating that the delamination defect of the composite material is more obvious during the tensile process. In the tensile process of ordinary bolts, the stiffness change is small after the failure of the laminate layer, indicating that the degree of damage to the composite material changes slowly during the load-bearing process, indicating that the pallet nut has a certain protective effect on the joint structure.

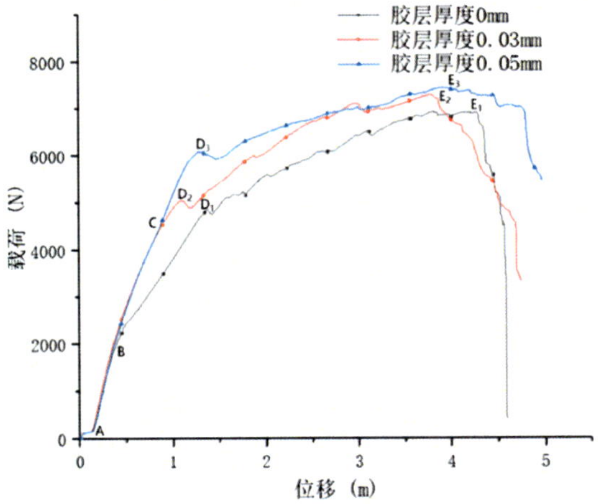

Fig. 5. Load-displacement curves of high-lock hybrid joints with different apertures

The average failure load of typical test specimens with different thicknesses of adhesive layer was taken for analysis. As shown in Table 3, the strength of the connector with glue between the nail holes is greater than that of the test piece without glue between the nail holes, and the failure load of 0.03 mm and 0.05 mm adhesive layer thickness is 5.9% and 12.5% higher than that of the non-adhesive test piece. It can be seen that the existence of the adhesive layer not only ensures its sealing performance, but also improves the bearing performance of the joint to a certain extent. The adhesive layer between the nail holes is within a small thickness range, as the thickness of the adhesive layer increases, the bearing capacity of the high-lock bolt connection is also gradually enhanced.

Table 3. Failure loads of high-lock connectors with different adhesive layer thicknesses.

Nail hole clearance (mm)	0.01	0.03	0.05
Failure load (N)	6546.2	6932.5	7365.6

4.2.2 Analysis of Damage Mechanism of Adhesive Layer

The macroscopic and microscopic morphologies of typical connection test pieces with different pore sizes after stretching are shown in Fig. 6. It can be clearly seen that after the tensile test, the typical connectors have different degrees of damage. From the scanning electron microscope image, it can be seen that the typical interlayer delamination phenomenon appears between the three types of typical test pieces during the tensile process, as shown in the red position in the figure. At the same time, it is found that when the initial interlayer delamination occurs, the degree of delamination without glue is lower, which explains the reason why the load-displacement curve without glue oscillates less at point D_1. After the stretching, the screw with glue layer is closely attached to the hole wall of the composite material. It is presumed that the adhesion of the adhesive layer hinders the separation of the screw and the hole wall, and also enhances the bearing capacity of the joint to a certain extent. The final failure mode of the three test pieces is the complete failure of the composite material due to the crushing of the bolt, as shown in the blue position in the figure. At the same time, it is found that the adhesive layer between the nail holes is not easy to be observed after the stretching. The main reason is that the thickness of the adhesive layer itself is low. As the stretching progresses, the composite material is crushed, and the adhesive layer is destroyed and accumulated in the composite debris.

Figure 7 shows the electron microscope pictures of the adhesive layer before and after stretching. It can be seen from the figure that the position of the adhesive layer is clear before stretching, the process is qualified, the adhesive layer is dense and has fewer defects. After the stretching, the adhesive layer between the nail holes suffered severe cohesive damage due to extrusion, only a small part of the adhesive layer was close to the bolt rod, and the adhesive layer on the surface of the composite hole wall was less. It shows that even if the pore wall of the composite material has the surface treatment

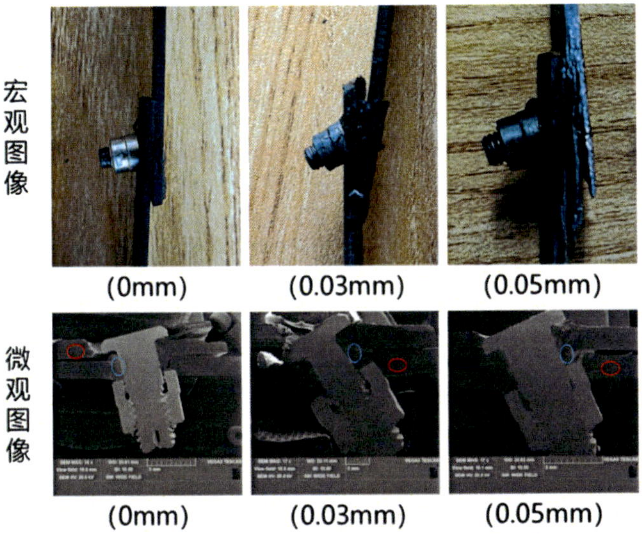

Fig. 6. Macroscopic and microscopic morphologies of test specimens with different thicknesses of adhesive layer after stretching

effect of the tackifier, its-adhesion performance is still poor, and the cohesive strength of the adhesive layer is low, which limits the improvement of the tensile strength of the structure.

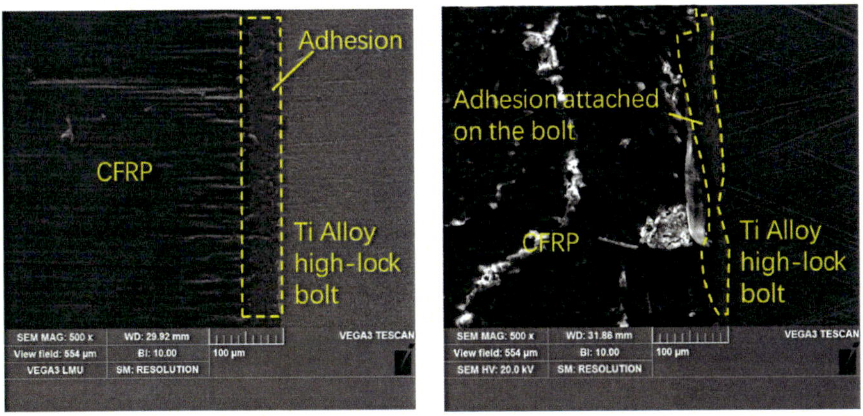

(a) The position of the rubber layer before stretching (b) The position of the adhesive layer after stretching

Fig. 7. Electron microscope pictures of the position distribution of the adhesive layer of the bonded-bolted hybrid joint before and after stretching

5 In Conclusion

In this paper, the tensile strength and failure behavior of a single lap joint with two kinds of bonded-bolted hybrid joint of ordinary bolts with pallet nuts and high-lock bolts under different pore diameters are studied. The following conclusions were obtained:

(1) For the two connection joints, the tensile strength of the connector with the nail hole adhesive layer is higher than that of the connector without the nail hole adhesive layer.
(2) In the small range of the thickness of the nail hole rubber layer (0–0.05 mm), the tensile strength of the ordinary bolt with pallet nut bonded-bolted hybrid joints increases first and then decreases with the thickness of the rubber layer; in the small range of the nail hole rubber layer thickness Inside (0–0.05 mm), the tensile strength of the high-lock bonded-bolted hybrid joints increases with the thickness of the adhesive layer.
(3) After stretching, the nail-hole adhesive layer showed the cohesion failure of the adhesive layer and the debonding failure of the adhesive layer and the composite hole wall, and the nail-hole adhesive layer and the bolt had good bonding performance.

References

1. Ding, L.: Structure Optimal Design for all Composite Wings of an Unmanned Aerial Vehicle. Changchun Institute of Optics, Fine Mechanics and Physics, Changchun (2014). (in Chinese)
2. Hart-Smith, L.J.: Bonded-bolted composite joints. J. Aircr. **22**(11), 993–1000 (1985)
3. Chan, W.S., Vedagiri, S.: Analysis of composite bolted/bonded joints used in repairing. J. Compos. Mater. **35**(12), 1045–1061 (2001)
4. Kelly, G.: Load transfer in hybrid (bonded/bolted) composite single-lap joints. Compos. Struct. **17**(69), 35–43 (2005)
5. Tan, S.C.: Evaluation of Composite Joints. US: Dayton Research Institute (1994)
6. Stewart, M.: An experimental investigation of composite bonded and/or bolted repairs using single lap joint designs. In: 38th Structures, Structural Dynamics, and Materials Conference. US: American Institute of Aeronautics and Astronautics, 2752 (1997)
7. Lopez-Cruz, P., Laliberté, J., Lessard, L.: Investigation of bolted/bonded composite joint behaviour using design of experiments. Compos. Struct. **170**, 192–201 (2017)
8. Chowdhury, N.: Static and fatigue testing bolted, bonded and hybrid step lap joints of thick carbon fibre/epoxy laminates used on aircraft structures. Compos. Struct. **142**, 96–106 (2016)
9. Peng, Z.O.U., Yongge, N.I., Xue, B.I., et al.: Research development on bonded-bolted hybrid joint in composite structure. Adv. Aeronaut. Sci. Eng. **12**(1), 1–12 (2021). (in Chinese)
10. Huang, W., Cheng, X., Wu, P., et al.: Analysis on tensile properties and influence factors of composite hybrid joints. J. Beijing Univ. Aeronaut. Astronaut. **39**(10), 1408–1413 (2013). (in Chinese)
11. Bodjona, K., Raju, K., Lim, G.H., et al.: Load sharing in single-lap bonded/bolted composite joints-Part I: model development and validation. Compos. Struct. **129**, 268–275 (2015)
12. Bodjona, K., Lessard, L.: Load sharing in single-lap bonded/bolted composite joints-Part II: global sensitivity analysis. Compos. Struct. **129**, 276–283 (2015)

13. Kelly, G.: Quasi-static strength and fatigue life of hybrid (bonded/bolted) composite single-lap joints. Compos. Struct. **72**, 119–129 (2006)
14. Yuan, H., Liu, P., Zhao, Q., et al.: Research on the influence factors of the bearing capacity for bonded-bolted hybrid joints. Fiber Reinf. Plast./Compos. **3**, 66–71 (2013). (in Chinese)

Check for updates

Simulation Research on Air Distribution of Electronic Equipment Bay Ventilation System of Civil Aircraft

Jin Rongjia(✉) and Yan Xudong

Shanghai Aircraft Design and Research Institute, Shanghai 201210, China
283912047@qq.com

Abstract. The electronic equipment bay ventilation system of civil aircraft is generally composed of a blowing subsystem and an extraction subsystem, which provides ventilation and cooling functions for the electronic equipment bay area. The quality of its design directly affects the reliability of the electronic equipment and thus directly affects the aircraft safety. In this paper, the electronic equipment bay ventilation system of a certain type of civil aircraft is taken as the research object, and the air distribution is investigated of the blowing subsystem and extraction subsystem. The Flowmaster is used to model the electronic equipment bay ventilation system, and the air distribution simulation of the blowing subsystem and extraction subsystem of the electronic equipment bay is carried out. The results show that, the air distribution simulation by using the restrictor ring size calculated by the flow balancing model can meet the air distribution design requirements of the electronic equipment bay ventilation system, and a reference for the design of electronic equipment bay ventilation system of civil aircraft is provided.

Keywords: Electronic equipment bay ventilation system · Air distribution · Simulation

1 Introduction

The function of the electronic equipment bay ventilation system is to provide ventilation and cooling functions for aircraft electronic equipment bay and onboard electronic equipment to ensure the normal and reliable operation of onboard electronic equipment [1]. At present, with the development of modern science and technology, the application of onboard electronic equipment is becoming more and more extensive, so the thermal load caused by electronic equipment increases sharply [2]. If the thermal cannot be dissipated in time, it will directly affect the normal operation of the electronic equipment, and even cause the damage of the electronic equipment, thereby greatly reducing the performance of the aircraft's onboard electronic equipment system. Therefore, the air distribution of the electronic equipment bay is a key issue in aircraft environmental control system.

In current studies, researches are mainly through experimental or simulation methods. For the experimental of air distribution, Liu et al. investigated the influence of air

Chinese Society of Aeronautics and Astronautics: CASTYSF 2022, LNEE 972, pp. 270–277, 2023.
https://doi.org/10.1007/978-981-19-7652-0_25

distribution on aerodynamic field and combustion performance in a 0.9 MW arch-fired furnace [3]. Nielsen et al. studied the air distribution which is in rooms or with the mixing ventilation and displacement ventilation [4]. Kumar et al. investigated the effect of tile air flow rate on the server air distribution at various locations in the rack [5]. Krajcik et al. studied the air distribution, ventilation effectiveness and thermal environment in a simulated room [6]. For the simulation of air distribution, Wang et al. used a multizone network program, and provided a theoretical basis for improving the accuracy for modeling air distribution in buildings [7]. Domanski et al. established a simulation model of a plate-fin, air-to-refrigerant heat exchanger based on a tube-by-tube approach [8]. Depypere et al. used the Fluent performed for a Glatt GPCG-1 fluidized bed coater and obtained a more homogeneous airflow towards the distributor [9]. Zhang et al. introduced the Flowmaster and showed the application in HVAC [10]. In conclusion, there is no studies on air distribution in the electronic equipment bay.

In this paper, the electronic equipment bay ventilation system of a certain type of civil aircraft is carried out as the research object, and the air distribution is investigated of the electronic equipment bay ventilation system, and a reference for the design of electronic equipment bay ventilation system of civil aircraft is provided. The Flowmaster is used to model the electronic equipment bay ventilation system, and the simulation of the blowing air and extraction air of the electronic equipment bay ventilation system is carried out.

2 Simulation Models and Boundary Condition

In this paper, the aerospace module of Flowmaster is used to model the ventilation system of the electronic equipment bay, and the physical model is converted into a simulation model. For the simulation of the electronic equipment cabin ventilation system, the incompressible steady-state analysis model is chosen.

For the modeling of the electronic equipment bay ventilation system, only the air of the working parameters of the electronic equipment blowing fan and extraction fan has an influence on the air distribution, so the blowing fan and the extraction fan are simplified as flow sources in the simulation The comparison table between the physical model and the simulation model of the electronic equipment bay ventilation system is shown in Table 1.

Table 1. The physical model and computational model comparison table

Boundary name	Boundary type
Blowing fan/Extraction fan	Flow source
Electronic equipment bay ducting	Pipe, bend, junction, transition
Restrictor ring	Orifice
Each blowing/extraction inlet or outlet	Pressure source

According to the above modeling principles, the modeling results of a certain type of civil aircraft electronic equipment bay ventilation system are shown in Figs. 1 and 2.

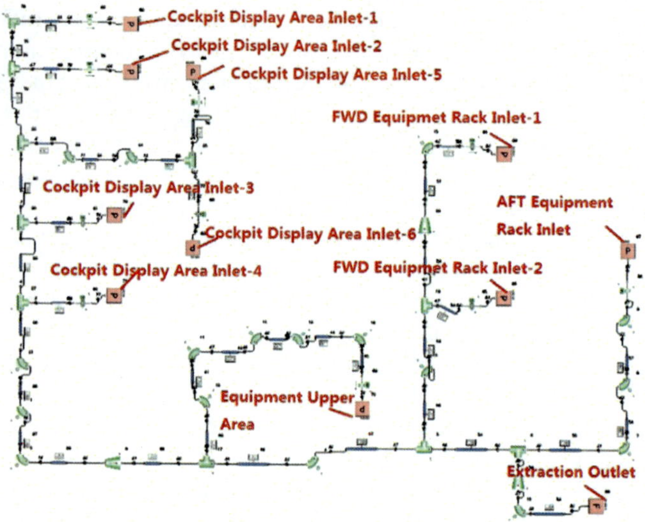

Fig. 1. Simulation model for extraction subsystem of a certain type of civil aircraft electronic equipment bay ventilation system

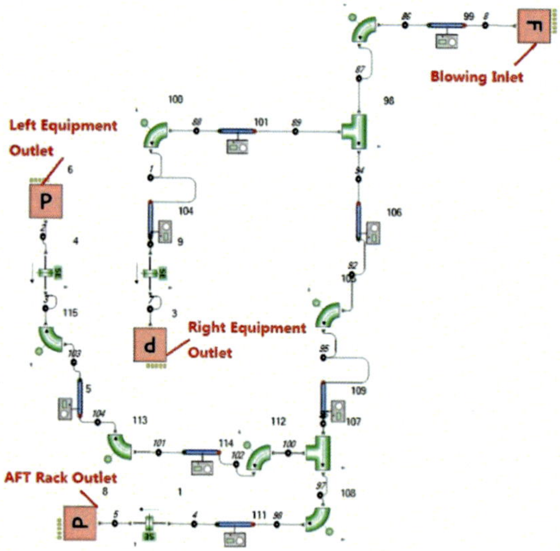

Fig. 2. Simulation model for blowing subsystem of a certain type of civil aircraft electronic equipment bay ventilation system

Figures 1 and 2 are respectively is the simulation model for the air distribution of extraction subsystem and blowing subsystem in the electronic equipment bay. In the extraction subsystem model, the inlet is each extraction inlet, which is expressed in the form of a pressure source; the air outlet is an extraction fan, expressed in the form of a flow source. In the blowing subsystem model, the inlet source is a blowing fan, which is represented in the form of a flow source; the outlet is the left electronic equipment inlet, the right electronic equipment inlet and the after equipment rack inlet, which are expressed in the form of pressure sources. Each extraction inlet or blowing outlet should be adjusted through the orifice of the restrictor ring on the ducting. The incompressible flow balancing model is used in the design calculation of the restrictor ring to obtain the diameter of the restrictor hole corresponding to each extraction inlet or blowing outlet.

The key to establish air distribution simulation model with the Flowmaster is to set the input parameters of the model. For components such as pipes, bends, transitions, junctions, restrictor rings, etc., the Flowmaster has a variety of mathematical models to choose from, and has certain engineering data support [11]. The parameter types of each simulation element are shown in Table 2.

Table 2. Air simulation parameter table of electronic equipment bay ventilation system

Component types	Parameter 1	Parameter 2	Parameter 3	Parameter 4
Pipe	Diameter	Length	Roughness	Friction coefficient
Restrictor ring	Pipe diameter	Orifice diameter		
Flow source	Flow	Temperature		
Pressure source	Pressure			

For the electronic equipment bay ventilation system simulated in this paper, the parameter settings of each component are designed according to the system parameters of a certain type of civil aircraft.

According to the development experience of other aircrafts, the total design air of the blowing subsystem is 320 L/s and the total design air of the extraction subsystem is 680 L/s. The boundary conditions of each inlets or outles of the electronic equipment bay ventilation system are provided in conjunction with the air distribution model of Figs. 1 and 2. Specifically, as shown in Table 3.

3 Simulation and Results Analysis

In this paper,the simulation steps of the air distribution of the electronic equipment bay ventilation system of a certain civil aircraft including the simulation of the orifice matching of the restrictor ring and the check of the air distribution results.

The flow balancing model is used to match the size of the restrictor ring on each branch, and the initial value of the restrictor ring orifice on each branch can be obtained. According to the development experience of other aircrafts, the actual current restrictor ring orifice is usually in units of 0.125in. After conversion, the rounded value of the current restrictor ring aperture can be obtained, as shown in Table 4.

The air distribution of the electronic equipment bay ventilation system according to the value of the restrictor ring orifice calculated in Table 4, compared with the design boundary conditions and the simulated flow value of each branch, and the flow deviation,

Table 3. The simulation boundary conditions of air distribution in the electronic equipment bay

Types	Inlet or Outlet	Air distribution (L/s)
Extraction air of the electronic equipment bay	Cockpit display area inlet-1	8
	Cockpit display area inlet-2	8
	Cockpit display area inlet-3	8
	Cockpit display area inlet-4	8
	Cockpit display area inlet-5	4
	Cockpit display area inlet-6	4
	FWD equipment rack inlet-1	20
	FWD equipment rack inlet-2	20
	AFT equipment rack inlet	280
	Equipment upper area	320
Blowing air of the electronic equipment bay	AFT rack outlet	220
	Left equipment outlet	50
	Right equipment outlet	50

and the air distribution of each branch in the electronic equipment bay. The results are shown in Table 5.

According to the simulation results in the table, it can be seen that the flow deviation between the calculated value of the air distribution in each temperature zone of the electronic equipment bay and the design value is within ±5%. Therefore, it can be considered that the air distribution by using the diameter of the restrictor orifice calculated by the flow balancing model and the converted diameter of the restrictor orifice can meet the air distribution requirements of the electronic equipment bay ventilation system.

4 Conclusion

In this paper, the electronic equipment bay ventilation system of a certain type of civil aircraft as the research object, and investigates the air distribution of the electronic equipment bay ventilation system. The electronic equipment bay ventilation system was modeled by the Flowmaster, and the air distribution of blowing subsystem and extraction subsystem in the electronic equipment bay was simulated.

The results show that the air distribution simulation by using the restrictor ring size calculated by the flow balancing model can meet the air distribution design requirements of the electronic equipment bay ventilation system.

Table 4. Simulation results of restrictor ring orifice

Types	Branch name	Initial value of restricting ring orifice (m)	Rounded value of restricting ring orifice (in)
Extraction subsystem	Cockpit Display Area Inlet-1	0.016	0.625
	Cockpit Display Area Inlet-2	0.016	0.625
	Cockpit Display Area Inlet-3	0.016	0.625
	Cockpit Display Area Inlet-4	0.016	0.625
	Cockpit Display Area Inlet-5	0.011	0.375
	Cockpit Display Area Inlet-6	0.011	0.375
	FWD Equipment Rack Inlet-1	0.022	0.875
	FWD Equipment Rack Inlet-2	0.022	0.875
	AFT Equipment Rack Inlet	0.079	3.125
	Equipment Upper Area	0.094	3.75
Blowing subsystem	AFT Rack Outlet	0.071	2.75
	Left Equipment Outlet	0.035	1.375
	Right Equipment Outlet	0.036	1.375

Table 5. Air distribution simulation results of the electronic equipment bay ventilation system

Types	Branch Name	Design boundary conditons (L/s)	Simulation value of flow (L/s)	Flow deviation (%)
Extraction system	Cockpit Display Area Inlet-1	8	8	0
	Cockpit Display Area Inlet-2	8	8	0
	Cockpit Display Area Inlet-3	8	8	0
	Cockpit Display Area Inlet-4	8	8	0
	Cockpit Display Area Inlet-5	4	4	0
	Cockpit Display Area Inlet-6	4	4	0
	FWD Equipment Rack Inlet-1	20	20	0
	FWD Equipment Rack Inlet-2	20	20	0
	AFT Equipment Rack Inlet	280	268	−4
	Equipment Upper Area	320	330	+3
Blowing system	AFT Rack Outlet	220	231	+5
	Left Equipment Outlet	50	49	−2
	Right Equipment Outlet	50	49	−2

References

1. Bruno, L.J., Hipsky, H.W., Jr. D.E.A, et al.: Aircraft Environmental Control System: US20190225343A1[P] (2019)
2. Shou, R.Z., He, H.S.: Aircraft Environment Control. Beihang University Press (2004)
3. Liu, R., Hui, S., Yu, Z., et al.: Effect of air distribution on aerodynamic field and coal combustion in an arch-fired furnace. Energy Fuels **24**, 5514–5523 (2010)
4. Nielsen, P.V., Topp, C., Sønnichsen, M., et al.: Air distribution in rooms generated by a textile terminal-comparison with mixing and displacement ventilation. ASHRAE Trans. **8**, 733–739; Ashrae Trans. **111**, 733–739 (2005)
5. Kumar, P., Joshi, Y.: Experimental investigations on the effect of perforated tile air jet velocity on server air distribution in a high density data center. In: IEEE. IEEE (2010)
6. Krajcik, M., Simone, A., Olesen, B.W.: Air distribution and ventilation effectiveness in an occupied room heated by warm air. Energy Build. **55**, 94–101 (2012)

7. Wang, L., Chen, Q.: Theoretical and numerical studies of coupling multizone and CFD models for building air distribution simulations. Indoor Air **17**(5), 348–361 (2010)
8. Domanski, P.A.: Simulation of an evaporator with nonuniform one-dimensional air distribution. ASHRAE Trans. **97**, 793–802 (1991)
9. Depypere, F., Pieters, J.G., Dewettinck, K.: CFD analysis of air distribution in fluidised bed equipment. Powder Technol. **145**(3), 176–189 (2004)
10. Zhang, G.P., X.N., Zhang, W.P., et al.: Application of Flowmaster in HVAC. Refrigeration Air-Condition (2006)
11. Wang, G.W., Wu, C.Y., et al.: Optimization design and experimental verification of cabin air distribution for civil aircraft. Civil Aircraft Des. Res. **4**, 4 (2015)

High Cycle Fatigue Properties and Fracture Behavior of TC4 Titanium Alloy at Room and Elevated Temperatures

Juan Li[1](✉), Jianming Cai[1], Min Zhou[2], Rui Duan[1], and Xu Huang[1]

[1] AECC Beijing Institute of Aeronautical Materials, Beijing 100095, China
lervly@sina.com
[2] Wuxi Hyatech Co., Ltd, Wuxi 214142, China

Abstract. The high cycle fatigue properties and fracture characteristics of TC4 titanium alloy at room temperature, 200 °C and 400 °C under the stress concentration factors $K_t = 1$ and $K_t = 3$ were studied. The *S-N* curves of rotary bending fatigue under different conditions were plotted by power function fitting of three parameters, and the fatigue limit of $N_f = 3 \times 10^7$ cycles was obtained. The results indicated that TC4 alloy had a high fatigue notch sensitivity at both room and elevated temperatures, which led to the significant decline of fatigue properties. It was found that the smooth specimen was more sensitive to the effect of temperature, whose high cycle fatigue performance decreased obviously with the increase of temperature. There was only one fatigue crack source region in the smooth specimen, and obvious fatigue bands and a small number of secondary cracks can be seen in the propagation region. With the propagation of cracks, instantaneous fracture occurred in the inner side of the specimen opposite the source region. Multiple fatigue crack sources were formed on the surface of the notched specimen, and then the fatigue cracks expanded from the circumference inward. Finally, instantaneous fracture occurred in the inner part of the specimen. TC4 alloy was characterized by ductile fatigue fracture.

Keywords: TC4 titanium alloy · Stress concentration factor · Temperature · High cycle fatigue · Fatigue fracture morphology

1 Introduction

Due to the excellent characteristics of low density, high specific strength, high temperature resistance and corrosion resistance, titanium alloys have been widely used in the fields of aerospace, automobile, ship, petrochemical, biomedical and marine engineering [1–3]. Among all the Ti alloys, TC4 alloy with the nominal composition of Ti–6Al–4V (wt.%) has the largest usage in the aviation industry at present [4–6]. This alloy is a typical $\alpha + \beta$ dual phase alloy, which exhibits both the advantages of α phase and β phase, and has good comprehensive mechanical properties, plasticity, weldability and corrosion resistance. The long-term service temperature of TC4 alloy can reach to

Chinese Society of Aeronautics and Astronautics: CASTYSF 2022, LNEE 972, pp. 278–286, 2023.
https://doi.org/10.1007/978-981-19-7652-0_26

350 °C. Therefore, it is often used in aero-engine discs, blades and other components as a high-temperature titanium alloy [7, 8].

The rotating parts of aero-engine are subjected to complex alternating loads during service. Under complex service conditions, fatigue failure is one of the main reasons for material fracture [9]. A large number of fatigue failure events of aero-engine turbines, rotors and blades show that the main failure causes are torsional and bending resonance fatigue. Among them, the accident rate caused by high cycle fatigue damage is as high as 24%, which is the largest among all accident causes [10, 11]. The high-cycle fatigue properties of Ti alloys are closely related to the service temperature and microstructure. In addition, since the fatigue properties of Ti alloys are sensitive to notches, the stress concentration factor also has a significant impact on its high-cycle fatigue properties [12–14].

At present, the research on high cycle fatigue of TC4 alloy mainly adopts the maximum stress of 1×10^7 without damage as the fatigue limit. In order to meet the development requirements of high reliability and long life of aero-engine, this work took the maximum stress of 3×10^7 without damage as the fatigue limit. The high-cycle fatigue properties at different temperatures were studied, the effect of stress concentration factor on the fatigue limit at different temperatures was analyzed, and the fracture characteristics of the fatigue specimens were analyzed.

2 Material and Experimental Procedures

2.1 Materials

The material used in this study was industrial ingot of TC4 alloy, which was melted through triple VAR melting, then processed into Φ 30 mm bar in $\alpha + \beta$ region by precision forging. The bar of equal length was cut and forged in $\alpha + \beta$ region to prepare a 60 × 40 × 15 mm flattened sample. The sample was subjected to ordinary annealing heat treatment at 700 °C for 2 h following by air cooling. The basic tensile properties obtained are shown in Table 1.

Table 1. Tensile properties tested at different temperatures for TC4 titanium alloy specimens.

T/°C	R_m/MPa	$R_{p0.2}$/MPa	A/%	Z/%
RT	1030	943	16.6	44.0
100	950	829	20.8	50.5
200	861	728	19.8	59.1
300	780	648	17.5	60.9
400	747	614	17.6	62.9

2.2 Experimental Procedures

Samples were taken along the longitudinal direction of the flattened sample and processed into the fatigue test sample shown in Fig. 1. The fatigue test was conducted on the QBWP type rotary bending fatigue testing machine, and the test preparation and process were strictly in accordance with the ISO 1143: 2010 metallic materials-rotating bar bending fatigue standard. In the air environment, the test temperatures were room temperature, 200 °C and 400 °C respectively, the stress concentration factor was $K_t = 1$ and $K_t = 3$, and the test speed was 5000r/min. The fatigue limit under the specified life of 3×10^7 was obtained by the lifting method, and each set of data met the 95% confidence requirement under the 50% survival rate.

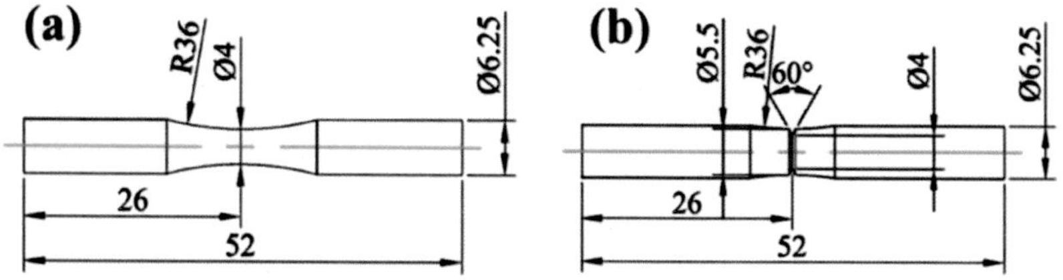

Fig. 1. Schematic diagram of high cycle fatigue **a** smooth and **b** notch specimens.

The fatigue fracture morphology and the microstructure of the samples were observed by JSM-7900F field emission scanning electron microscope (SEM) equipped with electron back-scattered diffraction (EBSD) detector. The diameter-axial surface of the sample was selected, and the microstructure characteristics were observed by SEM after mechanical polishing and chemical etching. The proportion of etching solution was 5 vol.%HF + 10vol.%HNO_3 + 85vol.%H_2O. The EBSD samples were prepared by electrolytic polishing with a mixed solution of 6% perchloric acid, 35 vol.% n-butanol and 59 vol.% methanol. The temperature of the polishing solution was controlled to about −25 °C and the voltage was about 20 V. The scanning area was 245 μm × 190 μm with a step size of 0.4 μm.

3 Results and Discussion

3.1 Microstructure

The microstructure of the TC4 alloy sample is shown in Fig. 2. It can be seen that after annealing, the TC4 alloy has a typical bi-modal structure, which is mainly composed of ellipsoid primary α phase and lath secondary α phase, and β phase is distributed among the secondary α phases. The content of primary α phase is about 40%, and the size is 5–15 μm (Fig. 2a). The crystal orientation distribution of the sample was obtained by EBSD analysis (Fig. 2b). It can be seen that the grain orientation of the primary α-phase is similar, and the grain orientation difference of the secondary α-phase varies greatly in comparison. Figure 2c is the pole figure corresponding to the orientation distribution. The relatively concentrated crystal orientations were <0001>//RD and <10$\bar{1}$0>//RD (ie, the sampling direction) deformation texture, which indicates that the sample still retains the texture formed by precision forging after flattening and annealing. However, the texture density was relatively low, and the maximum texture strength was 4.389.

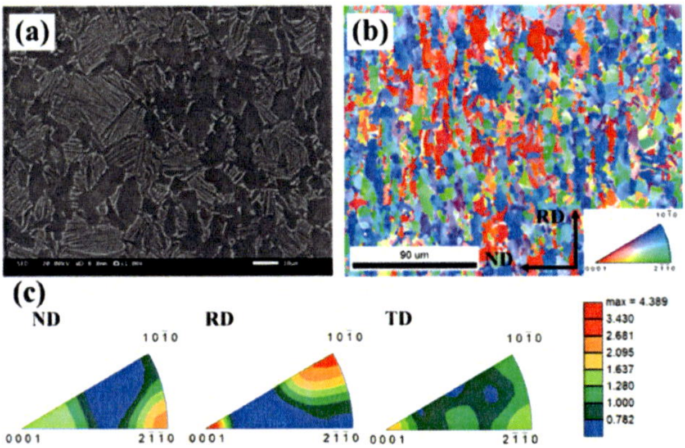

Fig. 2. Microstructure of TC4 alloy **a** SEM morphology; **b** inverse pole figure; **c** inverse pole figures parallel to ND, TD and RD.

3.2 High Cycle Fatigue Properties

The stress fatigue life curve (*S-N* curve) is the fatigue life curve measured under the condition that the maximum cyclic stress does not exceed the elastic limit or yield strength, and is one of the main basis for evaluating the fatigue performance of materials. Generally, there are three expressions of *S-N* curve, the power function expression, the exponential function expression and the three-parameter power function expression [15]. The three-parameter power function method was used to fit the rotational bending fatigue *S-N* curves of the smooth ($K_t = 1$) and notched ($K_t = 3$) specimens of the TC4 alloy at different temperatures, as shown in Fig. 3. It can be seen that with the increase of the test temperature, the fatigue life shows a downward trend, indicating that the temperature has a significant effect on the fatigue properties of the TC4 alloy. This is related to the change of the deformation mechanism of the material at high temperature. The resistance to crack initiation and propagation decreases at high temperature [16–18]. Comparing the fatigue curves with different stress concentration factors, it can be seen that for the smooth specimen, the fatigue performance is more sensitive to the change of temperature. The *S-N* (median value) fatigue curve equation and corresponding fatigue limit of TC4 alloy under different test conditions are shown in Table 2. It can be seen that under the same test conditions, the fatigue life of the notched specimens is significantly reduced, which is consistent with previous investigation [19]. Compared with the smooth specimens, the fatigue limit(σ_D) of the notched specimens is reduced by 65.6, 65.7 and 64.9% at room temperature, 200 °C and 400 °C, respectively, indicating that the high-cycle fatigue properties of TC4 alloys exhibit obvious notch-sensitive effects at room temperature and high temperature.

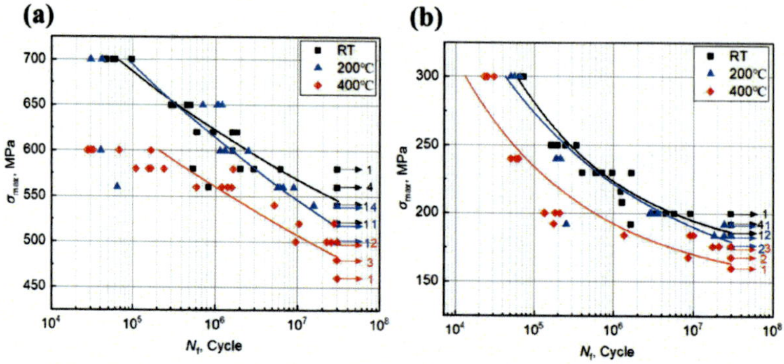

Fig. 3. *S-N* curves of TC4 alloy **a** smooth and **b** notched samples at different temperatures.

Table 2. Fatigue limit and S-N equation of TC4 alloy under different conditions.

Stress concentration	Temperature/°C	S-N equation	σ_D/MPa
$K_t = 1$	RT	$\lg N_f = 40.54–13.59\lg(\sigma_{max} - 275)$	570
	200	$\lg N_f = 47.53–15.41\lg(\sigma_{max} - 120)$	542
	400	$\lg N_f = 55.76–18.69\lg(\sigma_{max} - 100)$	498
$K_t = 3$	RT	$\lg N_f = 11.98–3.38\lg(\sigma_{max} - 165)$	196
	200	$\lg N_f = 14.95–4.68\lg(\sigma_{max} - 140)$	186
	400	$\lg N_f = 12.10–3.64\lg(\sigma_{max} - 145)$	175

3.3 High Cycle Fatigue Properties

Figure 4 shows the fatigue fracture morphologies of the smooth specimen at room temperature and 400 °C. It can be clearly seen that the fatigue fracture is divided into three typical regions, the fatigue crack source region, the crack propagation region and the instantaneous fracture region. Since the experiment adopts a rotating bending loading method, leading to the surface of the sample is subjected to a large stress, the crack source is usually formed on the surface or subsurface [20]. At room temperature, the crack of the smooth specimen started on the surface, with only one crack source area, accompanied by an obvious radial pattern (Fig. 4b). The crack propagation rate in this area is slow, and the cracks are repeatedly opened and closed under cyclic loads, causing friction on the surface, therefore they are generally flat and smooth. As the crack propagates into the crack propagation region, a large area of dense fatigue stripes can be seen, accompanied by a small number of secondary cracks (Fig. 4c). Fatigue stripes will increase the non-uniform deformation during crack propagation and hinder crack growth. The thinner the fatigue strip, the greater the resistance of the material to crack growth and the slower the crack propagation [21]. The instantaneous fracture region is composed of many small dimples, showing typical ductile fracture characteristics (Fig. 4d). However, at 400 °C, the cracks of the smooth specimens start from the interior of the specimens, which is also the origin of single-source cracks. Due to the weak resistance to crack propagation at high temperature, the crack source region is flatter, and the fatigue band in the propagation region is slightly wider than that at room temperature. As the crack propagates, an instantaneous fracture occurs on the inner side of the opposite side of the crack source area after instability occurs, forming a larger dimple.

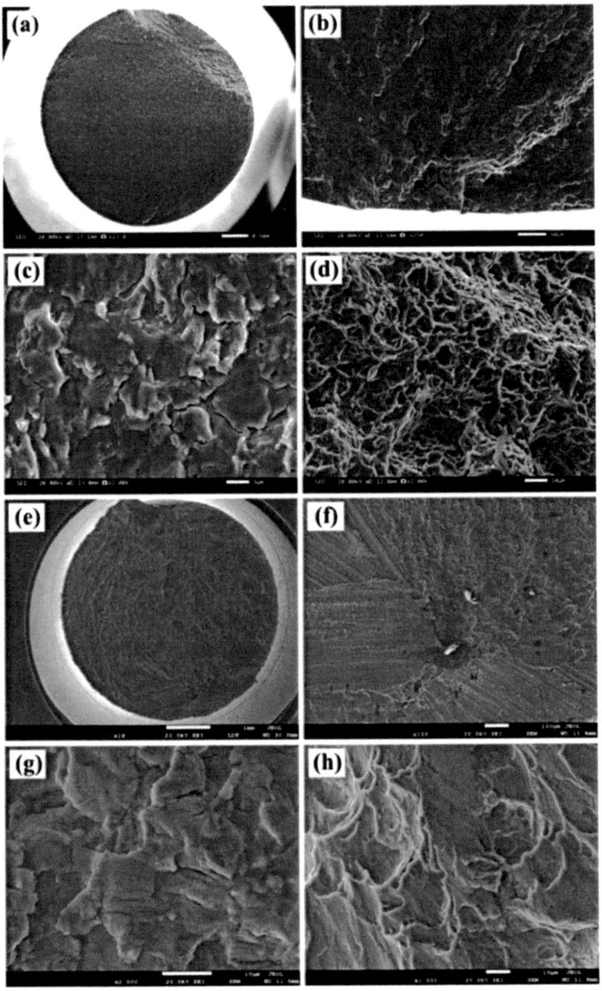

Fig. 4. Fatigue fracture morphology of TC4 alloy smooth specimens at room temperature (σ = 580 MPa, $N_f = 2.96 \times 10^6$): **a** macroscopic micrograph, **b** fatigue crack initial stage, **c** region of stable crack growth, **d** fast fracture zone; and 400 °C (σ = 500 MPa, $N_f = 2.57 \times 10^7$): **e** macroscopic micrograph, **f** fatigue crack initial stage, **g** region of stable crack growth, **h** fast fracture zone.

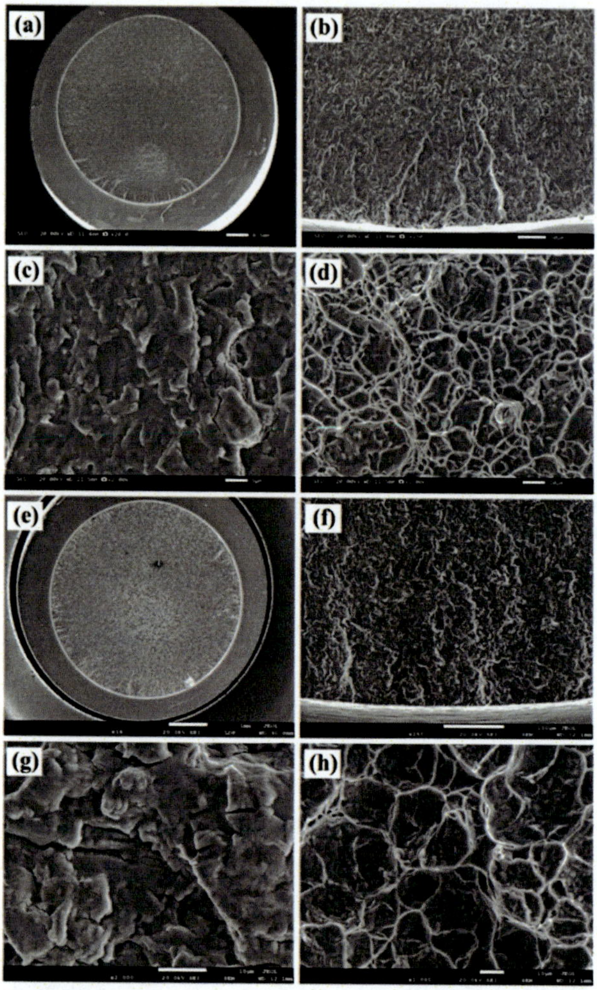

Fig. 5. Fatigue fracture morphology of TC4 alloy notched specimens at room temperature ($\sigma = 200$ MPa, $N_f = 9.19 \times 10^6$): **a** macroscopic micrograph, **b** fatigue crack initial stage, **c** region of stable crack growth, **d** fast fracture zone; and 400 °C ($\sigma = 176$ MPa, $N_f = 2.11 \times 10^7$): **e** macroscopic micrograph, **f** fatigue crack initial stage, **g** region of stable crack growth, **h** fast fracture zone.

Figure 5 shows the fatigue fracture morphologies of the notched specimens at room temperature and 400 °C. Different from the smooth specimens, the notched specimens exhibited multiple crack sources on the surface at both room temperature and high temperature. This is attributed to the stress concentration at the root of the notch is the largest in the process of fatigue loading. Fatigue failure starts in weak areas where stress concentrations are concentrated, resulting in a significant decrease in fatigue performance. With the crack propagation, obvious fatigue strips can also be observed in the steady-state propagation region, and more secondary cracks can be seen in the fatigue propagation region at high temperature. Unlike the smooth specimen, the fatigue crack of the notched specimen propagates inward from the circumference, and finally an instantaneous fracture occurs in a small area inside the specimen, which exhibits ductile fracture characteristics.

4 Conclusions

(1) The *S-N* curves of smooth and notched specimens of TC4 alloy were measured at room temperature, 200 °C and 400 °C. The high cycle fatigue of TC4 alloy shows obvious notch sensitive effect at room and high temperature. When $N_f = 3 \times 10^7$ cycles, the fatigue limits of smooth specimens are 574 MPa, 542 MPa and 543 MPa respectively, and the fatigue limits of notched specimens are 196MPa, 186MPa and 175MPa, respectively.

(2) With the increase of temperature, the high cycle fatigue properties of TC4 alloy were decreased. For the smooth specimen with $K_t = 1$, the effect of temperature on fatigue performance was more obvious.

(3) Both of room temperature and high temperature smooth samples had only one fatigue crack source. The room temperature fatigue crack originated from the surface of the smooth specimen, while the high temperature fatigue crack originated from the interior of the specimen. Obvious fatigue strips and a few secondary cracks can be seen in the propagation zone. With the crack propagation, instantaneous fracture occurred in the opposite and inner side of the crack source region, which was characterized by ductile fracture.

(4) Fatigue cracks of notched specimens originated in the notched root region with large stress concentration, and there were multiple fatigue crack sources. Fatigue bands can be seen in the propagation zone, accompanied by more secondary cracks at high temperature. The fatigue cracks extended from the circumference inward, and finally occurred instantaneous fracture in a small area inside the sample.

References

1. Banerjee, D., Williams, J.: Perspectives on titanium science and technology. Acta Mater. **61**(3), 844–879 (2013)
2. Leyens, C., Peters, M.: Titanium and Titanium Alloys: Fundamentals and Applications. Wiley (2003)
3. Cui, C., Hu, B., Zhao, L., et al.: Titanium alloy production technology, market prospects and industry development. Mater. Des. **32**(3), 1684–1691 (2011)
4. Seshacharyulu, T., Medeiros, S.C., Frazier, W.G., et al.: Hot working of commercial Ti-6Al-4V with an equiaxed $\alpha-\beta$ microstructure: materials modeling considerations. Mater. Sci. Eng. **284**(1–2), 184–194
5. Wang, X.D., Hao, B., Lu, F.S., et al.: The basic properties and application situation of titanium in China. Titan. Ind. Prog. **21**(1), 6–10 (2004)
6. Cheng, M., Zhao, J.G., Song, H.W., et al.: Effect of α grain size on deformation behaviors of TC4 titanium alloy with equiaxed microstructure. J. Aeronaut. Mater. **29**(1), 22–26 (2009)

7. Lütjering, G.: Influence of processing on microstructure and mechanical properties of (α+ β) titanium alloys. Mater. Sci. Eng. **243**(1–2), 32–45 (1998)
8. Evans, W.: Optimising mechanical properties in alpha + beta titanium alloys. Mater. Sci. Eng. **243**(1–2), 89–96 (1998)
9. Tao, C.H., Liu, Q.Q., Cao, C.X.: Failure and Prevention of Aeronautical Titanium Alloy, p. 39. National Defence Industry Press, Beijing (2002)
10. Zhong, P.D.: Failure and lesson of turbo rotation blades of aero-engines. J. Mater. Eng. **31**(z1), 30–33 (2003)
11. Cowles, B.A.: High cycle fatigue in aircraft gas turbines-an industry perspective. Int. J. Fract. **80**(2/3), 147–163 (1989)
12. Tokaji, K., Mito, N.: High cycle fatigue behaviour of Ti-6Al-4V alloy at elevated temperatures. Scripta Mater. **54**(12), 2143–2148 (2006)
13. Nalla, R.K., Boyce, B.L., Campbell, J.P., et al.: Influence of microstructure on high-cycle fatigue of Ti-6Al-4V: bimodal versus lamellar structures. Metall. Mater. Trans. **33**(13), 899–918 (2002)
14. Li, J., Cai, J.M., Duan, R., et al.: High cycle fatigue of TC6 alloy. **23**(z1), s277–s281 (2013)
15. Li, Y., Zhang, C.C., Gao, J.Y., et al.: High-cycle and giga-cycle fatigue properties of titanium alloy for aero-engine. Mech. Eng. **38**(3), 250–254 (2016)
16. Li, J.K., Liu, Y.J., Wang, Q.Y., et al.: High-temperature ultra-high cycle fatigue test of TC17 titanium alloy. Hangkong Dongli Xuebao/J. Aerosp. Power **29**(7), 1567–1573 (2014)
17. Li, G., Sun, C.Q.: High-temperature failure mechanism and defect sensitivity of TC17 titanium alloy in high cycle fatigue. J. Mater. Sci. Technol. **122**, 128–140 (2022)
18. Chen, Y.N., Xiong, Y., Fan, M.X., et al.: Fatigue facture analysis of TC11 titanium alloy at different temperatures. Trans. Mater. Heat Treat. **8**, 8 (2019)
19. Liu, Y., Le, Z., Shi, X., et al.: High cycle fatigue properties and fracture behavior of Ti–5Al–5Mo–5V–1Cr–1Fe titanium alloy. Rare Met. Mater. Eng. **47**(12), 3666–3671 (2018)
20. Sun, Y.L., Yu, J.J., Wang, Z.J., et al.: Rotary bending high cycle fatigue behavior of single crystal superalloy DD499 in <111> orientation. Rare Met. Mater. Eng. **40**(2), 239–242 (2011)
21. Morrissey, R.J., Mcdowell, D.L., Nichloas, T.: Frequency and stress ratio effects in high cycle fatigue of Ti-6Al-4V. Int. J. Fatigue **21**(7), 679–685 (1999)

Analysis on Architecture Design and Evaluation Technology of Civil Aircraft Fuel System

Li Zenghui(✉), Wang Yong, Cheng Gang, Zhang Jianlan, Wang Pengbo, and Cui Ze

AVIC Nanjing Engineering Institute of Aircraft Systems, Nanjing 211106, Jiangsu, China
lizenghui515@126.com

Abstract. Based on the fuel system architecture of civil aircraft, this paper forms the ability to establish the design and evaluation of aircraft fuel system architecture. Firstly, it is proposed to convert the stage results of demand development into the input of architecture design through demand analysis, then analyze the subsystems of the fuel system, including key design elements and trade-offs, and then form a structure. To further realize the mapping relationship between functions and components. In the process of architecture evaluation, based on the establishment of the system model and simulation analysis, the research on the comprehensive performance evaluation method of the fuel system is carried out, and the architecture trade-off evaluation is carried out, and finally a set of overall architecture design method of the civil aircraft fuel system is realized.

Keyword: Civil aircraft fuel system · Functional architecture · Physical architecture · Architecture design and evaluation

1 Overview

With the implementation of Special Research on Large Civil Aircraft of China, the fuel system of civil aircraft is an important subsystem in the aircraft system, and its research has very important engineering significance and application. The mission of the fuel system is to ensure that the fuel in the fuel tank is continuously and reliably delivered to the engine and APU under any flight conditions, and to ensure the normal flight of the aircraft. In the research of fuel system, its architecture design and evaluation are particularly important. A reasonable aircraft fuel system architecture design can ensure that the system meets the overall requirements and can avoid many potential errors and risks that may occur later. Whether the architectural design of the fuel system is reasonable and reliable directly affects the economy, safety and airworthiness of the aircraft, and directly determines the advanced technical solutions of the fuel system. In view of the pros and cons of the fuel system configuration, there is currently no complete evaluation method. To establish the system capability, it is necessary to establish an evaluation system to comprehensively evaluate the feasibility, economy, safety, maintainability and other factors of the system configuration and design [1, 2].

Foundation item: Special Research on Large Civil Aircraft of China (MJZ-2016-S-04).

Chinese Society of Aeronautics and Astronautics: CASTYSF 2022, LNEE 972, pp. 287–296, 2023.
https://doi.org/10.1007/978-981-19-7652-0_27

Compared with foreign civil aircraft fuel system design capabilities, domestic civil aircraft fuel system design is still in its infancy and cannot meet the safety and reliability requirements. The specific research on fuel system in China is relatively limited, such as the anti-strike research on military fuel tanks, fuel measurement computer, fuel measurement algorithm research, etc., and most of them are mainly theoretical research in colleges and universities, and engineering implementation is very small. Compared with foreign countries, there is a big gap in system design capabilities. Compared with foreign countries, the current domestic fuel system design capability gap is relatively large. Under the current demand for vigorously developing large civil aircraft, it is necessary to improve the overall strength of domestic equipment suppliers at all levels and master the ability of system-level architecture design and evaluation.

2 Fuel System Demand Capture and Confirmation

Civil aircraft fuel system requirements are the basis for system definition and form the basis for architecture, design, synthesis, and validation. According to the aircraft top-level document, analyze the original stakeholder requirements, and capture the corresponding fuel system requirements, including functional requirements, safety requirements and derived requirements. The security requirement is the minimum performance constraint on the availability and integrity of the function, and the functional requirement is to clearly stipulate the performance of the fuel system. These requirements are collated and analyzed to identify and define the required fuel system functions. A complete and correct requirements matrix is formed through the confirmation of requirements.

The top-level process of the fuel system includes the determination of fuel system functions and their associated requirements. The fuel system functions include functional interfaces and corresponding safety requirements, which form the basis for establishing the system architecture. The choice of architecture establishes the additional requirements needed to implement the architecture. At each stage of the requirements identification and allocation process, more details of current requirements and new derived requirements are obtained. The problems encountered and choices made in the implementation phase are the main reasons for the derived demand. Detailed design activities will continuously introduce new requirements or change current requirements.

The requirements validation process is to ensure that the requirements presented are sufficiently correct and complete to enable the product to meet the needs of customers, users, suppliers, maintainers, certifiers, aircraft, system and component developers, as shown in Fig. 1. The validation process runs through the fuel system development cycle, and at each development stage, validation activities increase confidence in the correctness and completeness of requirements.

3 Functional Architecture Design of Fuel System

The focus of functional analysis is to translate functional, presentation, interface, and other requirements into a clear description of system functionality, which is then used to guide subsequent design integration. A functional description and performance parameter description. Functional analysis is to analyze the functional process of each system

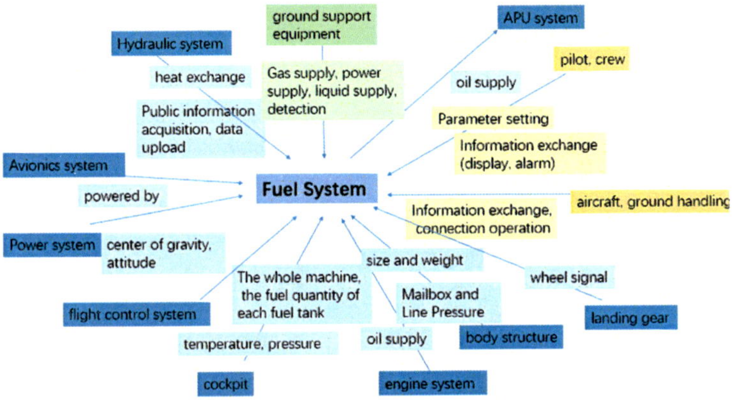

Fig. 1. Fuel system stakeholder identification

use case in a black box state, and then identify the interaction between the system and the outside world, and finally describe the state behavior of the system completely. A use case performs a functional analysis process, which mainly includes the following research contents: functional flow analysis, scene analysis, port and interface definition, system status behavior analysis, confirmation and verification through execution, etc. The process is to look at the system from the perspective of a black box, and how to design the architecture needs to consider the functional process of each subsystem, subsystem The internal and external interfaces, the state behavior of subsystems.

After the functional architecture is formed, when evaluating it, a fair and feasible trade-off method must be established to select the optimal architectural solution. It is ensured that the formed functional models of each subsystem can be used by system engineers at the subsystem level to perform demand analysis, function analysis and design synthesis at the subsystem level.

The establishment of the system function architecture is actually to sort out the system function levels, establish the logical relationship between the functions, determine the interface information between the functions, and formulate the relationship between the time sequence and state mode of each function. The division of functional levels includes both the decomposition of functions and the allocation of technical indicators. The function decomposition process involves the establishment of the state function diagram, the functional logic analysis, and the determination of the functional interface, which requires software such as Rhapsody to realize this process. The establishment of the logical relationship between functions can further realize the mapping relationship between functions and components, and determine the components or component combinations that each function implements.

Along with the function decomposition, there is also the allocation of technical indicators. This allocation process is subdivided and decomposed layer by layer according to the system, subsystem, and component levels, the total technical parameters of the fuel system are allocated downward to refueling, oil discharge, oil supply, and oil delivery, ventilation, measurement and other subsystem levels, then continue to decompose parameters down to the fuel pump, check valve, shut-off valve, pipeline and other component levels according to the subsystem. The allocation of technical indicators involves the simulation calculation of the system, such as using Flowmaster to build system and subsystem models, assign parameters to each component, and verify the rationality of parameter allocation, as shown in Fig. 2. This process needs to be based on certain design and development experience in system components. By decomposing the preliminary sub-functions of the fuel system and synthesizing the overall functions of the fuel system, the functional architecture as shown below is formed.

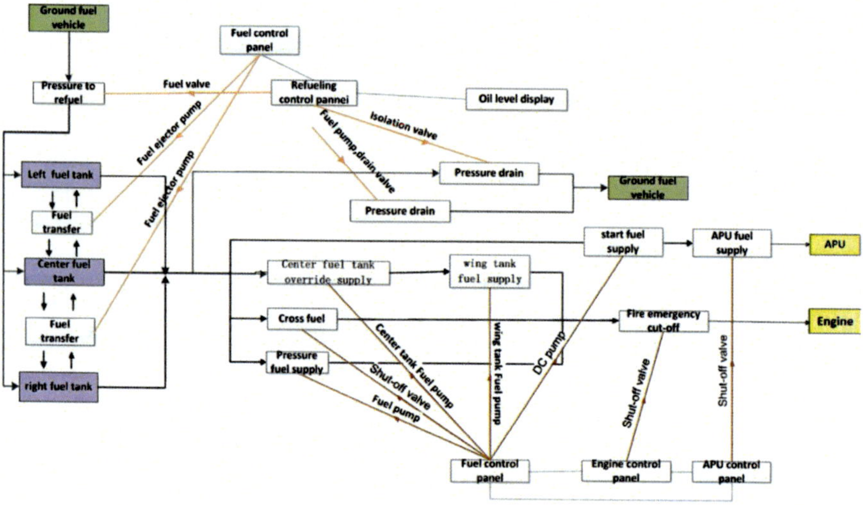

Fig. 2. Fuel system functional architecture example

4 Design of the Physical Architecture of the Fuel System

The physical architecture design of the fuel system of civil aircraft mainly includes the architecture design of storage subsystem, distribution subsystem and measurement subsystem.

4.1 Storage Subsystem

The main function of the civil aircraft fuel tank is to store the fuel required by the engine and APU. The storage subsystem mainly includes three functions: fuel storage, fuel tank ventilation, and discharge. In order to make the fuel supply unobstructed, the ventilation system ensures that the fuel tank is connected to the outside atmosphere to ensure the pressure balance inside and outside the fuel tank, and the discharge function is to meet the maintenance of the aircraft.

When designing the fuel tank ventilation system, the following four factors are mainly considered: (1) When the aircraft is cruising or ascending and descending, it must have air circulation capacity. Prevent the fuel tank from being damaged or even destroyed by the pressure difference caused by the fuel tank; (2) The ventilation system must ensure that the fuel tank is connected to the outside atmosphere within the range of various flight attitude changes of the aircraft; (3) the storage subsystem needs to have the function of over pressure protection to prevent the fuel tank structure from bearing excessive load; (4) Since the ventilation system is directly connected to the outside atmosphere, if a fire occurs on the ground or is struck by lightning during flight, there may be a danger of fuel gas.

Therefore, must have fire resistance. For the design and realization of fuel storage and discharge functions, the following factors should be considered: (1) The fuel tank needs to be structurally strengthened; (2) The discharge function should ensure that all the fuel in the fuel tank can be discharged.

4.2 Distribution Subsystem

The distribution system also includes refueling system, oil supply system, oil pumping system and water removal system. The key elements of the fuel distribution system architecture design mainly include the following aspects:

(1) Refueling system: Civil aircraft generally use pressure refueling, and only small aircraft still use gravity refueling. The refueling port of civil aircraft is generally located on the wing, and two refueling ports can also be used at the same time on large aircraft. The number of fuel tanks in the aircraft determines the number of refueling valves, and its capacity determines the refueling flow and the size of the fueling pipeline, pump, and valve. The composition of the fuel tank determines the location of the high-level fuel sensor and the fuel port. The pressure refueling interface can use two types of devices, the refueling header and the refueling connector. The refueling header has higher reliability than the refueling connector but takes up more space. When the wing fuel tank is composed of multiple semi-closed small fuel tanks, the fueling pipeline needs to use a balance connecting pipe to connect the inner fuel tank and the outer fuel tank. When the fuel tank reaches the balance pipe, the fuel flows into the outer fuel tank, and then passes through The semi-closed valve is gradually added to the inner fuel tank until the predetermined fuel value is reached or the high-level fuel sensor is encountered. A schematic diagram of the balance tube is shown in Fig. 3.

    On some planes with simple distribution of fuel tanks, there is also a "cascading" refueling method. As shown in Fig. 4, the fuel tanks are refueled in sequence. Only after the initial fuel tank is full, will the second fuel tank be refueled due to pressure, until the end.

    On some larger civil aircraft, the refueling system and the fuel supply system are a combined system, which avoids the repeated laying of pipelines and reduces the weight of the aircraft.

(2) Oil supply system:There are generally two oil supply methods: gravity oil supply and pump oil supply. The gravity oil supply system is the simplest, but its oil supply capacity is limited and the flow rate is small. The most common way is to pump oil, commonly used pumps include electric pumps and ejector pumps.

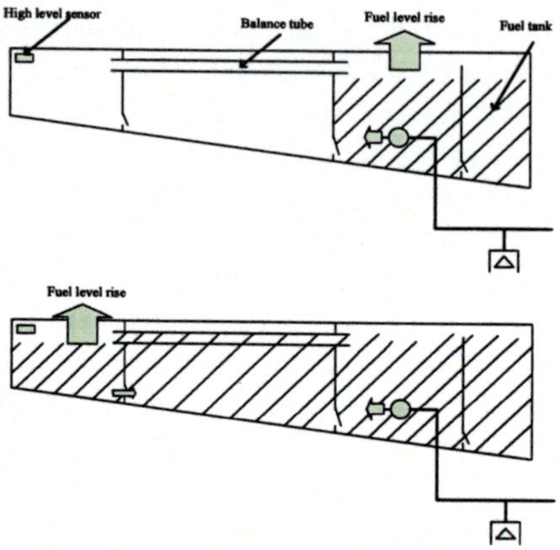

Fig. 3. Schematic diagram of the work of the balance tube

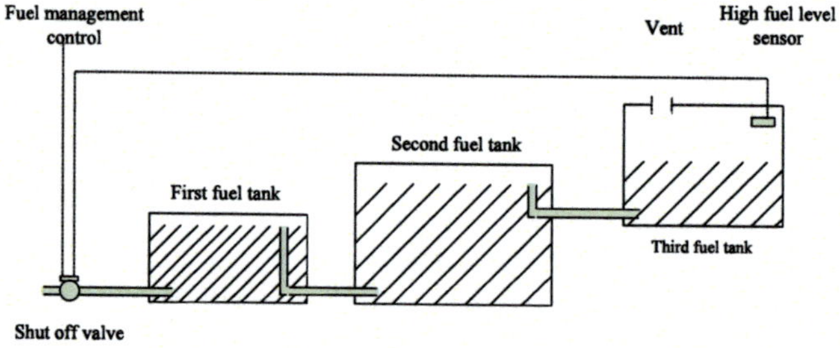

Fig. 4. Schematic diagram of stacked oiling

The electric pump can provide a large amount of fuel supply and pressure. When the power fuel cannot be provided, the electric pump can only be used to provide fuel supply power. The electric pump usually adopts two installation methods: the lower wall plate installation and the beam installation. The lower panel mount installs the fuel supply pump on the lowest level of the fuel supply tank. This installation method has good high-altitude performance of the pump and can reduce the amount of unusable fuel, but requires a large opening in the fuel tank, which is not conducive to the structural strength of the aircraft. The beam-type installation method is easy to install and disassemble, but this layout has an adverse effect on performance, and needs to increase the exhaust function, the structure is complex.

The ejector pump needs to be provided with power fuel. It has the advantages of higher reliability, easy installation and low cost. Because there is no electric facility involved, its stability and safety are also high, but due to the relatively high nozzle structure Precision, high quality requirements for fuel. The ejector pump is mainly used in some small aircraft that do not require high fuel flow.

In order to ensure flight safety and control the center of gravity of the aircraft, the fuel management system should control the distribution of fuel in each fuel tank to ensure the balance of the aircraft and the safe supply of fuel to the engine.

There are two main ways of oil supply sequence: over-control oil supply method and sequential oil delivery method. Most Boeing commercial aircraft generally use the over-control fuel supply method, which is characterized by not requiring a large number of pumps and having less valve opening and closing cycles. The sequential fuel delivery method needs to monitor the fuel quantity of the fuel supply tank. The engine is always directly supplied by the fuel supply tank. When the fuel in the fuel supply tank is less than a certain limit, the fuel supply system controls the fuel from the central fuel tank to the fuel supply tank. Until it is full, this kind of system will cause frequent movements of pumps and valves during the flight of the aircraft, which is easy to cause fatigue and wear of the equipment, the structure is complicated and the weight is increased.

4.3 Measurement Subsystem

The fuel subsystem includes the following functions: measuring fuel density, temperature, fuel position and detecting fault diagnosis. A typical architecture is shown in Fig. 5, which usually includes a fuel sensor, low fuel level sensor, density meter, temperature sensor, FRDC, and wiring harness.

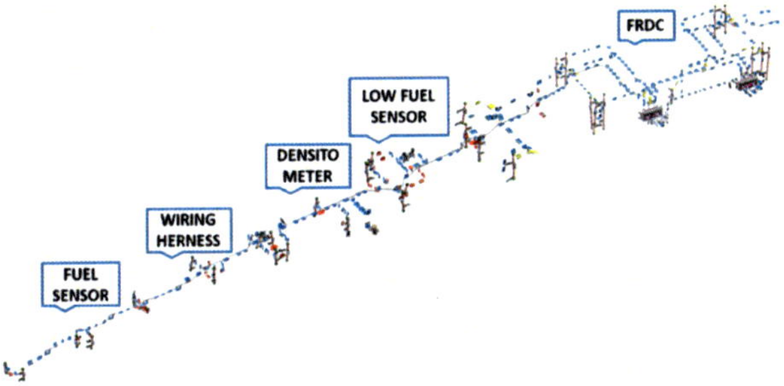

Fig. 5. Typical model fuel measurement and indication system

For the architectural design of measurement systems, reliability is a key factor. In the design process, only the part whose reliability cannot be improved is usually regarded as the redundant part, which is usually called backup parallel design. In the measurement system design, in order to make the output reliable, a control-monitoring architecture is usually adopted. This design calculates fuel quality by the main processor, which is then checked by the monitoring processor.

5 Simulation Evaluation of Fuel System

The physical architecture modeling of the fuel system is divided into three levels: system, subsystem, and component. Under the multiple operating conditions of the flight envelope, the initial fuel system architecture model, subsystem model, and component model are established in turn, finally, the subsystems and components are integrated into the fuel system architecture model [3, 4]. In the process of establishing the preliminary model, according to the physical model of the fuel system architecture, the framework of the system under the flight envelope is established, and the key parameters are determined. The fuel system can be divided into subsystems such as distribution, storage, and measurement. By establishing a preliminary model, the corresponding modules and connections between the subsystems are clarified, which lays the foundation for the establishment of subsequent subsystem models and component models.

In the subsystem modeling, the subsystem models are different under different working conditions in the flight envelope. For the distribution subsystem, it can be divided into: normal fuel supply and delivery conditions, APU starting fuel supply conditions, engine starting fuel supply conditions, Single pump failure oil supply conditions, etc. The establishment of the component model is the basis for the digital design and simulation analysis of the fuel system. The components can be sorted out, and corresponding models can be established for different series of components, which can be integrated into the fuel system through software. Finally, based on the preliminary model of the fuel system architecture, the subsystem model and component model are integrated into the fuel system architecture model, the corresponding model interface is established, the data is transmitted and exchanged, the detailed parameters of the system are determined, finally the fuel system architecture model is established [5, 6].

In the simulation analysis of the aircraft fuel system, it is necessary to consider the actual working environment of the aircraft under different flight states, and set the corresponding boundary conditions, so as to obtain the simulation operation data under multiple working conditions. The architecture simulation evaluation process is shown in Fig. 6.

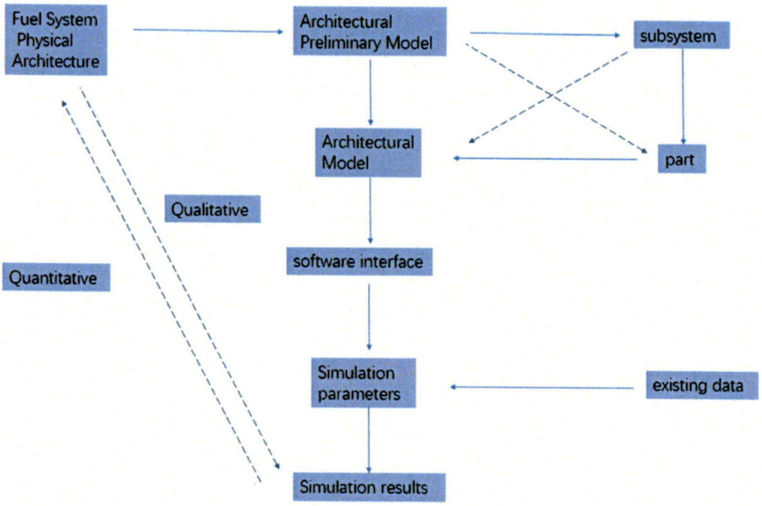

Fig. 6. Fuel system architecture simulation evaluation process

6 Fuel System Architecture Trade-Offs and Evaluation

The architecture trade-off assessment needs to comprehensively consider indicators such as weight, safety, maintainability, cost, airworthiness, etc., weigh and evaluate multiple solutions to determine the best solution and finally determine the system architecture.

The fuel system architecture evaluation should include: (1) Through functional simulation of different architecture schemes, confirm whether the selected architecture scheme can meet the corresponding functions and requirements, and verify that all functions defined in the operating concept can be implemented by the candidate architecture. (2) By simulating the performance of different architecture schemes, it is confirmed whether the architecture can meet the corresponding performance indicators, and the performance indicators defined in the verification requirements can be realized by the candidate architecture, such as the refueling rate, pressure, stress, etc. during the refueling process. (3) Evaluate the architectural design plan based on comprehensive indicators based on qualitative trade-off points such as weight, safety, maintainability, cost, and airworthiness, quantitative trade-off points such as pressure, stress, temperature, and fuel delivery rate, finally select the most suitable fuel. System structure. The details are shown in Table 1.

Table 1. Fuel system architecture trade-off guidelines

Judgment criteria	Detailed evaluation criteria	Scenario X			Scenario Y	
		Weights	Score	Weighted score	Score	Weighted score
Functional Performance						
Performance						
Reliability						
Maintainability						
Guaranteed						
Testability						
Safety						
Environmental Adaptability						
…						
Total						

7 Conclusion

This paper takes the civil fuel system architecture as the basis to form the architecture design and evaluation method, and mainly analyzes the civil aircraft fuel system requirement development, functional and physical architecture design, simulation analysis and architecture trade-off methods. Aiming at the evaluation and confirmation of the fuel system architecture of civil aircraft, this paper proposes that based on the establishment of the system model and simulation analysis, by integrating the models of various subsystems, components and structures of the fuel system, the matching and optimization are carried out by means of virtual simulation. Evaluate and iterate on the architecture design of the civil aircraft fuel system, and form the overall architecture design scheme of the civil aircraft fuel system.

The main research work of this paper includes:

(1) Research on the development method of civil aircraft fuel system requirements, including top-level demand capture, functional and non-functional requirements of civil aircraft fuel system, and fuel system safety requirements, to provide a design basis for the subsequent design of fuel system functional architecture and physical architecture and input in the security direction.
(2) By sorting out the functional levels of the fuel system, and establishing the logical relationship and system model between the functions based on Rhapsody and Flowmaster, the functional architecture of the fuel system is formed.
(3) Expounds the key elements of the civil aircraft fuel system and the physical architecture design of each subsystem. Aiming at the evaluation and confirmation of the fuel system architecture of civil aircraft, this paper proposes an evaluation based on the establishment of the system model and simulation analysis by integrating various trade-off points such as various subsystems, components and structures of the fuel system. Propose key trade-off elements for fuel system architecture trade-off assessment

References

1. Roy, L., et al.: Aircraft Fuel Systems. Wiley, Chichester, UK (2009)
2. Moyle, I., et al.: Introduction to Aircraft System Design and Development. Aviation Industry Press (2012)
3. Feng, Z., Gao, X., Liu, Y., et al.: Modeling and simulation analysis of an airplane fuel system. Aircr. Des. **27**(005), 65–67 (2007)
4. Bian, G., Zhong, Y., Liu, D.: Design and simulation on pipelines of aircraft fuel system based on CATIA/CAA. Comput. Aided Eng. **018**(003), 37–41 (2009)
5. Yaguo, L.: Research on the Calculation of Aircraft Fuel System. Northwestern Polytechnical University (2006)
6. Lv, Q.: The Design and Implementation of the Data Acquisition and management System of Aircraft Fuel System Test Bench. Nanjing University of Aeronautics and Astronautics (2014)

Research on Beidou Short Message Interference and Power Optimization Control Method

Jiuyuan Guo(✉), Zhan Zhang, Yaying Hu, and Boyuan Gong

COMAC Beijing Aircraft Technology Research Institute, Beijing, China
602539321@qq.com

Abstract. AS ground-based civil aviation airborne navigation system gradually develops to space-based system, global satellite navigation system has become an important part of airborne navigation system. Beidou short message, as a unique function of Beidou satellite navigation system, can provide tracking function for aircraft. With the release of relevant CTSO, the modification of aircraft will be gradually completed. However, flight test found it is possible that Beidou short message has potential interference with GPS. In view of the above problems, the interference was first analyzed and reproduced in this paper, power optimization control method was proposed and verified through experiments. The experiment result shows that this method can solve the problem that Beidou short message interferes with GPS and has little impact on the success rate of short message communication.

Keywords: Global satellite navigation system · Beidou short message · Aviation radio navigation

1 Introduction

Global navigation satellite system (GNSS) is proposed by the International Civil Aviation Organization (ICAO) to integrate various satellite navigation systems (GPS, GLONASS, Galileo, BDS), augmentation system and airborne navigation system (such as inertial navigation system, etc.), making it "the system of systems". On the basis of ensuring accuracy, availability, continuity and Integrity, the implementation of GNSS plan can accurately locate civil aviation aircraft, increase the flexibility of route design, improve the capacity of air traffic on the premise of safety, reduce the dependence of aircraft on ground radio navigation equipment and maintenance cost of ground navigation equipment, and finally realize global consistent operation.

Beidou short message communication is a unique function of Beidou satellite navigation system, it is also an important feature, differing from other satellite navigation systems. The communication process is not affected by terrain, environment, climate and other factors, it can be used for communication in remote areas and areas without other conventional means of communication. The Civil Aviation Administration of China issued the standard CTSo-2C604 in 2019, specifying that Beidou short message function is used to periodically send the obtained aircraft identification number, position,

Chinese Society of Aeronautics and Astronautics: CASTYSF 2022, LNEE 972, pp. 297–303, 2023.
https://doi.org/10.1007/978-981-19-7652-0_28

speed, time and other parameters to the ground platform to realize the tracking function of aircraft and ensure the safety of civil aviation operation.

2 Beidou Short Message Interference

Beidou short message has the capability of two-way communication between user terminals, and between user terminal and command aircraft. So that Beidou system can transmit its own location and other information without relying on other communication means, which is beneficial for monitoring and management of command center [30]. In addition, each Beidou user terminal has a unique user ID for user identification and encryption. The tracking process of Beidou short message to civil aircraft is shown in Fig. 1:

(1) The civil aviation aircraft uses airborne Beidou short message user terminal to apply for communication, encrypts the content of the short message and the user address of the receiver and sends it to the satellite.
(2) After receiving the communication application, the satellite will forward it to the ground center. The ground center will decrypt the signal and encrypt again, then send it to the satellite for continuous outbound broadcasting, which will be forwarded to the monitoring hall through the satellite.
(3) The monitoring hall receives the outgoing broadcast message, decrypts the communication content, and realizes civil aircraft tracking.

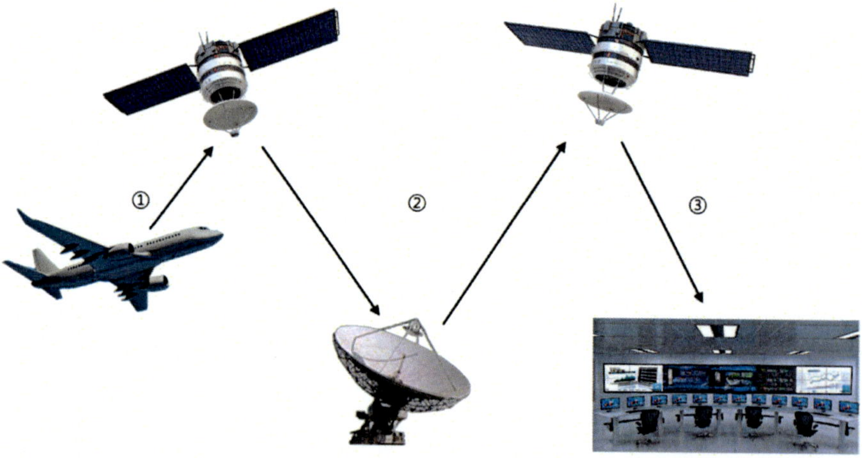

Fig. 1. Tracking process of Beidou short message to civil aircraft

The transmitting frequency of the airborne Beidou short message antenna is 1615 MHz, which is close to the receiving frequency of the Airborne GPS antenna 1575 MHz, and both are installed on the back of the fuselage. Through the flight test data, it is found that, as shown in Fig. 2, the DOP value and horizontal integrity limit of the aircraft body GPS receiver have increased significantly and regularly after Beidou is turned on, GPS altitude occurs step change and the time of DOP value increase overlaps with the time when the Beidou short message is transmitted. If the DOP value is too large, it indicates that the GPS positioning result is unreliable, which is easy to induce GPS related alarms.

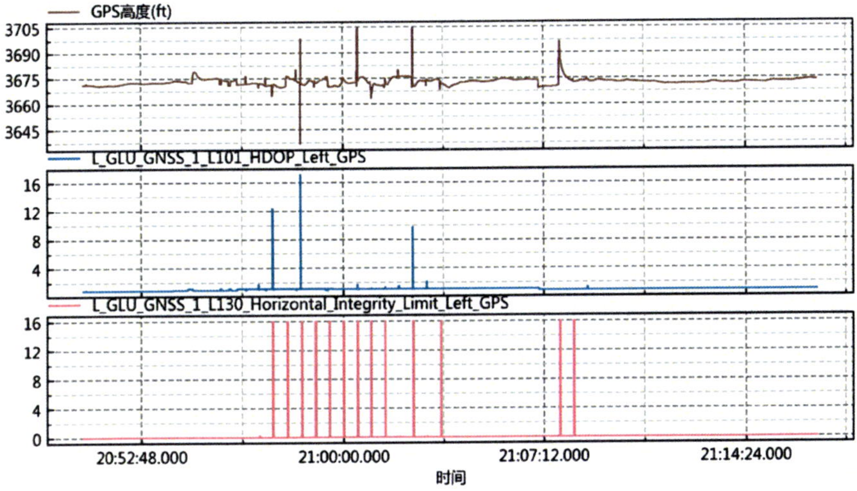

Fig. 2. Beidou short message interferes with GPS flight test data

3 Fault Reproducing

In view of the phenomenon that the airborne Beidou short message interferes with the Airborne GPS during flight test, the standby equipment is used in the laboratory to reproduce the interference. As shown in Fig. 3, the test process is carried out in a dark room. The GPS satellite signal is generated by satellite navigation simulator outside the dark room and transmitted by transmitting antenna in the dark room. Transmitting power of the airborne Beidou short message antenna is set to 10W and satellite navigation simulator is controlled to make the transmitting intensity of GPS signal gradually weaken from strong. The test results are shown in Table 1.

The test results show that the carrier noise ratio of GPS receiver decreases gradually in the process of GPS signal changes from strong to weak. When the carrier noise ratio becomes 34, the interference phenomenon in flight test will be reproduced, which is, at the moment of short message transmission, GPS receiver positioning data output is invalid lasting for 2 s. Keep GPS signal strength unchanged, make the carrier noise ratio of GPS receiver 34, and gradually reduce the transmission power of Beidou short message antenna until the GPS interference phenomenon is eliminated. The test results are shown in Table 2, when the change of Beidou short message transmission power is −2.9 dbm, the output positioning result of GPS receiver is normal, which means the interference disappears.

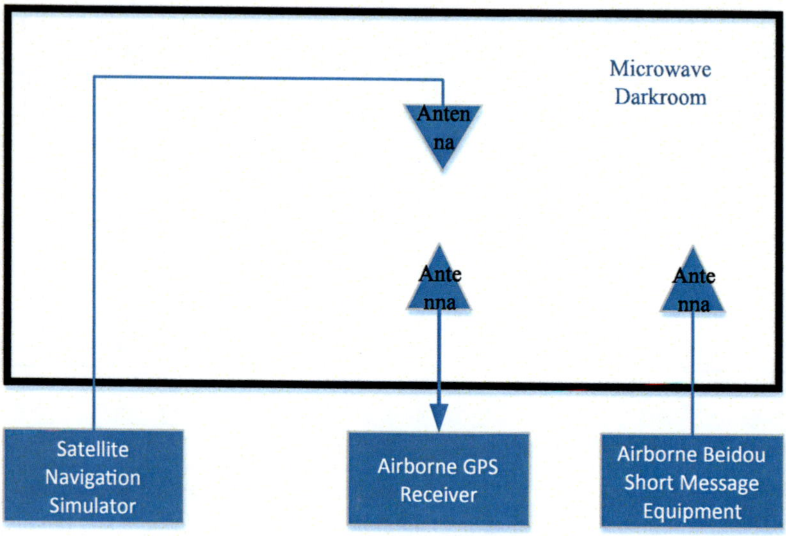

Fig. 3. Fault reproduing test block diagram

Table 1. Fault reproducing test results

No	GPS receiver carrier noise ratio	Whether reproduce GPS interference	Remarks
1	42	No	
2	39	No	
3	37	No	
4	34	Yes	At the moment of short message transmission, the positioning data output is invalid lasting for 2S

4 Power Optimization Control Method

looseness-1The fault reproduing test shows that the interference of Beidou short message transmission to GPS belongs to power suppression. When the GPS signal is weak, excessive Beidou short message transmission power will interfere with GPS and cause positioning failure. However, reducing the transmission power of Beidou short message will affect the communication success rate of Beidou short message. To solve the above problems, a Beidou short message power optimization control method is proposed in this paper to avoid GPS interference and ensure the communication success rate of Beidou short message. The airborne Beidou short message terminal obtains the carrier noise ratio output by GPS receiver through the 664 bus, and outputs different transmission

Table 2. Fault elimination test result

No	Power change of Beidou short message (dB)	Whether reproduce GPS interference	GPS Receiver carrier noise ratio	Remarks
1	0	Yes	34	
2	−0.9	Yes		
3	−1.9	Yes		
4	−2.9	Yes		At this time, the output positioning result of GPS receiver is normal

power according to the carrier noise ratio. Transmission power can be determined by the following formula.

$$P = \begin{cases} P_0 & , CN_0 > TH \\ P_0 + \Delta P & , CN_0 \leq TH \end{cases}$$

In the formula, P_0 is transmitting power of Beidou short message under normal conditions, ΔP is power value adjusted by airborne Beidou short message terminal, CN_0 is GPS output signal-to-noise ratio obtained from the bus, TH is carrier noise ratio threshold. The value of and TH can be determined by the method used in the fault reproducing test.

The power optimization control method is verified by flight test. The test results are shown in Fig. 4, as the power optimization control method is used in the flight test process, DOP value and Horizontal Integrity Limit do not occur step change, which means Beidou short message does not interfer with GPS. As a comparative test, it is found that both Dop value and Horizontal Integrity Limit occurs exception-step change, when turning off the power optimization control after flight. From the test results, it can be seen that the power optimization control method proposed in this paper can solve the interference of Beidou short message to GPS.

In order to evaluate the influence of power control method on short message communication success rate, this paper compares communication success rate of this flight verification test with the rate without applying this method. As shown in Table 3, the communication success rate of this test is 95.59%, which is equivalent to the communication success rate in previous tests. Therefore, the impact of this method on the communication success rate is limited and within acceptable range.

5 Conclusion

In this paper, the interference of Beidou short message to GPS is analyzed and reproduced in the laboratory. According to the analysis results, the power optimization control method is proposed. Experiments show that this method can avoid the interference of

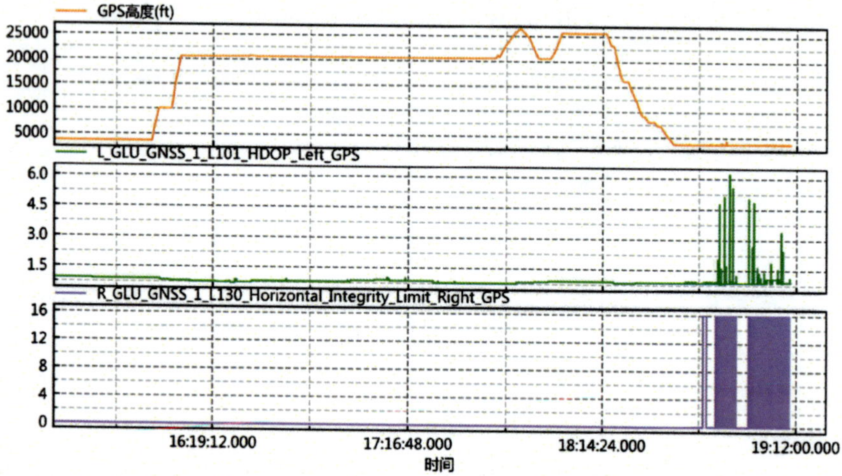

Fig.4. Flight test results of power optimization control method

Table 3. Comparison of communication success rate

	Communication times	Success times	Success rate
This experiment	318	306	95.59%
Previous experiment 1	246	238	96.75%
Previous experiment 2	382	371	97.12%

Beidou short message to GPS and has little impact on the success rate of short message communication.

References

1. Zhaohui, M.: Research on Key issues of Air Traffic System Based Upon Performance-Based Navigation. Nanjing University of Aeronautics and Astronautics, Nanjing (2017)
2. Yanqun, W.: Application Research and Simulation of Multi-satellites Integrated Navigation for PBN. Civil Aviation University of China, Tianjin (2016)
3. Weichuan, P.: Research on Realtime Airborne ARAIM for Precision Approach of Civil Aviation. Shanghai Jiao Tong University, Shanghai (2020)
4. Enge, P., Enge, N., Walter, T.: Aviation benefits from satellite navigation. New Space **3**(1), 19–35 (2015)
5. Haoyuan, Y.: General aviation command and surveillance system based on BDS. Civil Aviation University of China, Tianjin (2016)
6. Tianqi, X.: General Aviation Flights Conflict Detection and Traffic Alert Based on BeiDou Short Messages Servers. Civil Aviation University of China, Tianjin (2019)
7. Yilong, Z., Ruihua, L., Jian, W.: Design and implementation of general aviation surveillance system based on RDSS. Aeronaut. Sci. Technol. **29**(02), 55–60 (2018)

8. Xingjian, Z., Zhenhua, W., Yifei, Z.: Study on service performance test of Beidou RDSS in general aviation surveillance. J. Civ. Aviat. Univ. China **35**(3), 1–5 (2017)
9. Xiaoxing, Y., Ming, C., Xiaojuan, G.: Design of reliable remote communication system based on Beidou satellite. Comput. Eng. **43**(3), 62–68 (2017)
10. Ruihua, L., Haoyuan, Y.: Design of long-messages communication based on Beidou General Aviation command and monitor system. Comput. Eng. Appl. **51**(S1), 104–107 (2015)
11. Siliang, D., Feng, J., Dequan, W.: Application of beidou satellite navigation system in ground surveillance system. Ordnance Ind. Autom. **32**(10), 80–83 (2013)
12. Xianhua, Z., Yi, Z., Zhengrong, L., Feixue, W.: Method short-time pulse interference blanking satellite navigation receiver. J. Data Acquis. **28**(1) (2013)

Robust Flutter Analysis Considering Control Surface Aerodynamic Uncertainty

Haigang Wang(✉)

COMAC Flight Test Center, Pudong District, Shanghai 201233, China
wanghaigang1@comac.cc

Abstract. To reduce the conservation of robust flutter analysis results, only the uncertainty of control surface unsteady aerodynamic was modeled. The $\mu - k$ approach was used to compute the robust flutter speed. The results indicate that robust flutter speed considering only the control surfaces unsteady aerodynamic is 716% less than entire surface case, also show that the approach for robust flutter analysis is applicable to multiple flutter modes in engineering practice.

Keywords: Robust flutter · Control surface · Unsteady aerodynamic · Uncertainty · Structure singular value

1 Introduction

Traditional aeroelastic system stability can be modeled as linear system stability, flutter boundary is defined by solving system matrix single value. Lind firstly applied robust control theory to flutter problem which take uncertainty into account [1]. Structural singular value became the criteria of stability for robust flutter. From system equation point of view, the uncertainty sources including mass, stiffness, damping, and aerodynamic.

For stiffness uncertainty, Gu Yingsong proposed robust flutter margin is calculated for an airfoil with flap freeplay uncertainty with uncertain nonlinear operator method [2]. Sebastian Heinze proposed two useful methods for taking mode shape variation one is increasing the modal base by adding additional vectors, the other is a iterative approach to capture mode shape variations without the need of mode shape derivatives [3]. Dai Yuting using robust flutter analysis considered the mode shape variations that result due to variations of structural properties [4]. For mass uncertainty, Sebastian Heinze presented an approach to assess critical fuel configurations using robust flutter analysis, the worst-case flutter speed and the corresponding worst-case fuel configuration is found [5]. For aerodynamic uncertainty, general modeling method is introduced [6] but more useful description of uncertainty to engineering practice is limited. In this study, only aerodynamic uncertainty of concerned area is considered to reduce robust flutter conservation, μ-k approach [7] is applied to an aircraft vertical fin with control surface aerodynamic uncertainty.

Chinese Society of Aeronautics and Astronautics: CASTYSF 2022, LNEE 972, pp. 304–311, 2023.
https://doi.org/10.1007/978-981-19-7652-0_29

2 Uncertainty Modeling

There are two possible sourcex of the control surface aerodynamic uncertainty, one is from the error or assumption in mathematical modeling process, the other is physically existed but not considered or modeled accurately. For example, traditional flutter model use Doublet Lattice Method (DLM) which is based on linear assumption, the aerodynamic interpolation spline can be source of uncertainty. Physically it can be complex local aeroflow condition, flight condition uncertainty etc. Aeroflow separation normal starts form trailing edge, which can cause control surface aerodynamic uncertainty, analysis shows that uncertainty at trailing edge is significantly high than leading edge [8]. Besides, for aircraft flutter boundary in transonic regime, nonlinear phenomenon such as shock wave is difficulty to accurate modeled [9].

Flutter equation in Laplace domain can be written in the non-dimensional form [3]:

$$[Mp^2 + (L/V)Dp + (L^2/V^2)K - (\rho L^2/2)Q(p, m)]\eta(p) = 0 \tag{1}$$

where M, D, K and $Q(p, m)$ are the mass, damping, stiffness and the aerodynamic transfer matrix, these matrices are $f \times f$, where f is the number of modes used in the modal basis. $\eta(p)$ is the vector of modal coordinates. Further, V is the airspeed, L is the aerodynamic reference length, ρ is the air density and M is the Mach number. The non-dimensional Laplace variable is denoted $p = g + ik$, where g is the damping and k is the reduced frequency, to compute the critical flutter boundary let g = 0.

To define the aerodynamic uncertainty model, it is necessary to partition the aerodynamic matrix [10]:

$$Q_0(ik, m) = R_0 S_0(ik, m) \tag{2}$$

where the $S_0(ik, m)$ defines the mapping from modal coordinates to lifting surface pressure coefficients, and R_0 defines the mapping from lifting surface-pressure coefficients to generalized forces. Note that R_0 is independent of the reduced frequency and the Mach number because it is made up of information from the spline interface between structural and aerodynamic models and modal eigenvector information.

Then the uncertainty aerodynamic transfer function can be written:

$$\begin{aligned} Q(ik, m) &= R_0(I + w_j\delta_j E_j)S_0(ik, m). \\ &= Q_0(ik, m) + w_j\delta_j Q_j(ik + m) \end{aligned} \tag{3}$$

where, w_j defines uncertainty bound i.e. 0.1 means 10% uncertainty is assigned to the specified boxes in aerodynamic model, uncertainty parameter δ_j meets $|\delta_j| \le 1$, E_j is the matrix with $e_{jj} = 1$.

$$Q_j(ik, m) = R_0 E_j S_0(ik, m) \tag{4}$$

Generalized description can be written as:

$$Q(ik, m) = Q_0(ik, m) + \sum_{j=1}^{n} w_j\delta_j Q_j(ik, m) \tag{5}$$

where n is the total pressure coefficient number, rewrite as:

$$Q(ik, m) = Q_0(ik, m) + V_Q(ik, m)\Delta_Q W_Q \tag{6}$$

where, $V_Q = [Q_1 Q_2 \ldots Q_n]$

$$\Delta_Q = diag(\delta_1 I_{f\times f}, \delta_2 I_{f\times f}, \ldots, \delta_n I_{f\times f})$$

$$W_Q = \begin{bmatrix} w_1 I_{f\times f} \\ w_2 I_{f\times f} \\ \vdots \\ w_n I_{f\times f} \end{bmatrix}$$

3 Nominal Flutter Analysis

An aircraft vertical fin-rudder model was used for analysis, the NASTRAN structural model as shown in Fig. 1, ZAERO aerodynamic model as shown in Fig. 2, in which the dash blue line indicates the rudder area.

Nominal flutter speed was determined using ZAERO, results indicates two flutter modes with difference frequency, the lower frequency mode is the critical mode which was due to coupling between first bending and rudder rotating, flutter speed is 355kn, flutter frequency is 16 Hz, i.e. the low frequency mode. The other is coupling among first torsional, rudder rotating and second bending, flutter speed 438 kn and frequency 32 Hz i.e. the high frequency mode. The V-g and V-f plot of flutter results is shown in Figs. 3 and 4.

For robust flutter analysis, the matrix in flutter equation were generated with OUTPUT function of ZAERO [11].

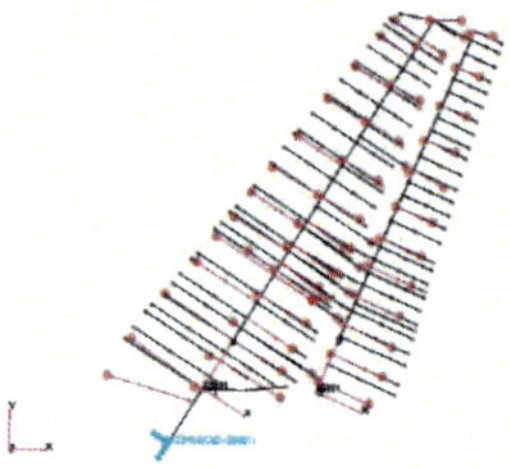

Fig. 1. Structure model.

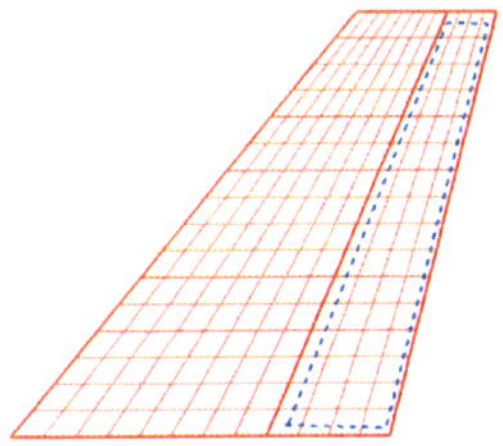

Fig. 2. Aerodynamic model.

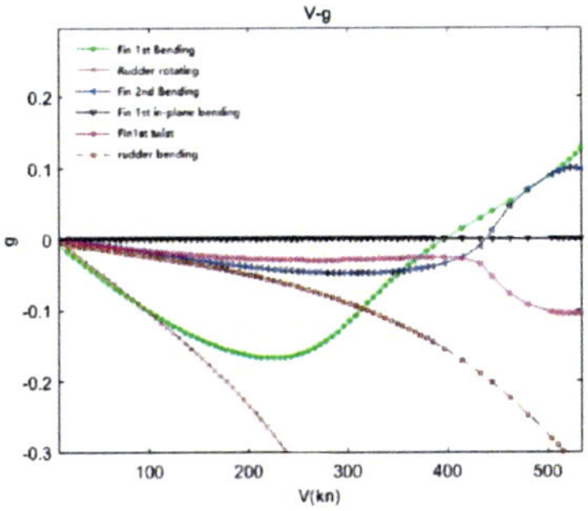

Fig. 3. V-g curve

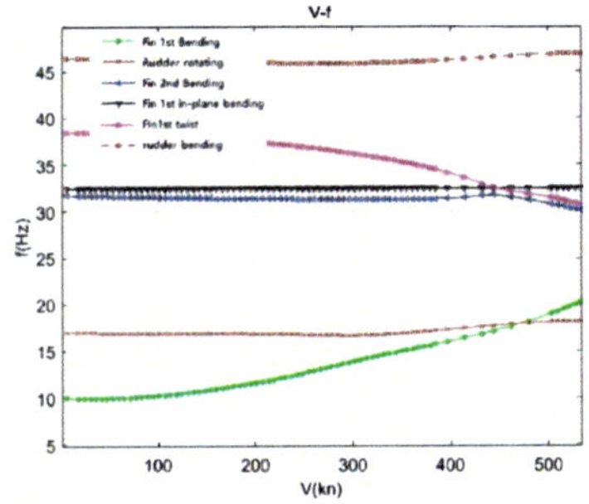

Fig. 4. V-f curve.

4 Robust Flutter Analysis

4.1 Robust Flutter Equation

Combing Eqs. (6) and (1) and uncertainty:

$$F_0(p, q, m)\eta = V(p, q, m)w \tag{7}$$

where, $F_0(p, q, m) = [Mp^2 + (L/V)Dp + (L^2/V^2)K - (\rho L^2/2)Q_0]$, $V(p, q, m) = (\rho L^2/2)V_Q(p, m)$

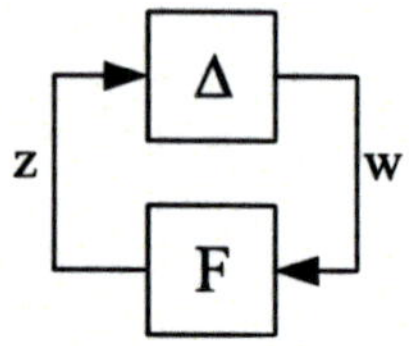

Fig. 5. Feedback model of uncertainty flutter equation

The uncertainty flutter equation feedback structure model as shown in Fig. 5: where, the input and output of uncertainty matrix being defined as:

$$z = W\eta \tag{8}$$

$$W = \Delta z \tag{9}$$

The transfer function F between input W and output z:

$$F(p, q, m) = WF_0^{-1}(p, q, m)V(p, q, m) \tag{10}$$

Rewrite using general aerodynamic force input ξ and general vector η in the modal base:

$$F_0(p, q, m)\eta = V(p, q, m)w + \xi \tag{11}$$

Transfer function between input ξ and output η which presents as Linear Fraction structure of uncertainty system as shown in Fig. 6:

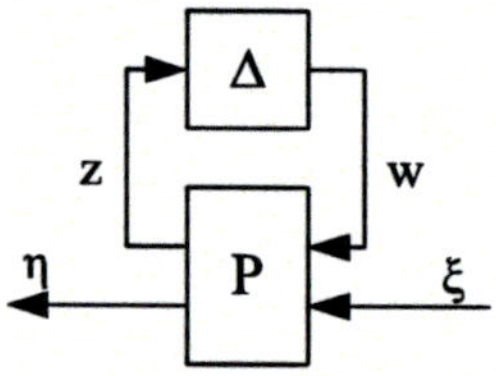

Fig. 6. Linear Fraction structure of uncertainty system

Then the uncertainty state space equation is as follow:

$$\begin{bmatrix} z \\ \eta \end{bmatrix} = \begin{bmatrix} P_{11} & P_{12} \\ P_{21} & P_{22} \end{bmatrix} \begin{bmatrix} w \\ \xi \end{bmatrix} = \begin{bmatrix} WF_0^{-1}V & WF_0^{-1} \\ F_0^{-1}V & F_0^{-1} \end{bmatrix} \begin{bmatrix} w \\ \xi \end{bmatrix} \tag{12}$$

Transfer function matrix between ξ and η:

$$\eta = f_u(\mathrm{P}, \Delta)\xi = [P_{22} + P_{21}\Delta(1 - P_{11}\Delta)^{-1}P_{12}]\xi \tag{13}$$

where $f_u(P, \Delta)$ is the linear fractional transformation, if system is nominal stable, the only instability term is $f_u(1 - P_{11}\Delta)^{-1}$ which is equivalent to $P_{11} - \Delta$ stability,with

$\|\Delta\| \leq 1$ and P_{11} is the flutter transfer function $WF_0^{-1}V$. The stability of input-output system is equivalent to flutter feedback stability.

The transfer function of flutter feedback system is:

$$F(p, q, m) = P_{11}(p, q, m) \tag{14}$$

4.2 $\eta - k$ Curve

If the system is stable at specified flight condition, and structural singular value meet:

$$\mu(k, q, m) = \mu\left[F(ik, q, m)\right] < 1 \tag{15}$$

Then the system is robust stable at specified condition. If all the structural singular value $\mu(k_j, q, m) < 1$, then system is robust stable. The critical robust flutter dynamic is determined at $\mu(k, q, m) < 1$, which is solved using μ toolbox of MATLAB [12]. A 10% aerodynamic uncertainty is used for all patches in control surface only, as the results shown in Figs. 7 and 8 is the result of whole fin 10% aerodynamic uncertainty case. Both shown two peak of μ which corresponding to the two flutter mode of nominal model. The robust flutter speed is shown in Table 1. Compare to nominal model, with only control surface 10% uncertainty, the low frequency robust flutter speed decreased by 10.6%, while high frequency mode decreased by 3.9%, which indicates low frequency flutter is more sensitive to control surface aerodynamic uncertainty, this also consists with the flutter mode tracking result of ZAERO. For the whole fin 10% aerodynamic uncertainty case, both low and high frequency robust flutter speed decreased by approximately 16%, which is obviously high than rudder case.

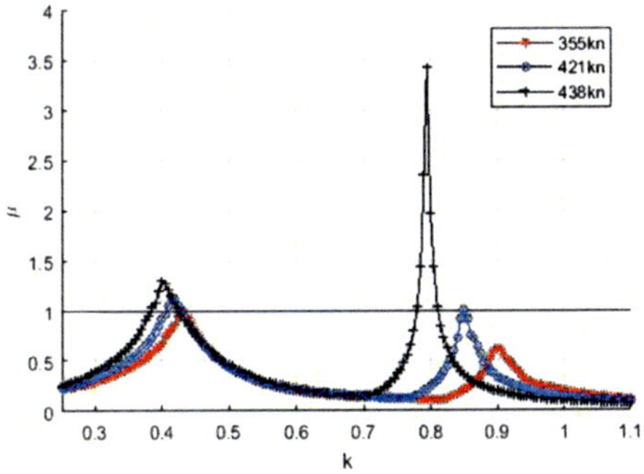

Fig. 7. $\mu - k$curve of control surface 10% aerodynamic uncertainty

5 Conclusion

For flutter problem with the control surface, the local aerodynamic uncertainty modeling of the control surface can effectively reduce the conservatism of robust flutter calculation results. At the same time, the calculation shows that the robust flutter analysis based on the approach is applicable to the presence of multiple flutter modes, which is of value for aircraft flutter design and flutter optimization in engineering practice.

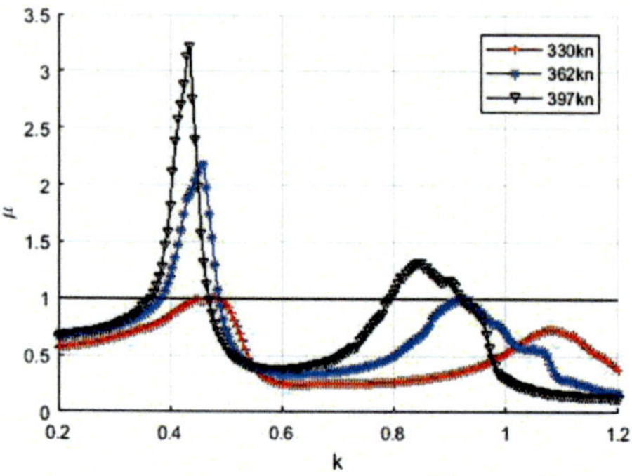

Fig. 8. $\mu - k$ curve of fin 10% aerodynamic uncertainty

Table 1. Robust flutter results

Model case	Low frequency mode	High frequency mode
Nominal model	397	438
Rudder aerodynamic 10% uncertainty	355	421
Fin aerodynamic 10% uncertainty	330	362

References

1. Lind, R., Brenner, M.: Robust Flutter Margin Analysis that Incorporates Flight Data. NASA,TM-19980206543 (1998)
2. Yingsong, G., Zhichun, Y.: Robust flutter analysis of an airfoil with flap freeplay uncertainty. In: 49th AIAA/ASME/ASCE/AHS/ASC Structures, Structural Dynamics, and Materials Conference. Schaumburg, IL (2008)
3. Heinze, S., Borglund, D.: Robust flutter analysis considering mode shape variations. J. Aircr. **44**(3), 1070–1074 (2007)
4. Yuting, D., Chao. Y.: Intrusive flutter solutions with stochastic uncertainty. Acta Aeronautica et Astronautica Sinica 2182–2189 (2014)
5. Heinze, S.: Assessment of critical fuel configurations using robust flutter analysis. J. Aircr. **44**(6), 2034–2039 (2007)
6. Yuting, D., Zhang, Z., et al.: Unsteady aerodynamic uncertainty estimation and robust flutter analysis. In: 29th AIAA Applied Aerodynamics Conference. Honolulu, Hawaii, AIAA-2011–3517
7. Borglund, D.: The μ-k method for robust flutter solutions. J. Aircr. **41**(5), 1209–1216 (2004)
8. Wright, J.R., Cooper, J.E.: Introduction to Aircraft Aeroelasticity and Load, vol. 194. Wiley, England (2007)
9. Talay, T.A.: Introduction to the Aerodynamics of Flight, pp. 54–55. NASA-SP-367, Washington D.C
10. Kun, Z.: Robust Flutter and Flutterometer Research, p. 62. Nanjing University of Aeronautics and Astronautics, Nanjing (2007)

11. ZONA Technology, Inc. ZAERO User's Manual, pp. 417–421. USA:ZONA Technology, Inc (2017)
12. Balas, G.J., Doyle, J.C.: μ-Analysis and Synthesis Toolbox for Use with Matlab. MathWorks Inc, Natick MA (1995)

Check for updates

Design and Performance Analysis of Aviation Axial Fan with Modified Variable Circulation Method

Tang Jincheng[1], Xie Yongqi[1](✉), Wu Hongwei[2], Gao Hongxia[1], and Yu Jianzu[1]

[1] School of Aeronautic Science and Engineering, Beihang University, Beijing 100191, China
xyq@buaa.edu.cn

[2] School of Physics, Engineering and Computer Science, University of Hertfordshire, Hatfield AL10 9AB, UK

Abstract. In aircraft oil cooling system, axial fan has been applied due to its characteristics of high rotation speed and pressure rise, as well as small flow rate and small volume. In this article, a modified variable circulation method for the aerodynamic design of axial fan with inlet guide vanes (IGVs) is newly developed. The design of aviation axial fan with advanced airfoil is performed based on a simplified three-dimensional (3D) flow theory. The effects of several different control parameters such as prewhirl parameters i.e. −0.75, −0.65, −0.55 and −0.45, circulation indexes i.e. 0.4, 0.45, 0.5, and number of IGVs and rotor blades i.e. R19S17, R17S15 and R15S13 on the fan performance are numerically analyzed in a systematic manner. The optimization design of axial fan with IGVs is proposed. Numerical study results indicate that the larger the prewhirl parameter and variable circulation index, the higher pressure rise and the wider effective working range of fan. The more the blade number, the higher pressure rise of fan, while the efficiency is normally kept the same. The effect mechanisms of above parameters on the fan performance are revealed. The current work can provide reference for the optimal design of advanced aviation axial fan.

Keywords: Axial fan · Inlet guide vane · Modified variable circulation · Prewhirl · Effective working range

1 Introduction

Axial fan is an important component of high performance aviation oil cooling system. Its main function is to provide cooling air with sufficient flow and pressure for system, overcome system resistance and remove the heat of lubricating oil in heat exchanger. The rotation speed and pressure rise of the fan used for aviation oil cooling system can exceed 8000 rpm and 5500 Pa, which is much higher than that of civil axial fans (≤2900 rpm and ≤4900 Pa). It has the characteristics of low flow rate and high pressure rise [1]. In addition, the volume and weight of the aviation axial fan should also need to meet the aviation equipment standards strictly. This becomes more challenging in designing high-performance aviation axial fan.

Chinese Society of Aeronautics and Astronautics: CASTYSF 2022, LNEE 972, pp. 312–323, 2023.
https://doi.org/10.1007/978-981-19-7652-0_30

In general, the volume flow rate, pressure rise and inlet airflow angle of aviation fan would be changed with the variation of aircraft's flight dimension, velocity and attitude [2]. Moreover, the heat exchanger behind fan in aviation oil cooling system may be blocked by dust and other objects during long-term operation, this could cause the pressure resistance of system to change and will also affect the operating condition of the fan. Therefore, the mass flow range of fan operating with high efficiency (called effective working range) is also an important evaluation indicator for the aviation fan used for oil cooling system. The axial fan with inlet guide vanes (IGVs) has the advantages of high pressure rise, wide effective working range, and small volume and weight [3]. According to the characteristics and requirements of aviation fan, it could be an ideal option to use axial fan with IGVs in aviation oil cooling system.

Conventional axial flow fan is often designed with equal circulation method [4]. Due to its characteristics of low flow rate and high pressure rise of aviation axial fan, using equal circulation method for designing could lead to excessive blade load. The aerodynamic performance at the blade root would be deteriorated. As a consequence, the variable circulation method which can reduce the load at blade boot, improve the efficiency and reduce the blade twist could be better for aviation axial fan design. Wang et al. [5] designed the twisted blade of micro axial fan with variable circulation method and analyzed the effects of variable circulation index on the fan aerodynamics performance. Wang et al. [6] and Liang et al. [7] also studied the axial fan with equal-variable circulation method. Their research improved the fan aerodynamic performance. At present, the twisted blade design of axial fan has achieved abundant research results [8]. About the study of fluid machinery with IGVs, Liu et al. [9] investigated the influence of IGVs on the performance characteristics of a small axial flow fan. They found the the IGVs can prove the static characteristics and reduce the noise of fan. The effects of different design parameters on the performance of axial fan with IGVs were numerically analyzed by Matai and Kharati-Koopaee [10, 11]. And Liu et al. [12] reviewed the previous researches of prewhirl regulation for pump and compressor with IGVs. Moreover, the parameter effect analysis and optimization design is also an important process for axial fan design. Wang et al. [13] and Li et al. [14] conducted related investigations. The optimization design of advance high performance axial fan involves a lot of design parameters such as hub ratio, tip clearance and others [15]. It is a complex process which needs a large amount of calculation to obtain the optimal result.

It appears from the previous investigations that the design and optimization method for axial fan has been quite mature. However, the variable circulation method for axial fan mainly focus on the fan with non IGVs. There is little study relevant to the high rotation speed and high pressure rise axial fan with IGVs, which is suitable for aviation oil cooling system. Due to the prewhirl airflow in the rotor inlet, the variable circulation method with the assumption that there is no torsion velocity of airflow in rotor inlet is no longer suitable for the axial fan with IGVs. In view of this problem, this paper modifies the formula of variable circulation method, and carries out the aerodynamic design of advanced aviation fan blade based on NACA65 airfoil. Moreover, the effects of several design parameters such as prewhirl parameter of IGVs, number of IGVs and rotor blades, and variable circulation index on fan performance are numerically studied.

2 Design of the Axial Fan with IGVs

2.1 Main Design Parameters

According to the working characteristics of aviation axial fan, the form of inlet guide vane and rotor that has wider effective working range is selected for the current fan design. Compared with other designs, the difference of the current one is mainly reflected in the guide vane. At the same time, the variable circulation design of rotor blade is also the focus and challenge of fan design. Therefore, the current work mainly focuses on studying three design parameters, namely the guide vane prewhirl parameter, number of IGVs and rotor blades, and variable circulation index.

The diameter of fan is derived from its space constraints, and the hub ratio is calculated by an optimization program through considering the efficiency and total pressure performance. The number of IGVs and rotor blades is studied according to the coprime principle within the empirical range. The recommended range of the prewhirl parameter of IGVs is −0.5 ~ −0.6 in ordinary axial fan. In this paper, a wider range of prewhirl parameters is studied due to the characteristics of aviation axial fan. Considering the influence of prewhirl airflow and the limitation of the equivalent diffusion coefficient at blade root, the variable circulation indexes of 0.4, 0.45 and 0.5 were selected for the fan design. The main design parameters of fan are shown in Table 1.

Table 1. Main design parameters of fan.

Design parameters	Value
Pressure rise (Pa)	5500
Design volume flow rate (m^3/min)	130
Rotation speed (rpm)	8873
Tip diameter (mm)	246
Hub ratio	0.64
Prewhirl parameter n	−0.75, −0.65, −0.55 and −0.45
Variable circulation index x	0.4, 0.45 and 0.5
Number of IGVs and rotor blades	R19S17, R17S15 and R15S13

2.2 Modified Variable Circulation Formula

A simplified three-dimensional flow theory is used for the rotor blade design, and the radial balance equation of airflow in the rotor [4] can be written as follows:

$$\frac{dP_{th}}{dr} = \frac{1}{2}\rho[\frac{1}{r^2}\frac{d(rc_u)^2}{dr} + \frac{dc_z^2}{dr}] + T\frac{ds}{dT} \tag{1}$$

where P_{th} represents the theoretical pressure rise; c_{u} and c_{z} are the torsional velocity and axial velocity of the airflow, respectively. r is the radius of blade.

The equation of variable circulation of the blade is define as [4]:

$$r^x c_u = \text{constant} \tag{2}$$

where x is the variable circulation index, and nomally its range is 0 ~ 1.

The airflow torsional velocity at rotor inlet is not zero in axial fan with IGVs. According to the relationship of velocity circulation between the IGVs and the rotor blade, the equation of the airflow torsion velocity at rotor inlet and the theoretical pressure rise of the fan can be expressed as follows:

$$P_{th} = \rho \omega r \Delta c_u = \rho \omega r \frac{1}{n} c_{1u} \tag{3}$$

where ρ is the density of airflow; ω is the rotation speed of fan; n is the prewhirl parameter of IGVs; Δc_u is the difference of airflow torsional velocity at the rotor inlet and outlet.

Due to the airflow velocity is different in the radial direction of fan, there must be a radius r_{m} where the airflow axial velocity c_{z} at this radius is equal to the average axial airflow velocity c_{zm} of fan. Ignoring the change of airflow temperature and combining Eq. (2) and Eq. (3), the following equation can be obtained by integrating Eq. (1):

$$\int_{r_m}^{r} \frac{dc_{1z}^2}{dr} dr = \int_{r_m}^{r} 2\omega \frac{1}{n} A_1 (1-x) r^{-x} dr - \int_{r_m}^{r} A_1^2 (2-2x) r^{-1-2x} dr \tag{4}$$

Thus, the relationship between the airflow axial velocity c_{1z}, c_{2z} at the rotor inlet and outlet with the radius r can be written as follows:

$$c_{1z} = \sqrt{c_{zm}^2 + 2\omega \frac{1}{n} A_1 r^{1-x} - r_m^{1-x} - A_1^2 \frac{x-1}{x} (r^{-2x} - r_m^{-2x})} \tag{5}$$

$$c_{2z} = \sqrt{c_{zm}^2 + 2\omega \frac{1}{1+n} A_2 r^{1-x} - r_m^{1-x} - A_2^2 \frac{x-1}{x} (r^{-2x} - r_m^{-2x})} \tag{6}$$

where A_1, A_2 is the circulation constant of Eq. (2) at the rotor inlet and outlet, respectively.

When the flow rate, pressure rise and other design parameters of fan are given, the modified variable circulation formula of the axial fan with IGVs with different indexes can be obtained through Eqs. (5) and (6). Compared with the variable circulation formula used for the fan with no IGV, the modified variable circulation formula includes the prewhirl parameter n. That is, the prewhirl effect of IGVs on airflow is considered.

3 Numerical Method

The flow will be analyzed using commercial software ANSYS CFX 17.1 [16]. As shown in Fig. 1, Turbo-Grid is used for single passage mesh discretization and the mesh was densified in the near wall area such as blade and hub, tip clearance and other areas, that can more accurately simulate the real flow field in these areas. The defining boundary conditions, solving, and post processing were performed in CFX-Pre, CFX-Solver and

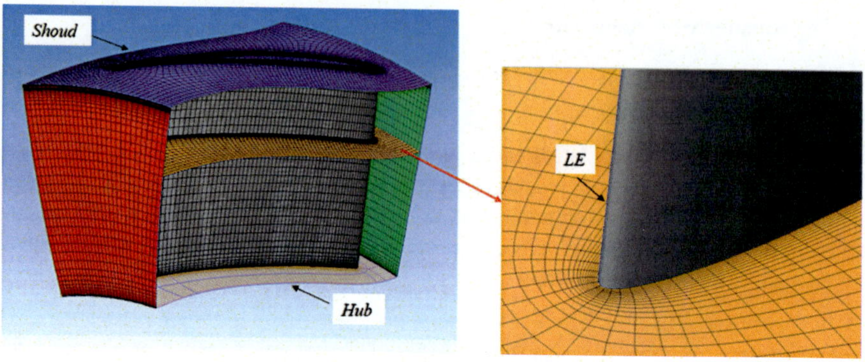

Fig. 1. The mesh discretization result at rotor blade.

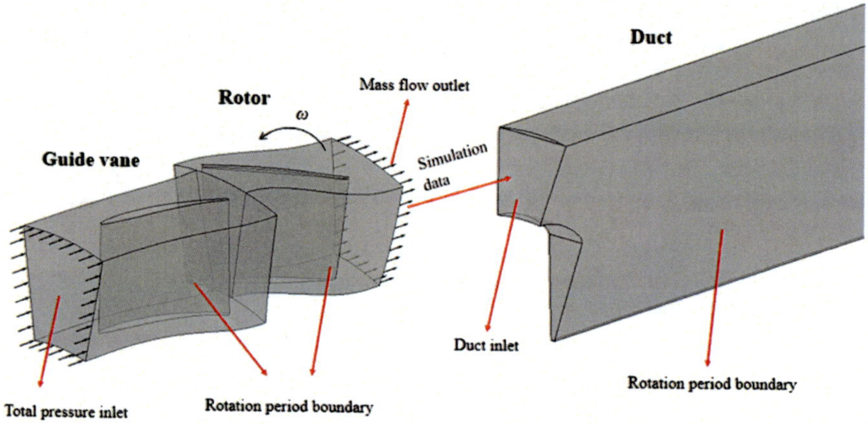

Fig. 2. The boundary conditions in simulation.

CFX-Post, respectively. The results of mesh independence analysis showed that when the mesh number was more than 300000, the simulation results are almost kept the same with mesh increases.

The boundary conditions set in simulation are shown in Fig. 2. The fan inlet and outlet were set as total pressure inlet with 0 Pa and mass flow outlet with variable mass flow, respectively. The working fluid was considered as air at 25 °C. The domain included rotor were set as rotation domain with the rotation speed of 8873 rpm. The mixing plane model was used at the interface between the domain of rotor and guide vane and the k-epsilon model was used to describe turbulence. The Simple algorithm was chose for steady solution and the residual of the solution was 10–5. Considering the influence of duct in fan performance experiment and in order to perform multiple simulations, this paper used the flow solution results of rotor outlet as the inlet condition of duct to calculate the pressure loss of duct. Finally, the total pressure of fan was the total pressure difference between rotor outlet and the pressure loss in duct.

4 Results and Discussion

4.1 Effects of the Prewhirl Parameter of IGVs on Fan Performance

For the case of blade number of R19S17 and variable circulation index of 0.5, the simulation results are illustrated in Fig. 3 with the prewhirl parameters of −0.45, −0.55, −0.65 and −0.75, respectively.

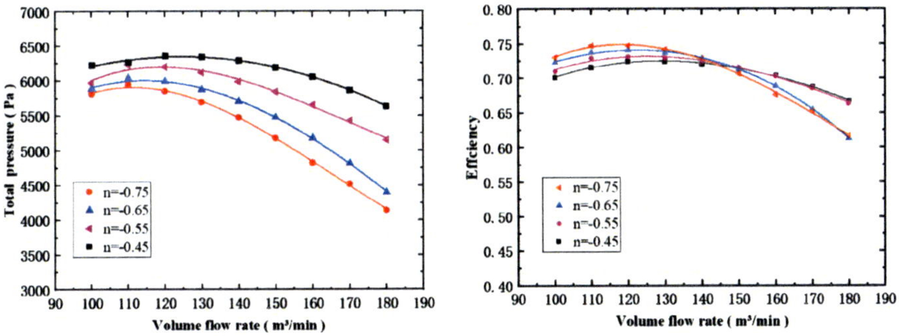

(a) Total pressure curves (b) Efficiency curves

Fig. 3. The simulated performance curves of fan with different prewhirl parameters.

At design flow rate of 130 m^3/min, Fig. 3a presents that the total pressure of the fan are 6350 Pa, 6123 Pa, 5878 Pa and 5700 Pa when the prewhirl parameters are −0.45, −0.55, −0.65 and −0.75, respectively. The total pressure decreases with the decrease of parameter. When the flow rate changed from the design point to 180 m^3/min, the total pressure decreased by 711 Pa, 964 Pa, 1178 Pa and 1555 Pa, respectively. It can be seen that the smaller the prewhirl parameter, the faster the total pressure decreases with increasing flow rate. Regarding the fan efficiency, it can be found from Fig. 3b that when the flow rate is less than 145 m^3/min, the smaller the prewhirl parameter, the higher the efficiency. This is opposite when flow rate is larger than 145 m^3/min. In general, the smaller the parameter, the faster the efficiency decrease. With prewhirl parameters of −0.75 and −0.45, the fan efficiency is 0.742 and 0.724 with the flow rate of 130 m^3/min, while the efficiency is 0.617 and 0.668 at 180 m^3/min. When flow rate changes from 130 m^3/min to 180 m^3/min, the fan efficiency with prewhirl parameters of −0.75 and −0.45 decreases by 0.125 and 0.056, respectively.

The influence mechanism of prewhirl parameter on the total pressure of fan can be explained as follows: from Eq. (3), it can be known that the theoretical pressure rise P_{th} is proportional to the rotor inlet airflow torsion velocity c_{1u}, whereas inversely proportional to the prewhirl parameter n. The IGVs does not impact on the airflow, and the inlet torsional velocity of airflow changes little with different prewhirl parameters. Therefore, the larger the absolute value of prewhirl parameters, the smaller the total pressure rise of fan. The influence mechanism of prewhirl parameters on efficiency of fan is very complicated. It should be noted that the efficiency of axial flow fan is strongly depends on the blade profile loss, annular wall loss, eddy current loss and tip clearance loss of fan. In the current work, the relative tip clearance of fan is kept the same at different prewhirl parameters, and the flow coefficients of all fans are relatively small (<0.6). As a consequence, the difference of the loss caused by annular wall loss and tip clearance loss is small. The influence of prewhirl parameter on the efficiency can be analyzed only from two aspects: blade profile loss and eddy current loss.

Figure 4 shows the pressure contour at 50% span of the rotor blade when the flow rates are 100 m^3/min and 180 m^3/min, respectively. It can be seen from Fig. 4a that the pressure rises faster when n = −0.45 with flow rate of 100 m^3/min. And the pressure difference between the suction surface and the pressure surface of blade is larger which leads to a larger blade profile loss. Meanwhile, the eddy current loss caused by tangential pressure gradient accounts for a larger proportion. The loss of fan is mainly eddy current loss. Therefore, the efficiency of fan at prewhirl parameter of −0.45 is lower than that at prewhirl parameter of −0.75. As shown in Fig. 4b, with the flow rate increase, the fan with larger prewhirl parameter has faster decrease of eddy current loss and slower blade profile loss. When the flow rate is 180 m^3/min, the loss of fan is mainly blade profile loss and it is smaller than that of the fan with larger prewhirl parameter. This would lead to a smaller loss and higher efficiency of the fan with prewhirl parameter of −0.45.

In general, for the fan design, the pressure rise performance of the fan with smaller prewhirl parameter would be better. Meanwhile, its total pressure and efficient have a smaller changes, that is, the effective working range is wider. However, the fan efficiency is lower at design volume flow rate. Considering the effective working range and efficiency of aviation fan, the current work takes the prewhirl parameter of −0.65 for the final design. And this parameter is beyond the application range of −0.5 ~ −0.6 for ordinary civil axial fans.

4.2 Effects of Variable Circulation Index on Fan Performance

When the fan with the variable circulation indexes of 0.4, 0.45 and 0.5, respectively, as well as with prewhirl parameter of −0.65 and blade number of R19S17, the airflow torsional velocity at the fan outlet is theoretically calculated, as demonstrated in Fig. 5.

Figure 5 shows the airflow torsional velocity at the fan outlet increase with the increase of variable circulation index, and decrease with the increase of the blade radius. At the blade root, the airflow torsional velocity achieves the largest, they are 22.8 m/s (x = 0.4), 26.3 m/s (x = 0.45) and 30.7 m/s (x = 0.5). The difference of airflow torsional velocity indicates the variable circulation index has an important impact on the fan aerodynamics performance.

The simulated performance curves of fan with different variable circulation indexes are presented in Fig. 6.

It can be seen from Fig. 6a that when the variable circulation indexes are 0.4, 0.45, and 0.5, the total pressure of fan at 130 m^3/min are 4728 Pa, 5269 Pa and 5878 Pa, respectively. The total pressure increases rapidly with the increase of variable circulation index. Figure 6b shows that the efficiency of fan is almost the same when the flow rate is less than the design point. However, when the flow rate is larger than the design point, the higher the circulation index and the wider the effective working range of fan. For example, with the indexes of 0.4, 0.45, and 0.5, the fan efficiency is 0.725, 0.734, 0.737 with the flow rate of 130 m^3/min, and 0.504, 0.563, 0.613 at 130 m^3/min, respectively.

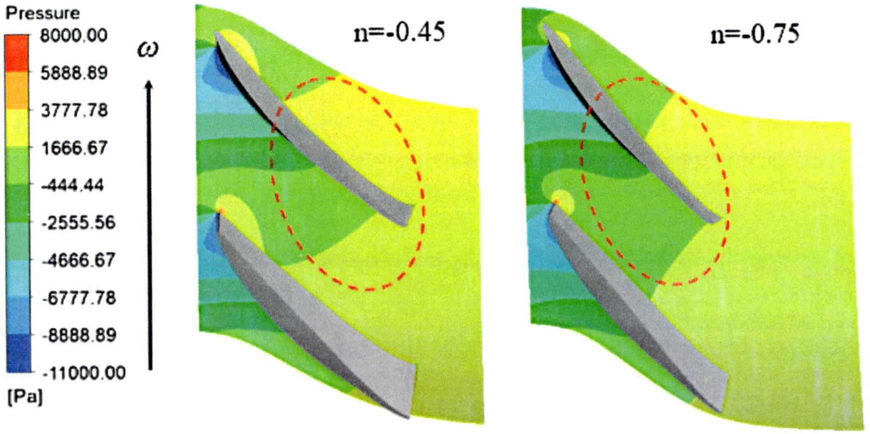

(a) Flow rate 100 m³/min

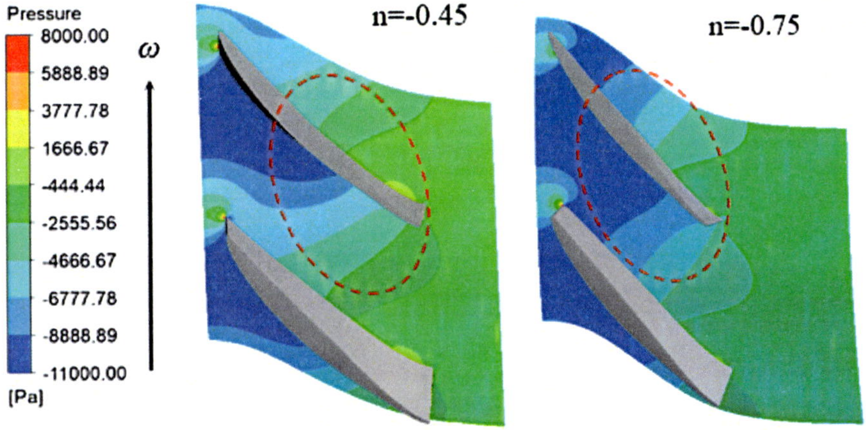

(b) Flow rate 180 m³/min

Fig. 4. Pressure contour at 50% span of the rotor blade with different prewhirl parameters.

When flow rate changes from 130 m^3/min to 180 m^3/min, the fan efficiency with the indexes of 0.4, 0.45, 0.5 decreases by 0.221, 0.171, and 0.124, respectively.

Figure 7 shows the total pressure contour on the blade pressure surface with different circulation indexes at the designed flow rate. It can be seen that the high pressure is distributed near the blade tip, and the larger the variable circulation index, the higher the pressure. It illustrates that the airflow can obtain more energy by utilizing the high peripheral velocity at blade tip region, thereby increasing the total pressure near blade tip. Thus, the higher the variable circulation index, the higher the total pressure and

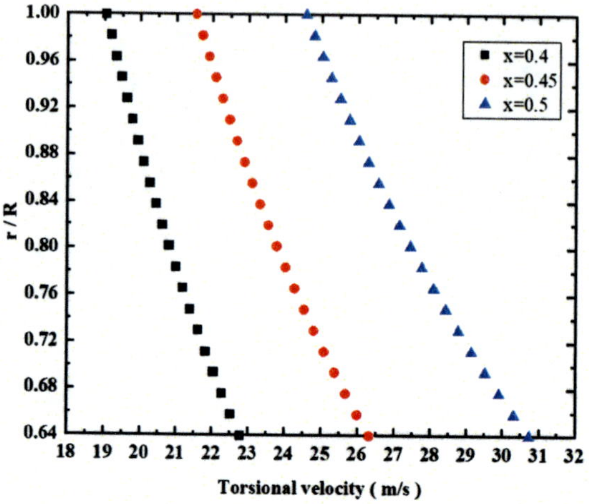

Fig. 5. The airflow torsional velocity at the fan outlet with different variable circulation indexes.

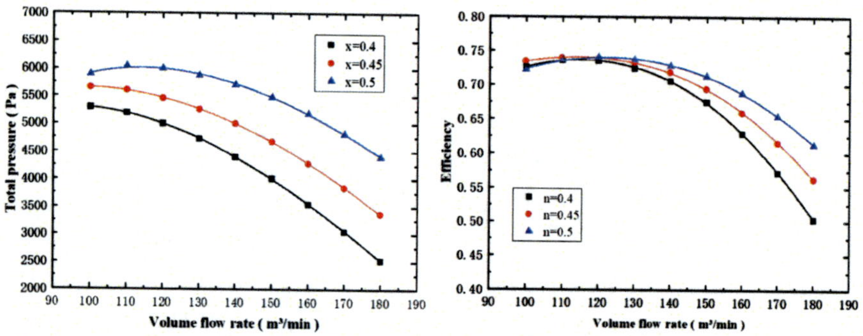

(a) Total pressure curves

(b) Efficiency curves

Fig. 6. The simulated performance curves of fan with different variable circulation indexes.

efficiency of the fan. However, when the variable circulation index is too large, the increase of the airflow torsional velocity will worsen the aerodynamic performance at the blade root. The equivalent diffusion coefficient would exceed the design limitation of it. Therefore, the variable circulation index should be reasonably selected according to the fan design parameters within a certain range through combining theoretical design and numerical results of fan.

4.3 Effects of the Number of IGVs and Rotor Blades on Fan Performance

With the prewhirl parameter of −0.65 and variable circulation index of 0.5, the simulated performance curves of fan with blade number of R19S17, R17S15 and R15S13 are shown in Fig. 8.

As can be seen from Fig. 8a, within a certain range, the total pressure performance of fan increases with the increase of the blade number. When the blade numbers are

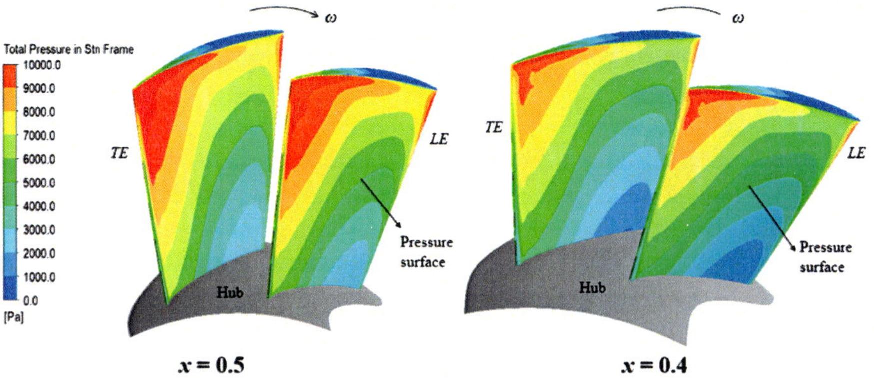

Fig. 7. The total pressure contour in pressure surface of rotor blade with different variable circulation indexes.

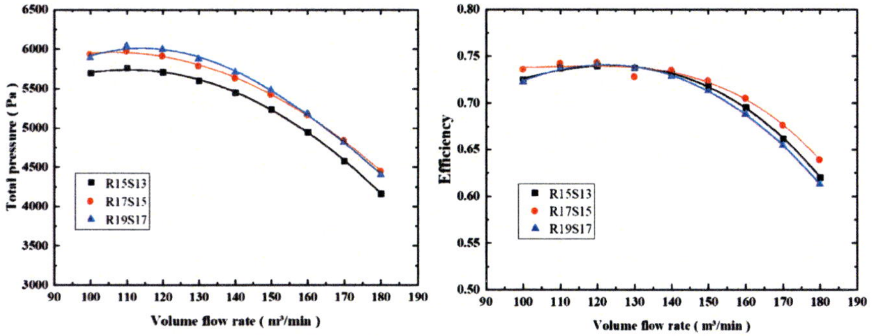

Fig. 8. The simulated performance curves of fan with different blade number.

R19S17, R17S15 and R15S13, the total pressures of fan at the design flow rate are 5878 Pa, 5791 Pa and 5601 Pa, respectively. Figure 8b shows that the fan efficiency is almost the same when the blade number changed, and it is also almost the same with the increases of the flow rate. The effect mechanism of the blade number on the fan performance is that the change of blade number leads to the change of blade pitch, which causes the change of the blade load factor. Within the blade load limit, the increase of the blade number will increase the blade load factor, resulting in the increase of the total pressure performance of fan.

4.4 Optimal Design Outcome with Modified Variable Method

According to the above effect analysis of different design parameters on fan performance, the fan can obtain the best overall performance with the prewhirl parameter of −0.65, variable circulation index of 0.5 and number of IGVs and rotor blades of R19S17. The performance curves of the fan design with modified variable method is shown in Fig. 9.

It can be seen from Fig. 9 that, with the rotation speed of 8873 rpm, the pressure rise and efficiency of the fan at the design flow rate of 130m^3/min are 5878 Pa and 0.74, respectively. Compared with the design pressure rise, the error of numerical result is 6.87%. The numerical results indicate that the fan design outcome in this paper has the characteristics of aviation axial fan with high pressure rise and low flow rate. This study could be beneficial to the design of axial fan used in aviation oil cooling system.

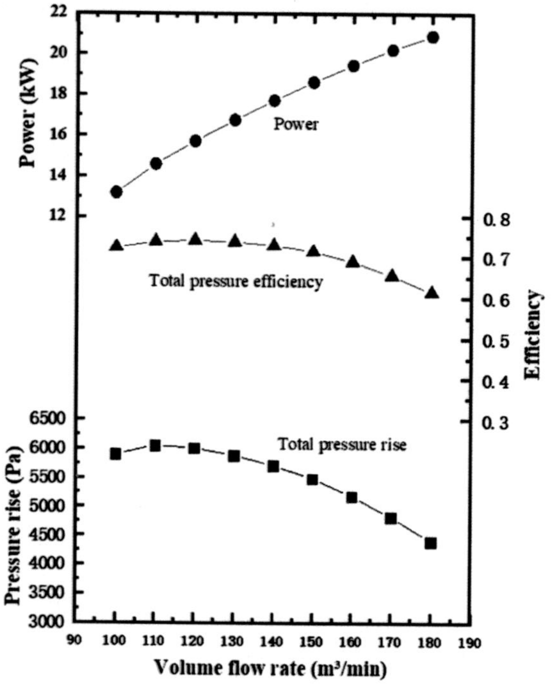

Fig. 9. The performance curves of the aviation axial fan in this paper

5 Conclusions

A high pressure rise, low flow rate axial fan with IGVs was designed using the modified variable circulation method. Numerical study on the effects of the prewhirl parameter of guide vane, variable circulation index, number of IGVs and rotor blades on the fan performance was performed. The main conclusions are as follows:

(1) The pressure rise of fan performance increases with the increase of prewhirl parameter. And the efficiency decreases with the increase of prewhirl parameter when the flow rate is less than 145 m^3/min. Compared with the large prewhirl parameter, the effective working range of the fan with small prewhirl parameter is narrower. The best fan performance is achieved when the prewhirl parameter is −0.65, although this value exceeds the recommended range of −0.5 ~ −0.6 for civil axial fans.

(2) The variable circulation index has obvious effect on the fan performance. The higher the index, the higher pressure rise performance and wider effective working range of fan. However, if the variable circulation index is too large, the equivalent diffusion factor at the blade root would exceed the limit value that would lead to the deterioration of the fan's aerodynamics.
(3) Within a certain range, the pressure rise of fan increases with the increase of blade number, but the efficiency is basically unchanged.

The numerical result shows that the fan design outcome in this paper has the characteristics of high pressure rise, low flow rate and small volume. This investigation can provide reference for the design and optimization of the axial fan used in aircraft oil cooling system.

References

1. Gao, H.X., Yu, J.Z., Xie, Y.Q.: Design of a high rotating speed, low flow rate axial flow fan used on helicopters. J. Aerosp. Power **01**, 119–124 (2006)
2. Liu, X., Liu, J., Wang, D., et al.: Experimental and numerical simulation investigations of an axial flow fan performance in high-altitude environments. Energy **234**, 121281 (2021)
3. Chang, Z.Z.: Practical Technology of Axial Flow Fan. Machinery Industry Press, Beijing (2005)
4. Cheng, X.D.: Pump·Fan·Compressor. Machinery Industry Press, Beijing (2011)
5. Wang, Q.K., Chen, K.M.: Influence of variable circulation index on aerodynamic performance of twisted blade in micro axial fan. J. Mech. Eng. **45**(04), 76–82 (2009)
6. Wang, S.J., et al.: Study on performance of axial fan blade with constant circulation design. Fluid Mach. **48**(07), 1–6+11 (2020)
7. Liang, D., Song, C., Liang, S., et al.: Design and performance analysis of blades based on equal-variable circulation method. Frontiers Energy Res. 730 (2021)
8. Castegnaro, S.: Aerodynamic design of low-speed axial-flow fans: a historical overview. Designs **2**(3) (2018)
9. Liu, Y., Lin, Z., Lin, P., et al.: Effect of inlet guide vanes on the performance of small axial flow fan. J. Therm. Sci. **26**(6), 504–513 (2017)
10. Matai, R., Yavuzkurt, S.: Evaluation of effects of different design parameters on axial fan performance using CFD. Turbo Expo: Power for Land, Sea, and Air. American Society of Mechanical Engineers 56628: V001T09A012 (2015)
11. Kharati-Koopaee, M., Moallemi, H.: Effect of blade tip grooving on the performance of an axial fan at different tip clearances in the absence and presence of inlet guide vanes. Proc. Inst. Mech. Eng., Part A: J. Power Energy **234**(1), 72–84 (2020)
12. Liu, M., Tan, L., Cao, S.: A review of prewhirl regulation by inlet guide vanes for compressor and pump. Proc. Inst. Mech. Eng., Part A: J. Power Energy **233**(6), 803–817 (2019)
13. Wang, Y.R., et al.: Research status and prospect of aerodynamic optimization design of fluid machinery based on machine learning method. Compress. Blower Fan Technol. **62**(05), 77–90 (2020)
14. Li, Z., et al.: Global aerodynamic optimization design of supersonic turbine blade. J. Propuls. Technol. **40**(05), 1051–1057 (2019)
15. Kim, J.H., Ovgor, B., Cha, K.H., et al.: Optimization of the aerodynamic and aeroacoustic performance of an axial-flow fan. AIAA J. **52**(9), 2032–2044 (2014)
16. Ansys, C.F.X.: 17.1, 2016, ANSYS CFX-Solver Theory Guide, ANSYS Inc.

Geometrically Nonlinear Static Aeroelastic Analysis Based on CFD/CSD Interaction Accelerated by Panel Method

Lan Yang[1](✉), Changchuan Xie[2], Deli Liang[1], and Chao An[2]

[1] Beijing Institute of Astronautical Systems Engineering, Beijing 100076, China
yanglan1105@sina.cn

[2] School of Aeronautic Science and Engineering, Beihang University, Beijing 100191, China

Abstract. High-fidelity computational aeroelasticity based on CFD/CSD coupling method can capture the nonlinear effects in both aerodynamic and structural aspects. In this paper, geometrically nonlinear static aeroelastic problems of very flexible wings are solved based on CFD/CSD interaction accelerated by panel method. The acceleration of convergence is realized by introducing aerodynamic stiffness matrix based on panel method to reduce the computational cost of CFD. Considering the large deformation case, aerodynamic loads are applied as follower forces which change in direction along with the structural deflection. Geometrically nonlinear finite element method is employed for structural analysis. The numerical research of a high aspect-ratio wing is provided and the results are verified by the wind tunnel test data. Compared with the conventional staggered strategy, the proposed accelerated algorithm can improve the calculation efficiency under the same accuracy requirements.

Keywords: Aeroelasticity · Geometrically nonlinear · Fluid-structure interaction · Aircraft design

1 Introduction

In recent years, high-fidelity computation of aerodynamic loads based on CFD (Computational Fluid Dynamics) technique has been applied as a powerful tool of performance evaluation in multidisciplinary design optimization. Because of the requirement of high-fidelity aeroelastic analysis, computational aeroelasticity based on CFD/CSD coupling method has been also developed and successfully used [1]. Especially for modern transport aircraft, the cruising speed is usually in transonic domain, which results in the highly nonlinearity of aerodynamic characteristics. Therefore, it is necessary to develop more accurate FSI (Fluid-Structure Interaction) analysis based on nonlinear aerodynamic models. Meanwhile, lighter structure is also pursued to improve the fuel economy. The very flexible structure is likely to encounter large deformations under flight loads, which causes a series of problems with geometric nonlinear effects [2].

High-fidelity computational aeroelasticity based on CFD/CSD coupling method has become an important research hotspot recently. Bendiksen reviewed the challenges in

Chinese Society of Aeronautics and Astronautics: CASTYSF 2022, LNEE 972, pp. 324–336, 2023.
https://doi.org/10.1007/978-981-19-7652-0_31

computational aeroelasticity of flutter analysis [3]. The main difficulty arises from basic differences in the numerical schemes used in fluid dynamics and in structural dynamics. Researchers have developed many coupling strategies to deal with the incompatibility. Farhat et al. proposed a second-order time-accurate loosely coupled solution algorithms for transient nonlinear aeroelastic analysis [4]. They proved that computationally efficient staggered solution procedures can be designed using simple mathematical constructs. Tallec et al. used an arbitrary Lagrangian Eulerian formulation to deal with FSI problems with large structural displacements [5]. They solved the kinematic compatibility problem at the fluid structure interface by consistent and energy conservative interface boundary conditions at the moving fluid structure interface.

Although CFD/CSD coupling method achieves high fidelity in aeroelastic analyses, the computational cost is rather expensive. In preliminary aircraft design and optimization, it is still impractical to conduct massive computations based on CFD. Therefore, researchers have been making great efforts to improve the efficiency of the aeroelastic computation as well as promise the high fidelity. One of the effective ways is to improve the iterative algorithm to speed up the convergence. Dang et al. introduced a golden section technique in loosely coupled computations for nonlinear static aeroelastic analysis [6]. By using this acceleration technique, the wing deformed monotonically from the original position to its static elastic equilibrium position. Feng et al.used nonlinear Krylov acceleration for CFD-based aeroelasticity [7]. They compared Gauss–Seidel and Schur-complement solution and declared the latter is more robust for relatively large oscillation amplitudes. Heil studied the fully coupled solution of FSI problems by GMRES algorithm. They presented efficient approximate preconditioners for the computationally expensive operations [8].

Panel methods (PM) are applied generally in aeroelastic research as a medium-fidelity tool of aerodynamic load computation [9,10]. PM provide efficient analyses for potential flow but cannot either consider nonlinear aerodynamic characteristics or estimate the drag accurately. It is worthy to combine CFD method with panel methods to improve the computational efficiency as well as provide high-fidelity results. Jovanov et al. tried to use low-fidelity aerodynamic tools to accelerate the convergence of static aeroelasticity [11, 12]. They formulated a defect-correction approach to reduce the total number of HiFi aerodynamic load calculations. Although the defect-correction approach is a low-fidelity solution, the results of equilibrium state is of high-fidelity.

This paper is aimed at achieving faster convergence of CFD/CSD coupling based static aeroelastic analysis by using panel method. High-fidelity CFD is used to compute the aerodynamic force, while the nonlinear aeroelastic equation is solved by a modified Newton iterative algorithm. As one of the most popular panel methods applied in aeroelasticity, vortex-lattice method (VLM) is used to provide the approximate incremental aerodynamic load during the Newton's iterations to accelerate the convergence of aeroelastic analysis. The static aeroelastic equilibrium still converges at the high-fidelity CFD solution while the number of CFD computations decreases by the proposed iterative algorithm, which achieves faster convergence and less cost of computation. To model the large structural deformations, geometrically nonlinear structural FEM is applied on a very flexible wing. Because the aerodynamic force direction changes with the large deformation of the wing structure, the aerodynamic loads are applied as follower loads.

Numerical static aeroelastic simulation of a very flexible clamped wing is conducted to verify the acceleration convergence using the proposed algorithm.

2 Theoretical Development

The loosely coupled CFD/CSD method is widely used in high-fidelity aeroelastic analysis. In this paper, the coupling process is improved to achieve faster convergence. In the section of theoretical development, the aerodynamic analysis, the geometrically nonlinear structural analysis, and the fluid-structure interaction framework will be presented, respectively.

2.1 Aerodynamic Analysis

In this paper, RANS solver based on Fluent is used to compute high-fidelity aerodynamic loads, while VLM is used as an efficient solver to provide the incremental aerodynamic force caused by structure deformations. Since CFD computation costs massive time, it is scarcely possible to obtained aerodynamic stiffness matrix based on CFD. Instead, a state-space model of VLM is adopted to calculate a low-fidelity solution for aerodynamic force caused by elastic deformation.

In subsonic cases, VLM is based on potential flow theory, which in steady case yields

$$\left(1 - M_\infty^2\right)\phi_{xx} + \phi_{yy} + \phi_{zz} = 0 \tag{1}$$

where ϕ is the velocity potential function of an arbitrary location in the flow field. Introduce the compressible factor $\beta = \sqrt{1 - M_\infty^2}$, and the potential equation can be simplified to Laplace's equation by the Prandtl–Glauert transformation [13]. The boundary conditions are satisfied on the wing camber surface, which includes the factors of the airfoil camber, sweep, twist, and structural deflections, shown in Fig. 1. Considering the compressive effect, the velocity induced by the vortex ring is calculated by the Biot-Savart law,

$$\boldsymbol{w} = \frac{\beta^2\Gamma}{4\pi}\oint \frac{d\boldsymbol{l} \times \boldsymbol{r}}{|\boldsymbol{r}\beta|^3} \tag{2}$$

where $\boldsymbol{r}_\beta = x\boldsymbol{i} + \beta \cdot y\boldsymbol{j} + \beta \cdot z\boldsymbol{k}$. And the strength of vortex rings is solved by applying Neumann boundary condition,

$$\boldsymbol{A} \cdot \boldsymbol{\Gamma} = -\boldsymbol{Q}_\infty \cdot (\boldsymbol{n}_1, \boldsymbol{n}_2, \ldots \boldsymbol{n}_N)^T \tag{3}$$

where $\boldsymbol{A}$ is the aerodynamic influence coefficient matrix including the information of normal wash; $\boldsymbol{\Gamma}$ is the column vector of vortex strength; $\boldsymbol{Q}_\infty$ is the freestream velocity and $\boldsymbol{n}$ is the normal vector of each element. The aerodynamic force can be obtained according to the Kutta-Joukowski theorem:

$$\Delta L_{ij} = \rho V_\infty\left(\Gamma_{ij} - \Gamma_{i-1,j}\right)\Delta y_{ij} \tag{4}$$

Set the state variable as the strength of the wake elements, the VLM can be as the state-space form:

$$\boldsymbol{A}_{aero}\boldsymbol{x}_{aero} + \boldsymbol{B}_{aero}\boldsymbol{w}_{aero} = 0 \quad (5)$$

$$\boldsymbol{f}_A = \boldsymbol{C}_{aero}\boldsymbol{x}_{aero} + \boldsymbol{D}_{aero}\boldsymbol{w}_{aero} \quad (6)$$

The state variable is the strength of wake vortex rings $\boldsymbol{x}_{aero} = \boldsymbol{\Gamma}_w$, with the input variables of the normal wash $\boldsymbol{w}_{aero}$ and the output variable of aerodynamic forces $\boldsymbol{f}_A$.

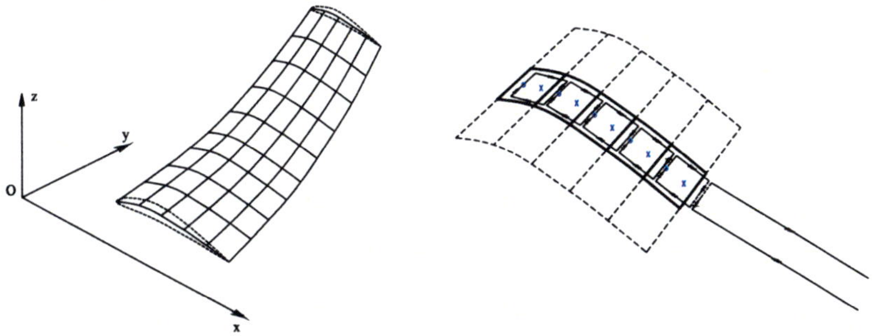

Fig. 1. Vortex-lattice model for a non-planar wing

2.2 Geometrically Nonlinear Structural Analysis

For the very flexible high aspect-ratio wing, geometrically nonlinear structural analysis is necessary to capture the large wing deformations. As a powerful structural solver, displacement-based finite element method (FEM) is used here with the updated Lagrange formulation (ULF):

$$\mathbf{M}^t\ddot{\boldsymbol{u}}^{t+\Delta t} + \left(\mathbf{K}_L^t + \mathbf{K}_{NL}^t\right)\boldsymbol{u}^{t+\Delta t} = \mathbf{F}_A^{t+\Delta t} - \mathbf{F}_I^t \quad (7)$$

where $\mathbf{M}$ is the inertial matrix; $\ddot{\boldsymbol{u}}$ represents the acceleration vector of the structural nodes [14]. The stiffness matrix includes the linear stiffness matrix $\mathbf{K}_L$ and the nonlinear part $\mathbf{K}_{NL}$. As a typical characteristic of nonlinear structural analysis, the stiffness matrix is changed in computational steps according to the different strain and stress levels. $\mathbf{F}_A$ includes all the outer forces such as the aerodynamic loads and thrust. $\mathbf{F}_I$ is the corresponding inertial force.

In the large deformation research, the aerodynamic forces are loaded as follower forces to capture the changing of force directions caused by the structural displacements. Taking the cantilever beam under the vertical follower force as an example, Fig. 2 shows that as the structure deflects, the changing of the force direction will lead to the lateral force component and therefore influence the displacement calculation results.

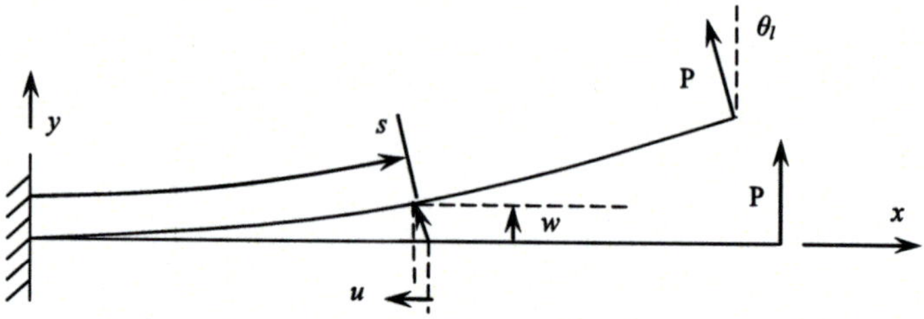

Fig. 2. A cantilever beam under vertical follower force

2.3 Surface Spline Interpolation

The generalized surface spline method is used in this paper to interpolate the displacements and forces between the aerodynamic and structural model [15]. The fundamental expression of the mapping relation is

$$w^k(\boldsymbol{X}) = c_1^k + \sum_{p=1}^{N} c_{p+1}^k x_p + \sum_{i=1}^{N} c_{N+1+i}^k r_i^2 ln\left(r_i^2 + \varepsilon\right) \tag{8}$$

where $\boldsymbol{X}_i = \{x_i^1, \cdots, x_i^N\}$, $(i = 1, 2, \cdots, n)$ is the given vector in N-dimension space and $\boldsymbol{W}_i = \{w_i^1, \cdots, w_i^N\}$, $(i = 1, 2, \cdots, n)$ is the corresponding image vector. The application of the surface spline interpolation method in the displacement vector can be written as

$$\boldsymbol{U}_A = \boldsymbol{G}\boldsymbol{U}_S \tag{9}$$

where $\boldsymbol{U}_S$ is the displacements of the structural nodes; $\boldsymbol{U}_A$ is the resulting displacements of the aerodynamic nodes; $\boldsymbol{G}$ is the interpolation matrix.

For the force interpolation between the structural model and the aerodynamic model, the virtual work principle must be satisfied:

$$\delta \boldsymbol{U}_A^T = \delta \boldsymbol{U}_S^T \boldsymbol{F}_S \tag{10}$$

Therefore, the force on the structural model $\boldsymbol{F}_S$ can be interpolated by the force on the aerodynamic model $\boldsymbol{F}_A$:

$$\boldsymbol{F}_S = \boldsymbol{G}^T \boldsymbol{F}_A \tag{11}$$

2.4 Modified Newton Iterative Algorithm for Nonlinear Aeroelastic Equation

For geometrically nonlinear aeroelastic analysis, it is necessary to update the aerodynamic model along with the structural deflections. The nonlinear static aeroelastic equation can be written as the following general form:

$$\boldsymbol{R}(\boldsymbol{u}) = \boldsymbol{F}_E(\boldsymbol{u}) - \boldsymbol{F}_A(\boldsymbol{u}) - \boldsymbol{F}_I = 0 \tag{12}$$

where $\boldsymbol{F}_E(\boldsymbol{u})$ is the structural internal force. For geometrically nonlinear problems, $\boldsymbol{F}_E(\boldsymbol{u})$ is a nonlinear function of the elastic deformation $\boldsymbol{u}$. When the aeroelastic system converges to the equilibrium state, the residual satisfies $\boldsymbol{R}(\boldsymbol{u}) = 0$. In the conventional algorithm of loosely coupled CFD/CSD solutions, the aerodynamic loads and the structural deformations are solved alternately. The CFD module and the mesh morphing procedure are conducted in each iteration, which consume massive computation cost and time. In this paper, we introduce VLM as a low-fidelity, high-efficiency aerodynamic method to the nonlinear aeroelastic iteration. Thus, the aerodynamic forces are calculated by two methods: the high-fidelity result $\boldsymbol{F}_{AH}(\boldsymbol{u})$ by CFD and the low-fidelity result $\boldsymbol{F}_{AL}(\boldsymbol{u})$ by VLM. The final aeroelastic solution is of high fidelity:

$$\boldsymbol{R}_H(\boldsymbol{u}) = \boldsymbol{F}_E(\boldsymbol{u}) - \boldsymbol{F}_{AH}(\boldsymbol{u}) - \boldsymbol{F}_I = 0 \tag{13}$$

Introduce the low fidelity aerodynamic force here:

$$\Delta \boldsymbol{F}_A = \boldsymbol{F}_{AH}(\boldsymbol{u}) - \boldsymbol{F}_{AL}(\boldsymbol{u}) \tag{14}$$

Conduct a first-order Taylor series expansion to Eq. (13), and we get

$$\boldsymbol{F}_E^n + \left(\frac{\partial \boldsymbol{F}_E}{\partial \boldsymbol{u}}\right)\Delta \boldsymbol{u} = \boldsymbol{F}_{AL}^n + \left(\frac{\partial \boldsymbol{F}_A}{\partial \boldsymbol{u}}\right)^n \Delta \boldsymbol{u} + \boldsymbol{F}_I + \Delta \boldsymbol{F}_A \tag{15}$$

Furthermore, Eq. (15) can be written as

$$\left(\boldsymbol{K}_T^n - \boldsymbol{K}_{AL}^n\right)\Delta \boldsymbol{u} = -\boldsymbol{R}_L^n(\boldsymbol{u}) + \Delta \boldsymbol{F}_A^k \tag{16}$$

$$\boldsymbol{u}^{n+1} = \boldsymbol{u}^n + \Delta \boldsymbol{u} \tag{17}$$

where $\boldsymbol{R}_L^n(\boldsymbol{u})$ is the residual based on the low fidelity aerodynamic force:

$$\boldsymbol{R}_L(\boldsymbol{u}) = \boldsymbol{F}_E(\boldsymbol{u}) - \boldsymbol{F}_{AL}(\boldsymbol{u}) - \boldsymbol{F}_I \tag{18}$$

The tangent stiffness matrix of the current equilibrium state $\boldsymbol{K}_T^n$ is relevant to the displacement $\boldsymbol{u}$ under the given load. It can be obtained by geometric nonlinear structural analysis. The aerodynamic stiffness matrix $\boldsymbol{K}_{AL}^n = \frac{\partial \boldsymbol{F}_{AL}(\boldsymbol{u})}{\partial \boldsymbol{u}}$ means the change of aerodynamic load due to the displacement $\boldsymbol{u}$, which is attained by VLM. According to Eqs. (5) and (6), the aerodynamic stiffness matrix is

$$\boldsymbol{K}_{AL}^n = \frac{\partial \boldsymbol{F}_{AL}(\boldsymbol{u})}{\partial \boldsymbol{u}} = \boldsymbol{G}^T\left(-\boldsymbol{A}_{aero}^{-1}\boldsymbol{B}_{aero}\boldsymbol{C}_{aero} + \boldsymbol{D}_{aero}\right)\frac{\partial \boldsymbol{w}_{aero}}{\partial \boldsymbol{u}} \tag{19}$$

Here the coefficient matrices $\boldsymbol{A}_{aero}, \boldsymbol{B}_{aero}, \boldsymbol{C}_{aero}, \boldsymbol{D}_{aero}$ should be updated according to the structural deformations during the iteration.

In Newton's method for nonlinear equations, the convergence $\boldsymbol{u}_*$ is reached only if the initial vector $\boldsymbol{u}_0$ is close to the solution. We use the original rigid model as the initial value, i.e. $\boldsymbol{u}_0 = 0$. It is obvious that $\boldsymbol{R}(\boldsymbol{u})$ is twice continuously differentiable, so the iteration is quadratic converged, which is much faster than the conventional staggered strategy (CSS). The flowchart of the iteration process is illustrated in Fig. 3.

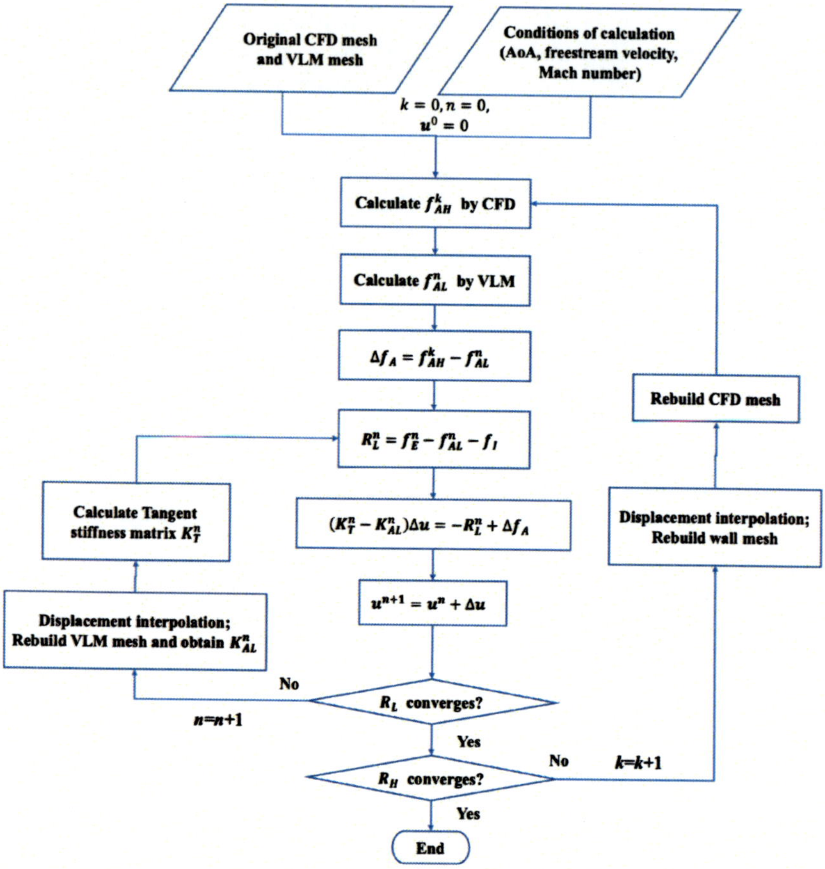

Fig. 3. Iteration process of nonlinear static aeroelastic analysis

The nonlinear iteration of aeroelastic analysis can be divided into two subcycles: On the one hand, VLM can provide an aerodynamic model for CFD computation which is closer to the convergent state, so that the iteration is accelerated. On the other hand, when the analysis converges, the aerodynamic loads are of high fidelity, which guarantees the computation accuracy.

3 Numerical Results

3.1 Model Description

Table 1 illustrates the main parameters of a high aspect-ratio wing. Figure 4 shows the structural model of the wing. In the aeroelastic analysis of this paper, the wing structure is clamped at the root and the flow field uses asymmetric boundary conditions. The wing model was tested in the wind tunnel for the aeroelastic research [16], shown in Fig. 5.

Table 1. Wing model parameters

Item	Value
Semi span (mm)	1542.1
Chord of the wing root (mm)	263
Chord of the wing tip (mm)	71
Aspect ratio	18.6
Reference area (mm^2)	2.6e + 5
Beam location	40% of the local chord
Weight (kg)	3.1

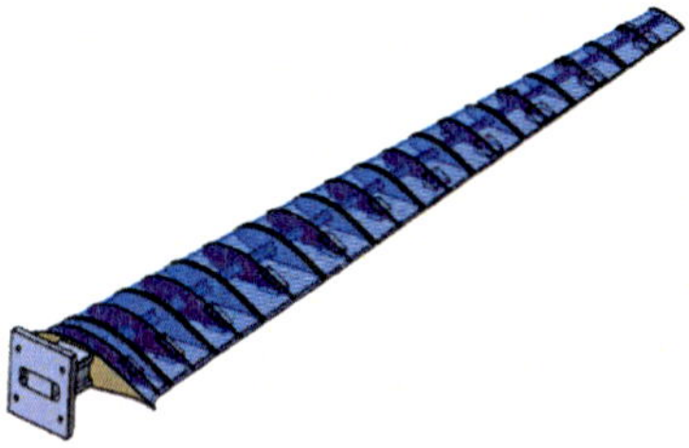

(a) CATIA model

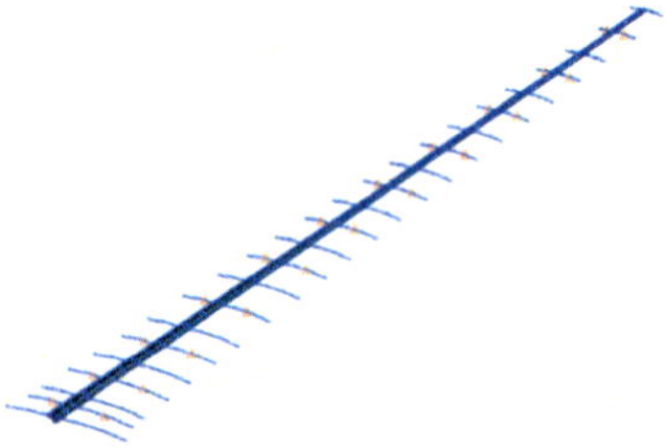

(b) Structural finite element model

Fig. 4. Wing model

Fig. 5. Wing model pictured in the wind tunnel test

Figure 6 shows the aerodynamic models for VLM and CFD analysis, respectively. Both aerodynamic models consider the camber and the preliminary twist of the wing. The coordinate system is defined as following: the origin is set on the root node of the main beam. The x axis points back following the air flow, and the z axis points up. The y axis is defined according to the right-hand rule.

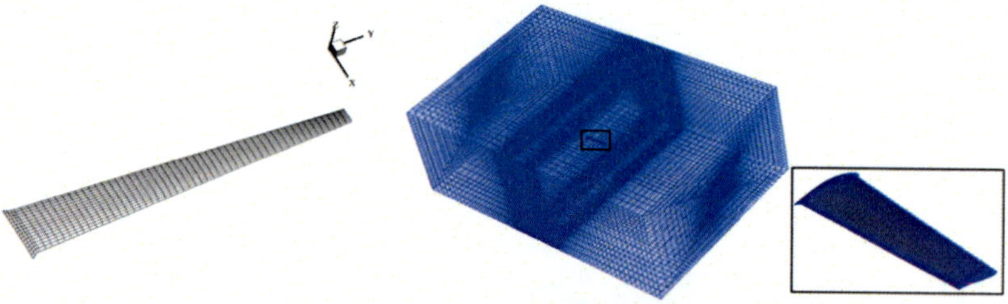

(a) Grid discretization for VLM (b) CFD mesh for the computational domain

Fig. 6. Aerodynamic models for VLM and CFD analysis

3.2 Nonlinear Static Aeroelastic Analysis

In this paper, comparative study of nonlinear static aeroelastic analysis based on CSS algorithm and the proposed modified Newton algorithm is conducted. Two cases of calculation are set: (a) the freestream velocity is $V_\infty = 30\,\text{m/s}$, and the angle of attack is $\alpha = 5°$; (b) the freestream velocity is $V_\infty = 40\,\text{m/s}$, and the angle of attack is $\alpha = 5°$.

Figure 7 shows the main beam deformation results by two methods. The identical results of both vertical and horizontal displacements verify that the proposed algorithm of nonlinear aeroelastic analysis achieves the same accuracy as the conventional staggered strategy (CSS). The clamped wing undergoes large deformations in both calculation conditions. In case (a), the maximal vertical deformation of the wing tip reaches more than 20% of the semi span, while the maximal horizontal deformation is about 4% of the semi span. In case (b), the maximal vertical deformation of the wing tip is approximately 39% of the semi span, while the maximal horizontal deformation is about 10% of the semi span. Figure 8 illustrates the contour of pressure coefficient on the wing surface and Fig. 9 shows the pictures taken during the wind tunnel test in corresponding conditions. The comparison between the simulation and the test results is presented in Table 2. The numerical results are consistent with the test data, which proves the accuracy of the nonlinear aeroelastic analysis.

The geometrically nonlinear effect is significant in such cases, which means the strong interaction between the large structural deformations and the aerodynamic loads. On the one hand, the aerodynamic configuration changes in pace with the wing structure deformations, which requires the update of the aerodynamic mesh. On the other hand, in the geometrically nonlinear structural analysis, the follower force effects of the aerodynamic force cause the horizontal deformations of the wing. Since the aerodynamic pressure is always perpendicular to the wing surface, when the wing structure deflects, the horizontal component of the aerodynamic force will result in the horizontal displacements. Meanwhile, the projected wing area and the affective angle of attack change significantly in geometrically nonlinear aeroelastic analysis, which is also very different from the conventional linear cases.

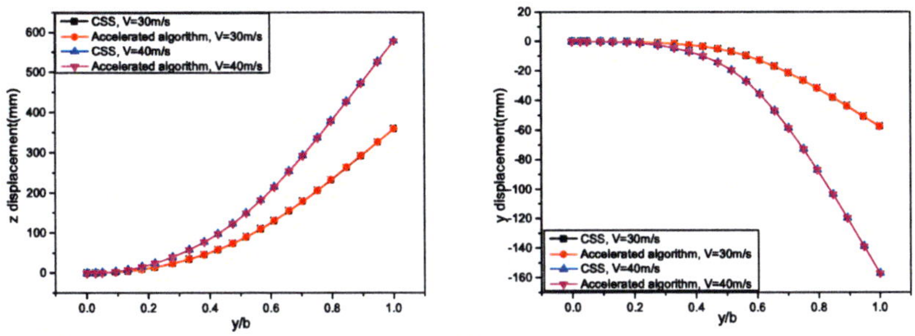

(a) Displacements in z direction (b) Displacements in y direction

Fig. 7. Comparative results of the main beam deformations by two methods

Table 2. Comparison of wing tip displacements between simulation and test data

	Case (a): $V_\infty = 30\,\text{m/s}$, $\alpha = 5°$		Case (b): $V_\infty = 40\,\text{m/s}$, $\alpha = 5°$	
	Simulation	Test data	Simulation	Test data
Vertical displacement (mm)	359.5	351.0	579.0	599.6
Horizontal displacement (mm)	−57.36	−61.6	−156.8	−184.0

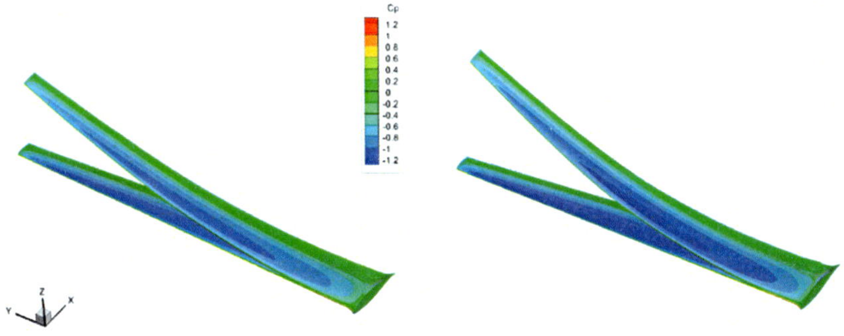

(a) $V_\infty = 30m/s, \alpha = 5°$ (b) $V_\infty = 40m/s, \alpha = 5°$

Fig. 8. Contour of pressure coefficient on the wing surface

The proposed modified Newton algorithm of nonlinear aeroelastic analysis achieves equivalent computational accuracy with CSS, and the computational efficiency can also

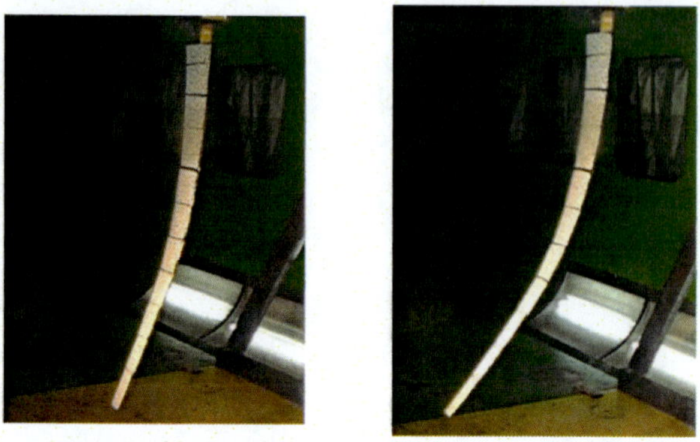

(a) $V_\infty = 30m/s, \alpha = 5°$ (b) $V_\infty = 40m/s, \alpha = 5°$

Fig. 9. Deformed wing in the wind tunnel test

be improved. Figures 10 and 11 present the comparison of convergence history by the two methods. First, the convergence curves of the accelerated algorithm are smoother than CSS algorithm. In the proposed accelerated algorithm, a low-fidelity solution is provided for CFD computation in each iteration step, which makes the wing deformations converge to the equilibrium state faster. In this paper, the convergence error is limited to 0.1mm. Figures 10b and 11b shows that when the accelerated algorithm is adopted, the number of iterations required for convergence is reduced from 9 to 7 times for the case $V_\infty = 30$ m/s, $\alpha = 5°$, increasing the calculation efficiency by 22%. For the case $V_\infty = 40$ m/s, $\alpha = 5°$, the number of iterations is reduced from 11 to 7 times, increasing the calculation efficiency by 36%. Therefore, compared with traditional CSS algorithm, the CFD/CSD coupling method accelerated by panel method can improve the calculation efficiency to a certain extent under the same accuracy requirements.

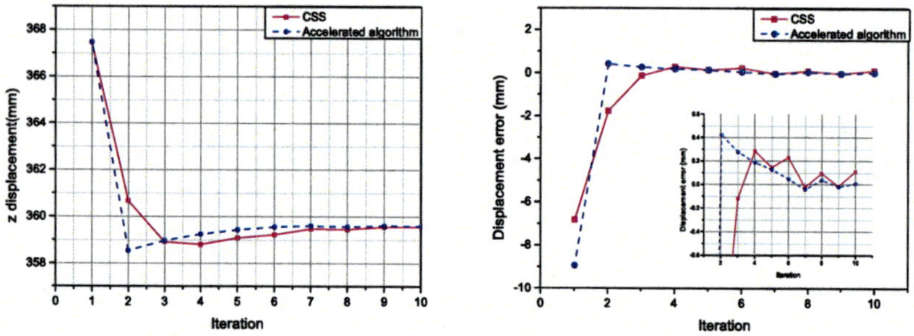

(a) Convergence of wing tip displacements (b) Convergence of displacement errors

Fig. 10. Convergence of nonlinear aeroelastic analysis ($V_\infty = 30$ m/s, $\alpha = 5°$)

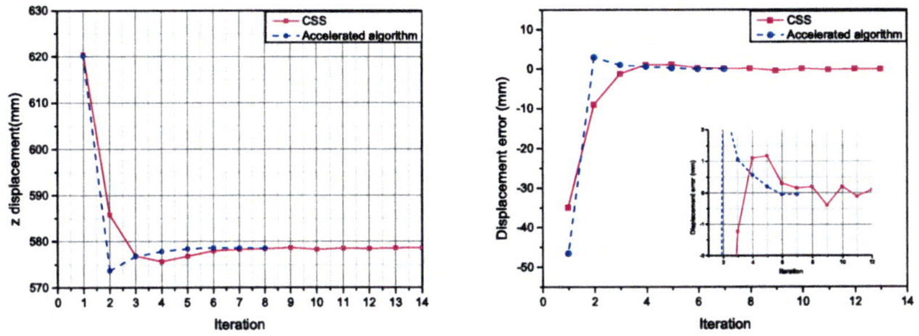

(a) Convergence of wing tip displacements (b) Convergence of displacement errors

Fig. 11. Convergence of nonlinear aeroelastic analysis ($V_{\infty} = 40\,\text{m/s}, \alpha = 5°$)

4 Conclusion

In this paper, an accelerated algorithm of CFD/CSD interaction for geometrically nonlinear static aeroelastic analysis is presented. Vortex-lattice method is introduced into the nonlinear Newton iteration as a low-fidelity, high-efficiency aerodynamic solver to provide the input for the CFD computation which is closer to the convergent state. Geometrically nonlinear FEM is used for structural analysis. It converges to the same equilibrium state as the conventional loosely-coupled algorithm, but the computational cost of CFD can be significantly reduced. The numerical case of a flexible high aspect-ratio wing is researched to verify the accuracy and efficiency of the proposed algorithm. The comparison between analysis and the wind tunnel test proves that the CFD/CSD interaction method achieves reliable results for geometrically nonlinear aeroelastic problems. The convergence history of aeroelastic iteration indicates that the CFD/CSD coupling method accelerated by panel method can improve the calculation efficiency by reducing the number of iteration loops while achieving the high-fidelity requirements.

References

1. Schuster, D.M., Liu, D.D., Huttsell, L.J.: Computational aeroelasticity: success, progress, challenge. J. Aircr. **40**, 843–856 (2003)
2. Hallissy, B., Cesnik, C.: High-fidelity aeroelastic analysis of very flexible aircraft. In: 52nd AIAA/ASME/ASCE/AHS/ASC Structures, Structural Dynamics, and Materials Conference (2011)
3. Bendiksen, O.O.: Modern developments in computational aeroelasticity. Proc. Inst. Mech. Eng., Part G: J. Aerosp. Eng. **218**, 157–177 (2004)
4. Farhat, C., van der Zee, K.G., Geuzaine, P.: Provably second-order time-accurate loosely-coupled solution algorithms for transient nonlinear computational aeroelasticity. Comput. Methods Appl. Mech. Eng. **195**, 1973–2001 (2006)
5. Le Tallec, P., Mouro, J.: Fluid structure interaction with large structural displacements. Comput. Methods Appl. Mech. Eng. **190**, 3039–3067 (2001)
6. Huixue, D., Zhichun, Y., Yi, L.: Accelerated loosely-coupled CFD/CSD method for nonlinear static aeroelasticity analysis. Aerosp. Sci. Technol. **14**, 250–258 (2010)

7. Feng, Z., Soulaïmani, A., Saad, Y.: Nonlinear Krylov acceleration for CFD-based aeroelasticity. J. Fluids Struct. **25**, 26–41 (2009)
8. Heil, M.: An efficient solver for the fully coupled solution of large-displacement fluid-structure interaction problems. Comput. Methods Appl. Mech. Eng. **193**, 1–23 (2004)
9. Murua, J., Palacios, R., Graham, J.M.R.: Applications of the unsteady vortex-lattice method in aircraft aeroelasticity and flight dynamics. Prog. Aerosp. Sci. **55**, 46–72 (2012)
10. Werter, N.P.M., De Breuker, R., Abdalla, M.M.: Continuous-time state-space unsteady aerodynamic modeling for efficient loads analysis. AIAA J. **56**, 905–916 (2017)
11. Jovanov, K., De Breuker, R.: Accelerated convergence of static aeroelasticity using low-fidelity aerodynamics. IFASD, pp. 1–10 (2015)
12. Jovanov, K., De Breuker, R.: Accelerated convergence of static aeroelasticity using low-fidelity aerodynamics. In: 56th AIAA/ASCE/AHS/ASC Structures, Structural Dynamics, and Materials Conference, pp. 1–10 (2015)
13. Dvorak, F.A., Woodward, F.A.: Viscous/Potential Flow Interaction Analysis Method for Multi-Element Infinite Swept Wings—V I. NASA Contractor Reports (1974)
14. Xie, C., An, C., Liu, Y., Yang, C.: Static aeroelastic analysis including geometric nonlinearities based on reduced order model. Chin. J. Aeronaut. **30**, 638–650 (2017)
15. Xie, C., Yang, C.: Surface splines generalization and large deflection interpolation. J. Aircr. **44**, 1024–1026 (2007)
16. Yang, L., Xie, C., et al.: Geometrically exact vortex lattice and panel methods in static aeroelasticity of very flexible wing. Proc. Inst. Mech. Eng., Part G: J. Aerosp. Eng. **234**(3), 742–759 (2019)

Research on Design of New Low Frequency Broadband Electromagnetic Stealth Structure

Zhang Qingdong(✉), Liu Kaiye, Fangliang, Zhang Wenwu, and Tian Junxia

The Research Institute for Special Structures of Aeronautical Composite AVIC, The Aeronautical Science Key Lab for High Performance Electromagnetic Windows, Ji'nan, Shandong 250023, China
544313918@qq.com

Abstract. Facing the demand of low-frequency broadband stealth technology is put forward for new generation weapon platform to improve the battlefield penetration and survival capability. Through scattering mechanism analysis, low-frequency broadband phase modulation microstructure design, broadband scattering angle control and travelling wave suppression fusion, the new low-frequency stealth structure is proposed. The suppression of backscattering of non-stealth shape and the goal of low-frequency broadband stealth are achieved, these technologies provide a technology path to improve low-frequency stealth performance of weapon equipment platform.

Keyword: Stealth structure · Low-frequency stealth · Angle controlling · Surface wave

With the development of technology, the battlefield scene is developing from a single confrontation mode to a system combat mode. In order to improve the aircraft battlefield survivability and reduce the probability of its detection, the application of stealth technology is very important and urgent. According to the existing radar countermeasure system, 10 dB reduction of the radar cross section (RCS) of the aircraft target can reduce the radar range by about 44% [1], which makes it one of the frontier technologies. Taking stealth fighters and warships as examples, special shapes and absorbant coatings can reduce RCS effectively, radar waves are scattered outside the threat angle through shapes, the coatings are used to further absorb radar wave energy and minimize the scattering, which can reduce the distance and probability of stealth fighters and warships being detected. However, with the emergence and application of new low-frequency radar technology, fighters and warships can be searched and found with high resolution, which will significantly weaken its stealth capability, mainly in S bands, C bands, X bands and Ku bands. Therefore, the research of low-frequency stealth technology is of great significance to make up for the deficiency of stealth performance of military equipment such as fighters, warships and missiles in low frequency band, which meets the military application requirements of full-frequency stealth, and improves the combat capability and penetration capability of military systems.

Chinese Society of Aeronautics and Astronautics: CASTYSF 2022, LNEE 972, pp. 337–347, 2023.
https://doi.org/10.1007/978-981-19-7652-0_32

1 Introduction of Stealth Technology

Stealth technologies mainly include absorbing material, absorbing structure, plasma, electromagnetic deflection, beam control, etc. [5, 6]. The absorbing material or structure mainly depends on the absorbing performance of the material or structure to improve the stealth performance, and it has obvious effects in the middle and high frequency bands, which is one of the most mature stealth technologies [7, 8]. However, the thickness of absorbing material or structure is directly related to the wavelength, which is limited by the space and profile height, resulting in poor stealth performance in low frequency band, so that it can't improve the low-frequency stealth capability.

Plasma stealth technology has been studied since 1960s, and it has been applied in MiG fighters in 1999. It refers to a technology that uses plasma to avoid radar wave detection. The plasma generator, generator or radioactive elements are used to form a plasma cloud on the surface of the aircraft. By controlling the energy, ionization degree, oscillation frequency and other characteristic parameters of the plasma, the radar wave impinging on the target plasma cloud can be absorbed, which mainly works at 4–14GHz. The theory of Plasma stealth technology is feasible, but its influence is difficult to eliminate due to the high power consumption and large volume of plasma generator [2], and there is no case of plasma stealth in low frequency band.

The metasurfaces can reduce RCS through electromagnetic deflection, so that it can improve the stealth performance of the platform. However, its shortcomings are narrow bandwidth and sensitive to the change of incident angle, which can't meet the low-frequency broadband stealth requirements.

In addition, active cancellation and beam control can also reduce low-frequency scattering. The active cancellation uses the transponder installed on the platform to transmit signals with the same amplitude, same opposite phase as the backscattering to the enemy radar, thus achieving the effect of cancellation stealth. This technology needs a large number of surface current distribution databases and instantaneous modeling for artificially manufacturing cancellation signals, which is difficult to apply. The beam control can scatter lobes in a wide angular range, which achieves low bistatic RCS, but the monostatic RCS may be not necessarily low.

2 Scattering Characteristics and Typical Structure Analysis

Generally, according to the ratio of the incident electromagnetic wave wavelength to the target size, the RCS characteristics of the target can be divided into three regions: Rayleigh region (low frequency region), resonance region and optical region (high frequency region), which respectively correspond to the fact that the wavelength is larger than the target size, the wavelength is equal to the target size and the wavelength is smaller than the target size.

The main scattering mechanism in Rayleigh region is induced dipole moment scattering, the RCS in this area increases rapidly with the increase of frequency, which is basically proportional to the fourth power of angular frequency. There is almost no phase change of the incident wave along the direction of the scatterer, each component of the scatterer is irradiated by almost the same incident site at a certain moment. Therefore, the scattering in Rayleigh region can be approximately equivalent to a static field induction problem. The influence of the overall shape of the target in this area exceeds the influence of some details in the target shape. In the resonance region, optical mechanism and surface wave effect dominate. The RCS in this area shows some oscillation behavior. Surface wave scattering often occurs at the discontinuity of the target surface, and the incident wave has obvious phase change along the direction of the scatter, so there is obvious interaction between scattered fields from various parts of the target. In the optical region, the surface wave effect is weakened, so the optical mechanism dominates. Because the wave length is smaller than the target size, the specific details of the target structure and shape have obvious influence on RCS. It can be seen that the three regions have their own dominant scattering mechanism, which leads to different RCS characteristics.

Taking a typical structure as an example, Fig. 1 shows the shape of the structure. The length direction of the structure is 1840 mm, the maximum length in the width direction is 1040 mm, and the maximum height is 380 mm. The front edge is blunt, and the rear edge is wedge-shaped. Generally speaking, the working wavelength of low frequency is 150–1000 mm, so it can be seen that the target size and wavelength size are the same order of magnitude. Therefore, the scattering of this structure is characterized by the resonance region, with optical mirror scattering and surface wave effect as the main features.

The method of moment (MoM) full-wave electromagnetic simulation method is used to analyze the low-frequency scattering source of the metal state with the front and rear edges mounted on the carrier. Figure 2 shows the simulation curve of 0–360° metal shape. It can be seen from the figure that the low-frequency scattering modes of typical structures mainly include mirror scattering, surface wave scattering, edge scattering, etc.

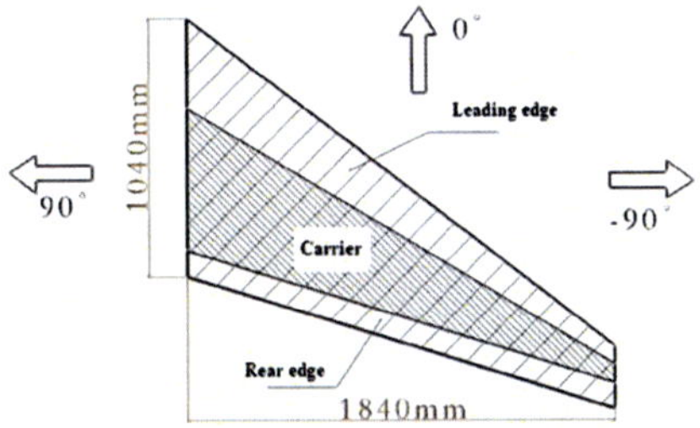

Fig. 1. Typical structure and azimuth

3 Analysis of Low Frequency Scattering Characteristics

Simulation model of triangular carrier loaded by edge structure is shown in Fig. 3, with the model size of 1200 mm in length direction, 400 mm in width and 450 mm in height., and the RCS simulation results of the leading edge structure in the direction of 0° in L band are shown in Fig. 4.

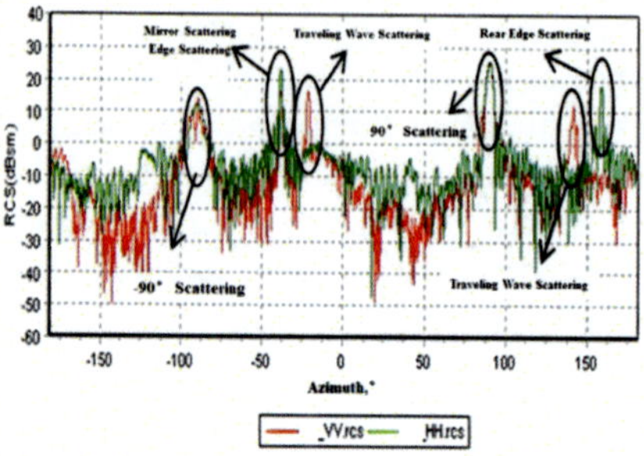

Fig. 2. RCS simulation results of typical metal structure

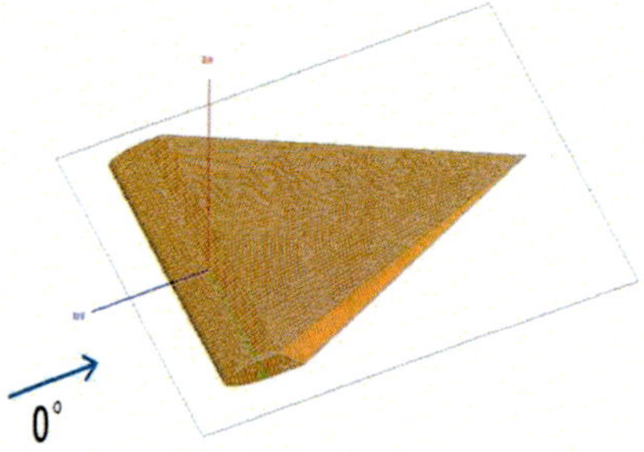

Fig. 3. Simulation model of edge structure

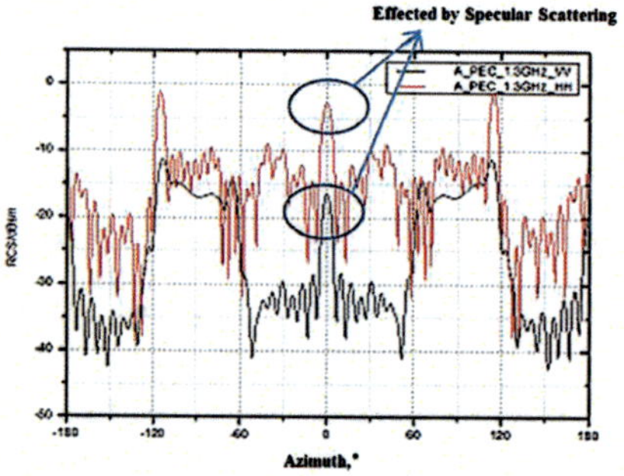

Fig. 4. RCS simulation results of leading edge structure

It can be seen from Fig. 4 that the scattering peaks on both sides are caused by the front edge specular scattering; The normal scattering peak of HH polarized wave is larger than that of VV polarized wave, and its low-frequency scattering mechanism mainly lies in:

(1) For HH polarization, the electric field vector has a component parallel to the edge, the current crawling along the leading edge will cause surface wave edge scattering. In addition, there may be some specular scattering. Therefore, the RCS of HH

polarized wave in the normal direction of leading edge is the superposition effect of specular scattering and edge scattering.

(2) For VV polarization, there is no electric field component parallel to the leading edge, the normal scattering is mainly specular scattering.

The simulation model of the trailing edge structure is shown in Fig. 5, with a length of 1200 mm, a width of 400 mm and a height of 200 mm (the trailing edge is a wedge-shaped structure, which is 250 mm thinner than the leading edge structure). The simulation results in L band are shown in Fig. 6.

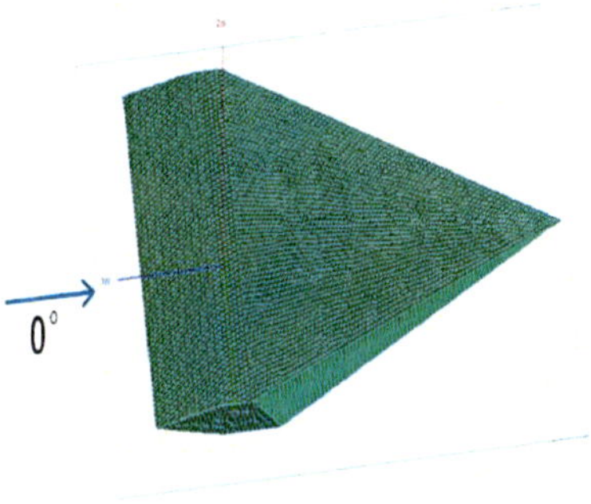

Fig. 5. Simulation model of rear edge

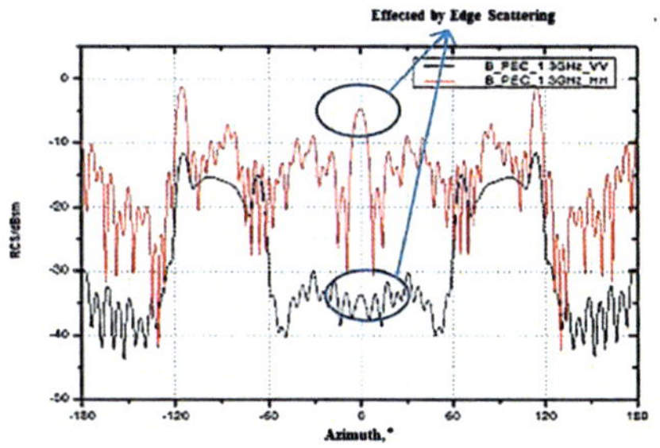

Fig. 6. Simulation results of trailing edge structure

The simulation results of the trailing edge structure show that the normal scattering peak of HH polarized wave is larger than that of VV polarized wave. Compared with the leading edge structure, this trailing edge is a wedge-shaped structure, the specular scattering is smaller from the normal direction, mainly the edge scattering. Figure 7 shows the surface current distribution of the trailing edge structure, the simulation results show that the main scattering is surface wave scattering under VV polarization.

From the above comprehensive analysis, it can be seen that the heights of the leading edge structure and the trailing edge structure are 450 and 200 mm, respectively, and other dimensions are the same, the scattering of the leading edge structure to the forward angular domain includes specular scattering and edge scattering, and the contribution of

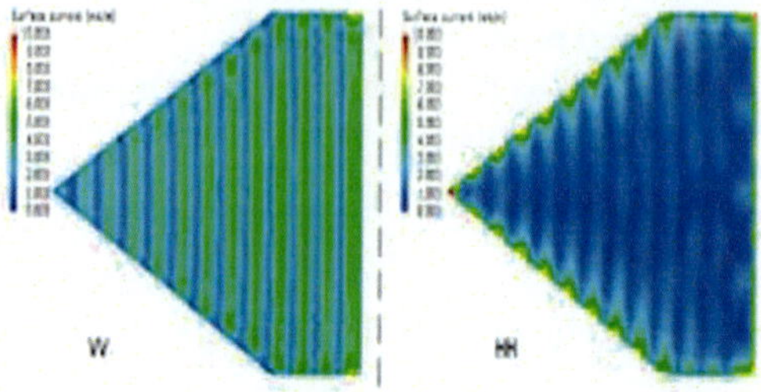

Fig. 7. Surface current distribution of the trailing edge structure

the trailing edge structure to the forward angular domain is mainly surface wave scattering. The scattering of the rear edge structure to the backward angle domain is mainly edge diffraction, and the contribution of the front edge structure to the backward angle domain is mainly surface wave scattering. The schematic diagram of typical structure low frequency scattering mode is shown in Fig. 8.

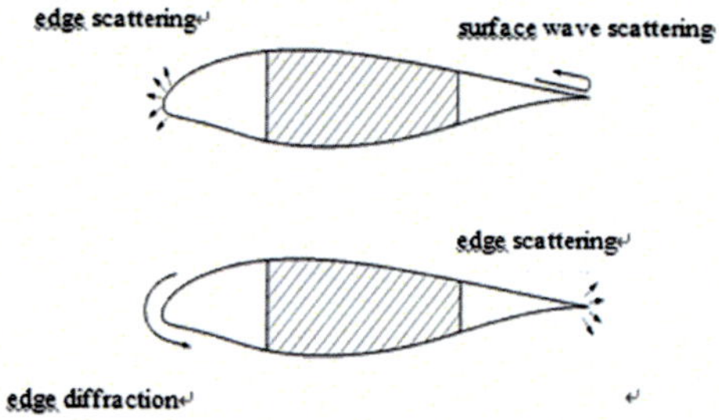

Fig. 8. Schematic diagram of typical low-frequency scattering modes of structures

4 Design of Low Frequency Stealth Structure

Based on the previous mechanism and analysis, the design of low-frequency stealth functional structure consists of two parts, one part is the design of low-frequency stealth structure, which is used to eliminate specular scattering and edge scattering; The other part is the design of surface wave suppression structure, which is used to eliminate surface wave scattering.

4.1 Low Frequency Stealth Structure Design

According to the shape characteristics of the structure, a stealth structure composed of impedance matching function layer, electromagnetic regulation function layer and absorption function layer is designed to achieve the low-frequency broadband scattering

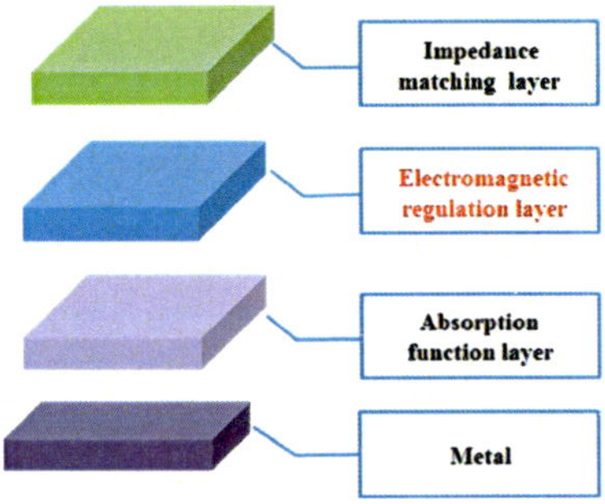

Fig. 9. The overall architecture of low frequency stealth structure

suppression effect. Figure 9 shows the overall architecture of low frequency stealth structure.

The impedance matching functional layer is usually a wave-transparent layer composed of multiple layers of low-loss materials, which can realize the non-reflection or low-reflection transmission of electromagnetic waves during the transmission of electromagnetic waves from optically-hydrophobic media to optically-dense media, so that as much energy as possible can reach the inside of the structure. In addition, it can also expand bandwidth through self-matching. The electromagnetic control function layer is the core of the stealth structure, which is used to control the scattering direction of electromagnetic waves to the non-threatening direction. In order to increase the bandwidth. The absorption functional lay mainly absorbs the electromagnetic wave entering the structure, and the distance between the electromagnetic regulation functional layer and the reflecting plate is controlled to meet the optimal thickness match. The reflective plate is usually the metal structure, which serves as the metal floor of the functional layer [3, 4].

The core of broadband impedance matching layer design is to make the reflection coefficient of each layer wall structure change with frequency to satisfy Chebyshev polynomial relation. It can be realized by low dielectric material in low frequency, and gradient structure can be considered in high frequency, which can be obtained by changing the dielectric constant and thickness gradient of the material. The stepped dielectric constant can meet the requirements of the following formula.

$$\square \tag{1}$$

$$\square \tag{2}$$

The optimal dielectric constant gradient structure is further optimized by using the 3D electromagnetic field simulation software, and the broadband matching between air and structure is well realized, as shown in Fig. 10.

The electromagnetic function layer modulates the transmission, reflection and phase characteristics of the structure in frequency domain by designing artificial electromagnetic structure units. For the structure with fixed shape, the direction control and peak suppression of scattering are realized by artificial electromagnetic structure. The absorption functional layer mainly adopts the lossy multilayer unitized film structure to expand the absorption bandwidth. Figure 11 shows the basic structure of the absorption functional layer.

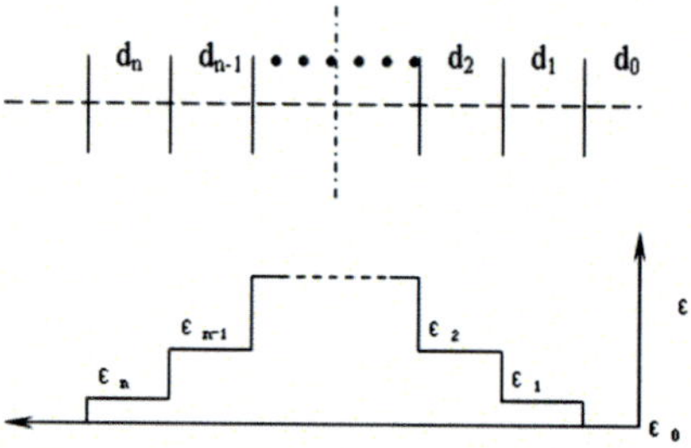

Fig. 10. Schematic diagram of dielectric constant gradient

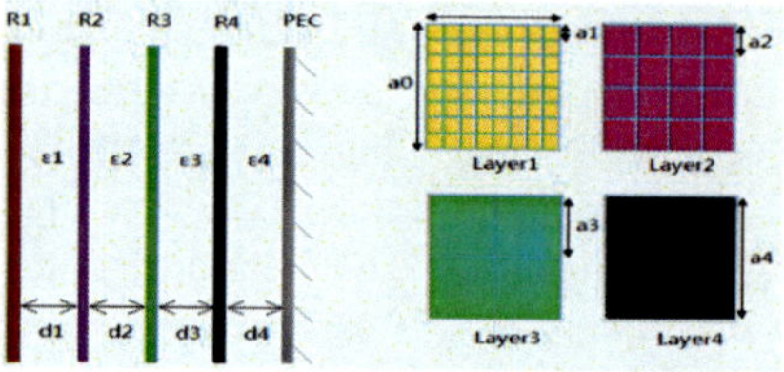

Fig. 11. Structure of absorption functional layer

4.2 Design of Surface Traveling Wave Suppression Structure

In order to suppress the forward traveling wave scattering of surface waves at the trailing edge and reduce the backward scattering of surface waves, a broadband traveling wave suppression structure is designed on a standard traveling wave plate with a size of 600 mm, as shown in Fig. 12, which is optimized and simulated. Figure 13 shows the RCS simulation results of the surface wave suppression structure. Compared with the metal structure, the surface wave suppression structure obviously reduces the scattering of surface waves at grazing incidence angle.

4.3 Performance Simulation of Low Frequency Stealth Integrated Structure

The stealth structure adopted in the leading edge structure shows in Fig. 9, the connecting area of the trailing edge structure adopts the surface traveling wave suppression structure, and the trailing edge adopts the stealth structure, which is installed on a metal carrier for comprehensive optimization simulation. The simulation results of the overall stealth structure and the metal structure are compared as shown in Fig. 14. It can be seen from Fig. 14 that the peak value of front edge scattering is obviously reduced, and the peak value of front edge surface wave scattering disappears, so that the RCS reduction is realized.

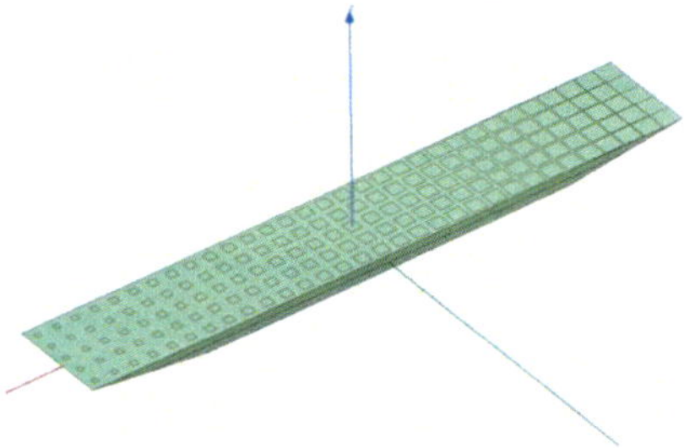

Fig. 12. Surface traveling wave suppression structure

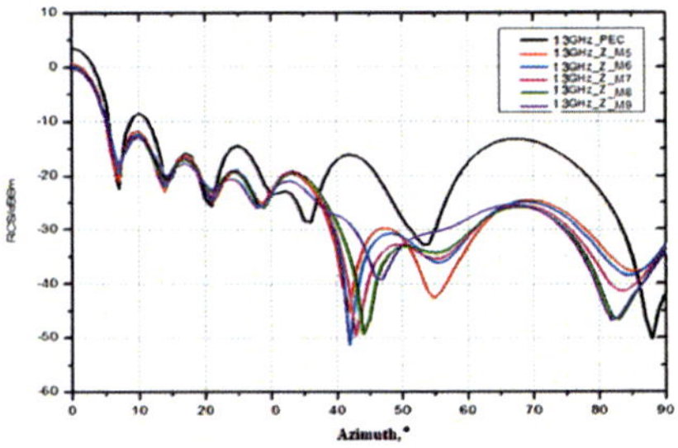

Fig. 13. RCS Simulation results in L band

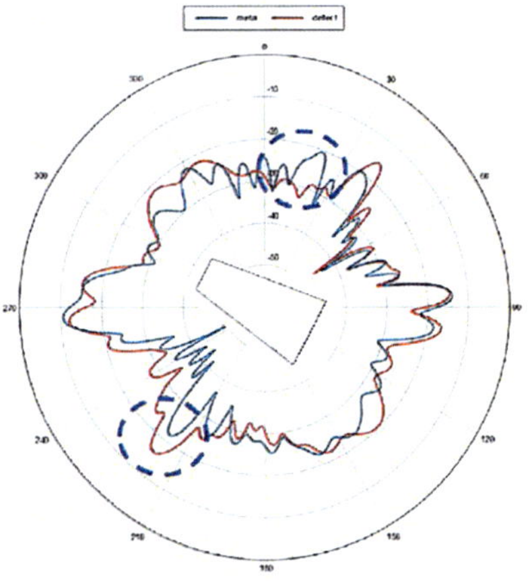

Fig. 14. Simulation results at L-band scattering of leading edge structure

5 Verification of Low Frequency Stealth Structure

The proposed structure is tested, and the experimental results at L-band are shown in Figs. 15 and 16. The results show that compared with the metal structure, the peak value of the typical structure decreases by 13 dB, and the scattering average value in the forward angle decreases by about 8 dB. It can be seen from Fig. 16 that the suppression effect of 21°scattering peak is obvious, and the number of forward scattering peaks is reduced from two peaks to one peak, and the scattering magnitude at the trough position in the angle range of 0–35° decreases by 3–5 dB.

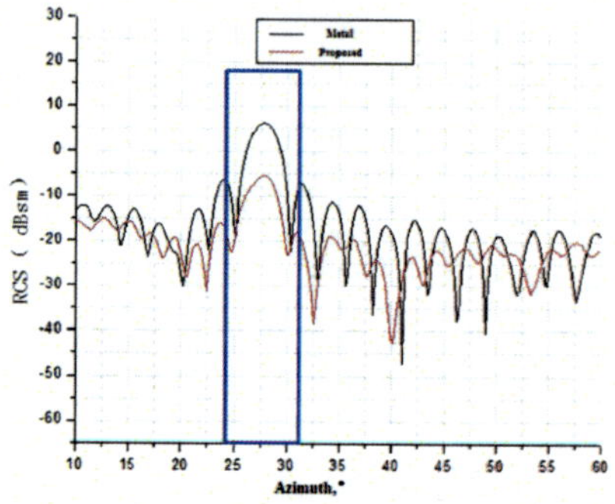

Fig. 15. L-band RCS of leading edge structure

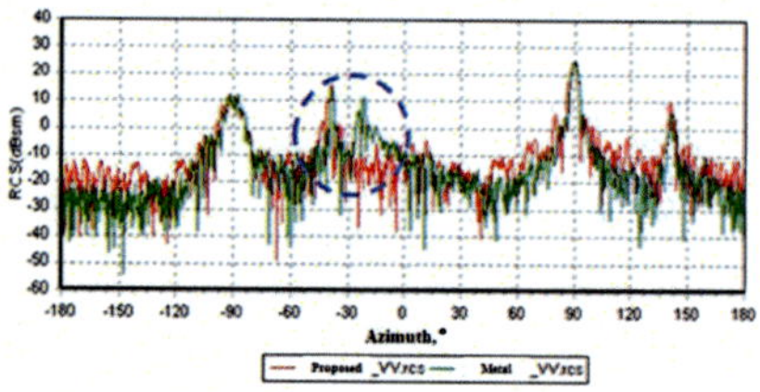

Fig. 16. L-band RCS of rear edge structure

6 Conclusion

In this paper, a new design method of electromagnetic stealth structure with low frequency and wide band scattering suppression is proposed, and the technical scheme of stealth structure with low frequency, wide band and thin thickness is proposed. Combining various mechanisms and methods, a typical structure is designed and tested. The results show that this method can be used to control and suppress the non-stealth shape scattering, and provides a technical way for weapon platform to realize low-frequency broadband stealth.

References

1. Jianhua, S.: Stealth Technology of Aircraft. Beijing: Aviation Industry Press
2. Zheng, R.: Radar Cross Section and Stealth Technology, 3rd edn. National Defense Industry Press, Beijing (2001)
3. Jianhua, S., Hai, Z., Yingwen, C.: Stealth design of advanced fighter. J. Aviation **38**(01) (2017)

4. Zhu, B., Wang, Z., Huang, C., et al.: Polarization insensitive metamaterial absorber with wide incident angle . Progress Electromag. Res. **101**, 231–239 (2010)
5. Zhu, B., Wang, Z., Huang, C., et al.: Polarization insensitive metamaterial absorber with wide incident angle. Progress Electromag. Res. **101**, 231–239 (2010)
6. Yanping, Z., Xiaopeng, Z., Shi, B., et al.: Dendritic metamaterial absorber based on impedance matching condition. J Phys **59**(9), 6078–6083 (2010)
7. Li, H., Yuan, L.H., Zhou, B., et al.: Ultrathin multiband gigahertz metamaterial absorbers. J. Appl. Phys. **110**, 014909 (2011)
8. Sun, J., Liu, L., Dong, G., Zhou, J.: An extremely broad band metamaterial absorber based on destructive interference. Opt. Express **19**(22), 21155–21162 (2011)

Optimization of Flight Procedure Method Based on Track Clustering

Cui Haiyang[1](✉), Zhang Zhaoning[2], Lu Tingting[2], Meng Qingyu[3], and Wang Jiechun[1]

[1] Airport Management College, Guangzhou Civil Aviation College, Guangzhou 510403, China
812329866@qq.com
[2] College of Air Traffic Management, Civil Aviation University of China, Tianjin 300300, China
[3] School of Electronic Engineering, Beijing University of Posts and Telecommunications, Beijing 100876, China

Abstract. Flight procedure devising is an important part to ensure flight safety, and the benchmark of aircraft safe operation. Therefore, if flight procedures are not suitable for the operating environment will also produce serious potential safety hazards. Based on the track information of the actual operation of an airport, combined with BP neural network and K-means algorithm, a new track clustering model is constructed. Through the cluster analysis of the departure track of the airport, the clustering center of the track is obtained. Combined with the inherent flight procedure of the airport, it is found that there are great differences between the actual operation track and the inherent flight procedure. After investigating and analyzing the causes of the differences, the program improvement scheme is put forward. Finally, the track clustering results are compared with the flight program optimization results, and the simulation analysis is carried out by using air top software. The results show that the new program can better adapt to the actual operation of the aircraft.

Keywords: Flight safety · Optimization of flight procedure · Track clustering · Pattern recognition · Track screening

Introduction: Flight procedure is an important link to ensure flight safety, but in the process of actual flight program design, some design results only meet the basic requirements of safety, but does not conform to the actual operating environment, thus cause serious problems such as track deviation, which caused great hidden trouble to flight safety. Therefore, we should explore the adaptability of the existing flight procedures, so as to increase the fluency of the flight process, reduce the difficulty of control, increase the comfort of passengers, and fundamentally eliminate the potential dangers affecting

Receipt date: 2xxx-xx-xx **Revision date**: 2xxx-xx-xx.
***Fund project**: Special funds for basic scientific research in central universities (3122017061): Guangdong University scientific research Platform and projects (21Z5221): National Key research and development plan (2020YFB1600100)
The first author: Cui Haiyang (1991-), Male, Schoolteacher, The main research direction is the flight program design optimization, Flight conflict detection and relief, 812329866@qq.com

Chinese Society of Aeronautics and Astronautics: CASTYSF 2022, LNEE 972, pp. 348–367, 2023.
https://doi.org/10.1007/978-981-19-7652-0_33

flight safety.Scholars at home and abroad often study the influence of noise, landform influence and environmental pollution influence. Literature [1] analyzed the influencing factors of flight programming in terminal area, including landform, radio equipment, and successively proposed the optimization of return flight program scheme and the comprehensive adjustment scheme of return flight parameters. Literature [2] evaluates the near-ground risk, macro-and micro-conflict risk of flight program operation. The adaptability of the flight program is also defined from the perspective of the aircraft safety interval. Using a hierarchical clustering model, the adaptability of the flight program was evaluated from the perspective of longitudinal, lateral deviation and unconventional track ratio. Based on the PBN principle of program design, literature [3] studied the adjustment strategy of potential conflict points, adjusted the route, departure route and approach route in the design stage, so as to achieve the goal of minimizing the long-term conflict, and optimized the original PBN program.After analyzing the causes of PM2.5 pollutants, literature [4, 5] established the emission and diffusion model of particulate matter in aircraft, and proposed the optimization scheme of flight program from the perspective of reducing pollutant emission.Literature [6–12] proposes the improvement scheme for the design results of the flight program from the perspective of noise pollution, using the noise impact area, based on which the noise reduction program model is used to reduce the impact of aircraft on residential areas.Literature [13] constructs the simulation model by comprehensively considering the lift characteristics, resistance characteristics, rising-resistance ratio characteristics, side-slip characteristics, rising and falling characteristics of the aircraft, and constructs the program adaptive model from the perspective of aircraft performance.Literature [14, 15] takes the goal of saving fuel consumption, established a fuel consumption calculation model, based on optimizing the flight program with large fuel consumption, and then realized the purpose of program optimization.Literature [16, 17] explores the method of air track clustering, combines the air track clustering technology with the system design of air traffic management, and improves the management efficiency of air traffic control.

In the above research results, the flight program is usually optimized based on environmental factors such as economy and pollution reduction, while rarely studied study from the perspective of track combined with controller control habits and pilot control habits, Considering the value of track data and the significance of flight procedure to flight safety guidance, this paper puts forward a model to optimize flight procedure by using the information reflected in track clustering center. Finally, important hazard information is found in track information, and the inherent flight procedure is optimized A flight trace classification model based on the BP neural network.

1 A Flight Trace Classification Model Based on the BP Neural Network

1.1 Build a Neural Network

BP neural network is a common classification method in pattern recognition, and a common means to use data features to double-classify or multiclassify data. As can be seen from the flight chart, there are four departure directions, corresponding to four types of flight procedures. To increase the accuracy of the k-means clustering algorithm, the BP neural network was used to classify the trace data into four categories.

First, the three-layer BP neural network shown in Fig. 1 is established, where the three-layer neurons of LA, LB and LC represent the three layers of input, implied and output, respectively. Considering the nonlinear characteristics of the trace data features, the logistic function was used as an intra-neuron activation function in both the LA and LB layers. The LC layer adopts the softmax function, which not only realizes the error function calculation, but also obtains various kinds of data gaps, which is more conducive to the evaluation of neural network performance.

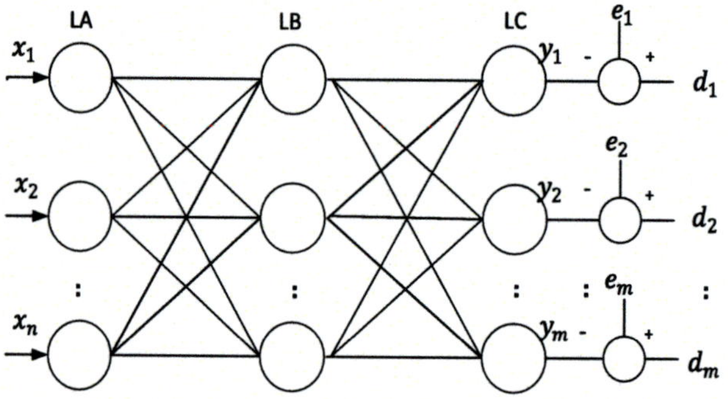

Fig. 1. Schematic diagram of the BP neural network

The BP neural network adopts the working mode of data forward transmission, error reverse transmission and modification of neuronal parameters. The mechanism of action is follows:

The LA layer corresponds to the n-dimension input $x_i(i = 1, 2 \ldots n)$, the number of LB neurons is the p, each neuron is connected to all the neurons in the upper layers, and from the formula (1) and (2) derive the output expressed as $out_j(j = 1, 2 \ldots p)$, the m neurons in the LC layer corresponds to the number of data species, the output from formula (3) and (4) is $y_k(k = 1, 2 \ldots m)$. The output vector and the real vector $D = [d_1, d_2, \ldots d_m]$ compare, error able by formula (5) $E = [e_1, e_2, \ldots e_m]$, then use formula (6), (7), (8), (9), (10) and (11) to reverse the optimization of parameters.

$$net_j = \sum_{i=1}^{n} w_{ij}x_i - b_j \tag{1}$$

$$out_j = f\left(net_j\right) = \frac{1}{1 + \exp(-net_j)} \tag{2}$$

$$sum_k = \sum_{j=1}^{p} w_{jk}out_j - b_k \tag{3}$$

$$y_k = p(sum_k) = \frac{e^{sum_k}}{\sum_{k=1}^{m} e^{sum_k}} \tag{4}$$

$$E_{total} = \frac{1}{2}\sum_{k=1}^{m}(d_k - y_k)^2 \tag{5}$$

$$\delta_k = |y_k - d_k| \tag{6}$$

$$\delta_j = \sum_{k=1}^{m} w_{kj} d_k \tag{7}$$

$$\Delta w_{ij} = \eta\delta_j \frac{\partial f_j(net_j)}{\partial(net_j)} x_i \tag{8}$$

$$\Delta w_{jk} = \eta\delta_k \frac{\partial f_k(net_k)}{\partial(net_k)} net_j \tag{9}$$

$$\Delta b_j = \eta\delta_j \frac{\partial f_j(net_j)}{\partial(net_j)} x_i \tag{10}$$

$$\Delta b_k = \eta\delta_k \frac{\partial f_k(net_k)}{\partial(net_k)} net_j \tag{11}$$

1.2 Data Processing and Model Construction

Different from the general single-dimensional feature vector data, the trace data is composed of multiple data points, and the data of each point is in 3 D data, so the data should be reduced in an appropriate way. The track data is composed of a set of points. The data volume of different track points is different. If all points are used as the input feature vector of the track, the dimension of the input vector is not unified, which seriously violates the requirements of BP neural network for the input vector. Therefore, a fixed length sampling method is proposed to maintain the unity of the dimension of the input data.

Take the time when the aircraft arrives at the designated point as the starting time, sample the track time series at a fixed time interval, and form the final sample feature set, and the process formula is shown in (12), (13), (14) and (15).

$$X_n = X[t_0 + (n-1)\tau] \tag{12}$$

$$Y_n = Y[t_0 + (n-1)\tau] \tag{13}$$

$$Z_n = Z[t_0 + (n-1)\tau] \tag{14}$$

$$[X_n, Y_n, Z_n] \tag{15}$$

The arrival and departure track information is transformed as follows: 15% is the test group, 15% is the verification group, and 70% is used to train the neural network. Figure 2 Confusion Matrix (confusion matrix) analysis chart, it can be seen that the accuracy of overall trace data classification is 95.8%, and the ROC (receiver operating characteristic curve) curve of classified data can be obtained. As can be seen from the ROC Fig. 3, with a small False Positive Rate (false positive rate), the True Positive Rate (true positive rate) can reach close to 1, indicating that the neural network has good sensitivity and accuracy.

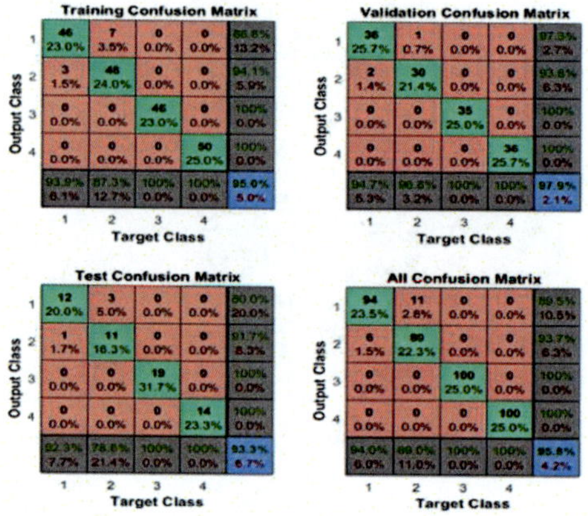

Fig. 2. Confusion Matrix (confusion matrix) analysis diagram

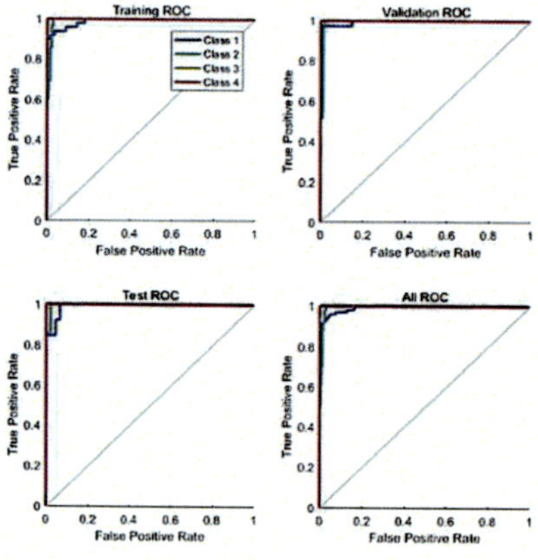

Fig. 3. ROC (receiver operating characteristic curve) graph

1.3 Division of Inbound and Departure Routes

The constructed track division model is used to divide the incoming and departure tracks respectively. Figure 4 departure procedures are TYN, HDH, DUBAG and UBDUN respectively, and the tracks of DUBAG are shown in Fig. 5.

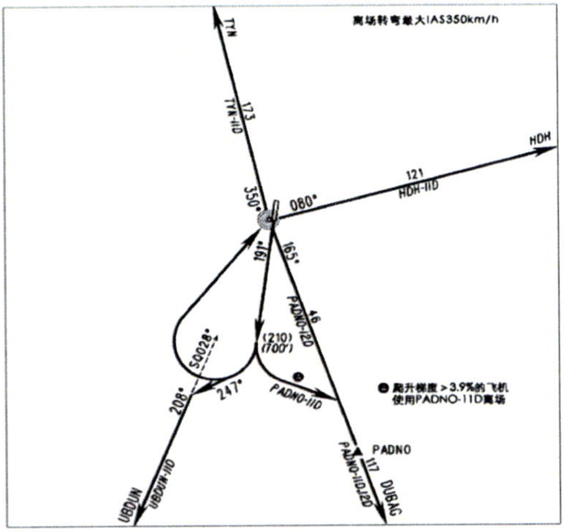

Fig. 4. Leaving the flight procedure diagram

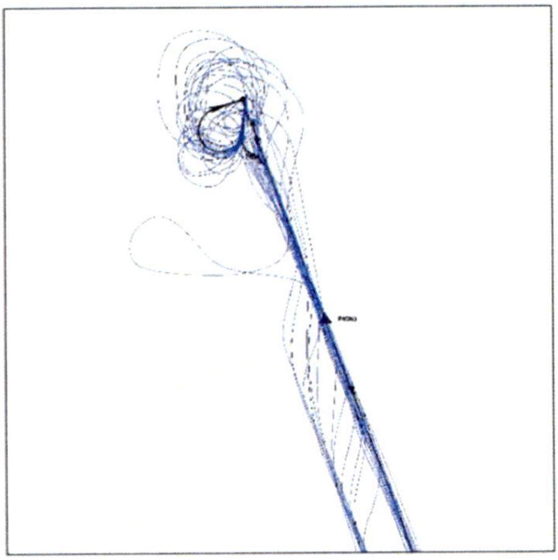

Fig. 5. DUBAG departure track classification results diagram

2 Trace Screening Together with k-Means-Based Trace Clustering

2.1 Abnormal Navigation Track Screening Model

The k-means clustering algorithm is a method of cluster analysis. After dividing the original data into k groups, k objects are randomly selected as the initial cluster center. By calculating the distance between the data unit and the cluster center, the point near the same cluster center is defined to realize the purpose of data clustering. After referring to the flight procedure and the corresponding protected area, it is found that some tracks exceed the protected area. This special track should be eliminated, and a track screening model is proposed.

Step 1: Determine the two endpoints of the line segment program where the track point is located, and obtain its plane coordinates as $x_{t1}y_{t1}$ and $x_{t2}y_{t2}$.

Step 2: Suppose that the coordinates of the track point are $x_d y_d$,bring the coordinates into formula (16) to obtain the deviation distanced.

Step 3: Comparing the size of the distance d to the $d_{\max}$ maximum distance, if there is a point with a distance greater than $d_{\max}$on a single track, the track is considered to be an abnormal track and needs to be removed, this article takes $d_{\max}$10 km in accordance with the regulations.

$$d = \frac{\left| y_d - y_{t1} - \frac{y_{t2}-y_{t1}}{x_{t2}-x_{t1}} x_d - x_{t1} \right|}{\sqrt{1 + \left(\frac{y_{t2}-y_{t1}}{x_{t2}-x_{t1}} \right)^2}} \tag{16}$$

2.2 Clustering Based on the k-means Algorithm

Through the track screening model, abnormal traces can be removed. The blue traces in Fig. 6, 7, 8 and 9 are normal traces, and other abnormal traces need to be removed. After removing the abnormal trace, the k-means model is used to cluster the value of k is mainly based on the survey data of the actual flight condition of the aircraft. If the same trace is based on n sets of flight instructions, the k value is n. Thus, the flight track clustering results coincide with the actual flight situation.The results of trace clustering are shown in Figs. 10, 11, 12 and 13.

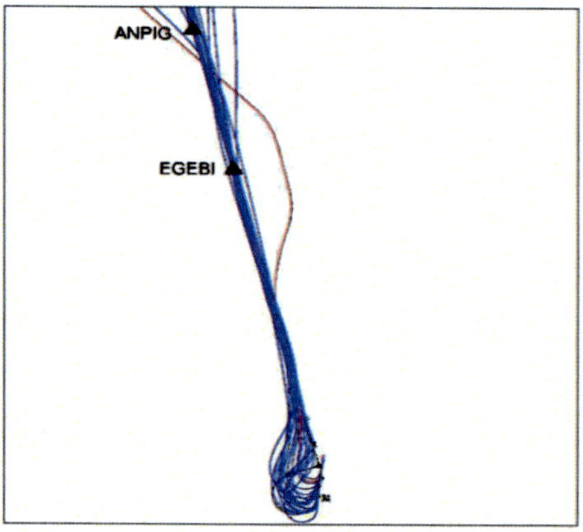

Fig. 6. Track of T Y N departure

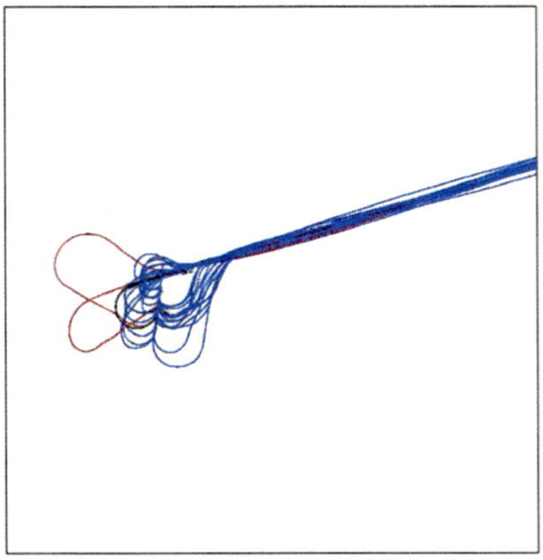

Fig. 7. HDH departure track

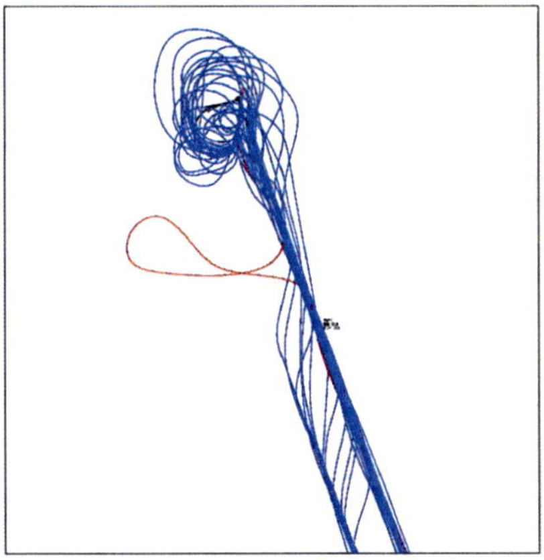

Fig. 8. Track of DUBAG departure

3 Analysis of Flight Track Clustering Results and Flight Program Optimization

By analyzing the relationship of cluster center track and inherent flight program, we can intuitively reflect the adaptability of inherent flight program in the actual working environment. For less adaptive flight programs, a more reasonable flight program design scheme is provided by combining the flight operation rules, and the new program can be evaluated to verify the availability of the program.

Figure 10 indicates the track cluster center of TYN departure, which is taken from 1 according to the flight manipulation situation k (blue line).The traditional program (black line) is to turn back to the platform, while the cluster center is to cut into the Changzhi NDB navigation station to the TYN VOR navigation station. According to the

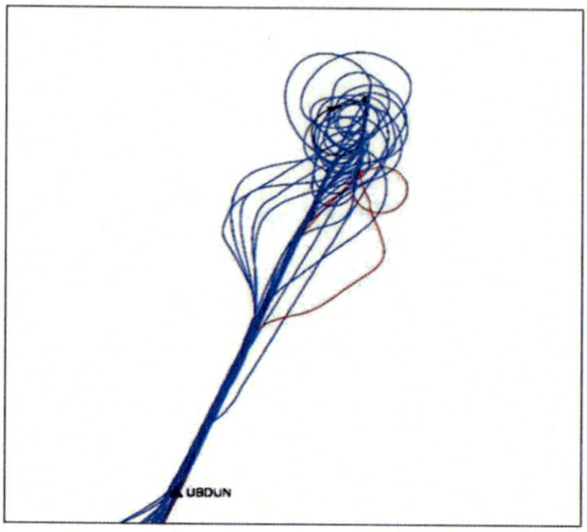

Fig. 9. UBDUN departure

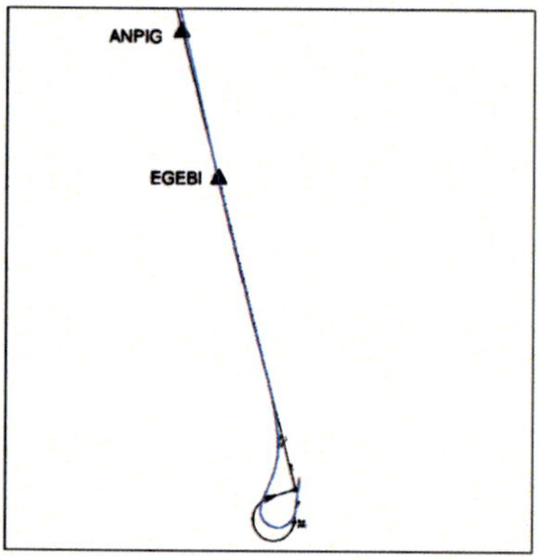

Fig. 10. TYN departure cluster center

existing program, the protection area can not protect the real flight track on the west side of the runway, and there are serious safety risks.

At the same time, the aircraft going to the TYN airport will not fly to the TYN navigation station, so the departure program termination point will be changed to EGEBI. Therefore, optimizing the optimized TYN departure segment flight procedure is shown in Fig. 11. On this basis, the flight reserve of Fig. 12 was redrawn and the obstacles were evaluated.

Figure 13 shows the track clustering center (blue line) of HDH departure, and the corresponding flight control situation K takes 2, that is, the clustering result is 2 tracks. Existing traditional program (black line) to take off right turn back to Taiwan, and track cluster center shows the actual flight situation for two kinds: leave left or turn right directly into the NDB navigation to HDH VOR navigation radial line, but the airport is no left track reserve, and left track for more than 50%, so there are serious safety risks. Therefore, the reserves were redesigned in Fig. 14, and the obstacles were evaluated in

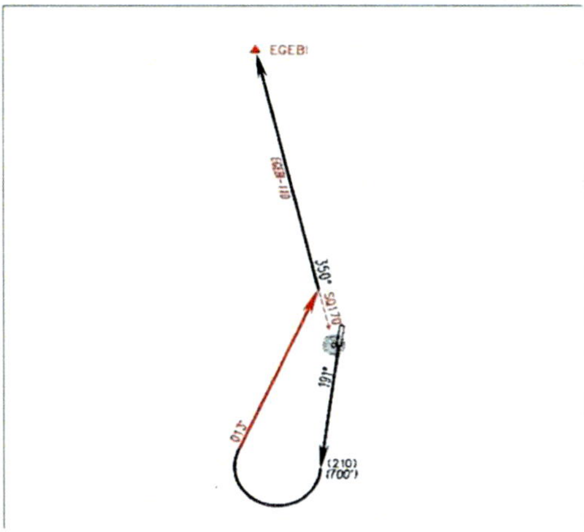

Fig. 11. Improved design of TYN direction

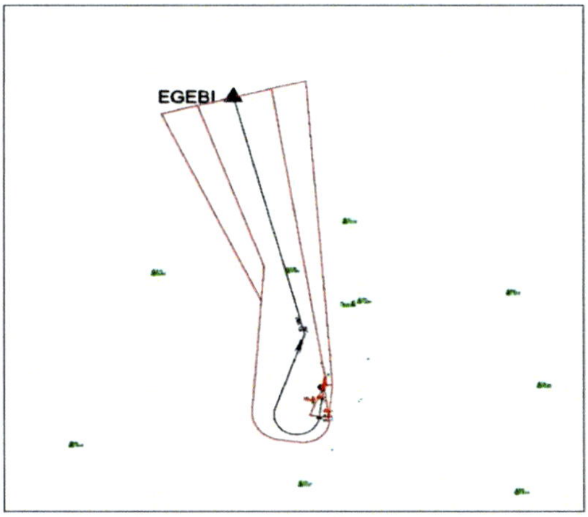

Fig. 12. Improved TYN departure flight program Reserve and obstacle diagram

Fig. 15. Considering that the east of the airport is high mountains, the drawing reserve assessment calculated that the runway 19 aircraft turning left at 210m field pressure requires at least 6.0% gradient departure.

Figure 16 shows the track clustering center (blue line) of HDH departure, and the corresponding flight control situation K takes 3, that is, the clustering result is 2 tracks. The inherent traditional procedure (black line of DUBAG) has two ways to leave runway 19: (1) turn right into the radial line of the navigation station to DUBAG with 3.9% gradient; (2) turn right back to the station with a 3.3% gradient of 210 m and fly directly to DUBAG via PADNO.

However, the track clustering center shows that the aircraft cannot directly enter the radial line of the airport NDB navigation station to DUBAG in the process of turning

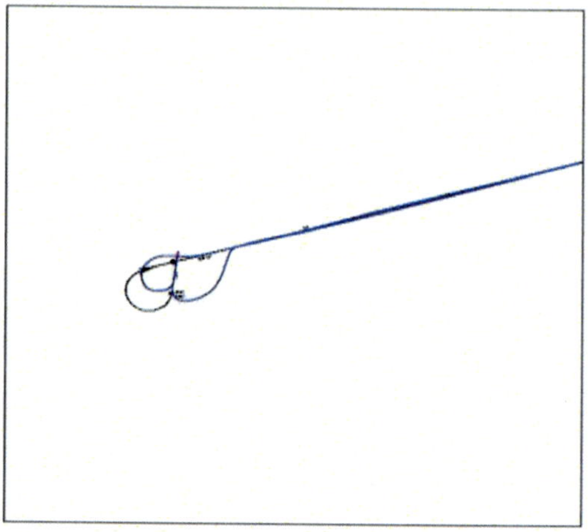

Fig. 13. HDH departure track cluster center

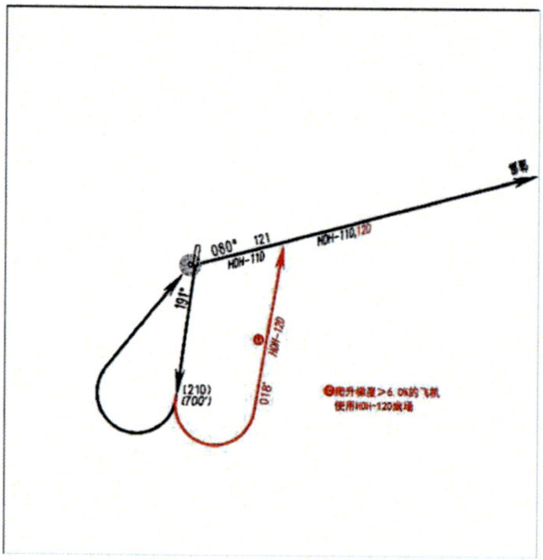

Fig. 14. Improved HDH direction departure program drawing

back to the station, and often needs to cut back to the station. In addition, the east side of the airport is a continuous mountain peak. It is not recommended that the aircraft operate at a low altitude on the east side of the runway. It is necessary to modify the departure turning procedure so that the aircraft can cut into the radial line before returning to the runway, as shown in Fig. 17. Finally the obstacle assessment in Fig. 18.

Figure 19 shows the track clustering center (blue line) of HDH departure, and the corresponding flight control situation K takes 2, that is, the clustering result is 2 tracks.The inherent traditional procedure for UBDUN departure (black line) is the only one way to leave runway 19, by turning right with a 3.3% gradient into the radial line of the airport NDB navigation station to UBDUN.

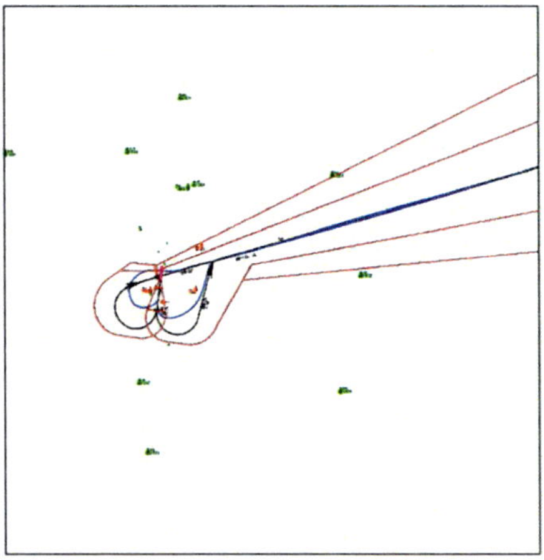

Fig. 15. Improved HDH departure flight program reserve and obstacle diagram

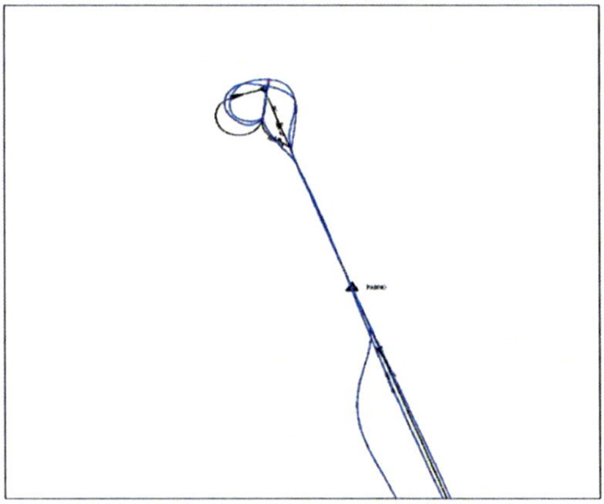

Fig. 16. DUBAG departure cluster center

The existing traditional program of the airport is a 3.3% gradient left turn into the NDB navigation station to the UBDUN radial line. By analyzing the causes of the cluster center, the survey results show that the actual control command requires the aircraft to enter with at least 7.5% departure gradient, or to turn back to the station with 3.3% departure gradient after entering the radial line from the field navigation station to UBDUN. However, the current procedures do not meet the requirements, so the departure turn back to the stage program is modified, and notes are added to adjust the UBDUN departure direction flight program, as shown in Fig. 20. Finally, the obstacle assessment as shown in Fig. 21 was conducted.

Finally, by comparing the track clustering center (blue line) in Figs. 22, 23, 24 and 25 with the improved flight program (black line), it is obvious that the improved program is more appropriate to the actual flight situation and more in line with the actual operation.

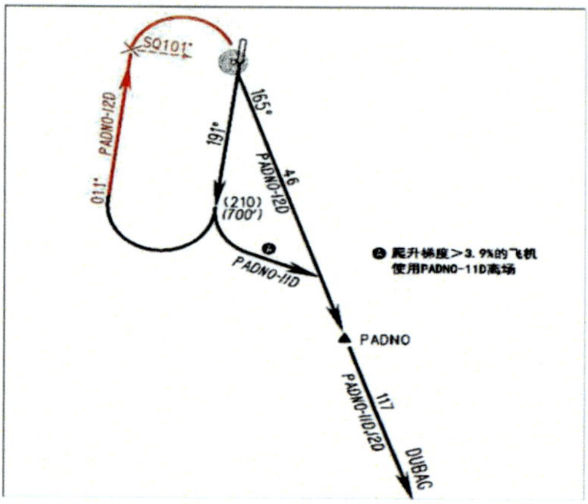

Fig. 17. Design design of DUBAG

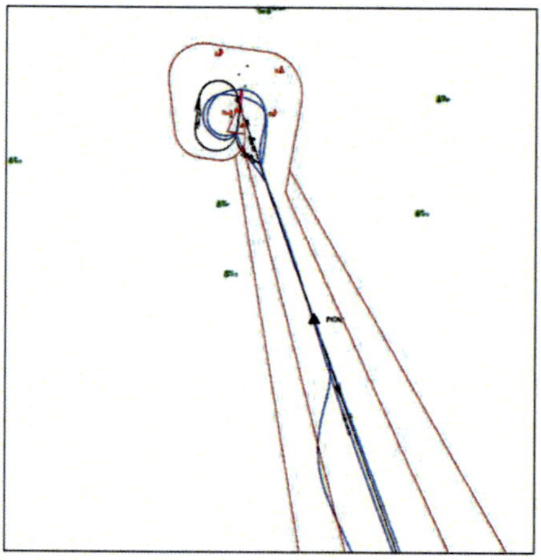

Fig. 18. Improved DUBAG departure flight program reserve and obstacle diagram

4 Program Performance Verification

The air top simulation software can be used to compare the average delay sorties and the average delay time information of the two flight programs to verify the nature of the optimization results. According to the data of Variflight, the existing flight plans at this airport are shown in Table 1.

Change the number of flights appropriately and based on the Preparation Guide for Simulation Research of Transportation Airport Master Plan, AC-158-CA-2020–02, the delay level is acceptable, the average all day (inbound and departure) flight delay is 8 min; the average delay of peak hour flight does not exceed 18 min.When there is a

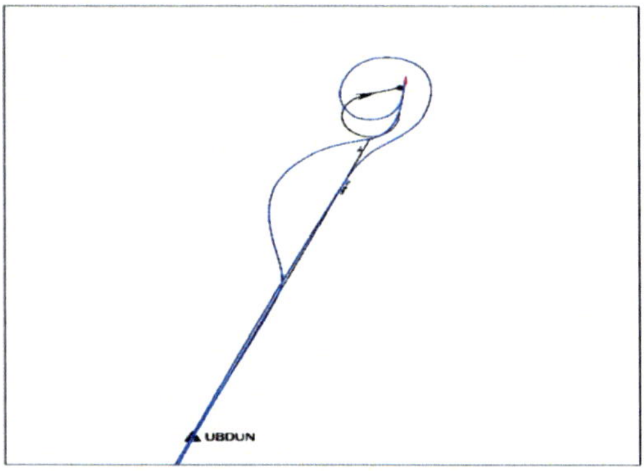

Fig. 19. UBDUN departure cluster center

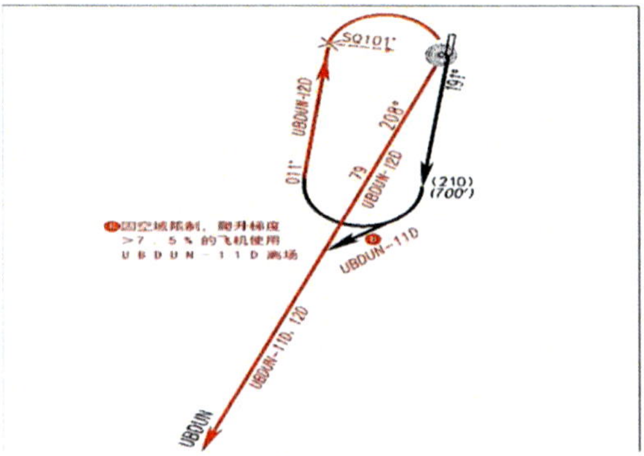

Fig. 20. Design design of UBDUN

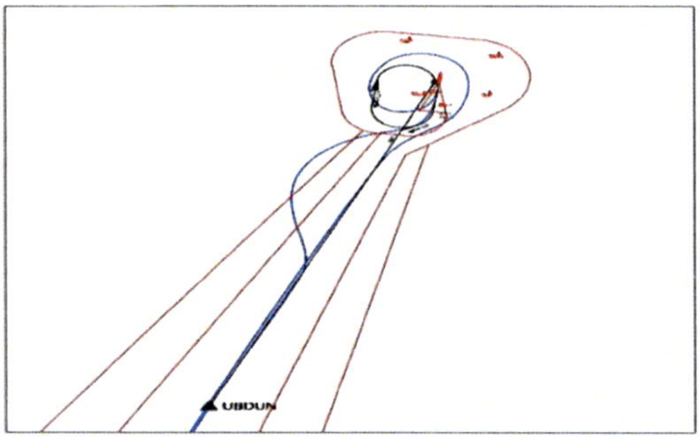

Fig. 21. Improved UBDUN departure flight program Reserve and obstacle diagram

breakthrough in the above standard, stop the increment. Based on the departure flight schedule, when the flight volume increases to 130 flights / day, the track heat map Figs. 26 and 27 are obtained, and the average delay data are shown in Table 2.

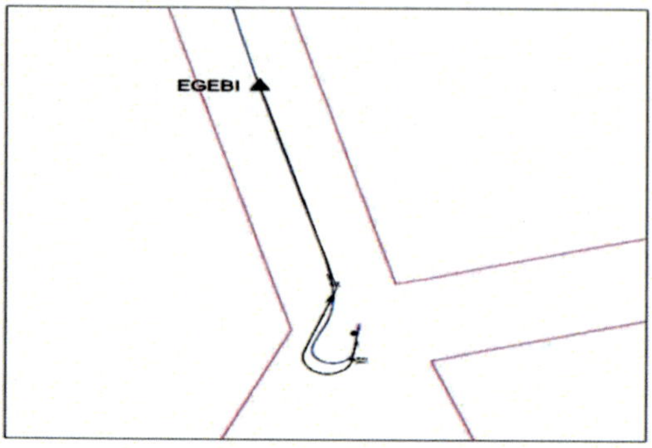

Fig. 22. Comparison of improved flight procedure (TYN departure) and track clustering center results

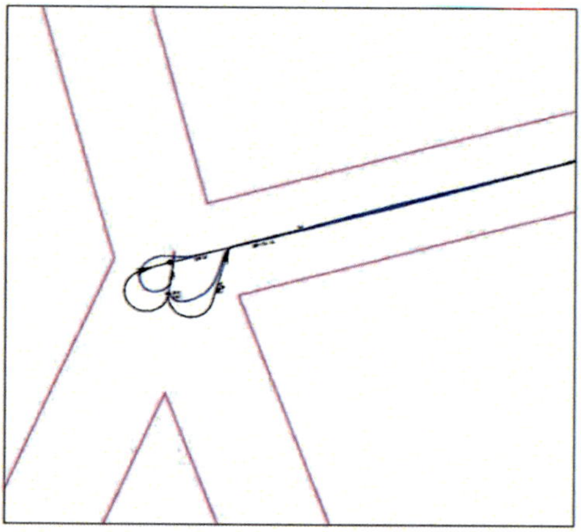

Fig. 23. Comparison of improved flight procedure (HDH departure) and track cluster center results

As shown in Table 3, compared with the traditional flight procedure, the improved flight procedure can reduce the average flight delay by about 2 min, and the average flight delay in peak hours can be greatly reduced; Table 5 and 6 shows the statistics of the two peak hour flights. Compared with the traditional flight procedure, the evening rush hour, and the flight volume in other periods is basically the same.

In summary, compared with the traditional flight procedures, the improved flight procedures increase the number of hourly flights during peak hours, but the flight delays are alleviated, indicating that the improved flight procedures reduce aircraft conflicts, alleviate the aggregation of aircraft, and improve the flight capacity in the airspace.

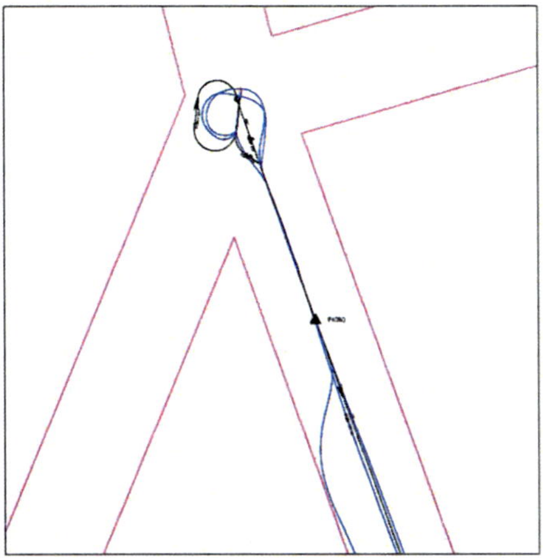

Fig. 24. Comparison of the improved flight procedure (DUBAG departure) and track cluster center results

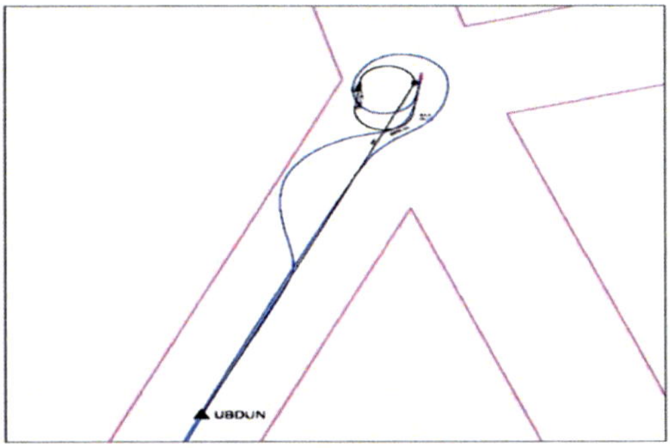

Fig. 25. Improved flight procedure (UBDUN departure) and track cluster center results comparison diagram

5 Conclusion

Considering that the flight procedure may not adapt to the actual work, a flight procedure optimization model is proposed in this paper we propose a flight program optimization model that combines BP neural network with k-means clustering algorithm to realize accurate clustering of trace information. After the original abnormal track removal model, the clustering results of the various flight tracks are obtained. By studying the difference between the inherent flight program and the clustering results and the actual operation environment, the final flight results show, the new flight program has better environmental adaptability.

Table 1. Airport departure flight plan

Take off	Destination	Flight number	Type	Between planned take-off	Plan between landing
Home course	Shenzhen	DZ6210	JET	9:05	11:55
Home course	Guangzhou	CZ3706	A320–214	14:30	15:55
Home course	Chongqing	G52720	A320–214	8:45	10:35
Home course	Nanchang	KN2927	B737–700	15:30	17:25
Home course	Daxing	KN2928	B737–700	20:40	22:15
Home course	Taiyuan	G54988	BOMBARDIER	18:05	18:45
Home course	Shenyang	A67314	B737–800	16:10	18:15
Home course	Capital	CA1400	B737–700	17:10	18:30
Home course	Guilin	A67313	JET	10:15	12:25
Home course	Wuhan	CZ3706	A320–214	13:25	14:50
Home course	Pudong	H01140	A320–214	11:15	13:20
Home course	Tianjin	G52719	A320-271N	22:15	23:35

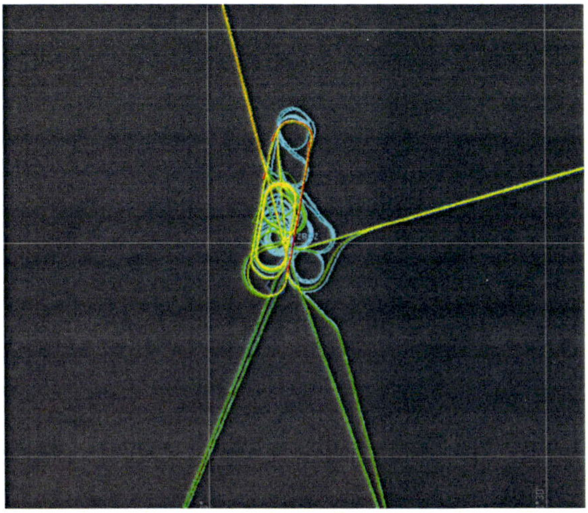

Fig. 26. Leaving the field with the original procedure

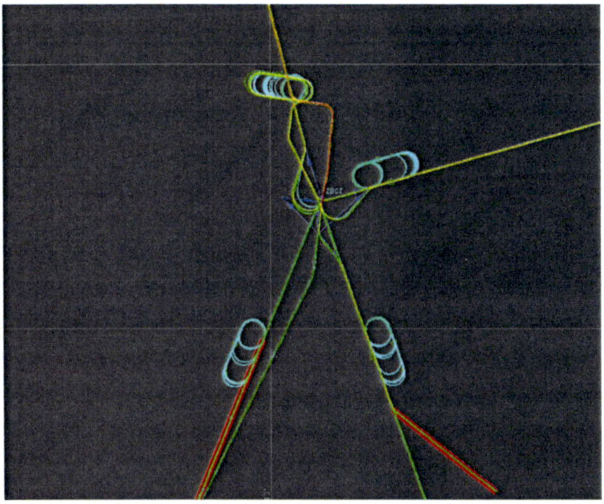

Fig. 27. Improvement procedure departure

Table 2. Average delay time

Time	Traditional program	Improvement program
At 7 o'clock	0:02:12	0:02:23
At 8 o'clock	0:07:41	0:05:33
At 9 o'clock	0:08:05	0:03:17
At 10 o'clock	0:02:32	0:02:05
At 11 o'clock	0:01:16	0:01:16
At 12 o'clock	0:08:19	0:05:45
At 13 o'clock	0:05:53	0:05:53
At 14 o'clock	0:00:01	0:00:01
At 15 o'clock	0:03:36	0:02:54
At 16 o'clock	0:06:58	0:07:28
At 17 o'clock	0:10:37	0:09:55
At 18 o'clock	0:17:54	0:05:29
At 19 o'clock	0:05:37	0:03:25
At 20 o'clock	0:03:30	0:03:29
At 21 o'clock	0:07:25	0:06:42
At 22 o'clock	0:09:06	0:05:41
Average value	0:07:39	0:05:44

Table 3. Peak hour sorties

Time	Traditional program	Improvement program
07:00–07:59	3	4
08:00–08:59	8	7
09:00–09:59	5	5
10:00–10:59	8	8
11:00–11:59	4	4
12:00–12:59	6	6
13:00–13:59	6	6
14:00–14:59	2	2
15:00–15:59	8	8
16:00–16:59	18	17
17:00–17:59	19	22
18:00–18:59	12	12
19:00–19:59	5	4
20:00–20:59	7	6
21:00–21:59	10	11
22:00–22:59	9	8

References

1. Chen, K.: Research on the layout optimization method of flight procedures and navigation facilities in the terminal area. Southwest Jiaotong University (2013)
2. Chao, W.: Theoretical Method and Simulation Application of Flight Program Operation Evaluation. Nanjing University of Aeronautics and Astronautics (2012)
3. Conflict evaluation and simulation. The Civil Aviation Flight Academy of China (2020)
4. Wenfei, W.: The Impact of Aircraft Emissions on Environmental PM2.5 in The Airport Area. Civil Aviation University of China (2020)
5. Wenjie, W.: Study on Emission Reduction Effect of Incoming Aircraft. Nanjing University of Aeronautics and Astronautics
6. Yuefeng, X.: Exit Noise Assessment and Control Based on Green Flight. Nanjing University of Aeronautics and Astronautics
7. Ting, W.: Study of Optimization Optimization for Reduce Noise Annoyance. Civil Aviation University of China (2017)
8. Zhen, W.: Effect of Implementing PBN Procedure on Airport Aviation Noise. Civil Aviation University of China (2017)
9. Dongya, L.: Study on Airport Noise Assessment and Control Based on the CDA Program. Nanjing University of Aeronautics and Astronautics (2017)
10. Yahang, Q.: Evaluation of the Noise Reduction Measures Based on the Flight Procedure. Civil Aviation University of China
11. Huiyu, H.: Airport Noise Prediction Study of Flight Delay Recovery at Multi-runway Airports. Civil Aviation University of China
12. Xuerui, C.: A Flight Program Optimization Study for Reducing the Impact of Airport Noise. Civil Aviation University of China
13. Hongxiang, L.: Quality Analysis of Flight Procedures Based on Flight Data. The Civil Aviation Flight Academy of China
14. Ge, Q.: Research on the Key Technology of Flight Program Optimization Design. Nanjing University of Aeronautics and Astronautics (2015)
15. Tsuchiya, T., Hasegawa, T., Mori, R.: Optimization of approach trajectory considering the constraints imposed on flight procedure design. In: 2014 Asia-Pacific International Symposium on Aerospace Technology (2014)
16. Wilson, S.R., Swieringa, K.A., Leonard, R.D., et al.: Statistical engineering approach to improve the realism of computer-simulated experiments with aircraft trajectory clustering. Qual. Eng. **29**(2), 167–180 (2016)
17. Mcfadyen, A., Martin, T.L., O'Flynn, M., et al.: Aircraft trajectory clustering techniques using circular statistics. IEEE Aerospace Conference. IEEE (2016)

Simulation and Test Study on Composite Honeycomb Sandwich Panel

Yang Kun(✉), Yan Qun, and Xu Fei

Aircraft Strength Research Institute, Xi'An 710068, People's Republic of China
254789945@qq.com

Abstract. Composite honeycomb sandwich panel is one of the ideal and most efficient structural forms for aircraft structures. In this paper, a finite element analysis model of honeycomb sandwich structure is established. Firstly, the natural vibration mode and frequency of the model are calculated and compared with the test results. In order to verify the calculated acceleration and strain results, a response test of the honeycomb sandwich structure under vibration is further carried out. The simulation of the model is in good agreement with the experimental results, which verifies the effectiveness and accuracy of the proposed analysis and method.

Keywords: Honeycomb sandwich panels · Vibration fatigue · Dynamic response · Experiment design

1 Introduction

Honeycomb sandwich structure with high specific strength and stiffness, light weight, energy saving and environmental protection is widely used in aerospace industry, automobile industry, marine shipbuilding industry. Since the constitutive relationship of honeycomb sandwich structure composites is complex and discontinuous, the analysis becomes more difficult. Therefore, the static and dynamic analysis and calculation of honeycomb sandwich structure have always been an important research topic [1–7].

The researches on the equivalent mechanical model characteristics of honeycomb sandwich structure mainly include The Reissner theory, The Hoff theory and The Kirchhoff theory [7], but the research on the vibration and acoustic coupling problems of honeycomb sandwich plate structure is not sufficient. In recent years, with the further application of honeycomb sandwich structure in the field of aerospace, the research on structural response characteristics and fatigue performance under vibration environment has become increasingly important. In 2000, Burman and Zenkert [8] first studied the structural fatigue failure and fatigue lifespan prediction of honeycomb sandwich structure. Belingardil et al. [9] carried out a four-point bending fatigue test of aluminum honeycomb sandwich structure in 2007, adopting the S-N curve proposed by Burman. For the core shear failure, it was found that the fatigue lifespan S-N curve of the structure can be well fitted by the Burman model. In 2019, Ma et al. [10] summarized the fatigue research

Chinese Society of Aeronautics and Astronautics: CASTYSF 2022, LNEE 972, pp. 368–379, 2023.
https://doi.org/10.1007/978-981-19-7652-0_34

progress of honeycomb sandwich structure, including the fatigue lifespan research test method, fatigue failure and life predicting model.

In this paper, the simulation analysis and vibration response test of a composite honeycomb sandwich structure are carried out. Firstly, a finite element analysis model is established based on the honeycomb sandwich structure. The natural frequency and vibration modes of the model are calculated and compared with the experimental frequency. On this basis, the acceleration and strain response distribution of the model are further calculated. Finally, the structural response test of the honeycomb sandwich structure under vibration load is carried out. The test results are compared with the simulation to verify the effectiveness of the finite element model, and the modeling method is summarized to optimize the calculation accuracy of the simulation. Finally, the modeling and calculation method will be applied to the response/lifespan analysis of the honeycomb sandwich structure under various loads.

2 Theoretical Model of Honeycomb Sandwich Structure

The upper and lower panels of the test piece are thin plates, so the stress is uniformly distributed along the thickness of the panel, which is, assuming the thickness of the panel only bears the axial force in the face and does not bear the transverse shear force. Since the core is relatively soft, the stress component parallel to the xy plane in the core is ignored ($\sigma_x = \sigma_y = \sigma_{xy} = 0$), meaning that the core only bears transverse shear force. In addition, it is assumed that the straight line segment perpendicular to the middle surface of the sandwich plate (z = 0) remains straight after deformation, but it is no longer perpendicular to the middle surface after bending. φ_x, φ_y Represents the rotation angle of the straight line segment perpendicular to the middle after deformation and w represents the deflection of the middle surface.

Based on Reissner's sandwich plate theory, the relationship between moment, shear force and generalized displacements can be expressed as follows:

$$\left.\begin{aligned} M_x &= -D(\frac{\partial \varphi_x}{\partial x} + \upsilon_f \frac{\partial \varphi_y}{\partial y}) \\ M_y &= -D(\upsilon_f \frac{\partial \varphi_x}{\partial x} + \frac{\partial \varphi_y}{\partial y}) \\ M_{xy} &= -\frac{1}{2} D(1 - \upsilon_f)(\frac{\partial \varphi_x}{\partial x} + \frac{\partial \varphi_y}{\partial y}) \\ Q_x &= C(\frac{\partial w}{\partial x} - \varphi_x) \\ Q_y &= C(\frac{\partial w}{\partial y} - \varphi_y) \end{aligned}\right\} \tag{1}$$

In the equation above, bending stiffness of the panel is $D = \frac{E_f(h+d)^2 d}{2(1-\upsilon_f^2)}$, and shear stiffness of the sandwich is $C = G_e h$. The equivalent shear modulus $G_e = G_x = G_y = \frac{1}{\sqrt{3}} \frac{t}{l} G_f$ can be obtained from the strain energy equivalent principle, where E_f, υ_f and

G_f represent the elastic modulus, Poisson's ratio and shear elastic modulus of panel materials, respectively.

Based on Reissner sandwich panel theory, the equilibrium equation of honeycomb sandwich structure along three coordinate directions is given below:

$$\left.\begin{array}{l} \dfrac{\partial M_x}{\partial x} + \dfrac{\partial M_{xy}}{\partial y} - Q_x = 0 \\ \dfrac{\partial M_{xy}}{\partial x} + \dfrac{\partial M_y}{\partial y} - Q_y = 0 \\ \dfrac{\partial Q_x}{\partial x} + \dfrac{\partial Q_y}{\partial y} - \rho \dfrac{\partial^2 w}{\partial t^2} + q = 0 \end{array}\right\} \tag{2}$$

In the equation, $\rho = h\rho_e + 2d\rho_f$ represents the mass of honeycomb sandwich structure per area, and ρ_e is the equivalent density of core, and ρ_f is the density of the panels.

Substituting Eq. (1) into Eq. (2), the coupled equations of three basic variables φ_x, φ_y and w can be obtained, but the process is complex and difficult to solve. In order to simplify the equation, the displacement function ϖ and f is introduced, and the generalized displacement φ_x, φ_y and w are expressed as follows.

$$\left.\begin{array}{l} \varphi_x = \dfrac{\partial \varpi}{\partial x} + \dfrac{\partial f}{\partial y} \\ \varphi_y = \dfrac{\partial \varpi}{\partial y} + \dfrac{\partial f}{\partial x} \\ w = \varpi - \dfrac{D}{C}\nabla^2 \varpi \end{array}\right\} \tag{3}$$

Substituting Eq. (3) into the coupled equations, the vibration control equation of honeycomb sandwich structure can be obtained.

$$\left.\begin{array}{l} D\nabla^2\nabla^2\varpi + \rho \dfrac{\partial^2}{\partial t^2}(\varpi - \dfrac{D}{C}\nabla^2\varpi) = q \\ \dfrac{D}{2}(1 - \upsilon_f)\nabla^2 f - Cf = 0 \end{array}\right\} \tag{4}$$

For a simply supported honeycomb sandwich structure with length a (direction x) and width b (direction y), the boundary conditions are expressed below.

$$\left.\begin{array}{l} x = 0, a : w = 0, M_x = 0, \varphi_y = 0 \\ y = 0, b : w = 0, M_y = 0, \varphi_x = 0 \end{array}\right\} \tag{5}$$

With the displacement function ϖ, f, the boundary condition are given below.

$$\left.\begin{array}{l} x = 0, a : w = \dfrac{\partial^2 \varpi}{\partial x^2} = \dfrac{\partial^4 \varpi}{\partial x^4} = 0, \dfrac{\partial f}{\partial x} = 0 \\ y = 0, b : w = \dfrac{\partial^2 \varpi}{\partial y^2} = \dfrac{\partial^4 \varpi}{\partial y^4} = 0, \dfrac{\partial f}{\partial y} = 0 \end{array}\right\} \tag{6}$$

Under the condition of all sides simply supported, f ≡ 0, and q = 0, the control Eq. (4) can be simplified as follows.

$$D\nabla^2\nabla^2\varpi + \rho\frac{\partial^2}{\partial t^2}(\varpi - \frac{D}{C}\nabla^2\varpi) = 0 \tag{7}$$

The function satisfying the boundary conditions (6) and the control Eq. (7) is assumed below.

$$\varpi = \sum_{m=1}^{\infty}\sum_{n=1}^{\infty} A_{mn} \sin\frac{m\pi x}{a} \sin\frac{n\pi y}{b} e^{jw_{mn}t} \tag{8}$$

A_{mn} is the undetermined function, and ω_{mn} is the circular frequency of the undamped vibration of the honeycomb sandwich structure, $j = \sqrt{-1}$. By substituting Eq. (8) into Eq. (7), the results can be obtained.

$$w_{mn}^2 = \frac{D\pi^4}{\rho a^4}\frac{(m^2 + \lambda^2 n^2)^2}{1 + \kappa^2(m^2 + \lambda^2 n^2)}, m, n = 1, 2, 3\ldots \tag{9}$$

κ is the dimensionless parameter of honeycomb sandwich structure, and λ is the aspect ratio, which are expressed as follows.

$$\kappa = \frac{\pi}{a}\sqrt{\frac{D}{C}}, \lambda = \frac{a}{b} \tag{10}$$

Thus, the natural frequency of the honeycomb sandwich structure is given below.

$$f_{mn} = \frac{1}{2\pi}\sqrt{\frac{D\pi^4}{\rho a^4}\frac{(m^2 + \lambda^2 n^2)^2}{1 + \kappa^2(m^2 + \lambda^2 n^2)}}, m, n = 1, 2, 3\ldots \tag{11}$$

3 Simulation Analysis

3.1 Finite Element Simulation Analysis Model

The size parameter of honeycomb sandwich structure is shown in Fig. 1. The thickness of upper and lower panels is the same. The sandwich core height is 15 mm and the panel thickness is 1.98 mm. One end is the clamping area, while the other is equipped with a counterweight block (496 g). The damping ratio of the structure is 0.05.

When using ABAQUS finite element analysis software to establish the model, the material properties of panel and sandwich equivalent structure establishing the finite element analysis model are shown in Table 1.

At present, there are two main finite element modeling methods for honeycomb sandwich structure. The equivalent plate element modeling method equivalents the entire structure to an integral element. The tensile and compressive stiffness is provided by the panel, while the shear stiffness is provided by the sandwich core. Although the

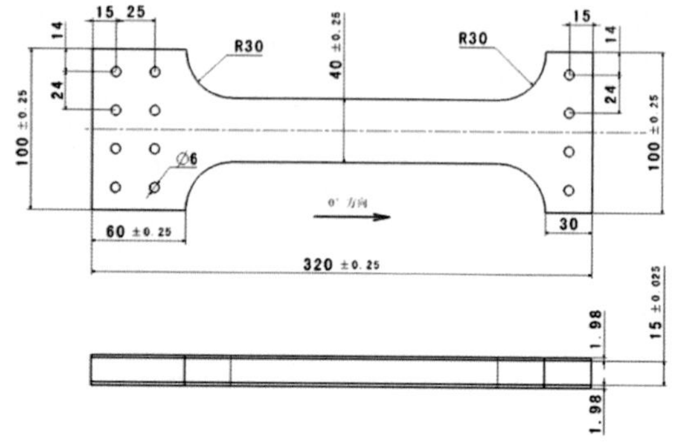

Fig. 1. Dimensions of honeycomb sandwich structure (mm).

Table 1. Material parameters of the model.

	E1 Mpa	*E2* Mpa	*E3* Mpa	ν12	ν_{13}	ν_{23}	*G*12 Mpa	*G* 13 Mpa	*G* 23 Mpa
Panels	36200	13100	13100	0.33	0.33	0.45	3890	3890	2900
Equivalent structure core	10.5	10.5	218.3	0.94	0.003	0.003	12.8	68.1	178.1

calculation efficiency of this method is high, however, the influence of the transverse shear deformation of the honeycomb sandwich is not considered, so it can only roughly evaluate the overall stiffness and stress level of the object. The cellular modeling method indicates that the upper and lower panels are equivalent to two-dimensional anisotropic layers, and the honeycomb sandwich is regarded as a single cell. The number of modeling nodes is too many and time-consuming, but the mechanic simulation results of the honeycomb core is more accurate.

On the basis of the above, combined with the advantages of the two modeling methods, the honeycomb sandwich structure is regarded as a composition of the upper, lower panels and sandwich core. The finite element model of the honeycomb sandwich structure with panel and paper honeycomb sandwich is established using equivalent anisotropic elements. After the equivalent processing of the sandwich, the boundary conditions are defined in the clamping area, and the basic excitation load is applied in the random vibration response analysis. The finite element analysis model is shown in Fig. 2.

3.2 Natural Frequencies of the Structure

Xu et al. [11] proposed that the natural frequency solved by the general honeycomb sandwich finite element model would be higher than the actual frequency of the structure. The model needs to be corrected to make it close to the real value. After processing the model, the honeycomb sandwich structure density given as 260kg/m^3. The first four

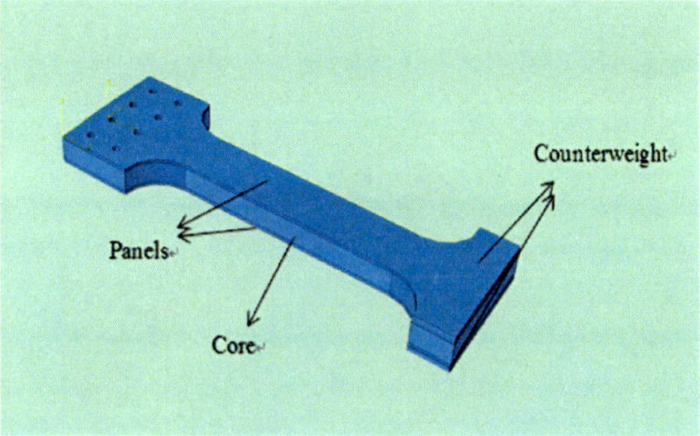

a) Composition of honeycomb sandwich structure.

b) Finite elements model.

Fig. 2. Finite elements model of honeycomb sandwich.

order resonance frequency of honeycomb sandwich structure is shown in Table 2, and the first four order vibration mode is shown in Fig. 3.

Table 2. Modal simulation results

Modes	First order	Second order	Third order	Fourth order
Frequency/Hz	53.6	81.9	84.6	587.5

3.3 Random Response of the Structure

It is found that the fatigue failure of the test piece is greatly affected by the first-order bending mode after experimental verification. Therefore, the narrow-band white noise

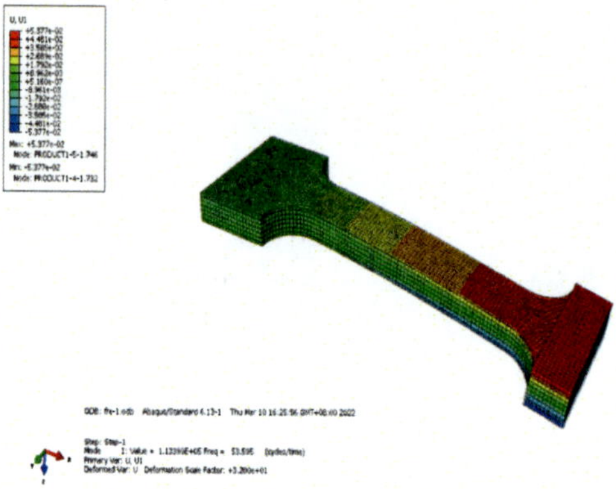

a) First order vibration mode.

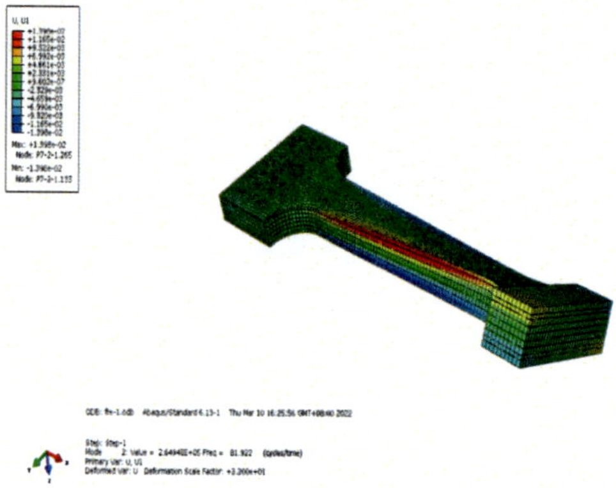

b) Second order vibration mode.

Fig. 3. Calculation results of first four modes.

excitation spectrum with the bandwidth of 45–55 Hz and level of 4.225 g^2/Hz is applied to the finite element model considering the first order natural frequency calculation result of the structure, as shown in Fig. 4. The modal damping of the structure is set to 0.05, and the random response of the structure is analyzed.

Figures 5 and 6 show the acceleration and strain responses of the finite element model under vibration loads.

After analysis, the position with the maximum acceleration response is at the center of the counterweight at the free end, and the root mean square value of the acceleration response at this position is about 68.9 g. At the same time, the analyses of the strain response results show that the maximum position is in the arc area of the clamping end, and the root mean square value of the calculated dynamic strain response is 3164 με.

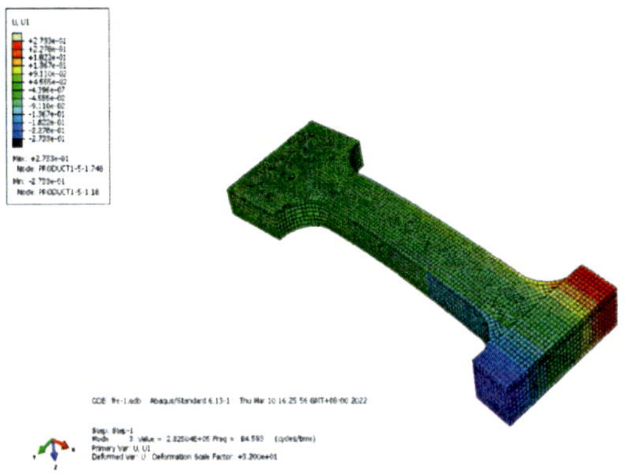

c) Third order vibration mode.

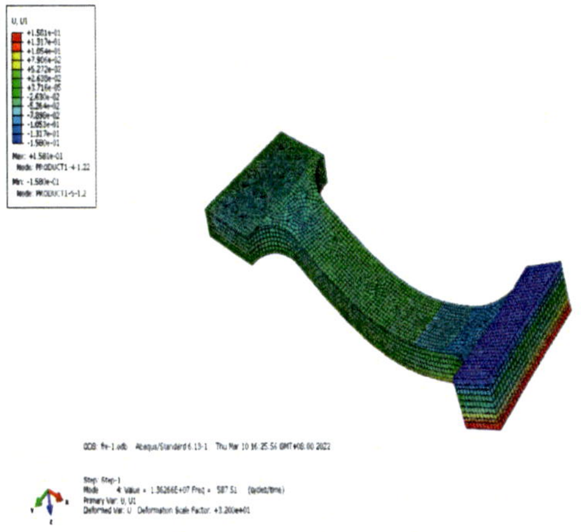

d) Fourth order vibration mode.

Fig. 3. (*continued*)

4 Test Verification Results

4.1 Test Method

In order to verify the effectiveness of the analysis model established in this paper, the honeycomb sandwich structure test pieces are designed according to the size given in Chap. 1, as shown in Fig. 7. The test pieces are installed on the vertical table of the vibrator through the fixture for test. The test load is consistent with the load in the simulation calculation. The test site photos are shown in Fig. 8.

In order to study the random vibration response of the test piece and verify the analysis model, four strain gauges were pasted on the arc region of the clamping end of the test piece to measure the vibration strain results.Meanwhile, an accelerometer is attached to the center of the counterweight block at the free end of the test piece to measure the acceleration response.

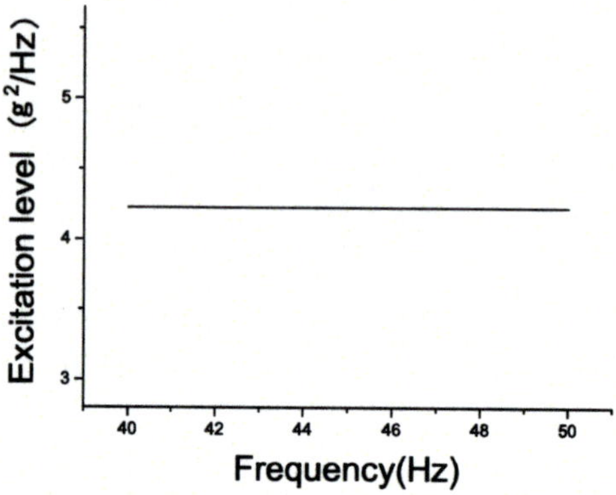

Fig. 4. Narrow-band white noise excitation spectrum.

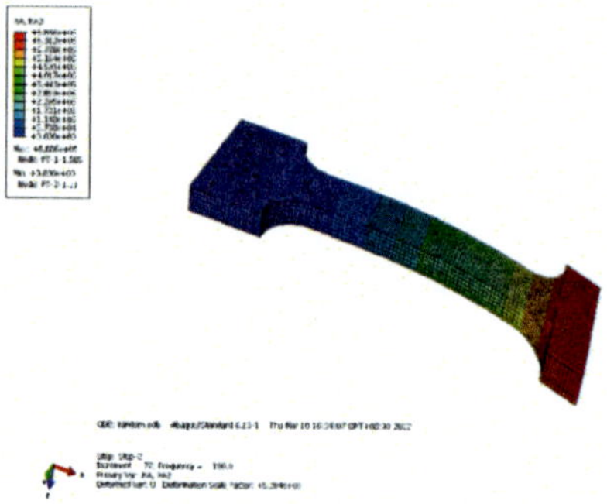

Fig. 5. Calculation results of the acceleration response of the model.

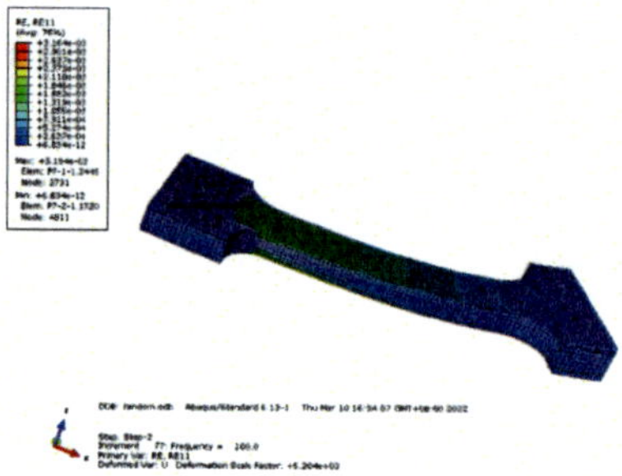

Fig. 6. Calculation results of the strain response of the model.

4.2 Comparison of Acceleration Response Results

Through the analysis of the vibration test results, the first-order frequency of the test piece is 49 Hz (slightly less than the simulation value), and the error is about 7.5% between the theoretical and actual value, which is considered that such error is within an acceptable range. Figure 7 shows that the root mean square value of the acceleration

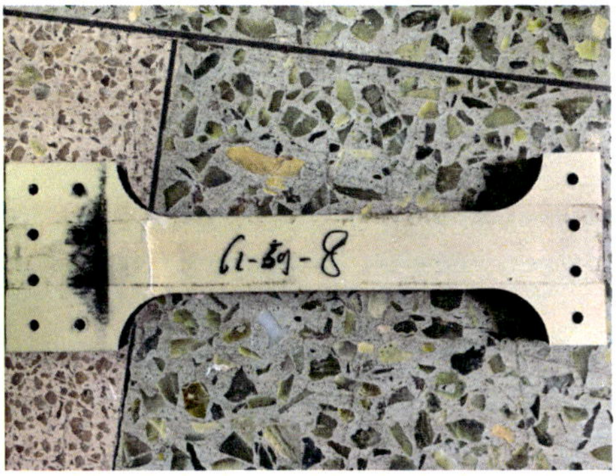

Fig. 7. Photo of the test piece

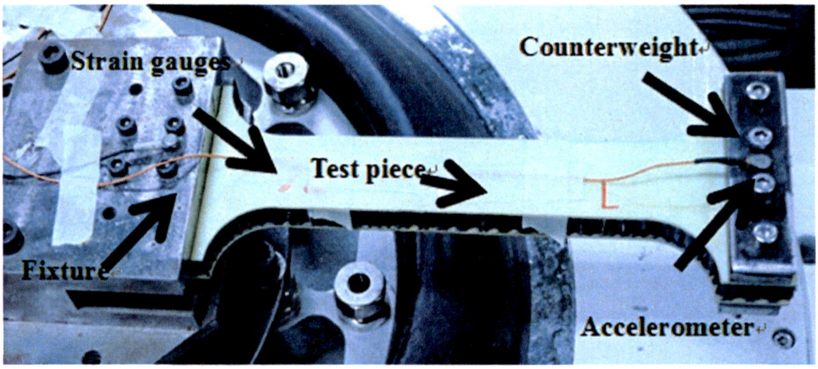

Fig. 8. Photo of the test site

response of the measuring point at the free end of the test piece is about 66.8 g, which is compared with the simulation analysis results given in Table 3. Figure 9 shows amplitude spectrum of the acceleration response in the frequency domain.

Table 3. Comparison of acceleration responses

	Test result	Simulation result	Error
Root mean square value of acceleration responses/g	66.8	68.9	3.1%

4.3 Comparison of Strain Response Results

Table 4 shows the results of the strain response measured by the test. The arithmetic mean of the results of the four measuring points is taken as the final measured strain result. The strain response after processing is 3095 με. The root mean square value of

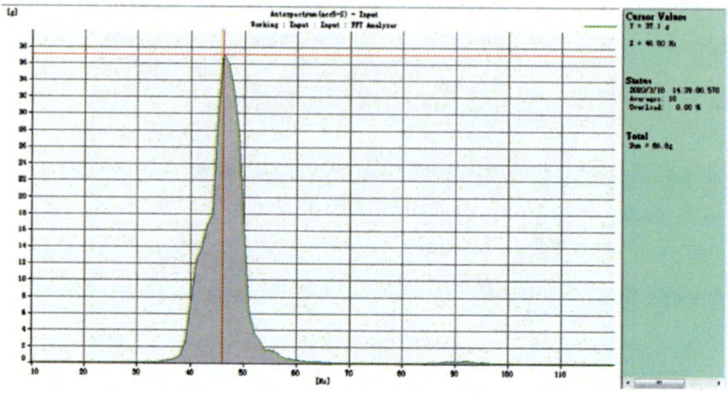

Fig. 9. Acceleration response diagram of power spectral density

the strain at the stress concentration position (arc region) given in Fig. 8 is 3164 με. After comparison, it is found that the error between the two is 2.2%, which is considered to be very consistent.

Table 4. Comparison of strain responses

Strain measuring points	1#	2#	3#	4#	Arithmetic mean
Root mean square value of test results/με	2700	1700	4680	3300	3095

It can be seen that the finite element analysis model of honeycomb sandwich structure proposed in this paper regards the honeycomb sandwich as a structure composed of upper, lower panels and honeycomb sandwich core. The finite element model of the structure with panel and paper honeycomb sandwich is established, and the vibration test is conducted for experimental verification. After calculation, it is found that the natural frequency of the first-order bending vibration mode which has a great influence on the vibration response characteristics of the test piece is in good agreement with the test results. Finally, the test verification is carried out, and the center of the counterweight at the free end of the test piece and the stress concentration position were selected respectively for further study. The acceleration and the strain response results of the arc region obtained by the simulation calculation and the test are compared. The errors are 3.1% and 2.2% respectively, and good consistency are obtained. It is verified that the established equivalent model and analysis method were reasonable and effective, which laid the good foundation for the subsequent response/fatigue life analysis under other loads.

5 Conclusions

In this paper, through the simulation analysis and experimental research on the honeycomb sandwich structure, the analysis and experimental results are compared, and the following conclusions are obtained:

(1) Firstly, a finite element model analysis model of equivalent honeycomb sandwich structure is established, and the resonance frequency of the first-order bending mode which has a great influence on the response characteristics is compared with test results. The calculation results are slightly higher comparing with the test, which conforms to the general law.
(2) Through the test verification, the acceleration response of the free end of the test piece and the strain response of the arc region (stress concentration position) under vibration load are obtained, and the results are compared with the calculation results. The error of root mean square results of the acceleration and strain responses are 3.1 and 2.2%, which indicate a great consistency. It is proved that the equivalent analysis model and analysis method proposed in this paper are reasonable and effective.
(3) On the basis of the analysis model proposed in this paper, the response/lifespan of honeycomb sandwich structure under various loads (noise, vibration, static force, high temperature, etc.) can be further studied in the future work, providing methods and support for the structural selection and optimization of key parts in aircraft design.

References

1. Gibson, A.: Porous Solid Structure and Performance, pp. 81–146. Tsinghua University Press, Bei Jing (2003)
2. Zhang, C.C.: Analytical Theory of Composite Plate and Shell Mechanics. National Defense Industry Press, Bei Jing (2009)
3. Wang, X.Y., Yang, F.L.: Design Principles and Applications of Sandwich Structure Composite Materials. Chemistry Industry Press, Bei Jing (2006)
4. Xia, L.J., Jin, X.D., Wang, Y.B.: Equivalent of satellite structure honeycomb sandwich panel. J. Shanghai Jiao Tong Univ. **37**(7), 999–1001 (2003)
5. Othman, A.R., Barton, D.C.: Failure initiation and propagation characteristics of honeycomb sandwich composites. Compos. Struct. **85**, 126–138 (2008)
6. Han, T.S., Ural, A., Chen, C.N.: Delamination buckling and propagation analysis of honeycomb panels using a cohesive element approach. Int. J. Fracture **155**, 101–123 (2005)
7. Abbadi, A., Koutsawa, Y., Carmasol, A., Belouettar, S., et al.: Experimental and numerical characterization of honeycomb sandwich composite panels. Simul. Model. Pract. Theory **17**, 1533–1547 (2009)
8. Plate and Shell Group, Mechanics of Solids Lab, Institute of Mechanics, Chinese Academy of Sciences. Bending Stability and Vibration of Sandwich Plate and Shell. Science Press, Beijing (1977) (in Chinese)
9. Burman, M., Zenkert, D.: Fatigue of undamaged and damaged honeycomb sandwich beams. J. Sadwich Struct. Mater. **2**, 50–74 (2000)
10. Belingardi, G., Martella, P., Peroni, L.: Fatigue analysis of honey-Comb composite Sandwich beams. Compos. A **38**(4), 1191 (2007)
11. Ma, M.J., Yao, W.X., Yan, C.: Research progress of honeycomb sandwich. Aeronaut. Eng. Progress **10**(02), 12–20 (2019)
12. Xu, D.F.: Research on Vibration Fatigue Failure of Composite Honeycomb Sandwich Structures. Nanjing University of Aeronautics and Astronautics (2011)

Guidance Law Design Against Unknown Maneuvering Target Based on Two ESOs

Chunhua Cheng(✉), Mingming Huo, Dongsheng Hao, and SiMin Bi

Naval Aeronautic University, Qingdao BranchQingdao 266041, China
chch715@126.com

Abstract. With the development of high maneuvering missiles and hypersonic maneuvering unmanned aerial vehicles, accurately intercepting such high-speed moving targets typically involves knowing the target's precise velocity and acceleration. Obtaining the velocity and acceleration of the target in real time is difficult and expensive in practice. On the other hand, developing a guidance law for maneuvering target with unknown velocity and acceleration is still a work in progress. This paper proposes three types of sliding-mode guidance rules based on two extend state observers (ESOs) for intercepting maneuvering targets with unpredictable velocity and acceleration, where the only information required is the LOS angle and distance, but not the LOS angle velocity or the relative velocity orthogonal to the LOS. To estimate the unknown things associated to the objective, two ESOs are built. The auxiliary signal system has been devised to compensate for the saturation constraint of the guidance command. To test the effectiveness and robustness of the proposed guidance legislation, numerical simulations are run with a non-moving target and a maneuvering target.

Keywords: Unknown maneuvering targets · Extend State Observer (ESO) · Sliding mode · Saturation

1 Introduction

Because of its simple design and ease of implementation, proportional navigation guidance (PNG) is the most preferred terminal homing guidance approach in unmanned craft guidance [1, 2]. In Refs. [3–5], it was shown that PNG is the best guidance for non-maneuvering targets. PNG's performance, on the other hand, will deteriorate and even fail against maneuvering targets.

With the advent of high maneuvering missiles and hypersonic maneuvering unmanned aerial vehicles in recent years, intercepting such targets now necessitates the ability to intercept high speed moving targets with greater precision. To deal with maneuvering targets, different PNG laws have been proposed to improve guidance precision, such as augmented PNG (APNG) law [6–8] and optimal guidance law (OGL) [9, 10], which both require correct target acceleration information or prediction. In practice, however, obtaining the goal acceleration or an exact estimate is difficult. As a result, nonlinear robust control approaches such as nonlinear geometric method [11], sliding

Chinese Society of Aeronautics and Astronautics: CASTYSF 2022, LNEE 972, pp. 380–391, 2023.
https://doi.org/10.1007/978-981-19-7652-0_35

mode control method [12–14], and others have been employed to create guidance rules. For maneuvering targets, a robust geometric guidance technique is provided, in which the terms relating to target acceleration are treated as disturbance [11]. In Ref. [15], an all-aspect guidance law based on backstepping is proposed to achieve the interception without knowing the target's acceleration.

Because the target's maneuvering acceleration is unknown, the variable structure control (VSC) guidance law is designed with the line of sight (LOS) normal velocity as the sliding surface [16]. However, when the angle between the line of sight and the missile flight path is equal to 90°, the VSC guidance law is no longer effective. An adaptive nonsingular terminal sliding mode control (SMC) with an extended state observer (ESO) to estimate the uncertain term is provided to handle the unknown target acceleration [17, 18]. However, it requires knowing the relative velocities orthogonal to the LOS.

Furthermore, all of the above guidance approaches imply that the target is stationary or moving at a constant velocity that must be known. The problem of designing a guidance law for a target with uncertain velocity and acceleration remains unsolved. It is quite difficult to collect the target's velocity information without delay in actual applications. As a result, designing a guidance law using the target's unknown acceleration and velocity is more practicable. To deal with the unknown target, two ESOs are used in this study. ESO has been shown to be a powerful technique for dealing with unfamiliar terms [18–23]. Sliding mode control is also used to improve the robustness of the guidance law. Furthermore, this guidance strategy just needs to know the LOS angle and distance, but neither the LOS angle velocity or the relative velocity orthogonal to the LOS.

2 Problem Formulation

The two-dimensional engagement geometry is described in Fig. 1, where the missile M is trying to intercept a ma-neuvering target T. Assuming that the target and missile are point masses, the dynamic equations of missile and target are given as follows

$$\begin{aligned} \dot{x}_M &= V_M \cos\varphi_M,\ \dot{y}_M = V_M \sin\varphi_M \\ \dot{\varphi}_M &= \frac{a_M}{V_M} \end{aligned} \tag{1}$$

$$\begin{aligned} \dot{x}_T &= V_T \cos\varphi_T,\ \dot{y}_T = V_T \sin\varphi_T \\ \dot{\varphi}_T &= \frac{a_T}{V_T} \end{aligned} \tag{2}$$

where φ_M and φ_T are the flight path angle of the missile and the flight path angle of the target, V_M and a_M are the velocity and acceleration of the missile, and V_T and a_T are the velocity and acceleration of the target.

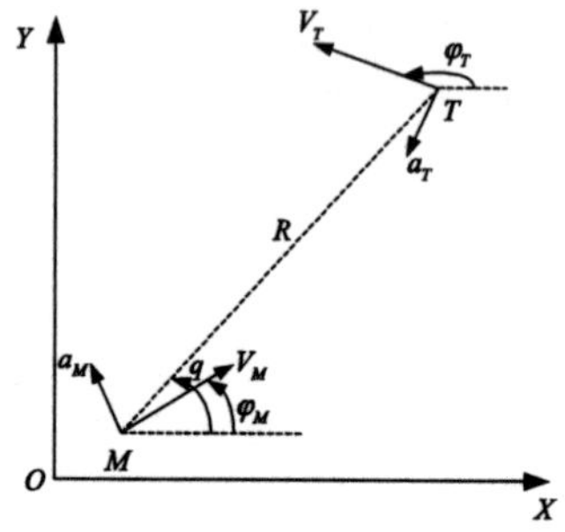

Fig. 1. Planar interception geometry

Combine Fig. 1 with Eq. (1) and Eq. (2), the dynamic of interception can be given by

$$\begin{aligned} \dot{R} &= -V_M \cos(q - \varphi_M) + V_T \cos(q - \varphi_T) \\ \dot{q} &= \frac{V_M \sin(q - \varphi_M)}{R} - \frac{V_T \sin(q - \varphi_T)}{R} \\ \dot{\varphi}_M &= \frac{a_M}{V_M} \\ \dot{\varphi}_T &= \frac{a_T}{V_T} \end{aligned} \tag{3}$$

where R and q represent the relative distance between the missile and the target and the line-of-sight (LOS) angle.

Assumption 1 both V_T and a_T can be treated as unknown term.

The reason of assumption 1 is that it is difficulty to obtain the accurate values of V_T and a_T without delay, and the sensors are affected by noise. Then the flight path angle of the target φ_T is also unknown. Therefore, $V_T \cos(q - \varphi_T)$ and $V_T \sin(q - \varphi_T)$ are unknown terms too, which means that $\dot{R}$ and $\dot{q}$ contain unknown terms. Therefore, $\dot{R}$ and $\dot{q}$ can not be directly used in the PNG law in this paper.

Assumption 2 both the relative distance R and the LOS angle q are available, and R doesn't equal zero.

It is reasonable for Assumption 2, because as long as $0 < R \leq r_0$ where r_0 is the sum of size of missile and target (see Ref. [1]), the hit-to-kill interception will be successful.

Assumption 3: V_T is less than V_M.

3 SMC Guidance Law with ESO

In order to design a guidance law for a homing missile against a high maneuvering target whose velocity and ac-celeration are unknown, SMC and ESO are applied. Meanwhile, the auxiliary signal is introduced to prevent the ef-fect of saturation.

3.1 ESO Design for Unknown Items

Obviously, the unknown terms $V_T \cos(q - \varphi_T)$ and $\frac{V_T \sin(q-\varphi_T)}{R}$ are differential and bounded, and their derivatives are also bounded in practical applications due to that the velocity and acceleration of target could not be infinite large. Define $V_{Tc} = V_T \cos(q - \varphi_T)$ and $V_{Ts} = -\frac{V_T \sin(q-\varphi_T)}{R}$, according to $0 < R \leq r_0$, then there exists two positive constant C_a and C_V, which satisfy that $\left|V'_{Tc}\right| \leq C_V$ and $\left|V'_{Ts}\right| \leq C_a$.

As in Refs. [18–21], in order to handle the unknown terms in $\dot{R}$ and $\dot{q}$, two ESOs for $\dot{R}$ and $\dot{q}$ can be constructed as follows.

$$\begin{cases} E_{1,1} = Z_{1,1} - R \\ \dot{Z}_{1,1} = Z_{1,2} - \beta_1 g_{c1}(E_{1,1}) - V_M \cos(q - \varphi_M) \\ \dot{Z}_{1,2} = -\beta_2 g_{c2}(E_{1,1}) \end{cases} \tag{4}$$

$$\begin{cases} E_{2,1} = Z_{2,1} - q \\ \dot{Z}_{2,1} = Z_{2,2} - \beta_1 g_{c1}(E_{2,1}) + \dfrac{V_M \sin(q - \varphi_M)}{R} \\ \dot{Z}_{2,2} = -\beta_2 g_{c2}(E_{2,1}) \end{cases} \tag{5}$$

with

$$g_{c1}(E_{i,1}) = E_{i,1},$$

$$g_{c2}(E_{i,1}) = \begin{cases} |E_{i,1}|^{\alpha} \operatorname{sign}(E_{i,1}), & |E_{i,1}| > \sigma \\ E_{i,1} \big/ \sigma^{1-\alpha}, & otherwise \end{cases} \quad 0 < \alpha < 1, i = 1, 2$$

where $E_{i,1}$ is the estimation error of the ESO, $Z_{i,1}$ and $Z_{i,2}$ are the observer output, $\beta_1 > 0$, $\beta_2 > 0$ are the observer gains. It has been proved [18–21] that for appropriate values of $\beta_1, \beta_2, \alpha, \sigma$, the observer output $Z_{1,1}$ and $Z_{2,1}$ can approaches to R and q respectively, meanwhile, $Z_{1,2}$ and $Z_{2,2}$ can approaches to V_{Tc} and V_{Ts} respectively. As well, $\dot{Z}_{1,1}$ and $\dot{Z}_{2,1}$ can approaches to $\dot{R}$ and $\dot{q}$ respectively. It had been proved that $E_{1,2} = Z_{1,2} - V_{Tc}$ and $E_{2,2} = Z_{2,2} - V_{Ts}$ are bounded [20]. In other words, we have that $\left|E_{1,2}\right| \leq \frac{\beta_1}{\beta_2}\sigma^{1-\alpha} C_V$ and $\left|E_{2,2}\right| \leq \frac{\beta_1}{\beta_2}\sigma^{1-\alpha} C_a$ [20], where C_V and C_a are positive constants need not to know the exact value.

3.2 SMC Guidance Law Design Procedure

As we all known, $V_r = -V_M \cos(q - \varphi_M) + V_T \cos(q - \varphi_T)$ denotes the velocity along LOS. Then it is necessary to keep $V_r < 0$ for a direct hit [14]. Also, it is well known that a direct hit can be achieved when $R = 0$ or $\theta_M = q - \varphi_M = 0$, where $\theta_M = 0$ mean that the velocity of missile is always in the LOS.

Then, the guidance task can be realized by stabilizing the following sliding mode surface with the missile acceleration

$$\begin{aligned} & S_1 = R + c_0 \dot{R} = 0, \\ or\ \ & S_2 = \theta_M = 0, \\ or\ \ & S_3 = \theta_M + c_1 R = 0 \end{aligned} \tag{6}$$

where c_0 and c_1 are positive designed constant parameters. Note that the units of variable θ_M and variable R are not the same, so the units of variable θ_M and variable R need to be consistent when selecting parameter c_1. Meanwhile, in order to hit the target, it is necessary to ensure that the missile's speed is along the LOS line firstly, that is to say it is necessary to ensure that the speed of convergence of θ_M is much faster than the speed of convergence of R. Therefore, we can choose $c_1 = \frac{\theta_M(0)}{100\times R(0)}$, where $\theta_M(0)$ and $R(0)$ are the initial values of variable θ_M and variable R, respectively.

Next, this article focuses on the analysis of the guide law when choosing the third sliding mode surface. The analysis of the other sliding mode surfaces is similar to this.

Derivate the sliding mode surface S_3, we can obtain

$$\begin{aligned}\dot{S}_3 = &\frac{V_M \sin(q-\varphi_M)}{R} + V_{Ts} - \frac{a_M}{V_M} \\ &+c_1\left[-V_M \cos(q-\varphi_M) + V_{Tc}\right]\end{aligned} \tag{7}$$

Obviously, the Eq. (7) includes the unknown terms V_{Tc} and V_{Ts}, which can be estimated by ESO. Then we can design guidance law as follows

$$a_M = V_M \begin{bmatrix} c_1\left(-V_M \cos(q-\varphi_M) + Z_{1,2}\right) \\ +\dfrac{V_M \sin(q-\varphi_M)}{R} + Z_{2,2} - \tau S_3 - \delta |S_3|^\gamma \operatorname{sgn}(S_3) \end{bmatrix} \tag{8}$$

where $Z_{1,2}$ and $Z_{2,2}$ are the states of the ESO Eqs. (4) and (5), τ, δ are positive designed parameters, and $\gamma \in (0, 1)$ is also positive designed parameter.

Note that, the acceleration of missile is usually constrained due the overload in practical applications. Therefore, the practical acceleration of missile with saturation constraint can be described as follows

$$a_M = sat(a_{M0}) = \begin{cases} a_{M0}, & |a_{M0}| \le a_{M\max} \\ a_{M\max}\operatorname{sgn}(a_{M0}), & |a_{M0}| > a_{M\max} \end{cases} \tag{9}$$

where $a_{M\max}$ is the maximum acceleration of missile and a_{M0} is the designed acceleration of missile without consideration of saturation.

In order to compensate the effect of saturation, the auxiliary signal system can be constructed as

$$\dot{\lambda} = -\tau\lambda + \frac{\Delta a_M}{V_M} \tag{10}$$

where $\Delta a_M = a_{M0} - a_M$, and τ is defined in Eq. (8).

Obviously, the auxiliary signal system is bounded input bounded output stable (BIBO). When $|a_{M0}| \le a_{M\max}$, the auxiliary signal λ will converge to zero.

Then the sliding mode surface S_3 in Eq. (6) can be compensated as followings:

$$\overline{S}_3 = S_3 - \lambda \tag{11}$$

Derivate the compensated sliding mode surface $\overline{S}_3$, we can obtain

$$\begin{aligned}\dot{\overline{S}}_3 &= \frac{V_M \sin(q - \varphi_M)}{R} + V_{Ts} - \frac{a_M}{V_M} \\ &+c_1\left[-V_M \cos(q - \varphi_M) + V_{Tc}\right] + \tau\lambda - \frac{\Delta a_M}{V_M} \\ &= \frac{V_M \sin(q - \varphi_M)}{R} + V_{Ts} \\ &+c_1\left[-V_M \cos(q - \varphi_M) + V_{Tc}\right] + \tau\lambda - \frac{a_{M0}}{V_M}\end{aligned} \tag{12}$$

Therefore, we can design guidance law as follows

$$\begin{aligned}a_M &= sat(a_{M0}) \\ a_{M0} &= V_M \begin{bmatrix} c_1\left(-V_M \cos(q - \varphi_M) + Z_{1,2}\right) \\ +\dfrac{V_M \sin(q - \varphi_M)}{R} + Z_{2,2} - \tau S_3 - \delta\left|\overline{S}_3\right|^{\gamma} \operatorname{sgn}\left(\overline{S}_3\right)\end{bmatrix}\end{aligned} \tag{13}$$

Remark 1 when the sliding mode surface is selected as S_1, the derivation of the unknown term $V_T \cos(q - \varphi_T)$ exists in the derivation of S_1 due to the target is maneuvering. Then the guidance law is

$$a_{M0} = \begin{cases} \dfrac{1}{c_0 \sin(q - \varphi_M)} \begin{bmatrix} c_0 \dot{Z}_{1,2} - V_M \cos(q - \varphi_M) \\ +Z_{2,2} + \tau S_1 + \delta\left|\overline{S}_1\right|^{\gamma} \operatorname{sgn}\left(\overline{S}_1\right)\end{bmatrix}, & |q - \varphi_M| > \varepsilon \\ 0, \quad |q - \varphi_M| < \varepsilon \end{cases} \tag{14}$$

where $\dot{Z}_{1,2}$ is the estimation of the derivation of $V_T \cos(q - \varphi_T)$, and ε is a small positive designed parameter, $\overline{S}_1 = S_1 - \lambda$.

Remark 2 when the sliding mode surface is selected as S_2, then the guidance law is

$$a_{M0} = V_M \left[\frac{V_M \sin(q - \varphi_M)}{R} - Z_{2,2} - \tau S_2 - \delta\left|\overline{S}_2\right|^{\gamma} \operatorname{sgn}\left(\overline{S}_2\right)\right] \tag{15}$$

where $\overline{S}_2 = S_2 - \lambda$ is the compensated sliding mode.

3.3 Stability Analysis

Substituting Eq. (13) into Eq. (12), we can obtain the follows

$$\dot{\overline{S}}_3 = -\tau \overline{S}_3 - \delta\left|\overline{S}_3\right|^{\gamma} \operatorname{sgn}\left(\overline{S}_3\right) - E_{2,2} - c_1 E_{1,2} \tag{16}$$

Choose lyapunov function as $V = \frac{1}{2}\overline{S}_3^2$, then we can obtain the follows

$$\begin{aligned}\dot{V} &= -\tau \overline{S}_3^2 - \delta\left|\overline{S}_3\right|^{\gamma+1} - E_{2,2}\overline{S}_3 - c_1 E_{1,2}\overline{S}_3 \\ &\le -\tau \overline{S}_3^2 - \delta\left|\overline{S}_3\right|^{\gamma+1} + \left|\overline{S}_3\right|\left(\left|E_{2,2}\right| + \left|E_{1,2}\right|\right)\end{aligned} \tag{17}$$

Then $|E_{2,2}| + |E_{1,2}|$ can be viewed as the disturbance input of the closed loop system. When $\|\overline{S}_3\| \geq \chi(|E_{2,2}| + |E_{1,2}|)$, with $\chi(|E_{2,2}| + |E_{1,2}|) = \frac{1}{\tau}(|E_{2,2}| + |E_{1,2}|)$, the Eq. (17) is smaller than zero. Therefore, the closed loop system is input to state stable (ISS). Base on input to state stability (see [19, 20]), it is easy to conclude that the estimation error $|E_{2,2}| + |E_{1,2}|$ by ESO will affect that whether the dynamic of the compensated sliding mode converge to $\overline{S}_3 = 0$.That is to say the dynamic of the compensated sliding mode will be restricted into the neighborhood of $\overline{S}_3 = 0$ as

$$\lim_{t\to\infty} \overline{S}_3 \in \left(|\overline{S}_3| \leq \frac{1}{\tau}(|E_{2,2}| + |E_{1,2}|) \right) \tag{18}$$

Fortunately, this neighborhood can be reduced to any small size by selecting the ESO parameters β_1, β_2, α, σ and the controller parameter τ. Furthermore, τ, δ, γ determines the speed of convergence and the final error. According to Eq. (17), the bigger the parameters τ and δ are, the faster speed converges to zero, and the smaller approaching error is. From the above analysis, a_{M0} is bounded, therefore, Δa_M is bounded too, then we can conclude that λ is bounded. When $|a_{M0}| \leq a_{M\max}$, we can obtain that $\lim_{t\to\infty} S_3 \in \left(|S_3| \leq \frac{1}{\tau}(|E_{2,2}| + |E_{1,2}|)\right)$.

Remark 3 From the above analysis, it is clear that S_3 will not converge to zero due to the estimation error of ESO. It implies that S_3 can only converge into a neighborhood of the desired sliding mode surface $S_3 = 0$ and remains within it.

4 Simulation Results

Numerical simulations are given to investigate the effectiveness of the proposed guidance law (13). It is assumed that the maximum acceleration value of the missile is $a_{M\max} = 50\,\text{m}/\text{s}^2$. The initial positions of the missile and the target are $x_M(0) = -20\,\text{km}$, $y_M(0) = -20\,\text{km}$ and $x_T(0) = 0\,\text{m}$, $y_T(0) = 0\,\text{m}$, respectively. The initial flight path angles of the missile and the target are $\varphi_M(0) = \frac{\pi}{2}$ rad and $\varphi_T(0) = \frac{\pi}{3}$ rad, respectively. The velocity of the missile is chosen as $V_M = 800\,\text{m/s}$.

The parameters for ESO are selected as $\beta_1 = 50$,$\beta_2 = 300$,$\alpha = 0.25$ and $\sigma = 0.15$. And the initial states of ESOs are set to $Z_{i,1}(0) = Z_{i,2}(0) = 0, i = 1, 2$. The parameters for the proposed guidance law are chosen as $\tau = 10$, $\delta = 1$ and $\gamma = 0.2$. The initial auxiliary signal is $\lambda(0) = 0$. The parameters for sliding mode are chosen as $c_0 = 0.1$ and $c_1 = \frac{\theta_M(0)}{100\times R(0)}$. And ε is selected as $\varepsilon = 0.01$, where it needs to note that the value of ε determines the stability of the slide mode S_1.

For comparison, the following PNG law is also considered.

$$a_M = NV_M\dot{q} = NV_M\left(\frac{V_M\sin(q - \varphi_M)}{R} + Z_{2,2}\right)$$

where $N = 3$ and $Z_{2,2}$ is defined above.

Define $a_M^{sum} = \int_0^{t_I} |a_M|dt$ as the total control effort of the required guidance command, where t_I is the interception time representing the smallest miss distance.

Case 1: It is assumed that the target is stationary. That is to say $V_T = 0\,\text{m/s}$ and $a_T = 0\,\text{m}^2/\text{s}$. The simulation time is set to 40 s and the simulation step is set to 0.0001 s. The engagement trajectory is shown in Fig. 2, and the relative distance is displayed in Fig. 3.

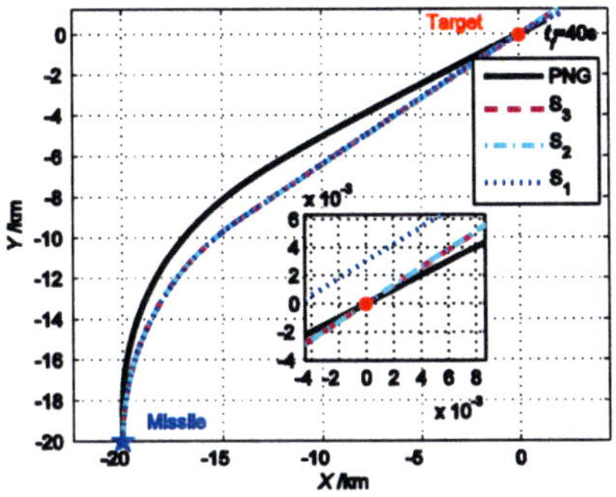

Fig. 2. Flight trajectories

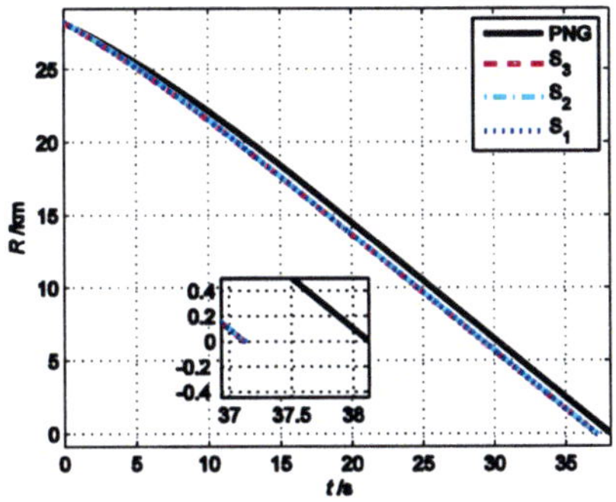

Fig. 3. Relative range

Table 1. Performance results of guidance laws for case 1

Methods	Miss distance (m)	Interception time (s)	Total control effort
PNG	0.0771	38.1296	9757110
S3	0.0402	37.1326	8087205
S2	0.0599	37.1325	10953274
S1	0.8956	37.1326	8027486

The performances at the interception time arerepresented in Table 1. By the simulation, the actual miss distances and corresponding interception times for four methods are respectively 0.0771 m and 38.1296 s, 0.0402 m and 37.1326 s, 0.0599 m and 37.1325 s,

0.8956 m and 37.1326 s. It can be clearly seen that when the target is stationary, the effect of the four guidance methods is similar. But the interception time of the three sliding mode guidance methods is shorter than that of PNG. Moreover, the engagement trajectories of the three sliding mode guidance methods are almost identical. The guidance command for the four methods is given in Fig. 4. As shown in Fig. 4 and Table 1, when the slide mode surface is selected as S_3, the proposed guidance law uses less guidance command than the other three guidance laws.

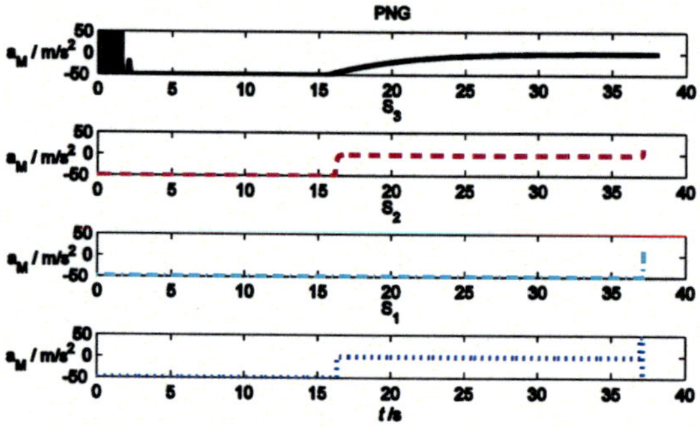

Fig. 4 Guidance command.

Case 2: It is assumed that the target is maneuvering. The velocity and acceleration of target is respectively set to $V_T = 450\,\text{m/s}$ and $a_T = 100\sin(t)\,\text{m}^2/\text{s}$. The simulation time is set to 100 s and the simulation step is set to 0.001s. The engagement trajectory is shown in Fig. 5, and the relative distance is displayed in Fig. 6.

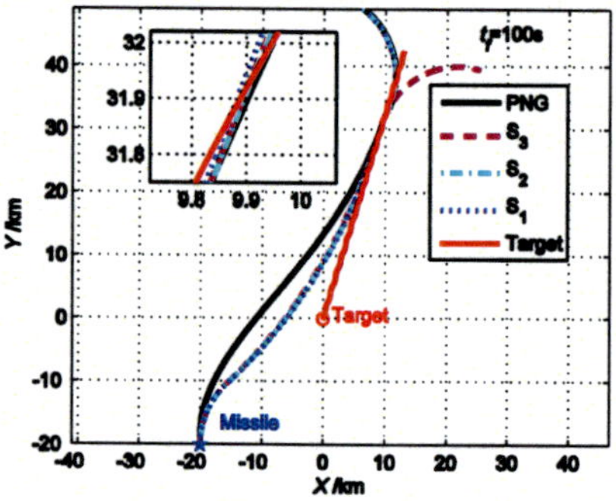

Fig. 5. Flight trajectories

The performances at the interception time are represented in Table 2. By the simulation, the actual miss distances and corresponding interception times for four methods are respectively 28.9759 m and 77.2706 s, 0.2217 m and 77.2920 s, 7.4823 m and 77.3372 s, 6.1301 m and 77.3396 s. It can be clearly seen that when the target is maneuvering, the effect of the three sliding mode guidance methods are better than that of PNG. The

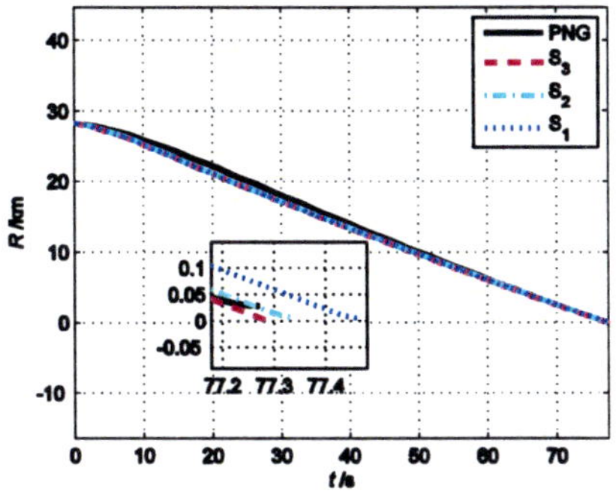

Fig. 6. Relative range

Table 2. Performance results of guidance laws for case 2

Methods	Miss distance (m)	Interception time (s)	Total control effort
PNG	28.9759	77.2706	13215847
S3	0.2217	77.2920	13873439
S2	7.4823	77.3372	13931688
S1	6.1301	77.3396	9626835

guidance command for the four methods is given in Fig. 7. As shown in Fig. 7 and Table 2, when the sliding surface is designed as S3, although the guidance command is not the least, the miss distance is the least compared to the other three methods.

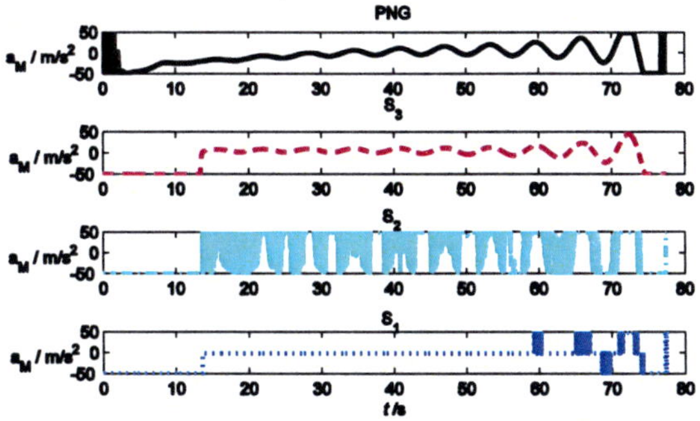

Fig. 7. Guidance command.

From the simulations in case 1 and case 2, it could be concluded that the proposed three slide mode guidance methods has better performance than PNG, especially when the slide mode surface is selected as S_3. The reason is that it can not only ensure that the velocity of the missile points to the target, but also ensure that the distance between the missile and the target is small, as the sliding surface S_3 converging to 0.

5 Conclusions

In this study, the problem of guidance law design against non-maneuvering and maneuvering targets with unknown velocity and acceleration is solved by designing three different sliding mode guidance methods. Compared to the traditional PNG method, the proposed guidance methods can ensures intercepting a maneuvering target. In addition, the proposed approach is novel in that the guidance law considers the uncertainties of target velocity as well as target acceleration. Simulation results, the stability and performance analysis show that the proposed schemes are effective against non-maneuvering and maneuvering targets.

Acknowledgements. This work was supported in part by the Natural Science Foundation of Shandong Province, China (Grant No. Zr2019mf065).

References

1. Zarchan, P., Zarchan.: Tactical and Strategic Missile Guidance, 3rd edn. AIAA, Washington, pp. 13–34 (1998)
2. Tyan, F.: Analysis of general ideal proportional navigation guidance laws. Asian J. Control **18**(3), 899–919 (2016)
3. Bryson, A.E., Ho, Y.C.: Applied Optimal Control: Optimization, Estimation and Control. Hemisphere Publishing Corporation, Washington, DC (1975)
4. Kreindler, E.: Optimality of proportional navigation. AIAA J. **11**(6), 878–880 (1973)
5. Tyan, F.: Analysis of 3D PPN guidance laws for nonmaneuvering target. IEEE Trans. Aerosp. Electron. Syst. **51**(4), 2932–2943 (2015)
6. Siouris, G.M.: Comparison between proportional and augmented proportional navigation. Nchriten technische Zeits chrift 278–280 (1974)
7. Ghosh, S., Ghose, D., Raha, S.: Capturability of augmented pure proportional navigation guidance against time-varying target maneuvers. J. Guid. Control. Dyn. **37**(5), 1446–1461 (2014)
8. Cho, N., Kim, Y.: Optimality of augmented ideal proportional navigation for maneuvering target intercep-tion. IEEE Trans. Aerosp. Electron. Syst. **52**(2), 948–954 (2016)
9. Cottrel, R.G.: Optimal intercept guidance for short range tactical missiles. AIAA J. **9**(7), 1414–1415 (1971)
10. Yang, C.D., Yeh, F.B.: Optimal proportional navigation. J. Guid. Control. Dyn. **11**(4), 375–377 (1988)
11. Youan, Z., Yunan, H., Tao, L.: Robust geometric approach to missile guidance. Control Theory Appl. **20**(1), 13–16 (2003) (in Chinese)
12. Hu, Q., Han, T., Xin, M.: Sliding-mode impact time guidance law design for various target motions. J. Guid. Control. Dyn. **42**(1), 136–148 (2019)
13. Moon, J., Kim, K., Kim, Y.: Design of missile guidance law via variable structure control. J. Guidance, Control, Dyn. **24**(4), 659–664 (2001)
14. Shtessel, Y.B., Shkolnikov, I.A., Levant, A.: Guidance and control of missile interceptor using second-order sliding modes. IEEE Trans. Aerosp. Electron. Syst. **45**(1), 110–124 (2009)
15. Kim, H.G., Kim, H.J.: All-aspect guidance with impact angle constraint against unknown target maneuver. IEEE Trans. Aerosp. Electron. Syst. **55**(2), 830–845 (2018)
16. Wen-Xue, S., Feng-Qi, Z., Jun, Z.: Nonlinear variable structure guidance law. J. Astronaut. **24**(6), 638–641(2003) (in Chinese)

17. Wang, W., Xiong, S., Liu, X., et al.: Adaptive nonsingular terminal sliding mode guidance law against maneuvering targets with impact angle constraint. Proc. Instit. Mech. Eng. Part G: J. Aerosp. Eng. **229**(5), 867–890 (2015)
18. Zhu, Z., Xu, D., Liu, J., et al.: Missile guidance law based on extended state observer. IEEE Trans. Indus. Electron. **60**(12), 5882–5891 (2012)
19. Cheng, C.H., Hu, Y., Wu, J.: Auto disturbance controller of non-affine nonlinear pure feedback systems. Acta Automatica Sinica **40**(7), 1528–1536 (2014) (in Chinese)
20. Chunhua, C., Yunan, H., Jinhua, W.: Auto disturbance rejection controller for non-affine nonlinear sys-tems with adaptive observers. Control Theory Appl. **31**(2), 148–158 (2014) (in Chinese)
21. Cheng, C., Di, Y., Xu, J., Yuan, T.: Advanced backstepping control based on ADR for non-affine non-strict feedback nonlinear systems. Automatika **59**(2), 220–230 (2018)
22. Liu, L., Wang, D., Peng, Z.: ESO-based line-of-sight guidance law for path following of underactuated marine sur-face vehicles with exact sideslip compensation. IEEE J. Oceanic Eng. **42**(2), 477–487 (2016)
23. Cheng C.-H., Wu J.-H., Hu Y.-A., Jing, L.: Adaptive control of non-affine nonlinear system with saturation constraint. Control Theory Appl. **31**(8), 1000–1008 (2014) (in Chinese)

Numerical Study on the Formation of Defects During Double Diaphragm Forming Using a Biaxial Non-crimp Fabric

Fei Yu[1](✉), Xiangming Chen[1], Shuai Chen[2](✉), and Lee T. Harper[2]

[1] Aircraft Strength Research Institute of China, Xi'an 710065, China
yuf078@avic.com
[2] Composites Research Group, University of Nottingham, Nottingham, UK
Shuai.chen@nottingham.ac.uk

Abstract. A macroscopic finite element model was employed to investigate the forming-induced macroscale defects during double diaphragm forming (DDF) using a bi-axial non-crimp fabric. The forming related wrinkling behaviour was predicted by the simulation considering the bending stiffness of the fabric. The location of wrinkles was influenced by the in-plane shear deformation and bridging force in the fabric-diaphragm assembly. The severity of wrinkles was influenced by the layup sequence and the frictional resistance at the fabric-fabric and fabric-diaphragm interfaces. The quad-axial layup [+45°/−45°, 0°/90°] was found to produce more wrinkles than the single-orientation layup, indicating that the dissimilar shear deformation between plies at different orientations can contribute to fabric wrinkling. A parametric analysis indicated that reducing the coefficients of friction at fabric-fabric and fabric-diaphragm interfaces can mitigate wrinkles. The deformation of fabric plies can be affected by the in-plane distance between preforms, when forming multiple components in one DDF operation. The component spacing therefore needs to be designed to ensure the quality of each preform is not adversely affected during multi-component forming.

Keywords: Fabric/Textile · Preforming · Finite element analysis · Defect

1 Introduction

Double Diaphragm Forming (DDF) utilises vacuum forming force to deep draw the fabric-diaphragm arrangement to conform to a male tool, which enables large composite preforms or components to be produced without using autoclaves or oversized hydraulic presses [1]. Thermoplastics [2] or thermoset prepregs [3] can be directly used in DDF to produce composite components, while dry fabric plies are typically used with binders in DDF to produce preforms for liquid moulding [4]. Two layers of deformable elastomeric sheets (diaphragms) are used to encapsulate the fabric plies and binders, whilst providing forming and clamping forces to the ply book during DDF. The air between the diaphragms is initially evacuated to generate some clamping force to mitigate out-of-plane ply buckling. The fabric-diaphragm assembly is typically constrained

Chinese Society of Aeronautics and Astronautics: CASTYSF 2022, LNEE 972, pp. 392–402, 2023.
https://doi.org/10.1007/978-981-19-7652-0_36

around the perimeter of the diaphragm using a rigid frame, which is then lowered to guide the fabric-diaphragm assembly to make contact with the tool. A second vacuum is subsequently generated between the lower diaphragm and the machine bed to deep draw the diaphragm over the male tool. This vacuum is then maintained to hold the shape of the fabric, with the preform heated to activate the binders in the fabric. The part is finally ejected from the tool once the system has cooled. Repeatable net-shape preforms can be produced via automated DDF process, but defects such as wrinkles and bridging can occur due to the limited out-of-plane constrains [4–6], which need to be mitigated in the design phrase of DDF.

It is therefore vital to use simulation tools to refine process parameters during the design phase, in order to ensure those defects do not evolve. Whilst several studies have focused on the forming behaviour of plies at the same orientation [4, 6], relatively little attention has been paid to the mechanisms of defect formation during forming multi-layered fabric plies at dissimilar orientations. The interfacial frictional behaviour was shown to dominate the wrinkling behaviour during forming multi-layered fabrics over a spar geometry with gentle changes in the cross-sectional area [7]. However, the influence of tool geometries was overlooked in that study. In addition, one of the advantages of DDF is that multiple components can be produced in one operation by introducing multiple tools and preforms for the same diaphragm arrangement. The influence of component spacing is therefore a critical parameter to ensure the quality of the preforms is not affected.

In this paper, a numerical model is employed to investigate the forming behaviour of multi-layered biaxial non-crimp fabrics (NCFs) during DDF using a hemispherical tool. The influence of ply orientation and inter-ply frictional resistance on the occurrence of wrinkling defects is investigated. The simulation tool is also employed to investigate the forming deformation of fabrics during multi-component forming.

2 Fabric Material Model

An existing non-orthogonal constitutive model was employed to model the forming behaviour of bi-axial fabrics during DDF, considering the fibre reorientation and fabric bending effects. The material model is briefly introduced herein, but a more detailed description of the numerical implementation can be found in [8, 9].

2.1 Fabric Bending Behaviour

The fabric bending behaviour is different from that of continuous material as fibre slippage and buckling can occur at mesoscale during the bending deformation. Consequently, it is not reasonable to derive the bending stiffness of fabric directly from the tensile modulus along primary yarns, indicating that the classic beam or shell finite element formulations are not applicable for modelling the fabric bending behaviour. In this study, the Composites Layup Toolset in Abaqus/Explicit was employed to independently define the membrane and bending behaviours of fabric material. As shown in Fig. 1, the in-plane membrane and out-of-plane bending behaviours are independently defined by modifying the thickness of sublayers within the layup and the integration points within each sublayer. A composite layup consists of three sublayers was assigned to a single layer of shell elements (S4R in Abaqus/Explicit) to model the fabric ply. Only one integration point was allocated to the central sublayer to remove the bending contribution of this layer to the shell element, while the bending stiffness of the element was defined by the modulus of two outer surface sublayers. The actual thickness of the fabric ply was used to define the thickness for the shell element in FE simulation.

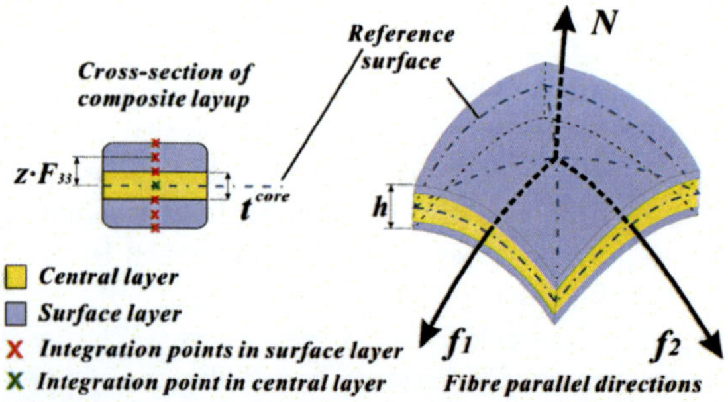

Fig. 1. Schematic diagram of the composite layup shell for decoupling the membrane and bending stiffnesses.

The bending moment in fibre directions, M_{fi}, can be calculated by integrating the moment related to the stresses along fibres over the thickness of the shell element.

$$M_{fi} = \int_{-h/2}^{+h/2} \left\{ \sigma_{fi}(z) \cdot \overline{F}_{33}^{2} \cdot z \right\} dz. \tag{1}$$

By assuming the shell element is incompressible, the thickness change of the shell element, $\overline{F}_{33}$, can be calculated from the in-plane components of the deformation gradient $\overline{F}_{ij}(i, j = 1, 2)$, i.e.

$$\overline{F}_{33} = \frac{1}{\overline{F}_{11}\overline{F}_{22} - \overline{F}_{12}\overline{F}_{21}}. \tag{2}$$

A single integration point was allocated to the central sublayer of the element, corresponding to the reference surface of the shell element where the thickness coordinate $z = 0$. Therefore, the bending moment of the shell element is irrelevant to the stresses at this integration point. Consequently, the bending behaviour of the ply is defined by the moduli of the two surface sublayers. The thickness of the fabric ply is much smaller than the in-plane dimension of the ply, the bending stiffness along the fibre direction can therefore be derived by assuming the ply as a thin-walled shell. An identical Young's modulus (i.e. E_{fi}^{surf}) was assigned to the top and bottom surface layers. In practice, three-point Simpson's integration rule was employed to calculate the bending moment on the shell cross-section, with only one integration point allocated to each sublayer to reduce the computational cost associated with the cross-sectional integration. This could result in reasonable accuracy as the stress-strain relationship at the integration points was assumed to be elastic. It should be noted that the inertial moment of each sublayer with respect to its centroid axis is eliminated when calculating the bending stiffness of the layup, which leads to expression for the bending stiffness parallel to the fibres, B'_{fi}

$$B'_{fi} = \frac{1}{16} E_{fi}^{surf} \left(h - t^{ctral}\right)\left(h + t^{ctral}\right)^2. \tag{3}$$

Whilst the expected value of the bending stiffness B'_{fi} can be achieved by modifying the value of either E_{fi}^{surf} or t^{ctral} or both according to Eq. (3), adjusting E_{fi}^{surf} is used herein to control the bending stiffness B'_{fi} due to a linear relationship between them. According to the Rule of Mixtures, the nominal Young's modulus in fibre directions, E_{fi}^{lam}, can be determined by the modulus value of each sublayer.

$$E_{fi}^{lam} = E_{fi}^{surf}\left(\frac{h - t^{ctral}}{h}\right) + E_{ctral}\left(\frac{t^{ctral}}{h}\right). \tag{4}$$

In practice, fibre modulus, E_{fi}^{lam}, and the bending stiffness, B_{fi} or B'_{fi}, are known parameters, which are used to derive the moduli values of each sublayer within the composite layup using the above equations.

2.2 Non-orthogonal Constitutive Model

An non-orthogonal constitutive model originally developed in [8, 9] was revised to incorporate the bending stiffness of each primary yarn into the modelling framework. The bending stiffness is defined in the fibre orientation system which updates as fabric shears in-plane, as shown in Fig. 2. The moduli values of the sublayers within the shell element were updated for each yarn direction during the analysis to incorporate a non-linear variation for bending stiffness into the constitutive model. The material model was implemented into Abaqus/Explicit via a user defined material subroutine (VUMAT).

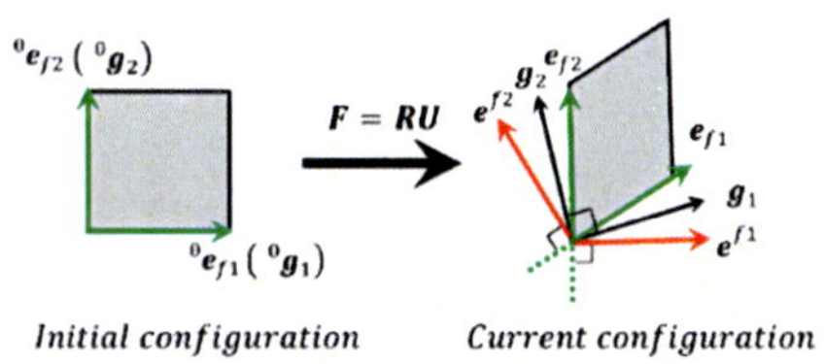

Fig. 2. Schematic of the in-plane shear deformation of a biaxial fabric unit cell.

2.3 Curvature Updating

As shown in Fig. 3, the bending moment and curvature along each fibre direction are updated during each time increment based on the stresses and strains of corresponding yarns determined via the non-orthogonal constitutive framework. In practice, the curvature along each primary yarn was calculated for each yarn direction using the strains through the thickness of the shell element. This procedure was implemented into VUMAT based on Abaqus/Explicit, enabling the bending stiffness to be updated according to the curvature of the primary yarns. The stresses and strains at all through-thickness integration points within the shell element are required to calculate the bending moment and curvature, a user defined subroutine VEXTERNALDB was therefore used to enable a communication mechanism between a global data array and VUMAT to update data at element level. The subroutine VEXTERNALDB is called once at the beginning of the analysis increment to access element information and date updated by the VUMAT subroutine. At the end of each time increment, the VEXTERNALDB is called once again to store the data (i.e. fibre curvature) into a global data array for the next increment. The moduli values can therefore be determined for each sublayer within the composite layup shell, updating the bending stiffness. The VUMAT subroutine is then be called multiple time to update stresses and strains for the integration points within the model, which are then stored in the global data container.

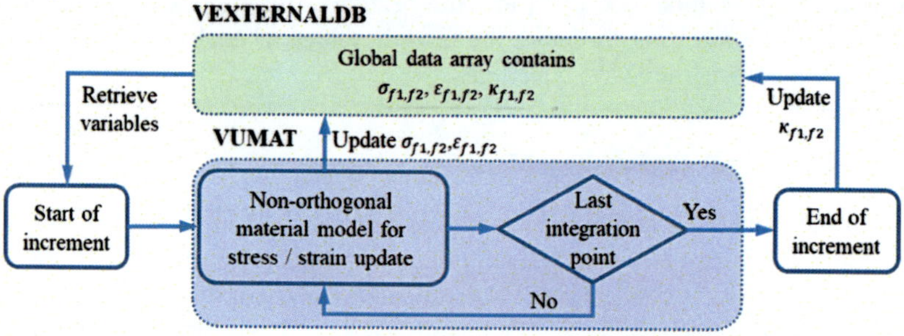

Fig. 3. Flowchart for data update and exchange between VUMAT and VEXTERNALDB.

3 Process Simulation

3.1 Material Modelling

The mechanical properties of the biaxial NCF, characterised in [8, 10], was employed to support the simulation in this study. The NCF ply is constructed by stacking two layers of 24 K non-crimped fibres oriented at ±45°, with the architecture fixed by pillar stitches at 0° to the warp direction. Hence, the NCF exhibits asymmetric in-plane shear behaviour under positive and negative shearing due to the constraining effect of the pillar stiches. The thickness (uncompressed) of the assembled ply is 0.4 mm, and the areal mass of the NCF ply is 440 gsm. The bending stiffness of the NCF was characterized to be nonlinear and asymmetric about the bending direction [8]. The diaphragm was modelled as a hyperplastic material using the data published in [11].

3.2 Process Modelling

As shown in Fig. 4a, the finite element mesh of the ply book is sandwiched between the top and bottom diaphragm meshes, which are constrained in the x-y plane around its perimeter to replicate constraints from the rigid frame. The male tool and the machine bed were modelled as discrete and analytic rigid body respectively. The frictional coefficients (CoFs) measured in [7] were used herein to model the frictional behaviour at contact interfaces in the process, based on the Coulomb friction model in Abaqus/Explicit. A pair of pressures of the same magnitude but opposite directions were applied to the top and bottom diaphragms to model the in-bag vacuum between time points A and B in Fig. 4b. A vertical displacement was applied to the frame to guide the fabric-diaphragm assembly to drape over the male tool until the perimeter of the lower diaphragm made contact with the machine bed. The pressure applied to the bottom diaphragm (denoted as red arrows) was then reduced to zero to generate a pressure differential (from time B to C in Fig. 4b) to deep draw the diaphragm assembly to conform to the tool surface.

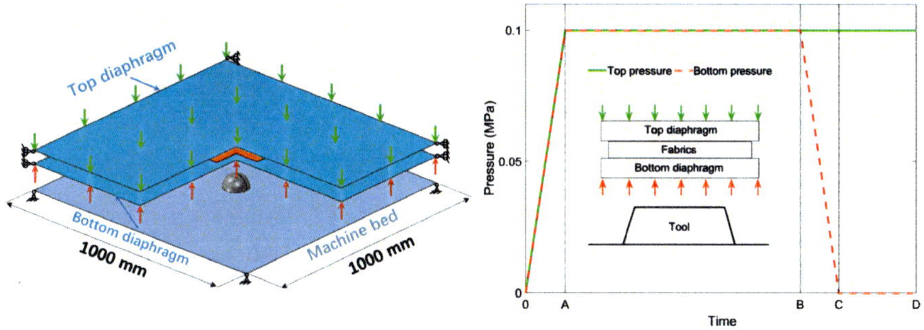

(a) FE model (b) Pressure variation for diaphragms

Fig. 4. FE model of the DDF process, and the variation of pressures applied to the diaphragm assembly as a function of time.

4 Results and Discussion

4.1 Influence of Ply Orientation

Forming simulations were performed using three different layups (see Table 1) to investigate the effect of ply-orientation sequence on the formation of defects. Figure 5 compares the shear angle distribution and the defect pattern predicted for the three layups. The location of all surface defect was visualised using the deviation distance between the deformed fabric mesh and the tool surface [9].

Wrinkle locations correspond with regions of positive shear deformation for layups L1 and L2, as shown in Fig. 5a, b, indicating a correlation between shear-locking and wrinkling. For layup L1, additional wrinkles occur in the area between the positive and negative shear regions, as a result of significant material draw-in (highlighted by dotted-lines in Fig. 5a). As shown in Fig. 5b, the negative shear regions are free from

Table 1. Layup sequences used for simulations.

Layup	Fibre orientation
L1	[0°/90°, 0°/90°]
L2	[+45°/45°, +45°/−45°]
L3	[0°/90°,+45°/−45°]

wrinkles for layups L1 and L2 due to the low in-plane shear stiffness in these regions. In comparison, more wrinkles are observed for layup L3 (see Fig. 5c) since plies at different orientations undergo different local in-plane shear deformations, hence compete against each other during forming. This indicates that the incompatible inter-ply movement can adversely affect the formability of stacked fabric plies with different ply-orientations.

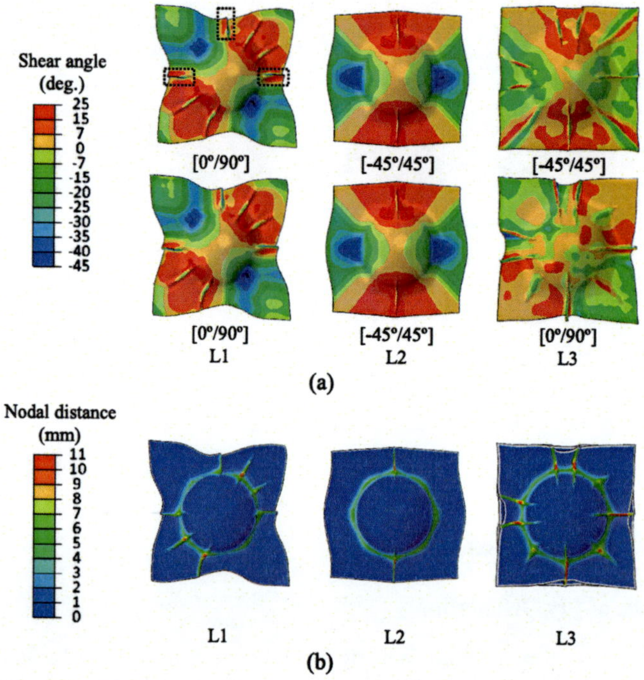

Fig. 5. Numerical predictions for layups L1-L3 in Table 1: **a** Shear angle and **b** Nodal deviation distance between the deformed fabric mesh and tool surface.

4.2 Influence of Inter-ply Frictional Resistance

A parametric study was performed for layup L3 using different coefficients of friction (CoF) to investigate the influence of inter-ply frictional resistance on the wrinkling behaviour. The fabric-fabric and fabric-diaphragm CoF were set to be identical and varied from 0 to 0.3 in increments of 0.1, as shown in Fig. 6. Increasing the interfacial CoF leads to higher frictional forces at the ply-ply interfaces, constraining the in-plane shear deformation of each ply. Consequently, the material draw-in of each ply decreases as the CoF increases, which can be seen from the changes in the perimeter shape of each ply in Fig. 6a. Additionally, the bridging area around the equator of the hemisphere increases due to the increase in the constraining effect at the ply-ply interfaces. Overall, the number of wrinkles decreases as the CoF is reduced, but wrinkles remain in the positive shear region for the case with no inter-ply friction (CoF = 0) where the location of these wrinkles is close to the area with significant ply draw-in. This indicates that reducing the inter-ply frictional resistance can mitigate wrinkles caused by ply-ply interactions, but this appears to be less effective for reducing the wrinkling defects related to fabric over-shearing or fabric bridging.

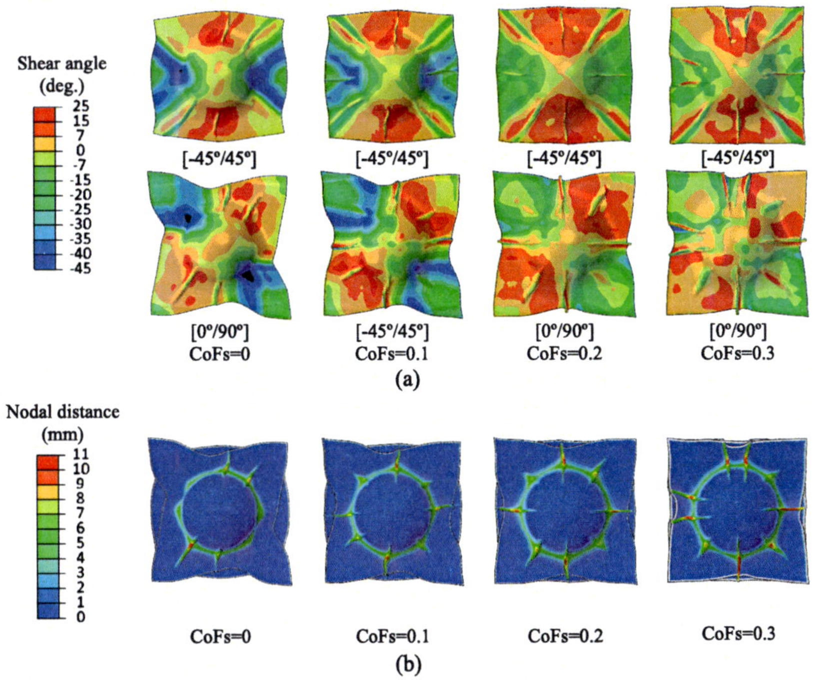

Fig. 6. Numerical predictions for layup L3 in Table 1 using different coefficients of friction (CoF). **a** Shear angle and **b** Nodal deviation distance between the deformed fabric mesh and tool surface.

4.3 Influence of Component Spacing

It is common for manufacturers to use DDF to produce multiple components in one forming operation, to increase production rate and to reduce material wastage. Therefore, it is important to determine the minimum distance between the tooling blocks to ensure the quality of each preform is not affected by neighbouring parts. A parametric analysis was performed by varying the distance between two preforms to investigate the influence of component spacing on the forming deformation of the fabric. Layup L2 (i.e. [+45°/−45°, + 45°/−45°]) was selected for this parametric analysis. The initial distance, *D*, between the centres of the two hemispherical tools varied from 200 mm to 300 mm in increments of 25 mm, as shown in Fig. 7a. Figure 7b presents the perimeter shape of the preform on the left side at different component spacing values. Compared with the results for the single-component case in Fig. 5b, additional wrinkles occurred between the tools when $D = 200$ mm, due to diaphragm pinching caused by the bridging effect, as shown in Fig. 7c. The wrinkle amplitude decreases as the distance *D* increases, since the bridging force reduces as the tools become further apart. Whilst the additional wrinkles disappear when $D = 250$ mm, the perimeter shape of the preform is still affected by the bridging force, as compared with the perimeter shape in Fig. 5b. The bridging force appears to have negligible influence on the perimeter shape when $D \geq 275$ mm, as shown in Fig. 7.

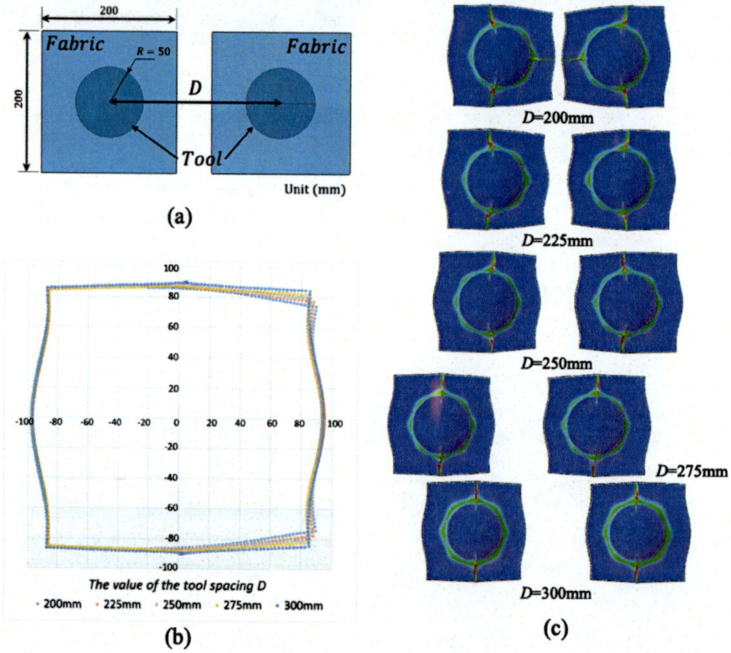

Fig. 7. **a** Schematic of the component layout. Numerical predictions for the layup L2 in Table 1 at different component spacing values: **b** Perimeter shape of the preform on the left side **c** Nodal deviation distance between the deformed fabric mesh and tool surface.

5 Conclusion

A macro-scale fabric model was employed to investigate the influence of process parameters during DDF of a biaxial NCF, including ply orientation, inter-ply friction and component spacing. The flowing conclusions can be drawn.

(1) Regions of high positive shear deformation are more prone to wrinkling than regions of negative shear, indicating that the constrains from the intra-ply stitching increase the likelihood of wrinkling for the NCF during DDF.
(2) The severity of wrinkles can be significantly influenced by the ply orientation within the layup. More wrinkles were found in the quad axial layup (i.e. [+45°/−45°, 0°/90°]) than layups with plies stacked at the same orientation.
(3) Reducing inter-ply friction can reduce the severity of wrinkles related to inter-ply slippage, but this method is less effective when wrinkles occur due to fabric bridging and fabric over-shear.
(4) Reducing component spacing when forming multiple preforms simultaneously can increase the tensile strain in the diaphragm-fabric assembly between adjacent tools, introducing additional wrinkles and affecting the material draw-in. Component spacing therefore needs to be optimised to ensure additional defects do not evolve when producing multiple preforms.

References

1. Krebs, J., Friedrich, K., Bhattacharyya, D.: A direct comparison of matched-die versus diaphragm forming. Compos. A Appl. Sci. Manuf. **29**(1), 183–188 (1998)
2. Bersee, H.E.N., Beukers, A.: Diaphragm forming of continuous fibre reinforced thermoplastics: influence of temperature, pressure and forming velocity on the forming of Upilex-R® diaphragms. Compos. A Appl. Sci. Manuf. **33**(7), 949–958 (2002)
3. Bian, X.X., Gu, Y.Z., Sun, J., Li, M., Liu, W.P., Zhang, Z.G.: Effects of processing parameters on the forming quality of c-shaped thermosetting composite laminates in hot diaphragm forming process. Appl. Compos. Mater. **20**(5), 927–945 (2012). https://doi.org/10.1007/s10443-012-9310-7
4. Chen, S., et al.: Double diaphragm forming simulation for complex composite structures. Compos. A Appl. Sci. Manuf. **95**, 346–358 (2017)
5. Alshahrani, H., Hojjati, M.: Optimum processing parameters for hot drape forming of out-of-autoclave prepreg over complex shape using a double diaphragm technique. In: 20th International Conference on Composite Materials. Copenhagen (2015)
6. Alshahrani, H., Hojjati, M.: Experimental and numerical investigations on formability of out-of-autoclave thermoset prepreg using a double diaphragm process. Compos. A Appl. Sci. Manuf. **101**, 199–214 (2017)
7. Yu, F., Chen, S., Harper, L.T., Warrior, N.A.: Investigation into the effects of inter-ply sliding during double diaphragm forming for multi-layered biaxial non-crimp fabrics. Compos. A Appl. Sci. Manuf. **150**, 106611 (2021)
8. Yu, F., Chen, S., Viisainen, J.V., Sutcliffe, M.P.F., Harper, L.T., Warrior, N.A.: A macroscale finite element approach for simulating the bending behaviour of biaxial fabrics. Compos. Sci. Technol. **191** (2020)

9. Yu, F., Chen, S., Harper, L.T., Warrior, N.A.: Simulating the effect of fabric bending stiffness on the wrinkling behaviour of biaxial fabrics during preforming. Compos. Part A: Appl. Sci. Manuf. 106308 (2021)
10. Chen, S., McGregor, O.P.L., Harper, L.T., Endruweit, A., Warrior, N.A.: Defect formation during preforming of a bi-axial non-crimp fabric with a pillar stitch pattern. Compos. Part A Appl. Sci. Manuf. **91**, 156–167 (2016)
11. Yu, F., Chen, S., Harper, L.T., Warrior, N.A.: Double diaphragm forming simulation using a global-to-local modelling strategy for detailed defect detection in large structures. Compos. A Appl. Sci. Manuf. **147**, 106457 (2021)

Research on Three-Proofing Technology Based on PCBA of Airborne Electronic Equipment

Han Yaguo(✉), Ding Lei, Wang Lu, Li Lanjun, Chen Xufan, and Sha Jing

AVIC Xi'an Aviation Brake Technology Co., Ltd, Xi'an, Shaanxi 710075, China
843129227@qq.com

Abstract. It is reported that 80% of airborne electric equipment failures have been linked to environmental factors. Adaptability to harsh environment has become an important tactical index for airborne electronic equipments. In order to improve the three-proofing properties (including moisture proofing property, salt spray proofing property and fungus proofing property),it is a must to improve the three-proofing effect in an all-round manner by means of rational structural layout, component design as well as PCBA design, process, production control, test verification. The three-proofing of airborne electronic equipment should be verified by tests in accordance with national military standards and other applicable standards so that requirements of environmental adaptability can be met. Meanwhile, data support can be provided for the upgrade of follow-up airborne electronic equipments.

Keywords: Three-proofing · Airborne electronic equipment · PCBA

1 Overview

With the progress of science and technology, airborne electronic equipments tend towards integration, modularization and miniaturization. The functions of these equipments are more and stronger and application becomes more and more broad. Operating environment and the characteristics should be taken into account during design of airborne electronic equipment. Airborne electronic equipments are very susceptible to environment conditions, especially for electromagnetic environment, moisture, salt spray and fungus and other environments. The environment where airborne electronic equipments work is mainly type B environment profile, in which condensation should be avoided and dust should be prevented from falling within the equipment during temperature change. In order to make airborne electronic equipment work reliably, new design methods, new technologies, new materials and new processes should be adopted based on existing mature theories and design experience. New design methods include modular design and miniaturization design. New technologies, such as SMT (surface mounting technology) and high-density packaging technique can be used. As existing theory and design experience, bracket or vibration damper can increase the vibration resistance and

Han Yaguo (1983)—male, M.S. degree, senior engineer.

Chinese Society of Aeronautics and Astronautics: CASTYSF 2022, LNEE 972, pp. 403–411, 2023.
https://doi.org/10.1007/978-981-19-7652-0_37

impact resistance. Electromagnetic shielding is very efficient in reducing external electromagnetic interference (EMI) and offering electromagnetic environment protection. Temperature control technology is also helpful. Additionally, three-proofing (including moisture proofing, salt spray proofing and fungus proofing) technology must be utilized.

Three-proofing are not limited to moisture proofing, salt spray proofing and fungus proofing. Airborne electronic equipments may be subject to dust, moisture, fungus, salt spray, Corrosive gas and so on. As a result, corrosion, softening, deformation, mildew and other problems begin to occur in PBCA and elements of airborne electronic equipments, which result in electric leakage, short circuit and device failure and product quality can be further affected. According to incomplete statistics, 80% of airborne electric equipments failures have been linked to environmental factors. Because moisture proofing, fungus proofing and salt spry proofing (can be referred to three-proofing for short) don't create effectiveness, they tend to be ignored at early stage. Once three-proofing problem is found, only remedial measures can be taken afterwards. The majority of three-proofing problems are found at late design stage or during operation. For delivered airborne electronic equipments, tactical index declines because of corrosion and their lives are reduced after used for a period of time.

Humid environment can dramatically reduce the insulativity between PCBA conductors, accelerate high-speed decomposition, decrease Q value and corrode conductors. It can also cause oxidation and electrochemical corrosion of metals and quicken chemical reactions. Condensation and free water may create short circuit. Heat conduction property can be changed. Or even worse, the copper on PBCA of airborne electromagnetic equipment may react with water vapor and oxygen, leading to the appearance of verdigris.

Fungus making conductive paths on insulating materials, can have a negative impact on the electric characteristics of precision circuits. Fungus growing on dust, grease and other pollutants deposited during manufacturing or operation process on the surface of PBCA can cause damage to substrate. Metabolites such as organic acid secreted by fungus can cause metal corrosion. Metabolite of fungus comes in contact with neighbouring antifungal materials and erosion is thereby produced, which can cause performance degradation of airborne electronic equipments as well as damaged electrical and electronic systems.

Salt spray environment makes electrochemical take place in airborne electronic equipment and stress corrosion accelerate. Salts ionize in water, from which acid solution or alkaline solution is produced. As a result, a conductive cover layer takes place, causing corrosion of insulating materials and metals. Mechanical components and movable parts of assemblies get blocked. Coatings of PBCA blister because of electrolysis effect. Salt deposits can make damage to airborne electronic equipments.

2 Standard Requirements

As the performance of airborne electronic equipments are continuously increased, the three-proofing requirements for airborne electronic equipments are also improved. Three-proofing capabilities of airborne electronic equipments must be raised if the equipments are to work normally. Adaptability to harsh environment (as is shown in Fig. 1) which has a direct impact on the performance of airborne electronic equipments even on their lives, has become an important tactical index for airborne electronic equipments.

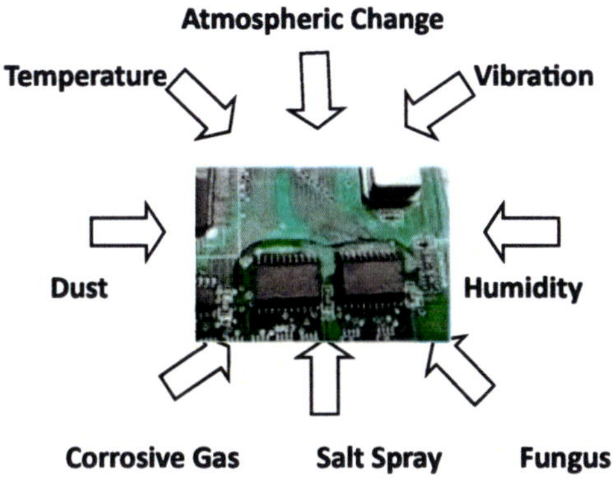

Fig. 1. PBCA is subject to various environments

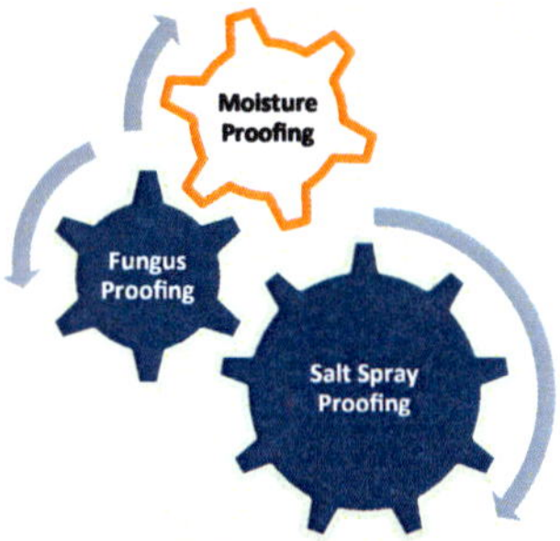

Fig. 2. Technical indexes of three-proofing coating of PBCA

Airborne electronic equipments are required to have good three-proofing (moisture proofing, salt spray proofing and fungus proofing) protection capabilities during operation and storage process, i.e. these equipments should endure the moisture, fungus and salt spray environment. Airborne electronic equipments must comply with RTCA/DO-160F Environmental conditions and Test Procedures for Airborne Equipment, GJB 150A.09-2009 Laboratory Environmental Test Methods for Military Materiel-Part 9:Damp Heat Test, GJB 150A.10-2009 Laboratory Environmental Test Methods for Military Materiel-Part 10: Fungus Test, GJB150A.11-2009 Laboratory Environmental Test Methods for Military Materiel-Part 11: Salt Fog Test and other requirements for test verification of airborne electromagnetic equipments.

3 Requirements for Three-Proofing Design

Three-proofing are not only a technological protection problem, but also related to design, process and manufacturing. Control measures should be taken in design, process and manufacturing. Additionally, staff's awareness on three-proofing should be raised and technologies should be improved. It is only in this way that three-proofing property of airborne electronic equipments can be obviously increased and that these equipments can work normally in moisture, fungus, salt spray as well as high shock and vibration environments.

3.1 Design Link

Three-proofing design is an important link in design and manufacturing of airborne electronic equipments and also a main approach of improving equipment lives. Three-proofing design is to develop design specification, incorporate three-proofing design into function design and property design of airborne electronic equipments, and adopt three-proofing design methods and protection measures for the equipments in accordance with environmental profile throughout the live cycle of airborne electronic equipments and requirements for three-proofing level. Randomness in design, simple technical protection and remedies made afterwards should be avoided. Three-proofing properties of airborne electronic equipments should be verified by test according to national military standards or related provisions specified in other standards. Meanwhile, three-proofing data should be tracked during the life cycle of airborne electronic equipments, providing data support for upgrade of follow-up equipments.

Structural layout and component design should be selected based on the operation environment of airborne electronic equipments and be carried out in overall layout, structural design and design in selection of material as below.

a. Overal layout. The design concept of three-proofing should be incorporated at design stage and three-proofing design should be integrated into structural design as the internal structure of airborne electronic equipments is rather complicated and the operating environment is harsh. The purpose is to increase the environmental adaptability and operating stability of airborne electronic equipments.
b. Structural design. Fillet should be adopted for the structure of parts for airborne electronic equipments, instead of sharp edges and acute angles. The surface of parts should be smooth, continuous so that the uniformity and cohesion of coating can be assured. On the basis of ventilation, heat rejection and electromagnetic compatibility (EMC), three-proofing should also be considered. Normally, the exposed portion of electronic equipments should be coated with three-proofing paint, which can be divided into two types: primer and finish. Typically, zinc chromate coating is selected for primer thanks to its excellent cohesion and corrosion resistance. Polyurethane coating is selected for finish because of its good weather resistance, chemical resistance and durability. Airborne electronic equipments should be designed into sealed structure so as to protect the printed board assemblies and make them work normally.
c. Design in selection of material. For the selection of materials of airborne electronic equipments, weight reduction, heat dissipation, ruggedization, three-proofing, shielding should be taken into synthetical account and specifications should be complied with. Airborne electronic equipments used under different environments, the surfaces of PCBA, discrete devices and integrated circuits of different airborne electronic equipments should be sprayed with three spoofing paints of different functions.

Three-proofing paints are coatings made of special formulas to protect PCBA of airborne electronic equipment from humid and high-temperature environment, prolong service life and ensure safety and reliability. Currently, three-proofing paint used for aerospace are S31-11 polyurethane insulating paint and H31-4 epoxy three-proofing insulating paint. Three-proofing can form a protective film with a certain thickness on the external surface of PCBA. Protective film can isolate the surface of printed board assembly from being in contact with the environment, playing a role in insulation, moisture proof, corrosion, occurrence of fungus, oxidation of solder joint, electric leakage, short circuit. In short, three-proofing paint can protect PCBA (as is shown in Fig. 2) from harsh environment, with the aim of long-term moisture proofing, fungus proofing and salt spray proofing.

3.2 Process Link

The typical process for spraying three-proofing paint on PCBA of airborne electronic equipments is shown in Fig. 3.

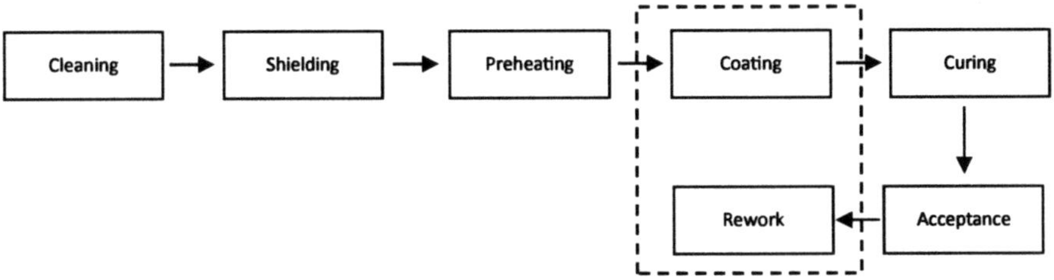

Fig. 3. Technical process for spraying three-proofing paint on PCBA assemblies

3.3 Cleaning

Cleaning should be performed on PCBA of airborne electronic equipment prior to coating in accordance with QJ3259-2005. The Technique Requirement for Preserving and Spreading in the Electrical Products of Aerospace. Dust, grease and residues of soldering flux should be cleaned. Cleaning solvent should be selected based on the cleaning objects, such as anhydrous ethanol and cleanser. Cleaning by hands or by machine is both allowed. For control of foreign object, GJB 5296-2004 Foreign Object Control Requirements and HB 7128-1994 Foreign Object Control Requirements should be followed before cleaning. For PCBA with ESD (electrostatic discharge) sensitive elements, tools and operators should comply with the antistatic requirements specified by GJB 3007A-2009 Requirements for Electrostatic Discharge Protected Area and SJ/T10694-2006 General Specification of Testing Methods for Electrostatic Protection in Electronic Production Manufacturing and Using System.

3.4 Shielding

Generally, solder joint doesn't need coating. The areas which don't need coating should be protected. For the small area, such as the plug of PCBA, duct tape 8204 and 3M adhesive tape can be used for sealing. For large area, it can be packed by antistatic bag and tied by cotton string. The non-coating area should comply with design documents and process documents.

3.5 Pre-heating

Oven should used for preheating before cleaned PCBA is coated in accordance with QJ3259-2005 The Technique Requirement for Preserving and Spreading in the Electrical Products of Aerospace. Preheating temperatures and preheating time are shown in Table 1, which can be adjusted based on the coating area and PCBA characteristics. Cut off the power supply and cool PCBA in the oven down to room temperature and then take the coated PCBA out of the oven when the parameters listed in Table 1 are reached. The PCBA which can't be put into oven or is forbidden to enter into oven for preheating and dehumidification, can be placed in a clean, dry and well-ventilated room and dried naturally for 24 h.

Table 1. Preheating temperatures and preheating time under different relative humidities

Relative humidity	Preheating temperature °C	Preheating time h
<60%	45 ~ 50	2
60% ~ 70%		2.5
70% ~ 80%		3
>80%		3.5

3.6 Coating

After appropriate three-proofing paint is selected based on the characteristics of airborne electronic equipments and operating environment, the effects of three-proofing of PCBA attach great importance to the right occasion of spraying. When the spraying occasion is selected, the secondary pollution prior to spraying should be avoided to the maximum.

The commonly used spraying methods are brush spraying, dip coating, manual spraying, machine spraying, selective coating and showering coating. The advantages and shortcomings of each coating methods are shown in Table 2. Based on production requirements, one method or combination method can be adopted. For example, use selective coating method and spraying method for PCBA. For the defective position, touch-up can be carried out by brush spraying or manual coating. For the position with sagging, corresponding measures should be taken. In this way, the paining quality can be improved on the basis of efficiency.

Table 2. Comparison between different spraying methods

No	Painting methods	Advantages	Shortcomings	Application
1	Brush spraying	Special fixture is not needed; coating material can be saved; small investment and broad application	Has the lowest efficiency; Shadowing effect when the whole board is coated; the consistency of coating is poor	Applicable to repair and maintenance of local position; Also applicable to lab environment or small batch trial production
2	Dip coating	Simple to operate, easy to carry out, no shadowing effect, can be applied to the whole product	Tend to have for impurities added and solvent volatilize; need covering for elements which isn't allowed to be coated	Applied to the occasion where full coating is needed
3	Manual spraying	Small investment; simple and convenient operation	Need covering during manufacturing; enormous waste of materials	Applicable to the products whose elements are not densely arrayed, products which doesn't need much local protection and products which don't have many demanding requirements
4	Machine spraying	High precision; good uniformity; high efficiency; automatic detection	Covering is needed; has shadowing effect; it is difficult for the lower part of narrowly spaced element to enter	Applicable to mass production or medium batch production
5	Selective coating	The uniformity of coating is good; The thickness of coating can be controlled; the consistency is good	The cost is high. It is applicable to short-term/small batch production The efficiency is lower, compared to automatic dip coating and automatic spraying	Applicable to high precision circuit board with high assembly density and many positions to be protected

3.7 Curing

Curing treatment should be performed after the first coating and the second coating of PCBA of airborne electric equipment or when multi-layers are coated in accordance with QJ 3259-2005 The Technique Requirement for Preserving and Spreading in the Electrical Products of Aerospace. Too long interval for coating multi-layer paint will result in weaker cohesion force between layers. After first coating, PCBA should be set at room temperature for 30 min ~ 60 min in order to let solvent volatilize and then the second coating should be applied. After the second coating, PCBA should be set at room temperature and then dried in infrared curing oven. When multiple layers are applied, coating of the other side can be carried out only when the organic solvent volatilizes to the extent that paint film is not sticky. For the drying temperature of paint film, paint processing plant's instructions can also be followed. The total thickness of coating can be decided on the basis on the design documents and process documents.

3.8 Touch-Up

For the position of PCBA which still can't meet requirements after coating and for the position where paint is missing, touch-up is needed. Touch-up can be carried out using brush method and paint spray gun method.

3.9 Acceptance After Coating

All PCBA of airborne electronic equipments should be 100% inspected for appearance. The PCBA, after coated with three-proofing paint, should be uniform in thickness, clean and tidy and free from element damage. The surface of boards and elements should be free from dirt, fingerprint, ripple, dust, shrinkage cavity, dust and other foreign objects. Pulverization, peeling, blister, pinhole and ripple are not allowed. Detection should be carried out in accordance with acceptance provisions for electronic assemblies, namely, free from loss of cohesion, void, blister, loss of wettability, powder points, wrinkles, cracks, ripples, fish eye, orange peel; free from imbedded or coerced foreign objects; free from discoloration, free from transparence; completely solid and uniformly distributed.

3.10 Production Link

3.11 Environmental Requirements

Spreading three-proofing paint should be conducted in an enclosed separate room where the ventilation facility should be equipped. The relative humidity should be controlled at 30% ~ 75% and temperature should be controlled at 23°C ± 5°C. Take electrostatic protection measures in accordance with GJB 3007A-2009 Requirements for Electrostatic Discharge Protected Area. Workbench, spray booth and three proof coating machine should be kept clean and floating dust should be avoided.

3.12 Process Control

The pollutants of PCBA, once coming in contact with moisture, can corrode conductors or even create irretrievable short circuit. The most common pollutants found in electrical system, may be chemicals from manufacturing process, such as flux, solvent release agent, metallic granules and marking ink, The pollutants may also be produced by personnel carelessly, such as human grease, fingerprint, cosmetics and food residue.

4 Conclusion

Beginning with the hazards of corrosive moisture, fungus and salt spray environment, shortcomings in environmental adaptability of airborne electric environment are put forward, i.e. these equipments have excellent three-proofing capabilities including moisture proofing, fungus proofing and salt spray proofing. This paper proposes that the effect of three-proofing should be promoted in an all-round way by three-proofing design, process, control during production and test verification on the basis of selecting structural layout and component design. In this way, the shortcomings of airborne electronic equipments in three-proofing design can be remedied, so that the airborne electronic equipments can work reliably for a long term in a moist, fungus and salt spray environment. Therefore, the failure risks of airborne electronic products are reduced. At the same time, as three-proofing data accumulate, data support can be provided for upgrade of follow-up airborne electronic equipments.

References

1. QJ 3259-2005 The Technical Requirement for Preserving and Spreading in the Electronical Products of Aerospace
2. QJ 2850A-2011 Prevention and Control for Foreign Object Debris (FOD) of Space Products
3. HB7262.1~2-1995 Installation of Elements of Aeronautical Printed Board
4. Sun, H.-L., Wang, X.-H.: Overview of sealing design of shipborne electronic equipment. Equip. Environ. Eng. **5**(5), 49–52 (2008)
5. HB7262.7-1995 Electronic Fitting Technology of Aeronautic Products—Conformal Coating of PCBA
6. Zou, S.-W., Xiao, K., Dong, C.-F., Li, H.-Y., Lixiao-Gang.: Effect of mold on corrosion behavior of ImAgfinished printed circuit board. Sci. Technol. Rev. **30**(11), 21–26 (2012)
7. Zhang, M.: Study on the Corrosion Behavior of Printed Board and Its Influencing Factors. Xiamen University, Xiamen (2008)

The Application and Research of Multi-core Processing in Avionics System

Zhan Wentao[1,2](✉), Niu Wei[2], Yuan Ji[2], Wang Xiaobo[2], and Yang Liu[2]

[1] School of Civil Aviation, Northwestern Polytechnical University, Xi'an, China
zhanwt1984@qq.com

[2] Xi'an Aeronautics Computing Technique Research Institute, Xi'an, China

Abstract. Multi-core processing is widely used for its high performance, but in avionics system, there are some time behavior interferences caused by cache sharing and bus sharing etc. To address application aspects of multi-core processing in avionics system, this paper made tests and analyses, which showed that dual-core processor can get doubled performance in some specified situations. Gathering service experience of multi-core processing in less critical system is also a usable way towards critical system. Then this paper presents a scheduling configuration of SMP mode multi-core AVIC OS to meet the needs of mixed critical applications. This envisioned that multi-core processing will support the development of next-generation avionics systems.

Keywords: Multi-core processing · Sharing interference · Application mode

1 Introduction

Avionics system is the core component of aircraft, which is essentially a complex real-time system, that is, the system must correctly respond to the changes of external physical processes within the specified time range [1]. The system is also considered as safety critical system because its failure can cause catastrophic consequences such as loss of lives [2]. In modern aircraft avionics, more and more functions traditionally transition from federated to integrated system architectures. In federated architectures, computers connected by buses usually host one application, while in integrated architectures, such as the Integrated Modular Avionics (IMA) [3], many applications are often integrated on powerful platforms of mixed criticality. This can improve systems in several dimensions, such as weight, space, energy consumption and maintainability etc. Now, next generation of new aircraft avionic programs will integrate more new functions, including safety functions, information services and comfort features, which will greatly increase the demand of avionics processing performance. The traditional approach to improve processing performance was to increase CPU clock frequency or increase pseudo-parallel processing on code level. However, with the development of semiconductor technology based on Moore's law, this method has reached its limit. The increase of CPU frequency will lead to sharp rise in power consumption, which will bring trouble to heat dissipation. In addition, increasing CPU frequency will also introduce more and more problems

Chinese Society of Aeronautics and Astronautics: CASTYSF 2022, LNEE 972, pp. 412–419, 2023.
https://doi.org/10.1007/978-981-19-7652-0_38

caused by chip internal and external crosstalk, and chip signal integrity. As one of the ways of parallel processing, multi-core processing has become an important measure to improve processing capacity of processors in recent years. In avionics system, multi-core processing is increasingly favored by system developers because of its advantages of high performance, high integration and low power consumption.

2 Challenges of the Application of Multi-core Processor

2.1 The Architecture of Multi-core Processor

Software designer of single core processor can directly improve the performance by deploying the existing software to the CPU with higher frequency and faster speed without modification. CPU is the only shared resources in single core system. Once a task obtains the execution right of CPU, it will be able to control other hardware resources of the system. But in multi-core system, as shown in Fig. 1, resource sharing exists in multiple dimensions. Parallel tasks not only need to share multiple processing cores, but also share hardware resources such as cache, bus, memory controller and main memory. The sharing among multiple dimension leads to interdependence and interference among tasks, which leads to the high unpredictability of the overall performance and time behavior of the system, and it is difficult to guarantee the logical correctness [4].

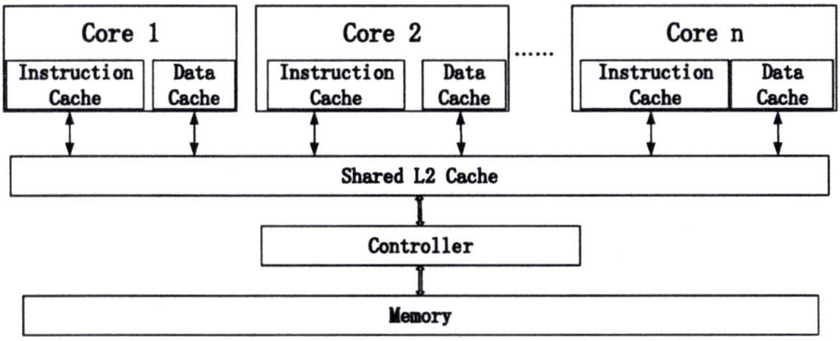

Fig. 1. Architecture of shared resource of multi-cores

2.2 The Problem of Shared Resources

Cache sharing interference. In multi-core system, a possible cache interference situation is that one core will continuously drive out cache blocks useful to another core, which will not bring significant performance improvement, but also increase the cache miss rate of the system and reduce system performance. Reference [5] shows that the shared cache interference between cores will increase the completion time of tasks by 40% compared with that running on a single core. In addition, the disordered preemption timing among cores makes it very difficult to analyze the system behavior. Inter tasks interference leads the WCET of a task not only rely on the task itself, but also affected by other tasks executed in parallel [6]. Moreover, the interaction among tasks explodes exponentially with the increase of the number of cores.

Bus sharing interference. Each core needs to access the off chip memory through the shared bus, which will cause the interference of the shared bus and increase the response time of the task on multi-core system. The bus access request of the task does not carry the priority attribute, so the bus will be reordered according to the access of the task, which may cause the access request service time of high priority task later than that of the low priority task. The delay of task bus access request, that is, determining the bus access service order, depends not only on the bus arbitration mechanism, but also the bus access volume of parallel tasks concurrently. So how to analyze the response time considering the interference of shared bus is still a challenging task.

Software interference. Software interference only occurs when the software is running concurrently. Interference severity mainly depends on the software architecture and the way the software utilizes shared resources of the cores. The existing operating system and software applied in avionics system comply with ARINC653 standard, which has been implemented and applied on a single core platform without considering the characteristics of multi-core platform. Previously, the system software only needed to consider avoiding shared resource conflict between partitions running concurrently. Now the system software must prevent unintended coupling and must not introduce additional interference due to concurrent access to shared resources among different cores and different partitions.

3 Research Development of Multi-core Processing in Avionics

Reference [7] proposed a static analysis method to solve the impact of shared cache on WCET to a limited extent. Considering the complexity of the problem, these solutions need to introduce some assumptions to simplify the problem. Such as only analyze the behavior of instruction cache and ignore the influence of data cache, and assuming that the data can always be accessed in data cache, this method cannot be verified on the actual multi-core computing platform. These simplified assumptions also limit the availability of the proposed solution. Therefore, some scholars turn to specific design to avoid or reduce the interference between cores, for example, time division multiple access (TDMA) arbitration mechanism [8]. Rosen described a scheme of TDMA bus scheduling on multi-core computing platform [9]. All tasks are scheduled based on the off-line scheduler and the scheduler is written to the memory of bus arbiter. When the system is running, the bus access request is coordinated and controlled according to the schedule, so as to avoid bus interference. However, these studies need system customization and re-development, which induced high cost and cannot meet the needs of generalization and marketization. Especially in avionic multi-core applications, if it cannot support incremental certification, the application will be greatly restricted. As described by Corradi [10], the semiconductor industry is facing technical and economic challenges, where fewer chip suppliers can afford the investments, safety-related and avionic semiconductor market is too small for semiconductor companies to provide economically competitive safety specific solutions compliant with defined certification standards. Therefore, it is expected that multi-core devices used in safety-related and avionics systems will have to select COTS chip. This is also in line with the goal of reducing life cycle costs for avionics system. Therefore, starting with the application mode, that is, the software architecture of the operating system, is also an optional method [11]. Multi core application modes include AMP (asymmetric multi-processing) and SMP (symmetric multi-processing) [12–14].

In AMP mode, each core is configured with an independent operating system, and the blueprint is configured to manage shared resources. The operating environment of AMP is similar to the traditional single core system, so it has good portability and is easy to inherit the software of single processor. AMP only needs to select an operating system that can provide distributed programming mode to maximize the use of multi-core. Applications on one core are allowed to communicate transparently with applications or system services on another core. In AMP mode, multiple partitions are executed in parallel, and all partitions run on their corresponding cores at the same time. When the number of processors increases, the system performance can be significantly improved.

In SMP mode, there is an operating system controls the simultaneous activities of all cores. Since these tightly coupled processors access a common memory area, they must synchronize with each other through a communication mechanism based on low latency shared memory. Since there is only one operating system, local communication between different cores is completed through process communication, which greatly improves the communication efficiency between processes. At the same time, it also allows the operating system to dynamically schedule any process to run on any core. SMP provides much more flexibility and a better load balancing than that of AMP. But The SMP software layer is more complex since it needs to protect its internal data and the internal data need to be arranged very carefully by system.

Incremental authentication is also a requirement that must be considered in multi-core integration, especially in civil avionics, traditional avionics software has completed certification and the process has been proved to be effective by the airworthiness authority. In multi-core processing system, how to inherit existing certification results and avoid providing additional interference related information to certification authority are also the key balancing factors for the system user.

4 The Promote Methods and Potential Solutions

4.1 The Test Result of AMP Mode Application

In AMP mode, each core runs its operating system instance, so the resources occupied by operating system instances increased with the number of cores. Therefore, the performance proportion of dual core is doubled in theory while the resource occupation is small, especially when there is no conflict in memory access. Based on a dual core processor platform, this paper introduces AVIC operating system conforming with ARINC653 standard, hosting the spec cpu2006 test cases which tests the integer performance and floating-point performance of the processor. These cases mainly tests the real computing performance of the processor with relatively few memory accesses. The test results are shown in Table 1 and Fig. 2. The smaller the test value, the better the performance. It can be seen that the measured performance of each core in dual core is the same as that of single core activated, which means that dual core can indeed achieve double performance improvements to a certain extent. It should be noted that in F_433_milc test case, the performance of dual core is not double in all, which may related to the fact that the test case uses more shared storage resources.

Table 1. Dual core processor spec 2006 integer and floating point benchmark results

Test case		Dual core		Single core
		Core0	Core1	
I_401_bzip2	Integer test case	1050.5	1050.6	1001
I_429_mcf		429.9	429.9	384.1
I_456_hmmer		1459.7	1459.7	1452.2
I_458_sjeng		3681.7	3682.1	3451.4
I_462_libquantum		112.8	112.9	103.1
F_433_milc	Float test case	255.0	255.3	137.9
F_470_lbm		764.2	764.2	764.2
F_482_sphinx3		276.2	276.2	267.4

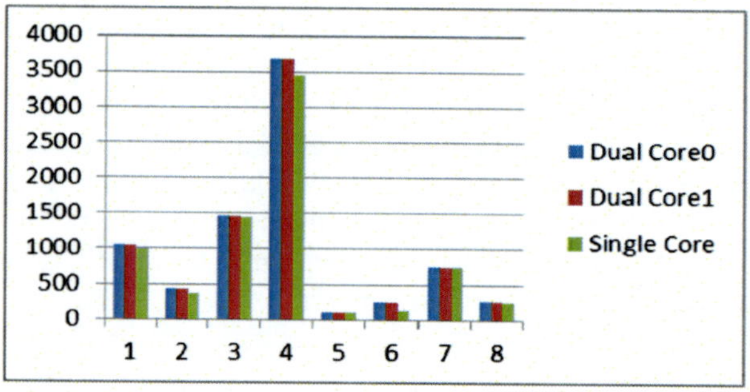

Fig. 2. Dual core processor spec 2006 integer and floating point benchmark results

4.2 Low Critical Application

Due to its good compatibility and inheritance, AMP mode is considered as the first step of multi-core processing application in avionics system. However, due to the limited number of cores and application scenarios, its performance improvement is also limited. In order to adapt more cores and significantly improve performance, a more effective way is to use SMP mode operating system with better flexibility and scalability. However, SMP mode needs to provide more additional evidence to airworthiness authority to ensure that the whole design process is approved. In fact, due to intellectual property restrictions, almost no suppliers are willing to provide detailed design documents of processor and operating system software to prove that it complies with the certification process of DO-254 or DO-178. An alternative way is to provide service experience to certification authority and prove that there is no problems happened with the operating system and multi-core processing when in service. If the highest IDAL of multi-core processing system is below C, the authentication process does not require equipment and software to conduct interference analysis, so as to avoid additional evidence collection process. Therefore, multi-core processing system can be firstly applied in the low critical aviation system to gather in-service data and problem reports as a basis for the application of mixed criticality in avionics system.

4.3 An Applicable Schedule Configuration for SMP

In fact, future IMA platform should not only meet high performance requirements, but also meet the requirements of lower criticality and critical applications running simultaneously. These applications may also be consistent with ARINC 653 standard. The operating system configuration must support exclusive access of platform resource for the most critical applications. The operating system AVIC OS is a good example that can be configured to meet the needs of mixed critical applications, from which less critical applications can get high performance and highly critical applications can get exclusive resources for determinism and isolation. Figure 1 shows a possible configuration of the AVIC OS partition scheduler.

The configuration is assumed to be based on a quad-core CPU, the major time frame is divided into three time partition windows. To avoid interference on hardware and software level, the critical application that is running on partition 3, will only execute on core '3' during 'T3' time slice. Although it seems to waste CPU time, it ensures critical application have exclusive access to a single core and have exclusive access to entire platform resource during its allocated time window, which guaranteed highest isolation (Fig. 3).

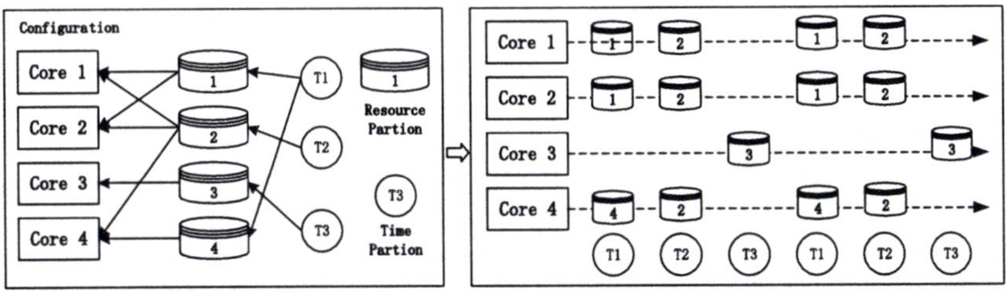

Fig. 3. A configuration using the AVIC OS Scheduler for mix-critical application

5 Conclusion

In addition to IMA platforms, the use of multi-core processing is necessary for future avionics system to meet high performance requirements. Dual-core CPUs using AMP mode are suitable for scenarios with less memory access and high computational requirements for performance improvement. Multi-core processing needs to be firstly introduced in less critical systems to gather in service experience and then be applied in critical systems with no problems. We also demonstrate a schedule configuration of multi-core platform with AVIC OS for critical applications with exclusive resources. Multi-core devices are envisioned to support the development of next-generation avionics systems.

References

1. Kopetz, H.: Real-Time Systems: Design Principles for Distributed Embedded Applications, 2nd edn. Springer-Verlag (2011)
2. J. Athavale, R.M., Paulitsch, M.: Flight safety certification implications for complex multi-core processor based avionics systems. In: Proceedings of the IEEE 25th International Symposium on On-Line Testing and Robust Syst. Design (IOLTS'19), pp. 38–39 (2019). https://doi.org/10.1109/IOLTS.2019.8854415
3. Watkins, C.B., Walter, R.: Transitioning from federated avionics architectures to integrated modular avionics. In: 26th AIAA/IEEE Digital Avionics Systems Conference (DASC) (2007)
4. Chen, G., Guan, N., Lü, M.S., Wang, Y.: State-of-the-art survey of real-time multicore system. Ruan Jian Xue Bao/J. Softw. **29**(7), 2152–2176 (in Chinese) (2018). http://www.jos.org.cn/1000-9825/5580.htm
5. Kim, H., Kandhalu, A., Rajkumar, R.: A coordinated approach for practical OS-level cache management in multi-core real-time systems. Real-Time Syst. **8114**, 80–89 (2013)
6. Lü, M.S., Guan, N., Wang, Y.: Survey of cache analysis for worst-case execution time estimation. Ruan Jian Xue Bao/J. Softw. **25**(2), 179−199 (in Chinese) (2014). http://www.jos.org.cn/1000-9825/4529.htm
7. Li, Y., Suhendra, V., Liang, Y., Mitra, T., Roychoudhury, A.: Timing analysis of concurrent programs running on shared cache multi-cores. In: Proceedings of the IEEE Real-Time Systems Symposium, pp. 56–57. IEEE, Washington (2009)
8. Kelter, T., Falk, H., Marwedel, P., Chattopadhyay, S., Roychoudhury, A.: Bus-aware multicore WCET analysis through TDMA offset bounds. Real-Time Syst. **6794**(29), 3–12 (2011)
9. Rosen, J., Andrei, A., Eles, P., Peng, Z.: Bus access optimization for predictable implementation of real-time applications on multiprocessor systems-on-chip. In: Proceedings of the 28th IEEE Int'l Real-Time Systems Symposium Tucson, pp. 48–60. IEEE (2007)
10. Corradi, G.: Tools, architectures and trends on industrial all programmable heterogeneous MPSoC (KeyNote). In: Proceedings of the 29th Euromicro Conference on Real-Time Systems (ECRTS'17) (2017)
11. Fuchsen.: How to address certification for multi-core based IMA platforms: current status and potential solutions. IEEE. IEEE (2010)
12. Agrou, H., Sainrat, P., Gatti, M, et al.: Mastering the behavior of multi-core system to match avionics requirements. In: Proceedings of the 31st Digital Avionics Systems Conference, pp. 6E5-1–6E5-12. IEEE/AIAA, Williamsburg (2012)

13. Nowotsch, J., Paulitsch, M.: Leveraging multi-core computing architectures in avionics. In: Proceedings of the 9th European Dependable Computing Conference, pp. 132–143. Sibiu IEEE (2012)
14. Crespo, A., Masnano, M., Coronel, J., et al.: Multi-core partitioned systems based on hypervisor. In: Cape Town. Proceedings of the 19th IFAC World Congress, pp. 12293–12298. IFCA (2014)

Research on Hierarchical Control Strategy of Aircraft Power Supply System

Liu Rui(✉), Ai Fengming, Wang He, Liu Jiao, and Wang Guan

Shenyang Aircraft Design & Research Institute, Liao Ning Shen Yang 110000, China
Liur_sy@163.com

Abstract. With the iteration and improvement of the combat capability of fighter jets, the power of airborne electrical equipment has been continuously improved. At the same time, various advanced high-power equipment has been equipped with new fighter jets, making the electrical load power of the aircraft continue to increase. Aircraft power supply systems have many problems by using traditional control strategies such as low system efficiency and high energy consumption when faced with high-power loads. And their limitations are becoming more and more obvious. In this paper, based on the F-35 fighter power supply architecture and the typical airborne power load profile, the hierarchical control method is used to optimize the power supply system control strategy. The important indicators, such as average power, energy consumption, average efficiency and heat dissipation of the aircraft power supply system are greatly optimized, which greatly eases the pressure on the onboard power supply.

Keywords: Hierarchical control strategy · Power supply system · High power load · Particle swarm optimization

1 Preface

Since 21st century, fighter jets have achieved the leap from the second and third generation aircraft to the fourth generation aircraft, and are moving towards a higher level. The continuous evolution of strategic combat requirements has led to the development of fighter jets.

One of the key technical points in the design of fighter aircraft is the power supply of the airborne power supply system and the heat dissipation of the whole aircraft. In the current fighter aircraft, the power is relatively low, and the energy requirements and heat dissipation requirements can also be satisfied under the current traditional design scheme. The use of high-power electrical equipment puts forward higher requirements on the power supply structure and heat dissipation capacity of the aircraft, while the traditional structure and its control strategy make the generator set work in the low-efficiency range for a long time. The overall efficiency is low which also means serious energy loss, and it increases the heat dissipation burden of airborne equipment. Therefore, the control strategy of the power supply system must be developed from uncontrolled to multi-source coordinated control to meet the bottleneck problems of power supply and heat dissipation caused by the above load changes, and to optimize the overall energy architecture within the boundary conditions of the airborne environment.

Chinese Society of Aeronautics and Astronautics: CASTYSF 2022, LNEE 972, pp. 420–432, 2023.
https://doi.org/10.1007/978-981-19-7652-0_39

2 Airborne Power Supply Architecture and Hierarchical Control Strategy

2.1 Aircraft Power Supply System

The early aircrafts were mainly based on AC parallel architecture [1], and the civil aviation passenger aircrafts were mainly based on AC power supply architecture. Boeing 787 mainly uses 235V and 115V alternating current for power supply, and then converts it into corresponding direct current through power electronic converter [2, 3].

Later, with the iteration of aircraft requirements, the problem of AC architecture gradually emerged [4], so AC power generation system is gradually replaced by high-voltage DC power supply system [5]. The DC power supply architecture built by scholars from the University of Nottingham has studied generator grid-connected characteristics and bus voltage characteristics of the DC power supply system in detail [6, 7]. At present, the power supply structure of the fourth-generation fighter jets in the world is dominated by high-voltage DC power supply systems. The earliest high-voltage DC power supply system design scheme has been fully tested in the United States in the F-22 and F- 35 [8]. Among them, the F-22 adopts dual independent channel 65 kW high-voltage DC generators to supply power to the whole machine load. The architecture also includes a variety of power electronic converters to convert the high-voltage DC into the required low-voltage DC.

Figure 1 is the architecture diagram of the power supply system of the F-35 fighter. The power system consists of two subsystems: the Electrical Power Generation System (EPGS) [9] and the Electrical Power Management System (EPMS).

The power generation system consists of a starter/generator (ES/G) which provides two independent 80 kW power outputs during normal operation. Two 160A rated voltage converters provide 28V DC for low power loads and critical loads. A 270V DC-115V AC inverter supplies the aircraft with 5.4 kVA of current (115V AC) for the F-35C's weapon bay and folding wings.

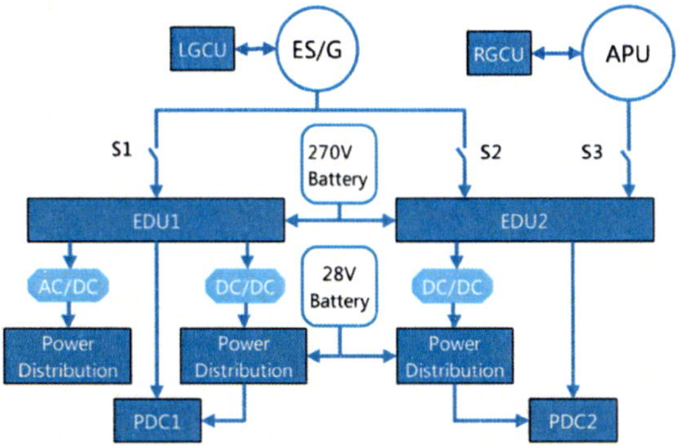

Fig. 1. Schematic diagram of the power supply system architecture of the American F-35 fighter jet

The 270V DC battery system consists of a Li-Ion battery and a charger/controller. The system provides auxiliary power for 270V flight critical loads and provides power for power and thermal management system self-starting.

The simulation of the power supply architecture requires the flight profile load usage. This paper sets a typical combat load profile with a high-power load as the core. This profile reflects the load usage of the airborne system in Figure below. The simulation and analysis in this chapter are based on this flight profile (Fig. 2).

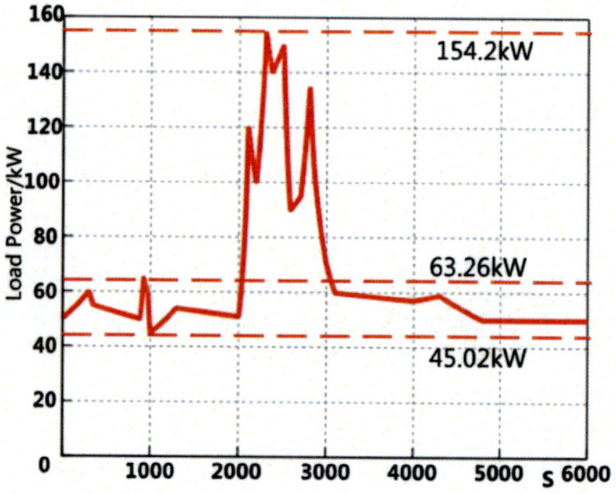

Fig. 2. Load variation curve of future fighter flight profile

The flight time of this mission profile is 6000 s. The peak power is 154.2 kW. The minimum power is about 45.02 kW and the average power is 63.26 kW.

2.2 Control Strategy Development and Hierarchical Control Strategy

At present, energy control strategies are mostly used in the hybrid energy power supply architecture of vehicles. The control strategies of the power supply architecture are mainly divided into two categories according to their characteristics. One is the rule-based energy management strategy, and the other is based on the optimization algorithm. Energy management strategy [10] and different control strategies have different scope of use and advantages and disadvantages. Rule-based energy management strategies include fixed rule control strategy, frequency division control strategy, fuzzy logic control strategy and predicted neural network control strategy. Energy management control strategies based on optimization algorithms include dynamic programming control strategies, transient optimization control strategies, Pontryagin's principle control strategies, swarm algorithm control strategies, and model predictive control strategies. Since the current sharing control strategy is mostly used for systerm control (hereinafter referred to as the "original strategy").

Hierarchical control strategy is a control scheme that is widely used in microgrids [11]. Scholars at home and abroad have done a lot of research and improvement work in power grid control optimization [12, 13], and continuously optimize the bus voltage operation quality from the control structure and control strategy [14, 15], so that the power grid maintains stable in different states. There are no specific requirements for the multi-layer control of the hierarchical control strategy. The hierarchical control strategy includes two to three control levels. Usually, the bottom layer is a component-level control layer, and the upper layer is a system-level control layer. There are also some layered control strategies in which the bottom layer is responsible for function implementation, such as droop control, the current sharing control mentioned above, and the master-slave control [16] method and so on. The upper layer test optimizes the algorithm of the underlying control strategy to make the underlying strategy more stable and responsive to run [17, 18].

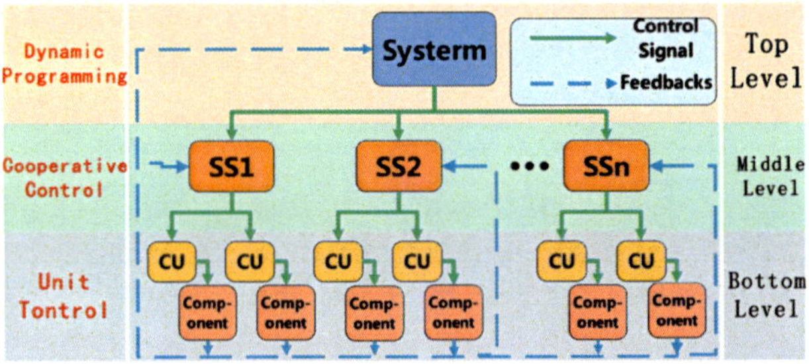

Fig. 3. Hierarchical control strategy of airborne power supply system

In this paper, the hierarchical control strategy controls power supply components such as generators and lithium batteries. The hierarchical control strategy is shown in Fig. 3. According to the system level, the control strategy is divided into system-level control strategy, subsystem-level control strategy and component-level control strategy. The component-level control strategy mainly analyzes the control strategy for the generator and lithium battery models, and ensures the stability of the bus power supply based on the combined power supply architecture of the generator and the lithium battery. The subsystem-level control strategy mainly studies the grid-connection of components, which mainly involves the grid-connection and power distribution of generators and the grid-connection and power distribution of DC/DC converters. Finally, the responsibility of the top layer, that is, the system-level control strategy, is to schedule the energy supply of the entire architecture. Based on the optimal efficiency of the entire architecture, the corresponding algorithm is used to provide the subsystem-level control strategy with specific instructions for grid-connected power distribution.

3 Top-Level Control Strategy Based on Particle Swarm Optimization

The top-level control strategy of the hierarchical control strategy mainly realizes the whole-system dynamic planning and overall decision-making, and guides the system in which state to run on the basis of the middle-level power distribution control strategy. Therefore, the core of the outermost control strategy is to perform calculation and optimization according to the operating state of the system, so that the system can run optimally based on a specific index or specific index. For the power supply architecture, system energy consumption or system operating efficiency is an important indicator when the parameters of the system components are determined.

Particle Swarm Optimization (PSO) [19] was proposed in 1995, and scholars at home and abroad have done a lot of research and improvement work on it.

The expression of the basic particle swarm algorithm is as follows:

$$v_{i,d}^{T+1} = v_{i,d}^{T} + c_1 r_1 (p_{ibest,d} - x_{i,d}^{T}) + c_2 r_2 (g_{best,d} - x_{i,d}^{T}) \tag{1}$$

$$x_{i,d}^{T+1} = x_{i,d}^{T} + v_{i,d}^{T+1} \tag{2}$$

in, $d = 1, \cdots, n$; x_i^T is the position of the i-th particle in the T-th iteration; v_i^T is the velocity of the i-th particle in the T-th iteration.; c_1 is the self-perception coefficient; c_2 is the social cognition coefficient; r_1 or r_2 is a random number from 0 to 1; p_{ibest} is the individual optimal position of the i-th particle in the T iterations; g_{best} is the global optimal position obtained by all particles in T iterations.

The top-level strategy uses particle swarm optimization based on the fact that the generator efficiency changes with the output power of the generator, which makes it possible to optimize based on the system efficiency. The abscissa in Fig. 4 is the ratio of the generator output power to the rated power of the generator, dimensionless; the ordinate is the generator efficiency, dimensionless. It can be seen from the generator efficiency curve that the generator has the highest efficiency under the rated power. And the generator efficiency on both sides of the rated point shows a downward trend. When the output power of the generator is close to 0, the efficiency of the generator is extremely small. With the increase of the output power, the efficiency of the generator increases rapidly. When the ratio of the output power of the generator to the rated power of the generator increases from 0 to about 0.15, the generator efficiency rises from a very low value to more than 60%. After that, with the increase of the output power of the generator, the increase of the generator efficiency gradually slows down to the highest level, and the peak value of the generator efficiency at this stage is generally about 90%. Then, as the degree of overload of the generator increases, the generator efficiency gradually decreases. In addition, when the output power of the generator is 0, the input power of the generator is not 0. At the same time, the generator still consumes energy although it does not supply power to loads. In the architecture design of this paper, a high-speed clutch device is used at the front end of the generator, which can be controlled to cut off when the output power of the generator is 0 in a necessary stage, so that the input power of the generator is reduced to 0 and energy consumption is reduced. At the same time, it can be controlled and connected quickly when generator power is required.

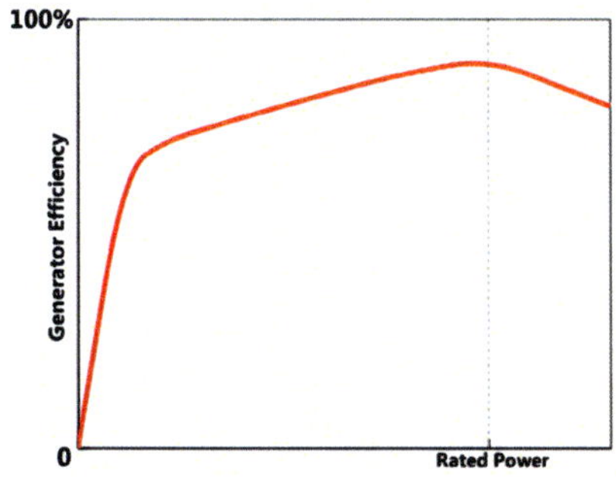

Fig. 4. Generator efficiency curve

Taking two permanent magnet synchronous generators with a rated power of 100 kW as an example, the particle swarm optimization simulation calculation is carried out. It is assumed that two generators need to provide electricity to a load of 130 kW at a certain time, and the optimization calculation is carried out through the particle swarm algorithm. In the quantum particle swarm algorithm of this paper, ten particles are initialized and the velocity and position information are randomly generated. In the constraints, the maximum power of the generator cannot exceed the load power and the generator cannot be overloaded by more than 2 times. Since the generator set uses two permanent magnet synchronous generators with the same rated power, the efficiency curves of the two generators are the same and are calculated using the line type in the above figure.

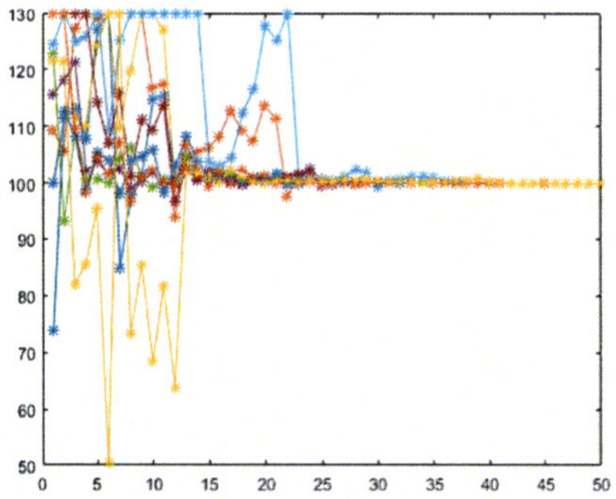

Fig. 5. Generator 1 particle swarm optimization process

Figure 5 shows the change curve of particle swarm when calculating the optimal power of generator 1 in the generator set. The particle position parameter cannot be higher than the injected load power value and greater than 0. The particles tend to aggregate after about 15 iterations, and the desired optimal solution is basically obtained after 35 iterations. The final power distribution particle swarm optimization result is that when the generator set supplies power to the load of 130 kW, the two generators output 100 and 30 kW of energy respectively, which can optimize the efficiency of the system.

4 Middle-Level Control Strategy and Bottom-Level Control Strategy

4.1 Middle-Level Control Strategy Based on Droop Control Strategy

In the power grid, the bus voltage is usually in a fluctuating state, and the fighter usually adopts a bus voltage 270V power supply scheme, so a control strategy based on the bus voltage is adopted. At the same time, in the case of multi-source grid connection, it is necessary to control the output power of each source. At present, the droop control method is relatively mature in practical application. The droop control mainly realizes the coordinated control of the parallel converter by introducing the droop coefficient and adjusting the output voltage reference value according to the magnitude of the output current.

When the droop control method is used to control the parallel system of DC/DC converters, the deviation of the output voltage will be caused. Therefore, the output voltage compensation control should be adopted in the secondary control to overcome the influence of the traditional droop control on the output voltage.

The mathematical model of droop control with output voltage compensation control is:

$$v_{refi} = v_{ref}^{*} - R_{di} i_{oi} + \left(k_{pv} + \frac{k_{iv}}{s}\right)\left(v_{ref}^{*} - v_{o}\right) \quad (3)$$

where v_{refi} is the actual output voltage reference value of the converter i; v_{ref}^{*} is the initial reference value of the output voltage; v_o is the actual bus voltage; k_{pv} and k_{iv} are the output voltage compensation PI adjustment parameters.

The droop control expression using voltage and current secondary compensation is:

$$v_{refi} = v_{refi}^{*} - \left(R_{di}^{*} - \left(k_{pi} + \frac{k_{ii}}{s}\right)\left(i_{oi} - \bar{i}_{o}\right)\right) i_{oi} + \left(k_{pv} + \frac{k_{iv}}{s}\right)\left(v_{ref}^{*} - v_{o}\right) \quad (4)$$

Comparing Eqs. (3) and (4), it can be found that the improved droop control is similar to the traditional droop control method. Because they both change the output voltage reference value of the converter to realize the coordinated control of the converter. However, different from the traditional droop control, when the output voltage reference value is given, the problem of output current unbalance and output voltage drop is considered in the improved droop control, and the output voltage command value of the basic converter unit is adjusted according to the unbalance degree of output voltage and the degree of bus voltage drop, so as to achieve Accurate distribution of output current and constant output voltage in parallel across the full range. The droop control simulation model of a single converter in the grid is shown in Fig. 6.

The lithium battery is connected to a DC/DC bidirectional converter, and then the power distribution can also be achieved by supplying such two groups in parallel. First, two lithium batteries are supplied with equal power, and the control strategy is changed in the first 0.5 s. The output power of the lithium battery pack supplies power to the load at a ratio of 2:1. The simulation results are shown in the Fig. 7 Simulation model output power variation curve. The functionality and stability of the droop control are both Guaranteed.

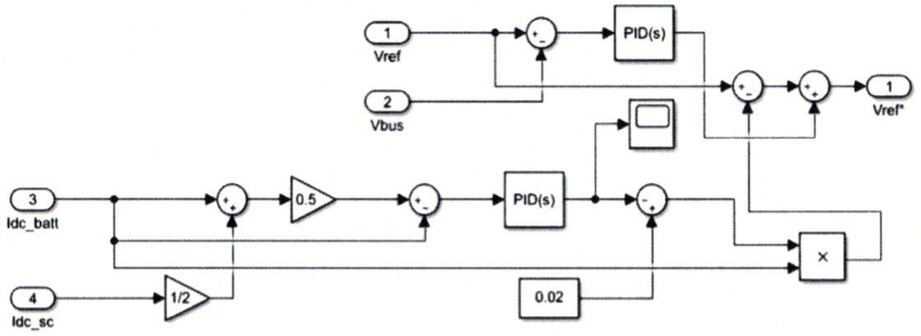

Fig. 6. Droop control simulation model

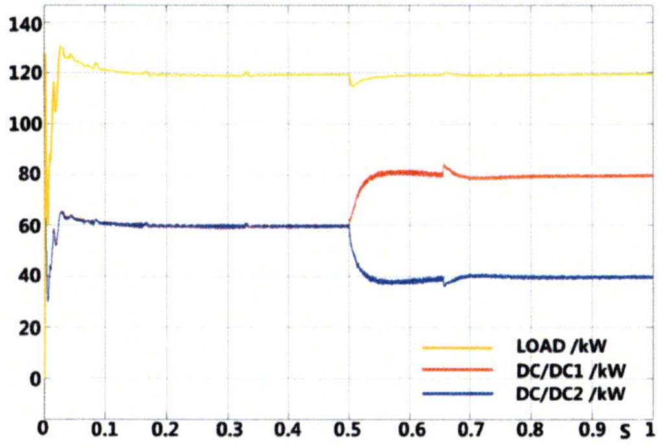

Fig. 7. Simulation model output power variation curve

4.2 The Underlying Control Strategy

The underlying control strategy mainly involves the control strategy of components such as generators and lithium batteries. The generator is generally equipped with a generator controller. For the simulation model, the dual-loop control strategy is a classic control strategy. Regardless of the control strategy used, the goal of the underlying control strategy is to make the generator work stably according to demand.

The underlying control strategy also includes a control strategy for lithium batteries. In the underlying control strategy, this paper set a threshold for the lithium battery. When the load power exceeds the threshold, the lithium battery discharges to reduce the power supply pressure of the system. When the system power is lower than a certain threshold or the SOC of the lithium battery is lower than a certain value, the lithium battery switches to the charging mode. Through the control of the lithium battery and the generator, the correct operation of each component in the system is realized, and the application of the upper-layer strategy is guaranteed.

5 Simulation Experiment and Result Data Analysis

According to the simulation load profile and layered control strategy set in the previous section, the simulation analysis is carried out based on the power supply architecture of the F-35 fighter. The simulation in this paper uses SimulationX software, and the simulation model is shown in Fig. 8.

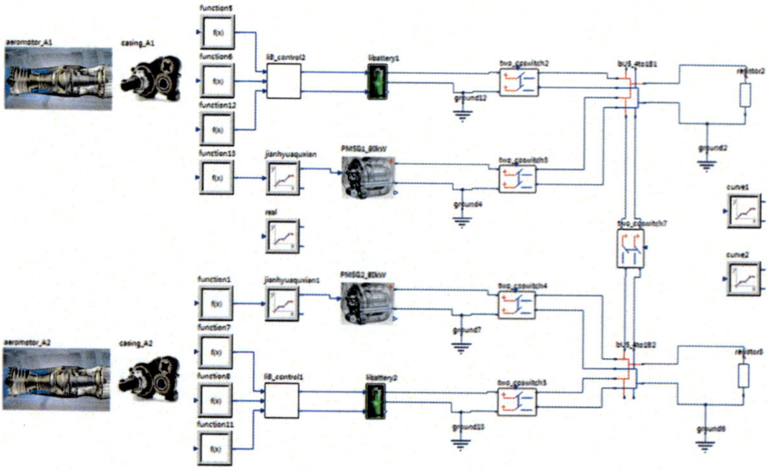

Fig. 8. Power supply system and its strategy simulation model

Through simulation and data analysis, the following results are obtained.

When the load is lower than a certain value, one of the generators in the generator set stops working (implemented by a special clutch device), and the other generator completely supplies power to the load according to the optimal solution criterion for system efficiency. The output power of the two generators is shown in Fig. 9. After the top-level control strategy calculates the optimal power distribution solution, the middle-level and bottom-level control strategies are implemented. It can be clearly seen in Figure that the output power of the two generators is no longer evenly distributed which is different from traditional architecture, whether in a low load power state or a high load power state. They are controlled in real time according to the load state. Regulation, in order to achieve the optimization of system energy consumption.

Figure 10 is a comparison of the efficiency curve of the system efficiency of the basic architecture and the system efficiency of the layered control architecture. The layered control strategy greatly improves the average efficiency of the system compared with the original strategy, from 76.82% to 84.84%. The improvement of efficiency means the reduction of energy consumption. In addition, it can be seen from Figure that the efficiency of the layered control architecture is relatively stable during the operation process, and the system efficiency is adjusted within a certain range.

The energy consumption comparison curves of the two architectures and loads are shown in Figure below. Both architectures use the same load profile, which has a total

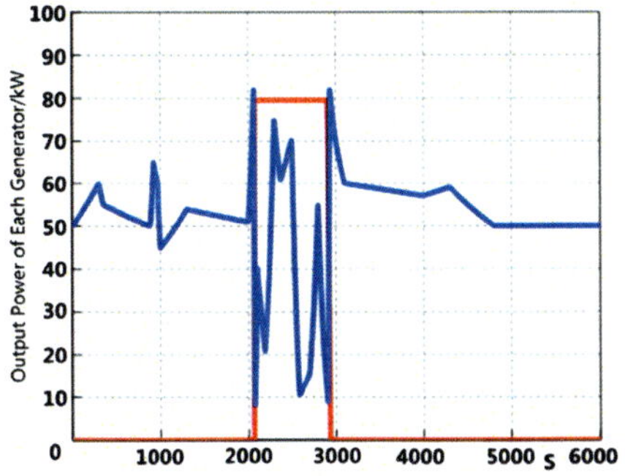

Fig. 9. The output power change curve of the two generators in the left generator set

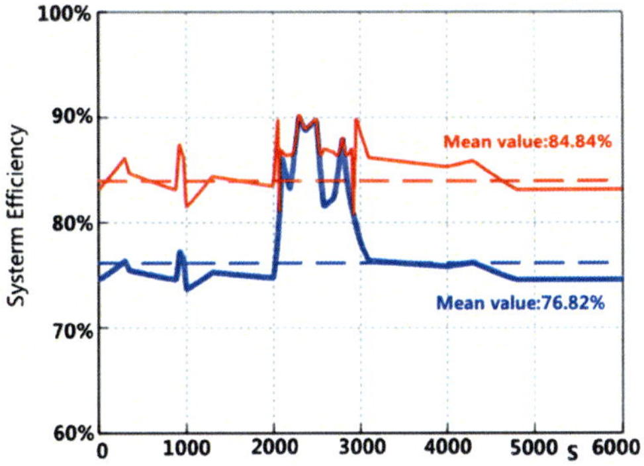

Fig. 10. Comparison of the efficiency change curves of the two-architecture systems

energy consumption of 379.55MJ. The energy consumption of the system under the original strategy and the hierarchical control strategy is 486.85MJ and 444.90MJ, respectively, and the energy consumption efficiency of the system is 77.96% and 85.31%, respectively. In this simulation, compared with the original strategy, the hierarchical control strategy achieves a reduction of more than 8% in system energy consumption (Fig. 11).

Table 1 summarizes the key parameters of the system simulation under the two strategies as follows.

The system load under the two strategies consumes 379.55MJ of energy, the energy consumption of the system using the original strategy is 486.85MJ, and the energy consumption efficiency is 77.96%; Compared with the conventional architecture, the data energy consumption is reduced by 8.62%, and the energy consumption efficiency is increased by 9.43%. From the point of view of power supply, the layered control strategy achieves a substantial optimization of the energy supply system of the system, and the average efficiency of the system is 10.44% higher than that of the system using the original strategy.

The reduction of the energy consumption of the whole machine not only means saving electric energy, weight and space cost, and improving the power supply capacity.

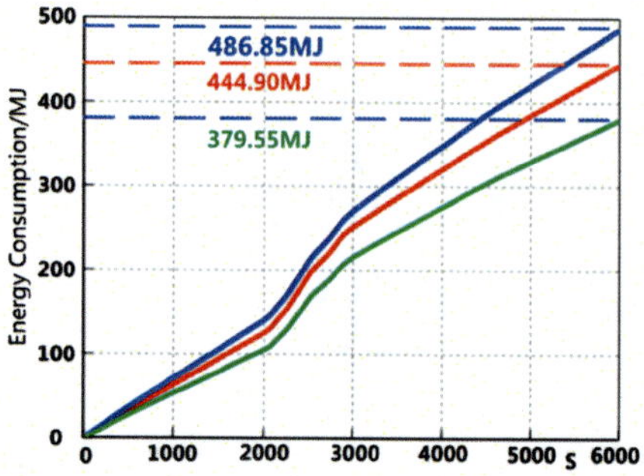

Fig. 11. The energy consumption curve of the two-type power supply system architecture

Table 1. Comparison of simulation results between two power supply architectures and control strategies

	Original strategy	Hierarchical control strategy
Total load energy consumption/MJ	379.55	379.55
Total system energy consumption/MJ	486.85	444.90
–	Benchmark	↓ 8.62%
System Energy Efficiency	77.96%	85.31%
—	Benchmark	↑ 9.43%
System calorific value/MJ	107.30	65.35
–	Benchmark	↓ 39.10%
Average efficiency	76.82%	84.84%
–	Benchmark	↑ 10.44%

Considering the full-function thermal system, the reduction of the energy consumption of the whole machine also means the reduction of the heat dissipation pressure of the whole machine. We assume that the energy dissipated by each component and unit in the power supply architecture due to the efficiency problem is radiated in the form of thermal energy, then it can be obtained by calculation that the calorific value of the system using the original strategy is 379.55MJ, and the calorific value of the system using the hierarchical control strategy is 444.90MJ. The heat is 65.35MJ, and the system heat production is reduced by 39.10%, which is a huge optimization for the airborne environment with extremely strict temperature control requirements, which greatly reduces the pressure on the airborne heat sink. It may bring about secondary energy and weight saving and from the system level.

6 Conclusion

Based on the use of high-power equipment on fighter aircraft, this paper adopts the layered control strategy of the power supply architecture to solve the problems of insufficient power supply capacity and low efficiency of the next-generation aircraft power supply architecture. The layered control strategy is studied and simulated in detail. The main conclusions are as follows:

(1) The layered control strategy is realized based on the characteristics of the relationship between the generator efficiency and its output power.
(2) The use of the hierarchical control strategy can optimize the system energy compared with the conventional current sharing control strategy. And the parameters such as system efficiency and energy consumption can be greatly optimized in the case of a unified load profile in the same architecture.
(3) The layered control strategy can reduce the heat dissipation of the system, so as to ease the design and operation pressure of the airborne heat dissipation system, and greatly reduce the airborne heat dissipation problem.

References

1. Larson, L.R.: Parallel operation of aircraft AC generators. Electr. Eng. **72**(11), 1021 (1953)
2. Wang, H., Sun, L.: Standard analysis of power supply characteristics of Boeing 787 aircraft. Aviat. Standard. Quality (05), 46–49 (2014). https://doi.org/10.13237/j.cnki.asq.2014.05.013
3. Whyattg, A., Chick, L.A.: Electrical generation for more-electric aircraft using solid oxide fuel cells. Oak Ridge: Office of Scientific and Technical Information (OS-TI) (2012)
4. Zhang, Z., Xu, Y., Yu, L., Li, J., Xia, Y.: Development status and key technologies of high-voltage DC parallel power supply system for multi-electric aircraft. Acta Aeronaut. Sinica **42**(06), 12–25 (2021)
5. Muehlbauer, K., Gerling, D.: Two-generator-conceptsor electric power generation in more electric aircraft engine. In: The XIX International Conference on Electrical Machines-ICEM 2010, pp. 1–5. IEEE Press, Piscataway
6. Gao, F., Bozhko, S., Costabeber, A., et al.: Control design and voltage stability analysis of a droop-controlled electrical power system for more electric aircraft. IEEE Trans. Ind. Electron. **64**(12), 9271–9281 (2017)
7. Liu, F.P., Xu, L., Li, Y.D., et al.: Permanent magnet synchronous machine starter/generators based high-voltage DC parallel electric power system for the more electric aircraft. J. Eng. **13**, 565–569 (2018)
8. Robbins, D., Bobalik, J., De Stena, D., et al.: F-35 subsystems design, development & verification. In: 2018 Aviation Technology, Integration, and Operations Conference, pp. 1–23. AIAA, Reston
9. Schettini, F., Denti, E., Di Rito, G.: Development of a simulation platform of all-electric aircraft on-board systems for energy management studies. Aeronaut. J. **121**(1239), 710–719 (2017)
10. Sankarkumar, R.S., Natarajan, R.: Energy management techniques and topologies suitable for hybrid energy storage system powered electric vehicles: an overview. Int. Trans. Electr. Energy Syst. **31**, e12819 (2021)

11. Bidram, A., Davoudi, A.: Hierarchical structure of microgrids control system. IEEE Trans. Smart Grid **3**(4), 1963–1976 (2012)
12. Feng, W., Sun, K., Guan, Y., Josep, M.G., Xiao, X.: An active suppression control strategy for grid-connected harmonic currents of microgrid based on hierarchical control. J. Electrotech. Technol. **33**(06), 1400–1409 (2018). https://doi.org/10.19595/j.cnki.1000-6753.tces.170691
13. He, J., Li, Y.W., Blaabjerg, F.: An enhanced islanding microgrid reactive power, imbalance power, and harmonic power sharing scheme. IEEE Trans. Power Electron. **30**(6), 3389–3401 (2014)
14. Savaghebi, M., Jalilian, A., Vasquez, J.C.: Secondary control scheme for voltage unbalance compensation in an islanded droop-controlled microgrid. IEEE Trans. Smart Grid **3**(2), 797–807 (2012)
15. Quanwei, L., Yong, T., Xunhao, L., et al.: Voltage unbalance and harmonics compensation for islanded microgrid inverters. IET Power Electron. **7**(5), 1055–1063 (2014)
16. Rajagopalan, J., Xing, K., Guo, Y., et al.: Modeling and dynamic analysis of paralleled dc/dc converters with master-slave current sharing control. In: Eleventh Annual IEEE Applied Power Electronics Conference and Exposition, pp. 678–684. IEEE, San Jose, CA, USA (1996)
17. He, J., Li, Y.W., Blaabjerg, F.: Flexible microgrid power quality enhancement using adaptive hybrid voltage and current controller. IEEE Trans. Ind. Electron. **61**(6), 2784–2794 (2013)
18. Chen, J., Zeng Guohong, W., Xuezhi, Y.J.: Hierarchical control strategy of bipolar DC microgrid. Power Grid Technol. **39**(08), 2217–2222 (2015). https://doi.org/10.13335/j.1000-3673.pst.2015.08.023
19. Kennedy, J., Eberhart, R.: Particle swarm optimization. In: Proceedings of ICNN'95-International Conference on Neural Networks (1995)

Research on Electromagnetic Compatibility Technology Based on Airborne Electronic Equipment

Zhi Qiang, Han Yaguo(✉), Wang Lu, Luo Kai, and Yang Yuan

AVIC Xi'an Aviation Brake Technology Co., Ltd, Xi'an, Shaanxi 710075, China
843129227@qq.com

Abstract. As more and more types and quantities of airborne electronic equipment are used, the frequency band becomes more and more wide, and the radiation power becomes more and more large, which makes the electromagnetic environment where airborne electromagnetic equipment work become increasingly complicated. The electromagnetic compatibility of airborne electronic equipments should be well designed so that they can work normally in complex electromagnetic environment. In this paper, the basic concept of electromagnetic compatibility (EMC) is pointed out first of all, the fundamental principle of EMC is then analyzed, and finally the three elements of electromagnetic interference (EMI) and the basic methods of EMC design are put forward, which provides guidance for design of the follow-up airborne electromagnetic equipments.

Keyword: Airborne electronic equipment · Electromagnetic compatibility · Design objectives

1 Summary

As informatization progresses in modern society, electrical and electronic equipments are widely used and increase rapidly in types and quantities. The trend for future electrical and electronic equipments is digitalization, miniaturization and integration. However, when these equipments work, electromagnetic energy will always be produced, which makes various electromagnetic waves available everywhere and every time. These electromagnetic energies make surrounding electromagnetic environment more complicated, causing electromagnetic interference (EMI) and electromagnetic radiation between high frequency communication equipments become increasingly serious. Electromagnetic wave now lives up to its reputation as a kind of environment pollution. However, the working environment of airborne electronic equipment is sensitive and special, which is characterized by the following. Many airborne electronic equipments of different functions are placed in a small space where cables are installed in a centralized manner; frequency band is wide, transmit power is large; receiver is highly sensitive and can

Zhi Qiang (1984)—male, M.S. degree, senior engineer.

Chinese Society of Aeronautics and Astronautics: CASTYSF 2022, LNEE 972, pp. 433–440, 2023.
https://doi.org/10.1007/978-981-19-7652-0_40

be interfered by many approaches. Electromagnetic interference deteriorates the performance of equipments and systems or causes damage to them, which have a severe impact on their reliability and performance. Consequently, it is an important issue for design of airborne electronic equipments and systems to improve their viability in complicated environment.

Electromagnetic compatibility is also referred to as EMC. The purpose of safety requirements for EMC is to make the signal amplitude generated by the airborne electronic equipments designed or produced by ourselves meets the requirements of others. And at the same time, the airborne electronic equipments designed or produced by us should work normally in the event of EMI caused by other airborne electronic equipments. As a result, EMC is part of airborne electronic equipment safety. One of the important researches related to military equipments conducted by developed countries is to improve the anti-electromagnetic interference resistance and electromagnetic compatibility of airborne electronic equipments and promote the defense capability by making equipments, systems and subsystems coexist for the same objective under the electromagnetic conditions of battlefield. Complete and accurate EMC analysis on airborne electronic equipments, cables as well as harnesses can provide important references for addressing the EMC problem of airborne electronic equipment during model research.

2 Standard Requirements

EMC problems were taken seriously step by step since the foundation of SAC/TC 79(Standardization Administration of the People's Republic of China/ Technical Committee) in 1968 and the issue of military standard GJB151-86. In 1997, national military standards on EMC, i.e. GJB151A-97 Electromagnetic Emission and Susceptibility Requirements for Military Equipment and Subsystems, GJB152A-97 Measurement of Electromagnetic Emission and Susceptibility for Military Equipment and Subsystems as well as other confidential committee standards were issued and enforced, EMC technologies made enormous progress.

In December 2001, Administration of Quality Supervision, Inspection and Quarantine (AQSIQ) announced Administrative Regulations for Certification of China Compulsory Products, replacing the previous Licensing System for Safety and Quality of Exported Products and Certification System for Safety of Electrical Products by compulsory certification system. China Compulsory Products, abbreviated with "CCC", is referred to 3C certification. 3C certification stipulates the safety, EMC, anti-electromagnetic radiation and other performance of electronic products in detail. Related departments of China attach creasing importance to EMC in recent years. Many units and colleges have set up or transformed EMC labs with various EMC measurement and test capabilities.

Normally, aeronautical (air force) products need going through the following EMC test items as seen in Table 1.

Table 1. EMC inspection items for aeronautical (air force) products

No	EMC inspection items
1	CE102 10kHz ~ 10MHz conducted emission of power leads
2	CE106 1MHz ~ 18GHz conducted emission of antenna terminal
3	CE107 conducted emission of power lead spike signal (time domain)
4	CS101 25 Hz ~ 50 kHz conducted susceptibility of power leads
5	CS106 conducted susceptibility of power lead spike signal
6	CS114 10kHz ~ 400MHz injection conducted susceptibility of cable bundle
7	CS115 conducted susceptibility of injection pulse excitation of cable bundle
8	CS116 10kHz ~ 100MHz conducted susceptibility of damped sinusoidal transients of cable and power leads
9	RE102 2MHz ~ 18GHz radiated emission of electric field
10	RS103 30MHz ~ 18GHz radiated susceptibility of electric field

3 Electromagnetic Compatibility Implementation and Design Principles

With the widely use of electronic equipments and information—based equipments, the electromagnetic environment in battlefield includes the signal environment of communication, radar, computer, photoelectric equipments as well as the signal environment of electronic countermeasures. The sources of electromagnetic radiation and electromagnetic signals in modern battlefield become highly intensive and their types become more complex, which make it difficult to precisely grasp the electromagnetic environment. The electromagnetic environment, which produces an effect on war process, has become an important part in battlefield.

Any EMI effect must have 3 fundamental conditions, namely, interference source, coupling path and sensitive equipment (Fig. 1).

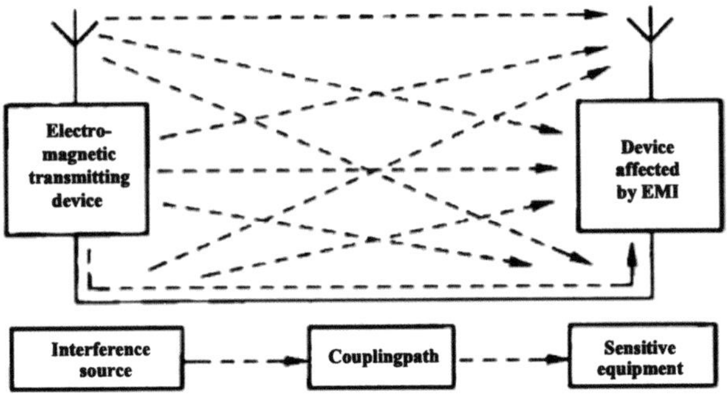

Fig. 1. F3undamental elements that cause EMI problem

Interference source: an object which can produce electromagnetic energy. The interference source can generate radiated electromagnetic interference having a certain strength, no matter the electromagnetic interference is intentionally or unintentionally emitted.

Coupling path: a channel or medium through which the energy of EMI is transmitted to sensitive equipment. Based on the transmission approach of interference source, coupling path is normally divided into radiative coupling and conductive coupling. It is through coupling path that the radiated electromagnetic energy transmits and furthermore has an influence on the normal operation of airborne electronic equipments (Fig. 2).

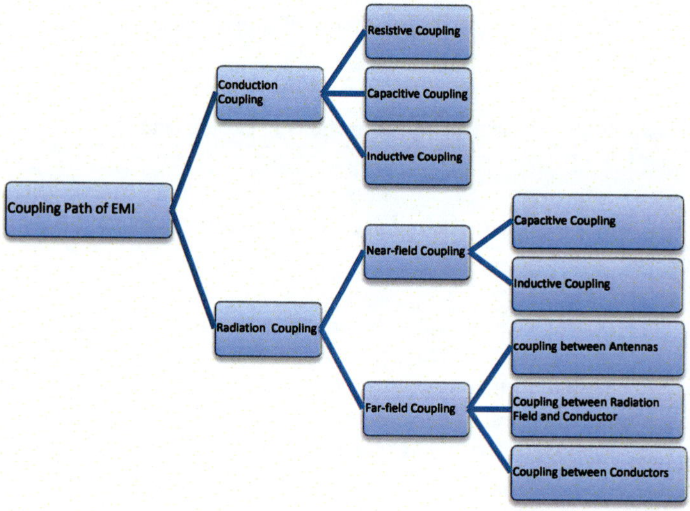

Fig. 2. Classification of coupling path of EMI

Sensitive equipment: The production of an EMI must have equipments that are susceptive to EMI. This kind of equipment is called sensitive equipment. Upon receiving the EMI signal through the coupling path, the sensitive equipment responds to the signal. The response normally manifests as performance degradation or loss of function. Sensitive equipment is the collective name for interfered objects which may be a component, an equipment or a set of a system.

For the 3 elements of EMI (interference source, coupling path and sensitive equipment), efforts can be made during EMC design by suppressing interference source, cutting the coupling path and increasing the immunity of sensitive equipment.

(a) Suppressing interference source: Suppress the electromagnetic energy of useless interference source. Analyze the electromagnetic environment where the airborne electronic equipment or system is located and relevant requirements. Select the correct design direction. Take a full account the electromagnetic environment of airborne electronic equipment during design stage.

(b) Cutting the coupling path. Increase the isolation between sensitive equipments. Meticulously select the frequency used by the airborne electronic equipment. Develop an EMC requirement and control plan.
(c) Increasing the immunity of sensitive equipment. Apply rational interference suppression and protection technologies to components, modules and circuits.

For the 3 elements of EMI, EMC control on one element is closely connected with the other two elements. For example, when good control is achieved in radiation energy of interference source, the requirement for isolation of sensitive equipments can be lowered. Once the isolation from sensitive is large enough, the requirement for suppressing electromagnetic energy of interference source can be lowered. In actual engineering, EMC design is concurrently performed in interference source, coupling path and sensitive equipment. Implementation of comprehensive and rational strategies and technical measures for EMI suppression is the prerequisite for EMC assurance.

4 EMC Design

EMC design can be carried out for airborne electronic equipments according to the following aspects based on the characteristics of equipments.

Firstly, conduct functional design. Select the components, devices and recruits which have low interference on each other. Make rational layout. On the basis that functional requirements are met, modify parameters including transmit power, operation frequency, receiving sensitivity and reselect components, in order to make EMC parameters meet standard requirements.

Secondly, conduct protective design. Design includes filtering, shielding, grounding and bonding. Time isolation, space isolation, frequency avoidance and other technical measures should also be taken. The purpose is to reduce the interference level and increase the attenuation of interference along the coupling path. Filtering is the most direct and most efficient method to suppress conducted interference. Good filtering is beneficial to resolve radiation interference problems. Shielding is an efficient approach in suppressing radiation interference. Shielding also includes relocation far from sensitive equipments and cutting interference path. It should be noted that shielding, only when used in conjunction with grounding, can play a role. Rational grounding is the most economical and most efficient EMC design technology.

When AC 115V and DC 28V are used at the same time in airborne electronic equipment, AC ground wire and DC ground wire should be designed separately for this system with hybrid power supply. Moreover, the two ground wires should be isolated reliably. The purpose is to reduce the interference of coupling between ground wires.

Thirdly, conduct layout design, including inspection of overall layout, cable layout and distribution, location inspection and adjustment of aperture, layout inspection and adjustment of assemblies and printed board.

Layout design includes the relative location and traces between units of airborne electronic equipments and internal modules of airborne electronic equipments. The fundamental principle is incompatibility and division, which means that PCB areas should be divided, based on the functions of circuits and that input/output elements should be placed as far from each other as possible. The connection between elements which are closely connected should be shortened so as to ensure that the signal flows smoothly and rationally. Sensitive elements should be placed as far from interference source as possible. Input port and output port should be isolated properly. High frequency and low frequency cables should be laid separately. The device within the case of airborne electronic equipment should be designed in layout, trying the best to make interference source be far away from sensitive elements and radiated electromagnetic source be far away from the hole of the case. The circuit elements and signal pathways should be designed to decrease the coupling with undesired signal to the maximum. The component with low level signal should be placed far from component with high level signal and can't be configured with power lead which isn't filtered. Avoid using long parallel line for the traces on PCB. When an element is laid out on a PCB, sensitive circuit should be separated from high level circuit and the circuit which can produce transient process. Take full advantage of the uncorroded part of PCB for grounding and use it as the shielded object. The shielded object should be connected to the main case as possible as we can. Use magnetic shielding around the sensitive circuit to reduce the magnetic coupling with power leads and to diminish low frequency interference.

Above all, airborne electronic equipment should be designed to meet the functional requirements, on the basis of which, combined methods such as grounding, shielding, filtering and rational layout should be adopted to carry out individually targeted and comprehensive design. The purpose is to reduce or even eliminate the effects caused by EMI, making ensure the airborne electronic equipment can operate normally.

5 Basic Methods for EMC Design

Normally, there are 3 basic methods for EMC design, which are problem solving method, specification method and system method.

Problem solving method is used for improvement when an airborne electric equipment is detected faulted. Problem solving method is to address a problem when it comes into being and tends to solve a problem by a specific but limited method. This method comes into being when EMC theory was incomplete and EMC design methods were unsystematic. Because this method is targeted, it is widely used by engineers. However, it works as a remedy for the EMC problems exposed at the late development stage of circuits, airborne electric equipments and systems. Therefore, this method is very passive. It is used at the late development stage of airborne electric equipment, so the change of technical status is very costly and considerably difficult.

Specification method is a design approach used at the development stage of airborne electronic equipments by complying with related EMC standards, to solve the problems of airborne electronic equipments which may appear at early stage. Because EMC standards and specifications reveal the common problems which exist in systems and of airborne electronic equipments and the rules of resolving problems to a certain extent, specification method provides prescient and comprehensive measures. Therefore, specification method is relatively rational and progressive compared to problem solving method.

System method is to build the model of elements, modules and circuits which may have an impact on the EMC of airborne electronic equipments and conduct EMC analysis, prediction, control and distribution by means of auxiliary tool, laying a foundation for airborne electronic equipments meeting requirements. This method can be used throughout the design of airborne electronic equipments. Electromagnetic coupling factors can be considered comprehensively. Design at all stages can be evaluated and corrected continuously.

In summary, EMC design of a system should be carried out at the beginning of development of circuits, airborne electronic equipments as well as systems by EMC prediction, analysis and design. We can gain the initiative to make overall plan for EMC measures for circuits and systems, test and calculate the EMC, and deploy EMC design together with system reliability, maintainability, supportability, testability, safety and environment adaptability. To accomplish design goals, problem solving method, specification method and system method should be combined when it is necessary. The effectiveness of methods should be based on the actual operation and test results of airborne electronic equipments.

6 Summary

EMC technology is an integrated discipline which is rapidly developing and also an application technology which is characterized by highly engineering. It is playing an increasingly important role in fields related to electronics. A comprehensive method should be adopted for EMC design. On one hand, noise source should be suppressed and on the other hand, the immunity of sensitive equipments should be increased. The two sides both are of importance. Any single measure taken won't achieve the desired result. For this reason, the EMC should be considered at the beginning of system design on a system level and improved during equipment manufacturing, operation and maintenance, ensuring that the design of airborne electronic equipments meet EMC standard requirements and that these equipments can work reliably. EMI and EMC are the common thread that run through design, finalization, production and use of aircrafts and have an important role in the safety and reliability of aircrafts. Approaches for resolving EMC problems and fundamental methods for EMC design are put forward, which provides guidance for EMC design of follow-up airborne electronic equipments.

References

1. EMC Society of Hubei Province: Principles and Application of Electromagnetic Compatibility. National Defense Industry Press, Beijing (1996)
2. Pengcheng, L.: Principles and Technologies of Electromagnetic Compatibility. Higher Education Press, Beijing (1993)
3. Dinghua, W., Jiasheng, Z.: Principle and Design of Electromagnetic Compatibility. University Press of Electronic Science and Technology of China, Chengdu (1995)
4. Dequan, L.: Design of Circuit Board Level Electromagnetic Compatibility Board. Microcontrol Department of Motorola, Hong Kong (2001)
5. Bai, T., Lv, X.: Design of Electromagnetic Compatibility. Beijing University Press of Posts and Telecommunications, Beijing, pp. 21–22 (2001)

6. Rengang, C.: Principle, Design and Prediction Technology of Electromagnetic Compatibility. Beijing University Press of Aeronautics and Astronautics, Beijing (1997)
7. GJB 152A-1997.: Measurement of Electromagnetic Emission and Susceptibility for Military Equipment and Subsystems, Administration of Quality Supervision, Inspection and Quarantine (1997)
8. GJB 17626.: Technical Standards for Electromagnetic Compatibility Test and Measurement, Administration of Quality Supervision, Inspection and Quarantine (1998)
9. The European Parliament and The Council of The European, Machinery Directive 2006/42/EC (2006)
10. The European Parliament and The Council of The European, Low Voltage Directive 2006/95/EC (2006)
11. Zhu Liwen etc: Design of Electromagnetic Compatibility. Publishing House of Electronics Industry, Correction Strategies and Case Analysis (2012)

Research on Lift Error of Horizontal Tail Based on Component Disassembly Method

Qing Wen(✉), Kangzhi Yang, Zhihang Cheng, and Yuanzhe Zhang

China Aviation Industry General Aircraft Institute Co, Ltd, Zhuhai 519040, China
wenqing0601@163.com

Abstract. It is very important for calculating aerodynamic load and strength design to obtain aerodynamic force of aircraft components accurately. The aerodynamic force of horizontal tail is usually obtained by means of component disassembly in wind tunnel test, That is, subtract the value of non horizontal tail configuration from the value of whole plane configuration to obtain the value of horizontal tail. Using CFD method, through the comparative study of component disassembly method and direct measurement results, it is found that for low Horizontal tail aircraft, the slope of Horizontal tail lift line obtained by component disassembly method is more than 25%. The main reason for the error is that the component disassembly method assumes that the aerodynamic characteristics of the wing and fuselage with/without Horizontal tail configuration remain unchanged, so the interference data of the whole plane Horizontal tail on the fuselage and wing are accumulated on the Horizontal tail components. The error change of component disassembly method when the Horizontal tail is in different positions is studied. When the Horizontal tail is installed on the fuselage, the error is large, and when it is installed on the vertical tail, the error decreases gradually with the increase of height. By using the moment instead of the lift difference method that subtracted from the value of the whole plane configuration, the interference of the Horizontal tail to the wing can be basically eliminated and the interference of the Horizontal tail to the fuselage can be reduced. For the aircraft with T-tail layout, the results obtained by using the component disassembly method combined with the moment inverse calculation are the same as the direct measurement results.

Keywords: Low horizontal tail · Component disassembly method · Wind tunnel tests · Aerodynamic force of horizontal tail

1 Introduction

Among many tail layout forms, low Horizontal tail layout has been widely used in large passenger aircraft and small general-purpose aircraft [1–5]. The aerodynamic characteristics of the Horizontal tail, especially the lift characteristics, are not only related to the shape design and aeroelastic deformation of the Horizontal tail [6, 7], but also directly determine the structural layout, structural stiffness and strength design of the Horizontal tail [8–11]. Therefore, it is of great significance to obtain the accurate lift characteristics of the Horizontal tail [12–14].

Chinese Society of Aeronautics and Astronautics: CASTYSF 2022, LNEE 972, pp. 441–454, 2023.
https://doi.org/10.1007/978-981-19-7652-0_41

In the process of aircraft design, the lift characteristics of the Horizontal tail are generally obtained through the component disassembly method in the wind tunnel test [15, 16], that is, the lift coefficient of the Horizontal tail is obtained by subtracting the lift coefficient of the non Horizontal tail configuration from the lift coefficient of the whole plane. This method assumes that the aerodynamic forces of the wing, fuselage and other components remain unchanged with or without a Horizontal tail. In fact, the installation of the Horizontal tail will inevitably produce aerodynamic interference to the fuselage, wing and other components, and the value of the interference will affect the accuracy of the Horizontal tail lift characteristics obtained by the component disassembly method. When the interference caused by the installation of the Horizontal tail is large, the lifting error of the Horizontal tail will be significantly increased by using the component disassembly method.

The lift of the horizontal tail of a small general-purpose aircraft with low Horizontal tail layout under research in China was initially obtained by component disassembly method. In the aerodynamic load calculation of Horizontal tail in the limit state, it was found that the Horizontal tail load was abnormally large, which was obviously unreasonable. The lift of the horizontal tail of this model is analyzed and studied. Through CFD Research, it is found that at the installation position of the Horizontal tail, the aerodynamic interference of the Horizontal tail to the fuselage and wing cannot be ignored, and the lift characteristic error of the horizontal tail obtained by component disassembly method is large. By moving the position of the Horizontal tail, the error of the Horizontal tail using the component disassembly method with the horizontal tail in different positions is studied.

2 Comparison Between Simulation and Wind Tunnel Test

2.1 Wind Tunnel and Model

The wind tunnel test with the whole plane and component disassembly force measurement was carried out in the fl-12 wind tunnel of the low speed Institute of China Aerodynamic Research and development center. The size of the wind tunnel test section is 4 M $\times$ 3 m. Due to the limitation of test conditions, the Reynolds number based on the average aerodynamic chord length of the wing is about 1.42 million and the wind speed is 70 m/s. The wind tunnel test model adopts a 1:5 all metal model. The installation of the whole machine test model in the wind tunnel is shown in Fig. 1.

2.2 Simulation Method

The governing equation of the flow field is Reynolds averaged N-S equation. Because the mainstream velocity in the flow field is incompressible flow with low Mach number, the pressure based separation method is selected and the SIMPLEC algorithm is adopted. The pressure term is discretized by the 2nd-order upwind scheme, the convection term is discretized by the 3rd-order MUSCL scheme, and the viscous term is discretized by the 1st-order upwind scheme. The turbulence model adopts transition SST model [17, 18].

Fig. 1. Wind tunnel test model

The mesh type is a cut volume mesh similar to Cartesian mesh. Since the state of the flow separation near the wing surface is not studied, the wall function model is adopted. The value of the y plus of the first grid layer is set as 30, the boundary layer grid has more than 20 layers, the height of the first layer is about 0.1mm, the growth rate is 1.2, and the total number of cells is 12 million.

The calculation model is a half mode model, the calculation far field is rectangular, and the model ratio is 1:5. The symmetry plane is the symmetry plane boundary condition, the far-field boundary behind the model is the pressure outlet boundary condition, and the other boundaries are the velocity inlet boundary condition.

The value of atmospheric environment calculation condition is 0 km in standard atmosphere, the incoming wind speed is 70 m/s, and the calculated Reynolds number is the same as that in wind tunnel test (Fig. 2).

Fig. 2. CFD model

2.3 Comparison Between Simulation and Wind tunnel Test

The CFD simulation state is the same as the wind tunnel test, that is, the aerodynamic characteristics of whole plane and non Horizontal tail configuration are calculated respectively. The cruise configuration is selected as the contrast state, and the angle of elevator is 0°.

The simulation results show that the CFD and wind tunnel test results coincide well, the slope error of lift line is within 1%, and the absolute error of lift coefficient at zero angle of attack is no more than 0.01 (Fig. 3).

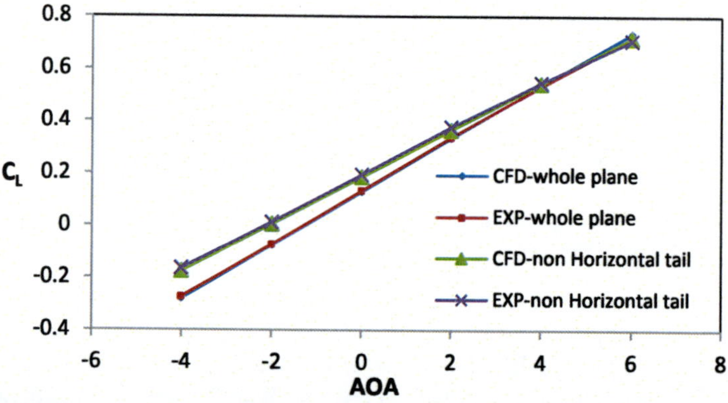

Fig. 3. Comparison of lift curves of the whole plane and non Horizontal tail

The disassembly method obtain the lift of the Horizontal tail by subtracting the lift of the non Horizontal tail configuration from the lift of the whole plane to obtain the lift of the Horizontal tail.

$$C_{L\_Pw\_zc} = C_{L\_All} - C_{L\_YS}$$

where: $C_{L\_\mathrm{Pw\_zc}}$ is Horizontal tail lift obtained by component disassembly method, C_{L_All} is the lift of the whole plane,and C_{L_YS} is the Lift of non Horizontal tail configuration.

According to the lift data of the whole plane and the non Horizontal tail configuration with wind tunnel test and CFD method respectively, the Horizontal tail lift comparison curve obtained by the component assembly method is shown in Fig. 4. Due to the small amount of Horizontal tail lift, the percentage error of Horizontal tail lift characteristic data calculated by disassembly method is increased compared with the wing configuration, and the slope error of lift line is 4.24%.

3 Study on the Error of Component Disassembly Method

3.1 Aerodynamic Analysis of Components Obtained by Component Disassembly Method

Component disassembly method: subtracting the lift of non horizontal tail configuration from the lift of whole plane configuration.

The lift of non Horizontal tail configuration is:

$$C_{L\_NHT} = C_{L\_Wing} + C_{L\_Body}$$

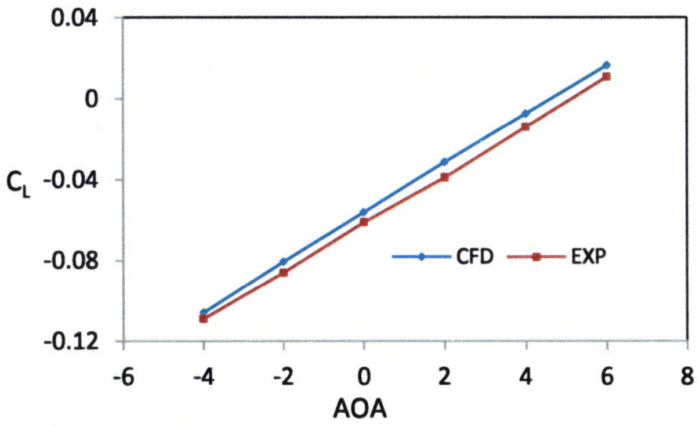

Fig. 4. Comparison of Horizontal Tail lift curves obtained by disassembly method

where C_{L_Wing} is the lift of wing of non Horizontal tail configuration, and C_{L_Body} is the lift of Fuselage of non Horizontal tail configuration (including vertical tail).

The lift of whole planen is:

$$C_{L\_\mathrm{WP}} = C_{L\_\mathrm{Wing\_2}} + C_{L\_\mathrm{Body\_2}} + C_{L\_\mathrm{Pw}}$$

where $C_{L\_\mathrm{Wing}_2}$ is Wing lift of whole planen, $C_{L\_\mathrm{Body\_2}}$ is Fuselage lift of the whole plane configuration(including vertical tail),and C_{L_Pw} is Horizontal tail lift of the whole plane configuration.

Since the lift of the wing in the whole plane state is different from that in the non Horizontal tail state, considering the interference effect of the Horizontal tail, then:

$$C_{L\_Wing\_2} = C_{L_{Wing}} + \Delta C_{L_{Wing\_PW}}$$

$$C_{L\_Body\_2} = C_{L\_Body} + \Delta C_{L_{Body\_PW}}$$

where $\Delta C_{L_{\mathrm{WingPW}}}$ is interference amount of Horizontal tail of whole plane configuration to wing, and $\Delta C_{L_{\mathrm{Body\_PW}}}$ is interference amount of Horizontal tail of whole plane configuration to fuselage (including vertical tail).

The Horizontal tail lift coefficient calculated by the component disassembly method is:

$$C_{L\_Pw\_zc} = C_{L\_WP} - C_{L\_NHT} = C_{L\_Pw} + \Delta C_{L_{Wing\_PW}} + \Delta C_{L_{Body\_PW}}$$

It can be seen from the above formula that the Horizontal tail lift obtained by the component disassembly method not only includes the lift of the Horizontal tail itself, but also includes the interference of the Horizontal tail to the wing, fuselage and other components. The magnitude of the interference is directly related to the accuracy of the Horizontal tail lift obtained by the component disassembly method.

3.2 Comparison of Component Disassembly Method and Direct Measurement Results

The comparison between the Horizontal tail lift characteristics obtained by component disassembly method and direct measurement is shown in Fig. 5. Direct measurement is to measure the lift characteristics of the Horizontal tail on the whole plane. This method is accurate. It can be seen from the calculation results that there is a great difference between the Horizontal tail lift coefficient obtained by the component disassembly method and the direct measurement. At the angle of attack of 0°, the Horizontal tail negative lift obtained by the component assembly method is 41.8% higher than that obtained by the direct measurement, and the lift line slope is 28.3% higher than that.

The Horizontal tail lift near 0° angle of attack is closely related to the Horizontal tail load during high-speed cruise, and the slope of the Horizontal tail lift line is directly related to the sudden wind load. This error has a decisive influence on the load evaluation of the Horizontal tail.

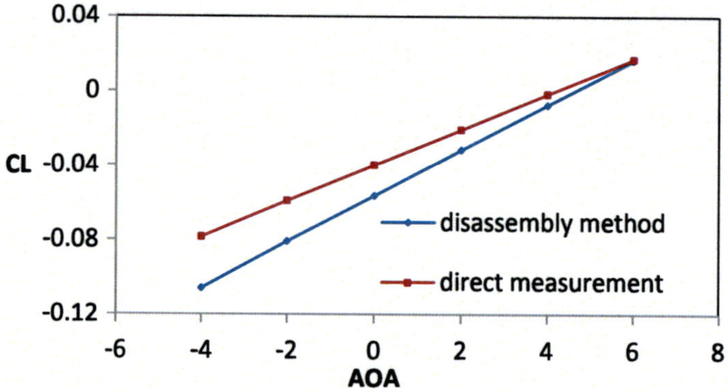

Fig. 5. Comparison of component disassembly method and direct measurement method with CFD

3.3 Analysis on the Error of Component Disassembly Method

According to the comparison of CFD results with/without Horizontal tail configuration, the interference amount of Horizontal tail to aircraft wing and fuselage is shown in Fig. 6. It can be seen from the calculation results that after the Horizontal tail is installed, the lift coefficients of the wing and fuselage are reduced by 0.0052 and 0.0099 respectively at the state of 0° of attack, totaling 0.0151. This part of the force originally acting on the wing and fuselage caused by the interference of the Horizontal tail is calculated on the Horizontal tail when the component disassembly method is adopted, because the aerodynamic forces of the fuselage, wing and other components are assumed to be unchanged, resulting in the increase of the negative lift of the Horizontal tail at zero angle of attack.

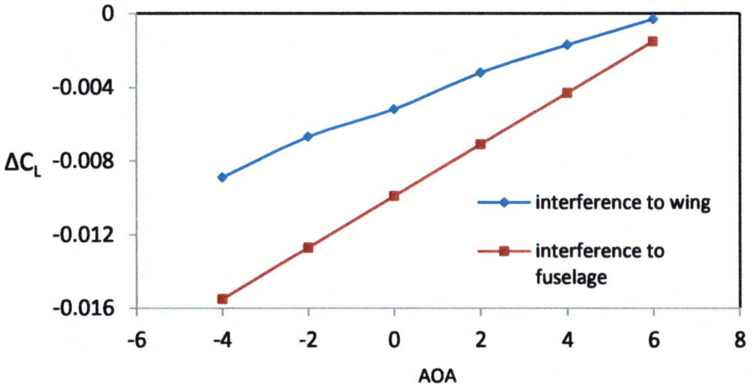

Fig. 6. Lift interference of horizontal tail to fuselage and wing

It can be seen from Fig. 6 that the interference of the Horizontal tail on the fuselage is relatively large. At the angle of attack of 0°, the magnitude of the influence on the wing is about 1/3 and the magnitude of the influence on the fuselage is about 2/3.

The influence of the Horizontal tail on the wing is mainly due to the change of the wing angle of attack caused by the wash flow generated by the lift force of the Horizontal tail in front of the Horizontal tail. In the state of angle of attack 0 degrees, the Horizontal tail causes negative lift, and there is a counterclockwise circulation on the Horizontal tail from the left side of the fuselage, which reduces the effective angle of attack and lift coefficient of the wing.

The interference effect of the Horizontal tail on the fuselage is more complex and has the effect of up washing. More importantly, the Horizontal tail is installed on the fuselage, and the Horizontal tail has an obvious influence on the pressure distribution of the fuselage. Figures 7 and 8 are side views of the pressure distribution in the rear section of the aircraft with and non Horizontal tail configuration at zero angle of attack. It can be seen from the Fig that the pressure distribution at the root of the Horizontal tail obviously affects the pressure distribution of the fuselage. A high-pressure area is formed on the upper side of the leading edge of the Horizontal tail wing root and a low-pressure area is formed on the lower side. Figure 9 shows the comparison of the pressure distribution of the fuselage profile at the leading edge of the Horizontal tail between the whole plane and the non Horizontal tail configuration at zero angle of attack. The area enclosed by the pressure distribution curve of the whole plane configuration is significantly larger than that of the non Horizontal tail configuration, indicating that there is an obvious lift change in the fuselage profile due to the installation of the Horizontal tail.

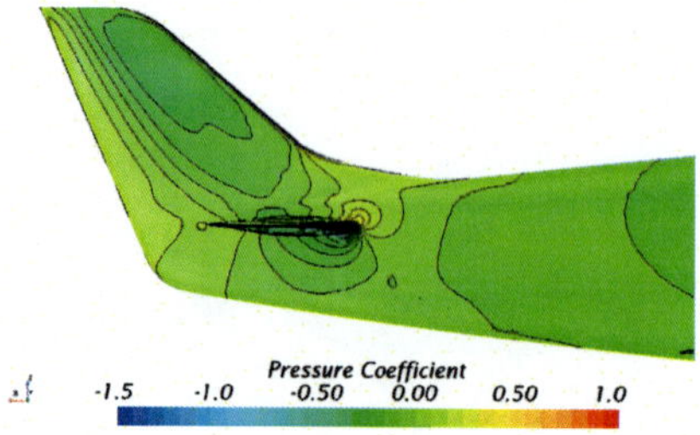

Fig. 7. Pressure distribution in the rear section of aircraft

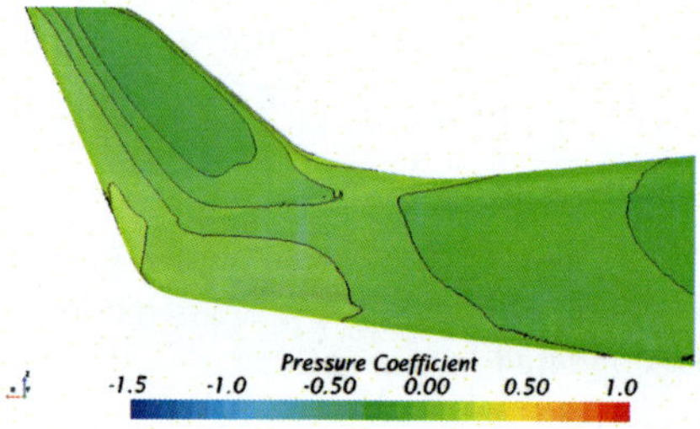

Fig. 8. Pressure distribution in the rear section of tailless aircraft

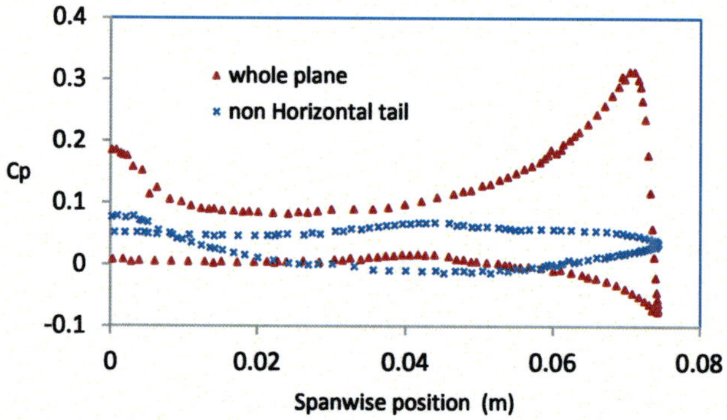

Fig. 9. Comparison of pressure distribution in fuselage section with/without horizontal tail configuration

4 Influence of Horizontal Tail Position on the Accuracy of Disassembly Method

Because the installation of the Horizontal tail has a great influence on the aerodynamic force of the fuselage, the influence of the Horizontal tail at different installation positions on the accuracy of the Horizontal tail aerodynamic force obtained by the component disassembly method is studied.

(1) The Horizontal tail is installed in the upper position of the fuselage, so the Horizontal tail moves downward in the vertical direction on the fuselage. According to the height limit of the rear fuselage, the Horizontal tail moves down 100 ~ 400 mm on the fuselage at an interval of 100 mm (the corresponding size of the full-size aircraft, the same below).

(2) Because the vertical tail has a certain sweep angle, when moving on the vertical tail, the Horizontal tail moves along the 45° angle direction, that is, moving upward and moving backward at the same time, and the moving amount is the same. The Horizontal tail moves up/back 400 mm, 600 mm, 800 mm, 1500 mm on the vertical tail.

The moving direction and partial position of the Horizontal tail on the fuselage and vertical tail are shown in Fig. 10.

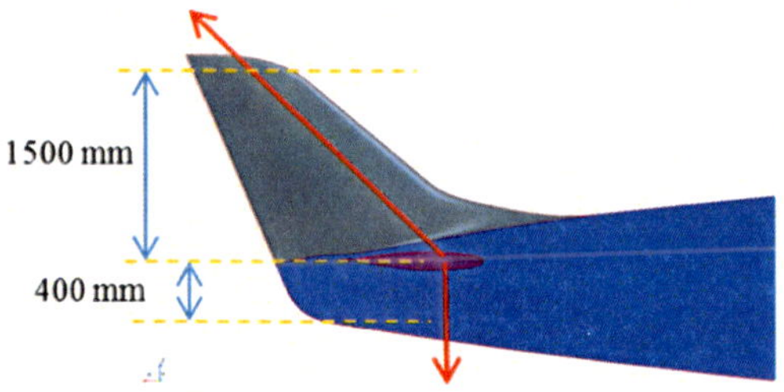

Fig. 10. Horizontal tail position diagram

4.1 The Horizontal Tail Moves on the Fuselage

When the Horizontal tail moves downward on the fuselage, the lift characteristic results of the Horizontal tail obtained by CFD method based on component disassembly method and direct measurement method are shown in Table 1. The calculation results show that when the Horizontal tail moves on the fuselage, due to the certain differences in the downwash characteristics of the wing at different heights and the different exposed areas of the Horizontal tail, the CLA and CL0 of the Horizontal tail are obviously different. However, at all Horizontal tail positions, the slope of the Horizontal tail lift line obtained by the component disassembly method is more than 25% greater than that obtained by the direct measurement method.

4.2 The Horizontal Tail Moves on the Vertical Tail

When the Horizontal tail moves on the vertical tail, the lifting characteristic results of the Horizontal tail obtained by CFD method based on component disassembly method and direct measurement method are shown in Table 2. The results show that the difference between the two method comparing that on the fuselage obviously decrease, the difference can be controlled below 10%, and gradually decrease by the moving up of the horizontal tail.

Table 1. Comparison of lift between component force measurement and component disassembly with different horizontal tail positions

The position of the horizontal tail	Component disassembly method		Direct measurement		Error	
	C_{L0}	C_{La}	C_{L0}	C_{La}	C_{L0}	C_{La} (%)
Designed position	−0.0563	0.01230	−0.0397	0.00959	−0.01660	28.3
Down 100	−0.04262	0.01189	−0.03048	0.00920	−0.01214	29.2
Down 200	−0.03004	0.01169	−0.02217	0.00906	−0.00787	29.0
Down 300	−0.01731	0.01151	−0.01364	0.00902	−0.00367	27.6
Down 400	−0.00633	0.01131	−0.00615	0.00905	−0.00018	25.0

After the horizontal tail is moved up by 1500 mm, the overall layout of the aircraft is changed from the conventional layout to the T-tail layout. At this time, the error of the component disassembly method is significantly reduced, and the slope error of the horizontal tail lift line is only 3.7%. Since the Horizontal tail is far away from the fuselage, the main reason for this error is the interference effect of the Horizontal tail on the wing, as shown in Fig. 11.

Table 2. Comparison of lift between direct measurement and component disassembly with different horizontal tail positions

The position of the horizontal tail	Component disassembly method		Direct measurement		Error	
	C_{L0}	C_{La}	C_{L0}	C_{La}	C_{L0}	C_{La} (%)
Designed position	−0.0563	0.0123	−0.0397	0.00959	−0.0166	28.3
Up 400	−0.0581	0.0126	−0.0496	0.0114	−0.0085	10.5
Up 600	−0.0523	0.0122	−0.0455	0.0111	−0.0068	9.9
Up 800	−0.0483	0.0124	−0.0428	0.0115	−0.0055	7.8
Up 1500	−0.0396	0.0140	−0.03657	0.0135	−0.00303	3.7

5 The Lift Obtained by Moment Inverse Calcultion Method

The lift coefficient can also be inversely calculated according to the pitching moment coefficient of the Horizontal tail, but the following two assumptions need to be met:

(1) When the horizontal tail lift is 0, the pitching moment coefficient is also 0;
(2) The lift of the Horizontal tail acts on 25% of the average aerodynamic chord point of the Horizontal tail.

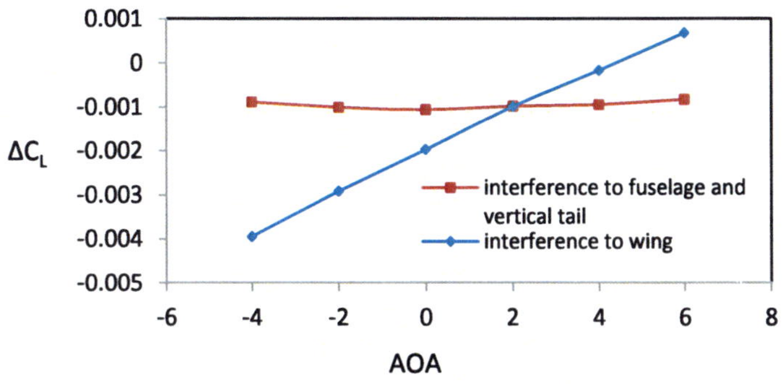

Fig. 11. Inverse calculation of lift by pitching moment

The above conditions are basically satisfied when the Horizontal tail is a symmetrical airfoil.

Ignoring the influence of the drag of the horizontal tail on the pitching moment, the lift and pitching moment coefficients of the Horizontal tail meet the following relationship:

$$C_L \frac{L_{pw} cos(\alpha)}{C_A} = C_m$$

where: L_{pw} is the tail force arm with Horizontal tail; C_A is the average aerodynamic chord length of wing; α is the fuselage angle of attack.

When the Horizontal tail is at the designed position, according to the CFD results, the lift characteristics of the Horizontal tail are obtained by using lift difference and the reverse calculation method of pitching moment according to the component disassembly method and, and compared with the results obtained by the direct measurement method. The results are shown in Fig. 12.

Using the component disassembly method, the aerodynamic force of the Horizontal tail is obtained by subtracting the non Horizontal tail configuration from the whole plane configuration. There is an obvious difference between the value obtained by lift difference method and the value obtained by moment inverse calculation method, and the result of moment inverse calculation method is closer to the result of direct measurement. The moment inverse algorithm helps to reduce the error, but compared with the direct measurement method, the slope of the Horizontal tail lift line obtained by moment inverse calculation is still 14.3% more than that using direct measurement.

There is an obvious difference between the component disassembly method, the direct measurement and the moment inverse calculation. The main reason is that the Horizontal tail aerodynamic force obtained by the component disassembly method includes three parts: the Horizontal tail aerodynamic force, the interference of the Horizontal tail to the wing and the interference of the Horizontal tail to the fuselage. The force arms of these three parts relative to the moment reference point are different, and the contribution of the same lift to the pitching moment is different. For example, the interference part of the wing, although the installation of the Horizontal tail has obvious interference to the lift of the wing, because the moment reference point is located on the wing, The force arm of the wing interfering with the lift is small and has little influence on the

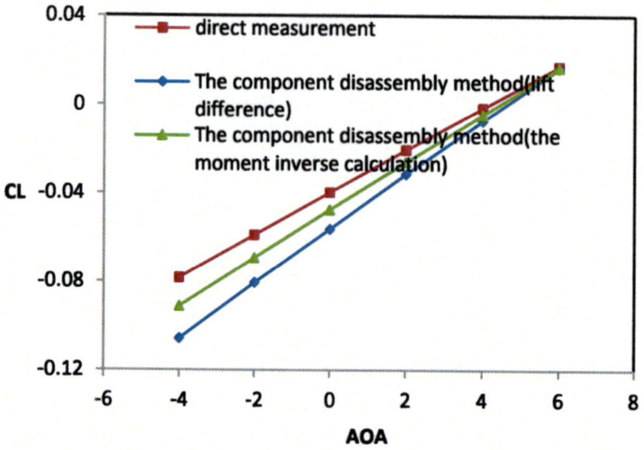

Fig. 12. Comparison of moment inverse method and lift difference method

pitching moment, and the result of rhe lift inverse calculation is close to 0. It can be seen from Fig. 13 that the moment inverse calculation basically eliminates the interference of the Horizontal tail to the wing and reduces the interference of the Horizontal tail to the fuselage.

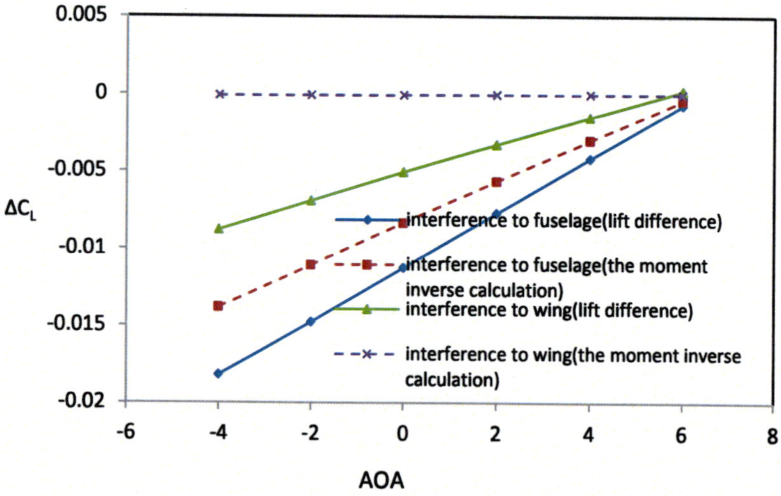

Fig. 13. Comparison of interference lift corresponding to lift difference and reverse from moment

When the Horizontal tail moves on the vertical tail, the research shows that after the Horizontal tail moves up 1500 mm, the overall layout of the aircraft changes from the conventional layout to the T-tail layout. At this time, the error of the component disassembly method is significantly reduced, and the slope error of the Horizontal tail lift line is only 3.7%. The main reason for this error is the interference effect of the Horizontal tail on the wing, as shown in Fig. 13. Because the interference of the Horizontal tail to the wing can be eliminated by using the moment inverse calculation, for the aircraft with T-tail layout, according to the component disassembly method and combined with the moment inverse calculation, the result is basically the same as that of the component disassembly method, and the error is only—0.3% (Table 3).

Table 3. Lift of Horizontal tail with different methods for T-tail

Method	C_{L0}	C_{La}	C_{L0} error	C_{La} error
The component disassembly method-using lift difference	−0.0396	0.01397	−0.003	3.5%
The component disassembly method-the moment inverse calculation	−0.0375	0.01346	−0.0009	−0.3%
Direct measurement	−0.0366	0.01350	–	–

6 Conclusion

The component disassembly method theoretically assumes that the aerodynamic characteristics of the fuselage and wing remain unchanged with or without a Horizontal tail. In fact, for the low Horizontal tail layout with a Horizontal tail installed on the fuselage, the installation of a Horizontal tail will significantly affect the aerodynamic characteristics of the fuselage and wing, which is mainly reflected in two aspects: after adding a Horizontal tail, the lift of the Horizontal tail will change the flow field of the wing, and the pressure distribution of the aircraft which can cause to change of the lift of the wing. The installation of the Horizontal tail on the fuselage changes the pressure distribution of the fuselage. Therefore, for aircraft with low Horizontal tail layout, the Horizontal tail aerodynamic data obtained by component disassembly method has large error.

The differences between the component disassembly method and the direct measurement method when the Horizontal tail is installed at different positions of the fuselage and the vertical tail are studied. When the Horizontal tail is on the fuselage, the error of the component disassembly method is large, and when the Horizontal tail is on the vertical tail, the error decreases gradually with the increase of the installation height.

The moment inverse calculation method can basically eliminate the interference of the Horizontal tail to the wing and reduce the interference of the Horizontal tail to the fuselage. Compared with the method using the substration of the lift, it can effectively reduce the error of the component disassembly method.

For the T-tail layout of high Horizontal tail, the error of Horizontal tail aerodynamic data obtained by component disassembly method is relatively small. With the moment inverse claculation, the results are basically the same as that of direct measurement method.

References

1. Haeni, S.: Horizontal Tail Sizing Pesawat Sport Ringan (LSA) Kapasitas 4 Orang Penumpang. J. Ind. Elektro dan Penerbangan **3**(1), 81–103 (2013)
2. Altunkaya, E.C., Ozkol, I.: Multi-parameter aerodynamic design of a horizontal tail using an optimization approach. Aerosp. Sci. Technol. **121**, 107310 (2022)
3. Schmitt, V., Reneaux, J., Thibert, J.: Design and experimental investigation of a laminar horizontal tail. In: AIAA 8th Applied Aerodynamics Conference, pp. 1–10. AIAA-90-3042, France (1990)
4. Ferrero-Guillén, R., Díez-González, J., Alija, J. M., Martínez-Gutiérrez, A., Verde, P., Perez, H.: Optimal Wing and Horizontal Tail plane design for maximizing the aircraft performance in cruise flight. Cybern. Syst. 1–17(2022)
5. Gomec, F.S., Unver, E.C., Arisoy, M.Z.: The effects of leading edge flap and trailing edge flaperon deflections on horizontal tail optimization. In: AIAA Scitech 2020 Forum, p. 0278 (2020)
6. van Arnhem, N., de Vries, R., Vos, R., Veldhuis, L.L.: Aerodynamic performance of an aircraft equipped with Horizontal Tail mounted propellers. In: AIAA Aviation 2019 Forum, p. 3036 (2019)
7. Benyamen, H., Mckinnis, A., Keshmiri, S.: Effects of propwash on horizontal tail aerodynamics of pusher UASS. In: 2020 IEEE Aerospace Conference, pp. 1–9 (2020)
8. Gao, D., Wang, Y., Wu, Z., Rahim, G., Bai, S.: Design of a sensor network for structural health monitoring of a full-scale composite horizontal tail. Smart Mater. Struct. **23**(5), 055011 (2014)
9. Tang, D., Dowell, E.H.: Computational/experimental aeroelastic study for a horizontal-tail model with free play. AIAA J. **51**(2), 341–352 (2013)
10. Chen, P.C., Lee, D.H.: Flight-loads effects on horizontal tail free-play-induced limit cycle oscillation. J. Aircr. **45**(2), 478–485 (2008)
11. Calderon, R., Aupoix, B., Calmels, B., David, C.: Modelling aerodynamics unsteady loads on the horizontal tail plane of a civil aircraft. Appl. Mech. Mater. **232**, 543–547 (2012)
12. Hoang, N.T.B., Bui, B.V.: Experimental and numerical studies of wingtip and downwash effects on horizontal tail. J. Mech. Sci. Technol. **33**(2), 649–659 (2019). https://doi.org/10.1007/s12206-019-0120-9
13. Ni, K., Hu, P., Zhao, H., Dowell, E.: Flutter and LCO of an all-movable Horizontal Tail with freeplay. In: 53rd AIAA/ASME/ASCE/AHS/ASC Structures, Structural Dynamics and Materials Conference 20th AIAA/ASME/AHS Adaptive Structures Conference 14th AIAA (2012)
14. Qiu, J., Sun, Q.: Research of aerodynamic load of horizontal tail. In: 2009 4th IEEE Conference on Industrial Electronics and Applications, pp. 1503–1506. IEEE (2009)
15. Paziresh, A., Nikseresht, A.H., Moradi, H.: Wing-body and vertical tail interference effects on downwash rate of the horizontal tail in subsonic flow. J. Aerosp. Eng. **30**(4), 04017001 (2017)
16. Hoang, N.T., Bui, B.V.: Investigation of wind tunnel wall effect and wing-fuselage interference regarding the prediction of wing aerodynamics and its influence on the horizontal tail. J. Mech. Sci. Technol. **33**(6), 2737–2746 (2019)
17. Spalart, P., Allmaras, S.: A one-equation turbulence model for aerodynamic flows. AIAA Paper 1992-439 (1992)
18. Spalart, P., Shur, M.: On the sensitization of turbulence models to rotation and curvature. Aerosp. Sci. Technol. **5**(1), 297–302 (1997)

Influence of Temperature on the Ability of Flying Over Obstacles

Zhu Wenlong(✉) and Wang Haozhi

Aviation College, Yantai Nanshan University, Longkou 265700, China
827374802@qq.com

Abstract. Take off stage is a key stage in aircraft flight mission, and it is also a frequent stage of aviation safety accidents, so take-off performance has become one of the main contents of flight performance research. This paper mainly studies the influence of temperature on aircraft take-off performance, and makes take-off analysis for different types of airports, and draws the final conclusion that temperature has great influence on aircraft load and taxiing distance during take-off and landing, and explores the influence of temperature at different stages on aircraft performance during take-off.

Keywords: Performance software · Emergency procedure design · Take off analysis

1 Introduction

With the development of China from a big civil aviation country to a powerful civil aviation country, people pay more attention to flight safety. The premise for airlines to pursue economic benefits is to ensure the safety of aircraft at all stages and meet the requirements of operation regulations. However, the take-off stage is a key stage of aircraft flight mission, and it is also the stage of frequent aviation safety accidents, so the take-off performance has become one of the main contents of flight performance research. It is very important to study the take-off performance of the aircraft in the take-off stage for ensuring the safety of the aircraft, optimizing the take-off performance of the aircraft and improving the economy of the aircraft operation. This paper aims to explore the law of the influence of different interval temperature on the aircraft take-off obstacle surmounting, and put forward corresponding suggestions for the take-off analysis of aviation companies according to the law.

2 The Influence of Temperature on Aircraft Take off Obstacle Surmounting

There are many limiting factors in the process of flight take-off, such as the limitation of aircraft performance and the limitation of meteorology. The atmospheric temperature in meteorology is a typical influencing factor, which has a direct impact on the obstacle

Chinese Society of Aeronautics and Astronautics: CASTYSF 2022, LNEE 972, pp. 455–463, 2023.
https://doi.org/10.1007/978-981-19-7652-0_42

surmounting of aircraft after take-off. Air temperature is the air temperature at the height specified by surface meteorological observation (i.e. 1.25–2.00 m, 1.5 m in the distribution map of temperature forecast in China). Air temperature records can characterize the thermal condition of a place. Air temperature is one of the conventional elements to be measured in surface meteorological observation. Air temperature is expressed in centigrade (°C), some in Fahrenheit (f), all to one decimal place, and negative values below zero. With the change of air temperature, the output thrust of the engine also changes. For example: analysis of the impact of temperature on aircraft take-off obstacle surmounting when the temperature rises, the air density will decrease. On the one hand, the engine thrust will decrease, the aircraft will accelerate slowly, and it will not reach the take-off speed quickly, so that the take-off distance and take-off distance will increase.

3 Impact Analysis

In this paper, B737-700/7b24 is used in Jiuzhai Huanglong Airport and B737-800/7b26 is used in Chongqing Jiangbei Airport for performance calculation. BPS carries the data obtained from various variable calculation into BCOP for obstacle surmounting judgment. The operation interface is shown in the Fig. 1.

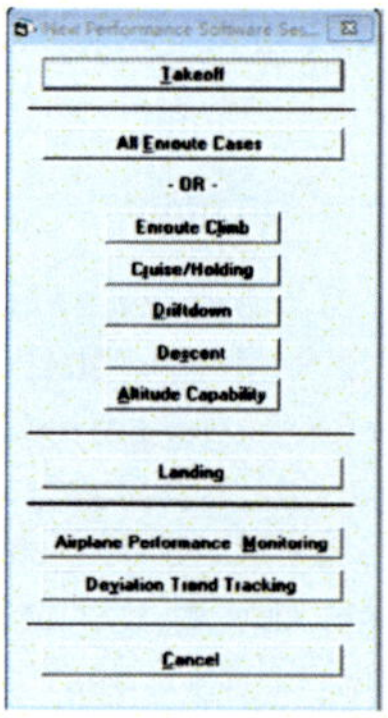

Fig. 1. BPS calculation interface

Different data are applied to different temperature conditions to study the existing factors that affect the take-off obstacle surmounting, and the influence of different temperatures on the aircraft take-off obstacle surmounting is obtained, so as to improve the economy of aircraft operation and provide guidance for operators.

4 Example Analysis

Airport is the place for aircraft to take off and land. Its various "states" naturally have a considerable impact on the take-off distance of aircraft. Here we mainly choose two typical airports to analyze.

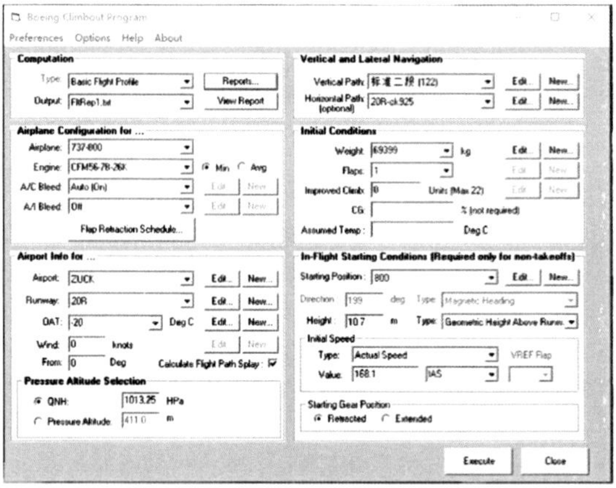

Fig. 2. BCOP calculation interface

Taking the data of Chongqing Jiangbei Airport (ordinary Airport) and Jiuzhai Huanglong Airport (High Plateau Airport) as examples, Jiuzhai Huanglong Airport uses B737-700/7b24 aircraft, Chongqing Jiangbei Airport uses B737-800/7b26 aircraft for performance calculation, and the take-off flight path results of BPS software under the same condition are compared.

Jiangbei Airport and meteorological data: the runway elevation is 411 m, the external atmospheric temperature is the monthly average maximum temperature of 32 °C, static wind, standard sea pressure. The available distance of take-off taxiing, take-off and acceleration stop are 3200 m. The main road surface and runway slope are 0%, and there are no obstacles.

Jiuhuang airport and meteorological data: the runway elevation is 3447.7m, the external atmospheric temperature is the monthly average maximum temperature of 26 °C, the static wind, and the standard sea pressure. The available distance of take-off taxiing, take-off and acceleration stop are all 3370 m. On the main road, the gradient of runway is 0%. The selection of obstacles is shown in Table 1.

Table 1. Obstacles in Jiuhuang airport

name	Altitude (m)	Lateral offset (m)	Correct height (m)	Horizontal distance (m)
Obstacle 1	3514	310	66	5130
Obstacle 2	3600	977	152	11190

Model data: (1) 737-700/cfm56-7b24, structure weight limit 69399 kg, optimal configuration flap 1, air conditioning automatic, anti icing off. Standard two-stage mode, balanced V1 strategy (2) 737-800/cfm56-7b26, the weight limit of structure is 79010 kg, the optimal configuration is flap 1, and the air conditioner is automatic and anti icing. Standard two-stage mode, balanced V1 strategy.

4.1 Jiangbei Airport, 737-800/CFM56-7B26 Model

The total track height range of BPS is 122 m to 292 m. The following calculation uses 122 m Leveling Height for leveling; The takeoff weight within the maximum takeoff weight range is used to simulate the takeoff of the aircraft. Different temperatures are used to simulate the aircraft take-off, only changing the air temperature of the airport, calculating the altitude of the aircraft when it reaches the designated point, quantitatively analyzing the influence of temperature on the aircraft take-off performance, and exploring its law.

As shown in Fig. 2, the vertical and horizontal tracks of the aircraft take-off simulation are as follows.

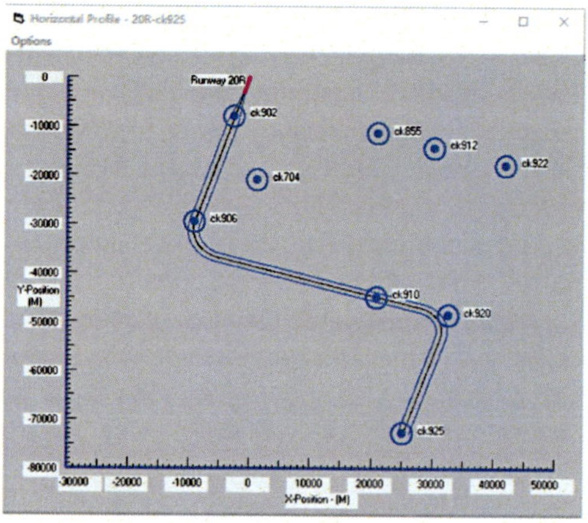

Fig. 3. BCOP horizontal interface

The KML file, which is the abbreviation of keyhole markup language, was originally developed by keyhole company. It is a coding specification based on XML syntax and format and used to describe and save geographic information (such as points, lines, images, polygons and models), It can be recognized and displayed by Google Earth and Google maps. As shown in Fig. 3, the track is displayed in Google Earth.

Different temperatures are used to simulate the take-off of the aircraft, and only the air temperature of the airport is changed to calculate the altitude of the aircraft when it reaches the designated point, as shown in Table 2.

According to the change rule of temperature and air pressure height at the designated point in Table 2, draw a line chart, and visually analyze the change rule, as shown in Fig. 4

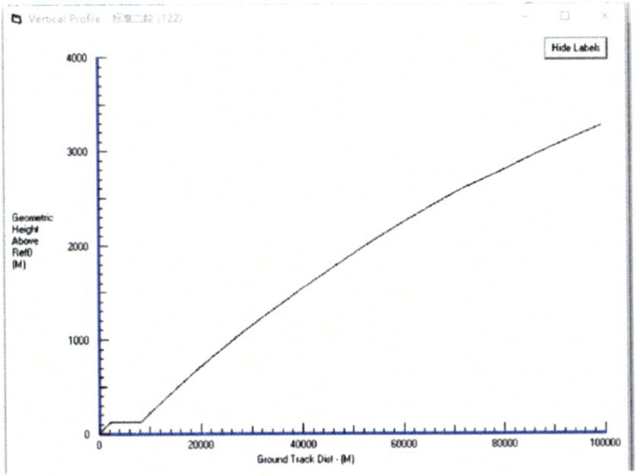

Fig. 4. BCOP vertical interface

Fig. 5. Google Earth track view

Table 2. The relationship between the temperature of Chongqing Jiangbei airport and the height of reaching the designated point

Temperature (°C)		−20	−15	−10	−5	0
Height (m)		4119	4067	4016	3966	3917
5	10	15	20	24	32.4	
3868	3820	3774	3727	3635	3162	

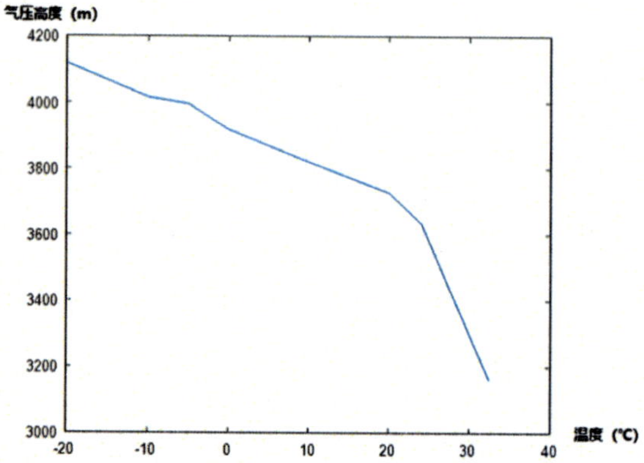

Fig. 6. Broken line chart of the relationship between temperature and the height at the designated point in Chongqing Jiangbei airport

It can be seen from Fig. 4 that the relationship between the airport temperature and the height at the designated point is basically linear in the range of—20 °C to 20 °C (the slope is about—9.8), and it is also basically linear when it is higher than 20 °C (the slope is about—45.6), but the slopes of the two stages are different and very different, with the slope nearly four times. It can be seen that when the temperature is high, the influence of temperature on the obstacle surmounting ability of the aircraft is far greater than that of the relatively low temperature operating environment.

4.2 Jiuhuang Airport, 737-700/CFM56-7B24 Model

The total track height range of BPS is 192 m–464 m; The following calculation uses 192 m Leveling Height leveling; The takeoff weight within the maximum takeoff weight range is used to simulate the takeoff of the aircraft. Different temperatures are used to simulate the aircraft take-off, only changing the air temperature of the airport, calculating the altitude of the aircraft when it reaches the designated point, quantitatively analyzing the influence of temperature on the aircraft take-off performance, and exploring its law. As shown in Fig. 5, the vertical and horizontal tracks of the aircraft take-off simulation are as follows.

The vertical and horizontal tracks of the aircraft are simulated by BCOP tracks, as shown in Fig. 6.

Different temperatures are used to simulate the take-off of the aircraft, and only the air temperature of the airport is changed to calculate the altitude of the aircraft when it reaches the designated point, as shown in Table 3.

According to the variation law of temperature and air pressure height at the designated point in Table 3, draw a broken line chart and visually analyze its variation law, as shown in Fig. 7

It can be seen from Fig. 7 that within the range of—20 °C to 20 °C, the relationship between the airport temperature and the height reaching the designated point is basically linear (the slope is about—27.8), and it is also basically linear when it is higher than 20 °C (the slope is about—68), but the slopes of the two stages are different and very

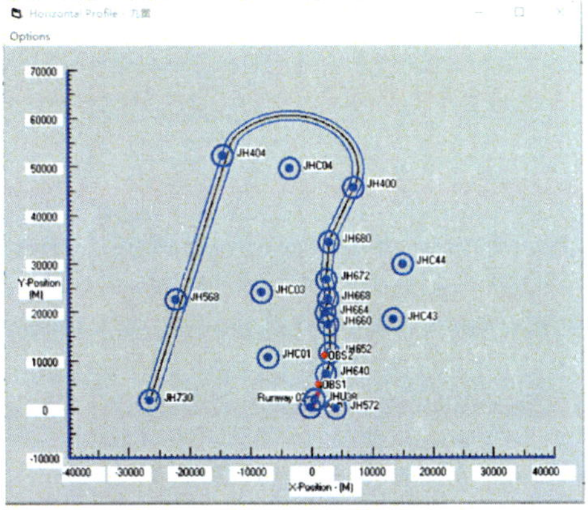

Fig. 7. BCOP horizontal interface

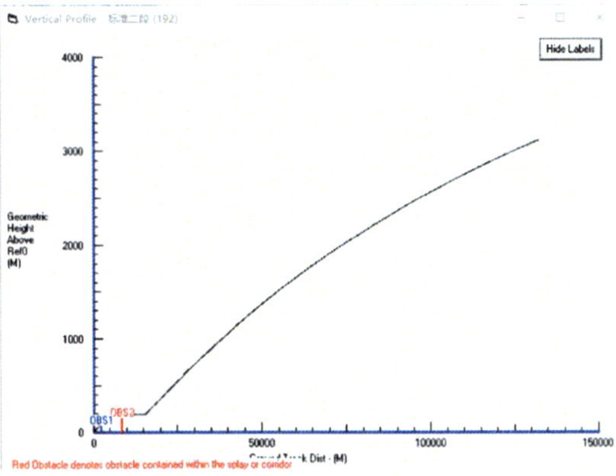

Fig. 8. BCOP vertical interface

different, and the slope is nearly three times. It can be seen that when the temperature is high, the influence of temperature on the obstacle surmounting ability of the aircraft is far greater than that of the relatively low temperature operating environment (Figs. 8, 9 and 10).

5 Conclusion

Through the data analysis of the two selected airports, combined with the above data curve, the following conclusion is drawn: with the increase of temperature, the obstacle surmounting ability of the aircraft after take-off is also declining. Moreover, it can be seen from the curve in the table that with the increase of air temperature, the takeoff obstacle crossing height basically increases linearly. Within the range of—20 °C to 20 °C, the relationship between the airport temperature and the height at the designated point is basically linear, and it is also basically linear when it is greater than 20 °C. However,

Fig. 9. Google Earth track view

Table 3. The relationship between the temperature of Jiuzhai Huanglong Airport and the altitude at the designated point

Temperature (°C)		−20	−15	−10
Height (m)		6731	6686	6641
15	20	24	26	32.4
5852	5532	5462	5122	4669

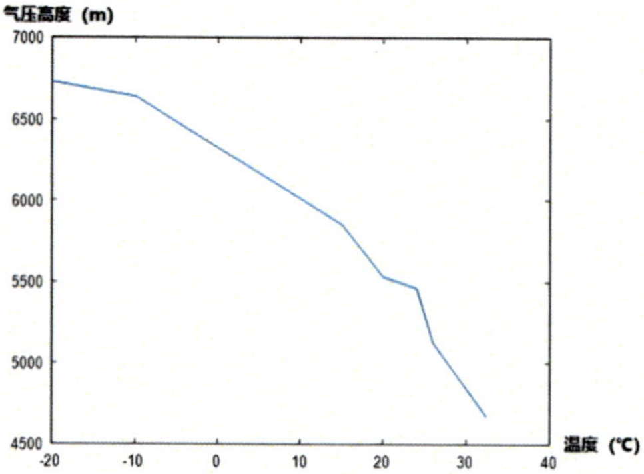

Fig. 10. A line chart of the relationship between the temperature of Jiuzhai Huanglong airport and the height at the designated point

the slopes of the two stages are different and have great differences. That is to say, the temperature change of ordinary airport and High Plateau Airport is more sensitive to the impact of taking off obstacle surmounting, and the higher the temperature is, the greater the impact on the obstacle surmounting ability after taking off. It can be considered that the temperature is a more sensitive factor for taking off obstacle surmounting, which is consistent with the theoretical analysis.

Analysis reason: the temperature has a great influence on the aircraft take-off obstacle. With the increase of air temperature, the air density becomes smaller, the lift is smaller, the aircraft load is smaller, and the ability to take off obstacle is weaker.

In this paper, the influence of temperature on obstacle surmounting is discussed in detail under two different conditions of ordinary airport and High Plateau Airport, and the corresponding specific values are given through the calculation of performance software. On the premise of ensuring flight safety, different temperatures are selected according to the meteorological conditions of the airport at that time to optimize the take-off performance of the aircraft and obtain better economic benefits. In this paper, the research on aircraft take-off obstacle surmounting is only a preliminary theoretical study. The above data are calculated by using the performance software, and there is a certain error with the actual situation in the practical application. However, the software is released by the original aircraft manufacturer, and the research goal is qualitative analysis, so the analysis conclusion obtained by using the software is still of guiding significance.

References

1. CAAC.: Ccar-25-r3 airworthiness standard for transport aircraft. Civil Aviation Administration, Beijing (2001)
2. CAAC.: Ccar-121-r4 large aircraft public air transport carrier operation certification rules. Civil Aviation Administration, Beijing (2006)
3. Jiang, Y.: aircraft performance analysis of plateau/complex terrain airports and routes. Southwest Jiaotong University Press, Chengdu. ISBN 978-7-5643-3838-1 (2015)
4. Chen, Z., Gu, R., Liu, J.: Aircraft performance engineering. Ordnance Industry Press, Beijing (2006)
5. Fu, Z.Z.: Load balancing and flight planning. China Civil Aviation press, Beijing (2007)
6. Liu, X., Su, B.: Flight performance and planning. Southwest Jiaotong University Press, Chengdu (2003)
7. Yu, H.: Meteorological factors affecting flight. Air Traff. Manag. (2) (1998)

Check for updates

A Theoretical Fatigue Acceleration Method for Polymer Composite Materials

Xu Fei(✉), Huang Guangqi, and Xiong Huafeng

Aircraft Strength Research Institute, Xi'an 710065, China
xufeiasri@hotmail.com

Abstract. A theoretical fatigue acceleration method for polymer composite materials is proposed in this paper, based on continuum damage mechanics and time-temperature superposition. By way of demonstration, it has been applied to problems involving constant strain rates and fatigue loading conditions. The damage evolution law for the matrix of composites was constructed using the Weibull distribution of defects which will develop into cracks as a result of deformation. In conjunction with the Wiechert model, the theoretical framework formulated here is capable of capturing the mechanical behaviour of UD composites. Demonstration and initial verification of this acceleration method were carried out through experiments.

Keywords: Polymer composites · Fatigue · Test acceleration · Continuum damage mechanics · Time-temperature superposition

1 Introduction

Fibre-reinforced composite structures, which have better stiffness-to-weight ratio, strength-to-weight ratio and significantly higher tolerance to fatigue loading than metallic materials, have been used widely in aerospace engineering to reduce the weight of structures. Composite structures started to be used as fairings and radomes in the 1970s on the Airbus A300-B2. By the time of writing, more than 45% composites by weight are used in the A350, including composite main wing and composite fuselage. The past design method was based on a no-damage growth criterion (it is assumed that no damage occurs during the life time) and this was achieved by conservative strain allowable and verification testing on components [1]. Fatigue was not an issue at that stage since the operational strain levels were kept safely below the fatigue threshold for graphite/epoxy composite materials.

As polymer composite materials began to be employed in the primary structures of aircraft, and therefore were subjected to higher strain levels, the improvement in these properties and a greater understanding of the static properties of composite materials have raised the allowable design strain to the critical point at which zero growth of defects due to fatigue cannot be guaranteed. Furthermore, the fatigue properties of helicopter rotor blades, ship propellers and wind turbines are a major concern of the designers. Thus

Chinese Society of Aeronautics and Astronautics: CASTYSF 2022, LNEE 972, pp. 464–478, 2023.
https://doi.org/10.1007/978-981-19-7652-0_43

the fatigue of composites has now become an issue that requires further investigation and predictive tools.

Due to the complex damage mechanisms for fibre-reinforced composite materials during fatigue such as matrix cracking, matrix crazing, debonding, fibre-matrix interface failure, delamination, fibre fracture, etc. and their interactions and their different growth rates, it is hard to construct a fatigue damage model which can cover all of the damage modes. However, some efforts have been made by a number of researchers to establish a fatigue model for composite materials. Existing composites fatigue models can be classified into three categories [2]. The fatigue life models [3–6] expressed fatigue life in terms of stress levels. Whilst this was in a desirable format in terms of applications, their applicability was confined within a limited scope in terms of the types of structures and the loading conditions, because these models rest heavily on the test data from which curves were fitted. It is therefore impractical to envisage the fatigue life prediction of composite materials and structures such as aircraft using these models. The constant-life models [7–11] shared much of the curve-fitting nature of the fatigue life models and hence their limitations. The semi-empirical models such as residual strength models and residual stiffness models [12–18] allowed damage accumulation but still relied on fitting of curves to experimental data. The progressive damage model directly predicts the growth of damage (e.g. transverse matrix crack density, size of the delaminated area), and correlates the damage growth with the residual mechanical properties (stiffness and strength) [19, 20] using damage variables which describe the degradation of the composite materials. Its applicability is limited to the specific damage types incorporated, such as matrix cracks and delamination.

Most of the fatigue models mentioned above are only directly applicable to composite coupons, and significant difficulties have to be overcome when modelling composite structure fatigue under general loading conditions for the following reasons. Firstly, most experiments are carried out under uniaxial stress conditions, although this stress state is not typical for real structures. Secondly, the governing damage mechanism varies with stress/strain level states [21, 22], and damage modes vary with cyclic stress level and even with the number of cycles to failure. The load history, in the form of different block loading sequences, will result in differences in damage growth [23]. Thirdly, the fatigue properties of composites such as residual strength and fatigue life have been found to decrease more quickly when the loading sequence is repeatedly changed [24]. This is the so-called ‘cycle-mix effect’ which refers to the observations that laminates experiencing small cycle blocks tend to have reduced average fatigue lives when compared with those subjected to large cycle blocks, even though the blocks are of similar structure but involving fewer cycles. Thus, in the authors’ opinion, the aim of formulating a comprehensive fatigue model for use in aircraft design is still too ambitious to achieve at the present time.

Experimental data from fatigue tests performed directly on composite materials and structures remain the primary means of evaluating their fatigue behaviour under a specific set of test loading conditions. However, aircrafts have long service lives (up to 50 years) during which they will suffer complex loading cycles. Direct replication of the load spectrum for fatigue tests is clearly not feasible due to the timescales involved. In practice, tests have to be accelerated and relatively established procedures are available for fatigue tests of metallic materials, typically by increasing the loading frequency. However, in the case of polymeric composite materials, an increase in the loading frequency generates a significant amount of heat owing to their high hysteretic losses and poor thermal conductivity, which lead to a temperature rise within the composite specimen. Moreover, the properties of the matrix or the interface of composites are typically sensitive to temperature, leading to a physically coupled problem amongst damage, viscoelasticity, temperature, frequency and the fatigue life. Consequently, a change of frequency will result in a different fatigue life from that obtained from testing at the normal loading frequency. Any reasonable acceleration scheme for composites will have to incorporate the effects of temperature change caused by the change in frequency.

In previous work, the authors presented a continuum damage model for transverse cracking in UD composite of linear viscoelastic behaviour. In that paper, it was demonstrated that the properties of the matrix or the properties in transverse direction of UD composites were affected by the combination of the damage and viscoelasticity [25].

The present paper describes a theoretical framework for accelerated fatigue testing of UD composite, derived by modelling the mechanical behaviour of the composite based on a full understanding of properties of its constituent fibres and resin. The temperature effect and frequency effect on the fatigue life and the time-temperature superposition principle are explained in this paper. This will lay a rational basis for the development of accelerated testing methods for fatigue testing for a specific UD composite. This paper will also present some simple demonstrations of the theory involving tests under simple fatigue loading conditions.

The paper is structured as follows. The damage model for the matrix of UD composites is introduced in Sect. 2. By taking advantage of the Wiechert model, the mechanical behavior model of UD composites is built in Sect. 3. Based on the mechanical behaviour model, the test acceleration methods for fatigue test are built in Sect. 4. Then the applications of these two methods are elaborated and discussed in Sect. 5. At last, Sect. 6 ends the paper with some conclusions and perspectives.

2 The Damage Model for the Matrix of UD Composite Material

The mechanical behaviour of cured polymeric resins exhibiting viscoelastic characteristics shows a degree of time and temperature dependence under certain conditions. This will further affect the mechanical properties of composites when the temperature of specimen rises due to the heat generated by the high frequency fatigue test.

When this occurs, neither of the effects of frequency nor the temperature on the composites can validly be ignored. From the authors' perspective, the effects of frequency and temperature on the fatigue life of the composites are mainly caused by the viscoelastic behaviour of the matrix and its tendency to undergo damage. In order to accelerate fatigue test of polymer composite materials, a damage model for the viscoelastic matrix material of the composites will need to be constructed in order to obtain a full understanding of their viscoelasticity, damage and their interaction.

2.1 The Constitutive Equation of Viscoelastic Materials

For a linear viscoelastic material, the constitutive relations can be written in an integral form using the Boltzman superposition principle [26] as

$$\sigma_{ij} = C_{ijkl}(t)\epsilon_{kl}^{0} + \int_{0}^{t} C_{ijkl}(t-\tau)\frac{\partial \epsilon_{kl}(\tau)}{\partial \tau}d\tau \tag{1}$$

or, inversely

$$\epsilon_{ij} = S_{ijkl}(t)\sigma_{kl}^{0} + \int_{0}^{t} S_{ijkl}(t-\tau)\frac{\partial \sigma_{kl}(\tau)}{\partial \tau}d\tau \tag{2}$$

Here $C_{ijkl}(t)$ and $S_{ijkl}(t)$ are the relaxation modulus and creep compliance, respectively.

2.2 Damage Evolution Law for Viscoelastic Materials

The Wiechert model is the most common mathematical model to describe viscoelasticity in a one-dimensional form (Fig. 1), which describes viscoelasticity using a series of springs (elasticity) and dashpots (viscosity) connected in parallel.

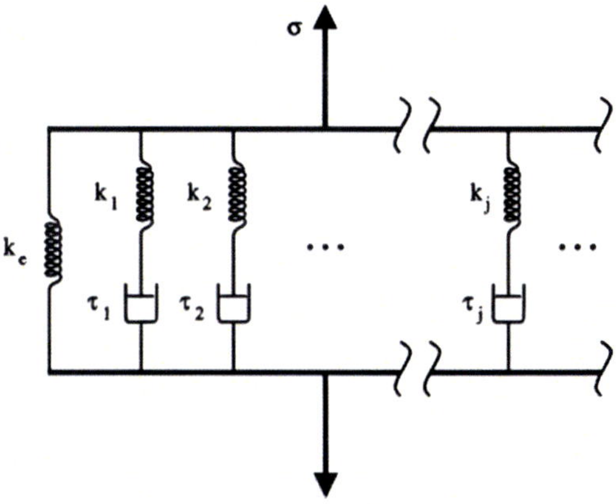

Fig. 1. The Wiechert model.

The relaxation modulus is

$$E_{rel}(t) = k_e + \sum_{j} k_j exp\left(-\frac{t}{\tau_j}\right) \tag{3}$$

where k_e is the elastic modulus of the main spring, k_j is the elastic modulus of a spring connected with a dashpot, τ_j is their relaxation time and

$$\tau_j = \frac{\eta_j}{k_j} \tag{4}$$

η_j is the viscosity of the dashpot.

In the continuum damage mechanics (CDM) theory, the damage effects are often represented in terms of a reduction in stiffness modulus [27–29]. For viscoelastic materials represented by the Wiechert model, it will be assumed that the damage affects all of its parts equally. Then the relaxation modulus with damage can be defined as

$$E_D(t) = (1 - D)\left[k_e + \sum_j k_j exp\left(-\frac{t}{\tau_j}\right)\right] \tag{5}$$

To describe cracking of the viscoelastic materials in terms of the CDM, a damage evolution law will be proposed here, based on the assumption that the damage evolution is controlled by the strain in line with Miyano and Nakada's work [30]. The Weibull distribution will be employed to characterise the defects in the viscoelastic material, in which damage (cracking) is expressed as a function of the strain. Tensile tests will be conducted on polymer resin to obtain the material properties.

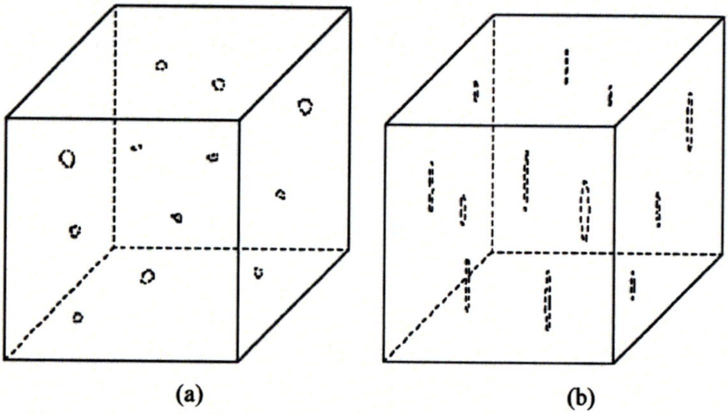

Fig. 2. The RVE showing, **a** microscopic defects and **b** the cracks which they develop into

The concept of a representative volume element (RVE) in a viscoelastic material will be employed to bring in damage. It will be supposed that the RVE contains many microscopic defects of different sizes as shown in Fig. 2a. These defects develop into cracks due to the strain increases as illustrated in Fig. 2b. The effect of these cracks is expressed via the damage parameter D, the value of which is in the range from 0 to 1. The probability density of these defects can be represented by the Weibull distribution which is defined as a function of strain.

$$\rho(\epsilon; \lambda, \mathrm{h}) = \begin{cases} \frac{h}{\lambda}\left(\frac{\epsilon}{\lambda}\right)^{h-1} e^{-\left(\frac{\epsilon}{\lambda}\right)^h} & \sigma \geq 0 \\ 0 & \sigma < 0 \end{cases} \tag{6}$$

where ϵ is the strain under uniaxial stress condition, and h and λ are material constants. Then

$$D = \int_0^{\epsilon} \rho \mathrm{d}x = 1 - e^{-\left(\frac{\epsilon}{\lambda}\right)^h} \tag{7}$$

There are various possible definitions or physical interpretations for the damage parameter D. Here it is defined as a ratio, specifically, the relative reduction in stiffness E due to damage as

$$D = (E^0 - E)/E^0 = 1 - E/E^0 \tag{8}$$

Once the damage D is obtained from the experimental data as a function of ϵ, parameters such as h and λ can be obtained by fitting the data to function (7).

Using the viscoelastic constitutive equations in (1) or (2) and treating the relaxation modulus as a function of damage as shown in (5), the constitutive relation for a damaged viscoelastic material is written as

$$\begin{aligned}\sigma(t) =& E_D(t)\epsilon_0 + \int_0^t \mathrm{E_D}(\mathrm{t}-\tau)\frac{\partial\epsilon(\tau)}{\partial\tau}\mathrm{d}\tau = (1-\mathrm{D})\mathrm{E_{rel}}(\mathrm{t})\epsilon_0\\ &+(1-D)\int_0^t E_{rel}(t-\tau)\frac{\partial\epsilon(\tau)}{\partial\tau}\mathrm{d}\tau = e^{-\left(\frac{\epsilon}{\lambda}\right)^h}\left[k_e + \sum_j k_j e^{\left(-\frac{t}{\tau_j}\right)}\right]\epsilon_0\\ &+e^{-\left(\frac{\epsilon}{\lambda}\right)^h}\int_0^t\left[k_e + \sum_j k_j e^{\left(-\frac{t-\tau}{\tau_j}\right)}\right]\frac{\partial\epsilon(\tau)}{\partial\tau}\mathrm{d}\tau\end{aligned} \tag{9}$$

where $\epsilon{=}\epsilon_0 + \int_0^t \frac{\partial\epsilon(\tau)}{\partial\tau}\mathrm{d}\tau$ and $\epsilon = \epsilon_0$ when $t = 0$. Then (9) describes the damage evolution law of the viscoelastic material.

In the case of a constant strain rate test at a strain rate a, (9) will be simplified as

$$\begin{aligned}\sigma(t) =& ae^{-\left(\frac{at}{\lambda}\right)^h}\int_0^t\left[k_e + \sum_j k_j e^{\left(-\frac{t-\tau}{\tau_j}\right)}\right]\mathrm{d}\tau\\ =& ae^{-\left(\frac{at}{\lambda}\right)^h}\left\{k_e t + \sum_j\left[k_j\tau_j - k_j\tau_j e^{\left(-\frac{t}{\tau_j}\right)}\right]\right\}\end{aligned} \tag{10}$$

The elastic part of the Wiechert model is absent for the cases of resin or the transverse properties of the UD composite according to Kumar and Talreja [26] and Koyanagi et al. [31], then (10) can be further simplified as

$$\sigma(t) = ae^{-\left(\frac{at}{\lambda}\right)^h}\cdot\sum_j\left[k_j\tau_j - k_j\tau_j e^{\left(-\frac{t}{\tau_j}\right)}\right] \tag{11}$$

The extremum values of a function can be determined when its first derivative vanishes. Then the strength under constant strain rate can be obtained when the maximum value of (11) corresponding to its first derivative with respect to time becoming zero as

$$\frac{\mathrm{d}\sigma(\mathrm{t})}{\mathrm{dt}} = \mathrm{a}e^{-\left(\frac{\mathrm{at}}{\lambda}\right)^{\mathrm{h}}}\sum_{\mathrm{j}}\mathrm{k_j}\left[\mathrm{e}^{-\frac{\mathrm{t}}{\tau_\mathrm{j}}} - \frac{\mathrm{ah}}{\lambda}\tau_\mathrm{j}\left(\frac{\mathrm{at}}{\lambda}\right)^{\mathrm{h}-1}\left(1-\mathrm{e}^{-\frac{\mathrm{t}}{\tau_\mathrm{j}}}\right)\right] = 0 \tag{12}$$

3 Mechanical Behaviour Model of UD Composites

The Wiechert model is used to describe the properties of UD composite materials as illustrated in Fig. 3. As mentioned above, there is no purely elastic part k_e in the matrix (resin). Thus part A is used to describe the properties of the fibre and part B is used to describe the properties of the resin.

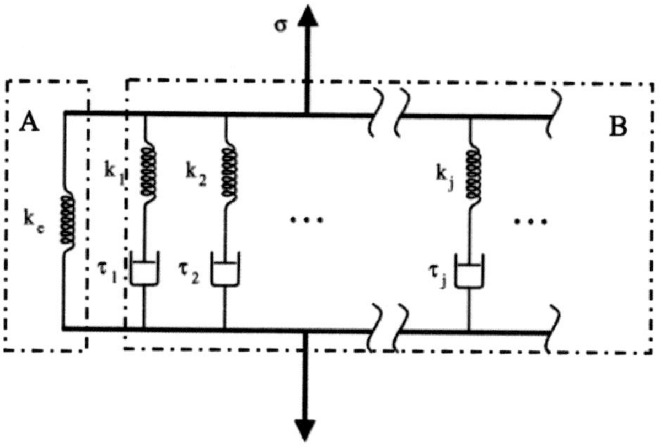

Fig. 3. The mathematic model for UD composites in 1 dimension.

The relaxation modulus including the consideration of damage evolution of the matrix for UD composites is obtained as

$$E_{rel}(t) = k_e + (1 - D)\sum_j k_j e^{\left(-\frac{t}{\tau_j}\right)} = k_e + e^{-\left(\frac{\epsilon}{\lambda}\right)^h}\sum_j k_j e^{\left(-\frac{t}{\tau_j}\right)} \quad (13)$$

As listed previously, there are several damage modes during the whole process of failure of UD composites such as matrix cracks, debonding and fibre breakage. If it is desired to predict the properties of UD composite during deformation, the damage model should be formulated to take account of all the damage modes as well as the interactions between them.

It is noted that a damage evolution law for the fibre has not been proposed. Due to the randomness of the fibre strength and the nonprogressive fibre breakage mechanism mentioned in Talreja's work [21, 32, 33], the fibre failure of UD composites during fatigue cannot be predicted. Also, to the best of the authors' knowledge, no damage model is currently available to describe the modes of the damage such as debonding (interface damage) which commonly occurs during the failure of UD composites. Whilst waiting for an acceptable predictive theoretical model to become available, the determination of fatigue strength of composites will have to rely mainly on experimental means. Fatigue experiments are notoriously time-consuming and an appropriate method to accelerate such experiments is essential. Even in the presence of applicable theoretical model, to support the model in terms of necessary input data to the model and the validation of the model will still require experimental data. This underlines the objective of the present paper.

4 The Fatigue Acceleration of UD Composites

The ultimate objective of this paper is the acceleration of fatigue tests for UD composite materials. To achieve this objective, it is not necessary to build a damage model for composite fatigue to predict fatigue life which would be an unrealistic aim when considering the complex damage mechanisms which occur during fatigue failure. Instead, the aim is to propose a rational method to accelerate fatigue tests through the mechanical behaviour model since the traditional fatigue acceleration method such as increasing the frequency cannot be used directly due to the viscoelasticity of the resin.

As illustrated in the previous section, the loading condition affects the process and result of monotonic loading tests such as creep tests, stress relaxation tests and constant strain rate tests. In order to accelerate fatigue tests, the viscoelastic behaviour under dynamic loading needs to be characterised. When a viscoelastic material is suffered to a sinusoidally stress, a steady state will be reached in which the resulting strain is also sinusoidal, having the same angular frequency but retarded in phase by an angle θ [34]. This is also true even if the strain is the controlled variable.

Suppose a sinusoidal load is applied on this model as follows

$$\sigma(t) = \sigma_0 \cos \omega t \tag{14}$$

The strain function should be

$$\epsilon(t) = \epsilon_0 \cos(\omega t - \theta) \tag{15}$$

Advantage can be taken of the complex modulus to calculate the strain of the UD composite. This calculation makes use of the composite mechanical behaviour model (13) proposed in Sect. 3 which is the relaxation modulus described by the Wiechert model with the consideration of damage evolution of the resin controlled by the strain [35].

$$\epsilon^* = S^*\sigma^* = \frac{k_e + \sum_j \frac{e^{-\left(\frac{\epsilon_0}{\lambda}\right)^h} k_j\omega^2\tau_j^2}{1+\omega^2\tau_j^2} - i\sum_j \frac{e^{-\left(\frac{\epsilon_0}{\lambda}\right)^h} k_j\omega\tau_j}{1+\omega^2\tau_j^2}}{\left(k_e + \sum_j \frac{e^{-\left(\frac{\epsilon_0}{\lambda}\right)^h} k_j\omega^2\tau_j^2}{1+\omega^2\tau_j^2}\right)^2 + \left(\sum_j \frac{e^{-\left(\frac{\epsilon_0}{\lambda}\right)^h} k_j\omega\tau_j}{1+\omega^2\tau_j^2}\right)^2}(\sigma_0 \cos \omega t + i\sigma_0 \sin\omega t)$$

$$= \sigma_0 \frac{\left(k_e + \sum_j \frac{e^{-\left(\frac{\epsilon_0}{\lambda}\right)^h} k_j\omega^2\tau_j^2}{1+\omega^2\tau_j^2}\right)\cos \omega t + \left(\sum_j \frac{e^{-\left(\frac{\epsilon_0}{\lambda}\right)^h} k_j\omega\tau_j}{1+\omega^2\tau_j^2}\right)\sin\omega t}{\left(k_e + \sum_j \frac{e^{-\left(\frac{\epsilon_0}{\lambda}\right)^h} k_j\omega^2\tau_j^2}{1+\omega^2\tau_j^2}\right)^2 + \left(\sum_j \frac{e^{-\left(\frac{\epsilon_0}{\lambda}\right)^h} k_j\omega\tau_j}{1+\omega^2\tau_j^2}\right)^2}$$

$$+i\sigma_0\frac{\left(k_e+\sum_j\frac{e^{-\left(\frac{\epsilon_0}{\lambda}\right)^h}k_j\omega^2\tau_j^2}{1+\omega^2\tau_j^2}\right)\sin\omega t-\cos\omega t\left(\sum_j\frac{e^{-\left(\frac{\epsilon_0}{\lambda}\right)^h}k_j\omega\tau_j}{1+\omega^2\tau_j^2}\right)}{\left(k_e+\sum_j\frac{e^{-\left(\frac{\epsilon_0}{\lambda}\right)^h}k_j\omega^2\tau_j^2}{1+\omega^2\tau_j^2}\right)^2+\left(\sum_j\frac{e^{-\left(\frac{\epsilon_0}{\lambda}\right)^h}k_j\omega\tau_j}{1+\omega^2\tau_j^2}\right)^2} \tag{16}$$

where

$$S^*=1/E^* \tag{17}$$

$$E^*=k_e+e^{-\left(\frac{\epsilon_0}{\lambda}\right)^h}\sum_j\frac{k_j\omega^2{\tau_j}^2}{1+\omega^2{\tau_j}^2}+ie^{-\left(\frac{\epsilon_0}{\lambda}\right)^h}\sum_j\frac{k_j\omega\tau_j}{1+\omega^2{\tau_j}^2} \tag{18}$$

$$\sigma^*=\sigma_0\cos\omega t+i\sigma_0\sin\omega t \tag{19}$$

and

$$\epsilon^*=\epsilon_0\cos(\omega t-\theta)+i\epsilon_0\sin(\omega t-\theta) \tag{20}$$

Then the strain $\epsilon(t)$ is obtained as the real part of (16)

$$\epsilon(t)=\sigma_0\frac{\left(k_e+\sum_j\frac{e^{-\left(\frac{\epsilon_0}{\lambda}\right)^h}k_j\omega^2{\tau_j}^2}{1+\omega^2{\tau_j}^2}\right)\cos\omega t+\left(\sum_j\frac{e^{-\left(\frac{\epsilon_0}{\lambda}\right)^h}k_j\omega\tau_j}{1+\omega^2{\tau_j}^2}\right)\sin\omega t}{\left(k_e+\sum_j\frac{e^{-\left(\frac{\epsilon_0}{\lambda}\right)^h}k_j\omega^2{\tau_j}^2}{1+\omega^2{\tau_j}^2}\right)^2+\left(\sum_j\frac{e^{-\left(\frac{\epsilon_0}{\lambda}\right)^h}k_j\omega\tau_j}{1+\omega^2{\tau_j}^2}\right)^2} \tag{21}$$

Substituting (4) into (21), the maximum strain ϵ_{max} and the phase angle θ are obtained as

$$\epsilon_{max}=\frac{\sigma_0}{\sqrt[2]{\left(k_e+e^{-\left(\frac{\epsilon_0}{\lambda}\right)^h}\sum_j\frac{k_j\omega^2{\eta_j}^2}{{k_j}^2+\omega^2{\eta_j}^2}\right)^2+\left(e^{-\left(\frac{\epsilon_0}{\lambda}\right)^h}\sum_j\frac{{k_j}^2\omega\eta_j}{{k_j}^2+\omega^2{\eta_j}^2}\right)^2}} \tag{22}$$

and

$$\theta=\tan^{-1}\left[\frac{k_e+e^{-\left(\frac{\epsilon_0}{\lambda}\right)^h}\sum_j\frac{k_j\omega^2{\eta_j}^2}{{k_j}^2+\omega^2{\eta_j}^2}}{e^{-\left(\frac{\epsilon_0}{\lambda}\right)^h}\sum_j\frac{{k_j}^2\omega\eta_j}{{k_j}^2+\omega^2{\eta_j}^2}}\right] \tag{23}$$

As shown in (22), the maximum strain is affected by the frequency. If the frequency is increased to accelerate the fatigue test, the maximum strain will decrease. According to Talreja's theory [21], the fatigue life of a UD composite is dominated by the maximum strain. When the maximum strain is decreased, the fatigue life will increase. Thus, if the frequency is increased directly to accelerate the fatigue test, an over-estimated fatigue life will be obtained. Furthermore, the phase angle will be changed (the changes of the phase angle can be calculated from the value of relaxation modulus) along with changes in frequency according to (23). The stress distributed on the fibre can be expressed by

$$\sigma_{k_e}(t) = k_e \epsilon_{max} \cos(\omega t + \theta) \tag{24}$$

It can be inferred from (24) that the load applied on the fibre at the same phase position will change when the frequency is changed since the phase angle θ is affected by frequency. This means that the stress distribution on fibre and resin will be changed if the frequency is changed.

Directly increasing the frequency to accelerate the fatigue test of UD composite will not only cause the maximum strain to decrease, but will also cause the stress distribution to change at a given phase position within the loading cycle. In other words, directly increasing the frequency does not validly accelerate the fatigue test on UD composites. The over-estimate of fatigue life will pose a hazard when using the results to evaluate a design.

Fortunately, the viscosity η_j and frequency ω always comes in pairs. If we can find a way to change the viscosity η_j properly, then the effects caused by increasing frequency should be eliminated.

The viscosity η_j is temperature dependent. From Ojovan's work [36, 37], if the temperature is serious lower than the glass transition temperature, $T \ll T_g$, then the viscosity can be described via an Arrhenius relationship as

$$\eta = A_H T \cdot e^{Q/RT} \tag{25}$$

with

$$Q = H_d + H_m \tag{26}$$

where H_d is the enthalpy of formation of broken bonds, H_m is the enthalpy of their motion and R is the gas constant, 8.314×10^{-3} [kJ/s(K $\times$ mol)].

The viscosity η_j can be reduced by increasing the temperature (25) by a suitable value in order to counteract the effect of increasing frequency. Then the maximum strain and the stress state in the fibres remains unchanged during each cycle and this means the stress state in the resin also remains unchanged. Thus, the fatigue test can be accelerated by increasing the frequency and the temperature at the same time. The viscosity-temperature shift factor b_T is defined as

$$b_T = b_{T_0}(T) = \frac{\omega_0}{\omega} = \frac{\eta}{\eta_0} = \frac{T}{T_0} e^{\frac{Q}{R}\left(\frac{1}{T} - \frac{1}{T_0}\right)} \tag{27}$$

and the value of Q can be obtained from stress relaxation tests.

5 Application

The methods of accelerating fatigue tests have been demonstrated experimentally. The specimens are made from T700 carbon fibre and VTM 264-1 resin provided by Cytec Industries Inc. The volume fractions of the fibre and resin are 65% and 35% respectively. The curing condition is 120 °C for 1 h followed by 80 °C for 5 h. The specimen design for the tests is illustrated in Fig. 4.

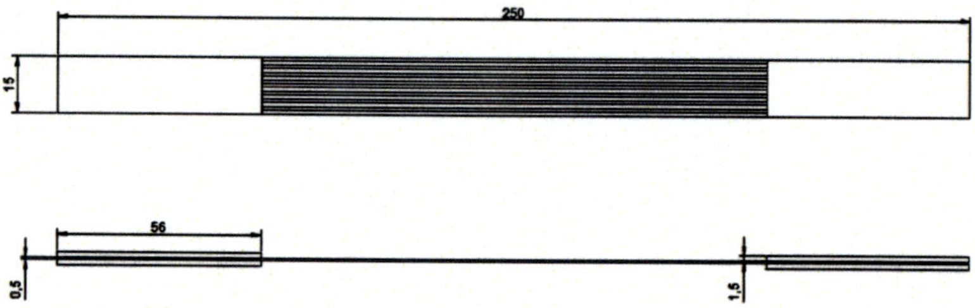

Fig. 4. Specimen drawing.

In previous work [25], the properties of the resin VTM 264-1 described by the Wiechert model in terms of values of parameters k_j and τ_j were obtained at the reference temperature 50 °C as (Table 1).

Meanwhile, the activation energy Q was 298.4 kJ/mol and the damage evolution parameters h and λ were 2.7381 and 0.0117 respectively [25].

The tensile fatigue tests were carried out under three different loading conditions using an Instron 8801 servohydraulic fatigue testing machine. The stress ratio R is 0.1. The original fatigue tests were carried out at 70 °C and 0.5 Hz. If the frequency needs to be increased to 5 Hz to accelerate the tests, the temperature should be shifted to 78 °C according to (28) which is a rearrangement of (27) noting that $\omega = \frac{f}{2\pi}$.

$$\frac{T}{T_0}e^{\frac{Q}{R}\left(\frac{1}{T}-\frac{1}{T_0}\right)} - \frac{f_0}{f} = 0 \tag{28}$$

Tests carried out at 70 °C and 5 Hz were also undertaken to provide a comparison. Testing results are shown in Fig. 5.

As shown in Fig. 5, when the original fatigue test (black line) is accelerated by increasing the frequency from 0.5Hz to 5Hz while maintaining the temperature at 70 °C, the fatigue life increases significantly (red line). This result is as would be expected, as simply increasing the frequency will reduce the maximum strain thus leading to a longer fatigue life. According to the calculation by the present theory, the fatigue properties when the frequency is 0.5 Hz and the temperature is 70 °C are equivalent to the fatigue properties when the frequency is 5 Hz and the temperature is 78 °C (blue line). Then the S-N curve of 0.5 Hz at 70 °C should overlap on the S-N curve for 5 Hz at 78 °C. However, the fatigue life at 5 Hz and 78 °C is less than the fatigue life at 0.5 Hz and 70 °C. There are several reasons which might cause this result, which will be discussed later. Although the present fatigue acceleration method leads to a shorter fatigue life compared with the

Table 1. Values of k_j and τ_j.

	k_j (MPa)	τ_j (s)
1	166.1564	10^{-3}
2	134.4521	10^{-2}
3	0	10^{-1}
4	127.5879	1
5	60.3306	10
6	270.0444	10^{2}
7	125.2148	10^{3}
8	278.8006	10^{4}
9	140.8153	10^{5}
10	725.0293	10^{6}
11	586.3370	10^{7}
12	1116.9276	10^{8}
13	1687.6311	10^{9}
14	1190.2172	10^{10}
15	921.5302	10^{11}

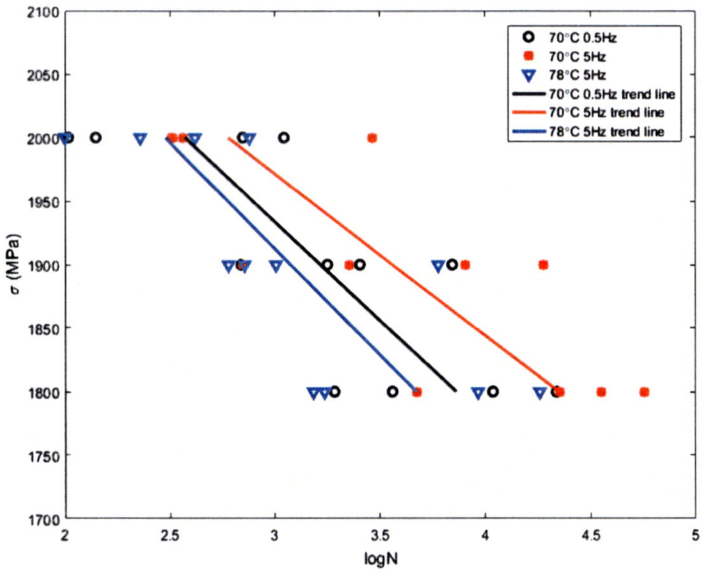

Fig. 5. The comparison of the S-N curves.

fatigue life without acceleration, this method still can be used to accelerate fatigue test since the accelerated result is conservative and therefore safe compared with directly increasing the frequency, for which the resulting fatigue lives are overestimated and unsafe.

The fatigue acceleration method is depended on the assumption that the fatigue life is dominated by the maximum strain and the stress distribution between fibre and resin during fatigue tests. Actually, there is another factor, the interfacial shear strength, which also affects the strength and fatigue life of UD composite materials. According to Koyanagi's theory [31], the interfacial shear strength is affected by the initial interfacial compressive stress which consists of thermal residual stress and will be affected by the temperature. When the temperature is increased, the interfacial compressive stress decreases. This will cause a decrease in the interfacial shear strength which will lead to a reduction of the strength or the fatigue life.

The described methods for accelerated fatigue testing methods for UD composites were demonstrated on a proof-of-concept basis, and were verified through a limited number of experiments which were carried out at representative values of loading rate and reference temperature. It is demonstrated that the present theoretical model is consistent with the experimental results well and the present acceleration method provides a much more accurate prediction for fatigue test of composite materials. These accelerated testing methods can of course be used for other loading rates, frequencies, loading ratios, temperatures or materials as long as the material exhibits linear viscoelasticity under those testing condition. The key parameter of these accelerated methods is the time-temperature shift factor which can be obtained from stress relaxation tests at different temperature. In order to apply these methods to the acceleration of fatigue tests of other composites at a different temperature, loading rate or frequency, the time-temperature shift factor for the relevant material at that temperature range will need to be determined. While the results presented here are promising and appear to be slightly conservative, the limited validation conducted within this proof-of-concept study means that further validation of the newly-proposed methods will be needed before they can be used for accelerating safety-critical tests.

6 Conclusion and Perspectives

A fatigue acceleration method for UD composite materials have been proposed and verified in this paper. A damage evolution law for viscoelastic materials (the composite matrix) has been proposed based on CDM. Then a mechanical behaviour model for UD composites has been formulated using the Wiechert model and the damage evolution law for the matrix. Based on this mechanical behaviour model, a rational method with a physical basis for acceleration of fatigue tests on a UD composite has been proposed and demonstrated on a proof-of-concept basis.

Both of the testing acceleration methods have been verified through tests. Compared with increasing frequency only for fatigue tests, which lead to an over-estimate of the UD composite properties, the testing acceleration method provides more accurate and conservative predictions.

The present testing acceleration method can be applied to a greater extent. Firstly, it can improve the testing efficiency of the durability verification of aircraft composite structure. Moreover, it can be extended to high frequency fatigue test and then helps evaluate the high cycle fatigue performance of composite blades used in aeroengine, wind turbine, etc.

References

1. Nyman, T.: Fatigue and residual strength of composite aircraft structures. Institutionen för flygteknik (1999)
2. Degrieck, J., Van Paepegem, W.: Fatigue damage modeling of fibre-reinforced composite materials: review. Appl. Mech. Rev. **54**(4), 279–300 (2001)
3. Hashin, Z., Rotem, A.: A fatigue failure criterion for fiber reinforced materials. J. Compos. Mater. **7**(4), 448–464 (1973)
4. Reifsnider, K., Gao, Z.: A micromechanics model for composites under fatigue loading. Int. J. Fatigue **13**(2), 149–156 (1991)
5. Fawaz, Z., Ellyin, F.: Fatigue failure model for fibre-reinforced materials under general loading conditions. J. Compos. Mater. **28**(15), 1432–1451 (1994)
6. Yao, L., et al.: A novel total fatigue life model for delamination growth in composite laminates under generic loading. Compos. Struct. **258** (2021)
7. Harris, B.: Fatigue behaviour of polymer-based composites and life prediction methods. In: Cardon, A.H. (ed.) Durability Analysis of Structural Composite Systems, pp. 49–84, Balkema, Rotterdam (1996)
8. Adam, T., et al.: Life prediction for fatigue of T800/5245 carbob-fibre composites: II Variable-amplitude loading. Int. J. Fatigue **16**(8), 533–547 (1994)
9. Gathercole, N., et al.: Life prediction for fatigue of T800/5245 carbon-fibre composites: I. Constant-amplitude loading. Int. J. Fatigue **16**(8), 523–532 (1994)
10. Beheshty, M., Harris, B.: A constant-life model of fatigue behaviour for carbon-fibre composites: the effect of impact damage. Compos. Sci. Technol. **58**(1), 9–18 (1998)
11. Beheshty, M., Harris, B., Adam, T.: An empirical fatigue-life model for high-performance fibre composites with and without impact damage. Compos. A Appl. Sci. Manuf. **30**(8), 971–987 (1999)
12. Chou, P.C., Croman, R.: Residual strength in fatigue based on the strength-life equal rank assumption. J. Compos. Mater. **12**(2), 177–194 (1978)
13. Chou, P., Croman, R.: Degradation and sudden-death models of fatigue of graphite/epoxy composites. In: Composite Materials: testing and Design (Fifth Conference). ASTM International (1979)
14. Hahn, H., Kim, R.: Proof testing of composite materials. J. Compos. Mater. **9**(3), 297–311 (1975)
15. Xiong, J.-J., et al.: Fatigue-driven failure criterion for progressive damage modelling and fatigue life prediction of composite structures. Int. J. Fatigue **145** (2021)
16. Yadav, I.N., Thapa, K.B.: Fatigue damage model of woven glass-epoxy fabric composite materials. J. Mater. Res. Technol.-JMR&T **9**(1), 301–306 (2020)
17. Wan, A., Xiong, J., Xu, Y.: Fatigue life prediction of woven composite laminates with initial delamination. Fatigue Fract. Eng. Mater. Struct. **43**(9), 2130–2146 (2020)
18. Hu, J., et al.: Novel mathematical-statistical models for the distribution of fatigue life and residual strength for fiber reinforced polymer composites. Int. J. Appl. Mech. **12**(9) (2020)
19. Akshantala, N.V., Talreja, R.: A micromechanics based model for predicting fatigue life of composite laminates. Mater. Sci. Eng. A **285**, 303–313 (2000)
20. Zhou, S., et al.: Progressive fatigue damage modelling of fibre-reinforced composite based on fatigue master curves. Thin-Walled Struct. **158** (2021)

21. Talreja, R.: A conceptual framework for studies of durability in composite materials. Fatigue Text. Compos. 1(2015)
22. Daniel, I., Charewicz, A.: Fatigue damage mechanisms and residual properties of graphite/epoxy laminates. Eng. Fract. Mech. **25**(5–6), 793–808 (1986)
23. Hwang, W., Han, K.: Cumulative damage models and multi-stress fatigue life prediction. J. Compos. Mater. **20**(2), 125–153 (1986)
24. Farrow, I.: Damage accumulation and degradation of composite laminates under aircraft service loading: assessment and prediction. Cranfield Institute of Technology (1989)
25. Xu, F., Jones, A., Li, S.: A continuum damage model for transverse cracking in UD composites of linear viscoelastic behaviour. Compos. Struct. **225** (2019)
26. Kumar, R.S., Talreja, R.: A continuum damage model for linear viscoelastic composite materials. Mech. Mater. **35**, 463–480 (2003)
27. Lemaitre, J.: A Course on Damage Mechanics, 2nd edn. Springer (1996)
28. Lemaitre, J., Chaboche, J.: Aspect phenomenologique de la rupture par endommagement. Journal de Mecanique appliquee **2**, 317–365 (1978)
29. Aboudi, J.: Stiffness reduction of cracked solids. Eng. Fract. Mech. **26**, 637–650 (1987)
30. Miyano, Y., Nakada, M.: Time and temperature dependent fatigue strengths for three directions of unidirectional CFRP. Exp. Mech. **46**(2), 155–162 (2006). https://doi.org/10.1007/s11340-006-6834-5
31. Koyanagi, J., Nakada, M., Miyano, Y.: Prediction of long-term durability of unidirectional CFRP. J. Reinfor. Plast. Compos. 1305–1313 (2011)
32. Talreja, R.: Fatigue of composite materials: damage mechanisms and fatigue-life diagrams. In: Proceedings of the Royal Society of London A: Mathematical, Physical and Engineering Sciences. The Royal Society (1981)
33. Talreja, R., Singh, C.V.: Damage and Failure of Composite Materials, pp. 141–155. Cambridge University Press, United States of America (2012)
34. Christensen, R.M.: Theory of Viscoelasticity, 2nd edn. Academic Press, New York (1982)
35. Emri, I.: Time-dependent phenomena related to the durability analysis of composite structures. In: Cardon, A.H. (ed.) Durability Analysis of Structureal Composite Systems: reliability, Risk Analysis and Prediction of Safe Residual Structural Integrity-Lectures of the Special Chair AIB-Vincotte 1995, pp. 85–122. CRC/Balkema, Rotterdam (1996)
36. Ojovan, M.I., Lee, W.E.: Viscosity of network liquids within Doremus approach. J. Appl. Phys. **95**(7), 3803–3810 (2004)
37. Ojovan, M.I., Travis, K.P., Hand, R.J.: Thermodynamic parameters of bonds in glassy materials from viscosity—Temperature relationships. J. Phys.: Condens. Matter **19**(41), 415107 (2007)

Research on CBB Mechanism Construction of Technical Service Organization

Li Huijuan(✉) and Feng Bing

China Aero Polytechnical Establishment (CAPE), Beijing 100028, China
Lhj1900@126.com

Abstract. In the process of knowledge productization, the demand for modularized design of comprehensive knowledge service solutions is increasingly urgent in technical service institutions. Establish CBB definition, classification and hierarchical design for products, services and solutions. Put forward the design concept of "equal responsibility and right", inspire the enthusiasm for the commercialization of scientific and research findings, form a return and a virtuous circle. Clarify the construction path, design a sharing platform, realize the module sharing and reuse of products and projects, improve efficiency, reduce costs and reduce risks.

Keywords: Knowledge services · Modularization · CBB

1 Introduction

With the rising demand for technical services in the whole life cycle of complex products, the rapid technological innovation and the development of the industrial chain, it is no longer possible for technical service institutions to rely on a single professional technical service. The long-term coexistence of products, services and solutions will become a typical service form—that is, comprehensive knowledge service solutions. Among them, products generally refer to commodity-off-the-shelf sold by the piece and with after-sales maintenance service in a narrow sense; services generally refer to timing, metering and pricing services, such as testing, consulting services, and training; solutions generally refer to technical services customized for specific products, specific fields or specific problems [1]. The knowledge service solution has the characteristics of comprehensiveness, totality and integration. Two urgent needs for the research results of technical service institutions should be satisfied: first, the components or modules that make up the knowledge service solution have the characteristics of high reuse and low coupling, which means that the achievements accumulated by the enterprise can be standardized, modularized and shared; second, the achievement modules can exist for a long time in a tradable mechanism, forming stable support and output for knowledge service solutions, and forming benefits feedback and a virtuous circle for scientific research project achievements. Therefore, the modularization design of technical service products has been paid more and more attention by institutions at home and abroad.

Companies with software and hardware as the main product form gradually establish an efficient R&D system based on integrated product development in order to achieve

Chinese Society of Aeronautics and Astronautics: CASTYSF 2022, LNEE 972, pp. 479–491, 2023.
https://doi.org/10.1007/978-981-19-7652-0_44

large-scale development [2]. They have built a product platform and select shared modules that support different types of products in the platform [3, 4]. Intel Corporation, the world's largest CPU manufacturer, promotes the construction of common foundation, aiming to unify the specifications of laptop accessories [5]. By building a shared information platform, the 16th Research Institute of the 9th Academy of China Aerospace Science and Technology Corporation (CASC) has preliminarily realized the coordinated transformation of IT-based scientific research and production models [6]. The 38th Research Institute of China Electronics Technology Group Corporation has realized the reuse of knowledge in design and development through the management of common basic modules [7]. Southwest China Research Institute of Electronic Equipment has used the concept of shared module management to carry out the PCB parallel design based on Cadence [8]. The No. 289 Factory of the Ninth Academy of Aerospace Science and Technology Group has promoted the research and development of general software modules through the development of software modular construction, which have been applied in many models [9]. Under the background of military-civilian integration, AVIC Qing'an Group has carried out a new model of research and development of aviation electromechanical products, and proposed the idea of modularized research and development of CBB for general-purpose military and civilian products [10].

Most of these management mechanisms are aimed at the design, development and production of software or hardware products with physical properties. From the perspective of physical modularization, through the analysis and decomposition of products, the service or solution products such as test, detection, analysis, consultation and training are rarely involved in each stage [11]. In view of the modularization requirements of comprehensive technical service solutions, this paper proposes the modularization and classification of project knowledge, the establishment of sharable technologies, services, and product shelves, to realize the reuse of design and development knowledge, reduce costs and improve efficiency; the equal design of "responsibilities and rights" promotes the transformation of achievements and benefits feedback, forming a virtuous circle.

2 Knowledge Modularization Connotation

2.1 Definition of Knowledge Modularization

Knowledge modules are common building blocks (CBB), supporting products, services and solutions, software, hardware, data, tools, forms and other finished products or finished components that can be reused, reduced costs or improved efficiency.

It includes the following four aspects:

(1) Product attributes. It is clear that CBB is an integral part of the product in the internal research and development process, and can also be a product with independent sales or use characteristics.
(2) Reusability. CBB is a physical module, knowledge module or technical module that can be reused in multiple products, services and solutions.
(3) Use effect. It can clearly achieve the ultimate goal of improving efficiency and reducing cost.

(4) Expression form. According to the product characteristics of technical service institutions, the expression form of CBB include software and hardware products, various data, forms and practical tools.

2.2 CBB Reuse Rate in Products

Comprehensive knowledge service products can be divided into three types: software and hardware products in a narrow sense, technical services, and single solutions. These three types of products are different in the form of existence, composition, service, etc. The specific classification characteristics are shown in Fig. 1.

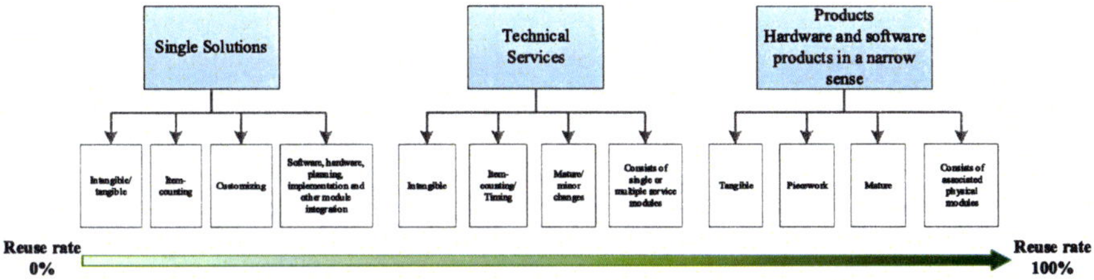

Fig. 1. Classification characteristics of comprehensive knowledge service products

In the narrow sense, software and hardware products are generally tangible value outputs, which are composed of mature modules through physical or logical connection to form a complete whole. They are produced in batch and serve users in the form of piecework. The main demand for its knowledge modularization is the physical module. In the narrow sense of software and hardware products, knowledge modules have the highest reuse rate. The modules that make up a mature product are fully mature with no changes to the modules allowed during mass production, so the reuse rate is 100%.

Technical service is a service output based on a product or a certain technology, consisting of relatively mature single or multiple services. It can provide services in batches or partially customized, or in the form of timing or item-counting. The main demand of its knowledge modularization is technical modularization and substantiality. The technical service relies on a certain technology. The basic service process and procedures remain the same, but certain adjustments will be made according to the differences in the service objects so that the knowledge modules that make up the technical service will be changed accordingly with the decreasing reuse rate.

The focus of the solution is integration, so as to provide integrated value output after integrating multiple products (multiple physical products or multiple technical services) into a whole. Most of them are customized services, providing services to customers in the form of items. The main requirements of its knowledge modularization include physical modules as well as technical modules. The solution focuses more on using existing products or technologies to solve the actual problems of customers, and provides more customized products or services. The proportion of direct use of knowledge modules is not high so that the reuse rate of this way is the lowest.

2.3 Mapping Relationship Between Products and CBB

CBB supports the production of enterprises as a product component, while it supports the production of upstream enterprises as a product in external sales. The specific mapping model is shown in Fig. 2.

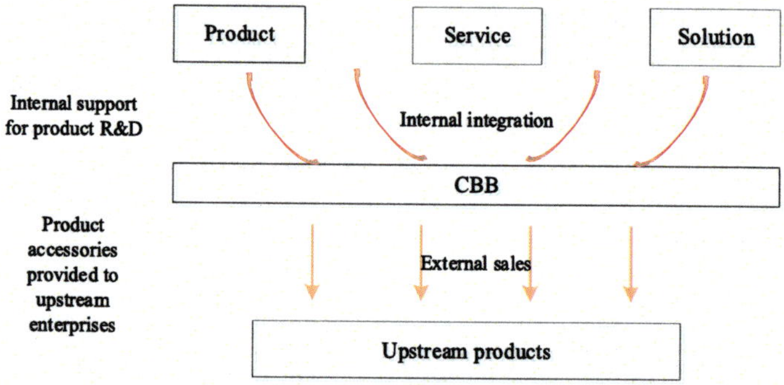

Fig. 2. Classification characteristics of comprehensive knowledge service products

Products, services and solutions are composed of multiple modules. In the research and development of new products, existing modules can be selected as product components. The selected CBB may be a component module of the previous generation product or a component module of other types of products. At this time, CBB is an essential component of a new enterprise product, service or solution, and it is a product component.

In order to maximize the technology and use value, CBB can also be sold as an independent product and used as a supporting component for products produced by other companies. At this time, CBB can be considered as a product of the enterprise.

3 Classification and Hierarchical Design

3.1 Classification of CBB

Considering the comprehensiveness of knowledge service solution products, CBB is divided into five categories: software, hardware, data, tools, and forms.

(1) Software CBB: a software module with independent functions that can be transplanted, updated and used independently, or a software composed of multiple functions that can be packaged and used independently. A Software CBB can be a software product or a module that composes a software product and has independent functions. The most important demand is that it must have source code, independent functions and external interfaces, which can be directly called during the development of other software.

(2) Hardware CBB: Hardware components with independent functions that can be combined, integrated and used, or hardware that can be packaged, used and delivered independently. A Hardware CBB can be a small hardware product, or a module or subsystem that composes a hardware system. The most important feature is the physical property. It can be an integral part of a specific product or a universal tool and accessory used in the test process.
(3) Tool CBB: a method carrier with specific functions, properties and uses. The Tool CBB is an electronic tool, calculation model and other tools that developed to realize a specific work. It can be a calculation tool that embeds cumbersome calculation formulas to simplify a certain work, or it can be a statistical tool to standardize the work process.
(4) Data CBB: a data package with specific purpose, which is not limited by the data carrier, has a certain amount of data, and can be updated, filled and used. The core of the Data CBB is data, so a large amount of data must be equipped and can be updated for a long time.
(5) Form CBB: templates, forms, processes, samples or cases with specific purposes. In addition to the software, hardware, tools, and data that can be made visible, there are a large number of technical methods, test methods, solution cases, work guidelines, work procedures, process plans, test specifications, report templates, checklists, consulting and training courseware, etc. in the research results of technical service institutions, these results are displayed and explained in the form of text, so they are uniformly classified as Form CBB. Form CBB is the main component in the CBB system of technical service institutions.

3.2 Hierarchy of CBB

Given the differentiated and customized design requirements of knowledge service solutions, CBB is divided into four levels: A, B, C, and D according to the use value and use performance of CBB in practical applications.

(1) A-level CBB can be independently packaged and sold. With the highest maturity, it can be sold directly as a finished product and can be used directly without any modification. A-level CBB mainly includes software modules, hardware components and other directly-formed products, and they can be directly called or selected for use among multiple products, such as manufacturing parts or software modules of a manufacturing company.
(2) B-level CBB is difficult to package and sell independently, but can be reused after minor or general modifications. The maturity is high, but it is seldom to sell directly as a finished product.
(3) C-level CBB is difficult to package and sold independently, and can be reused only after major modifications or adjustments. The maturity is not too high, but it has high reusability.
(4) D-level CBB is that which has not reached the A, B, and C maturity levels, but has use value or reference within a certain range. This type of CBB is used in a relatively small range, and needs to be modified during the use process.

4 Construction Path

Enterprises can build CBB through forward-planning and backward-sorting methods. This paper focuses on building CBB with backward-sorting.

The forward-planning method adopts a slim waist type structure. In the product development/project research planning, design, and demonstration stages, the module is divided according to the product planning and structure, and the product function and technology are decomposed according to the core and key technologies determined in the Functional Flow Block Diagram (FFBD). These are to find shared technologies, analyze and form modules with technological leadership and general to similar products/projects. The formation and path of forward-planning CBB are shown in Fig. 3.

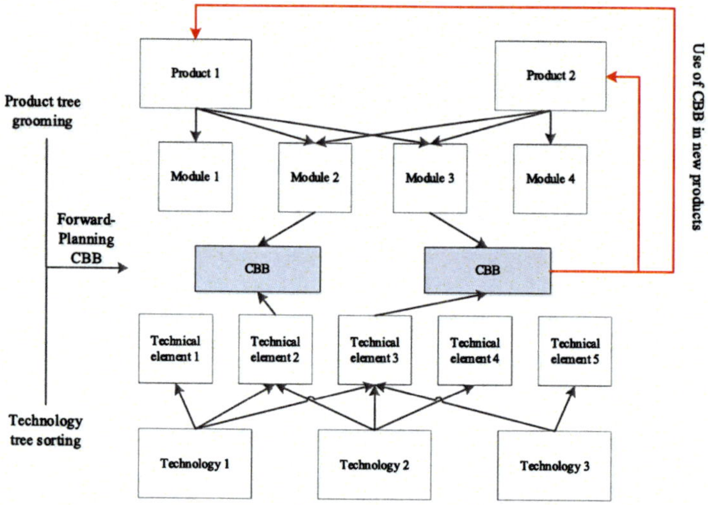

Fig. 3. Forward-planning CBB construction

Forward-planning is the basis for promoting products to continuously occupy the market. However, it is generally difficult to form a CBB based on market demand, product lineage planning, and technical direction planning because it requires companies to have strong market analysis capabilities and a clear product structure. When constructing the CBB mechanism, most enterprises use the backward-sorting method to form the CBB. This paper adopts the backward-sorting method to construct the CBB management mechanism.

The backward-sorting method relies on the research results of previous projects. After the product/project research is completed, the results of multiple products/projects are shared, analyzed and extracted, and the software and hardware modules and technical modules that can be shared and participated are sorted out. Figure 4 shows the formation and use path of CBB in backward-sorting method.

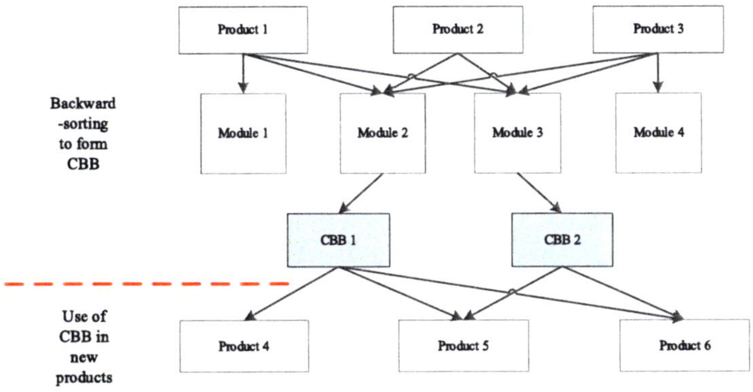

Fig. 4. CBB construction in backward-sorting method

5 Design of "EQUAl Responsibility and Right"

5.1 Stakeholders of CBB

As a management mechanism of an enterprise, CBB involves many stakeholders such as decision-making, management, marketing, design, production, sales, etc. The purposes and demands of each party are quite different. Therefore, how to identify the relevant parties and realize the balance of interests of the parties is the top priority to ensure the effective implementation of the CBB mechanism. The model of each stakeholder is shown in Fig. 5.

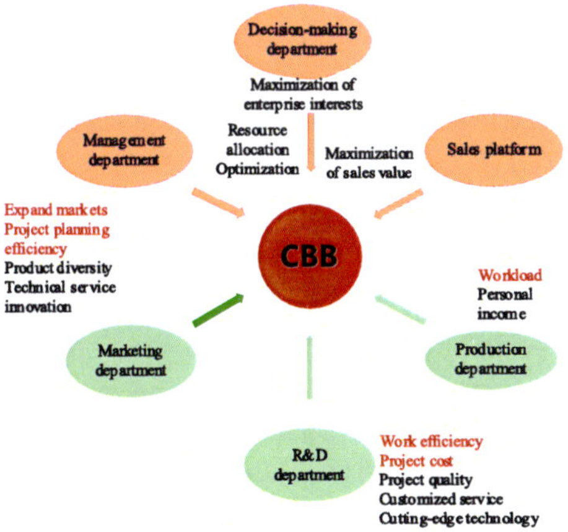

Fig. 5. CBB stakeholder model

The decision-making department, management department and sales department are the supporters of the construction of the CBB mechanism. The marketing department, R&D department and production department have positive and negative perceptions of CBB, which are closely related to their management demands:

(1) The decision-making department is most concerned about maximizing the overall benefits of the enterprise. CBB can achieve efficient reuse of past results, avoid repeated investment, reduce labor costs, and improve production efficiency, which has a significant effect on improving the overall efficiency of the enterprise.
(2) The management department is the executive department of various management mechanisms. The most concern is whether the overall resource allocation of the enterprise is optimal. CBB mechanism can fully revitalize the previous product modules or technical modules to realize the sharing among multiple products, and greatly improve the utilization of resources.
(3) The sales department pursues the maximization of sales output value. The CBB mechanism promotes the upgrading of products to a certain extent, which is more conducive to the realization of output value growth.
(4) The marketing department has the function of developing the business market, and tends to carry out efficient product development, making it the supporter of the CBB mechanism. On the other hand, it has the function of planning and launching new products, which is more inclined to technological innovation and inhibits its use of CBB.
(5) The R&D department tends to use CBB in order to improve R&D efficiency and reduce project costs. But when conducting cutting-edge technology R&D or developing customized solutions, it focuses more on innovation and uniqueness rather than the use of CBB.
(6) When the production department uses the existing CBB, it has the advantages of low cost and high efficiency. Based on the consideration of reducing workload, CBB will be used. However, when product production, technical services or external solutions are provided based on existing basic modules, personal creative work will be reduced with lower income, so CBB will not be used all the time.

Based on the analysis of the CBB stakeholder model, the main stakeholders are identified as decision-making and management, planning and implementation, and CBB holders. The specific responsibilities and rights of each stakeholder are shown in Table 1.

5.2 Ownership and Protection of Intellectual Property Rights

Considering that CBB consists of condensed and shared modules of multiple products, it is generally formed by the joint design and development of multiple persons. In order to avoid intellectual property disputes and benefits feedback disputes, CBB application adopts the method of joint declaration of applicants, who need to negotiate the contribution ratio of CBB and sign to confirm, as the basis for determining the CBB holders and holding ratio.

After accepting the CBB declaration, a certain range of announcements will be made. The main function is to avoid any disagreement among employees on the value, composition and contribution ratio of the CBB during the application process. This can prevent unfair competition and avoid intellectual property risks. When there is an objection, the employee can file an appeal and submit supporting materials. At the same time, it is necessary to review the confidentiality and loss of state-owned assets through technical organizations to ensure legality.

Table 1. Responsibilities and rights of stakeholders

	Responsibility	Right
Decision making and management	Policy guidance and resource guarantee Improve the CBB mechanism, eliminate policy barriers, and strictly control management links Promote the generation and use of CBB	Maximize the benefits of the organization Organizational resource optimization
Planning and implementation	Analyze, select and accurately use CBB Propose CBB requirements and carry out CBB planning CBB use value or effect analysis and evaluation	Use project funds independently Benefits feedback from cost savings
CBB holder	Modify and maintain CBB Promotion of CBB holdings Priority use of CBB held	CBB benefits feedback of use CBB benefits feedback of sale

5.3 CBB Use and Maintenance

In the process of product/project R&D, CBB can be selected at each stage of planning, implementation and delivery according to the development and production requirements. According to CBB's security level, technical value, application scope, etc., it is used in the form of external sales, internal free use, application for paid use, approved use, and restricted use. When loopholes and defects in design, production and use are found during use, CBB holders have the rights and obligations to maintain them. Maintenance should not cause subversive changes to the main functions and performance, but should maintain the common functional interface of CBB as much as possible.

5.4 Benefits Feedback Mechanism

In order to promote the establishment of the CBB mechanism and establish a virtuous cycle of benefits feedback, certain material incentives are given to the teams and individuals that generate CBB to ensure that CBB can be continuously generated. According to the level and use form of the CBB, a certain percentage (such as 5–15% of the gross profit from sales) will be commissioned and given out to the CBB holder. In the initial stage of the establishment of CBB mechanism, in order to promote the use and gradually improve the performance and versatility of CBB, it is advisable to appropriately reward teams that actively use CBB.

6 Shared and Participating Platform Design

In order to promote the operation of the CBB mechanism, realize the explicitness of the CBB shelf, and facilitate CBB sharing and participating, it is necessary to build a sharing and participating CBB platform. The platform includes basic information, functional performance introduction, statistical analysis and other information by access of browsing or downloading.

(1) CBB basic information. In order to ensure the CBBs to be unique and identifiable, the platform includes basic information such as the number, name, type, maturity level, security level, holder and other basic information of each CBB. The sharing and participating platform structure is shown in Fig. 6.
(2) Function and performance introduction. In order to fully demonstrate the characteristics of CBB and facilitate its selection and use, the platform includes a brief introduction of CBB, description of main functions and performance, possible application fields and usage methods.
(3) Browse and download. In order to promote the wide access of CBB, the direct download is permitted for some CBBs that do not involve state secrets, trade secrets, and core competitive technologies.

Numbering	Name	Type	Classified	Maturity	Holder	Main Function Introduction	Listing Status	Browsing/Downloading Person	Browsing/Downloading Time	Download Filename
CBB unique number	CBB Chinese name	Software/Hardware/Data/Tools/Forms	Fill in based on whether it involves state secrets or commercial secrets	Class A/Class B/Class C/Class D	All holders (including the contribution ratio of holders)	The technology that CBB relies on, the functions it realizes, the products it can support, etc.	Officially put on the shelf/stay on the shelf/off the shelf	The browsing/downloading usage of each CBB recorded in the information system.		

Fig. 6. Sharing and participating CBB platform data structure

7 Typical Applications

7.1 The Predicament of Technical Service Institutions in the Military Industry

The demand for modularization of technical service products in military research institutes is becoming more and more prominent. Military industrial research institutes are mainly responsible for technical research work such as national defense, basic scientific research projects of military industry, key technology research, and equipment pre-research. They take the completion of the research and production of weapons and equipment as the core, and focus on technical research, technical services in specific fields and small batch production.

In recent years, with the increase in model development tasks, the amount and complexity of projects undertaken by military research institutes continues to grow, and at the same time, the design and development cycle is shortened but the engineering quality requirements are gradually improved. In order to improve R&D efficiency, reduce

R&D costs, and relatively adapt to the diverse development needs of equipment, military research institutes are gradually exploring the CBB management mechanism during the R&D process of their projects.

7.2 Exploration of an Aviation Research Institute

The CBB mechanism was established and operated in an aviation research institute, the standard of General Requirements for CBB Management was issued, and all staff were trained for publicity and implementation, and a CBB management system was built to form a CBB shelf. Since the CBB mechanism has been in operation for one year, nearly 50% of the CBB in the institute have been submitted by the researchers, and a total of 254 items have been approved and put on the shelves. The CBB on the shelf is used in the browsing/downloading, application in projects, and online sales, and so on.

With the continuous advancement of the CBB mechanism and the increasing number of CBBs on the shelves, the number of browsing/downloading, as a primary form of use, continues to grow steadily, which reflects the use value of CBB to a certain extent. Taking the browsing/downloading situation from August 2019 to December 2021 as an example, the CBB on the shelf has a total of 12,473 browsing/downloading times, of which 5669 are downloaded. The distribution trend of CBB shelf browsing/downloading is shown in Fig. 7.

The number of CBB browsing and downloading are growing at an average rate of nearly 7% per month, with 7.8 browsing/downloading per capita across the institute. During the product/project planning stage from March to June 2020, the number of CBB browsing increased significantly.

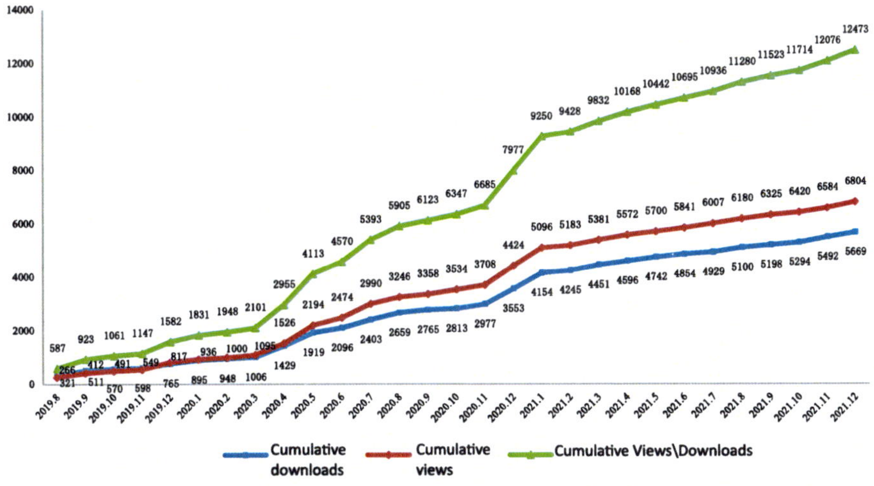

Fig. 7. CBB shelf browsing/downloading trend graph

7.3 Initial Results

In view of the strong relationship between the generation and use of CBB and the project/product R& D process, CBB generation is a necessary condition for project delivery. A positive cycle is formed, from CBB generated by the project to CBB used by the project. At the same time, the generation and use of CBB can be included in the relevant requirements of employee assessment and evaluation, which plays a role of policy-promotion in the early stage of the establishment of CBB mechanism.

CBB has been applied in hundreds of projects/products planning and implementation, and the project R&D utilization rate of CBB on the shelf is nearly 60%. Some projects have reduced the field service time by nearly 30%, some project costs have been reduced by more than 20%, some projects have realized the replacement of senior employees by new recruits, and some projects have realized multi-functional composite development.

Some CBBs have achieved Internet sales revenue, and some templates, workflows, cases have achieved hundreds of repeated sales, exploring a new growth pole for online sales of knowledge service solutions. A virtuous circle has been initially realized, that is the research results continued to generate benefits with just one-time investment in research projects.

7.4 Benefits Feedback

Internet sales of CBB have been fed back to dozens of CBB holders according to the specified proportions based on the online sales profits; CBB used in the project has been fed back according to the specified proportions of project cost saving contribution.

8 Conclusion

The construction of the CBB mechanism is a useful exploration of the management transformation of technical service institutions, a beneficial discovery of the redistribution of project results for knowledge service solutions, and a practice under the principle of ownership by enterprises, income-right enterprises and individuals sharing. It contributes to realize the modularization of products, knowledge and technology and the sharing and participating of research results. It will significantly improve the efficiency of scientific research and production, reduce costs, and shorten the research and development cycle, and realize one-time R&D investment of products/projects, reuse of products and modules, and multiple outputs.

An aviation research institute has proved through practice that the CBB management mechanism can maximize the reuse of knowledge, technology, and product modules, improve work efficiency, reduce R&D costs, and achieve "once input, multiple outputs".

The research direction and focus of the future CBB management mechanism will focus on: (1) CBB technical value assessment, including CBB pricing, use value evaluation and so on; (2) Exploration of the reform of disposal rights, including CBB holder conversion, CBB inter-enterprise transactions, and price-based shareholding 3) Embedded application of knowledge-based CBB in projects/products, including software-based knowledge-based CBB, interface standardization of knowledge-based CBB, and intelligent notification and invocation of knowledge-based CBB.

References

1. Li, W.: Study on the Performance Evaluation System in the Technical Service Organization. Jilin University (2003)
2. Hui, Z.: Product Research and Development Management. Electronic Industry Press, Beijing (2012)
3. Xiaobin, Y.: Design and implementation of CBB platform. Modern Electron. Technol. **15**, 104–106 (2015)
4. Shaojie, C., Wei, S.: Analysis of CBB construction of military scientific research enterprises. Enterp. Sci. Technol. Dev. **7**, 190–194 (2018)
5. Yi, A., Cao, G., Xu, X.: The pilot study of portable computer DIY under the CBB standard. Comput. Knowl. Technol. 1054–1176 (2007)
6. Yanhui, W.: Transformation practice of scientific research and production mode based on aerospace product engineering. Aerosp. Ind. Manag. **8**, 39–45 (2016)
7. Zubing, W., Ning, S.: Research on the problems in the CBB management of military electronic enterprises and the countermeasures. Stand. Pract. **8**, 73–76 (2019)
8. Jinwang, X., Yan, Z., Renlun, H.: The electronic products concurrent design methodology and CBB reuse based on cadence designer. Electron. Sci. Technol. **4**, 33–36 (2017)
9. Su, Q., Liu, S.: Practice and exploration of software CBB construction. In: Proceedings of Quality Management System Evaluation Practice, pp. 219–223 (2017)
10. Chao, Y., Xingmin, D., Yanping, W.: Study on the research and development model of military-civilian integrated aviation electromechanical products based on CBB module. Enterp. Sci. Technol. Dev. **6**, 48–51 (2019)
11. Yuntong, L., Renzhong, T., Jun, Z.: A knowledge organization for product design approach model based on modular method. J. Zhejiang Univ. **45**(11), 1900–1907 (2011)

Check for updates

Service-Oriented Intelligent Avionics System Architecture

Zhao Yan[1,2](✉), Wen Pengcheng[1,2], Bai Linting[1,2], and Wu Na[1,2]

[1] Xi'an Aeronautics Computing Technique Research Institute, AVIC, Xi'an 710068, China
zhaoyan@ftcl.hit.edu.cn

[2] Aviation Key Laboratory of Science and Technology On Airborne and Missile-Borne Computer, Xi'an 710065, China

Abstract. As the core component of the aircraft, the avionics system has gone through the independent, federated, and integrated architecture to deal with the increasingly complex air combat scenarios. With the development of artificial intelligence and integrated circuits, avionics systems should be prepared to equip a large number of intelligent applications, which contributes to more effectively completing the military missions. In addition, the next generation of avionics systems need to be more flexible, and thus can be conveniently updated and allow intelligent applications effectively communicate with each other. Therefore, to achieve a more intelligent and flexible avionics system, this paper proposes a novel Service-oriented Intelligent Avionics System Architecture, named as SIASA. With such a service-oriented architecture, SIASA provides a flexible software platform by treating intelligent applications as services. Specifically, the proposed SIASA consists of seven layers, i.e., airborne intelligent chip layer, abstraction layer, hashrate abstraction layer, equipment abstraction layer, airborne service capability layer, airborne intelligent application atomic service layer, and airborne intelligent application layer. With these layers, SIASA well supports intelligent applications and is able to make effective and efficient updates. Furthermore, SIASA also supports more advanced scenarios, including airborne OTA, drag-and-drop software development, and system intelligence.

Keywords: Avionics system · Service-oriented architecture · Artificial intelligence

1 Introduction

The concept of Artificial Intelligence (AI) was proposed at the Dartmouth Conference in 1956 [1]. After more than 60 years of continuous development, AI has become the core driver of a new round of industrial revolution [2]. In the defense industry, AI plays an important role in promoting the upgrade of combat equipment and the transformation of combat mode. However, general AI cannot fully support the needs of intelligence in the defense industry. The reason is that military scenarios contain the characteristics of highly complex environment, strong confrontation, and requiring real-time response. Therefore, multiple countries have put forward projects to support intelligence for military purposes,

Chinese Society of Aeronautics and Astronautics: CASTYSF 2022, LNEE 972, pp. 492–503, 2023.
https://doi.org/10.1007/978-981-19-7652-0_45

e.g., the Intelligence and Data Acceleration (ADA) program from U.S. Department of Defense and the "Future Defense 2030 Technology Strategy: Defense AI Technology Roadmap" released by the Korea Institute for Defense Technology Promotion.

As a key component of combat equipment, military aviation equipment needs to be comprehensively upgraded for improving the degree of intelligent defense industry [3]. Applying AI to avionics contributes to extending and enhancing human capabilities and efficiency [4]. To achieve the goal of intelligent aviation equipment, the avionics system architecture needs to be enhanced. Specifically, avionics system is the sum of all electronic systems on the aircraft [5], including fire control, display and control, radar, communication and navigation, external management, and many other subsystems. Currently, most military aviation systems have their own independent requirements, and each of them is provided by a single supplier with customized platforms and services. This circumstance results in a lack of hardware and software reusability between various aircraft platforms, long delivery times, and makes the improvement processes complicated. Especially, after the introduction of AI technology, the avionics system may consist of tens or even hundreds of intelligent applications inside, and the applications often need to be upgraded or even reconfigured online to cope with different operational tasks. To confront these circumstances, the characteristics of openness, dynamism, integration, modularity, and generalization might be essential for the future avionics system architecture [6].

To more effectively manage the increasingly complex software system and provide a more flexible and dynamic platform, Service-Oriented Architecture (SOA) has been adopted by multiple industrial areas. Specifically, SOA is an architectural design method that abstracts the capabilities of a system into services with standard interfaces that can be accessed, thus an application can be implemented by invoking one service or composing multiple services [7]. By defining the logical service interfaces, the service-oriented architecture clarifies the division of functions and the agreement of information communication. Therefore, SOA can be utilized to shorten the development cycle of the intelligent OODA loop for each scenario, enhances the reusability of software, and reduces the cost of system integration.

To make the avionics system more intelligent and flexible, this paper proposes a service-oriented intelligent avionics system architecture, named as SIASA. Specifically, SIASA consists of seven layers, i.e., airborne intelligent chip layer, hashrate abstraction layer, airborne equipment abstraction layer, airborne service capability layer, airborne intelligent application atomic service layer, service intelligent governance layer, and airborne intelligent application layer. In the following, this paper first introduces the related work on avionics system architecture and service-oriented architecture, then presents the proposed SIASA, after that makes an outlook, and finally summarizes the paper.

2 Related Work

2.1 Avionics System Architecture

Since the 1960s, the avionics system architecture has gone through four stages, i.e., independent, federated, integrated, and advanced integrated avionics system architecture [8].

In the 1940s-60s, the avionics system architecture was independent and consisted of multiple control systems [9]. Each of these control systems consists of a number of independent controllers connected point-to-point to dedicated sensors and peripherals. In the independent avionics system architecture, no interconnection mechanism exists among controllers.

In the 1970s and 1980s, as a result of the development of digital avionics information systems, the federate avionics system architecture interconnected the avionics subsystems through the 1553 bus [10], which enhanced the information interaction capability and made the control systems loosely coupled with each other.

In the 1980s and 1990s, with the rapid development of integrated circuits and network technology, the integrated avionics system further enhanced the data processing capability by utilizing high-performance microprocessors over fiber-optic data buses [11]. Later, the advanced integrated avionics system architecture has been proposed based on the integrated avionics system architecture to further enhance the integration of various system processing and sensor signal processing capabilities. At the same time, a number of new technologies related to AI are introduced to assist the aviator to operate military operations.

Currently, with the application of emerging technologies such as deep learning and big data in aviation systems, the performance requirements of avionics systems continue to rise. To support the intelligent applications over big data, heterogeneous intelligent processing architectures are required to fuse multiple intelligent chips, such as GPUs, FPGAs, and ASICs [8]. In addition, with the further development of system decoupling and service-oriented concepts, the top-level applications may face a considerable number of atomic services, which require a service-oriented architecture to support flexible and efficient intelligent management and scheduling.

2.2 SOA Architecture Research and Application Status

Service-Oriented Architecture (SOA) [12] is an architectural design approach that abstracts the capabilities of a system into services with standard interfaces that can be invoked or composed to meet the specific needs of the users. It is recognized that SOA has the characteristics of high cohesion within services, loose coupling between services and standardization of service communication, etc. There is also a consensus that SOA can build a flexible and variable platform system, which contributes to effectively maintaining and extending the increasingly complex software systems.

SOA has been widely utilized in many fields, e.g., the automotive industry, IoT, and banking industry. Taking the automotive industry as an example, SOA software architecture technology for in-vehicle systems has received increasing attention from global vehicle manufacturers due to the demand for intelligent and networked vehicles, especially for the development and application of autonomous driving systems. Specifically, Volkswagen, Audi, BMW, Ford, and other automotive giants have formed their own alliances to conduct research on the application of SOA technologies and specifications, which are expected to start being applied to mass production models around 2023.

3 Service-Oriented Intelligent Avionics System Architecture

3.1 Overall Framework

Figure 1 demonstrates our proposed Service-oriented Intelligent Avionics System Architecture, named as SIASA. Specifically, SIASA contains 7 layers, i.e., airborne intelligent chip layer, abstraction layer, hashrate abstraction layer, equipment abstraction layer, airborne service capability layer, airborne intelligent application atomic service layer, and airborne intelligent application layer.

The above-mentioned layers are briefly introduced in this paragraph and more details are presented in Sects. 3.2–3.8. Airborne intelligent chip layer contains intelligent chips such as CPU, GPU, FPGA, ASIC, and neuromorphic chip, which are suitable to support the computational task from intelligent applications. These chips are abstracted by arithmetic abstraction layer through resource virtualization and heterogeneous chip difference shielding. Similarly, the equipment, including sensors and weapons, are abstracted by the equipment abstraction layer to provide a formalized interface. Then, the airborne service capability layer is provided to support service description, service registration, service discovery and service communication, which are the fundamental elements of a service-oriented architecture. After that, we devised the airborne intelligent application atomic service layer, which contains observe services, orientation services, decision services, and action services. Finally, based on these atomic services, the airborne intelligent applications including autonomous flight applications and air combat applications are implemented in the airborne intelligent application layer.

3.2 Airborne Intelligent Chip Layer

Despite of high performance, intelligent applications require a large number of calculations and can hardly be fulfilled by the conventional CPU. Thus, intelligent chips have been studied and manufactured to fill this gap. Intelligent chips generally refer to all hardware employed to accelerate intelligent applications, especially applications based on deep learning techniques. Specifically, based on the technical architecture, intelligent chips can be classified into GPU, FPGA, ASIC, and neuromorphic chips. Different classification methods exist for intelligent chips. For example, based on the location, intelligent chips can be classified into cloud chips, edge chips, and terminal chips; and based on the usage stage, intelligent chips can be classified into training chips and inference chips. In the airborne scenario, the intelligent chip is usually used as inference chips at the edge end or terminal end, while the training process is executed in the ground-based cloud end.

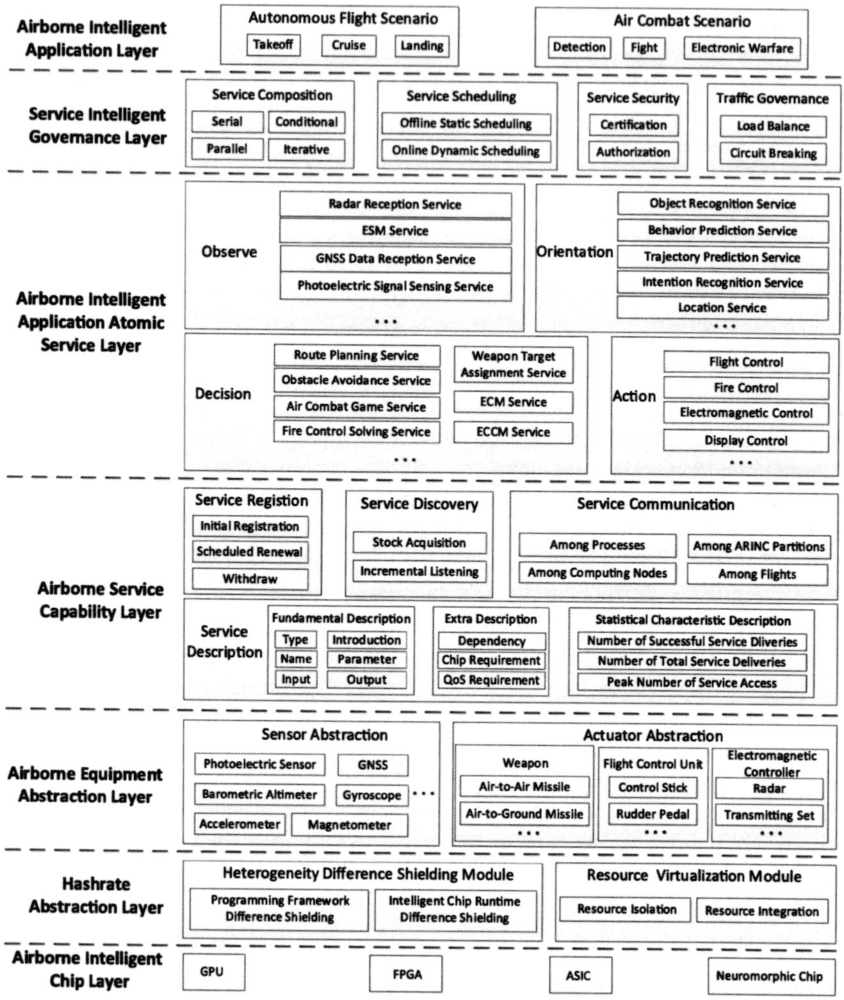

Fig. 1. The overall framework of the service-oriented intelligent avionics system architecture

3.3 Hashrate Abstraction Layer

The demand for intelligent applications in airborne scenarios is various, and a single type of chip cannot support intelligent applications in all airborne scenarios. Therefore, multiple types of chips are essential for the airborne computing platform to improve the performance of intelligent applications. In order to achieve flexible deployment of intelligent applications across multiple chip types and improve the utilization rate of computing resources, it is necessary to mask the differences in runtime and programming frameworks between different chip types. Moreover, in order to increase the utilization ratio, the virtualization technique is also essential to allow one chip to be utilized by multiple applications. Therefore, this paper devises a heterogeneous difference masking module and a resource virtualization module. The heterogeneous difference masking module includes deep learning programming framework difference masking and intelligent chip runtime difference masking, and the resource virtualization module includes resource isolation and resource integration.

3.4 Airborne Equipment Abstraction Layer

This layer is dedicated to the abstraction of the access control interface of relevant sensors, actuators, 11 and other physical devices to shield their physical differences, and thus realize the decoupling of hardware and software. Through this way, SIASA shortens the research and development cycle, enhances portability, and achieves the purpose of plug-and-play. Focusing on intelligent avionics systems, this paper abstracts the sensors and actuators related to intelligent applications. Specifically, typical sensors include photoelectric sensors, barometric altimeters, acceleration instruments, GNSS, gyroscopes and magnetometers, etc.; and typical actuators include weapons, e.g., air-to-air missiles and air-to-ground missiles, flight controllers, e.g., control sticks and rudder pedals, and electromagnetic controllers, e.g., radars and radio transmitters.

3.5 Airborne Service Capability Layer

This layer provides the essential components of a service-oriented architecture, and is a key role of SIASA. As presented in Fig. 2, the service-oriented architecture contains three key roles, i.e., service consumer, service provider, and service registry; and contains four main functions, i.e., service description, service registration, service discovery, and service communication.

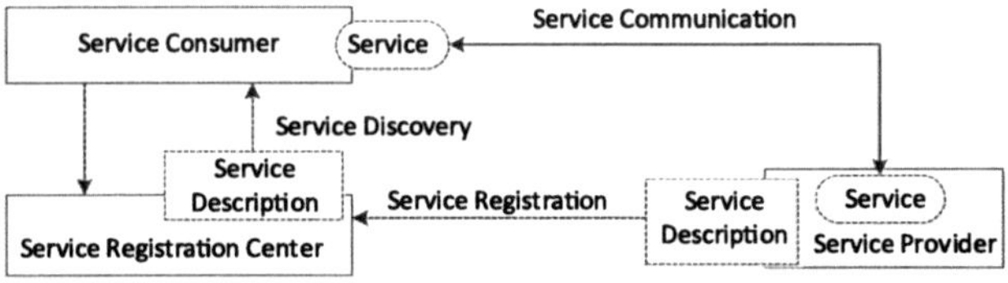

Fig. 2. The key roles and main functions of the service-oriented intelligent avionics

Brief descriptions of service provider, service consumer, and service registry are as follows.

- Service provider. The entity that provides services. Typical functions needed to be implemented include service development, service description (i.e., the standardized definition of the service interface, which will be introduced in the following), service registration, and other functions.
- Service consumer. The entity that consumes the service. Typical functions needed to be implemented includes querying the service description and communicating with the service.

- Service registry. The entity that supports the registration, publication, search, and evaluation of services. The typical functions needed to be implemented include publishing, adding, deleting, modifying, querying and classifying service description, verifying whether the service meets the standards, and notifying service consumers when the registration information updates.

A detailed functional description of the service-oriented architecture is as follows.

Service Description. Service description refers to a standardized definition of the service interface with a unified syntax that describes the basic information, functions, requirements, and statistical information of services. According to the characteristics of airborne services, the service description defined in this paper contains three parts, i.e., basic description, additional description, and statistical characteristics description. Specifically, the basic description includes attributes such as type, name, parameters, input format, output format, and service introduction; additional description includes attributes such as service dependency, intelligent chip requirements and QoS requirements; and the statistical characteristic description includes attributes including the number of successful service deliveries, and the number of total service deliveries.

Service Registration. Service registration means that the service provider registers the description information of the service to the service registry. After the initial registration, which allows service consumers to query the service description. Service registration is divided into three stages, i.e., initial registration, scheduled renewal, and withdrawal.

- Initial registration. The service needs to be registered to the registration center once it starts up and is ready to accept requests, which allows it to be discovered by other services or programs.
- Scheduled renewal. Scheduled renewal is equivalent to keeping alive, which periodically notifies the service registry that the service is still available and can continue to provide the service.
- Withdrawal: The registration information should be actively withdrawn when the service stops.

Service Discovery. Service discovery refers to the process by which a service gets access to services that can meet specific functional and non-functional requirements. This process involves the matching between the service requirement from the service consumer and the service description from the service provider. Service discovery is generally divided into two phases, i.e., stock acquisition and incremental listening.

- Stock acquisition: The service registry should provide a list of registered services to the new service or program.
- Incremental Listening: When a new service is registered to the service registry, the service registry shall notify this information to existing services.

Service Communication. Service communication refers to the cross-platform and cross-programming language data transfer between services through a unified and standardized communication mechanism. In order to improve the flexibility and efficiency of communication between services in the airborne environment, it is necessary to realize multi-channel communication, such as communication between processes, ARINC partitions, computing nodes, between flights, etc.

3.6 Airborne Intelligent Application Atomic Service Layer

Atomic services are the services with the smallest granularity and are generally designed as stand-alone logic modules and algorithms. Note that device-oriented functional atomic services are based on the device abstraction layer, and provide input or executed the instructions from logic or algorithmic services. Mapped to the field of airborne intelligence applications, atomic services can be classified into two types, i.e., (1) stand-alone intelligent algorithms, and (2) sensor-oriented or actuator-oriented functions associated with intelligent algorithms. As demonstrated in Fig. 3, this paper splits intelligent applications into four types of atomic services, i.e., observe services, orientation services, decision services, and action services based on the OODA loop theory.

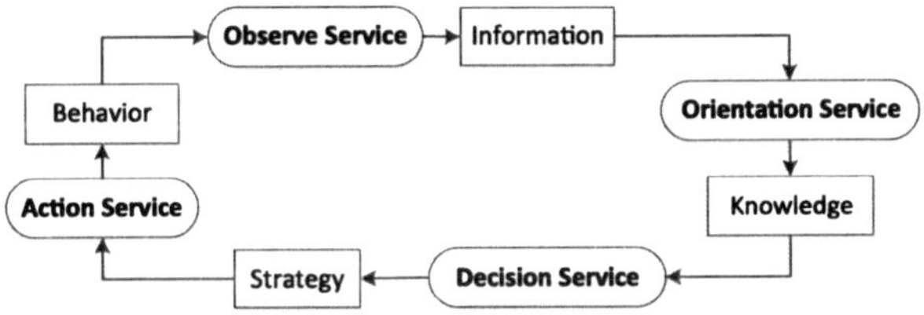

Fig. 3. Illustration of airborne atomic services in an OODA loop

In this paper, atomic services are devised for the autonomous flight scenario and air combat scenario. Specifically, the typical applications for the autonomous flight scenario include takeoff, cruise, and landing; and the typical applications for the combat scenario include detection, fight, and electronic warfare. To compose the applications mentioned above, a total of 18 atomic services have been devised. These atomic services contain 4 types of observe services, i.e., radar reception service, GNSS data reception service, photoelectric signal sensing service, ESM service; 5 types of orientation services, i.e., object recognition service, behavior prediction service, trajectory prediction service, intention recognition service, and location service; 7 types of decision services, i.e., route planning service, autonomous obstacle avoidance service, air combat game service, fire control solving service, weapon target assignment service, ECM service, and ECCM service; and 4 types of action services, i.e., flight control service, fire control service, electromagnetic control service, and display control service.

3.7 Service Intelligent Governance Layer

This layer focuses on the intelligent governance of services and is responsible for the intelligent composition of atomic services and their effectivity, security and reliability. Specifically, this layer consists of four modules, i.e., service composition, service scheduling, service security, and traffic governance. More details about these modules are introduced in the following.

Service Composition Module. The service composition module defines the logical execution order of atomic services and is devised specifically for composing intelligent applications. In this paper, 4 types of execution order are defined, i.e., serial order, parallel order, conditional order, and iterative order. The service composition design for the typical intelligent application, i.e., autonomous cruise, is presented in Fig. 4. In this case, serial order exists between radar reception service and location service, parallel order exists between radar reception service and GNSS data reception service, conditional order exists between route planning service and location service as route planning service is not always called, and the whole OODA loop is an iterative order.

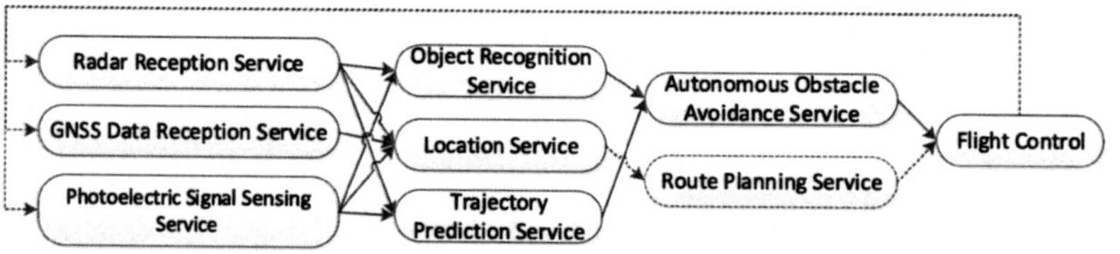

Fig. 4. Atomic service design and combination for autonomous cruise

Service Scheduling Module. This module is responsible for the mapping from atomic services to the computational resources by intelligent techniques. Such a mapping process should consider the matching degree between atomic services and resources as well as the matching degree between communication requirements among atomic services and the communication efficiency. Generally, service scheduling can be divided into two stages, i.e., offline static scheduling and online dynamic scheduling.

- Offline Static Scheduling refers to the scheduling process before taking off and is generally conducted on the ground center. The goal of this scheduling process is to provide an optimal or near-optimal accurate mapping relation from atomic services to the computational resources given enough scheduling time. The typical techniques suitable for this scheduling process are generally from the field of operation research, e.g., mixed-integer programming.
- Online Dynamic Scheduling refers to the scheduling process when the flights are in the air. For example, such online dynamic scheduling is essential when some computational resources are not available because of malfunction or being destroyed. The goal of this scheduling process is to provide a feasible solution given limited scheduling time. The typical techniques suitable for this scheduling process are evolutionary algorithms, e.g., the ant colony algorithm.

Service Security Module. This module is responsible for the security of service communication. To achieve this goal, the generally employed techniques are authentication and authorization.

- Authentication verifies whether a request has the permission to the called service, and thus enforces access control to the services.
- Authorization is the process that asserts the identity of the service. Its typical application is to secure service-to-service communication from the man-in-the-middle attack.

Traffic Governance Module. This module is responsible for the governance of traffic-related issues. Traffic governance contributes to the stability and reliability of service-oriented architecture, Specifically, the typical techniques employed for traffic governance include load balance and circuit breaking.

- Load Balance is a mechanism that utilizes multiple replicas of services on different computing nodes, and then intelligently distributes the requests to different service replicas. This mechanism effectively avoids the single-point failure and increases the availability of the system. Typical load balance algorithms include round-robin, random, and least requests.
- Circuit Breaking is a mechanism to limit calls to a service, such as the number of concurrent connections or the number of failed calls to record the number of call errors. Once the limit has been reached within a certain service, calling this service returns the error code immediately. Such a process allows users to limit the impact of failures, latency spikes, and other undesirable effects of network peculiarities.

3.8 Airborne Intelligent Application Layer

Focusing on two main airborne scenarios, i.e., autonomous flight scenario and air combat scenario, this paper lists six typical airborne intelligent applications. Specifically, for the autonomous flight scenario, three typical intelligent applications have been considered, i.e., takeoff, cruise, and landing. As for the air combat scenario, this paper lists detection, fight, and electronic warfare as three typical intelligent applications.

4 Outlook

4.1 Airborne OTA

Over-The-Air (OTA) [13] is a remote wireless upgrade technology, which has been widely employed in multiple fields, e.g., automotive and Internet of Things. Applying OTA technology to avionics contribute to rapid updates and upgrade of intelligent applications while aircraft are flying, which are of vital importance in the highly dynamic and competitive military environment. With the service-oriented avionics system architecture, based on its unified and formatted service interface, the intelligent applications can be more easily updated or upgraded with the OTA process.

4.2 Drag-and-Drop Software Development

The drag-and-drop application development platform [14] enables developers to quickly complete simple application development for new scenarios through drag-and-drop operations or a small amount of code. Such a platform is especially helpful when developing avionics intelligent applications, which are commonly composed of several functions and might be updated frequently. The drag-and-drop application development platform could be conveniently implemented with a service-oriented architecture. Specifically, first devise each of the required functions as an atomic services, then define the unified and formatted service interfaces, and finally implement the application by service composition with a drag-and-drop user interface.

4.3 System Intelligence

System intelligence [3] is an advanced stage in the development of military intelligence, which aims to intelligently integrate all combat units for taking effective military operations. Specifically, system intelligence requires intelligent control, on-demand combat resource extraction and scheduling. The information communication and sharing as well as interoperability are the essence of system intelligence. The service-oriented architecture is a potential solution to realize the system intelligence with the unified service interface, service registration, and service discovery.

5 Conclusions

In order to provide a more intelligent and flexible avionics system, a Service-oriented Intelligent Avionics System Architecture, named as SIASA, has been proposed. In this paper, firstly, the development trend of avionics system architecture and the related work on service-oriented architecture is introduced. Then, the proposed SIASA is presented. Specifically, SIASA consists of seven services, i.e., airborne intelligent chip layer, abstraction layer, hashrate abstraction layer, equipment abstraction layer, airborne service capability layer, airborne intelligent application atomic service layer, and airborne intelligent application layer. With these layers, SIASA is able to make effective and efficient updates and well support the intelligent applications. After that, some advanced scenarios supported by SIASA have been presented, including airborne OTA, drag-and-drop software development, and system intelligence.

Acknowledgement. This work was partially supported by Aeronautical Science Foundation of China—Aerospace Artificial Intelligence Special (Grant No. 2018ZC31003).

References

1. Moor, J.: The dartmouth college artificial intelligence conference: the next fifty years. AI Mag. **27**(4), 87–91 (2006)
2. Zhou, J., Li, P., Zhou, Y., Wang, B., Zang, J., Meng, L.: Toward new-generation intelligent manufacturing. Engineering **4**(1), 11–20 (2018)
3. Lei, H., Yao, C.: Technical architecture of aviation artificial intelligence for military application. Navig. Position. Timing **7**(1), 1–11 (2020)
4. Zhou, Q., Li, B., Chen C., Zheng, X.: Research on avionics system architecture design technology of integrated AI. In: Proceedings of the Chinese Society of Aeronautics and Astronautics Youth Science and Technology Forum'2020, pp. 1054–1061 (2020)

5. Collinson, R.P: Introduction to Avionics Systems, 3rd edn. Springer (2013)
6. Jin, Y., et al.: The Characteristics and technical approaches of avionics system in the future from the new concept of combat. Aircr. Des. **41**(3), 49–52 (2021)
7. Perrey, R., Lycett, M.: Service-oriented architecture. In: Proceedings on Symposium on Applications and the Internet Workshops'2003, pp. 116–119 (2003)
8. Yan, L., Zhu, G., Qu, G., Peng, Q.: Research on the architecture development of avionics system core processing platform. Avion. Technol. **52**(3), 23–29 (2021)
9. Blasch, E., Kostek, P., Pačes, P., Kramer, K.: Summary of avionics technologies. IEEE Aerosp. Electron. Syst. Mag. **30**(9), 6–11 (2015)
10. Watkins, C.B., Walter, R.: Transitioning from federated avionics system architectures to integrated modular avionics. In: Proceedings on Digital Avionics Systems Conference, pp. 1–10 (2007)
11. Gaska, T., Watkin, C., Chen, Y.: Integrated modular avionics-past, present, and future. IEEE Aerosp. Electron. Syst. Mag. **30**(9), 12–23 (2015)
12. Krafzig, D., Banke, K., Slama, D.: Enterprise SOA: service-Oriented Architecture Best Practices, 1st edn. Prentice Hall (2004)
13. Halder, S., Ghosal, A., Conti, M.: Secure over-the-air software updates in connected vehicles: a survey. Comput. Netw. **178**, 107343 (2020)
14. Joy, J.: Review on different types of drag and drop mobile app development platforms. Int. J. Comput. Sci. Eng. (2018)

A Guided-Wave Based Damage Diagnosing Method with Energy Spectrum and Multi-scale Network

Lyu Shuaishuai(✉), Yang Yu, and Wang Binwen

Aircraft Strength Research Institute of China, Xi'an 710065, China
647817545@qq.com

Abstract. As a promising approach of structural health monitoring (SHM), Guided-wave is being widely used for damage diagnosis of carbon fiber reinforced composite (CFRP) in aeronautics. The traditional methods usually employ threshold for damage detection. However, as Guided-wave is easily influenced by structural and environmental factors, the man-made thresholds may vary significantly from case to case, which deteriorates its universal application. To overcome this weakness, a deep-learning-based method is proposed. Firstly, the sample is produced in the form of grey scale, so that its label is determined objectively rather than by human judgement. Secondly, it employs a multi-scale convolutional neural network with dedicated settings of kernel size and number of convolutional layers, which can effectively extract damage features from multi-dimensional signals while keeping the negative influences of increased network width low. Additionally, Gradient-class activation mapping (GRAD-CAM) technique is used to reveal the sample areas where the deep learning model concerns, which helps to interpret the physical significance of damage features. The experimental results demonstrate that the accuracy of damage identification and localization reaches 92.6%, which is superior to the traditional methods.

Keywords: Composite material · Damage diagnosis · Guided wave · Convolutional neural network · Grey scale

1 Introduction

Carbon fiber reinforced polymer composite (CFRP) have been widely used in the primary structures of aircrafts [1]. However, they are susceptible to the low-velocity impact of blunt objects, such as tools and runway debris. This kind of impact usually degrades as much as 40% of the laminate's compressive strength by inducing large areas of delamination inside with no observable dent from the surface [2]. Apparently, this is of severe safe hazard, and requires reliable approach of structural health monitoring (SHM) [3].

Guided wave is a very potential candidate of SHM, as it has the nature of long-range propagation and sensitivity to typical damages (such as crack, debonding and delamination) of aeronautic thin-walled structures [4, 5]. The mechanism of damage

Chinese Society of Aeronautics and Astronautics: CASTYSF 2022, LNEE 972, pp. 504–517, 2023.
https://doi.org/10.1007/978-981-19-7652-0_46

recognition is by differentiating the signal variations of two guided waves before and after the damage event, and its general method is to quantify this variation as Damage Index (DI) and compare it with the DI threshold [6–8]. However, as the guided wave is so susceptible to geometrical configurations of structure and environmental factors such as temperature and loading conditions that the DI threshold is usually determined by individual experience, and the quality of damage recognition is heavily expertise-dependent and not fitted for general application.

As the intrinsic work of damage diagnosis is to identify and extract damage-related signals from a large amount of overwhelming noise signals, deep learning method is spontaneously adopted by many researchers in the hope of enabling the damage recognition more accurate and less individual-experience-dependent [9–14]. Tian Guo [15], et al., presented a model for crack identification of metallic beam. It used the vibrating waves at diverse modules as input, and employed multi-scale CNN、ResNet、Fully Connected Layer to alleviate the side effect induced by noise and data missing. Its accuracy of crack recognition reaches 90% with a good manifest of generalization. However, this good result is obtained partly due to the relative simplicity in this case, and a large number of numerical samples collected through finite element simulation, which, however, is a general obstacle for many cases. For instance, Sbarufatti [16] used ANN model to identify crack and quantify its size in aluminum plates. Merely 38 physical cracks and 410 numerical ones are generated. Therefore, some particular measure such as data augmentation has to be adopted to deal with the problem of inadequate samples. Liang Xu [17] investigated the influential factor of structural dispersion specimen after specimen in crack detection. 6 aluminum hinges with identical structural configuration were loaded to collect guided-wave signals during the period of crack initiation and propagation. The deep learning model consisted of 3 convolutional layer, 1 pooling layer, 2 fully connected layers and 1 output layer.

Although all these learning models presented above claimed that they have a good performance in accuracy, the accuracy is reported to achieve from internal testing only, which means that the models have not been verified by completely new samples, and the over-fitting situation exists somewhat.

Therefore, inadequate samples and poor sample-labelling strategy relying on individual experience, are the two major challenges for the application of deep learning method in SHM.

The study in this paper endeavors to address these two problems. Firstly, it improves the form of sample from a single piece of wave signals to a grey scale by aligning all pieces of signals over the monitoring area side by side. This enables the updated sample to be labelled by objective truth instead of individual judgement. Secondly, a multi-scale convolutional neural network was proposed to extract damage features from multi-dimensions to avoid over-fitting problems caused by insufficient data. All these approaches together serve to improve the accuracy and reliability of the deep learning model, especially when no individual expertise is involved and only limited samples are available.

2 Sample and Model Design

2.1 Sample of Grey Scale

The stiffened panel tested is made of CCF300/BA9916, and its geometric configuration is demonstrated in Fig. 1. A grid of 9 piezoelectricity (PZT) sensors forms a rectangle monitoring network, which is partitioned into 4 regions as A, B, C and D. Region A and C are on the skin, while Region B and D over stiffeners. PZT sensor could work as either guided-wave generator or receiver, and any two sensors apart would generate a wave-propagating channel (namely monitoring path). Therefore, every sensor in the network in turn activates wave signals for other sensors to receive, and, as to say, 6 monitoring paths for each region in Fig. 1 are established.

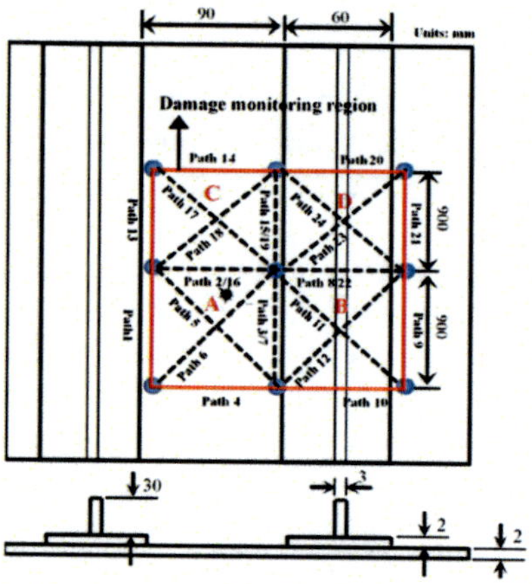

Fig. 1. Structure diagram of Lamb wave monitoring net-work

The fundamental mechanism of damage recognition is by differentiating the signal variations of two guided waves along the same path before and after the susceptive damage event, and its general method is to quantify this variation as Damage Index (DI) and to compare it with the DI threshold [18, 19]. Since guided wave is so susceptible to many factors, such as the distance from the damage to the path, this plausible method faces an uncompromising embarrassment. For example, if the damage occurs at point P in Region A (seen Fig. 1), the DI value of respective paths in Region A varies significantly. How to determine the DI threshold? It is quite difficult to establish a general rule, and, in most cases, it ends up with individual experiences [20].

By far, this practice is equally used for sample labelling in deep learning model. As there is not a fixed DI threshold, the sample state (damaged or undamaged) also has to be determined by individual experiences. However, for deep learning model, sample labelling is one of the most critical work. The operation like this will have significantly negative impact and largely mislead model training.

In contrast to a single monitoring path, the integration of all monitoring paths will reverse this situation completely [21]. When 24 pieces of signal variations obtained from the monitoring network in Fig. 1 (the signals from paths on the region boundaries are repeated) are aligned side by side (seen Fig. 2 in terms of energy distribution of scattered wave), the sample in this new pattern represents the state of the whole area instead of a single path, and consequently the sample could be labelled according to objective truth (damage occurs in this area or not). This practice completely avoids the artificial intervention in sample labelling. Additionally, the new sample presents in a form of image (Fig. 3), which could take full advantage of the latest developments of the deep learning model which is particularly good at image recognition and disposition.

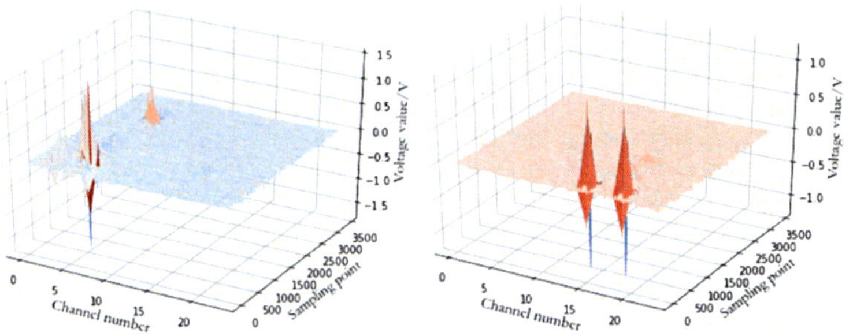

(a) The damage in Region A (b) The damage in Region C

Fig. 2. The energy distribution of the scattered wave in 24 monitored paths

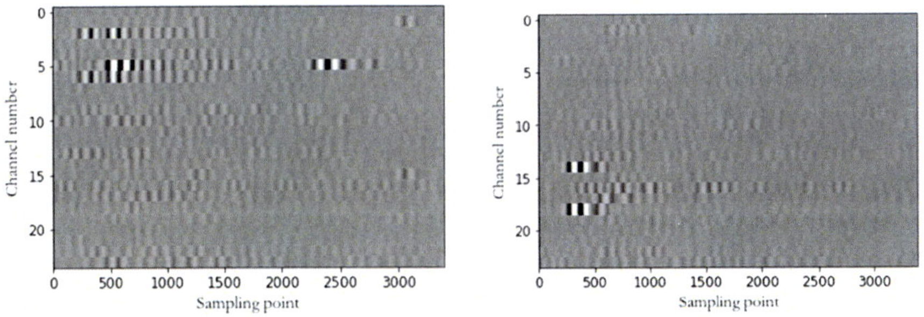

(a) The damage is at position 1 (b) The damage is at position 2

Fig. 3. The corresponding grey scales of Fig. 2

For the convenience of image disposition, the 24 pieces of signal variations are transferred to integrated grey scale map (IGM). Each IGM has 5 types of labels: 0, 1, 2, 3, 4 and 5, representing undamaged, damaged in Region A, B, C and D, respectively.

2.2 Data Augment

The shortage of sample is the leading hazard for the application of deep learning model in SHM. The deep learning model is expected to manifest the capability of reliable damage recognition while being immune to other disturbing factors such as structural dispersion. This capability relies upon a large amount of samples, which, however, is exactly the weak point in SHM. Expensive experiment cannot be afforded, and the solution is using virtual samples at one hand, while developing the adaptive deep learning model at the other hand. This paper focuses on the latter one and tries to improve it from several aspects.

Some practical problems are encountered when deep learning methods are applied:

- shortage of data samples. Limited by expensive and rare structures [24], the quantity of data sample usually cannot meet the requirement of model training;
- severe class skewness. In real application, the "damaged" signals generated are much less than the "undamaged" ones;
- Non-uniform data length. In practical application, the setting of data length varies from case to case. However, the uniform data length is essential as the input format for Convolutional Neural Network (CNN).

In order to tackle above problems, the module of data augment is introduced to increase the amount of data samples, as well as to alleviate class skewness.

- for the data, of which the length is shorter than the requirement, some meaning-less Fig. 4 are added to extend its length;
- for the data, of which the length is longer, a stride Ci of sliding window is defined in Eq. (1) to cut out segments from original data (schematic illustration in Fig. 4).

$$stride_{C_i} = \left\lceil MS * \frac{|samples\ labeled\ \ C_i|}{\max_{j=1}^{m}|samples\ \ labeled\ C_j|} \right\rceil \tag{1}$$

where, MS is the maximum stride length. It is a super parameter, and its value ranges between 50 and 300 empirically. C is the label set, C = [C1, C2,…, Cm], where m is the pattern of label. In this paper, C = [C1, C2, C3, C4, C5] = [undamaged, Region A, B, C, D]. The symbol of absolute value is to obtain the magnitude of that labelled samples.

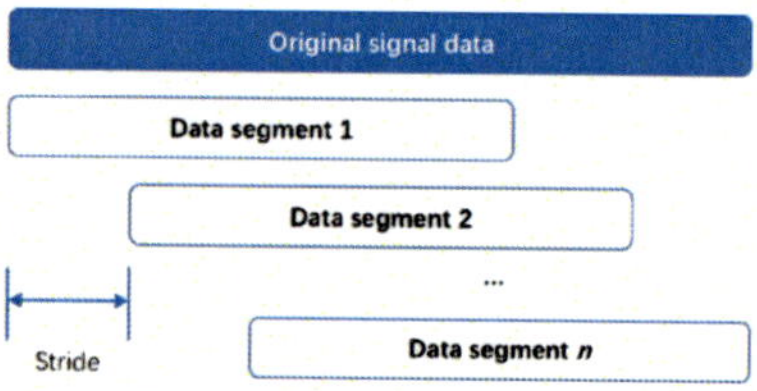

Fig. 4. Schematic diagram of using sliding window step to intercept the original data

2.3 Multi-scale Network Design

2.3.1 The Convolutional Kernel

Unlike general images, the IGM is extremely ill-proportional. Its dimensions are 24 * n, in which n is the sample rate of the wave-signal, usually 3,000–5,000 points. Therefore, the receptive field is specifically designed, of which the parameters m, n, f1 and f2 are the length, width, and the stepping interval in length and width direction, respectively (Fig. 5). There are 3 patterns of convolution kernel, 3 × 16, 6 × 16, and 3 × 32 at hand. As in each region there are 6 paths and usually no more than 3 paths of them are mostly affected by a damage, the width is set to be 3 or 6. The length also needs to be set properly. Over-long kernel will lead failure of characteristic extraction, while over-short kernel will undermine the generalization. In this study, 16 and 32 are preliminarily selected.

In order to acquire the optimal result, 3 patterns of kernel are tested separately or in a form of combination. The working mechanism of single and multi-scale kernels are demonstrated in Fig. 5a, b, respectively.

2.4 The Model Configuration

The model constitutes of 3 major modules, data augmentation, semantic feature extraction, and classification. The difference between Fig. 6a, b lies on the number of kernels, and the constitution of the parameter layer is presented in Fig. 6c.

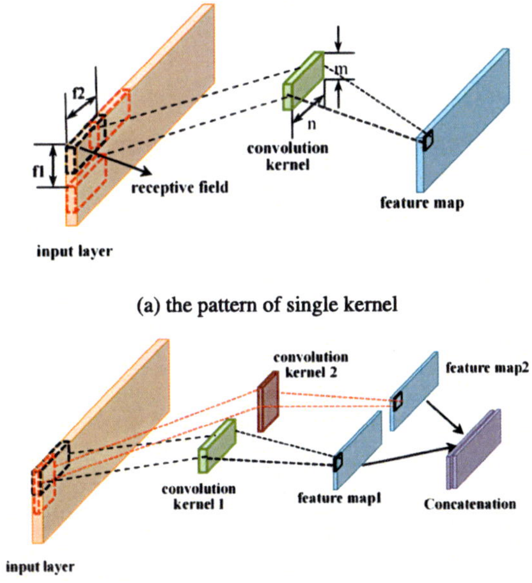

(a) the pattern of single kernel

(b) the pattern of multi-scale kernels

Fig. 5. The pattern of kernel

The semantic feature extraction is comprised of 3 convolutional layers, while for the pattern of multi-scale kernels an additional concatenation is adopted afterwards. This is because the feature map of the multi-scale kernels processed after 3 convolutional layers is significantly larger than that of the single one, and the additional convolutional layer helps to condense the size and reduce the parameter quantity at fully connected layer.

In the parameter layer, Relu is used as activation function in case of gradient explosion and gradient diffusion, while Normalization and "Dropout" for sparsely processing of the training parameters of each layer.

Two fully connected layers constitute of the classification. The first one extracts the top features and still uses Relu as activation function. The latter one finalizes the classification and uses SoftMax as activation function, which is presented as.

$$S_i = \frac{e^{V_i}}{\sum_j e^{V_j}} \tag{2}$$

where S is the output of SoftMax, *V is the output vector of the last fully connected layer,* V_i is the ith element in V, which has 5 elements is this paper. The position of the greatest element of the five is regarded as the classifying result.

The loss function is presented as

$$L(w) = -\sum_{i=1}^{m}\sum_{k=1}^{K} 1\{y^{(i)} = k\} \log S_k(w) + \frac{\lambda}{2}\|w\| \tag{3}$$

where m is the amount of samples in each batch, k is the sample type, y(i) is the real value of the ith sample, λ is the normalizing coefficient, w is the weighted matrix. $1\{y^{(i)} = k\}$ equals 1 when the sample is positive, otherwise equals 0.

Adam is used to update weight values. It is capable of taking the first and second orders of estimation into account, and consequently stabilizes the optimization process and avoids gradient diffusion.

3 Experiment

A group of 6 CFRP stiffened panels (S1-S6) with the same geometrical configuration as shown in Fig. 1 are prepared, in which S1-S4 and S5-S6 are used for model training and testing, respectively.

As confined by expensive specimens, false damages in the way of adhering clay is used. The size of clay is retained fluctuates from 8 mm * 8 mm to 15 mm * 15 mm. Each time when the clay is adhered, the specimen is scanned and corresponding guided-wave signal produced is specified as the damaging data. After every ten times of damaging data collection, one baseline data is collected, which is realized by scanning the specimen with no adhesion of clay. Following this way, 2614 damaging data and 270 baseline date are obtained from all 6 specimens. Random pairing of one damaging data and one baseline data collected from the same specimen is conducted, and consequently the signal variation is obtained, which is labelled as A, B, C or D according to its damaging position. Meanwhile, the signal variation produced by any pair of two baseline data from

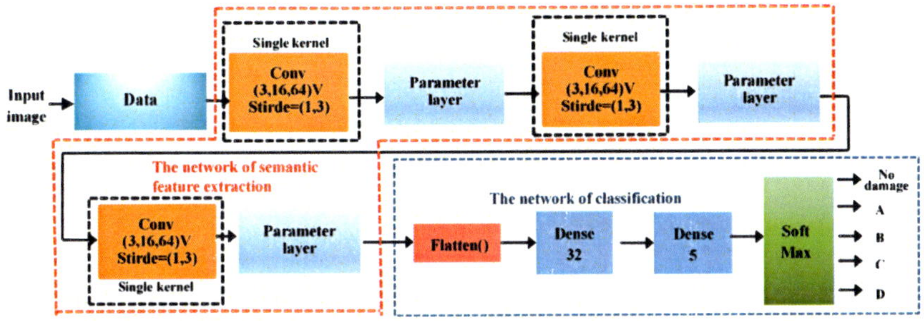

(a) Model with single-scale kernel

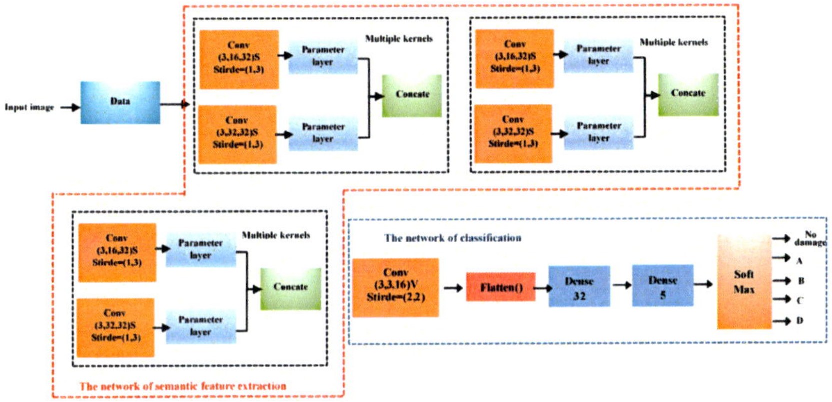

(b) Model with multi-scale kernels

(c) parameter layer

Fig. 6. Model configuration

the same specimen is labelled as undamaged. Following this way, total 5911 samples are produced, in which 5592 are collected from S1–S4 and 319 from S5 to S6 (Table 1). The producing process of experimental data lasts 21 days, during which temperature fluctuated within 3°, and 4 PZT sensors failed and were replaced.

At the process of data augment, the dimensions of sliding window are defined as 24 * 3250, and the stride as 50. The reason of setting the window length equal to 3250 is because almost all the important wave information lies between 150 and 3250. Such operation can secure that each piece of augmented data will always cover this section. Favored by data augment, as many as 22,368 pieces of samples from S1-S4 are produced. Among them 20,000 pieces are randomly selected for training. After every use of 128

Table 1. The number of collected data and generated samples of S1 ~ S6

Test article	Damage signal	Baseline signal	Sample (totality)	Sample (label = 0)	Sample (label = 1)	Sample (label = 2)	Sample (label = 3)	Sample (label = 4)
S1	650	66	1574	400	390	158	496	130
S2	650	66	1574	400	390	158	496	130
S3	595	61	1222	147	410	190	350	125
S4	595	61	1222	147	410	190	350	125
S5	62	8	160	35	29	28	49	19
S6	62	8	159	35	28	28	49	19
Sum	2614	270	5911	1164	1657	752	1790	548

samples for training, the model parameters are updated, and once over 20,000 sample are used, the model is validated by the left 2,368 samples. This process is repeated until convergence.

Once the training is finished, the model is verified by the samples generated from S5 and S6. The result is presented in Fig. 7, and the accuracies for the 5 patterns reach 97%, 96%, 93%, 85% and 92%, respectively. The average accuracy rate is 92.6%. The relatively lower rate occurs at Label 3, although, the root cause is the inadequate samples collected in Label 4, which accounts for merely 9% of the total samples and undermines the model training effect. Nevertheless, the general accuracies are excellent, especially at the case when no individual experience is involved.

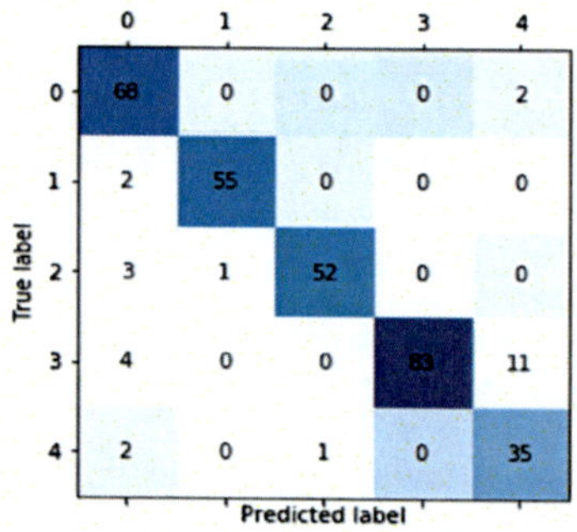

Fig. 7. Mixed matrix of verification

4 Discussion

4.1 The Selection of Kernel

3 patterns of kernel are tested separately or in a form of combination. For the pattern of single-scale kernel, all 3 models converge on the order of 10 iterations. The models

with the kernel of 3 × 16 and 3 × 32 have similar performance. Their loss functions stop at 0.10 and 0.093, respectively. With 6 × 16 kernel the loss function of model stops at 0.15, which does not perform as good as the other two. Therefore, in the pattern of double kernels, only 3 × 16 and 3 × 32 are adopted. The converging rate slows down, contrasted to that of the single-scale kernel. Nevertheless, the recognizing rate is observably improved. Its final loss function is 0.062 after around 40 iterations. For the model with 3 kernels, it doesn't converge at all. Therefore, only the pattern of double kernels of 3 × 16 and 3 × 32 are used in the following analysis.

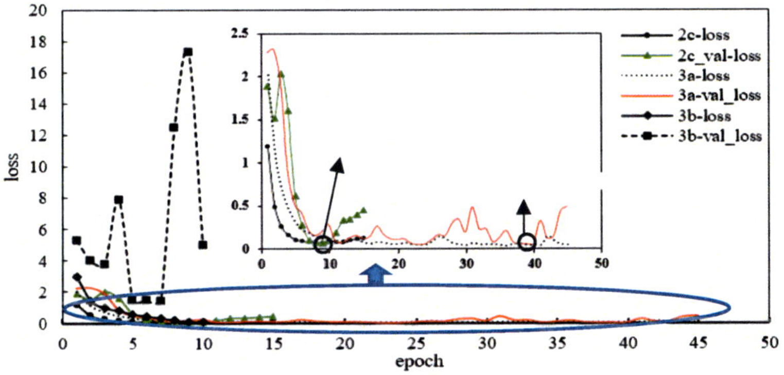

Fig. 8. The loss function curves of the training process of three models (2c represents single-scale kernel 3 × 16; 3a represents multi-scale kernels 3 × 16and3 × 32; 3b represents multi-scale kernels 3 × 16, 6 × 16 and 3 × 32; loss represents the loss function of training set; val_loss represents the loss function of validation set)

4.2 The Generalizing Capability

The nature of generalization, referred in this paper, is confined to the capability of overcoming the structural uncertainty when dealing with a group of specimens with same geometric configurations.

Here the Euclid distance of among extracted features is adopted as the judgement of generalization, of which the lower the value is, the greater similarity the extracted features share, and the better the generalization is [17]. The reason why the way of comparing the final damage-recognizing results between the deep-learning and traditional method is not adopted is that the recognition quality of traditional method is highly expertise-dependent (such as the determination of threshold value of Damage Index). Contrasted to the result comparison, the Euclid distance method is more objective and intrinsic.

For traditional method, the extracted feature matrix consists of 24 monitoring paths, each of which presents 8 patterns of Damage Index, including Cross correlation, Spatial phase difference, Spectrum loss、Central spectrum loss, Differential curve energy, Normalized Correlation Moment, Differential signal energy, Root mean square deviation, respectively. Therefore, its matrix dimensions are 24 × 8, and the specific matrix for specimen S5 and S6 are designated as $[X_5]_{24\times8}$ and $[X_6]_{24\times8}$, respectively. The Euclid distance (namely, the similarity) is expressed as

$$D^X_{56} = \frac{1}{192}\sqrt{\sum_{i=1}^{24}\sum_{j=1}^{8}(x'_{5ij} - x'_{6,ij})^2} \quad (3)$$

$$x'_{k,ij} = x_{k,ij}/\max([X_k]), k = 5, 6 \quad (4)$$

where, $x_{k,ij}$ is the element of $[X_k]_{24\times8}$ in the ith row and kth column.

For the deep learning method, the extracted feature is set to be the output of the first fully connected layer (Dense 32 in Fig. 6), which is one dimensional string of 32 elements. Following the naming rule, the specific matrix for specimen S5 and S6 are $[Y_5]_{1\times32}$ and $[Y_6]_{1\times32}$, respectively. Their Euclid distance is defined as

$$D^Y_{56} = \frac{1}{32}\sqrt{\sum_{i=1}^{32}\left(y'_{5i} - y'_{6,i}\right)^2} \quad (5)$$

$$y'_{k,i} = y_{5,i}/\max([Y_k]), k = 5, 6 \quad (6)$$

where, $y_{k,i}$ is the ith element of $[Y_k]_{1\times32}$. 10 "damages" are produced in either specimen, and each "damage" in S5 has an identical counterpart in S6 in terms of damaging location and size. Consequently, 10 Euclid distances are acquired from either specimen, and their respective average values are $\overline{D^X_{56}} = 0.0108$, and $\overline{D^Y_{56}} = 0.0053$. This proves that the deep learning method may extract features with greater similarity from different individuals, and is more likely to be immune to the geometric uncertainty. Furthermore, this immunity, with more samples, is believed to grow up more.

4.3 The Advantage of Multi-scale Kernels

Figure 9a is an example of the grey scale representing undamaged, which is diagnosed as damaged in Region A and undamaged (the correct diagnosis) by single and multi-scale kernels, respectively. Their corresponding heatmaps are presented in Fig. 9b, c. It is seen that the single-scale kernel concentrates on the greyest zone alone, and consequently submits the false diagnosis. As a contrast, the "multiple" pattern takes all fluctuations over the whole image into account. It is found that the undamaged state generally presents in the form of grey scale with a characteristic of irregular and widely spread fluctuation. Therefore, the "multiple" pattern, which is good at global scanning, is more favored.

4.4 The Denoising Capability

Figure 10 is an example of the grey scale representing the damaged in Region B. Due to poor attachment of Sensor 9, considerable noise is produced on the last monitoring path (seen at bottom right of Fig. 10a). From 10(b) it is seen that the deep learning model successfully differentiate it from other normal damaged signals. This proves the model with the capability of noise suppression to a degree.

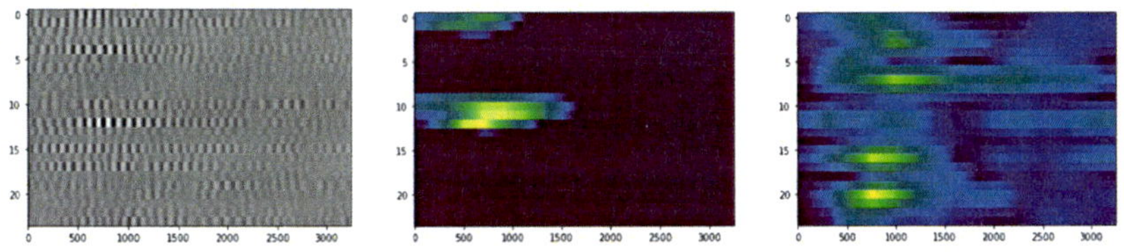

(a) Sample without damage

(b) Grad-CAM heatmap of single-scale kernel network

(c) Grad-CAM heatmap of multi-scale kernels network

Fig. 9. Comparison of heatmaps between single-scale kernel and multi-scale kernels networks

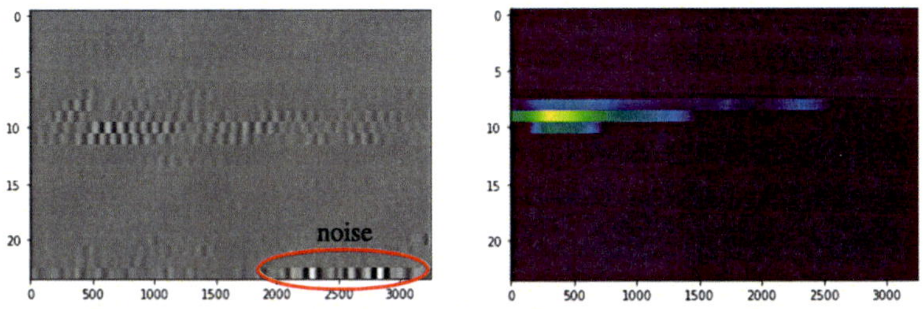

(a) grey scale

(b) Grad-CAM heatmap

Fig. 10. Sample of damage in Region B with noise generated in Region D

5 Summary

Guided wave is a very potential candidate of structural health monitoring. In this paper, an innovative deep learning model is proposed and it proves to help the damage recognition of carbon fiber reinforced composite using guided wave more accurately and reliably. The major contributions include:

- It uses the integrated grey scale of all guided-wave-propagating paths instead of a single path before as the sample for deep learning model. This practice enables the sample to be labelled according to objective truth rather than individual experience, which may largely mislead the model training;
- The model is elaborately designed with multiple convolutional kernels and the module of data augment, which alleviates the natural shortage of inadequate sample collection in the field of structural health monitoring.

Both measures above help the deep learning model to have a good performance in terms of generalization. Individual experience is eliminated during the labelling process, and the recognizing accuracy reaches 92.6% in the verification.

References

1. Shanyi, D.: Advanced composite materials and aerospace. Acta Mater Compos Sinica **24**(1), 12–24 (2007)
2. Craven, R., Iannucci, L., Olsson, R.: Delamination buckling: a finite element study with realistic delamination shapes, multiple delaminations and fibre fracture cracks. Compos. A Appl. Sci. Manuf. **41**(5), 684–692 (2010)
3. Xinlin, Q., Yishou, W.: Structural health monitoring technology and its Application in aerospace field. Exp. Mech. **27**(5), 517–526 (2012)
4. Purekar, A.S., Pines, D.J.: Damage detection in thin composite laminates using piezoelectric phased sensor arrays and guided Lamb wave interrogation. J. Intell. Mater. Syst. Struct. **21**(10), 995–1010 (2010)
5. Lei, Q., Shenfang, Y., Xiaoyue, Z., et al.: A time reversal focusing based impact imaging method and its evaluation on complex composite structures. Smart Mater. Struct. **20**(10), 105014 (2011)
6. Ng, C.T., Veidt, M.: A Lamb-wave-based technique for damage detection in composite laminates. Smart Mater. Struct. **18**(7), 074006 (2009)
7. Ong, W.H., Chiu, W.K.: Redirection of Lamb waves for structural health monitoring. Smart Mater. Res. **2012** (2012)
8. Memmolo, V., Monaco, E., Boffa, N.D., et al.: Guided wave propagation and scattering for structural health monitoring of stiffened composites. Compos. Struct. **184**, 568–580 (2018)
9. Sbarufatti, C., Manson, G., Worden, K.: A numerically-enhanced machine learning approach to damage diag-nosis using a Lamb wave sensing network. J. Sound Vib. **333**(19), 4499–4525
10. Worden, K., Manson, G.: The application of machine learning to structural health monitoring. Philo-Sophical Trans. **365**, 515–537 (1851)
11. Atashipour, S.A., Mirdamadi, H.R., Hemasian, M.H., et al.: An effective damage identification approach in thick steel beams based on guided ultrasonic waves for structural health monitoring applications. J Intell. Mater. Syst. Struct. SACE (2012)
12. Yan, G.: A Bayesian approach for damage localization in plate-like structure using Lamb waves. Smart Mater. Struct. **22**(3), 035012 (2013)
13. Su, Z., Lin, Y.: Lamb wave-based quantitative identifica-tion of delamination in CF/EP composite structures using artificial neural algorithm. Compos Struct **66**(1–4), 627–637
14. Agarwal, S., Mitra, M.: Lamb wave based auto-matic damage detection using matching pursuit and machine learning. Smart Mater. Struct. **23**(8), 085012
15. Guo, T., Wu, L., Wang, C., et al.: Damage detection in a novel deep-learning framework: a robust method for feature extraction. Struct. Health Monitor. Int. J. **19**(2), 424–442 (2020)
16. Sbarufatti, C., Manson, G., Worden, K., et al.: A numeri-cally-enhanced machine learning approach to damage diagnosis using a Lamb wave sensing network. J. Sound Vib. **333**(19), 4499–4525 (2014)
17. Xu, L., Yuan, S., Chen, J., et al.: Guided wave-convolutional neural network based fatigue crack diagnosis of aircraft structures. Sensors **19**(16), 3567 (2019)
18. Yao, P., Zheng, B., Dawood, M., et al.: Real time mon-itor-ing of spot-welded joints under service load using lead zirconate titanate (PZT) transducers. Smart Mater. Struct **26**(3), 35–59 (2017)
19. Guoqiang, L., Xiasheng, S., Yingchun, X.: T-type reinforcement damage monitoring for composite materials based on Lamb wave and Hilbert transform. Acta Mater Compos Sinica **31**(3), 818–823 (2014)

20. Dworakowski, Z., Ambrozinski, L., Packo, P., et al.: Appli-cation of artificial neural networks for com-pounding multiple damage indices in Lamb-wave-based damage detection. Struct. Control. Health Monit. **22**(1), 50–61 (2015)
21. Guoqiang, L., Yingchun, X., Hua, Z., et al.: Probabilistic imaging method for damage identification of composite Reinforced panels. Acta Mater Compos Sinica **35**(002), 311–319 (2018)

Research on Health Monitoring and Intelligent Diagnosis Technology of Large-Scale Low-Speed Wind Tunnel

Wang Jianfeng[1,2], Liu Boya[1,2](✉), and Liu Shi[1,2]

[1] AVIC Aerodynamics Research Institute, Harbin 150001, China
1748323030@qq.com
[2] Aviation Key Laboratory of Science and Technology On Low Speed and High Reynolds Number Aerodynamic Research, Harbin 150001, China

Abstract. The large-scale low-speed wind tunnel test system is complex, involving many equipment, parameters, and data. In order to monitor the state of the system, quickly determine and locate the cause of the fault when the system is abnormal. Based on advanced sensors and network technology, this paper monitors the components and parameters of the system in real time. Through the three-dimensional simulation method, the digital twin system construction of the wind tunnel system including the model is realized, and the data visualization method is combined with the machine learning technology to intelligently predict and diagnose the faults and causes of the test data. Through the research in this paper, the technical foundation is laid for the construction of the smart wind tunnel laboratory.

Keywords: Low-speed wind tunnel health monitoring · Fault diagnosis · Digital twin · Deep learning · Artificial intelligence

1 Introduction

The concept of health monitoring and intelligent diagnosis was first proposed by the United States in 1998 during the research and development of its F35 joint attack aircraft, with the purpose of improving the safety and survivability of fighters [1]. Since then, the failure prediction and health management system has played a big role in the military field [2], and it has attracted the attention of military powers in the world. Nowadays, Prognostic and Health Management (PHM) technology has been continuously developed from the original military field to private enterprises [3], and is widely used in the maintenance of new-generation equipment in the fields of transportation, equipment manufacturing, power generation, etc. [4]. Such as aircraft, ships, vehicles, CNC machine tools, etc. Boeing applied PHM technology to civil aviation and developed an aircraft health monitoring system [5]. The system collects flight status data in real time, and makes maintenance decisions after failure analysis [6], Maintenance personnel can log on the website to obtain fault information, and realize the integration of network system functions [7]. NASA is also studying PHM technology as an important

Chinese Society of Aeronautics and Astronautics: CASTYSF 2022, LNEE 972, pp. 518–536, 2023.
https://doi.org/10.1007/978-981-19-7652-0_47

part of aviation maintenance [8]. Bai et al. proposed a multi-objective CBM optimization method, using PHM technology to systematically balance the relationship between optimization objectives and find the optimal solution that best represents the decision maker's preference [9]. Moussa Hamadache, Joon Ha Jung, etc. based on REB prediction and PHM technology, introduced different bearing failure modes, outlined modern PHM technology, and explored deep learning methods for bearing fault detection and diagnosis [10].

The concept of failure prediction and health management has also attracted extensive attention of domestic scholars [11]. In 2015, based on the monitoring system and data acquisition, Huang Lili proposed the PHM technology of wind turbine operation and maintenance data, which is used for the condition assessment and fault diagnosis of wind turbines, and designed and developed the prototype control system of wind turbine PHM [10]. Wang Tianyu, Wang Hongdong and others used Bayesian network to establish a reliability model, and studied the application of this model in PHM, predicting the failure time of unmanned ships, and monitoring their health in real time [12]. Guo Yangming, Mi Qi et al. combed the technical connotation of PHM and the system architecture of avionics, and gave a dynamic reconstruction model of distributed integrated modular avionics based on PHM, which provided the foundation and guarantee for avionics [13]. Cui Liming applied PHM technology to bearings, and used AR model to extract the characteristics of bearing degradation trend and predict its service life [14].

The large-scale low-speed wind tunnel test system has a complex composition, involving many sections and their associated sensors; there are many types of wind tunnel tests, the test parameters are diverse, the data processing process is complex, and the test accuracy is extremely high. The use of health monitoring and intelligent diagnosis technology to carry out intelligent operation and maintenance management of wind tunnel has important research significance and application value for improving the stability of wind tunnel operation, wind tunnel test efficiency and test accuracy.

The FL-10 wind tunnel of the AVIC Aerodynaiviics Research Institute is currently the largest production-type low-speed wind tunnel in China. The size of the test section is 8 m × 6 m. In this paper, based on this wind tunnel, research work on health monitoring and intelligent diagnosis technology is carried out, and a large-scale low-speed wind tunnel health monitoring system is built. Its interior covers high-fidelity models such as factory areas, movable wind tunnel sections, test equipment, test models, and components based on 3D vision; digital mapping of test equipment and test status driven by multi-source data fusion such as real-time data and CFD simulation data; use multi-dimensional sensor network to realize the health monitoring of test equipment and parameters, realize equipment predictive maintenance, fault detection and diagnosis through artificial intelligence technology. Finally, a digital and intelligent wind tunnel management platform with multi-dimensional sensors as the skeleton, artificial intelligence and machine learning technology as the main body, and digital twin as the terminal embodiment will be realized.

2 Design of Health Monitoring System for Large-Scale Low-Speed Wind Tunnel

2.1 Identification of Wind Tunnel Monitoring Factors

Based on the expert knowledge base and the historical operation experience of the wind tunnel, the factors of each subsystem and equipment of the FL-10 wind tunnel are identified, and the key systems and equipment that affect the test efficiency or test data quality are determined. It mainly includes power system, cooling system, support system, wind tunnel body, etc. Health monitoring, fault diagnosis and predictive maintenance of the above subsystems and their ancillary equipmen, It will improve the reliability of equipment and facilities, ensure the long-term stable operation of the wind tunnel test environment to the greatest extent, and provide favorable quality assurance for obtaining effective and accurate test data.

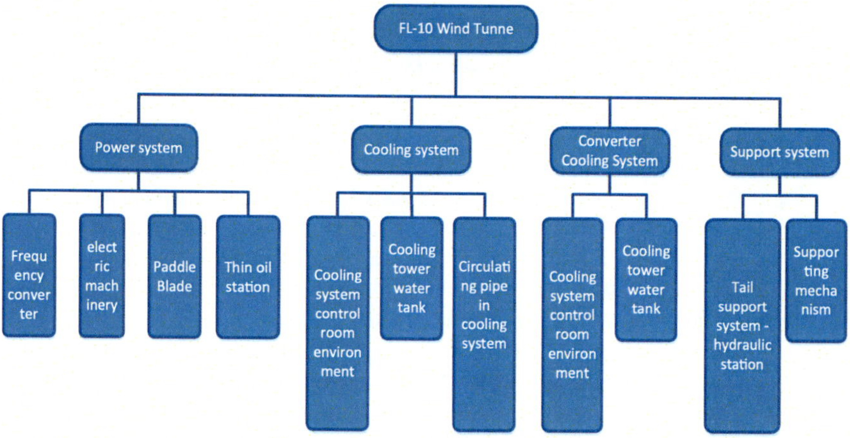

Fig. 1. Identification of monitoring factors in FL-10 wind tunnel

2.2 Sensor Arrangement and Networking

Divide the FL-10 wind tunnel according to the system identified in Fig. 1. Arrange sensors in the key sections of the wind tunnel and the necessary points of the auxiliary equipment of each subsystem and form a network. The sensor monitoring network covers all key systems and equipment of the entire FL-10 wind tunnel. Segments and equipment, so as to achieve a full range of health monitoring and fault diagnosis for the FL-10 wind tunnel system.

The signals of the power system monitoring objects are mainly inverter fault information, temperature information of the motor rotor and bearing bush, and shaft vibration information. The signals of the monitoring objects of the cooling system are mainly cooling tower water level and temperature information, cooling water pipeline pressure, flow and inlet and outlet temperature information. The signals of the monitoring object of the support system are mainly the vibration information of the support system, the pressure information of the hydraulic tail support pipeline, the liquid level of the fuel tank in the pump station, and the temperature information. The specific sensor arrangement is shown in the Table 1.

The data required for wind tunnel health monitoring and intelligent diagnosis algorithms are collected from the underlying data acquisition layer by PLC and field sensors, and then entered into the central switch of the test network through the data access layer switches of each subsystem and data acquisition system. The aggregated data is sent to the wind tunnel equipment operation database for storage to form a historical database, which is used as the basic training data for algorithm modeling. At the same time, the real-time data is transmitted to the health monitoring system server, and the health monitoring and intelligent diagnosis algorithms are used for real-time calculation and analysis. The data networking result is shown in the following Fig. 2.

2.3 Monitoring System Based on Sensor Network

Figure 3 is a monitoring system of a sensor network arranged in the health monitoring and intelligent diagnosis system of the system. The operation interface of the monitoring system takes the top view of the FL-10 wind tunnel as the main body and is divided into subsystems, and its content covers all the sensor parameters in Table 1. The key information such as the system, equipment running status, and test progress status can be visually viewed in the wind tunnel. When the sensor parameters are abnormal, the interface will flash and turn red to give an alarm.

3 Construction of Digital Twin System

3.1 3D Scene Digital Reconstruction

High-fidelity 3D scene reconstruction of large-scale low-speed wind tunnel health monitoring and intelligent diagnosis system is to use 3Dmax software to carry out high-fidelity and refined modeling and reconstruction of FL-10 wind tunnel scene and internal test section equipment. Built-in three-dimensional digital model including the overall structure of the test section, the floor and tail support linkage mechanism, the floor and turntable linkage mechanism, the open test scene, and the closed test scene; switchable test model groups including civil aircraft standard models, military aircraft standard models, missile models, and train models can be quickly imported through 3D scanning technology, It also includes the test support system model group including hydraulic tail support, single rod support, turntable support, etc. The system can freely switch the test type, support mechanism, and support form according to the actual test scene of the wind tunnel, and finally realize the 3D high-fidelity digital scene reconstruction based on the FL-10 wind tunnel (Fig. 4).

Table 1. Equipment status monitoring table

Monitoring system	Monitoring content	Monitoring variables	Number of sensors
Power system	Inverter	Inverter temperature	1
		Inverter voltage	1
		Inverter power	1
		Inverter current	1
		Inverter power-on indication	1
		Inverter fault indication	1
		Inverter running indication	1
	electric machinery	Motor stator temperature	5
		Motor noise	4
		Motor vibration	4
		Motor bearing bush temperature (thrust, radial)	4
		Axial movement of motor	1
	Paddle Blade	Blade stress	9
	Thin oil station	Low pressure state of thin oil station	1
		Operation indication of high pressure pump in thin oil station	2
		Operation indication of low pressure pump in thin oil station	2
Wind tunnel cooling system	Cooling tower water tank	Tank temperature	11
		Tank level	11
		Ambient temperature and humidity	2
	Circulating pipe in cooling system	Discharge	1
		Water temperature	1

(*continued*)

Table 1. (*continued*)

Monitoring system	Monitoring content	Monitoring variables	Number of sensors
		Hydraulic pressure	1
	Cooling system control room environment	Ambient temperature and humidity	2
Converter cooling system	Converter cooling system	Internal circulation cooling water pump shell temperature	2
		Internal circulation cooling water level	1
		Internal circulation cooling water flow	1
		Inlet temperature of internal circulation cooling water	1
	External circulation cooling system	External circulation cooling pressure	1
		External circulating cooling water temperature	1
Support system	Tail support system—hydraulic station	Oil tank temperature	1
		Tank level	1
		Fuel tank supply pressure	2
	Supporting mechanism	Voltage	1
		Electric current	1
		In cylinder displacement becomes (actual angle of attack,α)	1
		In cylinder displacement (sideslip angle, β1)	1
		Fault indication	1
		Operation indication	1
		Tail brace height	1

3.2 3D Digital Test Scene Mapping

The 3D model established by the fusion of real-time equipment operation data, CFD simulation data, test data and other multi-source data is driven to realize the digital

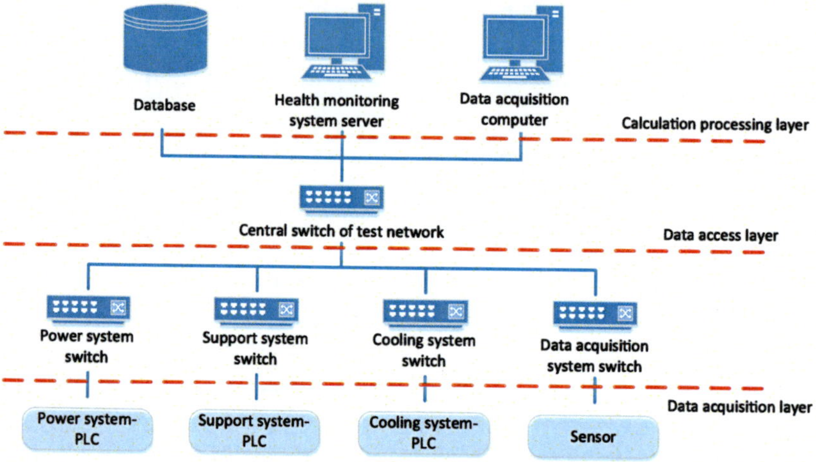

Fig. 2. Sensor network architecture diagram

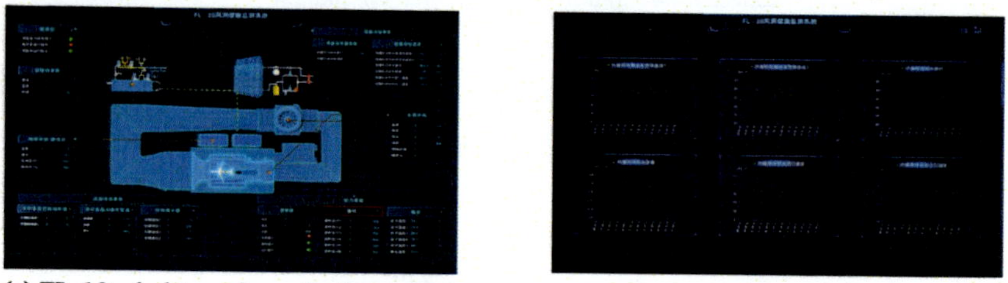

(a) FL-10 wind tunnel monitoring system (b) Sensor data monitoring

Fig. 3. Monitoring system based on sensor network

mapping of test equipment and test status. For the model attitude angle data, the system performs Inverse Kinematics (inverse dynamics) calculation after obtaining the target value, and controls the synchronous change of the movable joints of the support mechanism to achieve the effect that the dynamic attitude angle of the test target is consistent with the actual test linkage angle. For the change of the test target's own components, the system correspondingly obtains the names and positions of the detachable components, and performs simultaneous display and concealment operations according to the test plan, so as to realize the structure and the key components of the test target (such as wings, landing gear, vertical tail, horizontal tail, etc.) High consistency of actual trials (Fig. 5).

3.3 Flow Field Visualization Technology Based on CFD Simulation Data

The health monitoring and intelligent diagnosis system built in this study has a built-in flow field visualization module based on CFD simulation data. Its specific implementation is as follow.

(a) Factory scene selection

(b) Test configuration interface

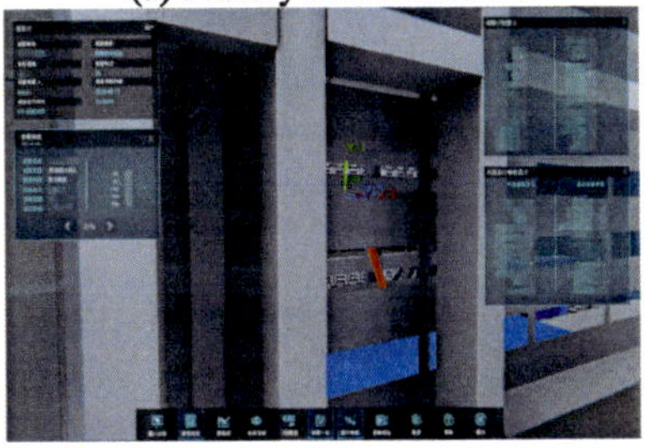

(c) Digital reproduction of equipment scene and parameters

(d) Built-in model library of the system

Fig. 4. Digital construction of high-fidelity 3D scene

The point cloud data calculated by CFD is extracted and processed, and each point data is positioned in three-dimensional space coordinates, and given specific meanings such as three-dimensional space coordinates, time series, and pressure coefficients. It adopts the development idea of Unity physics engine + distributed loading method, and has a data point reduction algorithm, which automatically filters some points according to the set rules and optimizes the point information, effectively solve the problem of slow loading and freezing of 3D visualization due to large amount of data, it can perform distributed fusion loading processing on millions of point cloud data, and display the effect in real time. The point cloud data is given different chromatographic colors according to the size of the CFD calculation data, and the calculation is comprehensively summarized, so as to realize the three-dimensional flow field visualization based on the CFD simulation data (Fig. 6).

4 Research on Data Intelligent Diagnosis Technology Based on Equipment Parameters

4.1 Key Equipment Prediction Learning Machine Based on Machine Learning

At present, conventional equipment health monitoring algorithms use a machine learning or deep learning algorithm to predict key equipment characterization parameters. Although this method can achieve a certain level of accuracy, due to the different spatial complexity of different algorithms and the possible selection of Not the best algorithm for the current data. As a result, although the accuracy meets the requirements, the

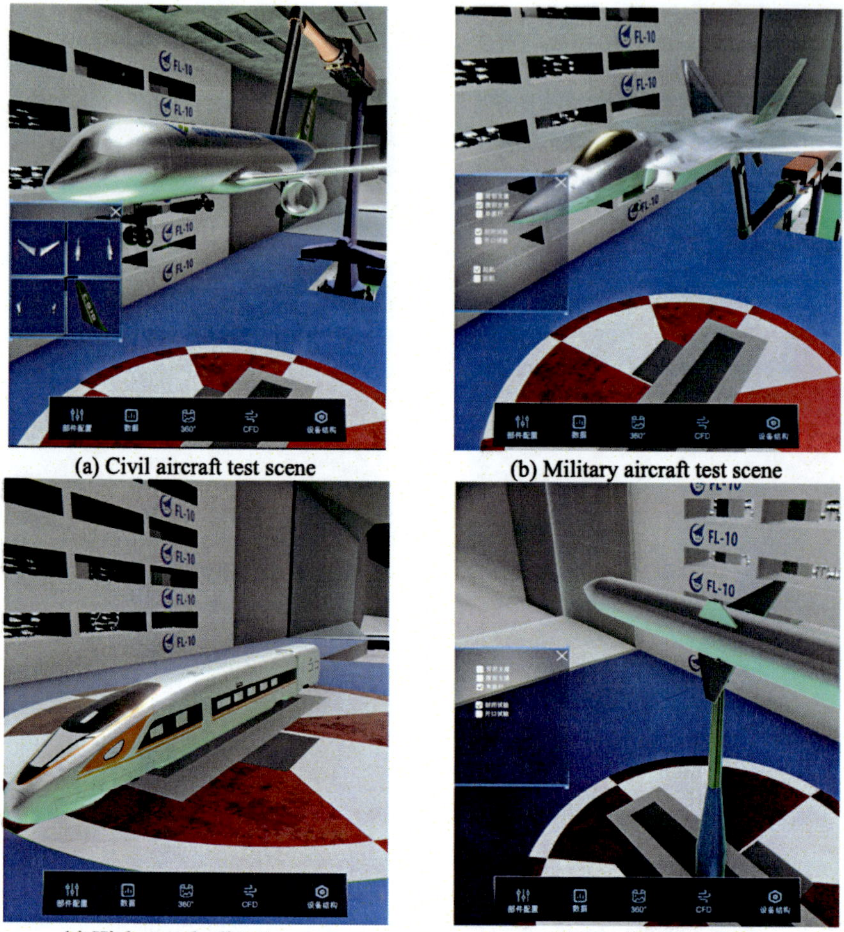

(a) Civil aircraft test scene (b) Military aircraft test scene

(c) High-speed rail test scene (d) Missile test scene

Fig. 5. Three dimensional digital mapping of test scene

selected algorithm is not the optimal algorithm, and it takes a lot of time to select the algorithm. In response to this situation, this paper proposes a key equipment health monitoring learning machine, which combines several algorithms suitable for wind tunnel health monitoring, and the learning machine compares the accuracy of each algorithm and various statistical indicators. Pick the optimal algorithm that fits the current running data. In this way, a more accurate prediction algorithm can be found clearly and quickly.

According to the historical operation data of the equipment, four machine learning algorithms including SVR (Support Vector Machine Regression), ElasticNet (Elastic Network Regression), Gauss (Gaussian Mixture), and GRU (Gated Neural Network) are combined to form a large-scale low-speed wind tunnel key equipment health monitoring learning machine. During algorithm training, the predictive learning machine forms four different algorithm models, and the learning machine independently judges statistical

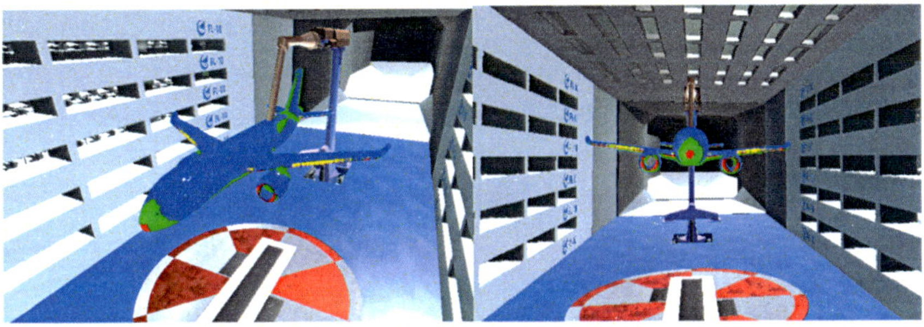

Fig. 6. Flow field visualization technology based on CFD simulation data

indicators such as fit, root mean square error, etc., and selects the algorithm with the highest fit and the smallest error as the final prediction algorithm.

4.2 Health Monitoring and Intelligent Diagnosis Algorithm

Data Processing

In this paper, six months of historical operating data of the wind tunnel were selected as the modeling data for the predictive learning machine. Based on the identification of key equipment in the wind tunnel based on expert experience, the motor speed is finally determined as the output parameter of the health monitoring algorithm; with 39 types of characteristic variables deployed by subsystems such as motors, frequency converters, propellers, and thin oil stations, a total of 86 characteristic values are the key factors affecting the motor speed. The algorithm structure is shown in the following Fig. 7.

Standardization

Since the variable data vector dimension and unit required to be monitored by the wind tunnel monitoring system are quite different [15], in order to improve the convergence speed of the predictive learning machine, the input data is standardized and preprocessed [16]. The formula is as follows:

$$S_n^k = \frac{X_n^k - \frac{1}{K}\sum_{K=1}^{N} X_K^n}{\sqrt{\frac{1}{K-1}\sum_{K=1}^{N}(X_K^n - \frac{1}{K}\sum_{K=1}^{N} X_K^n)^2}} \tag{1}$$

In the formula: X_n^k represents the production parameters of the Kth sample under n-dimensional data standardization, S_n^k represents the n-dimensional metadata of the Kth sample arranged in a time series, and K is the number of data sets.

Health Monitoring Algorithm of Key Equipment

The health monitoring prediction learning machine proposed in Sect. 3.1 is used to predict the historical operation data of the wind tunnel. The learning machine predicts the output curve as follows.

Table 2 lists the fit, MAE, MSE and RMES errors of the four regression algorithms SVR, ElasticNet, Gauss, and GRU to the wind tunnel historical data.

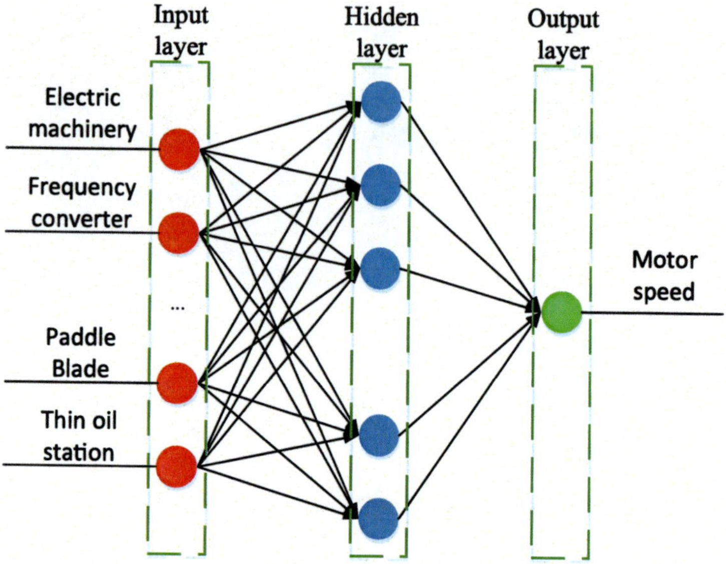

Fig. 7. Health monitoring algorithm structure

The following conclusions can be drawn from Fig. 8 and Table 2: the predicted value of the GRU algorithm has the best fit, and the errors of MAE, MSE and RMES are the smallest. Therefore, the model established by the GRU algorithm is selected as the prediction algorithm for the health monitoring of this system.

Failure prediction algorithm of key equipment

This study is based on a large amount of historical data with time series properties running in the wind tunnel, and uses the Bi-LSTM (Bi-directional Long Short-Term Memory network) algorithm to model it. The Bi-directional Long Short-Term Memory network is based on LSTM and combines the input The information of the sequence in both forward and backward directions, for the output at time t, the forward LSTM layer has the information at time t and before in the input sequence, and the backward LSTM layer has the information at time t and later in the input sequence. Information. The output of the forward LSTM layer at time t and the output of the backward LSTM layer at time t are processed together as the final algorithm prediction output. The structure of the Bi-directional Long Short-Term Memory network algorithm is shown Fig. 9.

In the Bi-directional Long Short-Term Memory network, the neural network model not only uses the data information before the current moment in the historical data, but also fully considers the data information that occurs in the future [17]. The purpose of the fault prediction algorithm established for the key equipment of the wind tunnel is to predict the operation status of the equipment for a period of time in the future. Therefore, when training the algorithm model, it is more suitable to choose the Bi-directional Long Short-Term Memory network, and fully learn the historical operation

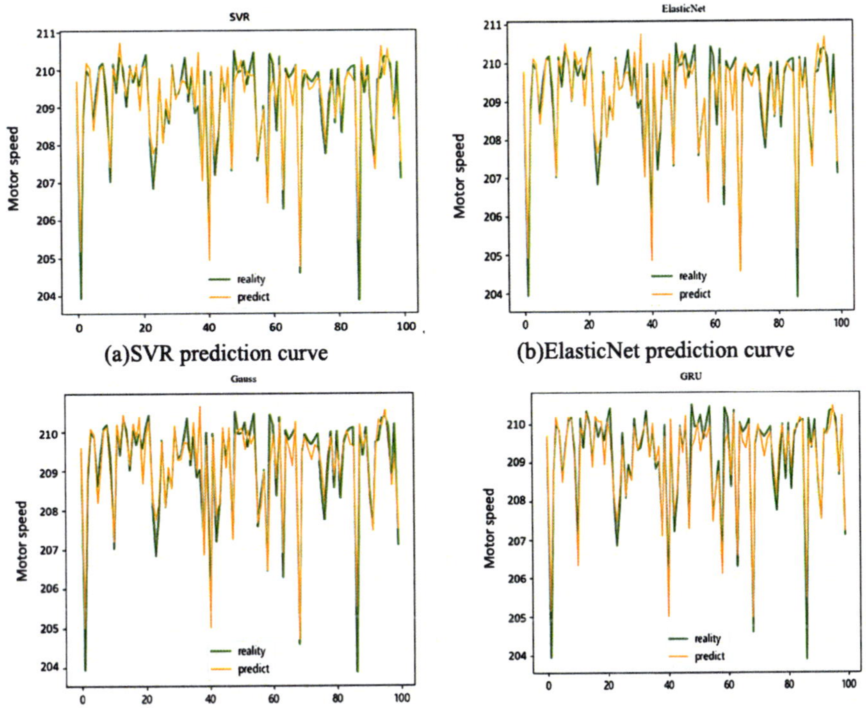

(a)SVR prediction curve (b)ElasticNet prediction curve

(c)Gauss prediction curve (d)GRU prediction curve

Fig. 8. Predictive learning machine prediction output curve

Table 2. Predictive learning machine output evaluation indicators

	R^2	MAE	MSE	RMSE
SVR	0.9009	1.5922	4.9379	2.2221
ElasticNet	0.9054	1.5497	4.7181	2.1721
Gauss	0.8991	1.6110	5.0318	2.2432
GRU	0.9314	1.2050	3.5232	1.8770

process of the equipment from the forward and reverse directions, which can greatly improve the model's ability to cope with sudden changes and abnormal moments.

For the input data of the Bi-LSTM network, the construction of feature engineering needs to be added on the basis of the data processing process described in Sect. 3.2.1. In this paper, the sliding window algorithm is used to construct feature engineering for historical data.

The sliding window algorithm is to predict the future data of the device, expand the value range of the current moment to an interval including this point, and use the

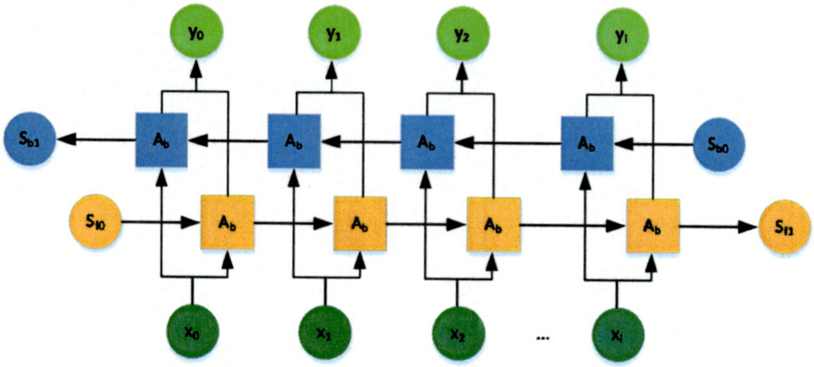

Fig. 9. Bi-LSTM algorithm structure diagram

interval to judge, this interval is the sliding window. The sliding window is to select the corresponding time series according to the specified unit length, so as to calculate the statistical indicators in the window. Its principle is shown in the Fig. 10.

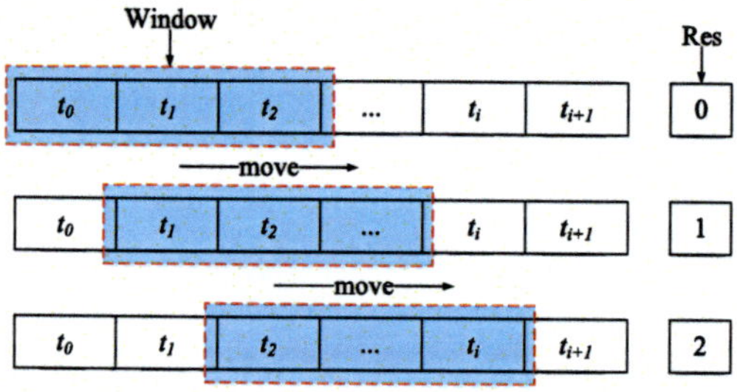

Fig. 10. Example of sliding window.

The following figure shows the error curve and accuracy of the mathematical model built by the Bi-LSTM network based on the above historical data. It can be seen from the figure that the algorithm reaches the optimum at the 125th iteration, the mean square error reaches e-4 level, the accuracy rate is close to 1, and the training error and test error are converged well, no overfitting or underfitting in the established model (Fig. 11).

In this paper, the sliding window size is selected as 30, that is, the data points of the past 30 moments are considered to predict the equipment operation trend of the next 3 moments. The figure below shows the prediction curve of the fault prediction algorithm based on the Bi-LSTM network. The blue curve in the figure is the prediction curve of the fault prediction algorithm for the running state of the motor speed at the next three moments (Fig. 12).

Intelligent fault diagnosis algorithm of key equipment

Due to the wide variety of equipment in the wind tunnel, the failure modes and failure mechanisms are very different, and the procurement time of each equipment is not

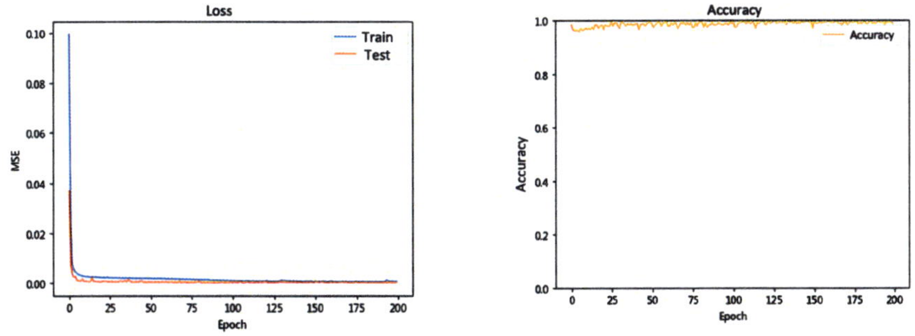

(a) Bi-LSTM algorithm error curve (b) Bi-LSTM algorithm accuracy

Fig. 11. Bi-LSTM algorithm error and accuracy curve

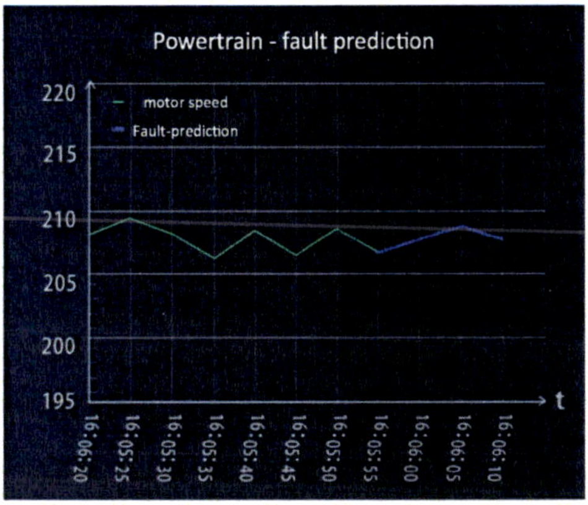

Fig. 12. Prediction output of fault prediction algorithm

uniform, resulting in different equipment operation data storage conditions, and it is difficult to use a fixed diagnosis method to meet the fault diagnosis requirements. Taking into account comprehensively, this system uses the combination of expert experience and principal component analysis (PCA) to diagnose the faults of the key equipment of the FL-10 wind tunnel.

PCA is to project high-dimensional process data into an orthogonal low-dimensional subspace and preserve the main process information [18]. Geometrically, the coordinate system formed by the sample is rotated to a new coordinate space through a certain linear combination, and the new coordinate axis represents the direction with the largest variance [19]. The statistical indicators of the intelligent diagnosis of the PCA algorithm are as follows:

The SPE indicator measures the change of the projection of the sample vector in the residual space [20].

$$SPE = \|(I - P \cdot P^T) \cdot x\|^2 \leq \delta_\alpha^2 \tag{2}$$

where δ_α^2 represents the control limit with confidence α. Its formula is as follows:

$$\delta_\alpha^2 = \theta_1 \left(\frac{C_\alpha \sqrt{2\theta_2 h_0^2}}{\theta_1} + 1 + \frac{\theta_2 h_0 (h_0 - 1)}{{\theta_1}^2} \right)^{1/h_0} \quad (3)$$

Finally, fault diagnosis is performed by calculating the contribution rate.
The SPE-based contribution graph is defined as:

$$Cont_i^{SPE} = (\xi_i^T (I - PP^T)x)^2, i = 1, \ldots, m \quad (4)$$

After the data is calculated and analyzed by the intelligent diagnosis algorithm, the three largest variables in the contribution graph are taken as the most likely cause of the fault, and combined with the expert knowledge base to make a comprehensive judgment. Finally, the precise cause of equipment failure and corresponding operation suggestions are given.

4.3 Implementation and Verification of Health Monitoring and Intelligent Diagnosis Technology

Implementation of Health Monitoring and Intelligent Diagnosis Technology

The working principle of the health monitoring and intelligent diagnosis system is as follows.

Through the established key equipment prediction learning machine, real-time prediction and analysis of the operating state of the key equipment in the wind tunnel are carried out. When the real-time acquisition value of the equipment operation and the predicted value of the learning machine have deviations beyond the allowable error range, the system sends a fault early warning signal; When the fault prediction curve exceeds the preset threshold, the system sends a fault early warning signal. The front-end interactive platform triggers the alarm information, the monitoring interface flashes and turns red and the alarm window pops up.

When the system receives an early warning signal from the health monitoring or fault prediction module, the system automatically enters the intelligent diagnosis module to perform fault diagnosis on the data at the current fault time. At the same time, a fault operation suggestion library is formulated based on expert experience, and finally gives the equipment fault cause and information. Corresponding action suggestions.

Verification of Health Monitoring and Intelligent Diagnosis System

In this paper, the failure of the large motor of the wind tunnel power system "Axial displacement of motor", which has a great impact on the operation of the wind tunnel, is selected to verify the system.

As shown in Table 1, for the axial movement of the motor, the system arranges a laser displacement sensor at the front end of the motor shaft section. The sensor is zeroed at the midpoint of its range, and the range is ±5mm. The normal fluctuation range threshold is set to ±3mm. In this paper, 100 groups of sensor monitoring network real acquisition values are selected when the wind tunnel system is running normally, and the "motor shaft" in each group of data is manually modified within the interval of [+3.5, +5] ∪ [−5, −3.5] The collected value of the "Axial displacement of motor". This data is input into the health monitoring and intelligent diagnosis system as a verification of the reliability of the system.

The figure shows the contribution of the 86 feature variables output by the intelligent diagnosis algorithm to the current fault data when one set of data is input. The intelligent diagnosis algorithm automatically selects the three characteristic variables with the largest contribution rate, and their histograms are displayed in red. The contribution rates of the remaining features are shown as blue histograms. In the figure, the feature variable with the highest contribution rate is the tag number 25, and the corresponding sensor in Table 1 is "Axial displacement of motor". The contribution rate of this feature is 73.7%, that is, 73.7% may be the fault cause of "excessive axial movement of the motor", and the contribution rates of other sensors are far lower than this cause, so they are not considered. It is confirmed that the intelligent diagnosis algorithm is accurate for the current fault judgment (Fig. 13).

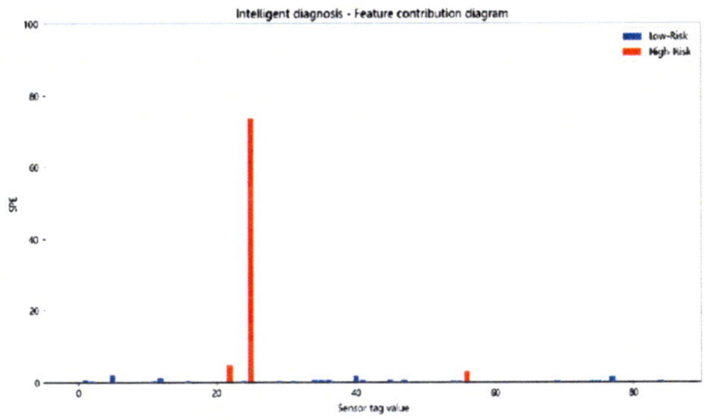

Fig. 13. Contribution rate of each sensor

The overall verification results are shown in the following table.

Table 3. Health monitoring and intelligent diagnosis system verification

Data quantity	Error data value range	Health monitoring alarm times	Accurate times of intelligent diagnosis
30	[+4.5, +5] ∪ [−5, −4.5]	28	27
30	[+4.0, +4.5] ∪ [−4.5, −4]	28	25
40	[+3.5, +4] ∪ [−3.5, −4]	36	33

It can be seen from Table 3 that among the 100 sets of artificially manufactured fault data for "excessive motor axial movement", 60 sets of data are in the interval [+4, +5]

∪ [−5, −4], the alarm rate of health monitoring can reach 93.3%, and the accuracy rate of intelligent diagnosis has also reached 86.7%. In [+3.5, +4] ∪ [−3.5, −4], which is 40 groups of data close to the correct data value range, the accuracy of health monitoring is 90%, and the accuracy of intelligent diagnosis is 82.5%. The combined accuracies are 92% and 85%, respectively. It has been confirmed that the system has good engineering practical value. The actual operation effect is shown in the Fig. 14.

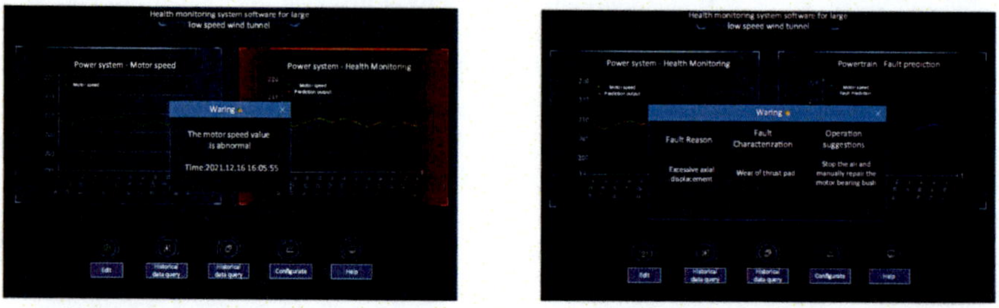

(a) Health monitoring and warning interface

(b) Fault diagnosis and operation suggestion interface

Fig. 14. Front-end interactive interface of fault diagnosis module

5 Conclusion

Identify the key sections and equipment of the FL-10 wind tunnel based on expert experience. Based on advanced sensors and networking technology, through the interactive interface with the overall view of the wind tunnel as the background, each subsystem and ancillary equipment of the wind tunnel are identified. Carry out real-time monitoring to realize the monitoring of the operating status covering the entire area of the wind tunnel.

Based on the high-fidelity 3D model driven by multi-source data fusion such as equipment real-time operation data, simulation data, test data, etc., the virtual-real combination of test equipment, test status and CFD point cloud is realized. Highly digital mapping of the internal test scene, test process and equipment operation status of the wind tunnel, so as to reduce the labor cost of wind tunnel management.

Based on the predictive learning machine based on machine learning proposed in this paper, an optimal algorithm for health monitoring of key equipment in wind tunnels can be quickly found, and the real-time operating status of the equipment can be accurately predicted; fault prediction based on Bi-LSTM algorithm The algorithm can predict the future time of the equipment, and realize the health management and predictive maintenance of the key equipment; Based on expert experience and principal component analysis, the intelligent fault diagnosis algorithm is developed to realize intelligent diagnosis of key equipment, reduce maintenance pressure, and improve the reliability of wind tunnel operation.

The large-scale low-speed wind tunnel health monitoring and intelligent diagnosis system drives the business management process with multi-source data fusion of wind tunnel operation, which improves the informatization and intelligence level of wind tunnel equipment operation management; at the same time, it is involved in advanced technologies such as the Internet of Things and machine learning. Technology, so that it has played an active role in the management and maintenance of key sections and equipment in the wind tunnel; improve the intelligent and automated fault diagnosis of key equipment in the wind tunnel, timely detect latent faults and predict development trends, play a role in The advantages of modern big data and artificial intelligence technology development; finally, it is visually presented in the form of a high-fidelity wind tunnel digital twin to improve the intuitiveness and convenience of wind tunnel management. This system greatly improves the work efficiency of the wind tunnel and the reliability of equipment operation, thereby ensuring the high-quality and high-precision operation of the wind tunnel test.

For the prospect of the research content of this paper, the wind tunnel digital twin operation and maintenance system built in this paper is currently only based on the key equipment and parameters of the wind tunnel to carry out the application of health monitoring and intelligent diagnosis technology. In the future, the small sample data characteristics of wind tunnel tests will be considered, and auxiliary analysis technologies such as intelligent prediction and intelligent diagnosis of wind tunnel test data based on small sample data will be added to the system to expand the application scope and engineering practicability of the system.

References

1. Hess, A., Fila, L.: The joint stirke fighter (JSF) PHM concept: potential impact on aging aircraft problems. Aerosp. Conf. Proc. **10**(3), 22–26 (2003)
2. Baozhen, Z.: Development and application of foreign comprehensive diagnosis, prediction and health management technology. Comput. Meas. Control **16**(5), 591–594 (2008)
3. Wu, B.R.: Condition based maintenance using proportional hazards model. Renew. Energy **204**, 7–19 (2009)
4. Hamadache, M., Jung, J.H., Park, J., et al.: A comprehensive review of artificial intelligence-based approaches for rolling element bearing PHM: shallow and deep learning. JMST Adv. **1**(3) (2019)
5. Baozhen, Z., Tianxiang, Z.: Advanced fault prediction and state management technology. Meas. Control Technol. **22**(11), 4–6 (2003)
6. Huang, L.: Research on wind turbine fault prediction and health management technology based on SCADA. University of Electronic Science and Technology (2015)
7. Wang, T., Wang, H., Liang, X.: PHM technology of unmanned ship propulsion system based on dynamic Bayesian network. Ship Sci. Technol. **41**(23), 80–86 (2019)
8. Yangming, G., Qi, M., Shuangkong, Z.: Overview of DIMA dynamic reconfiguration technology based on fault prediction and health management. Comput. Meas. Control **27**(10), 1–4 (2019)
9. Liming, C.: Research on Life Prediction Method and Life Test of Intermediate Bearing. Dalian University of Technology, Dalian (2016)
10. Yujun, X.: Application and practical analysis of health status monitoring in the whole life cycle of gas blower units. China Equip. Eng. **23**, 265–266 (2021)
11. Tian, N., Ji, X.: Research on health monitoring technology of underwater vehicle propulsion motor. Meas. Control Technol. **30**(01), 71–74+78 (2011). https://doi.org/10.19708/j.ckjs.2011.01.020

12. Hang, Y., Gao Haibo, F., Bo, L.Z., Qianming, S., Chenxing, S.: Fault diagnosis of permanent magnet motor based on PCA-RF. Appl. Sci. Technol. **48**(06), 90–96 (2021)
13. Zio, E.: Prognostics and health management (PHM): where are we and where do we (need to) go in theory and practice. Reliabil. Eng. Syst. Safety **218**(PA) (2022)
14. Wu, Q., Yuan, H., Yuan, H.: Development of ground special vehicle PHM with case-based reason model. Appl. Sci. **11**(10) (2021)
15. Omri, N., Al Masry, Z., Mairot, N., Giampiccolo, S., Zerhouni, N.: Towards an adapted PHM approach: data quality requirements methodology for fault detection applications. Comput. Ind. **127** (2021)
16. Weiss, B.A., Brundage, M.P.: Measurement and evaluation for prognostics and health management (PHM) for manufacturing operations—Summary of an interactive workshop highlighting PHM trends. Int. J. Progn. Health Manag. **12**(1) (2021)
17. An inspired machine-learning algorithm with a hybrid whale optimization for power transformer PHM. Mach. Learn.; Nanchang Univ. Res. Have Published New Data Mach. Learn. Robot. Mach. Learn. (2020)
18. Transfer learning strategies for deep learning-based PHM algorithms. Sci. Appl. Sci.; Res. Beihang Univ. Publ. Find. Appl. Sci.; Sci. Lett. (2020)
19. Fan, Y., Wenjin, Z., Laifa, T., Jian, M.: Transfer Learning strategies for deep learning-based PHM algorithms. Appl. Sci. **10**(7) (2020)
20. Fei, G., Wei-Wei, C., Lian-Feng, L., Jie, W.: A comprehensive evaluation method of sensor selection for PHM based on grey clustering. Sensors **20**(6) (2020)

Progress of Study on the Application of Actuators in Helicopter Airframe ACSRs

Hongyan Zhu(✉), Qingli Wang, Xinhao Yu, Wei Hu, Zheng Liu, Gongnan Li, and Yuqing Luan

AVIC Harbin Aircraft Industry Group Co., Ltd, Harbin 150066, China
zhuhy012@avic.com

Abstract. The severe vibration of helicopters have influenced the flight safety and the riding comfort. A good many technologies have been developed to suppress the helicopter fuselage vibration. They can be divided into two main types: active vibration control, passive vibration control. Now, the active vibration control technologies have been applied widely. The active control of structural response system (ACSRs) is one of the active vibration control technologies. It has many advantages, for example, no influence on rotor aerodynamics, and have been equipped in many helicopters. The actuator is a main component of ACSRs. There are many kinds of actuators, such as centrifugal inertial actuators, hydrostatic pressure actuators, electromagnetic actuators and piezoelectric actuators. Each kind of them have their own advantages and shortcomings, so they may be applicable to different helicopters. In this paper, the application of some kinds of actuators in helicopters, the basic principle, and the main performance parameters, such as output force are reviewed. A prospect of the development for the application of actuators in aviation for the future is made.

Keywords: Helicopter · Vibration · ACSRs · Actuator

1 Introduction

The vibration response of helicopter airframe will be led by the vibration load from the effect of excitation force caused by the main rotor blade, tail rotor blade and engine during the forward flight of helicopters. The reference shows that 40% of helicopter accidents are related to the vibration. Therefore, it is necessary to control the vibration level of the airframe during the forward flight of helicopters. At present, there are many passive and active vibration control techniques [1, 2]. The ACSRs is a widely used active vibration control method. It was developed by Westland Helicopters for the W-30 helicopter in the late 1980s [3] initially. ACSRs incorporating accelerometers, actuators, and so on, have been available commercially. It has many advantages, such as weight saving, vibration level control improving at a wider frequency band, and a more longer fatigue life of the component. ACSRs have been applied to many helicopters, such as S-92, V-22, UH-60M, S-76D, AW139, and EC225/725. These active methods have been proved effective over a range of configurations and helicopter frequencies. They have

Chinese Society of Aeronautics and Astronautics: CASTYSF 2022, LNEE 972, pp. 537–549, 2023.
https://doi.org/10.1007/978-981-19-7652-0_48

brought a significantly lower rotor-induced vibration level comparing to those passive approaches [4].

The actuator is a critical component in ACSRs in order to produce a controllable force for eliminating the vibration caused by the main rotor. The actual control effect of ACSRs on the airframe vibration is determined by its properties directly. It is used to introduce control force to the airframe to change the vibration response of the airframe. Now, though there are many kinds of actuators, but the centrifugal inertial, electromagnetic, hydraulic and piezoelectric actuators are used basically in helicopter ACSRs.

2 Actuators of Typical ACSRs

Currently, commercially available actuators are single-point. They apply inertial forces on the fuselage at their attached points. There are two main kinds of inertial actuators. The main difference is how the inertia is moved to produce force: rotating inertia and oscillating inertia. The principle of producing inertial force for the rotating inertia actuator is dependent on rotating an eccentric mass, while the principle of producing inertial force for the oscillating inertia is through oscillating a mass to produce a linear force.

2.1 Centrifugal Inertial Actuator

At present, force generators (FGs) of ACSRs applied already are usually centrifugal inertial actuators. The principle of centrifugal inertial actuators is using resonance of the mass to generate inertial force. But the magnitude of output force from the inertial actuator depends on the weight of the mass. To achieve a good controlling effect for the fuselage vibration, the weight of the inertial actuator used is 4–5% of the airframe weight in most cases while controlled by the inertial actuator [3]. The LORD active vibration control system (AVCs) on the Bell 429 helicopter is studied by Mahmood et al. [5]. Two different kinds of actuators were used. One of them is circular force generator (CFG) (Fig. 1).

The CFG is composed of two rotor disks. There is a mass attached on one side of each disk to form an imbalanced rotating mass. Every rotor is controlled by a single motor and rotate in the same direction at the rotor blade passing frequency. The angle between the disks changes to change the magnitude of the resultant force generated. The force generated by the two adjacent CFG's can be combined to generate a resultant linear or elliptical force as required. The maximum force output of the AVC system at rotor blade passing frequency for the Bell 429 is 288 lbf. The number of actuators installed is 4. The research shows that the weight of the active system with both actuator technology chosen can be reduced relative to the passive technology. Moreover, the vibration level is reduced below the target with the both active systems.

X2 Technology™ Demonstrator is a high speed helicopter developed by Sikorsky Aircraft [6, 7]. The flight speed arrives at 417 km/h. The vibration load transmitted to the airframe increases sharply. The helicopter is dependent on ACSRs completely without any passive control device in order to cancel the huge vibration load. There are six centrifugal inertial actuators in the ACSRs (Fig. 2). The actuator is the Moog Vibration Control Actuation System (VCAS). The VCAS is composed of a electronic unit (EU) and FGs. There are a pair of counter-rotating eccentric masses spinning at 4/rev. The flight test shows that the vibration level of the airframe meets the goal all through in the range of flight speed.

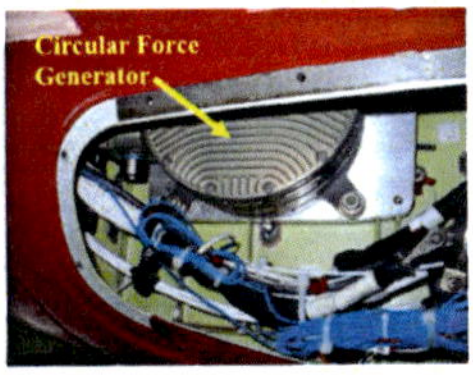

Fig. 1. CFG installation on keel beam [5].

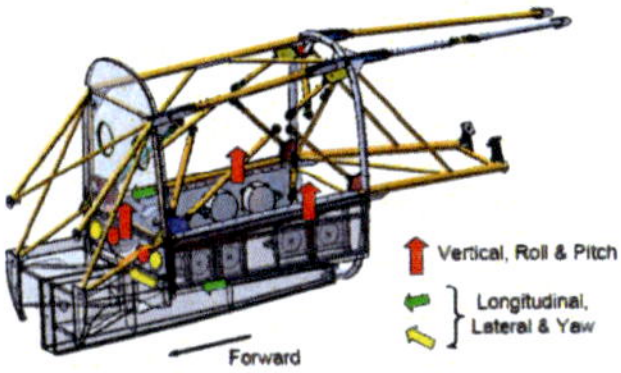

Fig. 2. Actuators in X2 helicopter [6, 7].

The circular rotating force generators (CRFGs) of Moog Inc. [8] have been used in several helicopters, such as S-92, UH-60M, and V-22. There are two eccentric masses in one CRFG. They rotate in opposite directions (counter-rotated) with a single electric motor and a gear system. The counter-rotating masses in one CRFG generate a sinusoidal inertial force along a single axis, as shown in Fig. 3. The force amplitude is directly proportional to the square of rotational frequency. For a given rotational frequency, the force amplitude generated is unchanged for a single CRFG. The result sinusoidal force amplitude and phase are controlled by the relative phase between a pair of CRFGs.

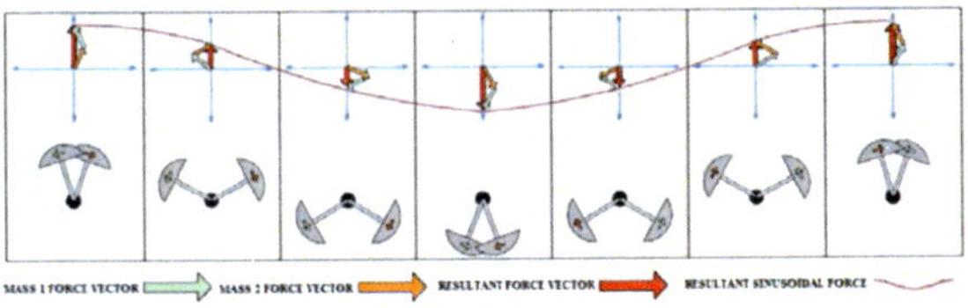

Fig. 3. Force vector diagram of a CRFG, Moog Inc. [8].

The CRFG and schematic diagram of a rotating eccentric mass in a pair of CRFGs are shown in Fig. 4 [8]. Moog Inc.'s CRFGs are used in the study on vibration control systems for a light civil helicopter (LCH) by Kim et al. [1] in the LCH Development Program funded by the Ministry of Trade, Industry and Energy (Korea). The key features of the two pairs of CRFGs used are listed in Table 1.

Table 1. Specifications of the CRFGs [1].

Model	Property	Value
CRFG1 (exciter pair)	Max. Force	398 N (18.0 Hz)
	Operating range	14.7–23.5 Hz
	Weight	14.6 kg
CRFG2 (actuator pair)	Max. Force	1,484 N (18.0 Hz)
	Operating range	14.7–23.5 Hz
	Weight	8.7 kg

An AVCs for the S-92 helicopter has been designed and flight-tested by Sikorsky Aircraft Corporation [9]. The inertial FGs were used. They are single point devices. A 1000 lb force could be produced by each inertial FG from a pair of counter-rotating eccentric disks. The first flight test of a S-92 helicopter with the AVCs was carried out in December of 1998. The vibration suppression of the AVCs was demonstrated in a large range of flight conditions and ballast configurations. The average level of cockpit and cabin vibration reduction was at least 50% (below 0.15 g) when speeds varied from hover to 150 knots.

Centrifugal inertial actuators have many advantages such as smaller volume, light weight, large output force, smaller power required, adapting a wide frequency band, and so forth. It is applicable to large helicopters, medium helicopters, and new helicopters with a changeable rotating speed.

2.2 Hydrostatic Pressure Actuator

The test verification on the ACSR was carried out by Millott for UH-60M at 2003 [10, 11]. The vibration transferred to the fuselage was controlled by three hydraulic servo actuators. The structure of the vibration control system is shown in Fig. 5. The experimental result shows that the controlling effect is better for different flight statuses.

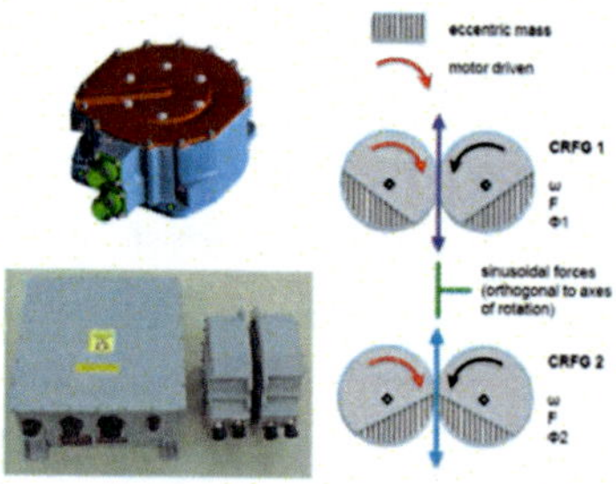

Fig. 4. Pairing two CRFGs, Moog Inc. [8].

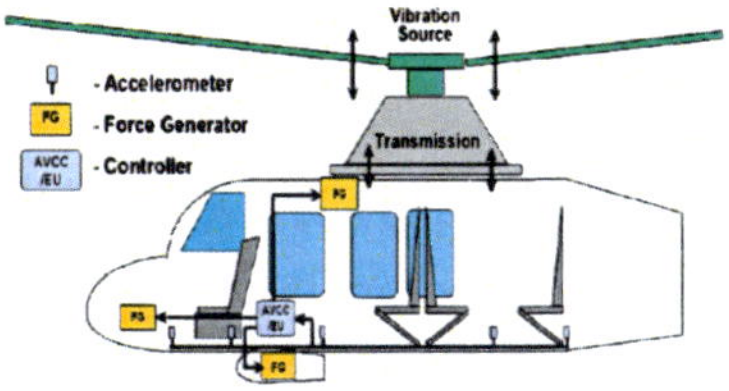

Fig. 5. The ACSR system of UH-60M [10, 11].

The experimental work was carried out over a total of five days in two visits to the Yeovil site during February and March 2011. The practical application of ACSRs developed initially on the Agusta Westland W30 helicopter is described by Mottershead et al. [12]. It was tested in the vibration test house at Yeovil. The experiment was carried out over a total of five days. In the experiment, the electro-hydraulic actuators was used. An electro-hydraulic actuator constitutes an part of the support system. It imparts a force between the raft and the airframe in parallel with each of the four elastomeric mounts. This configuration, described in Fig. 6a, b, makes up of the original Agusta Westland ACSRs. The flight experiment is successful. The result shows that the average vibration level of cockpit is reduced by 72–82% than the passenger cabin.

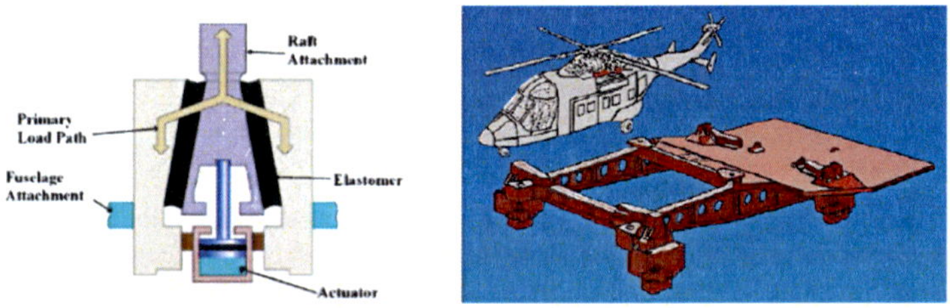

Fig. 6. Electro-hydraulic actuator and it's location [12].

The same active vibration control technology is applied in the EH-101 helicopter by Westland/Agusta [13]. The difference is that four small electro-hydraulic actuators are incorporated into struts of the main gearbox (Fig. 7). The structure is compact. The flight test shows that the average vibration level is reduced by 59–88% at ten observation points throughout the entire flight speeds.

The ACSRs for S-76B helicopter is also studied by Sikorsky Aircraft [14]. Six electro-hydraulic actuators are incorporated into the main rotor gearbox attachments in parallel with four stiff spring mounts (Fig. 8). The ground shake test showed a 86% of N/rev cockpit vibration reduction and a 50% of N-1/rev and 2 N/rev cockpit vibration reduction. The vibration at the pilot and copilot stations was controlled effectively by the ACSRs throughout the entire speed range. The intrusion index was kept well below the ADS-27 limit. The actuator force and power was predicted in the simulated flight. The maximum actuator force ranged from 2000 to 3200 lbs within the forward speed range. The maximum power of 9.5 horsepower was needed for operating four actuators.

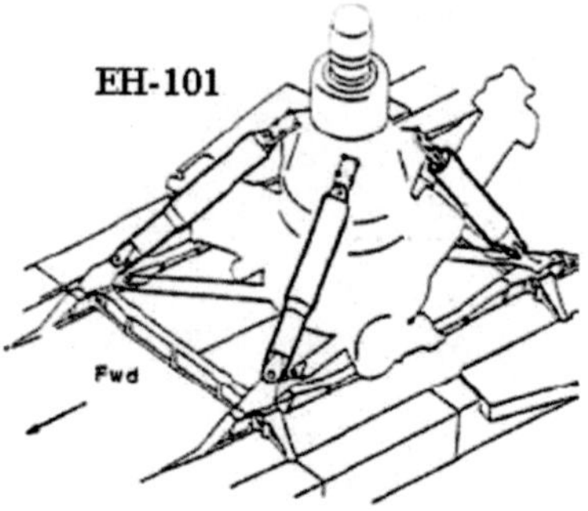

Fig. 7. EH101 helicopter actuator location [13].

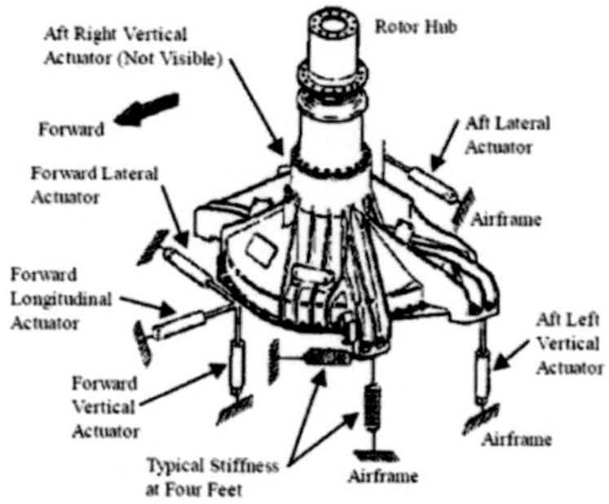

Fig. 8. S-76 helicopter actuator location [14].

An ACSRs like AVCs was demonstrated by Welsh et al. on a Sikorsky UH-60 Black Hawk helicopter [15]. Two inertial-type FGs were used. They apply force at a single point of the fuselage. The inertial FG is shown in Fig. 9. It is a hybrid resonant device. It incorporated the UH-60 passive vibration absorber with a hydraulic servo actuator. Therefore, the FG can be used as a passive vibration absorber, or a active actuator. The inertial mass of each force generator is 40 lbs, and the maximum travel distance is 0.2 inches. A 1200 lb force could be produced by each FG at the N/rev frequency (17.2 Hz). The AVCs is better continuously than the passive vibration control system throughout all forward speeds. A 65% vibration reduction is achieved at a forward speed of 145 knots, while a 45% reduction for the passive system. The maximum hydraulic power needed to manipulate the two force generators is in the range of 1.0–8.0 horsepower, in accordance with the main rotor speed and flight condition.

The Active Vibration Reduction system (AVRs) is an ACSR like system developed by Kawasaki Heavy Industries of Japan [16, 17]. Hydraulic actuators are incorporated in parallel with the main gearbox mounts to eliminate vibrations at N/rev and 2N/rev. At 1997, the AVRs was equipped in series production on the Kawasaki BK 117 helicopter. Flight tests of the AVRs were carried out on a series of production model BK 117 helicopters. The aim is to evaluate the system at a forward speed from hover to 150 knots. The N/rev and 2N/rev vertical vibration under the pilot seat was reduced below

0.1 g by the AVRs throughout the entire forward speed range. At least 45% reduction of N/rev vibration level at all forward speeds is achieved.

The Multipoint Adaptive Vibration Suppression System (MAVSS) is developed by Bell Helicopter. It is used to reduce the rotor-induced vibration over a broad area of the airframe [18]. Ground shake testing of the system for Model 412 helicopter was carried out. Four hydraulic control actuators were placed in parallel with elastomeric gearbox mounts. It is showed that the vibration is reduced by 94% at six sensor locations. The system has also been tested on a V-22 0.2-scale aeroelastic semispan tiltrotor model.

2.3 Electromagnetic Actuator

The electromagnetic actuator is used for the first ACSRs developed and certified by Eurocopter [19]. The actuator is composed of a magnetic inertial mass fixed on a spring. It is controlled by an electromagnetic field produced by a coil (Fig. 10). The magnet mass and the stiffness of the spring element is selected such that the resonance frequency of the system is nearby the vibration frequency range to be controlled. It may provide large dynamic effort with a small current. So, the power required is minimum.

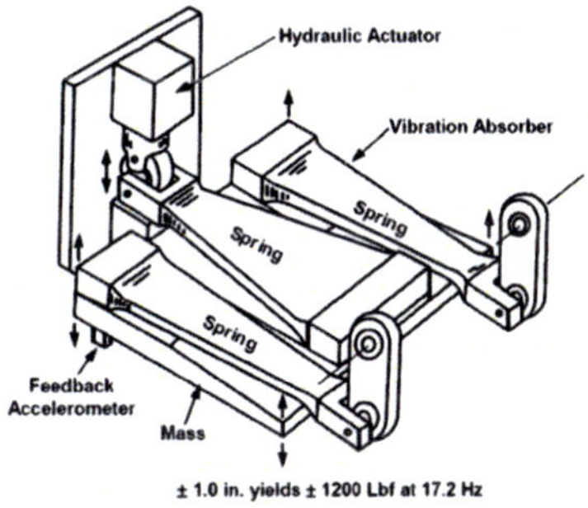

Fig. 9. UH-60 servo inertial force generator [15].

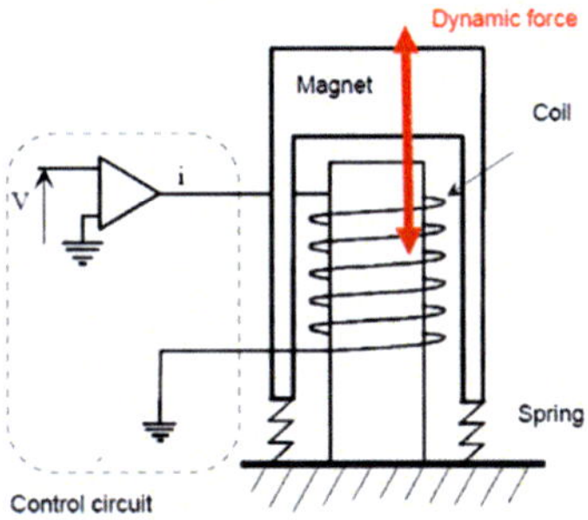

Fig. 10. Electromagnetic actuator [19].

EC225 is a 11 tons helicopter. A new generation of vibration control system is equipped with a electromagnetic actuator (Fig. 11). The system is composed of 3 actuators (one in the cockpit and others in the cabin rear). The ACSRs is the only vibration control system equipped on EC225. The mass is less than 0.8% of the whole mass of the helicopter. Furthermore, the electrical consumption keeps low [19].

The inertial electrodynamic linear actuator (IELA) used in the AVCs on the Bell 429 helicopter is shown in Fig. 12 [5]. The IELA utilizes a voice-coil type motor. A magnet assembly makes up a part of the moving mass at the end of a cantilevered flexure component. The flexural stiffness is chosen to have a resonant frequency at or near the passing frequency of the rotor blade. The mass and magnet assembly is actuated by the current flowing through a coil of copper wire. It is attached to the base of the setting. The setting is fixed firmly to the helicopter on keel beam. The maximum force output of the AVCs at rotor blade passing frequency is 250 lbf. The number of actuators installed is 2 [5].

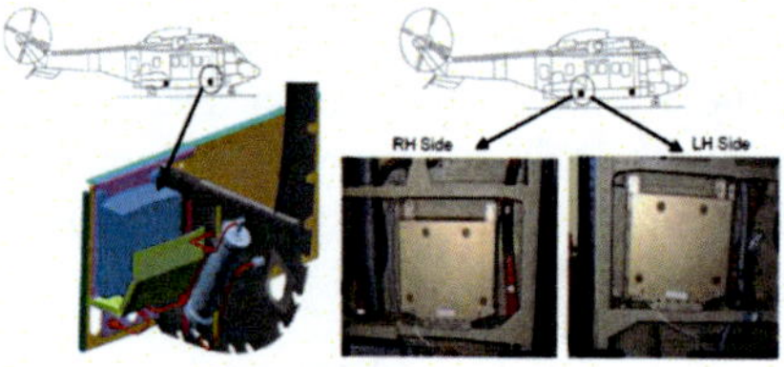

Fig. 11. The electromagnetic AVCs on EC 225 [19].

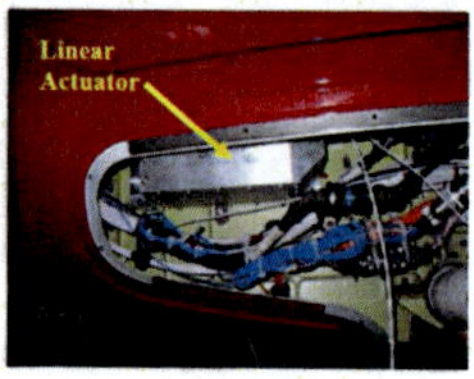

Fig. 12. Linear actuator on Bell 429 [5].

Boeing Helicopter's Active Vibration Suppression (AVS) utilizes electromagnetic resonant actuators. It is placed in the cockpit to eliminate vibration at N/rev and 2 N/rev [20]. The actuators generate structural force to cancel vibration at the sensor locations. There are a pair of electromagnetically controlled spring/mass assemblies. It is tuned to the N/rev and 2 N/rev frequencies. The ACSRs was installed on a CH-47D helicopter, and one N/rev actuator was placed under the pilot seat. The ground shake test shows that the N/rev vibration under the pilot seat reduced by a 97%. The peak of actuator force output was 2600 lbs at N/rev and 1500 lbs at 2 N/rev.

The flight test of ACSRs utilizing two self-developed electromagnetism inertial actuators for a light helicopter is carried out by Lu et al. [21]. The actuator is shown in Fig. 13. The weight of each actuator is 8kg. The operating frequency is in the range of 18.9–19.7 Hz. The maximum output force is 78kg. The result shows that a reduction of 33–66% is achieved through the ACSRs for each speed state.

2.4 Piezoelectricity Actuator

Piezoelectric stack actuators (PSAs) is effective for the control of helicopter vibration. There are many advantages such as large output force, rapid response, wide range of operating frequency, and lightweight, and so forth. The harmonic control of helicopter fuselage vibration through PSAs was studied by Song and Xia [22, 23]. The experiment was carried out. The result shows good vibration reduction.

The feasibility of utilizing a piezoceramic actuator for vibration reduction in the helicopter airframe has been investigated analytically by Hanagud and Babu [24]. The airframe is modeled through a simple beam finite-element-model with concentrated masses. It is composed of 10 elements, 11 nodes, and 33 degrees-of-freedom. A 60% vibration reduction is realized in the vertical direction at the sensor location. They came to the conclusion that a more sophisticated airframe and actuator model should be included.

The concept of the vibration control system based on piezo-actuator is show in Fig. 14. The vibration in the cabin induced by the rotor is eliminated by a secondary vibration field produced by piezo-actuator. This ACSRs based on piezo-actuator is realized by a new actuation technology, that is piezo-ceramic actuation [25]. The piezo active spring design is shown in Fig. 15. The actuator produces control forces by the attached mass inertia.

Fig. 13. Actuator in a light helicopter [21].

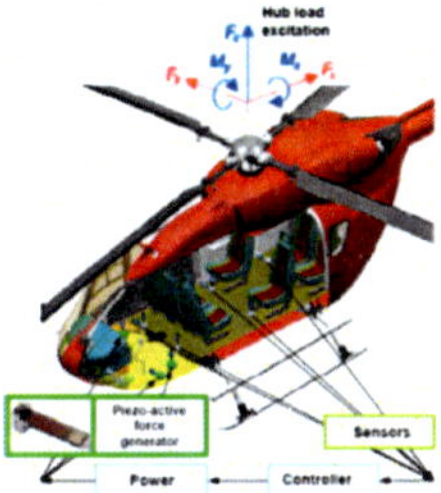

Fig. 14. Piezo-AVCS concept [25].

Piezo-ceramic elements are embedded into a GFRP structure allowing direct electrical control over the actuation forces. The actuation technology has some of advantages because of no moving components involving, that is no wear, no friction, high life time, silent actuation, direct electrical control, extremely accurate, very high bandwidth.

The test includes preliminary test, endurance test and full-scale test. The articles, the single-mass and dual-mass inertial force generators are showed in Fig. 16.

Fig. 15. Piezo-active inertial force generator [25].

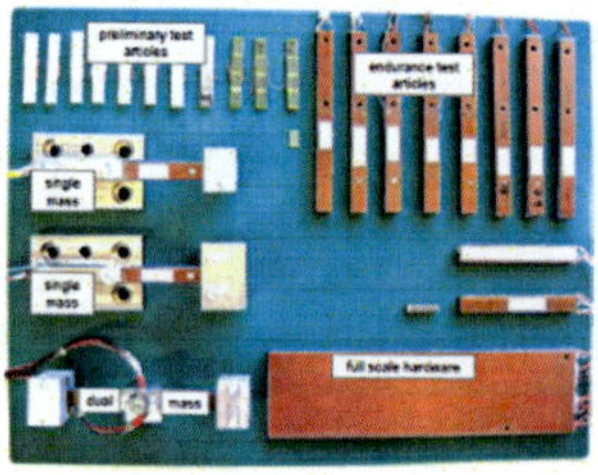

Fig. 16. Test articles [25].

The AVCs using piezoelectric stacks as actuators for Z-11 helicopter is studied by Meng et al. [26]. Four piezoelectric stack actuators (PSAs) are installed at four corners of the scaled Z-11 helicopter fuselage floor (Fig. 17). The PSA used is PSt 150/7/100 VS12. Its physical parameters are listed in Table 2 [26]. The experimental result shows that the effectiveness of the PSA-driven active vibration control is quite good. A reduction of 92 and 94% at two objective points is achieved when two dominant harmonics are controlled, and the vibration is reduced by 69 and 72% when controlling the first harmonic.

Table 2. Properties of PSt150/5/100 VS12 actuator [26].

Property	Value
Length, mm	107
Diameter, mm	12
Maximum free displacement, mm	0.1
Weight, g	68
Maximum force, N	1,400
Axial stiffness, N/mm	15,000
Voltage range, V	0–150

The simulation and test of active vibration control for AH64-4 helicopter tail beam model using piezostack actuators is studied by Walchko [27]. The vibration of tail beam model is controlled through four piezostack actuators fixed at the load paths of tail beam (Fig. 18). The result shows that the control effective is well.

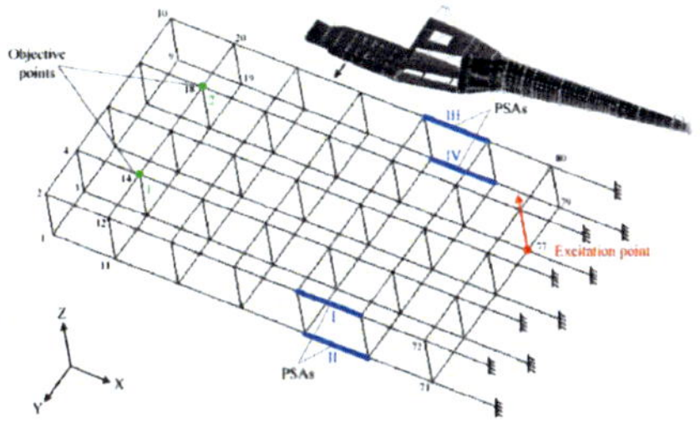

Fig. 17. Frame structure with PSAs [26].

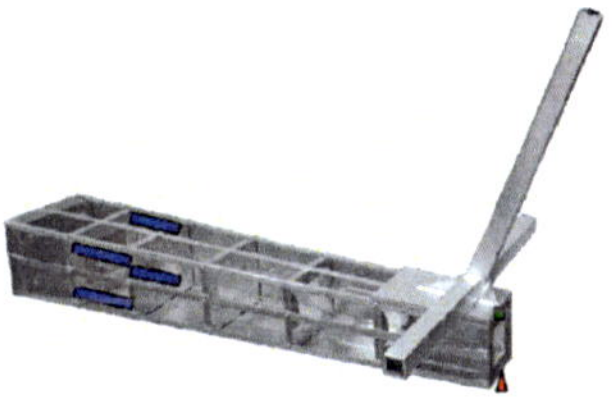

Fig. 18. Tail beam model using piezo actuators [27].

3 Conclusions and Prospect

In conclusion, the ACSRs can suppress vibration over the whole forward speed range effectively. At least 50% of cabin vibration could be reduced by a majority of the systems. The control force of single point actuation systems are produced by inertial devices. The maximum control force is in the range of 1000 and 3200 lbf. The weight of the ACSRs is less than the passive vibration control system. The power of ACSRs is lower. The application of ACSRs in helicopters will be more widely used with the development of Aviation Intelligence. Especially, piezoelectric actuator might be used widely.

References

1. Kim, D.H., Kim, T.J., Jung, S.U., et al.: Test and simulation of an active vibration control system for helicopter applications. Int. J. Aeronaut. Space Sci. **17**(3), 442–445 (2016)
2. Hu, M., Wang, L., Chen, K., Yan, W., et al.: Research and implementation of centrifugal harmonic force generator system for vibration control. In: Proceedings of the 5th China Aeronautical Science and Technology Conference, LNEE 821, pp. 385–397. Zhejiang, China (2022)
3. Staple, A.E.: An evaluation of active control of structural response as a means of reducing helicopter vibration. In: Proceedings of the 46th Annual Forum of the American Helicopter Society. Washington, DC (1990)

4. Swanson, D., Black, P., Girondin, V., et al.: Active vibration vontrol using circular force generators. In: 41st European Rotorcraft Forum, Munich, Germany (2015)
5. Mahmood, R.S., Heverly II, D.E.: In-flight demonstration of active vibration control technologies on the Bell 429 helicopter. In: Proceedings of the American Helicopter Society 68th Annual Forum, Fort Worth, Texas (2012)
6. Blackwell, R., Millott, T.: Dynamics design characteristics of the Sikorsky X2 technologytm demonstrator aircraft. In: Proceedings of the American Helicopter Society 64th Annual Forum. Montreal, Canada (2008)
7. David, E.H., et al.: Adaptive algorithms for rotorcraft active vibration control. In: Proceedings of the 65th Forum of the American Helicopter Society, Grapevine, Texas (2009)
8. Garnjost, K.D., Rey, G.J.: Modular vibratory force generator, and method of operating same, US Patent 5,903,077. Moog Inc, USA (1999)
9. Goodma, R.K., Millott, T.A.: Design, development, and flight testing of the active vibration control system for the Sikorsky S-92. In: Proceedings of the 56th Annual Forum of the American Helicopter Society, pp. 764–771. Virginia Beach, VA (2000)
10. Welsh, W.A., et al.: Flight test on an active vibration control system on the UH 60 Black Hawk helicopter. In: Proceedings of the 51st Annual Forum of the American Helicopter Society, pp. 393–402. Fort Worth, TX (1995)
11. Millott, T.A., Robert, G.K., Jonathan, K., et al.: Risk reduction flight test of a pre-production active vibration control system for the UH-60M. In: American Helicopter Society 59th Annual Forum. Phoenix, USA (2003)
12. Mottershead, J.E., Tehrani, M.G., James, S., et al.: Active vibration control experiments on an Agusta Westland W30 helicopter airframe. Arch. Proceed. Inst. Mechan. Eng. Part C J. Mech. Eng. Sci. 1989–1996 (203–210) **226**(C6), 1504–1516
13. Mei, Z.: Robust Structural Response Control for Helicopter. Nanjing University of Aeronautics and Astronautics, M. D. diss. (2007)
14. Welsh, W.A., Von Hardenberg, P.C., Von Hardenberg, P.W., et al.: Test and evaluation of fuselage vibration utilizing active control of structural response (ACSR) optimized to ADS-27. In: Proceedings of the 46th Annual Forum of the American Helicopter Society, pp. 21–37. Washington, DC (1990)
15. Welsh, W., Fredrickson, C., Rauch, C.: Flight test of an active vibration control system on the UH-60 Black Hawk helicopter. In: Proceedings of the 51st Annual Forum of the American Helicopter Society, pp. 393–402. Fort Worth, TX (1995)
16. Aso, M., Bandoh, S.: The development of the total vibration reduction (TVR) system. In: Proceedings of the 55th Annual Forum of the American Helicopter Society, pp. 202–208. Montreal, Quebec, Canada (1999)
17. Kawaguchi, H., Bandoh, S., Niwa, Y.: The test results of AVR (Active Vibration Reduction) system. In: Proceedings of the 52nd Annual Forum of the American Helicopter Society, pp. 123–136. Washington, DC (1996)
18. Settle, T.B., Nixon, M.W.: MAVSS control of an active flaperon for tiltrotor vibration reduction. In: Proceedings of the 53rd Annual Forum of the American Helicopter Society, pp. 1177–1193. Virginia Beach, VA (1997)
19. Konstanzer, P., Enenkl, B., Aubourg, P.A., et al.: Recent advances in Eurocopter's passive and active vibration control. In: American Helicopter Society 64th Annual Forum. Montréal, Canada (2008)
20. Teal, R.S., McCorvey, D.L., Malloy, D.: Active vibration suppression for the CH-47D. In: Proceedings of the 53rd Annual Forum of the American Helicopter Society, pp. 211–219. Virginia Beach, VA (1997)
21. Lu, Y., Gu, Z., Ling, A., et al.: Flight test of active control of structure response for helicopter. J. Vib. Eng. **25**(1), 24–29 (2012)

22. Song, L., Xia, P.: A harmonic synchronous identification-updating method for Active control of helicopter structural response driven by piezoelectric stack actuators. J. Am. Helicopter Soc. **60**(3), 032013-1-14 (2015)
23. Song, L., Xia, P.: Active control of helicopter structural response using piezoelectric stack actuators. J. Aircr. **50**(2), 659–663 (2013)
24. Hanagud, S., Babu, G.L.: Smart structures in the control of airframe vibrations. J. Am. Helicopter Soc. **39**(2), 69–72 (1994)
25. Konstanzer, P., Grünewald, M., Jänker, P., et al.: Piezo tunable vibration absorber system for aircraft interior noise reduction. In: Euronoise 2006. Finland (2006)
26. Meng, D., Xia, P.Q., Song, L.H., et al.: Experimental study on piezoelectric-stack-actuator-driven active vibration control of helicopter floor structure. J. Aircr. **57**(2), 377–382 (2019)
27. Walchko, J.C.: Hybrid Feedforward-Feedback Control for Active Helicopter Vibration Suppression. Department of Mechanical and Nuclear Engineering, Pennsylvania State University, Pennsylvania (2007)

Thermoelastic Analysis in Imperfect Rotating Functionally Graded Disks with Control Volume Finite Element Method

Qi Liu[1(✉)], Yanxia Du[1], Wenliang Qi[2,3], Lei Liu[1], and Yewei Gui[1]

[1] State Key Laboratory of Aerodynamics, Aerodynamics Research and Development Center, Mianyang 621000, China
liuqi202@hrbeu.edu.cn

[2] Xi'an Aeronautics Computing Technique Research Institute, AVIC, Xi'an 710068, China

[3] School of Mechanical Engineering, Northwestern Polytechnical University, Xi'an 710072, China

Abstract. A control volume finite element approach (CV-FEM) for the thermoelastic issue in functionally graded material (FG) spinning disks have been developed in this work. Because of a technological constraint during the FGM production, the FG disks have porosities. Two main forms of porosity distribution (uniform and ununiform porosity distributions) along the radial axis of the rotating disk are considered. Due to the FG is often served in extreme temperature environment, the temperature-dependency material properties are considered in this study. The modified mixed model is used to determine the FG disk material properties involving the existence of porosity and temperature-dependency property effects. The effects of material distribution, mechanical and temperature loads, such as boundary temperature, FG gradient index, porosities distribution, and rotating angular velocity on the thermoelastic performance have been analyzed.

Keywords: Thermoelastic modeling · Functionally graded materials · Rotating disk · Control volume finite element method

1 Introduction

Functionally gradient materials (FGMs) are novel types of materials whose material characteristics change in a continuous gradient direction. Due to their appealing physical and mechanical features, FG structures have been widely employed in a variety of industrial industries in recent years, including heat-engine components, nuclear power plants, gas turbines, etc. Many engineering structures such as centrifugal compressors, turbine rotors, gear wheels are modeled as rotating circular disks. The strength and safety of these mechanical parts are significantly affected by the high thermal loads and large centrifugal force. The temperature and the strength of the structural component can be optimized with functionally gradation of the material properties. Based on analytical and numerical methods, numerous thermoelastic analyses have been performed for FGM shells, disks, and plates. Attial [1] studied the stress and deformation distribution

Chinese Society of Aeronautics and Astronautics: CASTYSF 2022, LNEE 972, pp. 550–569, 2023.
https://doi.org/10.1007/978-981-19-7652-0_49

of the fiber-reinforced, FGM for rotating hollow disk based on finite element method (FEM), the results revealed that the choice of gradation has a less significant effect on deformation and stress in solid disks than it did in hollow disks. Bayat [2] studied thickness-variable FGM hollow rotating disks under plane stress condition, the results indicated that it may be more efficient to use a disk with parabolic or hyperbolic thickness profile compared to that with uniform-thickness. Under a steady-state temperature field, Bayat [3] investigated the effects of the grading index and thickness profile of an FMG rotating disk on stress and weight. Using the small and large deflection theories, Tanaka [4, 5] created a procedure to decrease the thermal stresses due to thermal and mechanical loads in an FGM annulus. Based on an improved Adomian decomposition method, Kutsal [6] studied the elastic field distribution of variable thickness rotating disk. Using FEM, sharma [7] investigated the deformation, stress, and strain distributions of a thin rotating FGM circular under thermal and mechanical load.

However, due to the manufacturing technology issues, porosity may exist inside the FG structures, and the inclusion of porosities greatly reduces the effectiveness of FGMs [8]. Furthermore, FGMs' intermediate zone exists more porosities than the top and bottom area of the FGMs due to the limitation of the multi-step sequential infiltration technique [9], which causes the uneven porosities to exist in the FGMs. Therefore, it is essential to consider the influence of the existence of uneven porosities on the thermoelastic response of FG structures. Babamiri [10] presented the thermoelastic analysis of roller-supported FGM rotating disks with and without imperfection, and with temperature-dependent properties. They pointed out that the stress and displacement of an imperfect FGM disk are greater than those of the perfect disk. Using the classical plate theory, Jabbari [11] studied the thermal buckling problems of a porosity FGM circular plate, the influence of porosity and temperature distributions on the critical buckling temperature were studied. Using the energy method and the higher-order shear deformation plate theory, Mojahedin [12] studied the buckling problem of a porosity FGM circular plate, assuming that the pores were soaked in fluid. Zhao [13] performed the dynamics characteristic of FG porous circular, annular, and sector plates based on general boundary conditions utilizing the first-order shear deformation theory. Dastjerdi [14] studied thick perfect and imperfect FGM rotating disks under a hydrothermal environment. The results showed that the presence of porous reduces the strength of the FGM rotating disk, and the type of the even porosity reduced the strength of the structure more than that of the uneven porosity. Using the Rayleigh-Ritz method, Wang [15] investigated the effects of even and uneven porous distributions on the thermal vibration characteristics of the FGM cylindrical shells. Taking into account the influence of porous on the structure's bulk density and moisture, Ting [16] carried out the magneto-electro-elastic field responses of an FGM rotating disk with varied thickness, the influences of material property, geometric structure, and external conditions on the multi-field mechanical responses of the FGM annular plates were investigated.

As can be seen from the above-mentioned literatures, few studies about thermoelastic problem in FG rotating disk with considering both efforts of temperature dependency material and properties are reported, therefore, in this paper, the influence of porosities with and temperature dependency material on the temperature, displacement, and stress fields of an FG rotating disk is investigated. The CV-FEM [17, 21] which has proven to be an efficient tool for the FG thermoelastic problem is extended to solve the FG rotating disk thermoelastic problem. The effects of some factors such as thermal effects, gradient index, porosity distribution, rotating velocity, and the thermoelastic field have been further studied. The effects of plane stress and plane strain assumptions on thermoelastic performance are also investigated in the study.

2 Formulation of the Problem

The problem to be studied is presented in Fig. 1, where a 2D FG isotropic porous rotating hollow disk with thermal loads on the inner and outer surface, the circular annulus is assumed to rotate about its center with a constant angular velocity ω. And the porosities inside the disk are assumed to distribute evenly or unevenly in the radial direction r. The thermal and mechanical parameters varying with both temperature T and radial direction r as

$$P = P(r, T) \qquad a \leq r \leq b \tag{1}$$

where $P\,(r, T)$ represents arbitrarily material cons such as thermal conductivity k, density ρ, specific heat capacity c, Young's modulus E, and Poisson's ratio ν, T represents time.

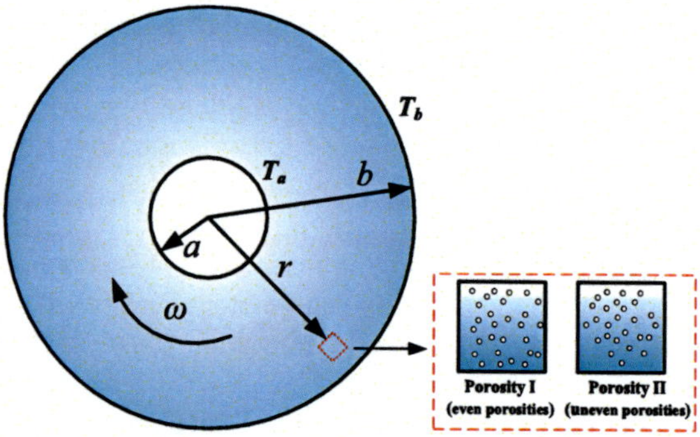

Fig. 1. A rotating FG hollow disk with even and uneven porosities under thermal loads.

The integral form of the transient heat conduction equation without internal heat sources can be given as

$$\int_S \frac{\partial(\rho c T)}{\partial t} dS = \oint_L k\left(\frac{\partial T}{\partial x}n_x + \frac{\partial T}{\partial y}n_y\right)dl \tag{2}$$

where t *is* the time, $\boldsymbol{n}(n_x, n_y)$ is the unit outer vector normal to the enclosed boundary L of the domain S.

$$\int_S \frac{\partial(\rho c T)}{\partial t} dS = \oint_L k\left(\frac{\partial T}{\partial x}n_x + \frac{\partial T}{\partial y}n_y\right)dl \tag{3}$$

Three types of thermal boundary conditions are used as

$$T = T_B \tag{4}$$

$$k_x \frac{\partial T}{\partial x} n_x + k_y \frac{\partial T}{\partial y} n_y = -q_B \tag{5}$$

$$k_x \frac{\partial T}{\partial x} + k_y \frac{\partial T}{\partial y} = -h_B (T - T_\infty) \tag{6}$$

where the subscript B denotes the variables on the boundary, T_B is the specified temperature on the Dirichlet boundary, q_B is the given normal heat flux on the Neumann boundary, and h_B and T_∞ are the given heat convection exchange coefficient and corresponding environment temperature on the Robin boundary condition.

The integral form of steady-state force equilibrium equation with body forces is given as

$$\int_S \nabla \cdot \boldsymbol{\sigma} dS + \int_S \rho \boldsymbol{f} dS = 0 \tag{7}$$

where $\boldsymbol{\sigma}, \boldsymbol{f}, \rho$, are the stress tensor, body force vector, density, respectively.

The centrifugal force in the radial axis in the rotating case is the body force, thus

$$\boldsymbol{f} = \omega^2 \boldsymbol{r} \tag{8}$$

where ω is the constant rotational angular velocity.

The radial displacement vector $\boldsymbol{r}(r_x, r_y)$ can be expressed as

$$\boldsymbol{r} = (r_x, r_y) = \frac{(x, y)}{\sqrt{x^2 + y^2}} \tag{9}$$

The constitutive equation for the isotropic linear elastic solid is

$$\boldsymbol{\sigma} = 2\mu\boldsymbol{\varepsilon} + \lambda \mathrm{tr}(\boldsymbol{\varepsilon})\mathbf{I} - \Gamma\alpha(T - T_r)\mathbf{I} \tag{10}$$

where

$$\boldsymbol{\varepsilon} = \frac{1}{2}\left[\nabla \boldsymbol{u} + (\nabla \boldsymbol{u})^{\mathrm{T}}\right] \tag{11}$$

is the linear strain tensor. $\boldsymbol{u}$, $\mathbf{I}$, α, and T_r are the displacement vector, identity tensor, thermal expansivity, and reference temperature, respectively. λ and μ are the Lame's constant, Γ is the thermal constant. λ. μ and Γ can be determined with $\boldsymbol{E}$ and ν. For the plane strain assumption, $\lambda = \nu E/(1 + \nu)(1\text{-}2\nu)$, $\mu = E/2(1 + \nu)$, $\Gamma = E/(1\text{-}2\nu)$; and for the plane stress assumption, $\lambda = \nu E/(1 + \nu)(1\text{-}\nu)$, $\mu = E/2(1 + \nu)$, $\Gamma = E/(1\text{-}\nu)$.

Using Eqs. (8), (10) and, and based on the Gauss divergence theorem, the force equilibrium Eq. (7) can be rearranged as

$$\oint_L \left[2\mu \frac{\partial u}{\partial x} n_x + \mu\left(\frac{\partial u}{\partial y} + \frac{\partial v}{\partial x}\right) n_y + \lambda\left(\frac{\partial u}{\partial x} + \frac{\partial v}{\partial y}\right) n_x - \Gamma a(T - T_r) n_x\right] dl$$

$$+\int_S \rho\omega^2 r_x dS = 0$$

$$\oint_L \left[2\mu\frac{\partial u}{\partial x}n_y + \mu\left(\frac{\partial u}{\partial y} + \frac{\partial v}{\partial x}\right)n_x + \lambda\left(\frac{\partial u}{\partial x} + \frac{\partial v}{\partial y}\right)n_y - \Gamma a(T - T_r)n_y\right]dl$$
$$quad + \int_S \rho\omega^2 r_y dS = 0 \tag{12}$$

where $\boldsymbol{u}(u, v)$ is the displacement vector.

Two types of mechanical boundary are used as Dirichlet boundary condition Eq. (13) and Neumann boundary condition Eq. (14).

$$\boldsymbol{u} = \boldsymbol{u_B} \text{ on } L_D \tag{13}$$

$$\boldsymbol{\sigma} \cdot \boldsymbol{n} = \boldsymbol{t_B} \text{ on } L_N \tag{14}$$

where $\boldsymbol{u}_\mathrm{B}$ and $\boldsymbol{t}_\mathrm{B}$ are the given displacement and the fixed traction on the boundary, respectively.

3 Numerical Methods

An applicable CV-FEM with considering the material properties temperature-dependent effect for the thermoelastic problem in FG porous rotating disks has been developed. 4-node quadrilateral (Q4) elements are used to discrete the spatial terms of the governing equation, and the back-difference method is used to discrete the time terms of the go. The material properties and unknown variables are defined at the cell center and the vertex, respectively. A control volume Ωn is constructed around a vertex by connecting the centers of each Q4 grid and the midpoints of its edge. See Fig. 2. The final discretized heat conduction governing is

$$\sum_{i=1}^{nc}\frac{\rho_i c_i s_i}{ncn_i}\frac{T - T^{t-\Delta t}}{\Delta t} = \sum_{i=1}^{nc}\sum_{j=1}^{ncn_i} T_{ij}\int_{l_i} k\left(\frac{\partial N_{ij}}{\partial x}n_x + \frac{\partial N_{ij}}{\partial y}n_y\right)dl$$
$$-\sum_{i=1}^{nNT} q_B|l_{Bi}| - \sum_{i=1}^{nRT} h_B(T - T_\infty)|l_{Bi}| \tag{15}$$

in which t and t-Δt represent the current and the last time steps, respectively. nc and $ncni$ are the number of the cells around the current vertex and the number of the nodes in the ith cell, respectively. n_{NT} and n_{RT} are the number of boundary edges with Neumann boundary condition and the number of boundary edges with Robin boundary condition. N_{ij} is the shape function of the jth node in the ith element, details about the discretization process of each term in governing equation can be found in Ref. [18].

The final discretized force equilibrium governing is

$$\sum_{i=1}^{nc}\sum_{j=1}^{ncn_i}\left(2u_{ij}^t\oint_{l_i}\mu_i\frac{\partial N_{ij}}{\partial x}n_x dl+u_{ij}^t\oint_{l_i}\mu_i\frac{\partial N_{ij}}{\partial y}n_y dl+v_{ij}^t\oint_{l_i}\mu_i\frac{\partial N_{ij}}{\partial x}n_y dl\right)$$
$$+\sum_{i=1}^{nc}\sum_{j=1}^{ncn_i}\left(u_{ij}^t\oint_{l_i}\lambda_i\frac{\partial N_{ij}}{\partial x}n_x dl+v_{ij}^t\oint_{l_i}\lambda_i\frac{\partial N_{ij}}{\partial y}n_x dl\right) \tag{16}$$
$$-\sum_{i=1}^{nc}\left[\left(\Gamma_i\alpha_i\left(\sum_{j=1}^{ncn_i}T_{ij}^t/ncn_i-T_r\right)\right)l_{xi}\right]+\sum_{i=1}^{n_{NS}}\sigma_{xB}l_{xBi}+\sum_{i=1}^{nc}\rho_i\omega^2 r_x=0$$

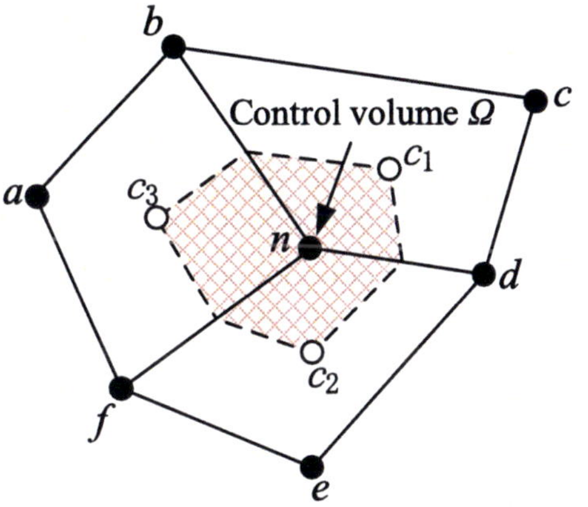

Fig. 2. A typical control volume with 4-node quadrilateral elements

$$\sum_{i=1}^{nc}\sum_{j=1}^{ncn_i}\left(2v_{ij}^t\oint_{l_i}\mu_i\frac{\partial N_{ij}}{\partial y}n_y dl+u_{ij}^t\oint_{l_i}\mu_i\frac{\partial N_{ij}}{\partial y}n_x dl+v_{ij}^t\oint_{l_i}\mu_i\frac{\partial N_{ij}}{\partial x}n_x dl\right)$$
$$+\sum_{i=1}^{nc}\sum_{j=1}^{ncn_i}\left(u_{ij}^t\oint_{l_i}\lambda_i\frac{\partial N_{ij}}{\partial x}n_y dl+v_{ij}^t\oint_{l_i}\lambda_i\frac{\partial N_{ij}}{\partial y}n_y dl\right) \tag{17}$$
$$-\sum_{i=1}^{nc}\left[\left(\Gamma_i\alpha_i\left(\sum_{j=1}^{ncn_i}T_{ij}^t/ncn_i-T_r\right)\right)l_{yi}\right]+\sum_{i=1}^{n_{NS}}\sigma_{yB}l_{yBi}+\sum_{i=1}^{nc}\rho_i\omega^2 r_y=0$$

Because the material properties are dependent on the solution temperature, the heat conduction equation and force equilibrium equation should be solved iteratively. In each time step, with the known initial conditions temperature ($T^{t-\Delta t}$) and the initial estimated values of unknown temperature (T^*), thermal coefficient properties are evaluated at T^* and Eq. (15) is solved iteratively until temperature T is converged within a given tolerance as Eq., after getting the convergent temperature solution, the mechanical properties are updated with the temperature field, and displacement field is obtained by solved Eqs. (16) and (17) under thermal loads and centrifugal force.

$$\left|\frac{T^* - T}{T^*}\right| \leq 10^{-6} \tag{18}$$

It's worth mentioning that the equilibrium equation is basing on the unknown node displacements, and stresses are derived using post-processing, using the external interpolation method based on the stresses on the cell center to obtain the stresses on the boundary will reduce the precision of the stresses close to the boundary, especially for the analysis in FGM [21]. The accuracy of the stress on the boundary can be effectively increased by first calculating the element node stresses and subsequently taking a weighted average of those stresses [21], in this paper, this technique is used to obtain the stress distribution.

4 Numerical Examples

4.1 Thermoelastic Analysis of FG Rotating Disk

The FG rotating disk form Ref. [22] subjected to thermal load is first carried out to verify the present method. The inner surface ($a = 0.2$ m) of the annulus is maintained at $T_a = 0$ °C, while the outer surface ($b = 1.0$ m) is maintained at $T_b = 1000$ °C, the constant angular velocity $\omega = 600$ rad/s. Poisson's ratio is constant to 0.3. Material temperature dependence is not considered in this study. The material properties of the FG annulus are given by the following equation

$$E(r) = E_0(\frac{r}{b})^n,\ \alpha(r) = \alpha_0(\frac{r}{b})^n,\ k(r) = k_0(\frac{r}{b})^n,\ \rho(r) = \rho_0(\frac{r}{b})^n \tag{19}$$

where $E_0 = 460$ GPa, $\alpha_0 = 10 \times 10^{-6}$ /K, $k_0 = 2.0$ W/m·K and $\rho_0 = 5700$ kg/m^3. n is the gradient index.

To compare the CV-FEM solutions with the results in Ref. [22], solutions are nondimensionalized by

$$\begin{Bmatrix} \overline{\sigma}_r \\ \overline{\sigma}_\theta \end{Bmatrix} = \frac{1 - \nu}{E_0\alpha_0 T_b + (1 - \nu)\rho_0\omega^2 b^2} \begin{Bmatrix} \sigma_r \\ \sigma_\theta \end{Bmatrix} \tag{20}$$

where E_0, α_0, ρ_0 are the material properties of the ceramic.

The predicted stresses present excellent agreement with those from the analytical method in Ref. [22], as shown in Fig. 3. Figures 4 and 5 plots the contours of the stress field of the FG hollow disk, the predicted hoop and radial stresses based on the CV-FEM distribute smoothly. This case indicates that CV-FEM is capable of thermoelastic analysis in FG disk subjected to centrifugal force.

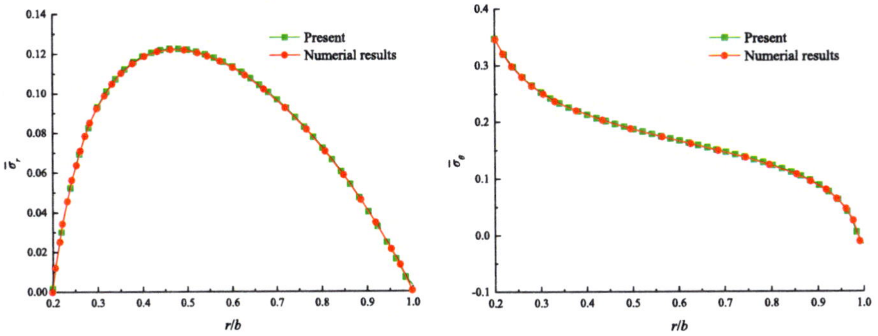

(a) $\overline{\sigma}_r$ (b) $\overline{\sigma}_\theta$

Fig. 3. Comparison of **a** dimensionless radial stress $\overline{\sigma}_r$ and **b** dimensionless hoop stress for the FG hollow disk with $\beta = 1$

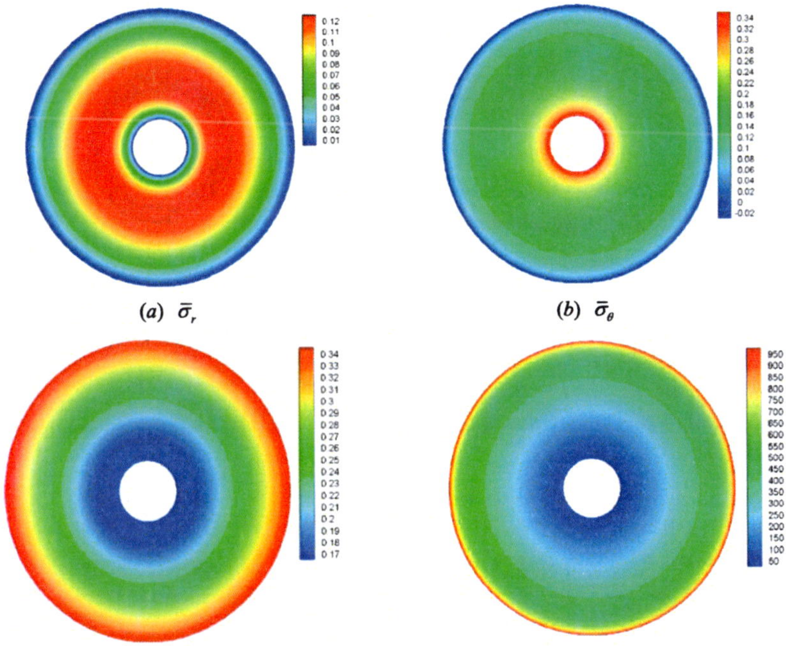

(a) $\overline{\sigma}_r$ (b) $\overline{\sigma}_\theta$

(c) $\overline{u}_r$ (d) T

Fig. 4. Contours of the thermoelastic field of FG hollow disk for $\beta = 1$: **a** dimensionless radial stress $\overline{\sigma}_r$ dimensionless hoop stress $\overline{\sigma}_\theta$, **b** dimensionless radial displacement $\overline{u}_r$ and **d** temperature T.

4.2 FG Annulus with Porosities

In this section, the effects of some factors such as thermal load, gradient index, porosity distribution, rotating velocity, and plane stress and plane strain assumptions on the steady

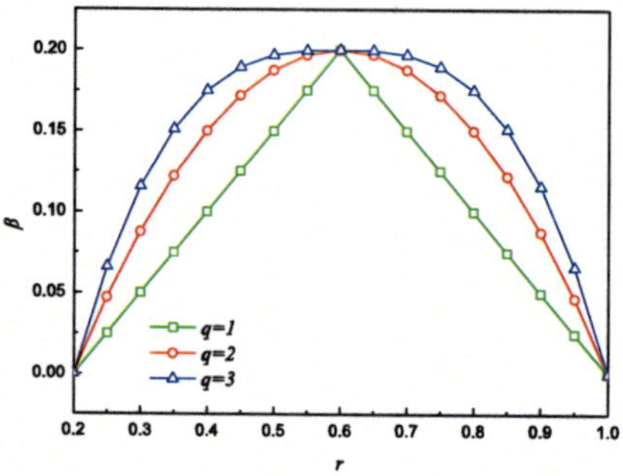

Fig. 5. Variation of the porosity for different values of *q*.

thermoelastic response of the FG disks have been studied. The geometry is as same as the FG disk in Sect. 4.1. The radial displacement of the inner surface is constrained to zero, and the outer surface is free from mechanical load. The material properties with the effects of property are estimated by the following equation

$$P(r) = P_M V_M + P_C V_C - \frac{\beta}{2}(P_M + P_C) \tag{21}$$

where the volume fractions of the metal and ceramic materials are $V_M = [(r\text{-}a)/(b\text{-}a)]^n$ and $V_C = 1\text{-}V_C$, respectively. β is the porosity coefficient. P_M and P_C represent the relevant properties of the metal and ceramic, their values are expressed as

$$P = P_0\left(\frac{P_{-1}}{T} + 1 + P_1T + P_2T^2 + P_3T^3\right) \tag{22}$$

where P_0, P_{-1} P_1, P_2, P_3 are material coefficients. Values of each coefficient for the metal and ceramic are shown in Table 1.

For porosity I β is constant, and for porosity II, β is given by the following equation

$$\beta{=}0.2 \times \left\{1 - \left[\frac{2|r-(b-a)/2|}{b-a}\right]^q\right\} \tag{23}$$

where q is the gradient index. In this study, q is set as 0.1, 0.2 and 0.3. The distributions of porosities in the radial direction with different values of q are shown in Fig. 5.

4.2.1 Effects of Thermal Load

When an FG annulus rotates at a certain angular velocity $\omega = 600$ rad/s, and the inner surface is maintained at $T_a = 0$ K, and only the temperature of the outer surface is changed. Effects of the temperature change at the outer surface on thermoelastic performance are examined. Results with and without porosity are presented. The variation of thermal properties and mechanical properties of the material with temperature is considered and the plane stress assumption is used in the analysis.

Figure 6 plots the distributions of the thermoelastic response along the radial direction. For this case, it can be observed that the existence of even and uneven porosities

Table 1. Coefficients of temperature dependent material property [23]

Material	Properties	P_0	P_{-1}	P_1	P_2	P_3
SU304	c(J/kg·K)	496.59	0	-1.151×10^{-3}	1.636×10^{-6}	-5.863×10^{-10}
	k (W/m·K)	15.379	0	-1.264×10^{-3}	2.092×10^{-6}	-7.223×10^{-10}
	α (10^{-6}/K)	12.330×10^{-6}	0	8.086×10^{-4}	0	0
	ν	0.3262	0	-2.002×10^{-4}	3.797×10^{-7}	0
	E (GPa)	201.04×10^9	0	3.079×10^{-4}	-6.534×10^{-7}	0
Si_3N_4	c(J/kg·K)	555.11	0	1.016×10^{-3}	2.920×10^{-7}	-1.670×10^{-10}
	k (W/m·K)	13.723	0	-1.032×10^{-3}	5.466×10^{-7}	-7.876×10^{-10}
	α (10^{-6}/K)	5.8723×10^{-6}	0	9.095×10^{-4}	0	0
	ν	0.2400	0	0	0	0
	E (GPa)	348.43×10^9	0	-3.070×10^{-4}	2.160×10^{-7}	-8.946×10^{-10}

(porosity I and porosity II) has a minor effect on the temperature and the displacement but has a great effect on the thermal stress, the maximum magnitudes of hoop and radial stresses increase with increasing the outer surface temperature. Due to the material variation is dependent on the temperature solution, the existence of porosity presents more influence on the thermoelastic field as boundary temperature T_b increases. Figure 6b plots the radial displacement subjected to different fixed temperatures on the outer boundary, it can be observed that the radial displacement with porosity is lower than that without porosity at $\Delta t = 400$ K, and the radial displacement with porosity is greater than that without porosity at $\Delta t = 900$ K and 1200 K. The reason for the displacement distribution is due to the existence of porosities have more influence on the expanded displacement caused by the mechanical load than the compressive displacement caused by the thermal load, the compressive displacement is smaller than the expansive displacement, and result in the displacement with the efforts of porosity is greater than that without porosity. As the thermal load increase, the compressive displacement increase, thus leading to the values of displacement with porosity efforts is smaller than the values without porosity. Compared to the results without porosities, the existence of porosities present lower maximum magnitudes of the radial and hoop stresses (Fig. 6c-d) and shift the locations of the maximum of hoop stress from the interior to the boundary of the annulus. Comparing the results with even and uneven porosities, it can be noted that the

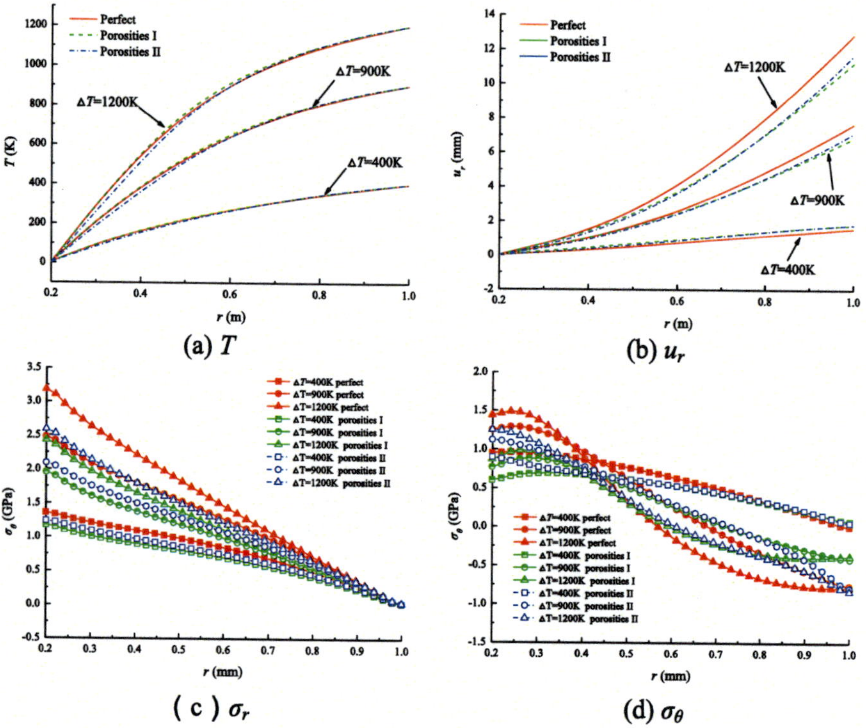

(a) T (b) u_r

(c) σ_r (d) σ_θ

Fig. 6. Effects of thermal loads on **a** temperature T **b** radial displacement u_r **c** radial stress σ_r and **d** hoop stress for $\omega = 600$ rad/s, $n = 1$, $\beta = 2$, $q = 3$, $T_a = 0$ K, $T_b = 400$ K, 900 K and 1200 K.

porosity has litter effect on the temperature and deformation (Fig. 6a-b), while has a significant effort on stresses, because of the variation of the material properties of porosity I is more continuous than that of porosity II, the hoop and radial stresses with porosity I distribute more smoothly than the porosity II. For this case, although the porosity weakens the structure stuffiness (see Eq. (21)), it helps reduce the maximum radial and hoop stresses and provide more smooth stress than that of the perfect structure. This test case indicates that the temperature, displacement, and thermal stress of the FG annulus have a strong dependence on the thermal load.

4.2.2 Effects of the Variation of Material Properties with Temperature

Effects of material temperature-dependent on the thermoelastic performance are discussed based on the perfect FG annulus. Gradient index n is set to 1 for this case, that is, the material properties distribute linear in the space when the material properties are not varied with temperature and the existence of porosity. The outer surface of the FG annulus is maintained at $T_b = 900$ K and the inner surface is maintained at $T_a = 0$ K. The angular velocity $\omega = 600$ rad/s. Plane stress assumption is adopted.

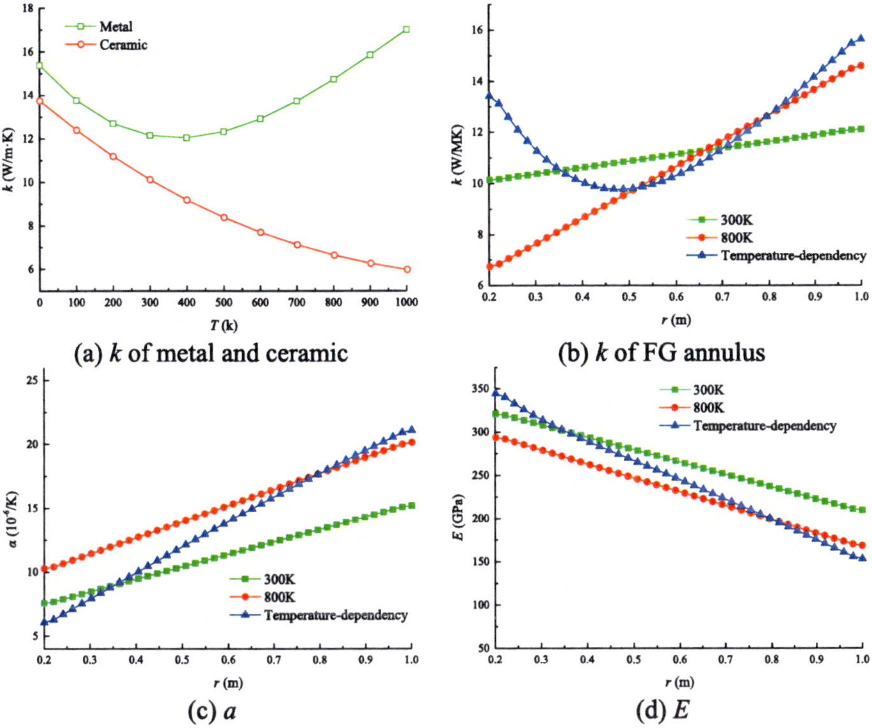

(a) k of metal and ceramic

(b) k of FG annulus

(c) a

(d) E

Fig. 7. Variation of material properties with temperature.

The material properties are evaluated by Eq. (21) at $T = 300$ K, 800 K and real temperature solution. The material properties distribution in the radial axis is presented in Fig. 7. Due to the nonlinear distributions of the conductivity of metal and ceramic caused by temperature-dependent, see Fig. 7a, the distribution of conductivity distributes strong nonlinearity for the real temperature solutions, see Fig. 7b, the properties at the inner surface of are constant to the values of ceramic at $T_a = 0$ K and the outer surface are constant to the values of metal of $T_b = 900$ K. Since the conductivity of the metal and ceramic at $T = 300$ K and 800 K is constant and the properties in the radial direction are assumed to distribute linear, the conductivity for a fixed temperature distribute linear. From Fig. 7c-d, it can be found that Young's module and thermal expansion coefficient change obviously for different temperature field assumptions. Figure 8 plots the thermoelastic response of the FG annulus, it can be observed that the material temperature-dependent have obvious influence on the temperature, displacement and stresses, in addition, it reduces the maximum magnitude of the radial and hoop stresses and shifts the location of the maximum of the hoop stress from the inner boundary to the interior of the FG annulus compared to the results with the properties evaluated at $T = 300$ K and 800 K. It should be pointed out that the concentration of stress in the interior of the structure is more easily to lead to the crack generation and then the low cycle fatigue failure of the structure under the periodic thermal and mechanical loads, even the maximum stress is not exceeded to the permissible values, therefore, the influence of material temperature-dependent on the thermoelastic field should be considered in the study of the FG rotating disk.

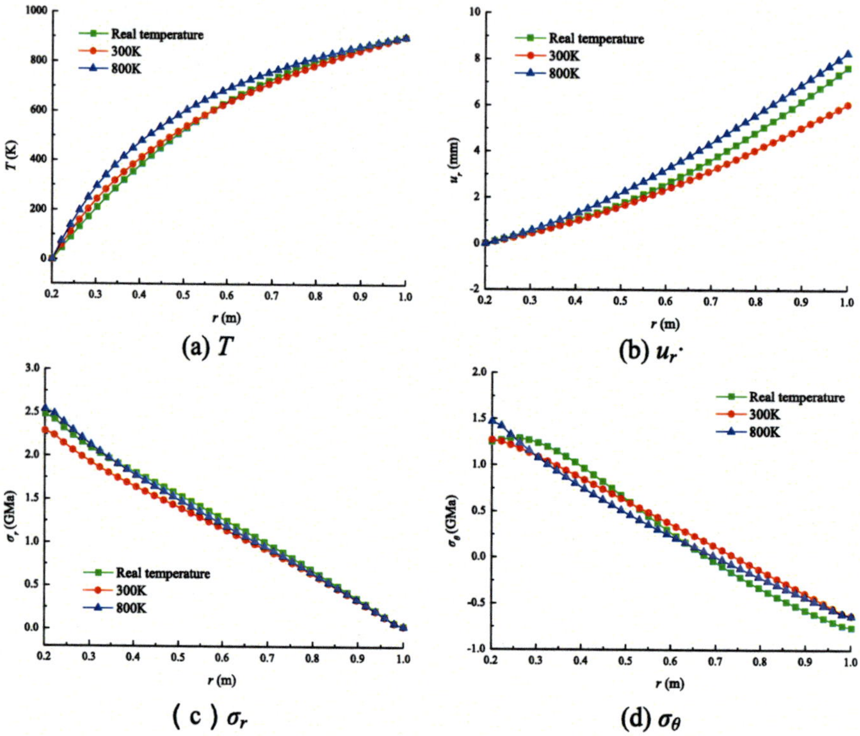

Fig. 8. Effects of material properties temperature-dependent on **a** temperature T, **b** radial displacement u_r, **c** radial stress σ_r and **d** hoop stress σ_θ for $n = 1$, $\beta = 0$, $\omega = 600$ rad/s, $T_a = 0$ K, $T_b = 900$ K.

4.2.3 Effects of Material Distribution

On the other hand, the influence of the FG gradient index and the porosity on the thermoelastic performance are investigated. The porosity parameter is set to $\beta = 0.1$ for the analysis of the effects of the gradient index, and the value of gradient index is set to be $n = 3$ for the analysis of the effects of the porosity. The results with even and uneven porosities are also compared. The material is varied with temperature. From Fig. 10, it can be observed that the gradient index has a significant influence on the thermoelastic response. As the gradient index n increases, the temperature gradient in the FG annulus increases (see Fig. 11a) since the FG annulus becomes ceramic rich, see Fig. 9. Increasing the gradient index n reduces the maximum of the stress and deformation and shifts the maximum of the hoop stress from the interior boundary to the inner surface of the FG annulus.

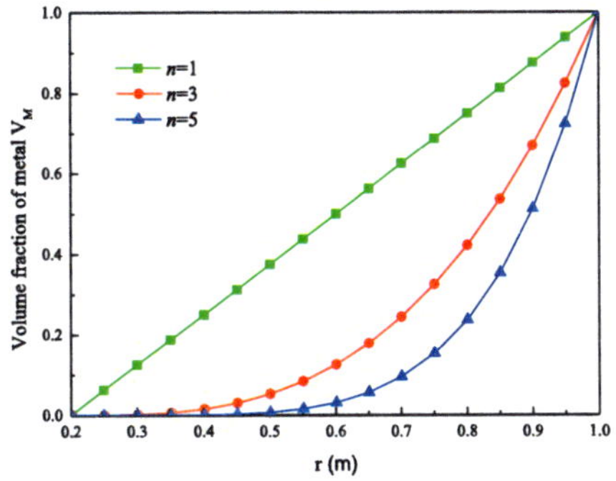

Fig. 9. Volume fraction of metal along the radial direction of the FG annulus.

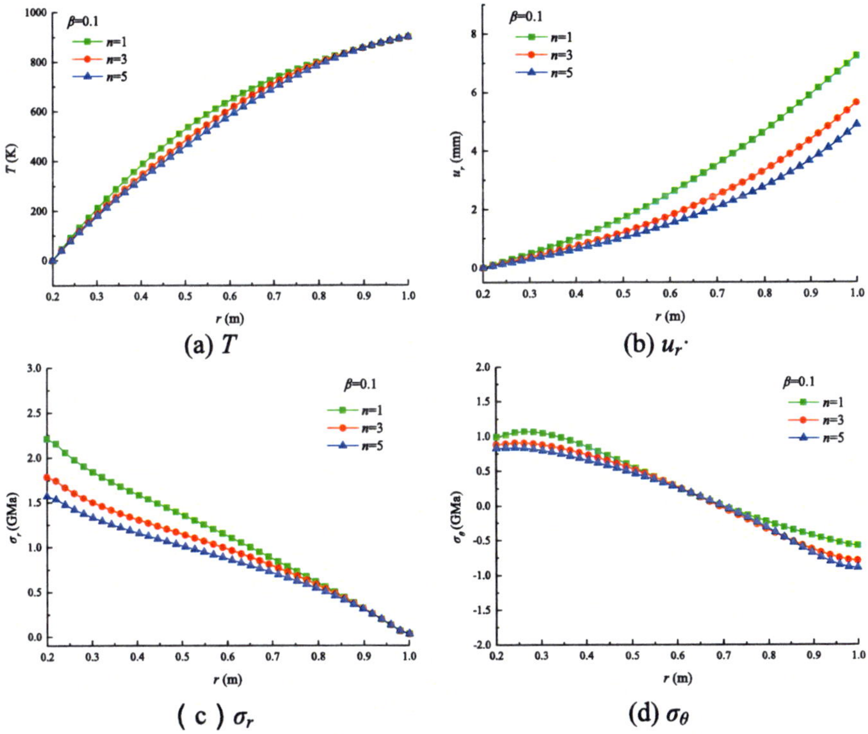

Fig. 10. Influence of FGM gradient index on **a** temperature T, **b** radial displacement u_r, **c** radial stress σ_r and **d** hoop stress σ_θ for $T_a = 0$ K, $T_b = 900$ K, $\omega = 600$ rad/s, $\beta = 0.1$.

Figures 11 and 12. Show the calculated results of the FG annulus with even and uneven porosities. It can be seen that the existence of porosities has small influence on the temperature and deformation while having significantly influence on the stresses, the stress decreases as increasing of the porosity coefficient β, see Fig. 11c, d. According to Fig. 12c, d, the deformation and stress decrease with the increase of the porosity index

q, compared to the results with even porosity for $\beta = 0.2$, the uneven porosity leads to the increase of the magnitude of stress. The variation of FG gradient index n and uneven porosity parameters index β performs a larger influence on the thermoelastic responses of FG annulus than the variation of uneven porosity parameters q. These results can give some advice in designing a rotating FG annulus to determine an appropriate material distribution to make the thermal stresses distribute more gently in the whole structure.

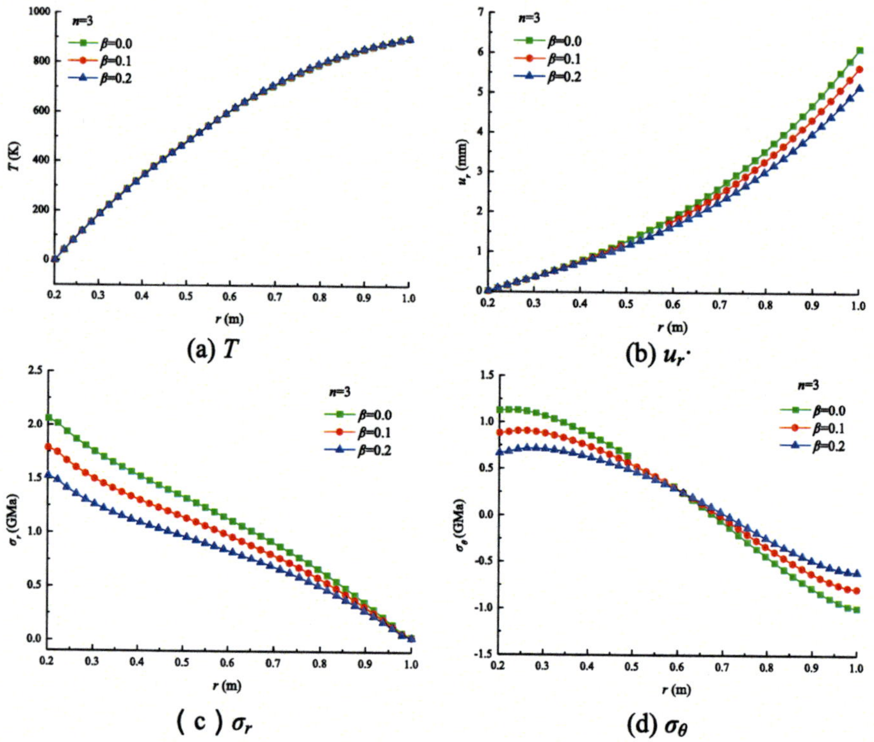

Fig. 11. Effects of even porosity on the thermoelastic performance of the FG rotating disk for $T_a = 0$ K, $T_b = 900$ K, $\omega = 600$ rad/s, $n = 3$.

4.2.4 Effects of Angular Velocity

Next, the effect of the angular velocity on thermoelastic performance is examined. The parameters as $T_a = 0$K, $T_b = 900$k, $\omega = 600$, $n = 1$, $\beta = 2$, $q = 3$ are used in this analysis. Due to the elastic solution having no influence on the temperature solution, for simplicity, the temperature results are not presented. Considering the material temperature-dependency. Figure 13 plots the displacement and stresses distributions for different angular velocities. It can be seen the displacement and stresses increase with increasing the angular velocity ω for the perfect and imperfect structure. The thermal stresses have a strong dependency on the angular velocity ω, the higher the angular

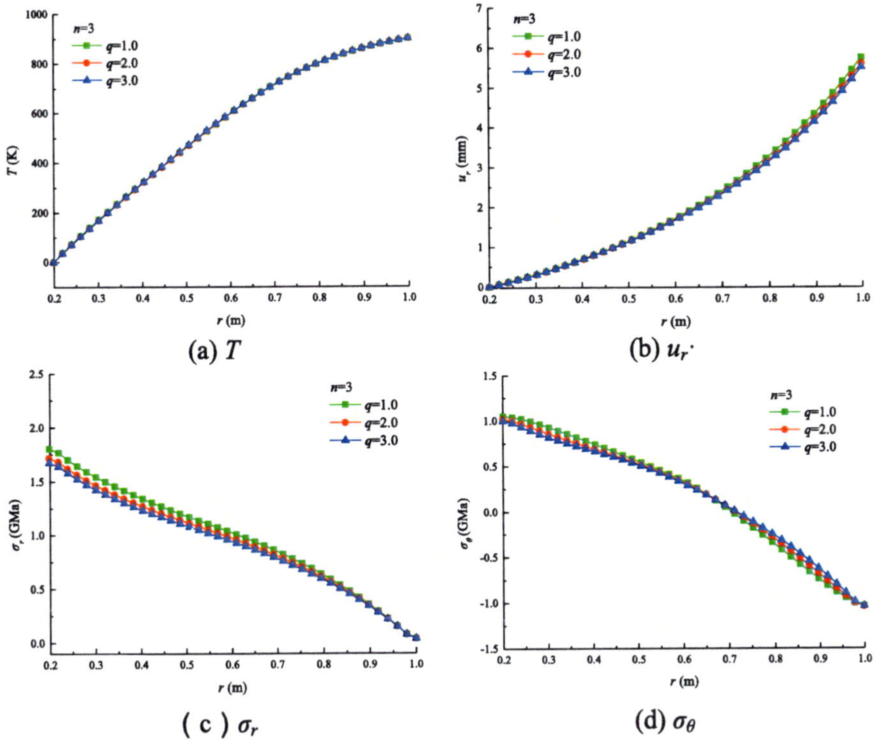

Fig. 12. Effects of uneven porosity on the thermoelastic field of the FG rotating disk for $T_a = 0$ K, $T_b = 900$ K, $\omega = 600$ rad/s, $n = 3$.

velocity, the higher stress on the inner surface. The existence of porosity (porosity I and porosity II) shifts the locations of the maximum hoop stress. There are obvious differences between the thermal stress of the perfect structure and the imperfect structure, both of the porosities can help reduce thermal stress under the mechanical load as similar as reducing the thermal stress under the thermal load. As respected, the stresses with even porosities distribute more gently than those with uneven porosities.

4.2.5 Effects of Plane Stress and Plane Strain Conditions

Finally, the effects of plane stress and plane strain conditions on the distribution of displacement and thermal stress of FG annulus are plotted in Fig. 14 for material coefficients $n = 1$, $\beta = 0.2$, $q = 3$ with $T_a = 0$ K and $T_b = 1000$ K. Considering the material temperature-dependency. For the perfect and imperfect FG annulus, the maximum radial displacements and stresses are the highest for the plane strain condition. The existence of even porosity helps to reduce the thermal stress obviously for plane strain conditions as it does for plane stress conditions.

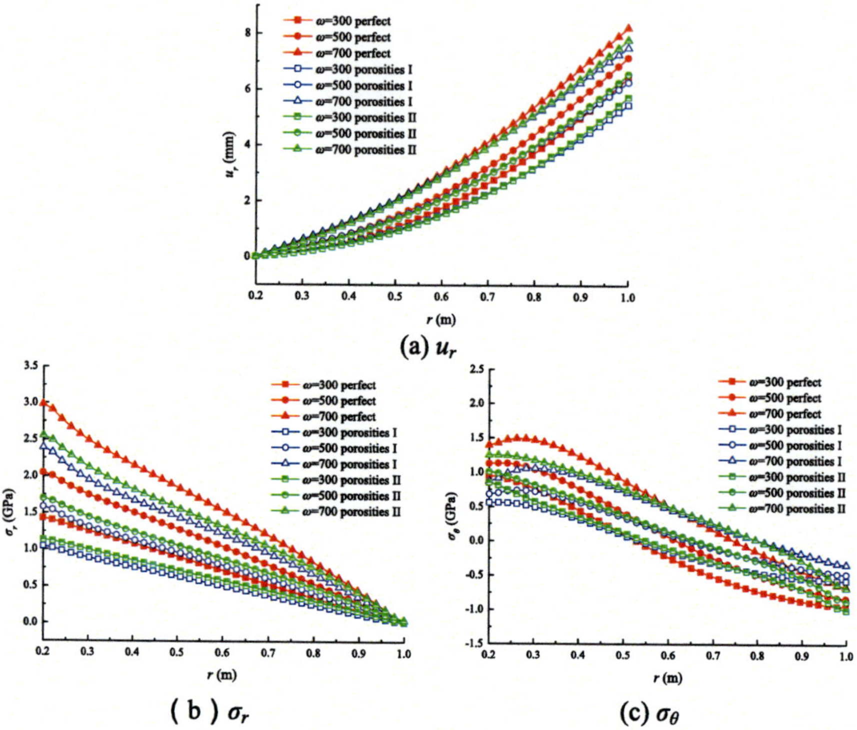

Fig. 13. Effects of angular velocity on the thermoelastic field of the FG rotating disk for $T_a = 0$ K, $T_b = 900$ K, $n = 1$, $\beta = 2$, $q = 3$.

5 Conclusions

The CV-FEM has been extended for the thermoelastic analysis for porous FGM rotating disk with material temperature-dependent. The method developed in this paper has proved to be effective for the analysis of the FG annulus with centrifugal force. For the present FG rotating annulus, the efforts of the thermal load, material distributions, rotating angular velocity, and plane stress and plane strain conditions have been investigated for the perfect and imperfect structures. The research shows that the temperature and displacement are much less sensitive to the thermal and mechanical loads or the material distribution or the plane strain and plane stress assumptions than the thermal stress. Although the porosity (even and uneven distributions) weakens the stuffiness of the structure, it helps reduce thermal stress. Because the variation of the material properties of even porosity is more continuous than that of uneven porosity, the stress with even porosity shows more gently than the uneven porosity under the same thermal and mechanical loads. The material temperature-dependent has a significant effort on the thermoelastic and it should be taken into account in the evaluation stage of the structure. The maximum stresses with plane strain conditions are greater than those with plane stress conditions. The gradient index and even porosity coefficient have more influence on the thermal performance than the uneven porosity index. The uneven porosity gradient index q has a smaller influence on the thermoelastic response compared with the

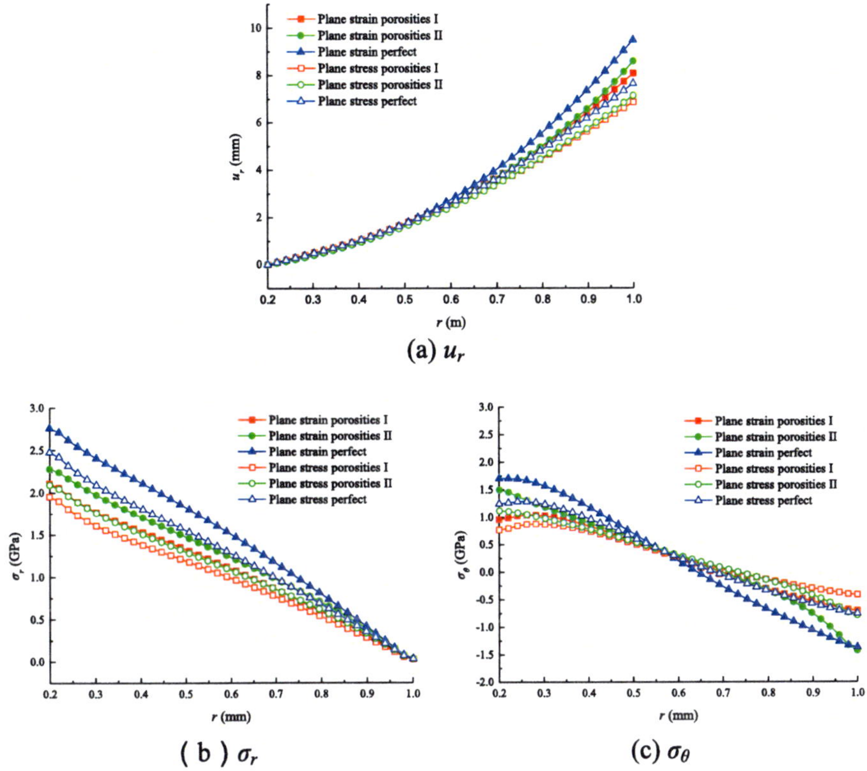

Fig. 14. Distributions of **a** temperature T, **b** radial displacement u_r, **c** radial stress σ_r and **d** hoop stress σ_θ for $T_a = 0$ K, $T_b = 900$ K, $\omega = 600$ rad/s, $n = 1$, $\beta = 2$, $q = 3$ for plane stress and plane strain assumptions

even porosity coefficients β and FG index n, and the thermal reduces obviously with increase of the gradient index and even porosity coefficient.

Acknowledgments. The authors gratefully acknowledge the financial support of the National Key Research and Development Program of China (Grant No. 2019YFA0405202), the National Natural Science Foundation of China (Grant No. 11972359), and the Natural Science Basic Research Plan in Shaanxi Province of China (Grant No. 2022JQ-564).

References

1. Durodola, J.F., Attia, O.: Deformation and stresses in functionally graded rotating disks. Compos. Sci. Technol. **60**, 987–995 (2000)
2. Bayat, M., Saleem, M., Sahari, B.B., et al.: Analysis of functionally graded rotating disks with variable thickness. Mech. Res. Commun. **35**(5), 283–309 (2008)
3. Bayat, M., Saleem, M., Sahari, B.B.: Mechanical and thermal stresses in a functionally graded rotating disk with variable thickness due to radially symmetry loads. Int. J. Press. Vessel. Pip. **86**(6), 357–372 (2009)

4. Tanaka, K., Tanaka, Y., Enomoto, K., et al.: Design of thermoelastic materials using direct sensitivity and optimization methods. Reduction of thermal stresses in functionally gradient materials. Comput. Methods Appl. Mech. Eng. **106**(1), 271–284 (1993)
5. Tanaka, K., Tanaka, Y., Watanabe, H., et al.: An improved solution to thermoelastic material design in functionally gradient materials: scheme to reduce thermal stresses. Comput. Methods Appl. Mech. Eng. **109**(3), 377–389 (1993)
6. Kutsal, S.M., Coskun, S.B.: Deformation analysis of variable thickness rotating disks using an improved adomian decomposition technique. Int. J. Appl. Mech. **12**(1), 4434–4440 (2020)
7. Sharma, J.N., Sharma, D., Kumar, S.: Stress and strain analysis of rotating FGM Thermo-elastic circular disk by using FEM. Int. J. Pure Appl. Math. **74**(3), 339–352 (2012)
8. Zhu, J., Lai, Z., Yin, Z., et al.: Fabrication of ZrO2–NiCr functionally graded material by powder metallurgy. Mater. Chem. Phys. **68**, 130–135 (2001)
9. Nuttawit, W.A., et al.: Free vibration analysis of layered functionally graded beams with experimental validation. Mater. Design **36**, 182–190 (2012)
10. Babamiri, B.B., Shahrjerdi, A., Bayat, M.: Effect of geometrical imperfection on the thermo-mechanical behavior of functionally graded material rotating disk. J. Braz. Soc. Mech. Sci. Eng. **42**(5), 271 (2020)
11. Jabbari, M., Hashemitaheri, M., Mojahedin, A., Eslami, M.R.: Thermal buckling analysis of functionally graded thin circular plate made of saturated porous materials. J. Therm. Stresses **37**(2), 202–220 (2014)
12. Mojahedin, A., Jabbari, M., Khorshidvand, A.R., Eslami, M.R.: Buckling analysis of functionally graded circular plates made of saturated porous materials based on higher order shear deformation theory. Thin-Walled Struct. **99**, 83–90 (2016)
13. Zhao, J., Xie, F., Wang, A.L., Shuai, C.J. et al.: Dynamics analysis of functionally graded porous (FGP) circular, annular and sector plates with general elastic restraints. Compos. Part B: Eng. **159**, 20–43 (2019)
14. Dastjerdi, S., Beni, Y.T., Malikan, M.: A comprehensive study on nonlinear hygro-thermo-mechanical analysis of thick functionally graded porous rotating disk based on two quasi-three-dimensional theories. Mech. Based Des. Struct. Mach. (2020). 1814812.11814812.31
15. Wang, Y., Ye, C., Zu, J.W.: Identifying the temperature effect on the vibrations of functionally graded cylindrical shells with porosities. Appl. Math. Mech. **39**(11), 1587–1604 (2018). https://doi.org/10.1007/s10483-018-2388-6
16. Ting, D., Hong, L.D., Zhong, Y.L.: Multi-field mechanical behavior of a rotating porous FGMEE circular disk with variable thickness under hygrothermal environment. Compos. Struct. **210**, 641–656 (2019)
17. Gong, J., Xuan, L., Ying, B., et al.: Thermoelastic analysis of functionally graded porous materials with temperature-dependent properties by a staggered finite volume method. Compos. Struct. **224**, 111071.1–111071.10 (2019)
18. Liu, Q., Peng, Q., Ming, P.J.: A control volume finite element method for the thermoelastic problem in functionally graded material with one relaxation time. In: Proceedings of the Institution of Mechanical Engineers Part C Journal of Mechanical Engineering Science 095440622097902 (2020)
19. Liu, Q., Ming, P.J.: A high-order control volume finite element method for 3-D transient heat conduction analysis of multilayer functionally graded materials. Numer. Heat Transf. Part B **73**(6), 363–385 (2018)
20. Liu, Q., Ming, P.J., Zhao, H.Y., Zhang, W.P.: A high order control volume finite element method for transient heat conduction analysis of multilayer functionally graded materials with mixed grids. J. Therm. Sci. **29**(1), 144–158 (2020)
21. Liu, Q., Yu, Y., Ming, P.J.: A high-order control volume finite element method for thermoelastic analysis of functionally graded solids with mixed grids. Proc. Inst. Mech. Eng. C J. Mech. Eng. Sci. **233**(11), 095440621881136 (2018)

22. Peng, X.L., Li, X.F.: Thermal stress in rotating functionally graded hollow circular disks. Compos. Struct. **92**(8), 1896–1904 (2010)
23. Reddy, J.N., Chin, C.D.: Thermomechanical analysis of functionally graded annulus and plates. J. Therm. Stresses **21**, 593–626 (1998)

Check for updates

A Survey of Few-Shot Learning for Image Classification of Aerial Objects

Haoxin Cai[1(✉)], Xuanyue Zhu[1], Pengcheng Wen[1], Wei Han[1], and Le Wu[2]

[1] AVIC Xi'an Aeronautics Computing Technique Research Institute, Xi'an 710065, China
18834140343@163.com
[2] Xi'an Thermal Power Research Institute Co. Ltd, Xi'an 710054, China

Abstract. The aircraft has high requirements for effective detection and rapid support in different situations, and in this process, the image classification of aerial objects plays a very important role. In view of the situation that many tasks in the field of image classification of aerial objects cannot get enough samples to learn, it is necessary to study few-shot learning technology, which is mainly divided into three aspects: data based, model based and optimization based. Considering the advantages and disadvantages of various few-shot learning methods and their contribution to the accuracy of few-shot image classification, the developing trend of few-shot learning technology in the field of image classification of aerial objects is summarized. The three research directions of data, model and optimization are not independent of each other, and it is the mainstream to combine them. The application of few-shot learning technology in aviation can realize the deployment of high-precision and high-stability models for aerial tasks with little available data, and reduce the high cost of data tags to a certain extent, which is meaningful. It is hoped that this paper can provide ideas for the application of few-shot learning in the field of aerial image, and then solve practical problems.

Keywords: Few-shot learning · Aerial objects · Image classification

1 Introduction

In the aerial background, the airplane has extremely high requirements for the ability to effectively detect emergencies and quickly support. In this process, the image classification of aerial targets plays a very important role. With the further improvement of safe flight and the development of airspace national defense, the development of image classification technology of aviation targets is also becoming more and more important in the military and civilian fields.

Aerial target images are easily affected by the environment. For example, different attitudes, different speeds, and different shooting angles will affect the attitude of the target. The traditional image classification method will have certain limitations. At present, the rapid development of traditional image classification methods is based on deep learning classifiers, and the target classifiers of deep learning need to adjust and optimize the network with a large number of sample data. However, in the aviation field,

Chinese Society of Aeronautics and Astronautics: CASTYSF 2022, LNEE 972, pp. 570–582, 2023.
https://doi.org/10.1007/978-981-19-7652-0_50

data is difficult to obtain under some special factors, such as data privacy and security, high data labeling costs, and rarely important data. In many application scenarios, there are insufficient conditions to obtain a large number of labeled samples. When the amount of data is very limited, the traditional deep learning target classifier is prone to over-fitting, which makes the image classification ability worse and the result is not ideal. Therefore, learning how to enable a system to effectively learn cognitive abilities from a small number of samples plays a very important role. In recent years, researchers have proposed few-shot learning to solve this kind of problem. For the aviation field, this kind of problem is more prominent, and it is very meaningful to study and summarize it.

This paper studies the few-shot learning technology in image classification, so as to provide some ideas for solving related problems in the aviation field.

In recent years, with the rapid development of deep learning, more few-shot learning methods based on deep learning models continue to emerge. These methods make use of the advantages of deep neural network in feature extraction and representation and provide ideas for solving few-shot problems from different perspectives (Fig. 1).

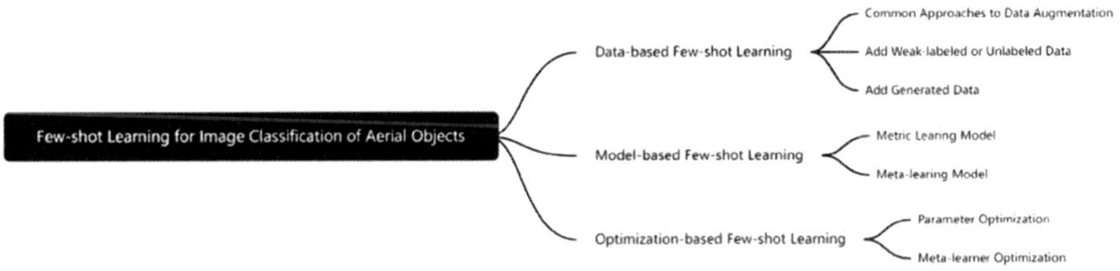

Fig. 1. Thinking diagram of few-shot learning technology research for image classification

This paper summarizes these few-shot learning technologies, which are mainly divided into three aspects: data-based, model-based and optimization-based. In the second, third and fourth sections of this paper, the few-shot image classification technology in the three research directions in the above figure is described respectively. The fifth section of the paper mainly describes the author's understanding and ideas about the development trend of few-shot learning technology in aerial image classification. Finally, in the sixth section, the research content and main ideas of this paper are summarized.

2 Few-Shot Learning Based on Data

Data enhancement is the most intuitive method to increase the number of training samples and enhance the diversity of data. To some extent, data enhancement can effectively avoid the over-fitting phenomenon of the network model and improve the robustness and generalization ability of the model.

2.1 Ordinary Data Enhancement Methods

For image data, the more commonly used geometric transformation methods are: flip, rotation, cropping, zooming, translation [1] and so on.

In addition, there are noise class transformation methods. Random noise is based on the original picture, randomly superimpose some noise, the main way is Gaussian noise. Coarse Dropout can lose information in a rectangular area of selected size at random positions. Simplex Noise Alpha can generate continuous noise masks and mix with the original image. Frequency Noise Alpha is capable of noise weighting random indices in the Frequency domain and then converting them to the spatial domain, so as to achieve speckle and cloudy effects on the image. In addition, there are fuzzy image processing methods by reducing the difference of each pixel value, including Gaussian blur, randomly erasing an area on the image, marking all edges in the image as black and white and many other methods.

These simple enhancement methods can not effectively improve the generalization ability of the few-shot learning model. Therefore, for some few-shot image classification problems, more complex and effective enhancement models, algorithms and networks may be needed.

2.2 Add Weak-Labeled or Unlabeled Data

The addition of weakly labeled or unlabeled data refers to the enhancement of training data by selecting samples identical or similar to the target image in the weakly labeled or unlabeled data set. Such real weak-labeled or unlabeled data is much easier to obtain in the real world, and costs are relatively low.

Wang [2] proposed a new unsupervised training module to enable the network to learn more unlabeled data in the real world. It improves the network's ability to express more general and rich data, and makes the network not only express specific data sets clearly, but also improve the generalization ability of network models. Boney [3] made Model-Agnostic Meta-Learning (MAML) model adjust the parameters of the embedding function and classifier in the presence of unlabeled data by adding unlabeled data, so as to adapt to the task of new classification and achieve good classification effect. Ren [4] expanded on the basis of the prototype network, increased the network's ability to utilize unlabeled data through training, improved the classifier algorithm and the parameterized model of the algorithm, and achieved better classification accuracy. Liu [5] designed a meta-learning strategy, introduced data sets without labels, and simulated the situation of few-shot data and unlabeled data in the real scene in the training stage, so as to make label prediction for unlabeled data in the case of small amount of data and improve the generalization ability of the model. Hou [6] proposed a novel and simple cross-attention network to solve the problem of small sample image classification, and used the unlabeled query set to expand the support set, so as to make the features of each class more representative. Xie [7] introduced unlabeled data into the classification model, evaluated the objective function of unlabeled data by using KL dispersion, and proposed a Training Signal rolling (TSA) method, which gradually released the training signal with labeled data. Therefore, when the model is trained on more and more unlabeled data, it will not lead to over-fitting, and finally improve the classification accuracy and generalization ability of the model performance. Bateni [8] proposed Transductive Conditional Neural Adaptive Processes, which combines unlabeled data to improve the classification accuracy during testing. This method has the advantages of simplicity and robustness.

By adding weakly labeled and unlabeled data, the network model for few-shot image classification can better adapt to the new type, and improve the generalization ability and classification accuracy of the network.

2.3 Add Generated Data

Adding generated data refers to synthesizing new image data for few-shot categories to expand training data, so that the subsequent network model can achieve more stable and accurate classification and further improve generalization ability. Recently, since Generative Adversarial Nets (GAN) [9] were proposed, research on image generation has developed rapidly (Fig. 2).

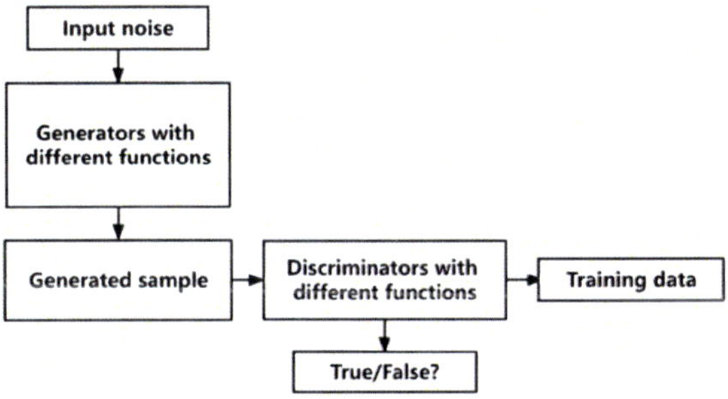

Fig. 2. Structure diagram of GAN

GAN is composed of two neural networks, generator and discriminator respectively. The generator turns the input noise into a picture that looks good enough to fool the discriminator. What the discriminator needs to do is to improve its ability to distinguish real samples from generated samples. In this structure of game confrontation, the two constantly improve each other, and finally reach a balance to achieve the goal of generating realistic images.

Mehrotra [10] proposed a new few-shot learning algorithm, which introduced strong regularization terms in the loss function. GAN network was introduced, generator loss and discriminator loss were used as regularization terms to enhance the similarity classification and matching task. The can effectively improve the classification effect. Aligombe [11] proposed a GAN-based few-shot image classification method, which classifies images into multiple fake or real classes. This method can classify fake classes and process labeled and unlabeled data simultaneously during training, which further achieves better classification performance. Xian [12] developed a conditional generation model that combined the advantages of variational auto-encoder and GAN to realize explainable image feature learning and improve the accuracy of image classification in the case of scarce training data. Nguyen [13] proposed a new self-supervised few-shot learning method, which uses GAN as the trunk model and increases reconstruction loss and triplet loss during training, so as to improve the performance of discriminator, improve the feature extraction ability of generator, and achieve better classification accuracy. LoFGAN [14] provides a way to generate more data for the few available samples. It randomly divides a small number of images into a basic image and several reference images, replaces the original local features with the closest related local features, and generates more fine-grained, more realistic, more diversified and higher quality images, and improves the stability of the network. Huang [15] proposed a data enhancement method for few-shot infrared airplane, which utilizes the pyramid multi-scale GAN structure to learn and fit the feature information of a single image at different scales. This method improves the generator structure of infrared aircraft image, improves the feature expression of small receptive field, and achieves the enrichment of image details. Liu [16] proposed a lightweight GAN, which consists of two main structures, a channel-level excitation module that skips layers and a self-supervised discriminator, to achieve the few-shot image synthesis task of GAN with the minimum computational cost.

GAN has become the most popular research direction in the field of image generation because of its continuously improving image generation ability in the game environment, which can finally realize the advantage of generating realistic images.

3 Few-Shot Learning Based on Model

3.1 Metric Learning Model

The goal of metric learning is learning similarity measurement, which is a comprehensive measure to evaluate the degree of similarity between two samples (or categories). Several typical few-shot learning methods based on metric learning include Siamese Network, Matching Network, Prototypical Network and Relational Network. These networks use different measure loss functions and various related similarity measures to solve the task of small sample image classification (Fig. 3).

Siamese Network [17] is composed of twin CNNS with the same weight. The two CNNS accept a pair of samples as input and achieve classification according to similarity. The main idea is to map the input to the target space by embedding function and calculate the similarity using simple distance function. Wang [18] proposed an attention-based Siamese Network, designed an efficient convolutional network to learn the embedding function, and used the simple attention function to calculate the similarity of the two feature vectors. This method improves the generalization ability of the classifier. Jadon [19] proposed a kernel-based non-parametric activation function to improve the accuracy and stability of the network by learning appropriate embedding. Shao [20] proposed Meta Siamese Network, which carries out interval training through a flexible framework. The network learns feature embedding and function of depth similarity measurement, and introduces two kinds of distance-based loss to assist optimization.

Matching Network [21] embedded image labels into vector space and mapped few-shot data to corresponding labels by constructing the attention mechanism between label semantic features and image features. Mai [22] proposed a feature-level attention mechanism, which enables the similarity function to better reflect the feature differences between classes. It improves the feature extraction capability and robustness of the network. Li [23] proposed a meta-network module that can learn transferable prior knowledge across tasks and directly generate network parameters for similar unknown tasks, so as to achieve fast learning and adaptation to new classes. Chen [24] proposed

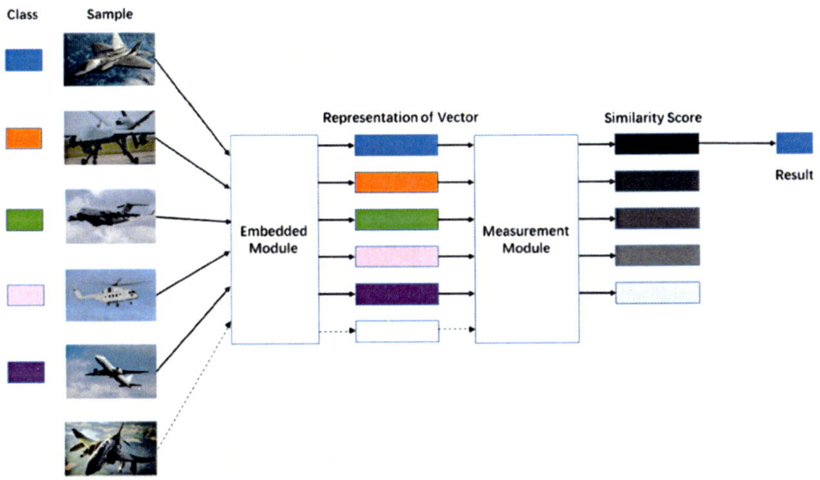

Fig. 3. Structure diagram of metric learning model

cascaded feature matching network, in which feature matching blocks can use multi-scale representation to correlate the high correlation between the object and the compared image, so as to achieve better classification effect.

Prototypical Network [25] assumes that there is a prototype for each category in the vector space. Deep neural Network is used to map images into vectors, and the average value of sample vectors belonging to the same category is used as the prototype of the category. The idea is to compare the Euclidean distance between the vector of the query set and the prototype of each category. Fort [26] proposed a novel architecture that uses the uncertainty of a single data point as a weight to carry out relevant distance measurement for categories in the embedding space, improving the accuracy of few-shot classification. Wang [27] proposed a novel architecture that added new modules to the Siamese Network for learning high-quality prototypical representations of each category, further improving classification accuracy. Lim [28] proposed Efficient-PrototypicalNet, which uses pre-trained models as feature extractors to reduce task complexity and also applies knowledge distillation to frameworks.

Relation Network [29] needs to be modeled.It does not use a single fixed distance measurement method, but trains a network to learn the distance measurement method. Hui [30] proposed a self-attention relation network, which introduced attention module to enhance learning features. Finally, the query set and support set were compared in the relational module to achieve accurate classification. Ashrafi [31] focused on the relation network and introduced embedded modules and attention mechanism to achieve the goal of faster generalization and better classification effect in short-term training. Abdelaziz [32] extended relational network to position perception, integrated multi-scale features, measured the similarity between combined features, and realized a model with more robust and better generalization ability.

The model based on metric learning can use a relatively simple way to achieve distance measurement. In recent years, it has developed more mature and achieved better image classification effect, and the technology is relatively stable. The core of measurement learning is to choose the appropriate distance measurement method.

3.2 Meta-learning Model

In few-shot learning, we often face many different new classification tasks. Meta-learning makes use of previous knowledge and experience to guide the learning of new tasks, so that the network has the ability to learn (Fig. 4).

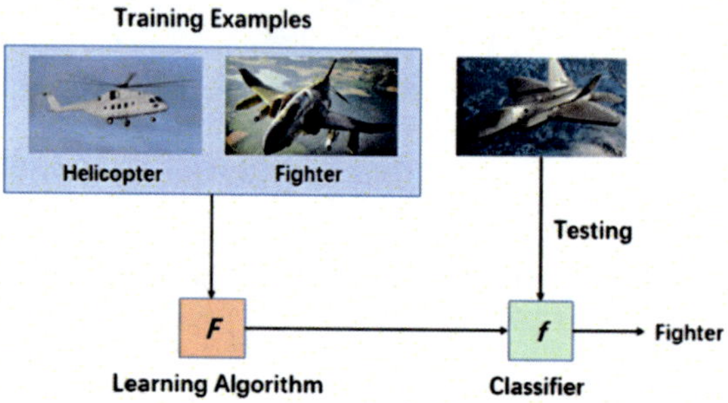

Fig. 4. Structure diagram of meta-learning model

Meta-learning mainly consists of two stages: meta-train and meta-test. In the meta-training stage, the model is exposed to many independent supervised tasks constructed from auxiliary data sets to learn how to adapt to future related tasks. The label space in the meta-test stage does not intersect with the label space in the meta-training period. Data sets are used to adjust the network in order to realize the task of few-shot learning.

Finn [33] proposed MAML framework. The key idea is to train the initial parameters of the model, so that the parameters of the model can only apply to a small number of samples in a new task and show good results after several steps of gradient descent update. Li [34] proposed a meta-learning-stochastic gradient descent model, which can be initialized and adjusted in a step in supervised learning. Compared with other models, it has the advantages of simplicity, easy execution and easy training. Jamal [35] proposed a general task-independent meta-learning model to conduct meta-learning on the unbiased initial model with the maximum uncertainty of the output label, so as to improve the generalization ability of the model. Xu [36] proposed an unsupervised meta-learning model, which uses clustering embedding and data enhancement functions to construct tasks and transfer data between internal and external cycles, so as to solve the problem of insufficient data diversity and achieve a good classification accuracy rate. Cao [37] proposed a new meta-learning method of concepts, which tried to learn new representations with explicable concept dimensions. The learner could learn the mapping of high-level concepts to semi-structured metric spaces, improving the generalization ability of models.

Meta-learning model goes through a lot of meta-training, and the tasks encountered in each training are different, so the model deals with different problems each time. After the training, the meta-learning model is naturally better able to deal with new tasks and ignore the characteristics of specific tasks. Therefore, when encountering a small sample of new tasks, the effect is also good.

4 Few-Shot Learning Based on Optimization

4.1 Parameter Optimization

In the problem of few-shot image classification, the parameters of feature extraction and classification network are very important to improve the accuracy of classification results, and appropriate parameters can make the network produce good classification results. For the research based on network parameter optimization, it is also a research direction of few-shot image classification to find appropriate strategies.

Lee [38] proposed an algorithm based on parameter optimization, which introduced task-related transformation matrix and mask matrix, so that the parameter updating process could be determined according to the requirements of the task. And parameters could be learned in the subspace corresponding to the task, so that the algorithm model could converge quickly and the classification effect was good. Kim [39] combined neural architecture search with meta-learning to train network parameters, and then searched for more complex structures based on good network structure. By repeating such steps, the optimal network structure and parameters were found to achieve better few-shot image classification effect. Dove [40] also used neural architecture search to find the optimal network and proposed a new structure for modifying parameters of different operations, which further improved the adaptability of the network to new tasks. Moreover, a two-level iterative optimization training method is adopted to alleviate the over-fitting problem of few-shot image classification. Elsken [41] proposed a few-shot learning algorithm based on parameter optimization, and introduced the neural architecture search into the few-shot learning algorithm, which realized the learning of initialization parameters and the parameters of network structure.

4.2 Meta-learner Optimization

For the problem of few-shot image classification, some common learners will overfit, because they need millions of times of training to converge, which will affect the correctness of few-shot image classification and the generalization ability of network model. Therefore, it is necessary to consider the problem of optimizing the learner, so that the model can learn efficiently, converge quickly and achieve accurate classification even with only a small number of samples.

Some researchers [42] proposed an accurate optimization algorithm based on Long Short Term Memory (LSTM) meta-learner, which can learn initialization parameters that can realize fast convergence of learner and learn how to select appropriate parameter to update, providing a good parameter update strategy and making the network more suitable for few-shot learning tasks. Antoniou [43] proposed an optimization strategy based on MAML, which made each step of the internal cycle unable to understand the learning rate and batch specification statistics, reduced hyperparameter sensitivity and generalization error, and improved the stability and convergence speed of MAML. Rusu

[44] introduced a low-dimensional hidden space on the basis of MAML, realized the inner layer cyclic updating of parameters, and obtained model parameters through random sampling, which could further adapt to the uncertainty in the case of few-shot learning. Bertinetto [45] optimized the classification method in the meta-learner and adopted Ridge Regression and Logistic Regression to improve the generalization ability of the model. Zhang [46] proposed a new optimization framework of prototype based on meta-learning, which regarded the gradient and its flow as meta-knowledge, and then proposed a new meta-optimizer based on Neural Ordinary Differential Equation to optimize the prototype and solve the problem of prototype deviation. Tian [47] proposed meta-contrastive loss as an appropriate meta-regularization to further improve the generalization ability of the meta-learning model.

5 Development Trend and Application for Few-Shot Learning in Image Classification of Aerial Objects

Firstly, the methods listed above for few-shot image classification are studied and summarized, and the comparison results are shown in the following table.

After comprehensively considering the advantages and disadvantages of various few-shot learning methods listed in the table above and their contributions to the accuracy of small sample image classification, the development trend of few-shot learning technology in the field of aerial image classification is summarized below (Table 1).

The major problem of the tasks in aviation few-shot image classification is insufficient data, so data enhancement is a very intuitive method. As for the trends of few-shot learning technology for aerial image classification in the future, at the data level, ordinary data enhancement is no longer the focus of research, and other prior knowledge will be used to train the model, or unlabeled data will be better utilized. Although the number of labeled samples in aviation is very small, a large amount of unlabeled data in the real world contains so much information, and the direction of using unlabeled data to train the model is worth further study. Secondly, GAN can generate images to be practically indistinguishable from the real image, and GAN is also suitable for data enhancement of aerial few-shot images. It is the focus of future research to find a stable GAN training strategy and a better feature extraction classification network to avoid pattern collapse and insufficient image richness.

Aiming at the research of model-based few-shot learning in aerial image classification, for the metric learning model, the feature extraction technology of a small number of samples and the appropriate measurement method are the most critical. And selecting the appropriate feature extraction network and comparison network can improve the accuracy of aerial image classification of the model. Secondly, It is also an important direction for further research to choose more effective neural network measurement method with unfixed measurement mode. Meta-learning model is a new trend in recent years, and further research is needed. So different network models are designed in view of different problems, such as introducing attention mechanism, designing appropriate loss function and punishment mechanism, so as to realize the strong generalization ability and the ability to adapt to new tasks proposed in concept. In addition, the explicable exploration of meta-learning network also needs to be continuously understood

Table 1. Summary of advantages and disadvantages of various few-shot learning methods

Class	Advantage	Disadvantage
Common approaches to data augmentation	Simple operation, easy control	Lack of feature-level enhancement, lack of data diversity
Add weak-labeled or unlabeled data	Unlabeled samples exist in large numbers Their acquisition costs are relatively low	It may appear that the enhancement effect is not obvious It is easy to increase the error rate in the early stage of training
Add generated data	Able to achieve feature enhancement to generate more realistic images	The training is unstable There may be a mode collapse problem
Metric learning model	Simple and feasible, good classification results	Strong dependence on data When the sampling strategy is too simple, it will not be able to learn complex samples
Meta-learning model	Able to learn at the task level and adapt well to new fields	Development is immature When the amount of data is small, the generalization ability of the model will decrease
Parameter optimization	Improve the accuracy of classification from network details	There is some inexplicability, which requires higher mathematical ability
Meta-learner optimization	Optimize the structure and initial parameters of the meta-learner, quickly converge in the face of few samples	It is prone to over-fitting Further research is needed on how to optimize the meta-learner

and studied, so as to better improve the network and training methods according to the internal mechanism.

Aiming at the trend of aerial few-shot image classification based on optimization, the research of parameter optimization for meta-learner will be more in recent years, and the current model is not mature enough, so better meta-learner need to be designed. It will be an important research direction in the future that how to combine parameter optimization with meta-learner optimization and how to design and improve the meta-learner to learn more or more effective meta-knowledge and better initialization parameters.

In general, the three research directions mentioned above are not independent of each other, and the combination of data, model and optimization is also an important trend of future development. The application of few-shot learning technology in the aviation field can achieve high-precision and high-stability model deployment for aviation tasks with little available data, and reduce the high cost of data tags in aviation applications to a certain extent, which is meaningful.

6 Conclusion

To sum up, there are many ways to solve the problem of few-shot image classification, but relevant researches in the aerial field are still very lacking. This paper studies and summarizes the existing few-shot image classification technology, and expounds the trends of few-shot learning technology in the field of aerial image classification. It is hoped that this paper can provide ideas for the application of few-shot learning in the field of aerial image, and then solve practical problems.

References

1. Shorten, C., Khoshgoftaar, T.M.: A survey on image data augmentation for deep learning. J. Big Data **6**(1) (2019)
2. Wang, Y., Hebert, M.H.: Learning from small sample sets by combining unsupervised meta-training with CNNs. 2016
3. Boney, R., Ilin, A.: Semi-supervised few-shot learning with MAML (2018)
4. Ren†, M., Triantafillou∗†, E., Ravi∗§, S., et al.: Meta-learning for semi-supervised few-shot classification[J] (2018)
5. Liu, Y, Lee, J., Park, M., et al.: Transductive propagation network for few-shot learning[J] (2018)
6. Hou, R., Chang, H., Ma, B., et al.: Cross attention network for few-shot classification[J] (2019)
7. Xie, Q., Dai, Z., Hovy, E., et al.: Unsupervised data augmentation for consistency training[J] (2019)
8. Bateni, P., Barber, J., Goyal, R., et al.: Beyond simple meta-learning: multi-purpose models for multi-domain, active and continual few-shot learning[J]. (2022)
9. Goodfellow, I., Pouget-Abadie, J., Mirza, M., et al.: Generative Adversarial Nets. MIT Press (2014)
10. Mehrotra, A., Dukkipati, A.: Generative adversarial residual pairwise networks for one shot learning[J] (2017)
11. Ali-Gombe, A.: Few-shot classifier GAN. [C]. In: 2018 International Joint Conference on Neural Networks (IJCNN) (2018)
12. Xian, Y., Sharma, S., Schiele, B., et al.: F-VAEGAN-D2: a feature generating framework for any-shot learning[J]. IEEE (2019)
13. Nguyen, K., Todorovic, S.: A self-supervised GAN for unsupervised few-shot object recognition[C]. In: 2020 25th International Conference on Pattern Recognition (ICPR) (2021)
14. LoFGAN: Fusing local representations for few-shot image generation
15. Pan, H., Yang Xiaogang, L., Ruitao, C.Z., Chuang, L.: Data augmentation method of infrared ship target based on spatial association[J]. Infrared Laser Eng. **50**(12), 20210281 (2021)
16. Liu, B., Zhu, Y., Song, K., et al.: Towards faster and stabilized GAN training for high-fidelity few-shot image synthesis[J]. 2021
17. Chopra, S., Had Sell R., Lecun, Y.: Learning a similarity metric discriminatively, with application to face verification[C]. In: 2005 IEEE Computer Society Conference on Computer Vision and Pattern Recognition (CVPR'05). IEEE (2005)

18. Wang, J., Zhu, Z., Li, J., et al.: Attention based siamese networks for few-shot learning[C]. In: 2018 IEEE 9th International Conference on Software Engineering and Service Science (ICSESS). IEEE (2019)
19. Jadon, S., Srinivasan, A.A.: Siamese networks for one shot learning using kernel-based activation functions[M] (2019)
20. Shao, H., Zhong, D., Du, X., et al.: Few-shot learning for palmprint recognition via meta-siamese network[J]. IEEE Trans. Instrum. Meas. **70**, 1–12 (2021)
21. Vinyals, O., Blundell, C., Lillicrap, T., et al.: Matching networks for one shot learning[J] (2016)
22. Mai, S., Hu, H., Xu, J.: Attentive matching network for few-shot learning[J]. Comput. Vis. Image Underst. **187**(Oct.), 102781.1–102781.13 (2019)
23. Li, H., Dong, W., Mei, X., et al.: LGM-Net: learning to generate matching networks for few-shot learning[J] (2019)
24. Chen, M., Wang, X., Luo, H., et al.: Learning to focus:cascaded feature matching network for few-shot image recognition[J]. Sci. China Inf. Sci. **64**(9), 86–98 (2021)
25. Snell, J., Swersky, K., Zemel, R.S.: Prototypical networks for few-shot learning[J] (2017)
26. Fort, S.: Gaussian prototypical networks for few-shot learning on omniglot[J] (2017)
27. Wang, J., Zhai, Y.: Prototypical siamese networks for few-shot learning[C]. In: 2020 IEEE 10th International Conference on Electronics Information and Emergency Communication (ICEIEC). IEEE (2020)
28. Lim, J.Y., Lim, K.M., Ooi, S.Y., et al.: Efficient-prototypicalnet with self knowledge distillation for few-shot learning[J]. Neurocomputing 459 (2021)
29. Sung, F., Yang, Y., Zhang, L., et al.: Learning to compare: relation network for few-shot learning[J] (2017)
30. Hui, B., Zhu, P., Hu, Q., et al.: Self-attention relation network for few-shot learning[C]. In: 2019 IEEE International Conference on Multimedia & Expo Workshops (ICMEW). IEEE (2019)
31. Ashrafi, I., Mohammad, M., Mauree, A.S., et al.: Attention guided relation network for few-shot image classification[C]. In: ICCCM (2019)
32. Abdelaziz, M., Zhang, Z.: Multi-scale kronecker-product relation networks for few-shot learning[J]. Multimed. Tools Appl. **81**(5), 6703–6722 (2022)
33. Finn, C., Abbeel, P., Levine, S.: Model-agnostic meta-learning for fast adaptation of deep networks[J] (2017)
34. Li, Z., Zhou, F., Fei, C., et al.: Meta-SGD: learning to learn quickly for few-shot learning[J] (2017)
35. Jamal, M.A., Qi, G.J., Shah, M.: Task-agnostic meta-learning for few-shot learning[J] (2018)
36. Xu, H., Wang, J., Li, H., et al.: Unsupervised meta-learning for few-shot learning[J]. Pattern Recogn. **116**(6), 107951 (2021)
37. Cao, K., Brbic, M., Leskovec, J.: Concept learners for few-shot learning[J]. ICLR (2021)
38. Lee, Y., Choi, S.: Gradient-based meta-learning with learned layerwise metric and subspace[J] (2018)
39. Kim, J., Choi, Y., Cha, M., et al.: Auto-meta: automated gradient based meta learner search[J] (2018)
40. Dove, H.S., Schwartz, E., Xue, C., et al.: MetAdapt: meta-learned task-adaptive architecture for few-shot classification[J] (2019)
41. Elsken, T., Staffler, B., Metzen, J.H., et al.: Meta-learning of neural architectures for few-shot learning[C]. In: 2020 IEEE/CVF Conference on Computer Vision and Pattern Recognition (CVPR). IEEE (2020)
42. Larochelle, S.: Optimization as a model for few-shot learning
43. Antoniou, A., Edwards, H., Storkey, A.: How to train your MAML[J] (2018)

44. Rusu, A.A., Rao, D., Sygnowski, J., et al.: Meta-learning with latent embedding optimization[J] (2018)
45. Bertinetto, L., Henriques, J.F., Torr, P., et al.: Meta-learning with differentiable closed-form solvers[J] (2018)
46. Zhang, B, Li, X, Ye, Y., et al.: MetaNODE: prototype optimization as a neural ODE for few-shot learning[J] (2021)
47. Tian, P., Gao, Y.: Improving meta-learning model via meta-contrastive loss[J]. Front. Comput. Sci. **16**(5), 1–7 (2022)

Remaining Useful Life Prediction Method of Aero-Engines Based on LSTM

Binghuan Duan(✉), Yukai Hao, and Yong Guo

Xi'an Aeronautics Computing Technique Research Institute, AVIC, Xi'an 710068, China
dbh_2010@126.com

Abstract. Fault diagnosis and remaining useful life prediction of aero-engines are important parts of its health management, which is of great significance for reducing maintenance costs and effectively preventing the occurrence of unexpected accidents. In order to improve the accuracy of engine fault prediction, this paper proposes a remaining useful life (RUL) prediction method of aero-engines, which is based on long short-term memory (LSTM) network. Firstly, in order to reduce noise and eliminate the influence caused by singular samples, the time-series in the dataset of aero-engines are processed by wavelet transform and normalization. Secondly, a LSTM prediction network is constructed, and the network is trained by clipping responses (i.e. RUL tag values) and adjusting network parameters. Finally, the RUL prediction is performed. A data set of aero-engines from practical engineering applications is used to validate the effectiveness of the proposed method. Compared with several other prediction algorithms, the proposed method in this paper effectively improves the prediction accuracy of RUL, and provides a decision-making basis for the maintenance of aero-engines.

Keywords: Remaining Useful Life (RUL) Prediction · Aero-engines · LSTM

1 Introduction

The structure of modern aero-engines is complex, and their working conditions are bad and changeable, which makes modern aero-engines a fault-prone system. The traditional periodic maintenance is costly and inefficient [1]. Furthermore, as a heart of an aircraft, the performance of the aero-engine is closely related to the safe navigation of the aircraft. Therefore, it is of great significance to predict the remaining useful life (RUL) and health status of aero-engines. On the one hand, the prediction of the RUL can help maintainers take protective measures before engine failures occur, which can provide a strong guarantee for the safety of a flight system. On the other hand, it can also help maintainers determine the best time for engine maintenance and reduce maintenance costs [2].

Many scholars at home and abroad have done a lot of research on RUL prediction. Wang et al. [3] applied a data mining model based on fuzzy clustering to predict the RUL of aero-engines. Wang et al. [4] proposed an intelligent fault prediction method that combines Hilbert-Huang transform (HHT) and support vector machine (SVM). Guo

Chinese Society of Aeronautics and Astronautics: CASTYSF 2022, LNEE 972, pp. 583–592, 2023.
https://doi.org/10.1007/978-981-19-7652-0_51

et al. [5] used grey rough set and BP neural network to predict equipment fault. Although this method has achieved good results, the prediction accuracy is still not high enough.

Compared with traditional methods, deep learning (DL) can extract more useful features from complex original data, and is better at learning and identification. Many subsequent approaches have adopted techniques from deep learning for RUL prediction, in order to solve the problem of model overfitting existing in traditional neural networks and SVM methods [6]. Yu et al. [7] proposed a multi-scale convolution neural network and Yu et al. [8] proposed a novel deep belief network (DBN) model in RUL prediction. The two networks have higher prediction accuracy compared with traditional DBNs, BP networks, and SVMs. Some scholars combined some approaches such as random forest (RF) with long and short-term memory (LSTM) network, and achieved improved prediction results [9, 10].

The prediction of the RUL of aero-engines is a typical time-series analysis problem. A big difficulty in accurate prediction is that the operating data of aero-engines may have problems such as inaccurate recording, noise pollution, complex time dependencies between multiple components. LSTM is a special type of recurrent neural network (RNN) capable of learning long-term dependencies. Its characteristics determine that it is suitable to deal with the time-series analysis problem of predicting the RUL. Many scholars' studies have also proved this [9, 10].

After the above analysis, a RUL prediction method of aero-engines based on LSTM is proposed in this paper. Firstly, the time-series data of aero-engines is preprocessed by wavelet transform and normalization to improve the quality of original data. Then, a LSTM prediction network is designed and a clipping responses (i.e. RUL tag values) mechanism is added when training, so that the network can learn more useful characteristics from samples where engines are about to break down. Furthermore, according to the actual applications, parameters of the network are adjusted in the training to obtain the target prediction model. And then, the RUL prediction is carried out.

The remaining sections are organized as follows. The theorem of LSTM is introduced in Sect. 2. The proposed RUL prediction method is described in Sect. 3. Validation is conducted by a data set of aero-engines from practical engineering applications in Sect. 4. Section 5 concludes the whole paper.

2 LSTM Principle

As a variant of the widely used RNN [11], LSTM is a neural network for processing time-series data [12]. LSTM introduces linear connections and three gate units of input gate, forget gate and output gate in the hidden layer cell to control the selective flow of information.

The forget gate f_t can determine the proportion of retained historical information (between 0 and 1), as shown in Eq. (1). The Input gate i_t determines the proportion of input information, as shown in Eq. (2). The forget gate and the input gate jointly control how much historical information C_{t-1} is forgotten and how much new information $\widetilde{C}_t$ is added to update the long-term memory unit C_t, as shown in Eq. (3) and Eq. (4). The output gate o_t determines how much information from the long-term memory unit C_t is output to the short-term memory unit h_t, as shown in Eq. (5) and Eq. (6). The calculations of the gates are listed as follows.

$$f_t = \sigma(W_f X_t + U_f h_{t-1} + b_f) \tag{1}$$

$$i_t = \sigma(W_i X_t + U_i h_{t-1} + b_i) \tag{2}$$

$$\widetilde{C}_t = \tanh(W_C X_t + U_C h_{t-1} + b_C) \tag{3}$$

$$C_t = f_t * C_{t-1} + i_t * \widetilde{C}_t \tag{4}$$

$$o_t = \sigma(W_o X_t + U_o h_{t-1} + b_o) \tag{5}$$

$$h_t = o_t * \tanh(C_t) \tag{6}$$

The notations W_f, W_i, W_C are the parameter matrices from the input layer to the hidden layer, connecting the input X_t to the three gates. U_f, U_i, U_C denote self-circular parameter matrices from the hidden layer to the hidden layer. The notations b_f, b_i and b_C denote bias parameter matrices of three gates and memory units. The notation σ is the sigmoid activation function that makes the output values of the three gates between 0 and 1. Figure 1 shows the flow of information in LSTM.

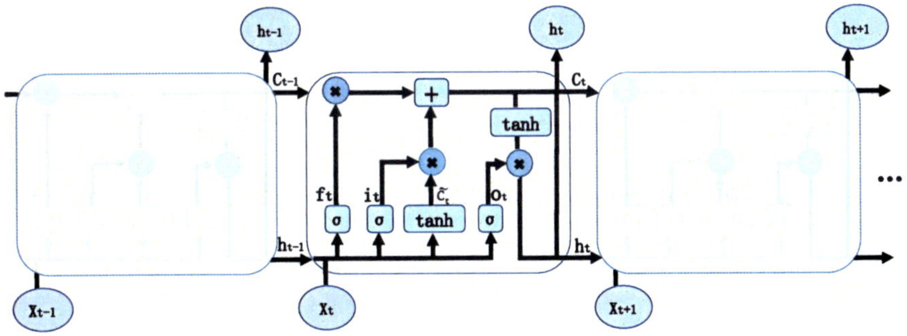

Fig. 1. The structure of LSTM.

The value of the long-term memory unit changes slowly, mainly to memorize the historical information, while the value of the short-term memory unit changes rapidly, mainly to memorize the newly input information. The two memory units together represent the cell state at a certain moment. This characteristic of LSTM determines that it is naturally suitable for strongly correlated time-series prediction.

3 The Proposed RUL Prediction Method Based on LSTM

In practical engineering applications, the collected raw operating data of aero-engines often contains a lot of noise, and the engine performance is affected by the cross-linking of multiple components. Therefore, in order to improve the accuracy of using LSTM to predict the RUL of aero-engines, data preprocessing and building an LSTM network with strong learning ability are the keys. The focus of the method in this paper is also placed in these two aspects, and the procedure of RUL prediction is shown in Fig. 2.

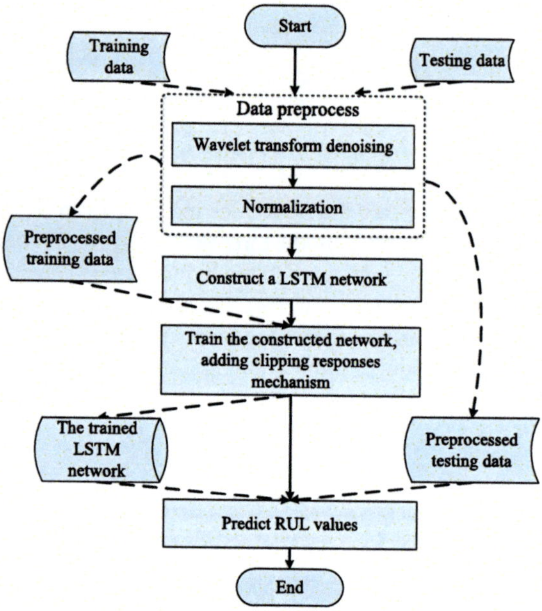

Fig. 2. The procedure of RUL prediction using the proposed method.

Firstly, the training data and testing data are preprocessed to improve data quality. The pre-processing includes wavelet transform and normalization. The wavelet transform can reduce noise. The normalization is conducive to improving the convergence speed and predictive accuracy of the network. After data pre-processing, the LSTM network is built and then trained. The number of network layers, the types of network layers, the number of hidden nodes and the selected training parameters will greatly affect the performance of the network. In addition, a clipping responses strategy is added during training. It can make the constructed network to learn more from the sequence data where aero-engines are about to fail, and then to further improve prediction accuracy. Finally, the trained network is used for RUL prediction.

3.1 Data Preprocessing

A. **Wavelet transform**

Due to the influence of temperature, air pressure, noise and other factors in the working environment, the collected time-series data of aero-engines contains a lot of noise. It is necessary to reduce noise in advance to improve data quality. Owing to the nonlinear and high noise characteristics of time-series, traditional Gaussian denoising and median filtering methods are difficult to achieve ideal results. Wavelet transform [13] can extract time-frequency features and is very suitable for processing non-stationary and nonlinear signals. In addition, the implementation process is simple and the amount of calculation is small. Therefore, this paper chooses the wavelet transform threshold technique to denoise the original data.

The composition of the collected signal *f(t)* is as follows,

$$f(t) = s(t) + e(t) \tag{7}$$

where *s(t)* denotes the real signal and *e(t)* denotes the noise.

The notation $W_e(a, t)$ is the wavelet transform function of *e(t)*, which is implemented as follows:

$$W_e(a, t) = \int_{-\infty}^{+\infty} e(u)\varphi_a(t - u)d_u \tag{8}$$

After doing wavelet transform on both sides of Eq. 7, Eq. 9 is obtained:

$$WT_f(a, b) = WT_s(a, b) + WT_e(a, b) \tag{9}$$

After wavelet transform, the correlation of the original signal *f(t)* is largely removed, and the effective part is mainly concentrated on a small number of wavelet coefficients with relatively large amplitudes. For the noise *e(t)*, the wavelet coefficients at each scale are greatly reduced after processing. Further, threshold processing is performed on all the obtained wavelet coefficients, and finally signal reconstruction is performed, so as to achieve the purpose of suppressing noise.

B. **Normalization**

When different features are listed together to represent time-series data, the small data in absolute value will be "eaten" by the big data in the process of network training. Normalization can prevent the above situation from appearing [14]. Normalization is to scale the data so that it falls into a small specific interval. After the denoised data is normalized, all features can be treated equally, which can speed up the convergence of the network. In this paper, Z-score normalization is performed on all the data processed by wavelet denoising. Normalization should be conducted as follows,

$$x^* = \frac{X - \mu}{\sigma} \tag{10}$$

where x^* denotes the normalized value of the input data X, μ denotes the mean of all input data, σ denotes the standard deviation of all input data.

3.2 The Constructed LSTM Network

Generally speaking, the more layers a LSTM network contains, the stronger learning ability of complex time-series it has. The number of layers of LSTM generally does not exceed three, otherwise it will be difficult to converge during training, and the prediction accuracy will be reduced instead [15]. Based on this, this paper constructs a LSTM prediction network, which contains two LSTM layers, two fully connected layers and three dropout layers, as shown in Table 1.

LSTM layers and fully connected layers are mainly used for feature extraction. The dropout layer after them invalidates some nodes of this layer with the probability of the drop rate *p*, which is used to prevent overfitting of the network and accelerate the convergence of the network. The empirical value of the *p* is generally between 0.2 and 0.5. The last fully connected layer is mainly used to change the output vector dimension of the network. From Table 1, the notation *nodeNum* denotes the number of hidden neurons in the LSTM layer and the fully connected layer, and the notation *classesNum* denotes the number of categories of aero-engines, that is the number of hidden neurons in the last fully connected layer.

Table 1. The composition of the constructed LSTM network.

Layer nums	Layer types	Parameters
1	lstmLayer	*nodeNum*(200)
2	dropoutLayer	*p*(0.2)
3	lstmLayer	*nodeNum* (100)
4	dropoutLayer	*p*(0.2)
5	fullyConnectedLayer	*nodeNum* (50)
6	dropoutLayer	*p*(0.4)
7	fullyConnectedLayer	*classesNum*

For the constructed LSTM network, a mechanism of clipping responses is added during training to improve the prediction accuracy. During training, if the response (i.e. RUL label value) is higher than a certain threshold, then the response is clipped to be equal to this threshold. Clipping responses makes the LSTM network treat instances with higher RULs as equivalent, which allows the engine to learn more from the time-series data near the time of failure, and results in a trained LSTM network with higher prediction accuracy.

4 Validation

4.1 Evaluation Indicators of Prediction Performance

For time-series models, or more broadly regression models, the scoring function and root mean square error (RMSE) are widely used evaluation metrics [16]. The formula of the scoring function is defined in Eq. (11), where *Score* is the scoring function value, N is the number of engines for testing, and $h =$ (Estimated RUL − True RUL).

$$Score = \begin{cases} \sum_{i=1}^{N} (e^{-\frac{h_i}{13}} - 1), h_i < 0 \\ \sum_{i=1}^{N} (e^{\frac{h_i}{10}} - 1), h_i \geq 0 \end{cases} \quad (11)$$

The scoring function shows different penalties for the error of the model predicting RUL. For the case of overestimating RUL (i.e. $h_i \geq 0$), the penalty is higher than the case of underestimating RUL (i.e. $h_i < 0$). This is because the cost of catastrophic consequences of not making timely repairs is far greater than the cost of overly wasting resources in repairs.

RMSE is used as the second evaluation indicator as follows,

$$RMSE = \sqrt{\frac{1}{N}\sum_{i=1}^{N} h_i^2} \quad (12)$$

For *Score* and *RMSE*, the smaller the values are, the better the prediction effect will get.

4.2 Data Set

The experimental data set is derived from the sensor data of aero-engines in engineering applications, including the time-series of 260 engines of the same type. Each engine was in normal use at the start of the test, and the test was continued under a combination of three different operating conditions until the engine degraded (i.e. reached the failure threshold). The 21 kinds of sensor data of each engine was monitored during the test.

The training set contains the whole life cycle monitoring data of each engine, from the time an engine can be used normally to the time when the engine degrades and cannot run, with a total of 53759 pieces of data. The testing set retains some data of normal operation before engines downgrade, with a total of 19482 pieces of data. The data set also includes the real RULs of these engines and can be used to evaluate the test results.

4.3 Experimental Results and Discussion

This paper adopts the wavelet transform fixed threshold technique to reduce noise. The parameter is set to soft threshold and the scale value is set to 'mln'. Wavelet basis function selects 'db3'.

The training parameters of the LSTM network are set as follows: the minibatch is set to 20, the maximum epoch is set to 80, the initial learning rate is set to 0.005 and is reduced by a factor of 0.1 after every 20 epochs. In addition, a clipping responses mechanism is added to the training, and the clipping threshold is generally set to no

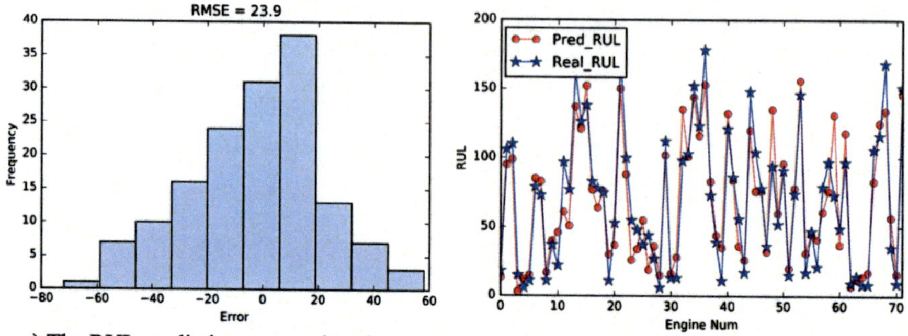

a) The RUL prediction errors of engines

b) The comparison of predicted and real RULs of engines

Fig. 3. The RUL prediction results of engines

more than 90% of the average life cycle of engines. In this way, all RUL label values in training are not higher than this threshold. Figure 3 shows the prediction effect.

The histogram of the RMSE of the 260 engines in the testing set as shown in Fig. 3a, where the abscissa is the difference between the predicted RUL and the real RUL, and the ordinate is the frequency of errors. The higher the frequencies in the area with smaller error values are, the more accurate the prediction result is. For a prediction, the comparison of the real and predicted RULs of engines extracted from the testing set are listed in Fig. 3b, where the predicted and real RULs of engines are represented in solid dots and solid stars, respectively. It can be seen from Fig. 3 that the predicted RUL values are relatively close to the real RUL values, and the method in this paper can accurately predict the RUL of the engines.

In order to further illustrate the feasibility and effectiveness of the prediction method in this paper, three other prediction methods were selected for comparative analysis on the experimental data set described in Sect. 4.2. The three methods are multilayer perceptron (MLP) [16], support vector regression (SVR) [16], convolutional neural network (CNN) [16]. The comparison results are shown in Table 2.

Table 2. *Scores* and *RMSE* values for various prediction methods

Indicators	MLP	SVR	CNN	Proposed method
Score	18652	2430	2376	1982
RMSE	50.71	28.34	25.85	23.90

From Table 2, the latter two deep learning methods achieve better results in both *Score* and *RMSE*, compared with traditional machine learning methods such as MLP and SVR. A longitudinal comparison of the two deep learning methods shows that the prediction method based on LSTM proposed in this paper has more advantages than CNN, when dealing with time-series problems with long-term dependencies. The *Score* and *RMSE* of the proposed method are 16.6% and 7.5% lower than that of CNN, respectively.

5 Conclusions

In this paper, a RUL prediction method of aero-engines based on LSTM is proposed. The method firstly performs preprocessing operations such as wavelet transform and normalization on the original time-series of engines to improve quality and credibility of the original data. Then a network is constructed combining LSTM layers, fully connected layers and dropout layers. Furthermore, the constructed network is trained through strategies such as clipping the responses (RUL tag values) and adjusting parameters. Finally, the RUL prediction is carried out, and the predicted RUL values are obtained. The experimental results on the data set from actual engineering applications show that the method in this paper has high prediction accuracy and feasibility, and has a positive role in improving equipment support capabilities and reducing maintenance costs.

References

1. Cai, G.Y., Gao, J., Miao, X.W.: Current situation and indexes of prognostics and health management system for aero-engines. Meas. Control. Technol. **35**(4), 5–10 (2016)
2. Dong, D., Li, X.Y., Sun, F.Q.: Life prediction of jet engines based on LSTM-recurrent neural networks. In: Prognostics & System Health Management Conference (PHM-Harbin), pp. 1–6 (2017)
3. Wang, Y.Y., Zhang, Z.H., Qian, W.W., et al.: Research on aero engine fault prediction based on fuzzy clustering. J. Nanchang Hangkong Univ. (Nat. Sci.) **32**(1), 23–28 (2018)
4. Wang, Y.S., Ma, Q.H., Zhu, Q., et al.: An intelligent approach for engine fault diagnosis based on Hilbert-Huang transform and support vector machine. Appl. Acoust. **75**, 1–9 (2014)
5. Guo, Y., Yang, Y.: Equipment fault prediction based on grey rough set and BP neural network. Appl. Res. Comput. **34**(9), 4–8 (2017)
6. Liu, H., Liu, Z.Y., Jia, W.Q.: Current research and challenges of deep learning for equipment remaining useful life prediction. Comput. Integr. Manuf. Syst. **27**(1), 34–52 (2021)
7. Yu, J., Peng, Y., Deng, Q.: Remaining useful life prediction based on multi-scale residual convolutional network for aero-engine. In: 2021 IEEE International Conference on Prognostics and Health Management (ICPHM). IEEE, pp. 1–6 (2021)
8. Yu, H., Tian, Z., Li, H., et al.: A novel deep belief network model constructed by improved conditional RBMs and its application in RUL prediction for hydraulic pumps. Int. J. Acoust. Vib. **25**(3), 373–383 (2020)
9. Tang, J., Xiao, L.: The improvement of remaining useful life prediction for aero-engines by classification and deep learning. In: 2020 11th International Conference on Prognostics and System Health Management (PHM-2020 Jinan). IEEE, pp. 130–136 (2020)
10. Tang, J., Wu, N., Xiao, L., et al.: The improvement of remaining useful life prediction for aero-engines by injecting noise. In: 2020 Global Reliability and Prognostics and Health Management (PHM-Shanghai). IEEE, pp. 1–6 (2020)
11. Li, S., Li, W., Cook, C., et al.: Independently recurrent neural network (indrnn): building a longer and deeper rnn. In: Proceedings of the IEEE Conference on Computer Vision and Pattern Recognition, pp. 5457–5466 (2018)
12. Smagulova, K., James, A.P.: A survey on LSTM memristive neural network architectures and applications. Eur. Phys. J. Spec. Top. **228**(10), 2313–2324 (2019). https://doi.org/10.1140/epjst/e2019-900046-x

13. Rhif, M., Ben, A.A., Farah, I.R., et al.: Wavelet transform application for/in non-stationary time-series analysis: a review. Appl. Sci. **9**(7), 1345–1352 (2019)
14. Zhang, X., Sugano, Y., Bulling, A.: Revisiting data normalization for appearance-based gaze estimation. In: Proceedings of the 2018 ACM Symposium On Eye Tracking Research & Applications. pp. 1–9 (2018)
15. Peng, Y., Liu, Y.H., Zhang, R.F.: Modeling and analysis of stock price forecast based on LSTM. Comput. Eng. Appl. **55**(11), 209–212 (2019)
16. Babu, G.S., Zhao, P., Li, X.L.: Deep convolutional neural network based regression approach for estimation of remaining useful life. In: International Conference on Database Systems for Advanced Applications. Springer, Cham, pp. 214–228 (2016)

The Optimization Design of Large-Camber Nozzle Based on Geometric Optimization

Yang Yuxin, Chen Yesi, Ma Shoudong, Yang Hua(✉), and Wu Changju

Zhejiang University, Hangzhou, Zhejiang, China
yhbuaa@163.com

Abstract. A large-camber nozzle is parametrically defined based on hyperelliptic equations and high-order continuous transition curves. Firstly, the transition curve that controls the geometric parameters of the nozzle is optimized, so that the nozzle has a smooth transition shape before aerodynamic optimization. Based on the geometric optimization solution, the optimization design of straight nozzle and large-camber nozzle is carried out with objectives of high thrust coefficient and high total pressure recovery coefficient. Compared with direct optimization, the work based on geometric optimization is more efficient and less time-consuming. The nozzle after geometric optimization with better thrust characteristics can be directly applied, which facilitates the rapid optimization design of nozzles in engineering.

Keywords: Transition curves · Large camber · Thrust coefficient · Optimization design

1 Introduction

Jet-based flight control technology without control surface is a new flight control scheme. The thrust vectorization is achieved by adding a large-camber nozzle after the outlet of engine to change flow direction, and the aircraft's pitch control is realized by thrust vectoring [1]. The design of nozzle profile is very important for thrust vectorization.

In the optimization design of nozzle, the main problem is how to efficiently and accurately generate nozzle profile and estimate its performance and efficiency [2]. Chen et al. [3] proposed that the geometric constraints should also be taken into account in nozzle design in addition to the optimization of aerodynamic performance. Zhou et al. [4] proposed a method of section shape generation and transition based on curvature control, which can quickly generate complex and variable cross-section profile that transitions from inlet section of arbitrary shape to circular outlet section. Wu et al. [5] used multi-segment spline curves to carry out finer parameterization of nozzle and took axial Mach number and root mean square error of the design parameters as optimization objectives. Yadav et al. [6] measured nozzle performance by the constant rate of momentum change, and designed the cross-section geometry. Schomberg et al. [7] introduced conjugate circular arcs into the design of rocket's nozzle profile, and investigated the nozzle performance by taking thrust coefficient, nozzle outlet pressure and a transition between flow separation regimes during the engine start-up phase as performance

Chinese Society of Aeronautics and Astronautics: CASTYSF 2022, LNEE 972, pp. 593–602, 2023.
https://doi.org/10.1007/978-981-19-7652-0_52

indicators. Wang et al. [8] analyzed the effect of nozzle profile parameters on performance by variable-controlling approach, and studied the coupling relationship between parameters by response surface method. Many parametric methods were introduced into nozzle profile design and different indicators were selected to measure nozzle performance according to the needs. However, it is time-consuming to optimize the nozzle design through aerodynamic calculations. If the geometric constraints of the nozzle can be used to carry out geometric optimization before aerodynamic optimization, the time cost of nozzle design will be greatly reduced.

In this paper, a rapid optimization method for large-camber nozzles based on geometric optimization is proposed. A large-camber nozzle is parametrically defined based on hyperelliptic equations and high-order continuous transition curves. Before the aerodynamic optimization, the geometric optimization on the transition curve is carried out with the optimization objectives of minimizing curvature radius and the standard deviation of slope of parameter *n*. Based on the geometric optimization solution, the optimization design of straight nozzle and large-camber nozzle is carried out with the optimization objectives of high thrust coefficient and high total pressure recovery coefficient. Compared with direct optimization, the optimization design of nozzles based on geometric optimization is more efficient and less time-consuming. The nozzle after geometric optimization with better thrust characteristics can be directly applied, which facilitates the rapid optimization design of nozzles in engineering.

2 Parametric Design and Geometric Optimization

The cross-section of nozzle is designed with hyperelliptic equation. The design function is shown in formula (1), where a is the semimajor axis of the super ellipse, b is the semiminor axis, and n is the index of the super ellipse. The cross-sections of different shapes are generated by changing parameters a, b and n. Based on the inlet cross-section parameters a_1, b_1, and n_1, the outlet cross-section area S_f can be calculated through the hyperelliptic area formula and the nozzle shrinkage ratio α in formula (2), and the outlet cross-section parameters a_f, b_f and n_f are calculated through the aspect ratio and area formula of the outlet cross-section.

$$\left(\frac{x}{a}\right)^n + \left(\frac{y}{b}\right)^n = 1 \tag{1}$$

$$\begin{gathered} S = \frac{4 \times \Gamma\left(\frac{1}{n}+1\right) \times \Gamma\left(\frac{1}{n}+1\right)}{\Gamma\left(\frac{2}{n}+1\right)} \times ab \\ S_f = \alpha \times S_1 \end{gathered} \tag{2}$$

After the inlet and outlet parameters of nozzle are obtained, the parameters of each cross-section along nozzle need to be determined, which make the cross-section from inlet to outlet smoothly transition. The transition curve adopts a fourth-order continuous curve, as shown in formula (3). *A*, *B*, *C* are the coefficients in the transition curve, *p* represents the parameters *a*, *b*, *S*, and *S* is the area of each cross-section along the nozzle. The transition curve of *a*, *b* or *S* is not identical. The transition curve is shown in Fig. 1, which is a continuous smooth curve with a slope of 0 at both ends. The cross-section parameters a, b and S along the nozzle can be calculated by formula (3), with given values of these parameters at nozzle inlet and outlet. Optimization design of nozzle profile can be transformed into optimization design of coefficient in the transition curve. The coefficient C can be changed to optimize the transition curve.

$$\begin{aligned} c(i) &= Ai^2 + Bi^3 + Ci^4 \\ p_i &= p_1 + c(i) \times (p_f - p_1) \end{aligned} \tag{3}$$

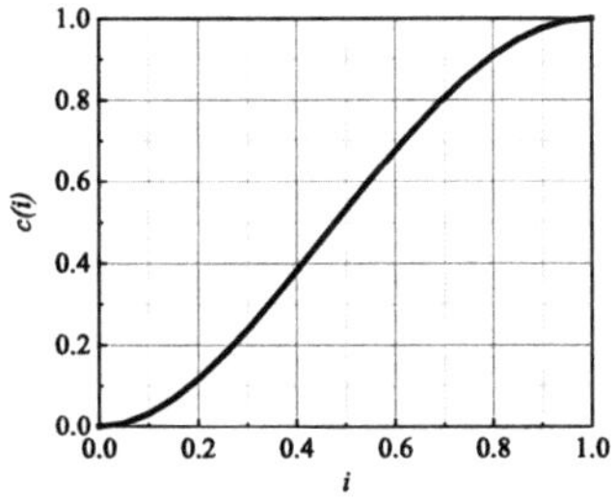

Fig. 1. The transition curve along the nozzle.

In the process of geometric optimization, the minimum curvature radius of the super ellipse on each cross-section along the nozzle will be calculated. Since the parameter n is obtained by the area formula, it is also necessary to calculate its change trend, that is, $k_{n_{i+1,i}} = n_{i+1}/n_i$. The optimization objective is to minimize the difference between the minimum curvature radius in the middle cross-section and the mean of these two values in the inlet and outlet cross-sections, and to minimize the standard deviation of $k_{n_{i+1,i}}$ along the nozzle. In other words, the parameter n has the most uniform transition.

The geometric optimization design of the nozzle is carried out based on the following optimization constraints: the horizontal axis length of each cross-section is greater than the vertical axis length; the area S_i of each cross-section calculated by the transition curve should be less than $4a_ib_i$, otherwise the super ellipse cannot be formed geometrically; these parameters a, b, S, n increase monotonically along the nozzle. The solution set is shown in Table 1. Finally, the parameters that can generate the nozzle with the highest thrust coefficient are selected as the basic values of subsequent aerodynamic optimization design.

3 Optimization Design of Non-camber Straight Nozzle

In this section, the aerodynamic calculation is carried out after straight nozzle sampling, and the influence of the parameters Ca, Cb and CS on the airflow density, static pressure, dynamic pressure and x-direction velocity at the nozzle outlet is investigated. In addition, a surrogate model is constructed based on the calculation results and design optimization is carried out.

The maximum prediction error and correlation coefficient (R^2) of the surrogate model are shown in Table 2. The values of R^2 are all greater than 0.9, indicating high accuracy

Table 1. The solution set of geometric optimization design.

Ca	*Cb*	*CS*	Thrust coefficient
−0.15056	2.39794	2.03713	0.91466
−2.86230	1.29252	0.80829	0.91197
−0.89883	2.17812	1.82298	0.91668
−2.38778	1.41913	2.03951	0.91516
−2.96493	1.23310	−2.38375	0.91298

of the surrogate model. The first-order linear responses of parameters are studied without considering the interaction between each parameter. It is shown by Figs. 2, 3 and 4 that the effects of various parameter changes on the airflow density, static pressure, dynamic pressure, and flow velocity in the *x* direction at the nozzle outlet.

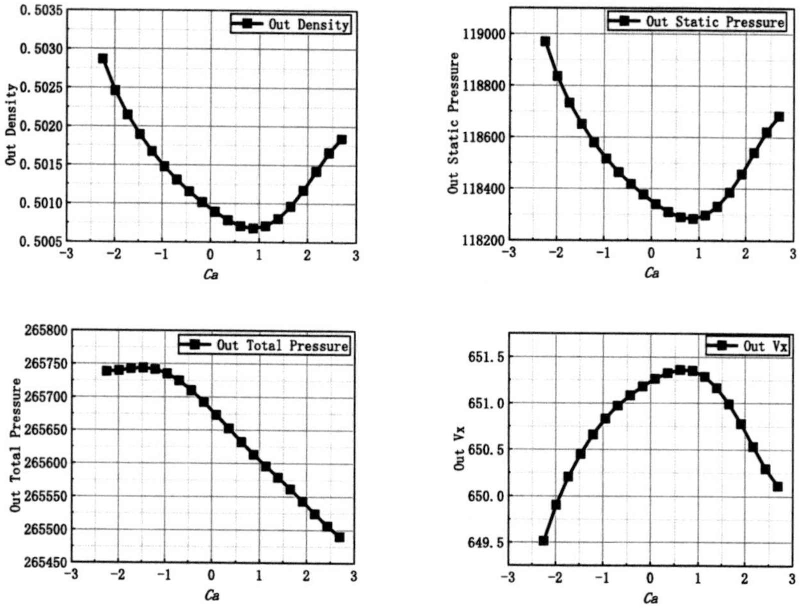

Fig. 2. First-order linear effect of design parameter *Ca* on response.

As shown in the figures, the outlet density and outlet static pressure will rise and then descend along *Ca* and *CS* increasing, while the outlet density will rise continuously to along *Cb* increasing. The increasement of each design variable will lead to the outlet total pressure decreasing. The *x*-direction velocity increases first and then decreases with the increase of *Ca* and *CS*, and decrease constantly with the increase of *Cb*. When the interaction of each parameter is not considered, the influence laws of each parameter on different responses are not the same, so it is difficult to analyze the optimization law of

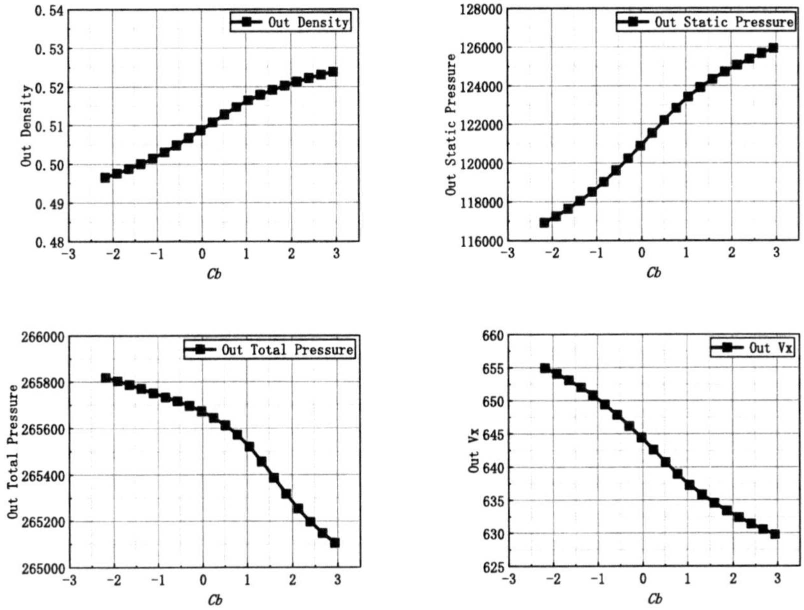

Fig. 3. First-order linear effect of design parameter *Cb* on response.

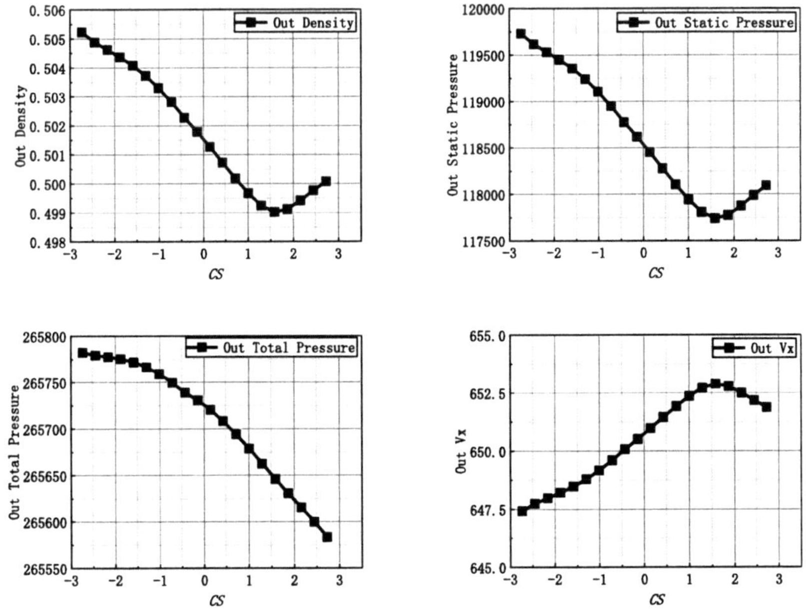

Fig. 4. First-order linear effect of design parameter *CS* on response

Table 2. The precision of surrogate model of straight nozzle.

Response	Maximum prediction error(%)	R^2
Outlet density	8.239	0.97195
Outlet static pressure	8.181	0.9729
Outlet total pressure	7.03	0.97557
Outlet *Vx*	8.055	0.97242

nozzle profile. Therefore, the interaction of each design variable should be investigated by second-order responses.

Figures 5, 6, 7 show the second-order response surfaces between the pairs of design variables and each response. Intercomparing these response surfaces, it can be seen that the interaction between *Ca* and *CS* is most significant, and there are obvious curved parts in all the four response surfaces. The interaction between *Cb* and *CS* is much weaker than that between *Ca* and *CS*. The interaction between *Ca* and *Cb* is weakest, and the second-order response surface is almost flat. It can be preliminarily inferred from the figure that the second-order interaction terms in response to outlet density, outlet total pressure, and outlet *Vx* are *Ca* × *CS*, *Cb* × *CS*, and *Ca* × *CS*, respectively.

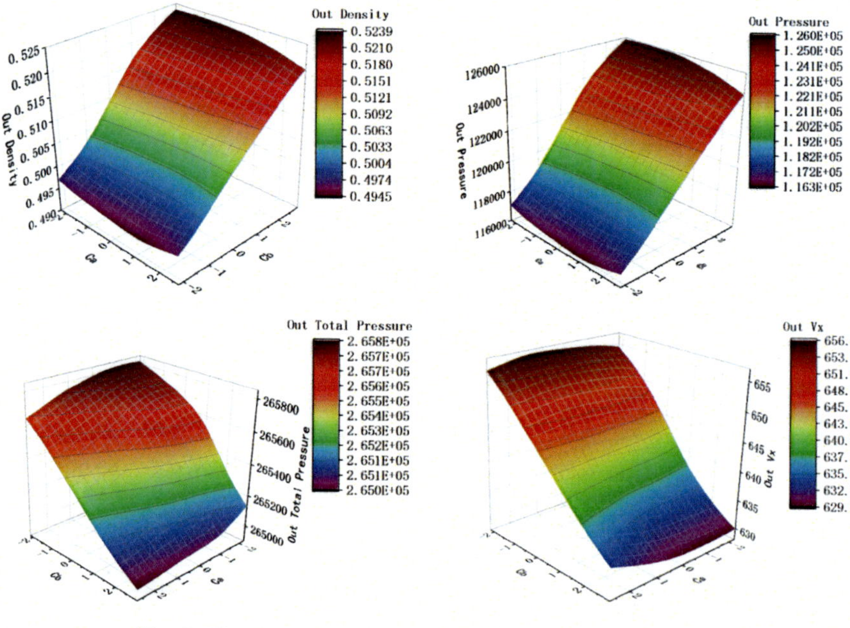

Fig. 5. The second-order response surfaces between *Ca* and *Cb*

A single-objective optimization for maximizing total pressure recovery coefficient and maximizing outlet thrust coefficient is carried out successively for a straight nozzle.

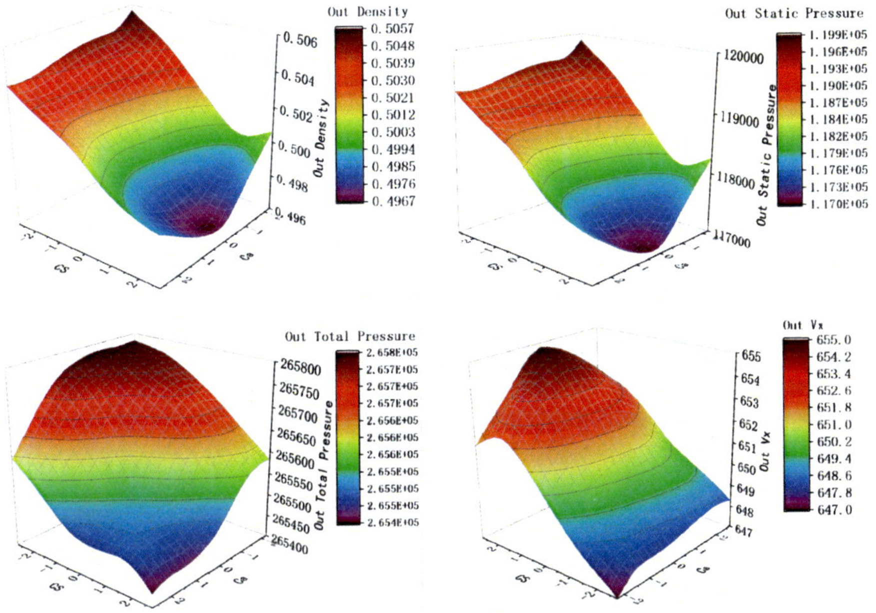

Fig. 6. The second-order response surfaces between *Ca* and *CS*

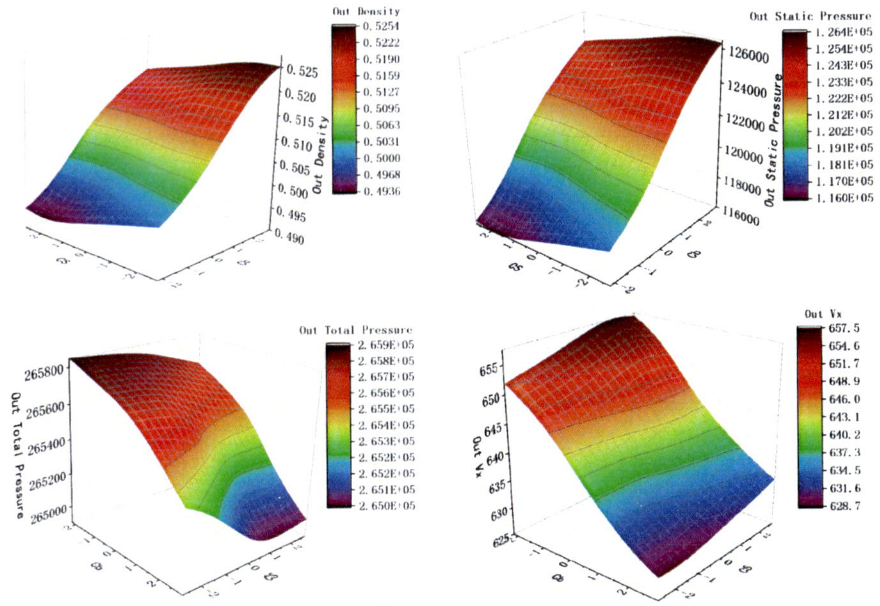

Fig. 7. The second-order response surfaces between *Cb* and *CS*

According to the sample, it is found that the outlet density with a larger thrust coefficient is larger, so the optimization objective of maximizing thrust coefficient is converted into maximizing outlet density, and then multi-objective optimization is performed. The solution set is shown in Table 3. When the optimization objective is to maximize the total pressure recovery coefficient, the obtained outlet total pressure is extremely high but the actual thrust decreases significantly. When the optimization objective is to maximize the thrust coefficient, the outlet total pressure does not change significantly compared with the previous optimization solution, but the thrust coefficient greatly increases, and the performance of nozzle is improved overall. Compared with the single-objective optimization results with the optimization objective of maximizing the thrust coefficient, the thrust obtained by the multi-objective optimization is not much different, and the total pressure recovery coefficient is slightly increased. But the improvement effect is not significant, therefore, the optimization solution with the optimization objective of maximizing the thrust coefficient is selected as the final optimization result. The optimized nozzle has both high thrust coefficient and high total pressure recovery coefficient, and the outlet thrust is up to 893*N*.

Table 3. The Optimization results of the non-camber straight nozzle.

Optimization objective	Design variables			Optimization results		
	Ca	*Cb*	*CS*	Pressure coefficient	Thrust coefficient	Thrust
Pressure coefficient	−1.82	−3	−3	0.98	0.91	878.29
Thrust coefficient	−1.15	3	−2.25	0.98	0.93	892.89
Multi-Objective	−1.99	1.29	−1.82	0.98	0.92	890.62

4 Optimization Design of Large-Camber Nozzle

The ridge of large-camber nozzle is constructed by conic curves, which expression is $y = k(x/L)^2$. The values of *Ca*, *Cb* and *CS* when the thrust coefficient is highest in the optimization of straight nozzle are selected to study the effect of nozzle camber change on the outlet thrust.

The precision of the surrogate model for the large-camber nozzle is shown in Table 4. R^2 is very close to 1, indicating very high accuracy of the surrogate model. Figure 8 shows the effect of camber change on *Fx* and *Fy* at nozzle outlet. It can be found that the larger the camber, the greater the *Fy*. Because the airflow direction will be twisted along the nozzle wall when the camber of nozzle changes greatly.

The optimization design of the large-camber nozzle is carried out. Firstly, vectoring angle is not fixed, and the optimization is performed with the objective to maximize thrust coefficient; then the optimization is carried out with vectoring angle as a constraint and maximizing thrust coefficient as optimization objective. The solution set is shown in

Table 4. The precision of surrogate model of large-camber flat outlet nozzle.

Response	Maximum prediction error (%)	R^2
Fx	2.47	0.99873
Fy	2.16	0.99904

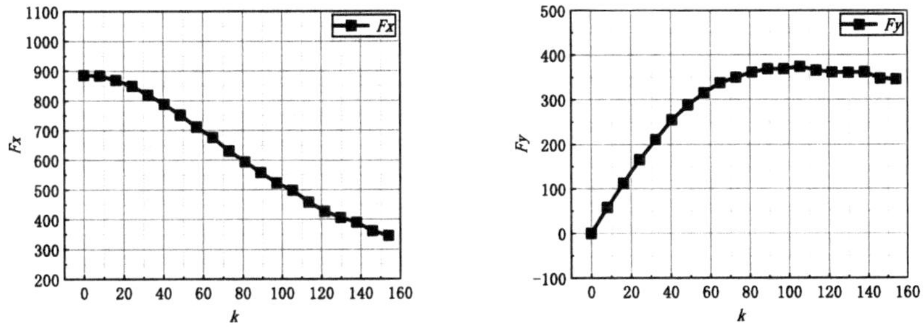

Fig. 8. The variation law of outlet thrust with the camber, *k*

Table 5. It can be found from the table that the larger the vectoring angle, the more curved the ridge, and the lower the thrust coefficient. When vectoring angle reaches 20°, thrust coefficient is still close to 0.9. Changing the camber can quickly adjust flow direction, which facilitate the rapid design of large-camber nozzle according to the requirements of vectoring angle.

Table 5. The optimization solution set for large-camber nozzle.

Vectoring angle (°)	Camber (mm)	*Fx*	*Fy*	Thrust coefficient
0	0	885.1613	0.0302	0.9188
10	23.7404	850.9576	161.6038	0.8990
20	34.9964	809.6082	223.7876	0.8719
30	79.7444	599.7578	357.7653	0.7249

5 Conclusion

In this paper, a rapid optimization method for large-camber nozzle based on geometric optimization is proposed. Firstly, the transition curve that controls the geometric parameters of the nozzle is optimized, so that the nozzle has a smooth transition shape. Based on geometric optimization solution, the optimization design of straight nozzle and large-camber nozzle is carried out with objectives of high thrust coefficient and high total pressure recovery coefficient. The main conclusions of this paper are as follows:

(1) Before the aerodynamic optimization, a smooth transition shape with high thrust coefficient above 0.9 is obtained through geometric optimization, which indicates the good thrust characteristic of nozzles.
(2) There are complex second-order interactions between the design variables of straight nozzle. The increasement of the coefficient *Cb* makes outlet density increase rapidly, resulting in greater thrust. The thrust coefficient of the optimized nozzle is up to 0.93.
(3) Changing the camber of large-camber nozzle can quickly adjust the thrust direction. The larger the vectoring angle, the more curved the ridge, the greater the thrust loss. The thrust coefficient of optimal large-camber nozzle based on straight nozzle is still close to 0.9 when vectoring angle is up to 20°.

References

1. Cao, Y.F.: Jet thrust vector control. Nanjing University of Aeronautics and Astronautics, Nanjing China, Master (2011)
2. Yang, F., Li, Z.H., Li, J.C.: Research progress on nozzle profile of solid rocket motor. Vacuum **57**(1), 40–47 (2020)
3. Chen, Y.L., Yu, K.K., X, J.L.: Design of thrust nozzle for scramjet with geometric dimension constraint. Acta Aeronaut. Astronaut. Sin. **42**(6), 124259 (2021)
4. Zhou, H.C., Han, H.J., Li, X.P.: A design method of complex variable section inlet. J. Aerosp. Power **24**(6), 1357–1363 (2009)
5. Wu, S.H., Liao, D.X., Chen, J.M., et al.: Optimization design of supersonic nozzle profile based on spline curve description. J. Northwest. Polytech. Univ. **036**(004), 785–791 (2018)
6. Yadav, S., Kumar, V., Pandey, K., et al.: Development of the constant rate of momentum change (CRMC) variable area nozzle. Mater. Today: Proc. **38**(2021), 2325–2331(2020)
7. Schomberg, K., Olsen, J., Neely, A., Doig, G.: Investigation of conjugate circular arcs in rocket nozzle contour design. Shock Waves **29**(3), 401–413 (2018). https://doi.org/10.1007/s00193-018-0834-0
8. Wang, Y.P., Xu, Y.H., Sun, H.J., et al.: Optimization design of nozzle profile of rocket motor based on response surface method. J. Aerosp. Power **37**(1), 214–224 (2022)

An Offline Inverse Model for Identification of Multiple Pollutant Sources in Aircraft Cabins

Yun Wei[1](✉) and Tengfei Zhang[2]

[1] Civil Aviation University of China, Tianjin 300300, China
y_wei@cauc.edu.cn
[2] Dalian University of Technology, Dalian 116024, China

Abstract. Gaseous pollutant source identification is important to protect passengers from infectious agents in aircraft cabins. Current identification method may either suffer from numerical instability or require extensive prior knowledge. This paper proposes an offline inverse model, which needs no iterations to identify the locations and release profiles of multiple gaseous pollutant sources. The model first uses airflow field in the domain to construct a transport probability matrix by Markov chain, which contains the probability of pollutant transports through faces in each cell. Then the transient concentration responses at the monitoring points of a unit impulse releases from each candidate source can quickly be predicted by the transport probability matrix, to construct another matrix describes the relation between the source releases and the monitored concentrations. Finally, Tikhonov regularization is used to inverse the matrix for identification. The above model was applied on a two-dimensional aircraft cabin with gaseous pollutant released from two passengers. Results showed that the proposed model could correctly reproduce the detailed travel path of pollutant transport, as well as identify the locations and release profiles of multiple pollutant sources. More than 90% decrease of the computing time as compared with the conventional CFD method showed that the proposed model had high efficiency.

Keywords: Markov chain · Tikhonov regularization · Offline inverse model

1 Introduction

The outbreak of Corona Virus Disease 2019 (COVID-19) in recent years increases the demand for gaseous pollutant source identification in aircraft cabins, to protect inside passengers from being infected. Pollutant source identification is an inverse problem that determines causal information from certain consequences, since outcomes such as continuing pollutant concentrations at the monitoring points, are always known as prior information to determine the pollutant sources.

Currently, a number of CFD based inverse models have concentrated on indoor gaseous pollutant source identification, which can be divided into forward methods and backward methods. Forward methods construct a number of candidate sources to find the one that makes the pollutant concentrations at typical points coincide best with that

Chinese Society of Aeronautics and Astronautics: CASTYSF 2022, LNEE 972, pp. 603–610, 2023.
https://doi.org/10.1007/978-981-19-7652-0_53

from the actual sensors, while backward methods directly reverse the flow to determine the source, by solving the N-S equations with a negative time step or a reversed airflow field. For the first strategy, Azimi and Daneshgar [1] proposed an inverse zonal method to optimize a defined objective function by iterations; Sohn et al. [2] applied Bayesian approach, which first builds numbers of scenarios contained hypothetical pollutant sources in the pre-event stage, then compares concentrations between the model simulations and actual sensor monitoring data by Bayesian statistics to determine the actual source in the second stage. Similar methods such as Neural Network [3], probability-based adjoint method [4] and that proposed by Cai et al. [5] were also used to identify the sources. Generally, forward methods have moderate computing efficiency, since all the candidate source locations and release strengths as well as the corresponding monitored concentrations are required to be known before the identification.

To further increase the source identification speed, some researchers turn to the backward method. For example, Zhang and Chen proposed quasi-reversibility (QR) method [6] and pseudo-reversibility (PR) method [7] for pollutant source identification. The former method reverses the time-marching direction of the pollutant transport equation, while the latter one reverses the airflow instead. Results showed that although PR method had less prior knowledge and was more stable, the accuracy was slightly worse than QR method. To balance the computing efficiency and stability, we raised an inverse model to identify multiple sources in indoor environments, which builds a matrix that describes the relation between the release rates of all the candidate sources and the monitored concentrations for inversion [8, 9]. However, to construct the matrix, a large number of pollutant transport equations need to be solved by CFD for preparation.

Several methods were also proposed to accelerate the solution speed for pollutant transport equations, such as FFD [10], which approximately replaces the diffusion and advection terms in N-S equations with other schemes that can decouple the pressure and velocity in CFD iterations to improve the computing efficiency. But the method still needs online iterations to obtain a convergent solution. Markov chain model has developed in indoor environment recently. The method constructs a transport probability matrix by determining the possibility of pollutant transports through faces in each cell, then the pollutant movement can be predicted by the matrix without iterations. Nicas [11] used the method to describe the gaseous pollutant dispersion in an indoor environment; Chen et al. [12] further applied the method on transient particle transport prediction. However, currently few studies have concentrated the method on inverse identification of the indoor pollutant sources. Although Zeng et al. [13] proposed a Markov-chain-based inverse modeling to identify the location and release rate of a pollutant source, the study only identified a single pollutant source, and the identification was applied in a multi-zone model, in which the detailed travel path of pollutant could not be obtained. This paper proposes a model that combined Markov chain and Tikhonov regularization, which is capable of reproducing the detailed travel path of pollutant transport, as well as identify the locations and release profiles of multiple pollutant sources in aircraft cabins. Solution speed and accuracy will be discussed in the paper for evaluation of the proposed model.

2 Methods

2.1 Construction of a Pollutant Transport Probability Matrix

The proposed model first uses Markov chain to construct a pollutant transport probability matrix for source identification. With a known airflow field, the probability of pollutant remains in the current cell can be obtained as [12]:

$$p_{i,i} = \exp\left(-\frac{\sum Q_{i,nbhd}}{V_i}\Delta t\right) \tag{1}$$

where $p_{i,i}$ is the probability of pollutant remains in the current ith cell, $\sum Q_{i,nbhd}$ is the sum of the airflow rate from the ith cell to the neighboring cells, which can be obtained from the airflow field in the domain, V_i is the cell volume, Δt is the time step. Equation (1) can be derived from a mass balance equation that describes variation of the pollutant amount in one cell in a certain time step must be equal to the sum of amount of pollutant that transports to other cells in this time step. Suppose that the probability of pollutant transports from one cell to another is proportional to the airflow rate through the corresponding adjacent face, then the probability of pollutant transports from the ith cell to the jth, denoted by $p_{i,j}$ in this study, can be solved as [12]:

$$p_{i,j} = \frac{Q_{i,j}}{\sum Q_{i,nbhd}}(1 - p_{i,i}) \tag{2}$$

where $Q_{i,j}$ means the airflow rate from the ith cell to the jth cell.

With the known $p_{i,i}$ and $p_{i,j}$, the transport probability matrix can be constructed, in which $p_{i,j}$ is the element in the ith line and jth column. Totally nn elements are contained, where n is the cell number in the domain used by CFD.

2.2 Identification of Multiple Pollutant Sources

The matrix P contains the probability of pollutant both remains in the current cell and transports to the adjacent cells, hence the pollutant amount field after k time steps M_k can easily be obtained by the matrix as:

$$M_k = M_0 P^k \tag{3}$$

where M_0 represents the initial pollutant amount field, which can be obtained by multiplying the initial pollutant concentration in each cell and the corresponding cell volume.

With Eq. (3), the concentration responses at the monitoring points of a unit impulse that releases from each candidate source can be predicted. For a candidate source, concentrations at a monitoring point that are contributed by this source can be written as:

$$\begin{bmatrix} c^j_{0,i} \\ c^j_{1,i} \\ \vdots \\ c^j_{k,i} \end{bmatrix} = \begin{bmatrix} c^j_{p,0,i} & 0 & \cdots & 0 \\ c^j_{p,1,i} & c^j_{p,0,i} & \ddots & \cdots \\ \vdots & \ddots & \ddots & 0 \\ c^j_{p,k,i} & \cdots & c^j_{p,1,i} & c^j_{p,0,i} \end{bmatrix} \begin{bmatrix} s_{0,i} \\ s_{1,i} \\ \vdots \\ s_{k,i} \end{bmatrix}, \text{ or } \mathbf{c}^j_i = \mathbf{A}^j_i \mathbf{s}_i, \tag{4}$$

where $c^j_{k,i}$ in the vector $\mathbf{c}^j_i$ on the left hand side of Eq. (4) represents the contributed concentrations at the jth monitoring point of the ith source at the kth time step, $c^j_{p,k,i}$ in the matrix $\mathbf{A}^j_i$ means the concentration responses at the jth monitoring point of a unit impulse that releases from the ith source at the kth time step, $s_{k,i}$ in the vector $\mathbf{s}_i$ means the release strength of the ith source at the kth time step. For all the candidate sources, the monitored concentrations at the jth point can be derived as:

$$\mathbf{c}^j = \sum_{i=1}^{m} \mathbf{c}^j_i = \sum_{i=1}^{m} \mathbf{A}^j_i \mathbf{s}_i \tag{5}$$

where $\mathbf{c}^j$ is the concentration responses at the jth point due to all the m candidate sources. To make all the matrices $\mathbf{A}^j_i$ construct a square matrix, concentrations of m monitoring points should be known, then Eq. (5) can be formulated into:

$$\mathbf{C}=\mathbf{AS} \tag{6}$$

where $\mathbf{C} = [\mathbf{c}^1, \cdots, \mathbf{c}^j, \cdots, \mathbf{c}^m]^{\mathrm{T}}$, which contains concentrations at m monitoring points, $\mathbf{S} = [\mathbf{s}_1, \cdots, \mathbf{s}_i, \cdots, \mathbf{s}_m]^{\mathrm{T}}$, which contains release strengths of m sources, $\mathbf{A}$ represents the whole square matrix on the right hand side of Eq. (5). Equation (6) illustrates that the monitoring points number should be equal to the candidate sources number for dimension balance.

To identify $\mathbf{S}$ in Eq. (6) with a known $\mathbf{C}$, $\mathbf{A}$ should be inversed. However, $\mathbf{A}$ is proved to be an ill-conditioned matrix for pollutant source identification. So, Tikhonov regularization is introduced to stabilize the ill-conditioned matrix inversion as [9]:

$$\mathbf{S} = \left(\mathbf{A}^{\mathrm{T}}\mathbf{A} + \lambda^2\mathbf{L}^{\mathrm{T}}\mathbf{L}\right)^{-1}\left(\mathbf{A}^{\mathrm{T}}\mathbf{C}\right) \tag{7}$$

where λ is the regularized parameter and L is the regularized matrix, both used to adjust the strength of regularization. Finally, locations and release strengths of the actual sources can be obtained by excluding the candidate sources that are determined to have approximate zeros of release strengths. With the monitored concentrations at several points and the airflow field as prior knowledge, no iterations are involved both in the transport probability matrix construction process and in the source identification process, hence the proposed model is an offline model.

3 Case Description

The proposed model was applied on half of a full-scale aircraft cabin for validation, as shown in Fig. 1a. The model has an inlet on the side wall near the ceiling, and an outlet near the floor, with three passengers seated inside. Three small circles above the passengers represent their nasal orifices, which are all supposed to be the candidate sources, denoted by S1, S2 and S3 from left. S1 and S2 are two actual sources, their release strengths are: S1 releases constantly with 100 ppmkg/(m^{2}s) for 2.5 s from the beginning, and S2 releases a sinusoidal wave for 2.5s with the peak value of 200 ppmkg/(m^{2}s) from the beginning. Both locations and release strengths of the actual sources are unknown before identification.

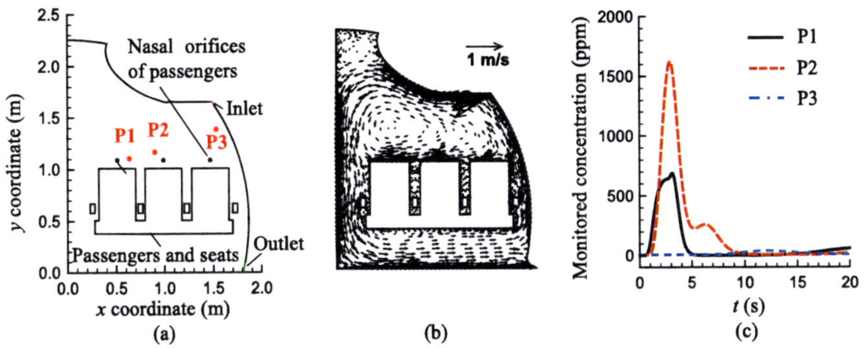

Fig. 1. Configuration of the cabin and prior knowledge for the proposed model: **a** the cabin; **b** the airflow field obtained by CFD; **c** the monitored concentrations at P1, P2 and P3.

Table 1. Boundary conditions for the aircraft cabin.

Item	Value (°C)	Item	Value
Ceiling temperature	22	Passenger surface temperature	30.3 °C
Side wall temperature	22	Air supply velocity	1.2 m/s
Luggage surface temperature	21	Air supply temperature	19.5 °C
Floor temperature	23	Turbulence intensity	10%

Prior knowledge for identification by the proposed model is the airflow field in the cabin and concentrations at the monitoring points whose number is equal to the candidate sources number, both were obtained by CFD in this study. Totally 5596 quadrilateral grids were generated in the cabin model, which has been verified by the grid independence. Boundary conditions are listed in Table 1, consistent with our previous study [14]. The turbulent flows were resolved with the renormalization group (RNG) *k-e* model with enhanced wall treatment, since y + for almost all the cells is less than 5. To solve the N-S equations, SIMPLE was selected as the pressure-velocity coupling algorithm, and PRESTO was selected for discretization of the pressure, Power Law for the momentum, while second-order upwind scheme was used for other variables. In addition, a time step size of 0.01 s was set to capture the transient features of pollutant transport with appropriate convergence criterion.

Figure 1b shows the airflow field obtained. The air supplies from the inlet, flows along the ceiling with the effect of thermal plume on passengers, then goes into the central seating region, and finally flows to the outlet. Monitored concentrations at three typical points are shown in Fig. 1c, and the monitored positions are marked in Fig. 1a with P1, P2 and P3, respectively.

4 Results

The transport probability matrix constructed in this case contains 5596 × 5596 elements, and the sum of elements in each line is verified to be 1, since the pollutant in each cell could only transport to other cells or remain in the current cell. With the obtained matrix, concentration responses at the monitoring points of a unit impulse that releases from each candidate source can be determined. Figure 2a shows the concentration responses at P2 of a unit impulse releases from S2 obtained by the proposed model and by CFD. The two profiles have quite similar trends, although the profile obtained by the proposed model seems to be a little behind that by CFD. The difference probably since that diffusion is not considered in construction of the transport probability matrix for the proposed model. Concentration fields at 5 s of the impulse releases from S2 determined by the two methods are further shown in Fig. 2b and c. Similar distributions can be seen, which shows that the high accuracy of the proposed model on prediction of the concentration responses of a unit impulse releases from the candidate sources.

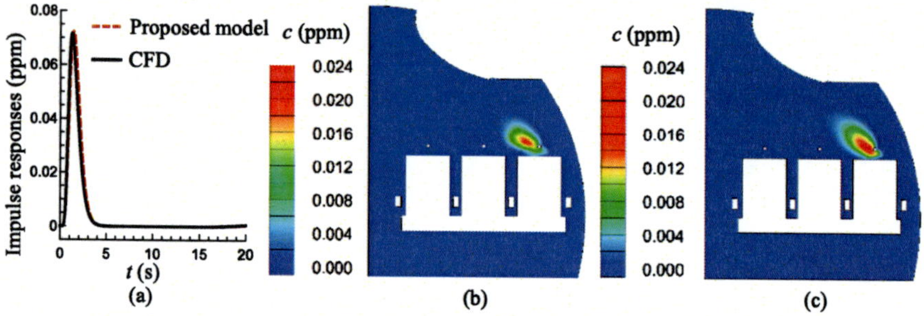

Fig. 2. Concentration responses of a unit impulse releases from S2: **a** concentration profiles at P2 obtained by the proposed model and by CFD; **b** the concentration field obtained by the proposed model at 5 s; **c** the concentration field obtained by CFD at 5 s.

Concentration responses for 20 s at 3 monitoring points of a unit impulse that releases from each candidate source were used to construct the matrix **A** in Eq. (6), which contains 3◊2000 lines and 3◊2000 columns for this case, since the time step size was set as 0.01 s. Then Tikhonov regularization in Eq. (7) was used for the matrix inversion, in which l is set as 80 to relieve the ill-condition of the matrix **A**, who has a huge condition number of 8.5210^{29}. Finally the release profiles of all the candidate sources can be determined, as shown in Fig. 3. S1 and S2 have obvious strengths, while S3 is almost zero, illustrates that S1 and S2 are two sources identified by our model. Source locations identified are consistent with the actual ones. Moreover, only slight differences can be observed when comparing the release profiles with the actual ones, shows that the proposed model could correctly identify the release strengths of multiple sources.

Computing time for the proposed model depends on the cells generated in the domain and the memory of the computer. For the validation case with 5596 cells worked on a computer with 3.7 GHz of CPU and 32 GB of memory, only less than one minute is required to construct the transport probability matrix as well as identify the multiple

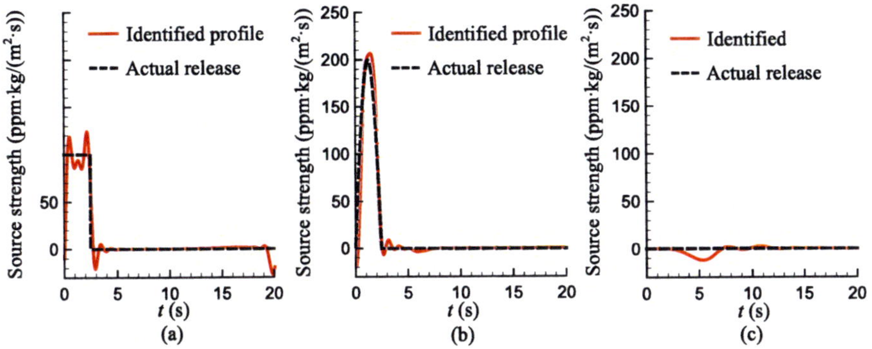

Fig. 3. Source strengths determined by the proposed model and its comparison with the actual release profiles: **a** S1; **b** S2; **c** S3.

sources. However, a few hours are required for CFD only to obtain the concentration responses of a unit impulse releases from each source, let alone the multiple sources identification. More than 90% of the computing time can be saved, which shows the high efficiency of the proposed model.

5 Conclusions

This study proposed an offline method combines Markov chain and Tikhonov regularization to rapidly and correctly identify multiple sources in aircraft cabins. By applying the model on an aircraft cabin, it is proved that the proposed model is capable of identifying the locations and release strengths of multiple sources, with the airflow field and concentrations at several monitored positions as prior knowledge. In addition, more than 90% of the computing time can be saved by the proposed model in comparison with the conventional CFD method, which shows its high efficiency.

Acknowledgements. This work was supported by the Scientific Research Foundation for Civil Aviation University of China (Grant No. 2020KYQD62).

References

1. Azimi, A., Daneshgar, E.: Indoor contaminant source identification by inverse zonal method: levenberg-Marquardt and conjugate gradient methods. Adv. Build. Energy Res. **12**(2), 250–273 (2018)
2. Sohn, M.D., Reynolds, P., Singh, N., et al.: Rapidly locating and characterizing pollutant releases in buildings. J. Air Waste Manag. Assoc. **52**(12), 1422–1432 (2002)
3. Vukovic, V., Tabares-Velasco, P.C., Srebric, J.: Real-time identification of indoor pollutant source positions based on neural network locator of contaminant sources and optimized sensor networks. J. Air Waste Manag. Assoc. **60**(9), 1034–1048 (2010)
4. Liu, X., Zhai, Z.: Prompt tracking of indoor airborne contaminant source location with probability-based inverse multi-zone modeling. Build. Environ. **44**(6), 1135–1143 (2009)

5. Cai, H., Li, X., Kong, L., et al.: Rapid identification of single constant contaminant source by considering characteristics of real sensors. J. Central South Univ. **19**, 593–599 (2012)
6. Zhang, T., Chen, Q.: Identification of contaminant sources in enclosed environments by inverse CFD modeling. Indoor Air **17**(3), 167–177 (2007)
7. Zhang, T., Chen, Q.: Identification of contaminant sources in enclosed spaces by a single sensor. Indoor Air **17**(6), 439–449 (2007)
8. Zhang, T., Yin, S., Wang, S.: An inverse method based on CFD to quantify the temporal release rate of a continuously released pollutant source. Atmos. Environ. **77**, 62–77 (2013)
9. Wei, Y., Zhou, H., Zhang, T., et al.: Inverse identification of multiple temporal sources releasing the same tracer gaseous pollutant. Build. Environ. **118**, 184–195 (2017)
10. Zuo, W., Chen, Q.: Real-time or faster-than-real-time simulation of airflow in buildings. Indoor Air **19**(1), 33–44 (2009)
11. Nicas, M.: Modeling turbulent diffusion and advection of indoor air contaminants by Markov chains. AIHAJ-Am. Ind. Hyg. Assoc. **62**(2), 149–158 (2001)
12. Chen, C., Liu, W., Lin, C.H., et al.: A Markov chain model for predicting transient particle transport in enclosed environments. Build. Environ. **90**, 30–36 (2015)
13. Zeng, L., Gao, J., Lv, L., et al.: Localization and characterization of intermittent pollutant source in buildings with ventilation systems: development and validation of an inverse model. Build. Simul. **14**(3), 841–855 (2021)
14. Wei, Y., Zhang, T., Wang, S., et al.: Prompt design of the air-supply opening size for a commercial airplane based on the proper orthogonal decomposition of flows. Build. Environ. **96**, 131–141 (2016)

Research on Data System and Platform Architecture of Complex Aeronautical Equipment Development

Li Rong-Qiang[1,3], Niu Shuang-Yue[2](✉), Guo Ke-Zhi[2], Long Bin[2], Lian Xiao-Feng[1], and Xu He-Yong[1]

[1] Northwestern Polytechnical University, Xi'an 710072, People's Republic of China
[2] AVIC Chengdu Aircraft Design & Research Institute, Chengdu 610041, People's Republic of China
937806396@qq.com
[3] AVIC (Chengdu) UAS Co Ltd, Chengdu 611731, People's Republic of China

Abstract. equipment is a typical representation of complex equipment, which is characterized by long development cycle, long service life, long industrial chain and many supporting contractors. In order to improve the internal and external cooperation and efficiency, the world's advanced aviation enterprises mostly adopt the single bill of materials (BOM), dual BOM and Data Lake configuration management mode; but there are still some problems such as lack of standard definition, one-way data transmission and difficulty in ensuring consistency. In order to solve the above problems, this paper proposes a multi-view xBOM architecture based on dual BOM infrastructure. On the basis of this multi-view xBOM architecture, a digital collaborative platform is established to implement standardized xBOM construction management and dynamic multi-view application. At present, the platform has been put into use in the development of a certain type of aircraft and has achieved the preliminary improvement of both quality and efficiency.

Keywords: Product life cycle management · xBOM · Architecture · Multi-view

Nomenclature

BOM	Bill of Materials
xBOM	Multiple Bill of Materials
EBOM	Engineering Bill of Materials
MBOM	Manufacturing Bill of Materials
ECI	Engineering Configuration Item
MCI	Manufacturing Configuration Item
BBOM	Build Bill of Materials
SBOM	Support Bill of Materials
CI	Configuration Item
DSI	Design Solution Instance
PBS	Product Breakdown Structure

Chinese Society of Aeronautics and Astronautics: CASTYSF 2022, LNEE 972, pp. 611–622, 2023.
https://doi.org/10.1007/978-981-19-7652-0_54

1 Introduction

BOM is the core data of manufacturing enterprise, which runs through the whole process of design, manufacturing and operation support. In the narrow sense, it refers to the list of parts and its structure required by the product. In the broad sense, it contains the collection of all the data information associated with it. It can effectively organize the data information through the product breakdown structure and provide data support for different business teams. Multi-view BOM (hereinafter referred to as xBOM) constructs and organizes the product structure and data based on the data management process of the whole product life cycle and different business perspectives, thus forming a BOM view oriented to different business teams (such as design, manufacturing, service), which meets the requirements of different users for the management, viewing and use of product data and information [1–14].

At present, the following problems exist in the whole life cycle data management of aircraft: (1) The requirement data management is difficult to verify; (2) Lack of view-based management of data, confused validity management and high management cost; (3) Lack of effective management of physical data, test data and support data, and it is difficult to trace the status of an individual aircraft; (4) It is difficult to accurately assess the impact scope of any change, and it is difficult to close the change process and thus leads to repeated changes. A unified data center and integrated data management system for aircraft product life cycle with xBOM as the core is constructed to track the configuration throughout the aircraft product life cycle, ensure the data uniqueness for a type of aircraft, and strengthen the fine management of aircraft physical status. It has caught more and more attention from the world advanced aviation enterprises or organizations and serves as an important direction for strategic development.

2 Analysis of Three BOM Management Modes

After years of research and practice, foreign aviation enterprises mainly adopt two management modes for the process of EBOM and MBOM, which embodies different configuration management ideas [15]. One is single BOM configuration management mode led by Airbus and Embraer; and the other is dual BOM mode represented by Boeing and Rolls-Royce. Single and dual BOM management modes are shown in Fig. 1 [16].

The essence of single BOM is that ECI and MCI are identical. Its advantages are good data consistency and traceability; High efficiency of communication and coordination can be achieved between designers, manufacturers and suppliers based on unified objects. The disadvantage is the tight coupling between design and manufacturing. Any change from the manufacturing organization at the manufacturing end needs to be changed from the design end, leading to mutual influence and high cost. To meet the requirements of single BOM, it needs to be constructed at the beginning of the project. For example, the development management mode of Airbus A380 model defines the unified configuration items between design and manufacturing. On the one hand, full communication is conducted between design and manufacturing in the early stage, the common reference is established for the division of CI items, and the smallest design unit is also the smallest manufacturing unit; On the other hand, when the design initiates

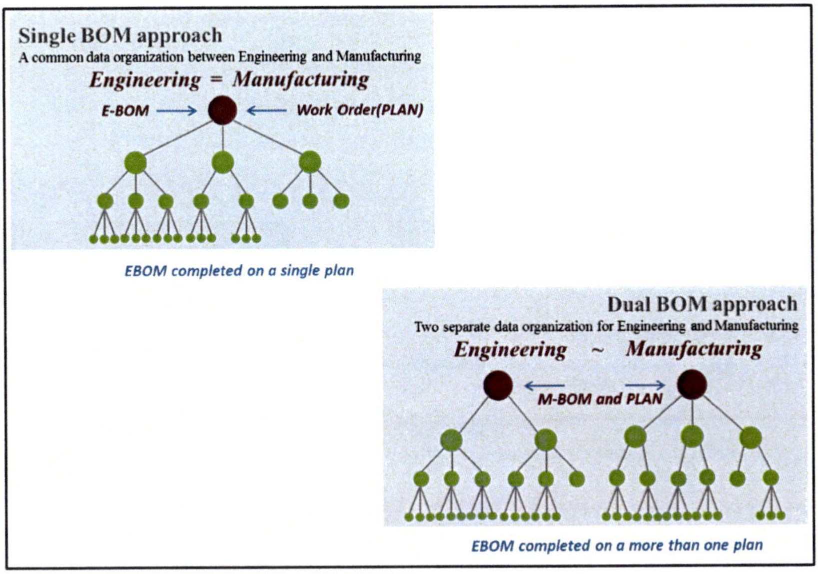

Fig. 1. Single and Dual BOM management modes

a CI change and produces a new solution, the manufacturing end simply evaluates how the internal components of the new solution are installed.

The essence of dual BOM is that ECI and MCI are independent of each other, and they establish two-way association through mapping. Its advantages are the decoupling of design and manufacturing, high flexibility; the material installation relationship can be flexibly adjusted according to the change of production organization, which will not affect the design. The disadvantages are the difficulty in achieving data consistency and traceability. Any reconfiguration of BOM needs a lot of computation to ensure consistency. In addition, the integration cost and PLM customization requirements are high as the conformity between EBOM and MBOM requires continuous verification [17, 18]. For example, when Boeing develops the B787 model, it performs configuration management based on business perspectives of design and manufacturing, and designs and manufactures with separate configuration management modes. On the one hand, there is no one-to-one correspondence between the designed minimum configuration unit defined by the design and the minimum assembly unit defined by the manufacturing; on the other hand, the manufacturing assembly unit needs to be constructed by obtaining and reorganizing parts from different design configurations [19].

The advantages and disadvantages of single and dual BOM management modes are shown in Fig. 2 [16].

Through the discussion and comparison, the two modes are analyzed deeply and the future management mode Data Lake is formed gradually. The multi-view management mode concept based on Data Lake is shown in Fig. 3 [16]. All the information related to the product life cycle is stored in a single Data Lake. The required views are generated in real time throughout the product life cycle, and multi-view tracing integrity is maintained

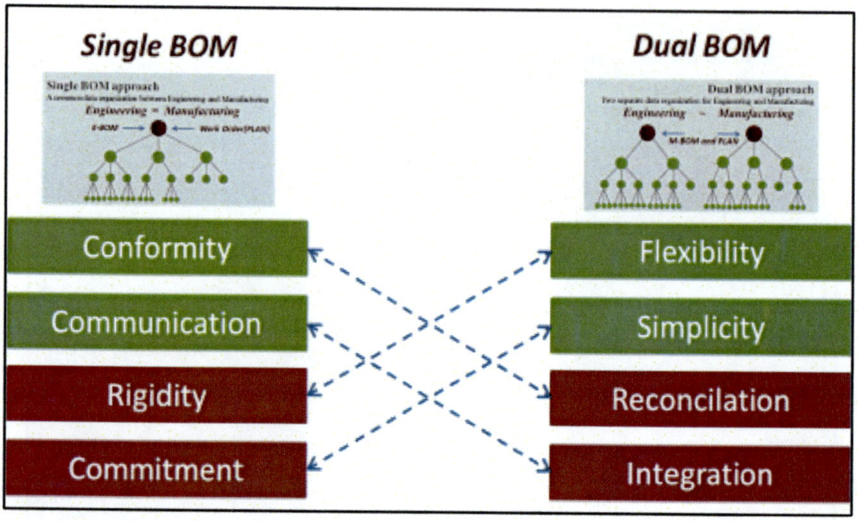

Fig. 2. Comparative analysis of advantages and disadvantages of single BOM and dual BOM

in the data pool to ensure the correctness of the views. Data Lake is still in the stage of basic theory and concept exploration at present, and the maturity is far from the actual model engineering application.

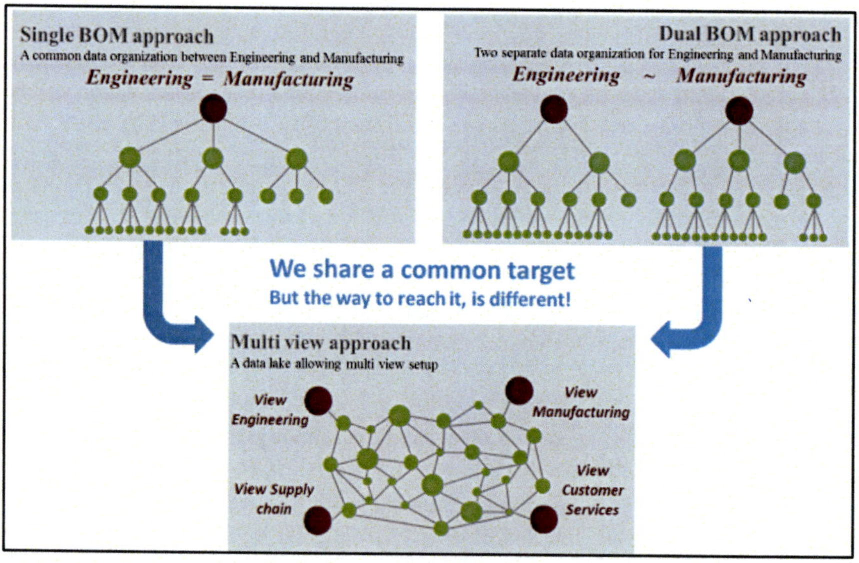

Fig. 3. Concept of multi-view management mode based on data lake.

Although the above three management modes are different in data organization, their core objectives are to improve the internal and external cooperation and efficiency. BOM architecture model definition and configuration consistency management of the BOM life cycle are the keys. This paper studies the xBOM architecture, which is oriented to the application of multiple view in aircraft design; it aims to improve the data standardization, the efficiency of cross-specialty design cooperation and design-manufacturing collaboration, so as to enhance the efficiency and quality of product development.

3 xBOM System Composition and Architecture

3.1 Composition of Full Life Cycle xBOM System

The characteristics of aeronautical equipment development are the continuous improvement of the capability in the life cycle of the aircraft, the continuous optimization of the design and manufacturing, and various enhancement suggestions due to failure and problems found after the aircraft enters service. Most of these improvements, optimizations, and troubleshooting solutions require engineering changes for implementation, resulting in ongoing engineering changes. In addition, the primary aircraft design organization is generally separated from the chief manufacturing organization in China.

Considering the separation of the manufacturing from the design and the frequent engineering changes, it is preferable to adopt a flexible, streamlined and mature dual-BOM infrastructure; that is, design-oriented EBOM and manufacturing-oriented MBOM adopt independent data organization infrastructure, and ECI and MCI are defined independently. At the same time, in order to reduce the complexity of EBOM-to-MBOM mapping, the following basic principles are introduced to solve the problems of insufficient EBOM-to-MBOM coordination, low efficiency of MBOM reconstruction and poor traceability to EBOM: (1) Uniformly define the components and sections in the top-level structure, and try to eliminate cross-section CI items; (2) The definition of CI layer shall be unified to the greatest extent, and the division and reconstruction from ECI to MCI shall be eliminated as far as possible; (3) Maximize uniform definition of part level below CI layer and eliminate combined parts for process as far as possible; (4) Establish the maturity definition and management rules of digital prototype, and standardize the parallel joint design process between manufacturing and design.

The xBOM system for the product life cycle consists of four BOMs [20], i.e. EBOM, MBOM, BBOM and SBOM, as shown in Fig. 4. Where EBOM refers to the bill of materials (including hardware, software or their combinations) formed during the design phase and reflecting the breakdown of the product. MBOM refers to the bill of materials (including tooling, rack, tools, equipment, raw materials/auxiliary materials and other data) organized according to the aircraft manufacturing process on the basis of EBOM. BBOM refers to the final physical manufacturing bill of materials (including plan and work order, production and manufacturing, quality inspection and other data) formed according to a single aircraft on the basis of MBOM. SBOM refers to the physical bill of materials (including delivery, operation, maintenance, repair and other data) for the whole process of operation support formed according to a single aircraft on the basis of EBOM and MBOM [21, 22].

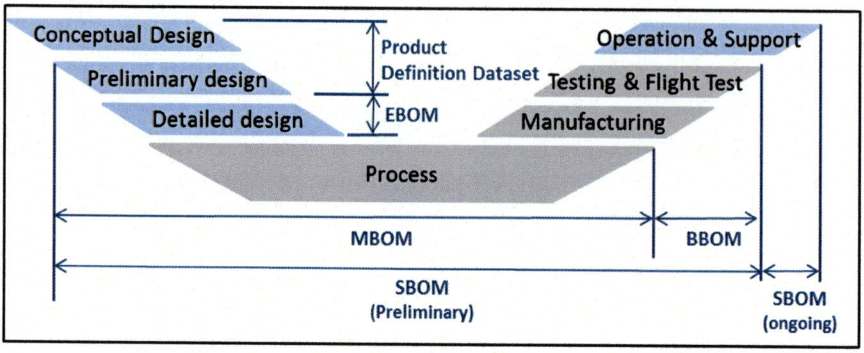

Fig. 4. xBOM Composition.

3.2 Multi-view Architecture Based on Dual BOM Infrastructure

The xBOM architecture uses a multi-view architecture based on a dual-BOM infrastructure, which can be divided into two parts: function-oriented and assembly-oriented, as shown in Fig. 5. Among them, EBOM and SBOM are in the multi-aircraft status based on the configuration; they are divided and organized by the aircraft and autonomous support from the function of a type of aircraft to form a three-layer structure. The airframe and related systems are defined as the top layer, which is further divided into the configuration layer based on the principles of CI division, including airframe CI, system installation structure CI, product CI, software CI, pipeline CI, wiring harness CI, equipment/tools and publications [23–25]. The above CIs continue to be divided into various DSIs based on the Manufacturing Attribute Set and Operation & Support Attribute Set, including Hardware DSI, Software DSI, Electrical Design DSI, Technical Publications DSI to form the bottom layer. In the other part, MBOM and BBOM for an individual aircraft are divided into five layers by considering the assembly. The aircraft type is defined as the type layer, and the manufacturing and the design jointly define 37 sections to form the section layer according to the aircraft structure. Any section can be divided into n sub-sections to form the sbu-section layer, and each Sub-section continues to be divided to form the configuration layer corresponding to the CI layer in the EBOM. The fifth layer is the bottom layer, and its part data is kept consistent with the EBOM bottom layer data.

The above construction process forms the association through data mapping, unified object and attribute definition; the attribute maintenance responsibility entity is clarified to achieve xBOM single data source management, design/manufacturing/support parallel collaboration, and two-way data tracing; it ensures data consistency, defines the work division of the manufacturing and the design, reduces the design and manufacturing coupling and improves the change efficiency and flexibility. The association of each BOM and multi-view application are shown in Fig. 6.

As shown in the figure, EBOM is the basis of the other three BOMs, including system view, product view, bay view, software view and file view. It provides design data for MBOM as input, guarantees BBOM manufacturing conformity, guarantees support for SBOM and status tracing. The MBOM covers production preparation view, plan view,

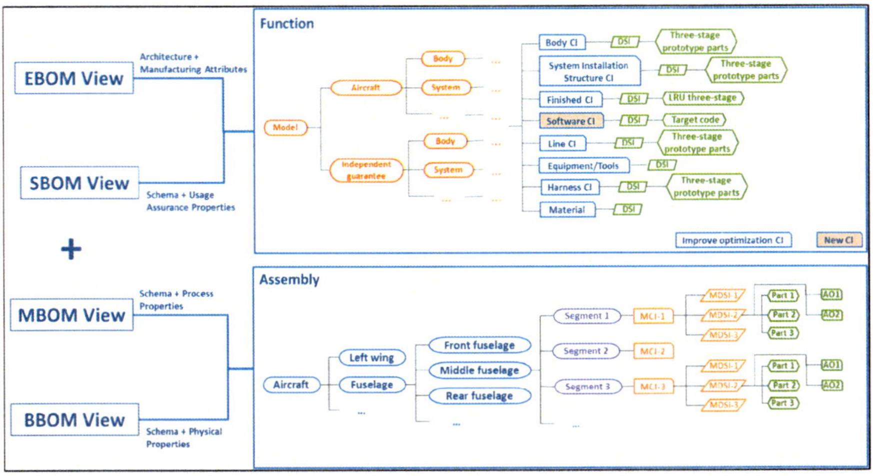

Fig. 5. xBOM Data Definition and View Construction Based on Dual BOM Infrastructure.

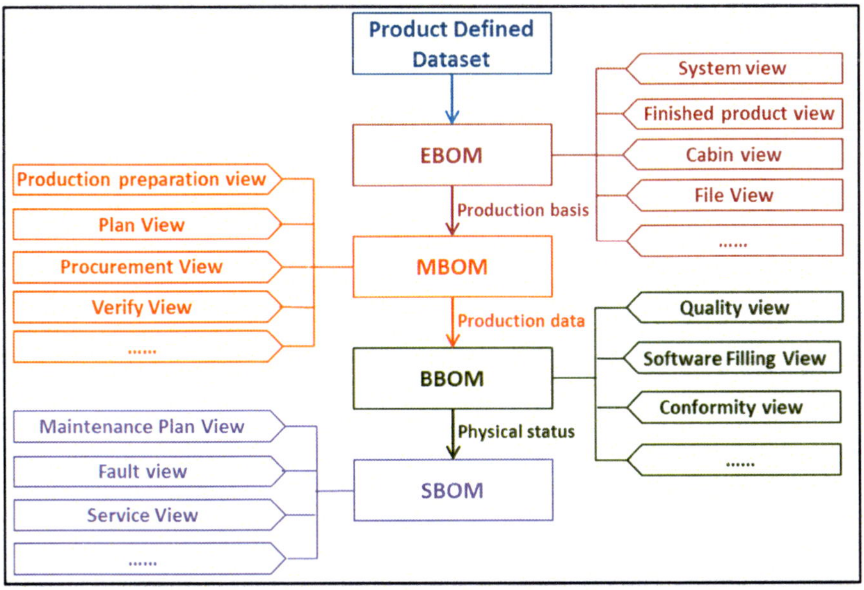

Fig. 6. Association of each BOM and multi-view application.

purchase view and inspection view; it transfers engineering status to the EBOM and provides process data for the BBOM. BBOM generates quality view, software encapsulation view and conformity view, to provide physical status for SBOM. SBOM works together with BBOM to document the physical status through failure view, repair view and maintenance plan view.

4 xBOM Management Platform Architecture

The multi-view xBOM architecture based on the dual BOM infrastructure is ultimately implemented on the platform, and its overall architecture is shown in Fig. 7. Considering the basic requirements of single product data source and the separation of design and manufacturing, the platform adopts the general architecture of logical unity and physical distribution, which is divided into physical layer, logical layer and multi-view layer.

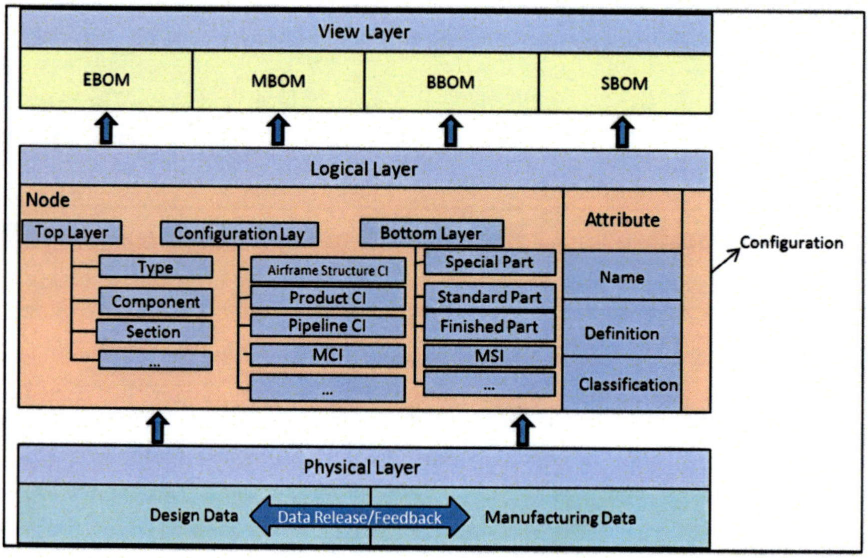

Fig. 7. Overall platform architecture.

The physical layer supports the unified storage of aircraft life cycle data, and achieves synchronization by the release and feedback of data between the manufacturing and the design. The logical layer supports the unified definition of each BOM node and its attributes. The node is divided into top layer, configuration layer and bottom layer. The attributes include special part, standard part, finished part and MSI. The multi-view layer supports dynamic multi-view applications of EBOM, MBOM, BBOM and SBOM.

In general, the scheme of the xBOM architecture is a multi-view architecture based on dual-BOM infrastructure. It introduces the joint definition rules of the design and manufacturing, and adopts the data storage platform architecture with logical unity and physical distribution. It is generally equivalent to dual-BOM +, that is, a step forward from dual-BOM mode to Data Lake mode.

5 Application and Discussion

Based on the research on the composition and architecture of xBOM system and the architecture of management platform, a digital collaborative platform supporting xBOM system is established to achieve standardized xBOM construction management and dynamic multi-view application. It has been put into use in the development of a certain type of aircraft. EBOM dataset management based on unified PBS enables unified and integrated management of aircraft, products, software, baseline files, and non-baseline files. The basic view of EBOM is shown in Fig. 8.

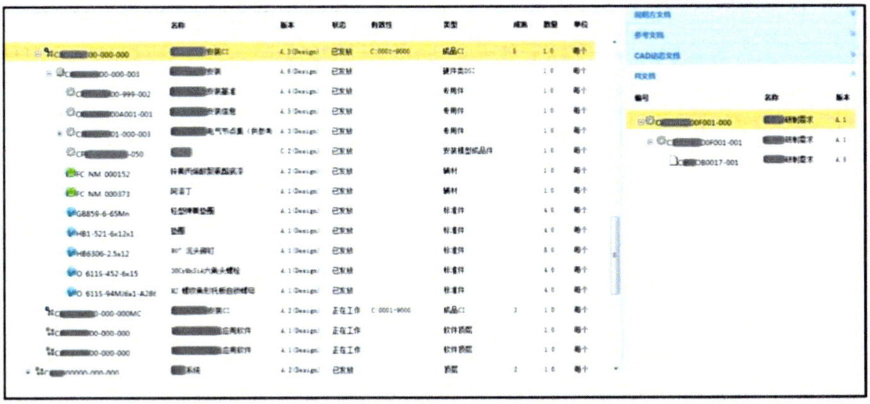

Fig. 8. Basic view of EBOM based on unified PBS.

Dynamic multi-view application is realized based on EBOM basic view, including bay view, section view and file view. An example of a bay sub-view application is shown in Fig. 9.

Ten history change tasks were randomly selected and their feedback days were counted. The results were verified and compared by the established digital collaborative platform, as shown in Fig. 10. It can be seen from the figure that the digital collaborative platform proposed in this paper can greatly reduce the change feedback time, improve the collaboration efficiency, and reduce the change task execution feedback time by 60.1% on average. Therefore, real-time and automatic generation of bay views based on single design data source can support the applications in different development phases and greatly improve the efficiency of digital prototype collaborative design.

6 Conclusion

THere is rare open literature on the research of the application of the dual BOM management mode at present. In this paper, the concept is further studied and detailed in a problem-oriented manner. Based on the forward design process, the overall scheme planning is carried out to closely combine the foreign advanced management ideas with the domestic characteristics of the manufacturing and design, to achieve the engineering practice application innovation. Through the actual engineering application, the aircraft development quality and efficiency are improved. In terms of data integrity, complete dataset management based on xBOM is achieved. In terms of data quality, the data standardization and consistency are significantly improved; In terms of collaboration efficiency, the synergistic research efficiency among design specialties and that between

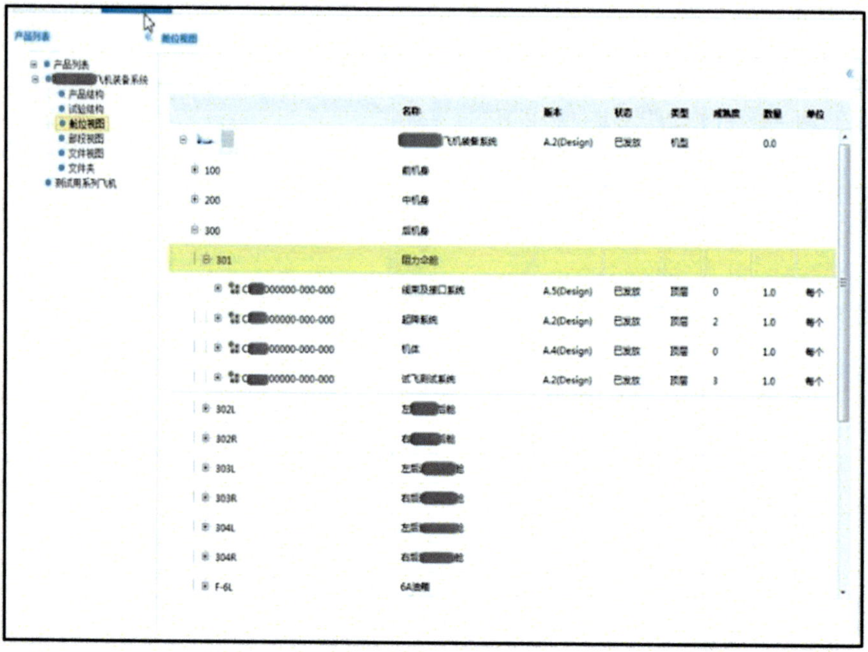

Fig. 9. Example of product structure filtered by bay view.

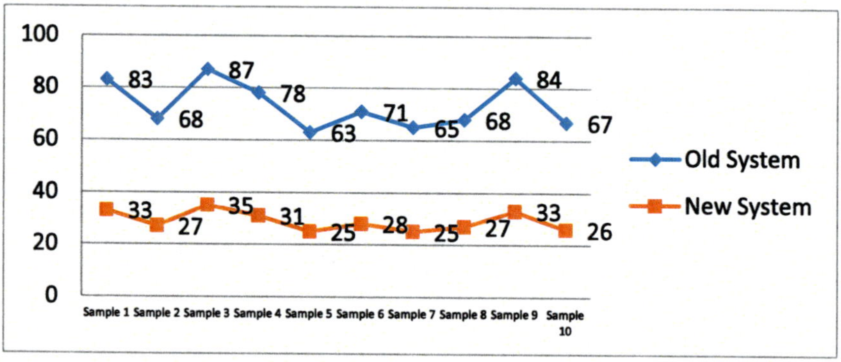

Fig. 10. Comparison of change task feedback days.

the manufacturing and the design are improved. The research also lays a good foundation for the future research and application of aircraft digital clues and digital twins.

References

1. Yan, J.J., Yang, Z.X., Gu, J.N.: Research on structure mapping method of BOM multi-view [J]. Modul. Mach. Tool Autom. Manuf. Tech. **2**, 22–25 (2020). (in Chinese)
2. Zhang, H., Wu, J.J., Fei, T.M., Zhang, S.W.: Research on assembly BOM conversion method for information integration[J]. J. Jiangsu Univ. Sci. Technol. (Nat. Sci. Ed.) **34**(01), 41–48 (2020)
3. Wang, J.T., Si, S.B.: The integrated modeling method for the maintenance BOM of complex equipments [J]. Mach. Des. Manuf. Eng. **40**(23), 50–54 (2011). (in Chinese)
4. Zhang, H., Shi, S.Y., Shi, Q.: PLM-oriented BOM data management and practice [J]. Manuf. Autom. **40**(8), 30–34 (2018). (in Chinese)
5. Geng, J.G., Cai, C.Z., Ni, Y.Q., Wen, X.L.: Design and implementation of multi-view dynamic management framework for product structure [J]. Mod. Manuf. Eng. **11**, 45–50 (2015). (in Chinese)
6. Kang, Y., Yan, H., Wang, Y.: Research and practice of multiple BOM management based on product life cycle [J]. China Manag. Inform. **22**(17), 95–98 (2019). (in Chinese)
7. Yuan, S.D., Wang, Y.: Function construction and application practice of manufacturing BOM in complex equipment industry under Domestic PLM platform[J]. Intell. Manuf. **06**, 93–99 (2021). (in Chinese)
8. INCOSE Systems Engineering Handbook, 4th ed. A guide for system life cycle processes and activities (2017)
9. Hu, J.: Study on BOM management and application for vehicle based on the SSPD [J]. Shanghai Auto **11**, 48–53 (2012). (in Chinese)
10. He, L.N., Ming, X.G., Kong, F.B., et al.: Research on airplane BOM management based on single source of product data [J]. Mech. Mater. **232**, 184–188 (2012)
11. Fan, Q.H., Zhou, R.Y., Wei, Q., Liu, B.X.: BOM multi-view mapping based on a single data source under PLM [C]. Int. Conf. Ind. Control. Electron. Eng. **45**(1):1705–1708 (2012)
12. Chen, B., Ji, J., Li, S., et al.: Single source of product data echnology in aero engine design and manufacture collaboration [C]. In: 2009 Second International Conference on Intelligent Computation Technology and Automation, vol. 819, pp. 430–433
13. Xie, X.C., Chun-Ming, Y.E.: Construction and maintenance of engineering machinery industry BOM [J]. Value Eng. **6**, 16 (2013)
14. Li, Y.: Aircraft Digital Manufacturing Technology and Application [M]. Aviation Industry Press, 2018. (in Chinese)
15. Lu, W.B.: Technology and Application of Product Structure Management Based on Lifecycle Data Relation Model [D]. Zheng Jiang University, 2015 (in Chinese)
16. Aerospace & Defense PLM Action Group: Multiple view bill of materials (BOM). (2019)
17. Yin, K.S.: Application of multidisciplinary BOM multi-view in PLM[J]. Urban Gas **04**, 36–41 (2019). (in Chinese)
18. Zou, B.W.: Reasearch on design and application of manufacturing-oriented EBOM architecture of tunnel boring machine [J]. Railw. Constr. Technol. **10**, 105–109 (2018). (in Chinese)
19. Zhang, X.M.: Definition and management of xBOM during aircraft model development[J]. Intell. Manuf. **06**, 21–28 (2021). (in Chinese)
20. Fan, Q.H., Zhou, R.Y., Wei, Q., Liu, B.X.: BOM multi-view mapping based on a single data source under PLM. In: 2012 International Conference on Industrial Control and Electronics Engineering, 2012, pp. 1705–1708
21. Wang, Y., Meng, M.: Integrated model and mapping technology of BOM multi-view for Hull block manufacture. Fifth Int. Jt. Conf. Comput. Sci. Optim. **2012**, 113–117 (2012)

22. Cai, S., Hu, D., Fei, Y., et al.: Research on BOM multi-view mapping method orient to aircraft in-service data management[C]. J. Phys.: Conf. Ser. IOP Publ. **1550**(3), 032124 (2020)
23. Han, Y., Li, C.R.: Research on aircraft configuration and effectiveness management based on modularization[J]. Aeronaut. Stand. Qual. **01**, 45–50 (2020). (in Chinese)
24. Qiao, W.X., Wang, Y.Q., Wei, R., Mei, Z.Y.: Research on the effectivity of modular configuration management of aircraft technical publications [J]. Aviat. Maint. Eng. **06**, 80–83 (2012). (in Chinese)
25. Wang, Q.L.: Aircraft Configuration Management Based on Systems Engineering [M]. Shanghai Science and Technology Press (2017) (in Chinese)

Simulation of Electro-Impulse De-Icing Process Based on an Improved Ice Shedding Criterion

Yongjie Huang[1], Xian Yi[2,3](✉), Qinglin Liu[2,3](✉), Zhangsong Ni[1], Pan Pan[1], and Ying Zhang[1]

[1] Chengdu Fluid Dynamics Innovation Center, Chengdu 610010, China
[2] Key Laboratory of Icing and de-Icing Research, Aerodynamics Research, Mianyang 621000, China
yixian0105@126.com, caso@163.com
[3] State Key Laboratory of Aerodynamics, Aerodynamics Research, Mianyang 621000, China

Abstract. , The electro-impulse de-icing (EIDI) system has wide development potential in the field of aircraft green de-icing. For the design and optimization of EIDI, it is quite necessary to study ice shedding criterion and simulate the de-icing process. Based on experimental observations and theoretical analysis, an improved ice shedding criterion considering ice/skin debonding and ice fracture is proposed in this paper. Nonlinear transient three-dimensional finite element models combined with the ice shedding criterion are established to simulate the de-icing process. The simulation method is verified by experiments and literature comparison. Compared with the traditional ice shedding criterion, the improved ice shedding criterion and de-icing simulation results in this paper are more reasonable and accurate. A parametric study is performed to determine the influence of ice strength parameters on de-icing effectiveness. The results of the parametric study can provide guidance for the design of a compound EIDI system with higher de-icing efficiency.

Keywords: Electro-impulse De-Icing · Finite element method · Ice shedding criterion · Ice fracture · Ice/skin debonding

1 Introduction

Since the first successful test flight of the Wright Brothers in 1903, the development of aircraft has always been accompanied by flight accidents [1]. According to statistics, one of the main causes of flight accidents is meteorological conditions, among which icing conditions are a very dangerous problem that aircraft designers must pay attention to [2].

For the in-flight ice accretion problems, the current ice protection technology mainly includes passive anti-icing technology and active de-icing technology [3]. As a mechanical de-icing technology, electro-impulse de-icing (EIDI) potentially meets the basic ice protection requirements, such as low power requirements, light weight, simple structure, high reliability and high efficiency [4]. Compared with the traditional electric/gas-heat

Chinese Society of Aeronautics and Astronautics: CASTYSF 2022, LNEE 972, pp. 623–634, 2023.
https://doi.org/10.1007/978-981-19-7652-0_55

de-icing system, the EIDI has low infrared characteristics, which can be more suitable for stealth combat aircraft with strict requirements on infrared stealth performance. EIDI system has a wide range of application scenarios and application potential in the future.

The working principle of EIDI system is quite simple as shown in Fig. 1. The discharging of electric capacitors triggers the coils generating an electromagnetic field. The magnetic field then induces a mechanical impulsive loading, which causes the mechanical vibration of the aircraft skin and the ice shedding [5].

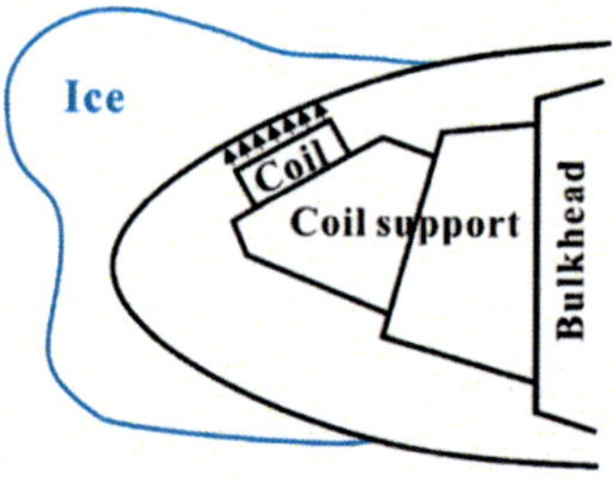

Fig. 1. Schematic diagram of EIDI system.

The earliest research on EIDI can be traced back to the 1930s [6]. The United States, the Soviet Union and the United Kingdom took the lead in carrying out the principal design and testing of EIDI system during this time. In the 1970s, the Soviet Union successfully applied the EIDI system to the Ilyshin series planes for the first time. However, due to various reasons, it was not widely promoted at that time [7]. Since the 1980s, the National Aeronautics and Space Administration (NASA) re-intensified the research of EIDI, and cooperated with Wichita State University and the Industry Consortium to carry out a series of icing wind tunnel experiments and flight tests [8]. Since 2010, Cox and Company of the United States successively developed Electro-Expulsive Separation System, Electro-Mechanical Expulsion De-icing System, Electro-Magnetic Expulsion De-icing System on the basis of EIDI system research, expanding and innovating EIDI technology [9].

The development of EIDI technology is inseparable from numerical simulation. Ice shedding process simulation and de-icing effectiveness prediction are of great significance to the engineering applications of EIDI system. The high accurate de-icing simulation model can help engineers quickly complete the design and analysis of the EIDI system. The finite element method (FEM) can discretize complex differential equations into linear equations to obtain stress and strain responses during EIDI process, and simulate the ice shedding based on damage mechanics.

Kandagal [10] and Palacios [11] believed that when the ice stress calculated by the dynamic model exceeded the bond strength between the ice layer and the aircraft skin, the ice layer would fall off. Khatkhate [12] used two-dimensional plane shell elements to model the aluminum skin and ice layer, and preliminarily predicted the ice shedding area on the skin surface under the impulsive load. The results indicated that shear stress between the skin and the ice layer was the main reason for the ice shedding. Considering the effect of interlayer normal tensile stress on ice shedding, Labeas [13] proposed an ice shedding failure criterion based on the coupling of ice/skin interfacial shear stress and normal tensile stress, and established a three-dimensional progressive damage ice shedding finite element model for EIDI system. The numerical results showed that the ice shedding failure criterion could be used to predict the de-icing process of EIDI system, but the prediction accuracy was limited. Zhang [14] analyzed the contribution ratio of ice/skin interfacial shear stress and normal tensile stress to ice shedding, improved the ice shedding criterion and compared the influence of ice shedding criterion with different stress ratios on the ice shedding prediction. Based on the re-development of ANASYS, Li [7] compared the difference between the maximum stress criterion and the ice-skin interface shear-tension criterion in predicting the de-icing effectiveness of EIDI. Li found that one of the reasons for the limited accuracy of ice shedding prediction was that the traditional ice shedding model only considered the failure of the ice/skin interface.

Although a lot of work has been done on the numerical simulation of EIDI, few studies involve ice shedding failure criterion and damage analysis. There is still a lack of refined ice shedding failure criteria and progressive damage failure de-icing models. Besides, traditional ice shedding criteria for EIDI only consider the effect of interfacial stress on ice/skin debonding, ignoring the effect of ice fracture.

Aiming at the above problems, considering ice/skin debonding and ice fracture, an improved ice shedding failure criterion for EIDI was proposed in this study. 3D nonlinear transient dynamic finite element models were developed to simulate the de-icing process of EIDI system. The simulation results were compared with the experimental results and literature simulation results. A parametric study was performed to determine the influence of the ice strength parameters on de-icing effectiveness. The numerical method can be used in the engineering field to quickly evaluate the de-icing effectiveness of the EIDI system.

2 Numerical Simulation Method

2.1 Ice Shedding Mechanism and Ice Shedding Criterion

As mentioned in the previous literature [10–14], current ice shedding failure criteria for EIDI are basically based on debonding failure of the ice/skin interface. The effect of ice fracture on ice shedding is ignored. However, according to a simple de-icing experiment observation of EIDI system (as shown in Fig. 2a), under the impulsive load, ice shedding typically involves two failure modes: (a) ice/skin debonding and (b) ice fracture.

From the perspective of structural failure analysis, the ice-covered skin can be regarded as a composite structure, and the ice layer, the ice/skin interface and the skin are substructures respectively. The ice/skin debonding corresponds to the structural interlayer failure, and the brittle fracture of the ice layer corresponds to the intralayer failure (as shown in Fig. 2b). The damage to the substructures eventually leads to the overall ice shedding.

Under the impulsive load, the instantaneous stress state of the ice layer and its interface is very complex. On the one hand, there are interfacial shear stress and normal tensile stress at the ice/skin interface. As the interfacial stress exceeds the adhesion strength of the ice/skin interface, the ice layer and the skin will be debonded. On the

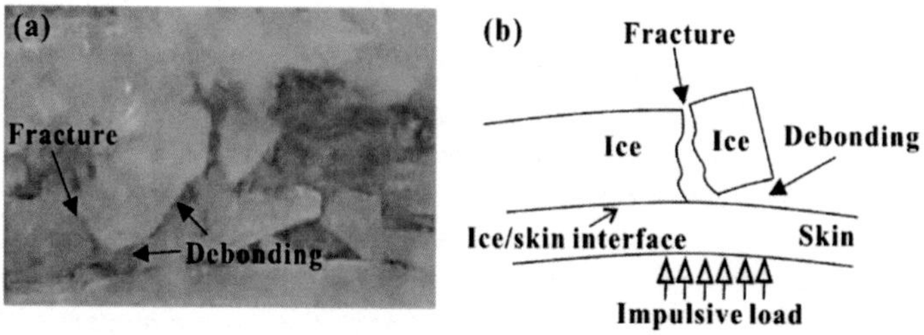

Fig. 2. Ice shedding mechanism for EIDI: **a** experimental phenomena, **b** failure analysis.

other hand, when the internal stress of the ice layer is large enough, it will promote the initiation of mode I crack in the ice layer, resulting in brittle fracture of ice.

Although it is difficult to capture the sequential relationship between ice/skin debonding and ice fracture in the EIDI experiment, it is obvious that the two failure modes can promote each other, and both eventually cause the ice shedding. Therefore, decomposing ice shedding into two failure modes: ice/skin debonding and brittle fracture of ice can more precisely and reasonably characterize the ice shedding phenomenon.

Based on the analysis of the ice shedding mechanism, an improved ice shedding criterion considering ice/skin debonding and ice fracture is proposed (as shown in Table 1), which is used for subsequent numerical simulation. When the stress state of the ice layer reaches the ice/skin debonding or ice fracture condition, it is determined that the ice layer falls off.

Table 1. Ice shedding criterion.

Ice shedding failure modes	Failure criterion
Ice/skin interface debonding	$f_{ID} = \begin{cases} \frac{\sigma_{int}}{\sigma_{iu}} + \left(\frac{\tau_{int}}{\tau_{iu}}\right)^2 \geq 1\ (\sigma_{int} \geq 0) \\ \left(\frac{\tau_{int}}{\tau_{iu}}\right)^2 \geq 1\ (\sigma_{int}0) \end{cases}$ (1)
Ice fracture	$f_{BF} = \frac{\sigma_1}{\sigma_{tu}^{I}} \geq 1$ (2)

The ice/skin debonding is determined based on the Labeas criterion [13], which comprehensively considers the effects of interfacial normal stress and shear stress on interface debonding failure. σ_{int} and τ_{int} represent ice/skin interface normal tensile stress and shear stress respectively, while σ_{iu} and τ_{iu} represent ice/skin interface normal tensile strength and shear strength respectively. σ_{iu} is of 1.44 MPa and τ_{iu} is of 0.4 MPa [13].

The fracture failure of the ice layer is simplified into the brittle fracture model, and the first strength theory is used to determine the failure. σ_1 and σI tu represent maximum principal stress of the ice layer and ice tensile strength respectively, where σI tu is of 1.6 MPa [15].

2.2 Finite Element Model and Ice Shedding Failure Analysis Process

The de-icing model of EIDI system is built by using the structural parameters of the ice-covered aluminum skin plate given in literature [2] (as shown in Fig. 3a). The size of the aluminum plate is 420 mm × 420 mm and the thickness is 2 mm. Two sides of the aluminum plate are fixed while the other two sides are free. The ice layer is evenly covered on the aluminum plate, and the thickness of the ice layer is 3 mm. The elastic modulus of the aluminum plate is 71.5 GPa, the Poisson's ratio is 0.33, and the density is 2780 kg/m^3; the elastic modulus of the ice layer is 5.5 GPa, the Poisson's ratio is 0.3, and the density is 897 kg/m^3.

The meshing of the model is shown in Fig. 3b. The ice layer is set with one element along the thickness direction while the aluminum plate is set with four elements. The meshes are optimized in the central electromagnetic impulsive loading area, and the total number of elements is 33060. In order to improve the computation efficiency and avoid the occurrence of abnormal computation phenomena such as shear locking and hourglass, the 3D eight-node incompatible elements (C3D8I) and the continuum shell elements (SC8R) are used to mesh the ice layer and the aluminum plate respectively.

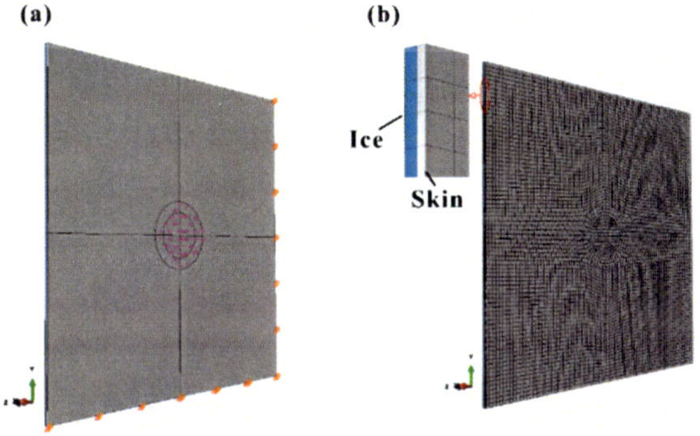

Fig. 3. Finite element model: **a** structure, **b** mesh.

The impulsive load generated by the EIDI system is applied to a 60 mm diameter area in the middle of the plate (as shown in Fig. 3a). Although the de-icing excitation acts on the skin for a very short time, the skin will still vibrate at its natural frequency after the driving force vanishes. As a result, the external load on the model can be described as shown in Fig. 4. The maximum impulse force is 130 N and the duration is 1 ms; the total de-icing process time is 80 ms.

De-icing simulation involves transient nonlinear dynamic numerical solution and ice shedding failure analysis. The whole process of EIDI de-icing simulation is shown in Fig. 5. t is the current time of the simulation process, and t_{max} is the total de-icing time. For each time increment Δt, the finite element dynamic equilibrium equation is established and solved, and the structural stress and strain responses are obtained. Afterwards, based on the improved ice shedding criterion, the ice shedding state is determined, and failed elements will be deleted.

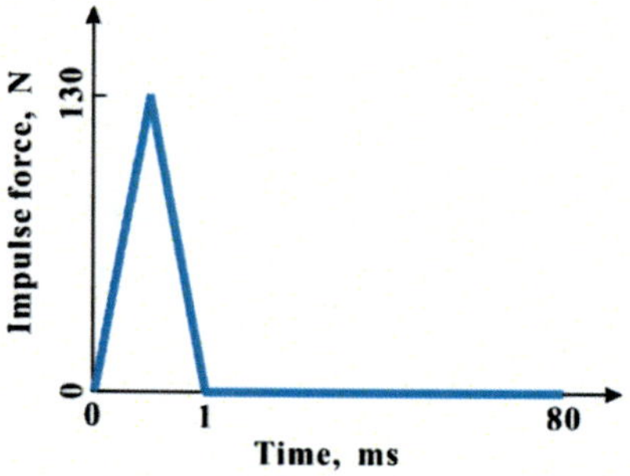

Fig. 4. The external load on the model.

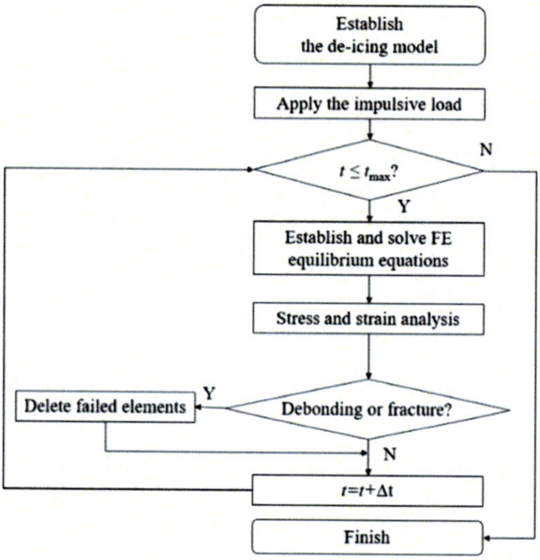

Fig. 5. Ice shedding failure analysis process.

The ABAQUS 2016/Explicit dynamics module is used to complete the transient dynamics solution of the finite element models, and the ABAQUS VUSDFLD is used for the failure analysis of ice shedding. The hardware platform for EIDI de-icing calculation is Intel Xeon Gold 6248R Server. The time to solve a single model is about 4 h.

3 Numerical Simulation Results and Analysis

3.1 Simulation Methodology Validation

To verify the accuracy of the de-icing model, the simulated results are compared with experimental and literature results (as shown in Fig. 6). The experimental conditions, structural geometry, material parameters and impulsive load are consistent with those in Sect. 2.2 for simulation models.

The de-icing experimental results [13] of the aluminum plate are shown in Fig. 6a. Due to the effect of impulsive loading induced by EIDI, most of the ice on the surface of the plate falls off, and only a small amount of ice is attached to the corner near the clamped boundary, forming an L-shaped region.

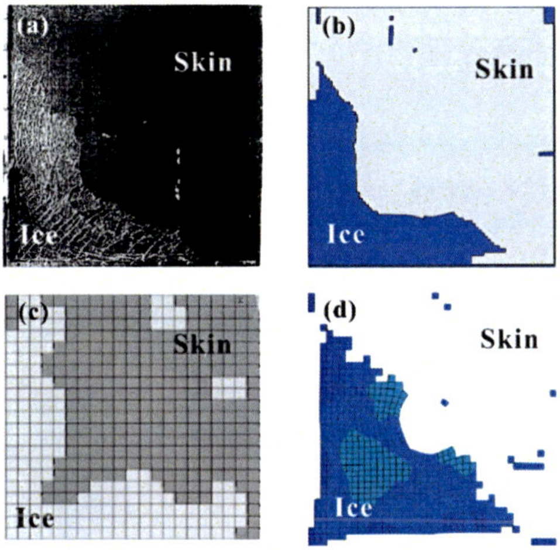

Fig. 6. Comparison of experimental and simulated results of de-icing process: **a** experimental result [13], **b** simulated result of this study, **c** simulated result of literature [13] and **d** simulated result of literature [14].

Figure 6b is the simulated results based on the improved ice shedding criterion proposed in this paper. Figure 6c and 6d are the simulated results of literature [13, 14], which are based on the traditional ice shedding criterion only considering ice/skin debonding failure.

From the comparison of the experimental and calculation results, it can be found that the predicted de-icing area and local characteristics of residual ice based on the improved ice shedding criterion are consistent with the de-icing experimental results, which proves the effectiveness of the simulation method. Moreover, the simulated results of the improved ice shedding criterion are more accurate than literature simulation results based on the traditional ice shedding criterion, indicating the improved ice shedding criterion are more reasonable.

3.2 Ice Shedding Failure Criterion Analysis

To analyze the effect of failure criteria on de-icing simulation, three kinds of ice shedding criteria were used for calculation: (a) ice shedding criterion considering both ice/skin debonding and ice fracture (as described in Eqs. (1) and (2)), (b) ice shedding criterion only considering ice/skin debonding (as described in Eq. (1)), (c) ice shedding criterion only considering ice fracture (as described in Eq. (2)). The simulated results of de-icing models corresponding to three ice shedding failure criteria are displayed in Fig. 7a, b, and c respectively.

The morphological characteristics of the residual ice calculated by ice shedding failure criteria (b) and (c) are quite different from the experimental result, which are also far inferior to the calculation result by ice shedding failure criterion (a). Moreover, the residual ice coverage areas predicted by ice shedding failure criteria (b) and (c) are much larger than the experimental result and the simulated result by ice shedding failure criterion (a), which also indicates that the failure criterion considering both ice/skin debonding and ice fracture is more reasonable than the failure criterion only considering ice/skin debonding or ice fracture.

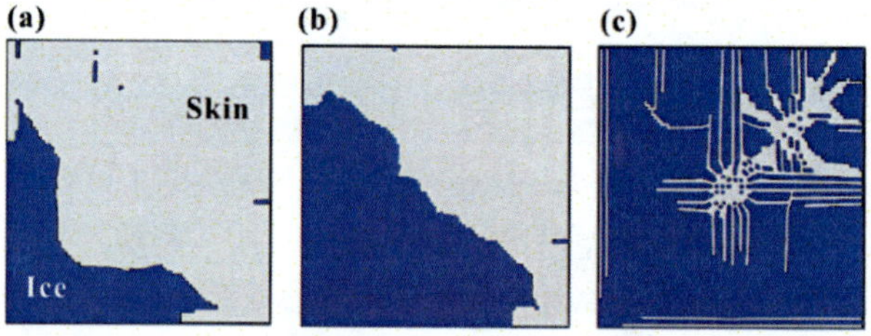

Fig. 7. Comparison of simulated results of ice shedding failure criteria: **a** criterion considering both ice/skin debonding and ice fracture, **b** criterion only considering ice/skin debonding, **c** criterion only considering ice fracture.

3.3 Parametric Study

Ice tensile strength, ice/skin interface normal tensile strength and ice/skin interface shear strength are three important strength parameters for the ice shedding criterion which affect both ice/skin debonding and ice fracture. In order to further clarify the influence of ice strength parameters on the de-icing process and provide a reference for the design of the EIDI system, a parameter analysis is carried out.

All investigated ice strength parameters for the de-icing simulation and their variations are presented in Table 2. The set of parameters corresponding to the benchmark is presented in the first column of the table. For each simulation, only one parameter changes, and the rest parameters keep the basic values. The variable range of the three strength parameters remains the same, which are 125%, 100%, and 75% of the basic values.

The de-icing effectiveness is quantitatively characterized by the de-icing rate, which is defined as the percentage of the de-icing area over the total plate area. The higher the de-icing rate, the better the de-icing effectiveness.

Ice tensile strength. Ice tensile strength is a physical parameter of the ice layer itself, which is affected by many factors such as the diameter of the water droplet, the air flow velocity, temperature, and ambient humidity. Ice tensile strength measured under different experimental conditions is quite different and it has a high dispersion.

The range of ice tensile strength investigated in this paper is 2.0 MPa (125%), 1.6 MPa (100%) and 1.2 MPa (75%). The simulated de-icing results of EIDI system are shown in Fig. 8a, b, and c, respectively. It can be seen that with the decrease of the ice tensile

Table 2. Strength parameters investigated and their variations.

Parameter	Basic set of parameters		
Ice tensile strength	1.6 MPa (100%)	2.0 MPa (125%)	1.2 MPa (75%)
Ice/skin interface normal tensile strength	1.44 MPa (100%)	1.8 MPa (125%)	1.08 MPa (75%)
Ice/skin interface shear strength	0.4 MPa (100%)	0.5 MPa (125%)	0.3 MPa (75%)

strength, the calculated de-icing rate gradually increases, indicating that the ice tensile strength is inversely proportional to the de-icing effectiveness within a certain range. Lower ice tensile strength is conducive to the ice shedding, which also reflects the rationality and necessity of considering ice fracture in the ice shedding failure criterion.

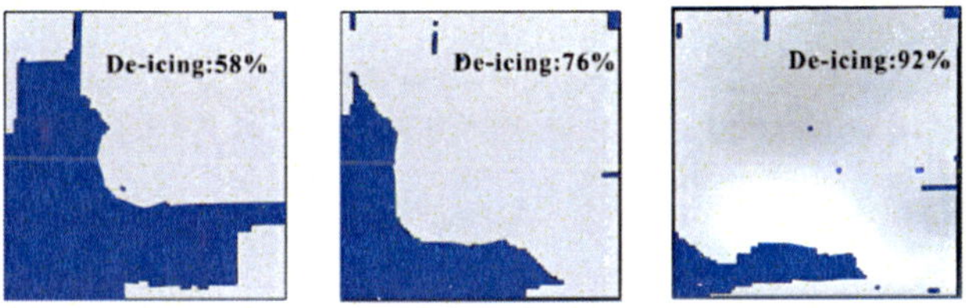

(a) Ice tensile strength:125% (b) Ice tensile strength:100% (c) Ice tensile strength:75%

Fig. 8. The influence of ice tensile strength on de-icing rate.

Ice/skin interface normal tensile strength. The adhesion strength of the ice/skin interface is not only related to the physical parameters of the ice layer, but also affected by the properties of the skin material, the coefficient of friction and the mesoscopic geometric characteristics of the interface. There are many experimental studies on the ice/skin interface shear strength, but few studies on the ice/skin interface normal tensile strength. The range of ice/skin interface normal tensile strength investigated in this paper is 1.8 MPa (125%), 1.44 MPa (100%) and 1.08 MPa (75%). The predicted de-icing results of EIDI system are shown in Fig. 9a, b, and c, respectively.

The results show that the de-icing rate gradually increases with the decrease of the ice/skin interface normal tensile strength. However, the increase of the de-icing rate is limited and the residual ice morphology characteristics does not change significantly. It indicates that under the same strength ratio range (75% ~ 125%), the effect of ice/skin interface normal tensile strength on ice shedding is less than that of ice tensile strength.

Ice-skin interface shear strength. The range of ice/skin interface shear strength investigated in this paper is 0.5 MPa (125%), 0.4 MPa (100%), and 0.3 MPa (75%). The predicted de-icing results of EIDI system are shown in Fig. 10a, b, and c, respectively.

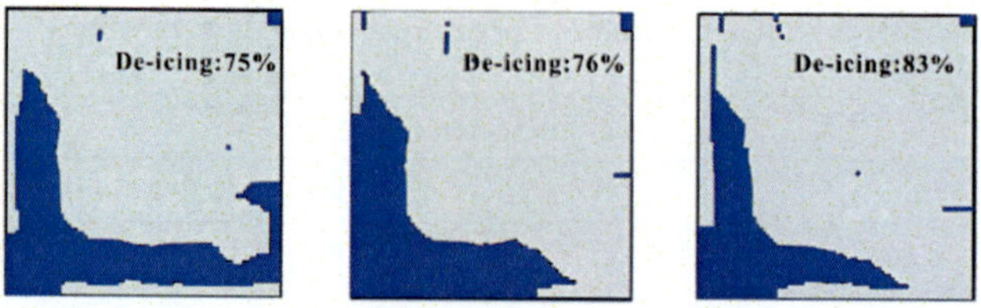

(a) Ice/skin interface tensile strength:125% (b) Ice/skin interface tensile strength:100% (c) Ice/skin interface tensile strength:75%

Fig. 9. The influence of ice/skin interface normal tensile strength on de-icing rate.

The simulated results show that the ice/skin interface shear strength has a significant effect on ice shedding. With the decrease of ice/skin interface shear strength, the de-icing rate increases from 47% to 98%. Moreover, compared with the ice tensile strength and ice/skin interface normal tensile strength, under the same conditions, the ice/skin interface shear strength has a more prominent influence on the de-icing process.

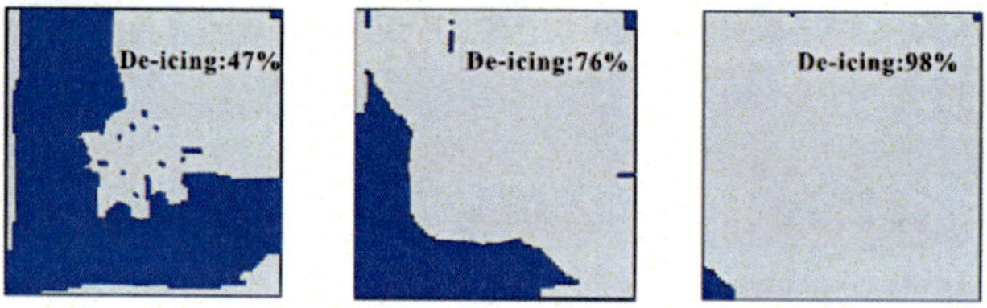

(a) Ice/skin interface shear strength:125% (b) Ice/skin interface shear strength:100% (c) Ice/skin interface shear strength:75%

Fig. 10. The influence of ice/skin interface shear strength on de-icing rate.

Therefore, in the design or simulation of the anti-icing system, it is necessary to pay more attention to the ice/skin interface shear strength. On the one hand, the accurate acquisition of the ice/skin interface shear strength value is helpful to improve the numerical simulation accuracy; on the other hand, through surface treatment technology, artificially reducing the ice/skin interface shear strength can significantly increase the effectiveness of EIDI system.

Furthermore, the comprehensive use of EIDI and surface materials with low ice adhesion strength (e.g. superhydrophobic materials) to construct a new compound EIDI system has a potential to further improve the anti-icing efficiency and reduce energy consumption, which is worthy of further research in related fields.

4 Conclusions

(1) Based on the experimental observation and theoretical analysis, the ice shedding process of EIDI can be refined and decomposed into two failure modes: ice/skin debonding and ice fracture.

(2) An improved ice shedding criterion considering ice/skin debonding and ice fracture is proposed in this paper. Compared with the experimental and simulated results in the literature, the de-icing simulation results based on the improved ice shedding criterion are more accurate and reasonable than traditional criteria which only consider ice/skin debonding failure.
(3) The de-icing rate of EIDI system increases with the decrease of ice tensile strength, ice/skin interface normal tensile strength and ice/skin interface shear strength. Compared with the other strength parameters, ice/skin interface shear strength has the most significant effect on the de-icing effectiveness of EIDI.
(4) A new compound de-icing system based on EIDI system and surface materials with low ice adhesion strength has a potential to further improve the de-icing efficiency and reduce energy consumption.

Acknowledgements. The authors would like to thank the National Natural Science Foundation of China (No.12132019) for their support in this research.

References

1. Tianxiang, Z.: Statistical analysis of aircraft accidents and their causes. Aviat. Stand. Qual. **12**(6), 37–43 (1998). (in Chinese)
2. Wei, Y., Xu, H., Xue, Y., Duan, X.,: Quantitative assessment and visualization of flight risk induced by coupled multi-factor under icing conditions. Chin. J. Aeronaut. **8**(173), 52–67 (2020)
3. Goraj, Z.: An overview of the de-icing and anti-icing technologies with prospects for the future. In: 24th Congress of International Council of the Aeronautical Sciences, pp. 34–42. Yokohama, Japan (2004)
4. Potapczuk, M.G.: Aircraft icing research at NASA Glenn research center. J. Aerosp. Eng. **26**, 260–276 (2013)
5. Fan, G., Shinan, C.: Design test of electro- impulse de-icing system of an aircraft. In: 2nd International Conference on Artificial Intelligence, Management Science and Electronic Commerce (AIMSEC), Zhengzhou, China, (2011)
6. Levin, I.: A: USSR electric impulse de-icing system design. Aircr. Eng. Aerosp. Technol. **44**(7), 7–10 (1972)
7. Li, Q., Zhu, C., Tian, B.: Numerical simulation and experimental verification of the electro-impulse de-icing system. In: 53rd AIAA/ASME/ASCE/AHS/ASC Structures, Structural Dynamics and Materials Conference, Honolulu, Hawaii (2012)
8. Zumwalt, G., Friedberg, R.: Designing an electro-impulse de-icing system. AIAA J. 1–9 (1986)
9. Al-Khalil, K.: Thermo-mechanical expulsive de-icing system – TMEDS. In: 45th AIAA Aerospace Sciences Meeting and Exhibit, Reno, USA (2007)
10. Kandagal, B.: Piezo-actuated vibratory de-icing of a flat plate. In: 46th AIAA/ASME/ASCE/AHS/ASC Structures, Structural Dynamics and Materials Conference, Austin, USA (2005)
11. Palacios, J.L.: Design, fabrication, and testing of an ultrasonic de-icing system for helicopter rotor blades. PhD thesis, The Pennsylvania State University (2008)

12. Khatkhate, A.A., Scavuzzo, R.J., Chu, M.L.: A finite element study of the EIDI system. AIAA J. (1988)
13. Labeas, G.N., Diamantakos, I.D., Sunaric, M.M.: Simulation of the electro-impulse de-icing process of aircraft wings. J. Aircr. **43**(6), 1876–1885 (2006)
14. Yongjie, Z., Wenjun, D., Bintuan, W.: Study on de-icing criterion of electro-impulse de-icing simulation. Comput. Eng. Appl. **48**(3), 232–233 (2012). (in Chinese)
15. Reich, A.: Ice property/structure variations across the glaze/rime transition. In: 30th Aerospace Sciences Meeting and Exhibit, Reno, USA (1992)

Resource Management in Cloud Computing Using Deep Reinforcement Learning: A Survey

Yuxin Feng(✉) and Feiyang Liu

AVIC, Xi'an 710068, China
609369689@qq.com

Abstract. Next generation aircrafts will not only require high-performance and intelligent computing capabilities, but also a fast design-developing-integration-update time cycle. Cloud computing technology provides a platform with large amounts of hardware resources and software services, making applications development and deployment much more convenient. Thus, airborne cloud computing becomes an important design methodology for next generation avionics system. However, the dynamic and uncertain cloud environment makes efficient resource management very complicated. Due to the characteristics of dynamic autonomous decision-making, deep reinforcement learning has become a promising resource scheduling algorithm. This paper firstly analyzes the requirements of airborne cloud computing systems, then studies the basic theory of cloud resources management and scheduling strategies. The resource management algorithms based on deep reinforcement learning (DRL), some common DRL models, experimental platforms, and evaluation parameters are introduced in details. Finally, some critical problems and challenges in the design of DRL-based resource management algorithm are summarized. This paper can provide some technique supports for the airborne cloud computing system.

Keywords: Airborne computing system · Cloud computing · Resource management · Deep reinforcement learning

1 Introduction

Avionics system is a common information processing platform in an aircraft. It can provide shared computation resources among different software applications in the form of central processing module. With the fast development of sensor systems, such as high-definition radio-frequency radar, optical radar, and distributed aperture system (DAS), next generation aircrafts must have high-performance and intelligent computing capabilities to processing the surged information. Meanwhile, the design time and full lifecycle cost of a new aircraft increase significantly recent years, next generation aircrafts should also improve the utilization of resources and speed up the design-developing-integration-update time cycle. Cloud computing is a widely used technology in commercial computing systems, it can provide an efficient platform with large amounts of hardware resources and software services, making applications development and deployment much more

Chinese Society of Aeronautics and Astronautics: CASTYSF 2022, LNEE 972, pp. 635–643, 2023.
https://doi.org/10.1007/978-981-19-7652-0_56

convenient. Thus, airborne cloud computing becomes an important design methodology for next generation avionics systems [1].

Cloud computing system provides a shared pool of configurable computing resources (*e.g.* networks, servers, storage, applications, services) that can be rapidly provisioned and released with the minimal management effort or interaction. Resource management is one of the key technologies in cloud computing. However, the dynamic and uncertain cloud environment makes efficient re-source management very complicated. With the continuous development of machine learning, researchers gradually use various algorithms to deal with the resource allocation. Reinforcement learning, is a special machine learning method that can be trained without a priori knowledge, and its policies are dynamically adaptive, which is very suitable in resource management. This paper studies the basic theory of cloud resources management and scheduling strategies, and analyzes the resource management algorithms based on deep reinforcement learning (DRL). It can provide some technique supports for the design of airborne cloud computing system.

2 Resource Management

2.1 Overview of Could Computing and Its Scheduling

In the cloud computing system, storage and computing resources are centrally placed in a public cloud resource pool, enabling customers to access them in a convenient, pay-as-you-go manner over the network. Cloud resource pools are the core of cloud services. Virtualization of cloud services is a major way to implement cloud computing to build resource pools [2]. Resource management requires the optimal utilization of virtual resources and efficient management of runtime resources to provide application-based services that meet the needs in a cost-effective manner.

Resource scheduling in cloud computing is divided into three layers: scheduling resources for applications, scheduling virtual resources (e.g. virtual machines) to physical resources, physical resource scheduling and real applications. Moreover, at each layer there can be several different objectives for optimization. In the case of resource management, for example, at the virtual resource layer, load balancing can be optimized, resource utilization such as CPU and memory shares can be improved, in addition to cost efficiency or energy savings. Therefore, the cloud resource management problem can be abstracted as a combinatorial optimization problem. A certain path or rule is used to obtain a solution that satisfies the user's requirements.

2.2 Cloud Resource Management Optimization Objectives

The cloud resource management scheduling problem is actually an optimization problem. The optimization objective, then, is closely related to the performance of the cloud computing resource scheduling system. Researchers often use SLA and QoS parameters as optimization parameters in resource scheduling, which can ensure that the requirements of SLA agreements are met, avoid violating the terms of the agreements, and efficiently integrate and allocate effective resources in the data center. From the cloud service provider's perspective, load balancing, resource utilization, and energy efficiency are the most important goals in task scheduling. Specifically, the optimization objectives of cloud resource management can be divided into four areas: energy optimization, time optimization, load balancing optimization, and other optimizations. Specifically, they are described as follows.

(1) Energy consumption optimization. It can be summarized into two categories, total energy consumption and energy utilization (energy efficiency). Energy consumption minimization is a common optimization objective in the literature.

(2) Time-related optimization. The time metrics for scheduling in cloud computing can be summarized into four types, including maximum completion time (execution time), delay time, response time, and waiting time. The optimization goals for cloud resource management are usually minimization of completion time, minimization of delay time (or delayed service), and minimization of response time.

(3) Load balancing optimization. Depending on the load object, the load balancing metrics are divided into two types: task volume load balancing and task number load balancing. The degree of load balancing can also be divided into two types of cumulative load balancing and real-time load balancing according to the time period. Maximizing load balancing is the usual optimization goal. In the literature, there are various functions used to calculate the degree of load balancing, such as variance or standard deviation, average success rate, coefficient of variance, load value and imbalance.

(4) Other optimization. It mainly includes resource utilization maximization, resource utilization minimization, etc.

2.3 Resource Management Strategies

Existing resource management methods can be divided into two main categories, traditional methods and intelligent methods. Traditional methods focus on adapting and extending traditional scheduling methods that rely almost entirely on manual computation, such as First In First Out (FIFO), shortest job first (SJF), First Come First Serve (FCFS), Round Robin (RR), Minimum-Minimum (Max-Max), and Maximum-Minimum (Max-Min). In [3], Xu et al. proposed a multi-workflow multi-QoS constrained scheduling policy (MQMW) for cloud computing. In [4], Li et al. developed a system cost function for jobs and a non-preemptive priority M/G/1 queuing model to help the policy and algorithm obtain the approximate optimal service value for each job. These scheduling algorithms are effective for a wide range of resource management problems. However, they can only support a limited number of parameters for optimization. In a cloud environment, many parameters need to be optimized simultaneously, intelligent algorithms should find the optimal solution.

Intelligent algorithms, on the other hand, are based on mimicking natural body algorithms, mainly ant colony algorithms, simulated annealing methods, particle swarm optimization, and genetic algorithms, which are capable of optimizing multiple parameters simultaneously. Reference [5] aims to minimize three conflicting objectives, i.e., completion time, resource utilization, and execution cost. For this purpose, a multi-objective optimization problem is proposed and then a composite discrete artificial bee colony technique based on epsilon fuzzy dominance is used to derive the Pareto optimal solution; a method based on multiple swarm GA is introduced in [6]. The method uses multiple swarm genetic algorithms to solve the load balancing problem, and the method achieves good results in terms of improving completion time, cost and load balancing.

However, similar to traditional methods, intelligent scheduling methods can lead to long optimization time and inefficient scheduling. In recent years, with the booming field of deep learning, cloud resource management through neural networks has become a major research hotspot. These methods use models such as neural networks to design scheduling strategies to avoid imbalanced resource allocation.

3 DRL Algorithm for Scheduling in Cloud Computing

Resource scheduling problems are NP-complete or NP-hard problems. Most of the deep learning-based resource scheduling algorithms are based on supervised learning. These works use the idea of recent workload to predict the current/future workload. Their main goal is to minimize the number of servers or virtual machines to save energy and reduce costs. After supervised learning, reinforcement learning is the second most prevalent approach to exploration. Most existing algorithms require hand-crafted rules that are difficult to adapt to complex and dynamic systems. In recent years, researchers have found that reinforcement learning (RL) ideas are very suitable for resource management. Reinforcement learning is a special machine learning method that can be trained without prior knowledge and is well suited for applications in cloud environments where labeled data is difficult to obtain (Fig. 1).

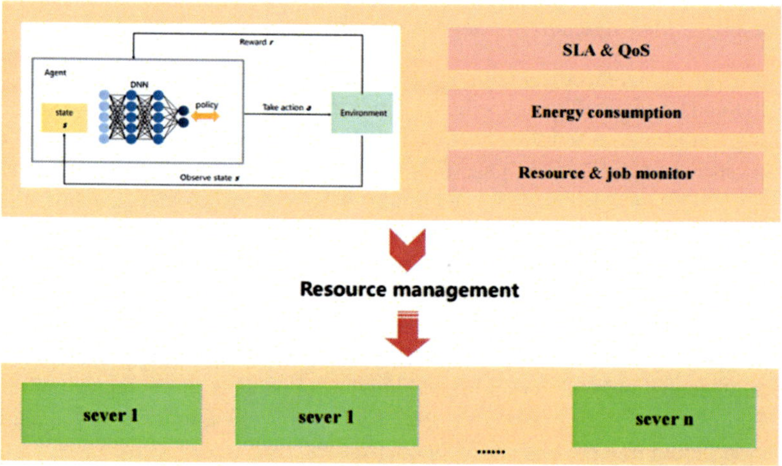

Fig. 1. Framework of resource management using deep reinforcement learning.

DeepRM [7] is the first DRL-based resource scheduler with a policy gradient reinforcement learning algorithm designed to schedule hierarchical jobs to heterogeneous resources and applied to a SoC chip scheduling simulator. However, resource management aims at minimizing the completion time of individual jobs, which can only correspond to a simple scenario. Later DeepRM_Plus [8], based on DeepRM, instead uses a network structure with six layers of convolutional neural networks to describe a decision mapper, which is based on the great success of DNNs for image processing. In DeepRM_Plus, data center clusters, waiting queues and to-do queues constitute the state of the environment, which uses hash codes as input state and reduces the size of the state space. The follow-up work of DeepRM Decima [9] in order to solve the scheduling problem of multiple Directed Acyclic Graph (DAG) tasks when running on multiple Executor. It uses a scalable GNN to represent the scheduling policy and can handle DAG-like tasks of arbitrarily large shape and size. However, the limitation is that it mainly provides performance for clustered applications on similar resource scheduling targets. DeepSoCS [10] extends the Decima architecture to schedule SoC jobs in DS3 by applying heuristic algorithms to map tasks to available resources. However, it has limited performance at faster job injection rates.

In recent years, DRL has shown superior performance in the current research on cloud computing resource management. QEEC [11] is a Q-learning based task scheduling framework for energy-efficient cloud computing, it uses Q-value tables to represent decision makers for actions. According to the operating state and task characteristics of the cluster environment, [3] uses deep Q learning to solve the CPU-GPU heterogeneous computing scheduling problem. RLPAS [12] introduced an algorithm based on the SARSA model, which was used for resource scheduling in cloud environments. RLPAS uses the idea of parallel multi-agent algorithm to learn from the environment and allocate resources to different tasks, thereby improving resource utilization, shortening response time and improving throughput.

At the core of RL is a policy mapper for cloud scheduling. In cloud scheduling using RL, the policy mapper is usually represented by a deep neural network or Q-table. Table 1 summarizes some of the applications of RL in cloud computing resource scheduling, including the resources scheduled, the task types, the RL model tuning used, and so on. It can be seen that the cloud resource scheduling environment is complex, heterogeneous and dynamic. In the use of resources, virtual machines, CPU, memory and other resources are used. DQN and Q-learning become the first choice of most researchers in algorithm selection.

4 Experiment and Evaluation

Experimentation and evaluation are important aspects to verify the effectiveness of algorithms. This section summarizes the direction of the selection of experimental datasets, experimental platforms and evaluation metrics for resource management algorithms based on reinforcement learning by categorizing and summarizing.

4.1 Experimental Datasets

Common datasets in the papers include GoogleCluster, Alibaba Cluster Data, Google-TaskEvents and Azure's scheduling dataset. The GoogleCluster dataset contains data on resource requirements and availability of tasks and virtual machines. The GoogleCluster dataset contains data about the resource requirements and availability of tasks and

Table 1. Summary of resource scheduling algorithms based on deep reinforcement learning.

Algorithm	Resource	Task type	DRL Model
W-FCFS [13]	VMs	Burst workload management	Q-learning
TSMRL [14]	VMs	Scheduling customers requests in dynamic cloud environments	NFQ-based
CDDQLS [15]	VMs	Task Scheduling in cloud computing environments	DQN
DeepRM2 [16]	CPU, Memory	Resource scheduling	DQN
SchedRL [17]	VMs, CPU, Mermory	VM scheduling in a Non Uniform Memory Access architecture (NUMA)	DQN
DRL-LSTM [18]	VMs	Automated task scheduling	Q-learning

virtual machines. Google Cluster consists of many machines connected by high-speed networks. The dataset includes 670,000 logging traces, which record about 40 million task events over 30 days for over 12,000 machines. Google traces contains production workload scheduling requests for 29 days. Alibaba Cluster Data contains production traces for 4k machines over 8 days. Both contain the CPU/memory numbers used by each workload at a granularity of 5 min, as well as scheduling details such as priority, class, and raw resource requests.

4.2 Experimental Platform

The right simulation platform has a very important impact on the effective verification of algorithms. The more commonly used cloud computing platform in the papers is CloudSim, in addition to PlanetLab, RapidMiner Studio, Pegasus Toolkit, and Xen hypervisor, which are also partially used.

CloudSim is an open-source framework, which is used to simulate cloud computing infrastructure and services. It is developed by the University of Melbourne, and is written entirely in Java. It is used for modelling and simulating a cloud computing environment. It can facilitate several cloud related processes such as data center definition, resource definition and scheduling method simulation. It also has several predefined examples and scenarios that can be modified based on algorithms. The PlanetLab project allows researchers to experiment with new services in real-world conditions and at large scale.

4.3 Experimental Metrics

Typically, energy consumption, time cost and load balancing are the main bases for evaluating the performance of scheduling algorithms in cloud computing. Resource scheduling based on intelligent algorithms can be measured in terms of cost, time, utilization, energy awareness, completion time, scalability, workload management, and profit. Specifically, [19] uses a combination of utilization metrics such as CPU utilization, average CPU utilization, RAM utilization, and average RAM utilization to evaluate the system performance, and [13] evaluates the scheduling strategy by calculating the standard deviation of the average resource utilization. in addition to this, [20] evaluates the effectiveness of the strategy by convergence speed. [14] evaluates the average load by load performance of the system. The SLA specifies the QoS content such as response time, reliability, and security of cloud services, so it can be said that SLA is a very important criterion for resource scheduling.

5 Research Challenges

The dynamic and uncertain characteristics of the cloud environment make the resource scheduling very complicated. Due to its characteristics of active learning and adaptive strategy, deep reinforcement learning is more and more applied in the design of cloud resource scheduling and management algorithms. However, the resource scheduling algorithm should also consider the following issues.

SLA violation. When designing the algorithm, the workload and energy efficiency balance should be fully considered to meet the requirements of SLA and QoS as much as possible.

Virtual machine management. Virtualization brings challenges to resource management, so the migration and placement of virtual machines and the creation of instances have a significant impact on algorithm performance.

Scalability. Resource can be scaled to manage demand extensively. Algorithms to meet different resource requirements as much as possible.

Fault tolerance. When one server fails, the algorithm should be able to shift the workload to another server to ensure the user's mission needs.

Complexity. The resource scheduling algorithm needs to optimize multiple objectives at the same time, so the execution time and computing power should be reduced as much as possible while meeting the needs of customers.

Energy consumption. Energy-efficient algorithms not only save costs, but also help data center communication systems reduce carbon emissions and protect the environment.

6 Conclusion

Next generation aircrafts will use airborne cloud computing system as an efficient information processing platform with large amounts of hardware resources and software services, making applications development and deployment much more convenient. Due to the characteristics of dynamic autonomous decision-making, deep reinforcement learning has become a promising resource scheduling algorithm to improve the utilization of resources, which can greatly improve the computing performance and shorten the design-developing-integration-update time cycle. This paper analyzes the requirements of airborne cloud computing systems, and summarizes some critical problems and challenges in the design of DRL-based resource management algorithm. It will provide some technique supports for the airborne cloud computing system.

References

1. Jianchun, X., Zhonghua, W., Yahui, L.: The distributed computing framework research for avionics cloud[C]. In: IEEE International Conference on Networking and Network Applications, pp. 390–393 (2018)
2. Qin, Y., Wang, H., Yi, S., et al.: Virtual machine placement based on multi-objective reinforcement learning[J]. Appl. Intell. **50**(8), 2370–2383 (2020)
3. Dasjupta, P., Leblanc, R.J., Apple, W.F.: The clouds distributed operating system: Functional description, implementation details and related work[C]. In: The 8th International Conference on Distributed, pp. 2–3 (1988)
4. Li, L.: An optimistic differentiated service job scheduling system for cloud computing service users and providers[C] 2009. In: Third International Conference on Multimedia and Ubiquitous Engineering, pp. 295–299 (2009)
5. Gomathi, B., Krishnasamy, K., Bslaji, B.S.: Epsilon-fuzzy dominance sort-based composite discrete artificial bee colony optimisation for multi-objective cloud task scheduling problem[J]. Int. J. Bus. Intell. Data Min. **13**(1–3), 247–266 (2018)
6. Wang, B., Li, J.: Load balancing task scheduling based on multi-population genetic algorithm in cloud computing[C]. In: 2016 35th Chinese Control Conference (CCC), pp. 5261–5266 (2016)
7. Mao, H., Alizadeh, M., Menache, I. et al.: Resource management with deep reinforcement learning[C]. In: Proceedings of the 15th ACM Workshop on Hot Topics in Networks, pp. 50–56 (2016)
8. Zhang, Y., Yao, J., Guan, H.: Intelligent cloud resource management with deep reinforcement learning[J]. IEEE Cloud Comput. **4**(6), 60–69 (2017)
9. Mao, H., Schwarzkopf, M., Venkatakrishnan, S.B., et al.: Learning scheduling algorithms for data processing clusters[G]. In: Proceedings of the ACM Special Interest Group on Data Communication, pp. 270–288 (2019)
10. Sung, T.T., Ha, J., Kim, J. et al.: Deepsocs: a neural scheduler for heterogeneous system-on-chip (soc) resource scheduling[J]. Electronics **9**(6), 936 (2020)
11. Ding, D., Fan, X., Zhao, Y., et al.: Q-learning based dynamic task scheduling for energy-efficient cloud computing[J]. Futur. Gener. Comput. Syst. **108**, 361–371 (2020)
12. Bibal Benifa, J., Dejey, D.: Rlpas: Reinforcement learning-based proactive auto-scaler for resource provisioning in cloud environment[J]. Mob. Netw. Appl. **24**(4), 1348–1363 (2019)
13. Balasubramanian, V., Aloqaily, M., Tunde-onadele, O., et al.: Reinforcing cloud environments via index policy for bursty workloads[C]. In: NOMS 2020–2020 IEEE/IFIP Network Operations and Management Symposium, pp. 1–7 (2020)
14. Balla, H.A., Sheng, C.G., Jing, W.: Reliability-aware: task scheduling in cloud computing using multi-agent reinforcement learning algorithm and neural fitted Q.[J]. Int. Arab J. Inf. Technol. **18**(1), 36–47 (2021)
15. Swarup, S., Shakshuki, E.M., Yasar, A.: Task scheduling in cloud using deep reinforcement learning[J]. Procedia Comput. Sci. **184**, 42–51 (2021)
16. Ye, Y., Ren, X., Wang, J. et al.: A new approach for resource scheduling with deep reinforcement learning[J] (2018). arXiv:1806.08122
17. Sheng, J., Hu, Y., Zhow, W. et al.: Learning to schedule multi-NUMA virtual machines via reinforcement learning[J]. Pattern Recognit. **121**, 108254 (2022)
18. Rjoub, G., Bentahar, J., Abdel Wahab, O., et al.: Deep and reinforcement learning for automated task scheduling in large-scale cloud computing systems[J]. Concurr. Comput.: Pract. Exp. **33**(23), e5919 (2021)

19. Yu, Z., Machado, P., Zahid, A., et al.: Energy and performance trade-off optimization in heterogeneous computing via reinforcement learning[J]. Electronics **9**(11), 1812 (2020)
20. Chou, Q., Fan, W., Zhang, J.: A reinforcement learning model for virtual machines consolidation in cloud data center [C]. In: 2021 6th international conference on automation, control and robotics engineering (CACRE), pp. 16–21 (2021)

Research on Analysis Architecture of Military Helicopter and Modeling Approach of Meta-Activity

Benchao Lou(✉) and Qiang Sun

Helicopter Research and Development Institute, Jingdezhen 333001, China
tjulbc@163.com

Abstract. In order to meet the architecture analysis requirements of military helicopter, the relationship of mission, system-of-system, system and measure is analyzed, and an analysis process is obtained. Then, based on DoDAF2.0 and *Mission Engineering Guide*, an analysis frame consists of 4 viewpoints and 14 models, is established. Meantime, the model expressions are given using UML diagram, mapping matrix or tabular form. Then, for the purpose of supporting basic and normal data for analysis architecture, the concept of metal-activity is proposed. Based on event-flow-chart, 7 modeling rules of metal-activities are established, including sequence rule, priority judgment rule, parallel rule, random rule, concentration rule, merge rule, and cycle rule. Finally, the conclusion is given that the analysis architecture is appropriate for military helicopter and can be referred by other weapon system.

Keywords: Military helicopter · Requirements demonstration · Architecture analysis · Viewpoint · Metal-activity

1 Introduction

With the development of modern warfare towards the direction of the systematism, informatization, and networking [1, 2], the architecture analysis method of equipment gets more and more attentions. In recent years, equipment military requirements demonstration experienced spontaneous formation phase, threats-based phase, and capability based phase [3–5]. Military helicopters are important equipment, and their systematic demonstration has great significance on development of satisfactory system-of-system.

In 2009, *the Department of Defense Architecture Framework (DoDAF)* version 2.0 was issued, which provides 8 viewpoints, including 52 models [6]. In 2020, *Mission Engineering Guide* was proposed by Office of the Under Secretary of Defense for Research and Engineering [7], which describes the main attributes of Mission Engineering (ME) and how to apply them, and enables practitioners to formulate problems. Many studies were carried out by scholars in China, including methodology of equipment requirements demonstration and engineering applications [8–10]. Despite some achievements, a complete demonstration system, from mission engineering to system-of-system to

Chinese Society of Aeronautics and Astronautics: CASTYSF 2022, LNEE 972, pp. 644–659, 2023.
https://doi.org/10.1007/978-981-19-7652-0_57

equipment, has not been formed. Especially for military helicopter, fewer related studies can be found. Although DoDAF is widely used in multiple areas of equipment demonstration, it is not completely applicable to military helicopters, because of its complex product types and lack of mission analysis and measure viewpoint. Furthermore, ME has no detailed engineering guidance, resulting to the less practical applications.

This paper aims to build a complete frame for military helicopter requirements demonstration, establishing an analysis system including mission, equipment system-of-system, equipment and measure, based on DoDAF and ME. Meantime, in order to improve analysis efficiency and ensure consistency of demonstration results, the concept and building method of meta-activity is proposed to provide data support for military helicopter requirements demonstration.

2 Architecture Analysis of Military Helicopter

3 Relationship of Mission, System-of-System, System and Measures

From the process of *problem posing-problem analyzing-solution posing-solution evaluating*, a frame of military helicopter requirements analysis is proposed (see Fig. 1).

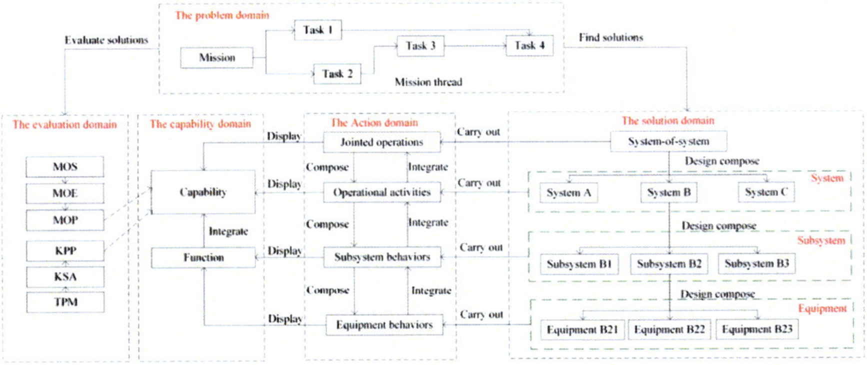

Fig. 1. The detailed relationship of mission, system-of-system, system and measures

Though the analysis of framework, there are 5 domains to support analysis work, including problem domain, solution domain, action domain, capability domain and evaluation domain.

Mission is the main responsibility or the main task for army to complete in a special time. In other words, mission, consisting of tasks, is the original requirements of the analysis architecture. In order to solve the question, an equipment system-of-system is required, which belongs to the solution domain. Furthermore, system-of-system consists of many systems, subsystems and equipment. Systems display various capabilities though kinds of activities, which consist of many behaviors carried out by subsystems and equipment. Meantime, activities aggregate to jointed operations. The jointed operations, activities, behaviors belong to action domain. Actions are the usage mode of systems-of-systems, systems and subsystems. The capability domain, including capabilities and functions displayed by actions, is the core links solution and problem domains.

Meantime, to evaluate the effective of accomplishing the tasks, several measures are proposed, including Measure of success (MOS), measures of effectiveness (MOE), measures of performance (MOP), key performance parameter (KPP), knowledge, skills, and abilities (KSA) and technical performance measure (TPM). MOS is the standard to judge the success of mission, and according to the system hierarchy, MOS can be distributed to MOE and MOP. Furthermore, TPM can integrate to KSA and KPP. Mission efficacy is the ability to produce a desired result or effect and is measured through criteria in the form of MOEs. MOEs help determine if a task is achieving its intended results. MOEs help to answer the question "Is the system doing the right things?" MOPs help measure the accomplishment of a system task. They answer questions such as "Was an action taken?" or "Is the system doing things right?" TPMs and KPPs are the "implementation" measures of success that should be traceable to the MOPs, MOEs and MOSs [11].

As the design process of system continues, KPPs, TPMs, MOPs and MOEs get more and more clear, and reach the requirements of higher hierarchy through continuous test.

4 Analysis Architecture of Military Helicopter Requirements

Through the above analysis, based on DoDAF2.0 and MEG, an analysis architecture frame is proposed to deal with requirements demonstration, including mission viewpoint (MV), system-of-system viewpoint (SoV), system viewpoint (SV) and measure viewpoint (MeV). Every viewpoint is composed of several models (see Table 1).

In analysis architecture, there is no fixed order, and many analysis procedures are iterative. Take Mev-1 for example, the data of measures of success derive from mission analysis, meantime, the data of measures of effectiveness may affect MeV-1. Therefore, an advice procedure is shown in Fig. 2.

5 Viewpoint of Military Helicopter Analysis Architecture

5.1 Mission Viewpoint (MV)

MVs are used to analyze the original requirements of the army, system-of-system and system, decompose the mission into tasks, and set detailed conditions for scenarios. The mission viewpoint consists of 3 models, including Mission Descriptions (MV-1), Mission Decomposition Tree (MV-2) and Mission Scenarios (MV-3).

Mission Descriptions (MV-1). MV-1 is usually a structured text product to describe the mission of army of interest. Definition of the mission starts with defining the strategic demand with linkage to the National Defense Strategy, Military Strategy, Joint War fighting Concept, etc. Based on the analysis, mission can be decomposed to several tasks.

Mission Decomposition Tree (MV-2). MV-2 is to break the mission into executable tasks, including the name, target, imaginary enemy, time frame of every task. Besides, the

Table 1. The viewpoint composition

Viewpoint	Model	Name	Data Type
Mission Viewpoint (MV)	MV-1	Mission Descriptions	Structured Text
	MV-2	Mission Decomposition Tree	Use Case Diagram
	MV-3	Mission Scenarios	Tabular Form
System-of-system Viewpoint (SoV)	SoV-1	Operational Relationships of Organization	Class Diagram
	SoV-2	Operational Activity Flow	Activity Diagram
	SoV-3	Event-Trace Description	Sequence Diagram
System Viewpoint (SV)	SV-1	System to Unit Mapping	Mapping Matrix
	SV-2	Systems Function Flow	Activity Diagram
	SV-3	Systems State Transition Description	State Machine Diagram
	SV-4	Systems Function to Operational Activity Traceability Matrix	Mapping Matrix
Measure Viewpoint (MeV)	MeV-1	Measures of Success	Tabular Form
	MeV-2	Measures of Effectiveness	Tabular Form
	MeV-3	Measures of Performance	Tabular Form
	MeV-4	Hierarchy of Measures	Use Case Diagram

completed standards are necessary to build measure viewpoint. MV-2 can be expressed using Use Case Diagram in UML (see Fig. 3).

Mission Scenarios (MV-3). MV-3 is to describe the detail of the possible scenarios. More attentions are focused on the time, environment, geographic area, conflict, threat laydown, red and blue forces, and order of battle. MV-3 can be expressed in tabular form (see Table 2).

5.2 System-of-System Viewpoint (SoV)

SoVs are used to analyze business information of system-of-system, including composition, activities of units, and information transmitted among units. There are 3 models in system-of-system viewpoint, including Operational Relationships of Organization (SoV-1), Operational Activity Flow (SoV-2) and Event-Trace Description (SoV-3).

Operational relationships of organization (SoV-1). SoV-1 describes the composition of participating units and their relationships. Lines with special information are used to

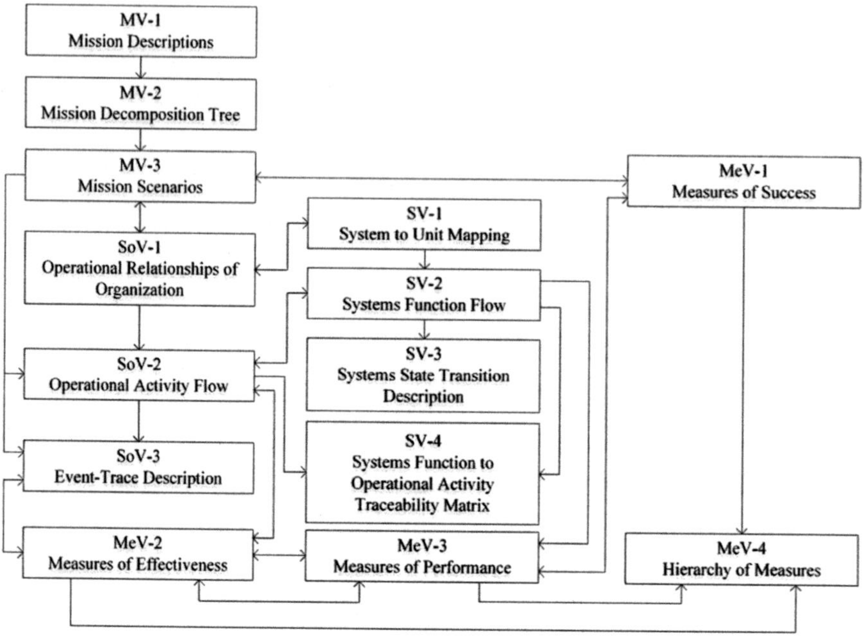

Fig. 2. An advice procedure

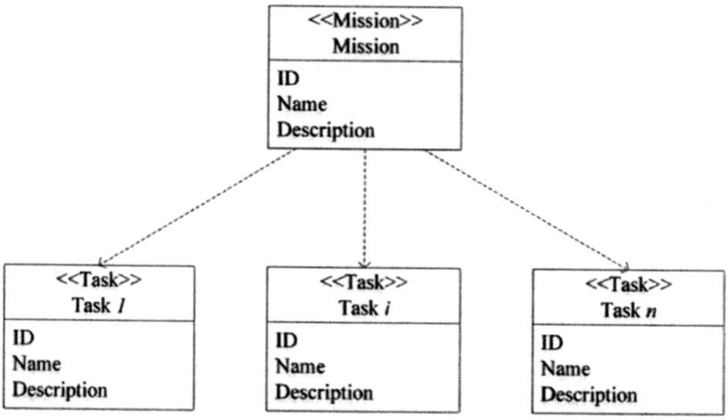

Fig. 3. MV-2 Model Expression

Table 2. MV-3 model expression

ID	Name	Time	Background	Imaginary enemy	Environment description	Objective	Red force	Blue force
.	.	.	.	.	.	.	.	.

describe the relations such as communication, interface, order, and coordination. SoV-1 can be expressed using Class Diagram in UML as Fig. 4.

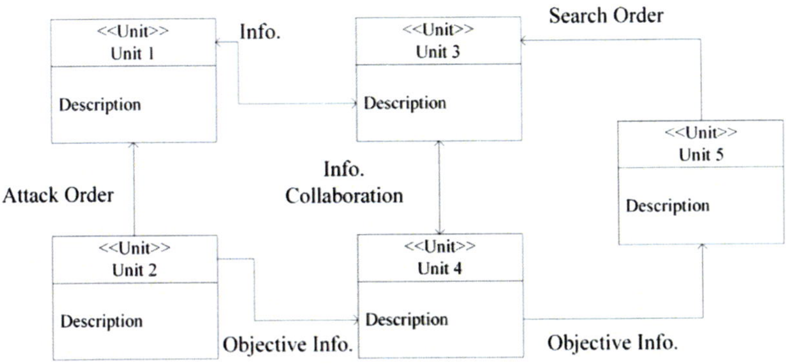

Fig. 4. SoV-1 model expression

Operational activity flow (SoV-2). SoV-2 is the same as OV-5b in DoDAF2.0, which describes operational activities of units, and their relationships among activities, inputs and outputs. Operational activities are closely related to operational rules, and can be considered standard actions according to operational rules. SoV-2 can be expressed using Activity Diagram in UML (see Fig. 5).

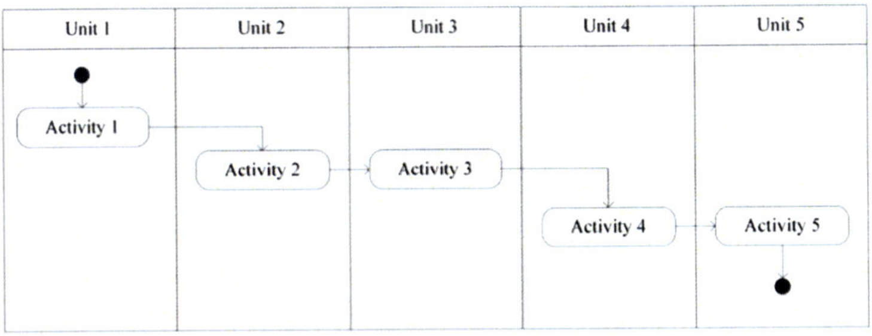

Fig. 5. SoV-2 model expression

Event-trace description (SoV-3). SoV-3 is the same as OV-6c in DoDAF2.0, which provides a time-ordered examination as a result of a particular scenario. Each event-trace diagram should have an accompanying description that defines the particular scenario or situation. SoV-3 can be expressed using Sequence Diagram in UML (see Fig. 6). What needs to be explained is dotted lines under participating units represent time continuing. The lines point to itself represent activities and lines point to other units represent information transmitted.

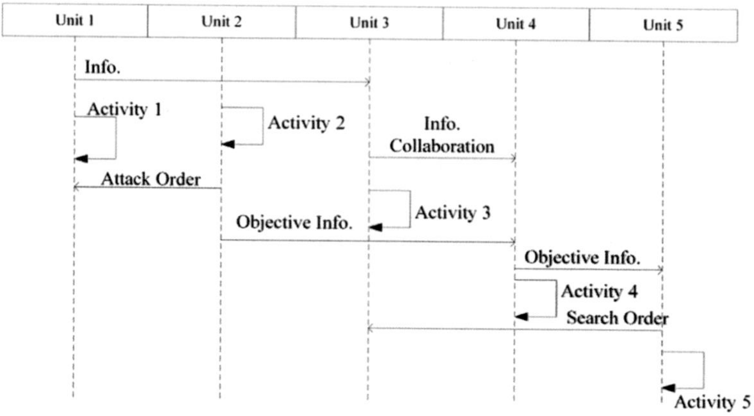

Fig. 6. SoV-3 model expression

5.3 System Viewpoint (SV)

The core elements of SoVs are participating units and their activities, which are used to draw the systems required and their functions. There are 4 models in system viewpoint, including System to Unit Mapping (SV-1), Systems Function Flow (SV-2), Systems State Transition Description (SV-3) and Systems Function to Operational Activity Traceability Matrix (SV-4).

System to unit mapping (SV-1). Based on the results of participating units, finding the corresponding systems are very important. SV-1 is used to analyze systems may be used according to current or future situation. One unit may own one or more systems.SV-1 can be expressed using mapping matrix (see Fig. 7).

	System 1	......	System *n*
Unit 1	1	0	1
Unit 2	0	1	0
Unit 3	1	0	0
Unit 4	0	1	0
Unit 5	0	0	1

Fig. 7. SV-1 model expression

Systems function flow (SV-2). SV-2 is the same as SV-4 in DoDAF2.0, which describes system functions, and their relationships among functions, inputs and outputs. SV-2 is similar to SoV-2 in use. Functions represent the characteristic of the systems and describe how to use the system. SV-2 can be expressed using Activity Diagram in UML (see Fig. 8).

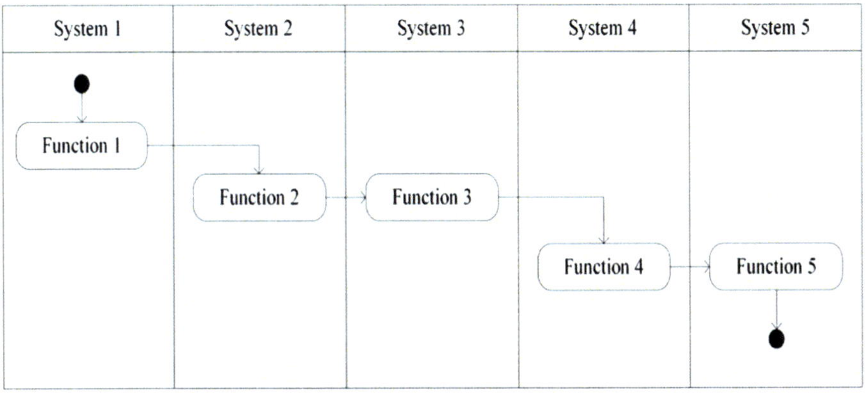

Fig. 8. SV-2 model expression

Systems state transition description (SV-3). SV-3 is the same as SV-10b in DoDAF2.0, which can describe the explicit sequencing of the functions. SV-3 can be expressed using State Machine Diagram in UML (see Fig. 9).

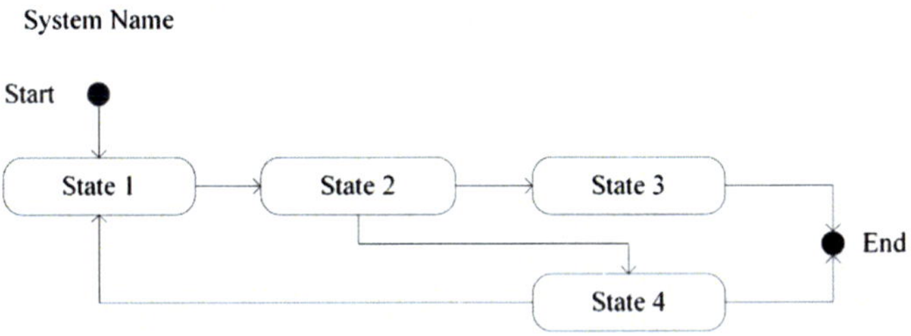

Fig. 9. SV-3 model expression

Systems function to operational activity traceability matrix (SV-4). SV-4 is the same as SV-5b in DoDAF2.0, which describes the mapping of system functions to operational activities to transform operational activity to system function. In SV-4, the relationship between operational activity and system function becomes clear though corresponding relation. SV-4 can be expressed using mapping matrix (see Fig. 10).

5.4 Measure Viewpoint (MeV)

Metrics are measures of quantitative assessment commonly used for assessing, comparing, and tracking performance of the mission or system. MeVs are used to describe the metrics. According to Fig. 1, several terms are used to refer to the metrics, including Measures of Success (MOS), Measures of Effectiveness (MOE), Measures of Performance (MOP), technical performance measure (TPM) and key performance parameter (KPP). For military helicopters architecture analysis, only MOSs, MOEs and MOPs are

	Function 1	......	Function *n*
Activity 1	1	0	1
Activity 2	0	1	0
Activity 3	1	0	0
Activity 4	0	1	0
Activity 5	0	0	1

Fig. 10. SV-4 model expression

required. For the reason of hierarchy of measures, MeV consists of 4 models, including Measures of Success (MeV-1), Measures of Effectiveness (MeV-2), Measures of Performance (MeV-3), and Hierarchy of Measures (MeV-4).

Measures of success (MeV-1). MeV-1 describes the ability to produce a desired result or effect, and help determine if a task is achieving its intended results. Therefore, the metrics data derive from MVs. In other words, MeV-1 measures the success of tasks in MV-2 with the scenarios in MV-3. In fact, one task may correspond to one or more metrics. MeV-1 can be expressed in tabular form (see Table 3).

Table 3. MeV-1 model expression

ID	Task	Scenarios	MOSs
1	Attack armored targets	Plateau counter-attack in self-defense	The probability of detecting and identifying enemy
			The number of destroying tanks with missiles
	......	......	......

Measures of effectiveness (MeV-2). MeV-2 describes indicate the target value of success or favorable result. The MOEs are closely linked to the MOPs in order to adequately reflect how the systems, system-of-system, or capabilities are used to achieve the objectives. In military helicopters architecture analysis, MeV-2 measures the capabilities mapping to activities from SoV-2 and interaction information from SoV-3. Normally, an MOE is considered a function of contributing MOPs and environment. MeV-2 can be expressed in tabular form (see Table 4).

Measures of performance (MeV-3). MeV-3 describes the basic characteristics of system. The characteristics are closely interrelated to system functions. Therefore, MeV-3 measures the specific performance of system function. Normally, the data of MeV-3 can obtain form SoV-3. MeV-3 can be expressed in tabular form (see Table 5).

Table 4. MeV-2 model expression

ID	Activities/Interaction	Capabilities	MOEs
1	Attack tanks	Attack capability	Kill probability The number of missiles used
2	Receive order	Information received capability	The rate of success of communication
	……	……	……

Table 5. MeV-3 model expression

ID	Functions	MOPs
1	Tracking objective	Tracking probability
2	Launch missile	The hit ratio of missile
	……	……

Hierarchy of measures (MeV-4). From above analysis, measures are divided into several hierarchies, from MOSs to MOEs to MOPs. MeV-4 is used to describe the relationships among above measures. MeV-3 can be expressed using Use Case Diagram in UML (see Fig. 11).

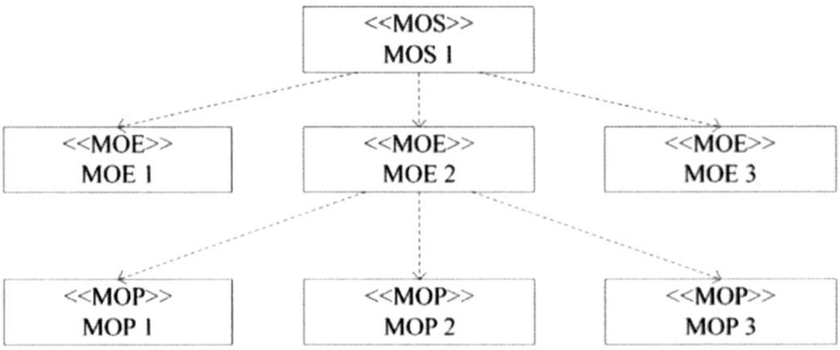

Fig. 11. MeV-4 model expression

6 Modeling Approach of Meta Activity

6.1 Concept of Meta Activity

Through the abov– that activity is an important element. Firstly, operational activities show how to use system to accomplish the tasks. Meantime, operational activities affect the combat results under given systems. Secondly, operational activities are the core of

SoVs, and draw the requirements of SVs. So, a good analysis of operational activities is the foundation for military helicopter architecture analysis. Therefore, it is necessary to build analysis rules of operational activities. In this section, the concept and rules of metal-activity is given to support base data and establish uniform standard for architecture analysis.

Metal-activity is minimum activity unit, which is independent, and can realize combat target [12]. Any metal-activity can not include others. Military helicopters complete different tasks, though different combinations to display different capabilities.

According to the above definition, based on event-flow-chart, a metal-activity model is established as follows:

$$Ma = \{N, Ca, Co\} \tag{1}$$

where *N* is the name of metal-activity, *Ca* is the capability driven by metal-activity and *Co* is the executive conditions of metal-activity such as environment, social conditions.

Furthermore, *Ma.t*, *Ma.T*, *Ma.p*, *Ma.pr* is respectively start time, duration, occurrence probability and priority of metal-activity. Then a metal-activity aggregation is defined as

$$MA = \{Ma_i | i = 1, 2, \ldots, n\} \tag{2}$$

The arc of metal-activity represents the connection of two metal-activities, which is defined as

$$A = \{a_j = (Ma_l, Ma_k) | j = 1, 2, \ldots, m; Ma_l, Ma_k \in MA, l \neq k\} \tag{3}$$

where Ma_l is start event, and Ma_k is end event.

Metal-activity input aggregation is defined as

$$Ma_i.in = \{Ma_l | Ma_l \in MA, (Ma_l, Ma_i \in A)\} \tag{4}$$

Metal-activity output aggregation is defined as

$$Ma_i.out = \{Ma_l | Ma_l \in MA, (Ma_i, Ma_l \in A)\} \tag{5}$$

A metal-activity model can be described as Fig. 12.

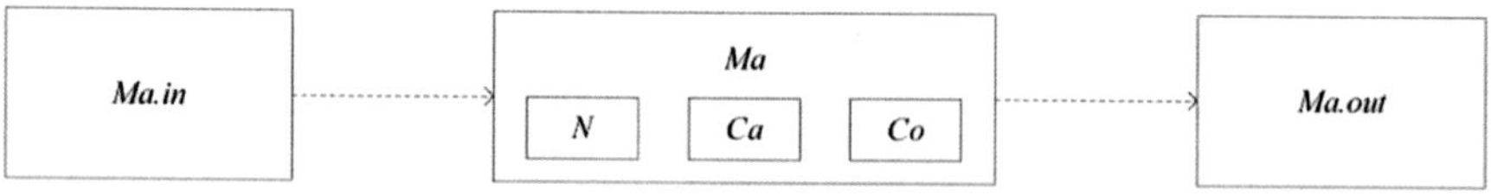

Fig. 12. A metal-activity model

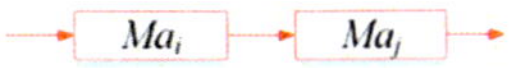

Fig. 13. Sequence rule

7 Modeling Rules of Meta Activity

According to the definition and model of metal-activities, the rules should be defined when analyzing architecture of military helicopters.

Sequence rule. $\exists\, Ma_i, Ma_j \in MA(i \neq j)$, make $Ma_i.out = \{Ma_j\}$, that is only one metal-activity output of Ma_i. Meantime, $Ma_j.in = \{Ma_i\}$, that is only one metal-activity input of Ma_j. The sequence rule is shown in Fig. 13.

Priority judgment rule. $\exists\, Ma_i, Ma_j, Ma_k \in MA(i \neq j \neq k)$, and meet.

(1) $Ma_j.in = Ma_k.in = \{Ma_i\}$.
(2) $Ma_j.t = Ma_k.t$
(3) $Ma_j.pr < Ma_k.pr$, that is $Ma_i.out = \{Ma_k\}$.

The priority judgment rule is shown in Fig. 14.

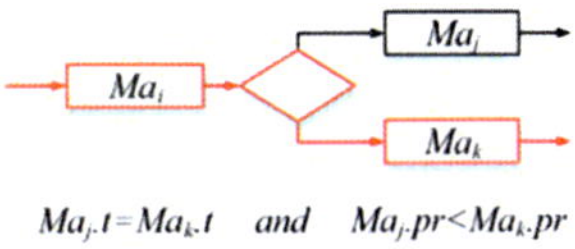

Fig. 14. Priority judgment rule

Parallel rule. $\exists\, Ma_i, Ma_{j1}, Ma_{j2}, \ldots Ma_{jn} \in MA(n \geq 2)$, and meet.

(1) $Ma_{jk}.in = \{Ma_i\}, k = 1,2,\ldots n$.
(2) $Ma_{j1}.t = Ma_{j2}.t = \ldots = Ma_{jn}.t$
(3) card $(\{a_k = (Ma_i, Ma_{jk}) | k = 1,2,\ldots n\}) = n$, that is every arc between Ma_i and Ma_{jk} exits.

The parallel rule is shown in Fig. 15.

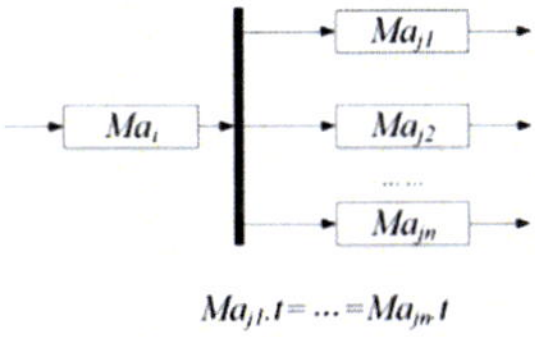

Fig. 15. Parallel rule

Random rule. $\exists$ Ma_i, Ma_{j1}, Ma_{j2},...Ma_{jn} $\in$ $MA(n \geq 2)$, and meet.

(1) $Ma_{jk}.in = \{Ma_i\}$, $k = 1,2,\ldots n$.
(2) $Ma_{j1}.t = Ma_{j2}.t = \ldots = Ma_{jn}.t$
(3) $Ma_{j1}.p + Ma_{j2}.p + \ldots + Ma_{jn}.p = 1$.
(4) $Ma_{jk}.pr = null$, $k = 1,2,\ldots n$.

The random rule is shown in Fig. 16.

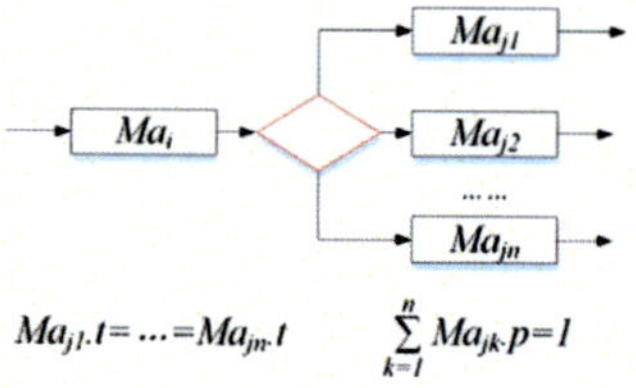

Fig. 16. Random rule

Concentration rule. $\exists$ Ma_i, Ma_{j1}, Ma_{j2},...Ma_{jn} $\in$ $MA(n \geq 2)$, and meet.

(1) $Ma_{jk}.out = \{Ma_i\}$, $k = 1,2,\ldots n$.
(2) $Ma_{j1}.t + Ma_{j1}.T = \ldots = Ma_{jn}.t + Ma_{jn}.T$
(3) $Ma_{jk}.pr = null$, $k = 1,2,\ldots n$.

The concentration rule is shown in Fig. 17.

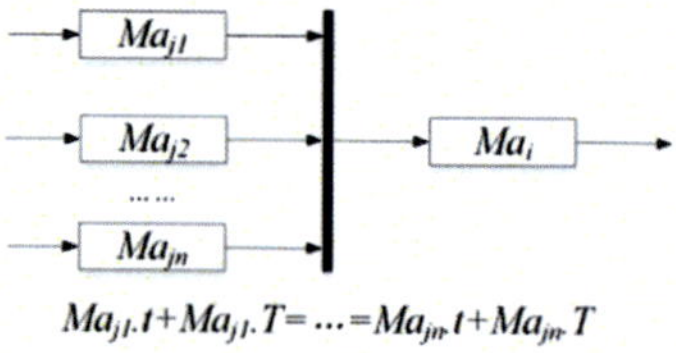

Fig. 17. Concentration rule

Merge rule. $\exists$ Ma_i, Ma_{j1}, Ma_{j2},...Ma_{jn} $\in$ $MA(n \geq 2)$, and meet.

(1) $Ma_{jk}.out = \{Ma_i\}$, $k = 1,2,\ldots n$.
(2) $Ma_{jk}.pr = null$, $k = 1,2,\ldots n$.

The merge rule is shown in Fig. 18.

Cycle rule. $\exists$ Ma_{i1}, Ma_{i1}, ...Ma_{in} $\in$ $MA(n \geq 2)$, and meet.

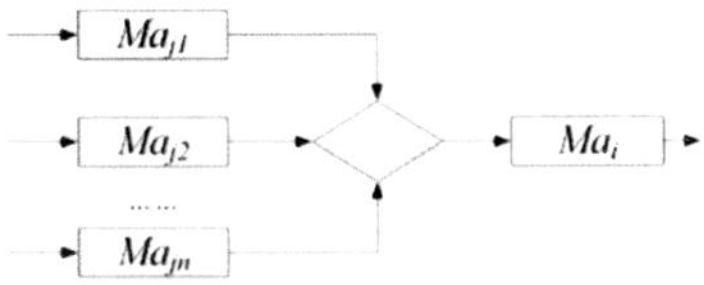

Fig. 18. Merge rule

(1) MA'⊂ MA, and MA' ≠ Ø.
(2) A'⊂ A, and A' ≠ Ø.
(3) card (MA') ≥ 2.

The cycle rule is shown in Fig. 19.

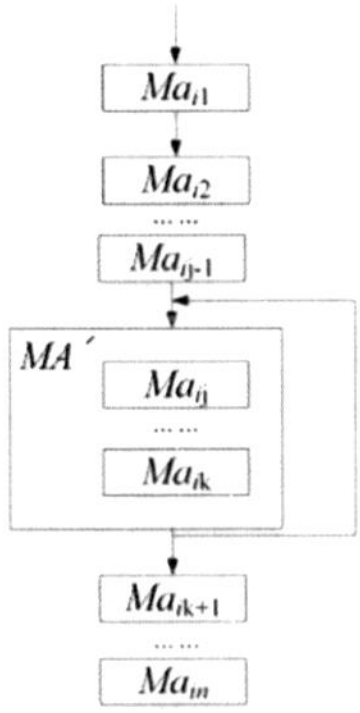

Fig. 19. Cycle rule

8 The Analysis of Capability and Executive Conditions of Metal-Activity

There is a mapping relationship between metal-activity and capability, which is conversion of executive level of mental-activity and capability. Meantime, capability is the original requirement drawing MOE. Therefore, capability is the bridge connecting activity and MOE. What is different from capability, *Ca* in mental-activity is minimum capability, corresponding to basic executive level. The relationship between mental-activity and *Ca* is called mental-activity-*Ca* rule. For different systems, mental-activity-*Ca* rule is different. There is no elaborate in this paper.

Co is the executive conditions of metal-activity, including natural conditions, military conditions and social conditions.

Natural conditions. For military helicopter, natural conditions are the relevant factors of land, air and sea environment involved in the tasks. Natural conditions have great influence on military helicopter during its mission, for example, different altitude has influence on the flight performance.

Military conditions. Military conditions are the military factors related to helicopter carrying out combat tasks, including preparation time, helicopter own combat ability, intelligence support ability and enemy situation threat ability, etc.

Social conditions. Social conditions are the political, economic, cultural and transportation conditions that influence military operations. For example, traffic in area can affect the rapid deployment of helicopter.

Through the above analysis, a database of mental-activity-*Ca* rule can be established. And the database can provide basic and uniform support for activity and capability analyzing, which can improve the analysis efficiency.

9 Conclusion

Based on DoDAF2.0 and MEG, a universal and simple analysis frame is proposed for military helicopter. On this basis, the concept and modeling rules of mental-activity is established. The advantages of the architecture are as follows:

(1) The universal analysis frame can be applied requirements demonstration of similar equipment, not only military helicopter. The process including mission analysis, system-of-system analysis, system analysis and measures analysis, is a complete forward design flow, which can solve general architecture analysis problems.
(2) The activity analysis is the core of equipment requirements demonstration, and mental-activity-*Ca* rule database can greatly improve the efficiency of demonstration.

Future research will be undertaken in 2 aspects, which are data transmission and measures evaluating method. Data transmission means to solve the problem of model-based data flowing, and measures evaluating method is used to carry out evaluating of MOSs, MOEs and MOPs. This could be useful for architecture analysis method practice.

References

1. Yang, K., Yang, Z., Tan, Y., et al.: Review of the evaluation methods of equipment system of system facing the contribution rate. Syst. Eng. Electron. **41**(2), 311–321 (2019)
2. Cheng, B., Tan, Y., Huang, W., et al.: Weapon system of system evaluation based on capability requirement perspective. Syst. Eng. Electron. **33**(2), 320–323 (2011)
3. Jianrong, C.H.E.N., Qisheng, G.U.O.: Method of determining requirements of operation capabilities based on effects-driven and exploratory. Analysis **24**(1), 8–13 (2010)
4. Wang, J., Guo, J., Zhou, Y.: Research on requirements analysis method of joint operations capability **23**(2), 30–34 (2009)
5. Wang, K., Wang, J., Fan, Y. et al.: Mode of capability-based equipment requirement demonstration. J. Acad. Armored Force Eng. **28**(3), 6–10 (2014)
6. DoD Architecture Framework Version 2.0. Volume 2: Architectural Data and Models Architect's Guide (2009)

7. Mission Engineering Guide: Office of the Under Secretary of Defense for Research and Engineering (2020)
8. Guo, Q, Li, T., Yang, X. et al.: Research on requirements generation process of army weapon equipment. J. Acad. Armored Force Eng. **22**(2), 6–9, 19 (2008)
9. Qisheng, G.U.O., Zhiming, G.U.O., Ge, M.U.: Research on the fundamental theory of equipment requirement demonstration engineerization. J. Acad. Armored Force Eng. **26**(1), 1–4 (2012)
10. Yepin, Q.U., Huiming, J.I.N., Zhaoxiong, H.E.: Carrier based aircraft equipment system-of-system and index demonstration method. Acta Aeronaut. Astronaut. Sin. **39**(5), 1–12 (2018)
11. System Engineering Handbook Version 4.0: A Guide for System Life Cycle Processes and Activities (2015)
12. Chen, Y., Dou, Y., Chen, B., et al.: Research on capability requirement generation of weapon system-of-systems based on operational activity decomposition. Syst. Eng.-Theory Oractice **31**(Sup.1), 154–163 (2011)

Spatial Temporal Characteristics and Driving Forces of CHINA's Aircraft Industry

Deng Yangu[1], Liu Guoliang She Ying[1(✉)], Jia Ling[1], and Dai Lu[2]

[1] School of Economics and Management, Nanchang Hangkong University, Nanchang 330061, People's Republic of China
sheying@nchu.edu.cn
[2] Business School, Ningbo University, Ningbo 315211, People's Republic of China

Abstract. The development of China's aircraft industry remained an increasing concern from the government and the public. Using the GIS method and the panel data from 2006 to 2019, this study visualized the spatial temporal characteristics of the aircraft industry in China from the following perspectives: the concentration ratio, the R&D intensity and the comprehensive development level. The characteristics were summarized as follows: (1) Across China's 23 provinces and municipal cities, the average agglomeration degree maintained a relatively low level during the study period, but there was an uneven trend among different regions. The agglomeration degree was decreasing in traditional western production centers and increasing in the emerging eastern production centers. (2) The R&D intensity went up and down from year to year and varied significantly among different regions. (3) The comprehensive development scores in western China kept decreasing and those in eastern China maintained increasing.

Keywords: Temporal and spatial characteristics · Agglomeration degree · R&D intensity · Development level · Aircraft industry

1 Introduction

For some time now, the spatial redistribution of the aircraft industry attracts many concerns from governments and the public. Observing the facts that the entry or the exit of the aircraft industry in the specific region will promote or depress the economy, this focus is well-grounded. It is therefore important to uncover the spatial temporal characters of the aircraft industry and identify the driving forces behind those characters.

From a historical perspective, the spatial temporal patterns in different countries presented different characteristics. In the US, the spatial diffusion of the aircraft industry was in line with its industrialization spatial expansion. The aircraft industry of the US began in the northeastern region and drifted toward the south and the east. Seattle got a lot of government financial support for military aircraft manufacturing during the World War II. Thus, it became the headquarters of Boeing and the production and the R&D centers of the aircraft industry in the world [1]. In 2001, the headquarters of Boeing moved from Seattle to Chicago for lower tax rate and geographical proximity to markets. In 2009,

Chinese Society of Aeronautics and Astronautics: CASTYSF 2022, LNEE 972, pp. 660–674, 2023.
https://doi.org/10.1007/978-981-19-7652-0_58

the second production line of Boeing 787 has been established in South Carolina. All those signs indicated that the spatial distribution evolution of the aircraft industry in the US.

The spatial temporal characteristics of the aircraft industry in China were quite different from those of the US, which was in reverse with its industrialization diffusion [2]. Generally speaking, the spatial distribution of the aircraft industry had experienced three stages. The first stage (1960–1980), China started to establish its aircraft industry initiated by the Third-line Construction Campaign[1]. For the sake of national security, the aviation industry was mainly distributed in key investment "Third-line" cities, such as Xi'an in Shaanxi Province, Chengdu in Sichuan Province, Nanchang in Jiangxi Province and so on. Due to the role of "path dependence", the aviation industry in these areas expanded towards the early of 1980s. The second stage (1980–2000), with the deepening of market economic reform, the aircraft industry became an "isolated island" which had little connection with the local economy. The reason lied in that those key investment "Third-line" cities were all western and central cities with relative scare capital and technology resources, which violated the capital-intensive and technology-intensive nature of the aircraft industry. Therefore, in this stage, the distribution of the aircraft industry commanded by the central government caused a low-efficient development. The third stage (2000-now), after 2000, in order to make use of the advantages of skilled labor, technology and capital resources in the eastern region, China tried to relocate the aircraft industry distribution by introducing new production lines in the eastern provinces. In this context, Tianjin, Zhuhai and other eastern cities have been chosen as new key aircraft industrial base. The aircraft industries in these cities have achieved rapid development by introducing aircraft assembly lines and establishing national aviation industry parks.

At present, the central government attached great importance to the aircraft industry. According to *the medium and long-term development plan of the civil aircraft industry (2013–2020)*, the annual revenue of the aircraft industry will exceed 1 trillion RMB Yuan by 2020. This huge revenue attracted local governments at the provincial level to make concerted efforts to develop their own aircraft industry. A province may try to preserve or improve its own enterprises' market share by providing financial supports or other local incentive policies. In fact, about 23 provinces have some aircraft capacity.

Would the government support cause a low level of repeated construction or a high level of agglomeration in aircraft industry distribution? In this context, what were the unique characteristics of spatial temporal evolution of China's aircraft industry? And what were the driving forces behind the spatial temporal agglomeration evolution? The answers to these questions will help us to understand the reason behind the spatial temporal evolution of the aircraft industry and provide useful policy implications.

Compare with other high technology and capital intensive industries, the aircraft industry required for even larger scale of production and much intensive R&D investment. Previous literatures documented that the comprehensive development level of the aircraft industries were influenced mainly by the production scale and R&D intensity [3, 5]. Therefore, this study will uncover the spatial temporal characteristics from

[1] Third Line Construction: A large scale infrastructure construction, spanning from 1964 to 1980, focused on national defense, and industrial transportation.

three aspects: the agglomeration degree, the R&D intensity and the comprehensive development level.

The civil aircraft industry agglomeration attracted many concerns in the literature [6–9] (Edward and Usha, 2016). As an industry that tends to agglomerate and requires huge R&D inputs, the aircraft industry has been studied in the previous literatures mainly from two aspects: the agglomeration and innovation patterns. Some literatures investigated how the agglomeration of the aircraft industry came into being? Or identifying influencing factors of the aircraft industry agglomeration, but no consensus had been achieved [9–11]. A hand of literatures emphasized the importance of geopolitical ambitions during the early stage of aircraft agglomeration. After investigated the aircraft agglomerations in Montreal, Seattle, Toulouse and Toronto, Niosi and Zhegu [11] demonstrated during 1960s, the trade protection policy implemented by the US government contributed a lot for the early formation of the US aircraft agglomeration. Controversially, other literatures demonstrated market force was even more important than government attention in the aircraft industry development. By survey from selected aircraft firms, Romero [3] identified the manufacturing advantages were the driving forces for aircraft agglomeration in Mexico, and incentive policies play little role. Spreen [4] emphasized the importance of the domestic market for civil aircraft agglomeration and pointed out the failure of Japan and South Korea in aircraft mainly came from their small domestic demand and improper international marketing strategy. Steenhuis & Kiefer [5] investigated three possible driving forces of the early-stage agglomeration development in the State of Washington and identified the main driver was manufacturers-led coordinating mechanism, the other two possible drivers: incentive policy and government organization contributed little to the agglomeration development.

Another branch of literatures discussed the aircraft industry from the perspective of R&D. McGuire [6] investigated the important role of R&D played and found in the UK, during 1990's R&D as a percentage of sales in the aircraft industry was about 10% which was much higher than the average level (2%). Vértesy [12] held that the international R&D cooperation in aircraft industry was well above the average of other high technology industries, but it had a decline trend after 1990. Some studies explored the technology diffusion patterns. After investigated the aircraft industry in Montreal, Seattle, Toulouse and Toronto, Wagner & Walton [13] found that compared with local technology spillovers, international spillovers contributed more to the aircraft industry development. Using a sample of 300 aircraft from 1945 to 2000 and logistic regression method, [14] found that when firm resources available was not matched with the product resource requirement, the firm would develop new aircraft products through horizontal alliances with competitors, which will lead to a technology spillover among aircraft plants.

Studies about China's aircraft industry mainly focused on the comprehensive development level evaluation. Based on Porter's Diamond Theory, most studies first established an evaluation index system, and then using global principal components methods or factor analysis method to evaluate the comprehensive development level [15–17]. In addition, using Czamanski's method, Chu (2010) first identified the aircraft industry agglomeration in China.

Although the existing literatures had explored the aircraft industry from three aspects: the agglomeration, the R&D input and the development level of the aircraft industry, some points still require further attention. (1) Existing literatures only investigated the spatial temporal evolution and its driving forces of the aircraft industry from the perspective of agglomeration, which failed to explore the aircraft industry from the perspective of the R&D intensity and the development level. Since the R&D intensity and the agglomeration is crucially important for the development of the aircraft industry, it is necessary to investigate the aircraft industry from the three aspects together. (2) In previous literatures, most studies were conducted by survey or through personal interviews, few literatures analyzed by quantitative methods for lack of data. (3) Countries studied in previous literatures about the aircraft industry, mostly were about the US and the EU. Some studies concerned about Asian countries only involved Japan and South Korea, few studies concerned the aircraft industry agglomeration in China.

To better understand the spatial temporal evolution pattern of the aircraft industry in China and its driving forces, we visualized the spatial temporal evolution pattern of the aircraft industry using the GIS method and the panel data from 2006 to 2016, and then identified the driving forces behind the evolution patter. This research contributed to the existing literature in the following two ways. (1) By illustrated the spatial temporal evolution pattern of the aircraft industry from three perspectives: the agglomeration degree, the R&D intensity and the development level, this study found that there was a separation of production centers from R&D centers and the development level of the aircraft industry in the eastern region were improved much faster than that in western region. (2) The different driving forces behind the spatial temporal evolution pattern of the aircraft industry had also been identified. Therefore, this paper can not only provide a more accurate and complete understanding of the spatial temporal evolution pattern of China's aircraft industry, but also present some policy implication for aviation industry in the near future.

2 Indicator Selection, Data Sources and Methodology

2.1 Indicator Selection

Considering the characteristics of the aircraft industry and the availability of data, this paper established an evaluation index system to describe the development of the aircraft industry, which included three aspects: product supply, product demand and technology innovation. As an industry with increasing returns to scale, the aircraft industry needed sustained investment to maximize its profit. To describe the supply of aircraft products, this study used the following index: the total fixed assets, the number of aircraft enterprises and the number of staff. The demand of aircraft products can be measured by the total industrial output value, sales revenue, total profit and export value. Technological innovation was a huge driving force for the development of the aviation industry, which was represented by: R&D intensity and proportion of R&D staff.

2.2 Data Sources

The data for aircraft industry used in this paper were collected from China Civil Aircraft Industry Statistical Yearbook 2007–2020, which covered the data from 2006 to 2019 and was the longest study period the data available. The provinces and municipal cities covered in this Yearbook were varied from year to year, reflecting the aircraft industries moved out or in some specific provinces. For example, the Yearbook didn't cover the data of Tianjin from 2006 to 2009, and it excluded Anhui, Fujian, Gansu, Hebei, Henan, He Longjiang, Jilin, Shaanxi, Yunnan Provinces in 2016. Therefore, the panel data used in this study was unbalanced.

2.3 Methodology

2.3.1 Concentration Ratio of Industry

Many approaches can be used to measure the industry agglomeration degree, such as Spatial Gini Coefficient, Herfindahl index, EG index, etc. Based on the availability of data and the characteristics of the aircraft industry, we use the Concentration ratio of industry in this paper to measure the industry agglomeration degree. The Concentration ratio of industry can be measured by the ratio of industrial output (or R&D, sales, employment, etc.) in a specific region to that of the whole market. The concentration ratio of industry (CR_i) was calculated for 2006–2019 as Eq. (1). The larger of the concentration ratio of industry represented the greater degree of agglomeration. Theoretically, the CR above 10% in the specific region indicated the agglomeration had been formed in this region.

$$CR_{\mathrm{i}} = \frac{X_i}{\sum_{i=1}^{N} X_i} \tag{1}$$

where CR_i represented the concentration ratio of the aircraft industry in province i; X_i indicated the industrial output or employment in the province i, $\sum_{i=1}^{N} X_i$ represented the industrial output of the total market. In this study, in order to explore the concentration ratio of the aircraft industry comprehensively, two indicators (industrial output and R&D intensity) have been used to calculate CR_i. More specifically, CR_{ii} will be used to represent the industrial output concentration ratio and CR_{ir} will be used to represent the R&D intensity concentration ratio of the aircraft industry.

2.3.2 Factor Analysis Method

Through combining a large number of related factors into a small number of unrelated comprehensive indicators, i.e. common factors, Factor Analysis Method can be used to evaluate the comprehensive development situation of the aircraft industry. It can eliminate the correlation between the original indicators, avoid the duplication of information and overcome the subjectivity of weight determination, thus making the evaluation results more accurate and objective. We established the model of Factor Analysis Method as Eq. (2):

$$X_{p\times 1} = A_{p\times m} F_{m\times 1} + \varepsilon_{p\times 1} \tag{2}$$

where X represented an observable random vector; F was an unobservable random vector with m dimensions, called a common factor of X; ε was a special factor, each component of which represents the part of the original variable that cannot be explained by the common factor; A was the correlation coefficient between X and F, which indicated the degree of dependence of X_i on F_j. Generally, the number of common factors was determined according to variance contribution rate and cumulative variance contribution rate. More specifically, when the cumulative contribution rate of several common factors exceeded 85%, the other common factors can be ignored. Thus, the number of indicators can be simplified. Finally, the comprehensive score of the aircraft industry development can be calculated according to the Eq. (3).

$$F = \frac{\sum_j^n \lambda_{ij} x_{ij}}{\sum_j^n \lambda_{ij}} \tag{3}$$

where F was the comprehensive score of the aircraft industry development; λ refered to the eigenvalues of common factors; x_{ij} represented the common factors. The higher the F indicated the higher level of aircraft industry development and vice versa.

2.3.3 Standard Deviational Ellipse (SDE)

Standard deviation ellipse was a commonly used method for quantitative analysis of the overall characteristics of spatial distribution. It consisted of three elements: rotation angle θ, standard deviation along the main axis (long axis) and standard deviation along the auxiliary axis (short axis). The rotation angle θ reflected the main direction trend of the geographical elements distribution, the long axis indicated the discrete degree of geographical elements along the main direction trend and the short axis represented the discrete degree of geographical elements along the minor direction trend. Their mathematical expressions were as follows:

$$\theta = \frac{\left(\sum_{i=1}^n w_i^2 x_i'^2 - \sum_{i=1}^n w_i^2 y_i'^2\right) + \sqrt{\left(\sum_{i=1}^n w_i^2 x_i'^2 - \sum_{i=1}^n w_i^2 y_i'^2\right)^2 + +4\left(\sum_{i=1}^n w_i^2 x_i'^2 y_i'^2\right)^2}}{2\sum_{i=1}^n w_i^2 x\prime y\prime} \tag{4}$$

$$\delta_x = \sqrt{\frac{\sum_{i=1}^n \left(w_i x_i' \cos\theta - w_i y_i' \sin\theta\right)^2}{\sum_{i=1}^n w_i^2}} \tag{5}$$

$$\delta_y = \sqrt{\frac{\sum_{i=1}^n \left(w_i x_i' \sin\theta - w_i y_i' \cos\theta\right)^2}{\sum_{i=1}^n w_i^2}} \tag{6}$$

$$\overline{X_w} = \frac{\sum_{i=1}^n w_i x_i}{\sum_{i=1}^n w_i} \tag{7}$$

$$\overline{Y_w} = \frac{\sum_{i=1}^n w_i y_i}{\sum_{i=1}^n w_i} \tag{8}$$

where θ was the rotation angle, representing the angle of the North direction clockwise rotation to the longitudinal axis of the ellipse. $(x_1, y_1) \ldots (x_i, y_i) \ldots (x_n, y_n)$ meant the coordinates of point elements. w_i represented the weight, x'_i, y'_i indicated the coordinate deviation of point elements x_i, y_i from the mean center respectively. δ_x, δ_y were the standard deviation of x axis and y axis. $(\overline{X_w}, \overline{Y_w})$ represented weighted mean center.

Table 1. Evaluation indexes of aircraft industry in cities of China

First rank index	Second rank index	Unit
Aircraft product supply	Number of enterprises	/
	Number of staff	/
	The total fixed assets	Hundred million RMB
Aircraft product demand	Total industrial output value	Hundred million RMB
	Sales revenue	Hundred million RMB
	Total profit	Hundred million RMB
	Export value	Hundred million RMB
Technology innovation	R&D intensity	%
	Proportion of R&D staff	%

3 Spatial Temporal Characteristics of CHINA's Aircraft Industry

Using the above method and the data described in Table 1, this paper calculated the CR_i and the comprehensive development score of aircraft industry in China from 2006 to 2019. We analyzed the spatial temporal characteristics of the aircraft industry from the following three aspects: agglomeration degree, R&D intensity as well as development level.

3.1 Time Series Characteristics of the Aircraft Industry

Promoted by the national strategy, the aircraft industry had developed rapidly since 2000. Figure 1 presented the agglomeration degree (CR_{ii}), the R&D intensity, the development level and the industrial output from 2006 to 2019, which showed although the trends of the four indexes in the study period were different from each other. The industrial output maintained a steady increasing trend during the study period. The annual average industrial output swelled from 2674.98 million RMB in 2006 to 14710.34 million RMB in 2019, with an average annual increase of 34.61%; the R&D intensity increased even more rapidly than the industry output, which increased from 3.22 in 2006 to 32.48 in 2019, with an average annual increase of 69.85%, but its increasing trend was not as smooth as the industry output. It resumed their upward trajectories after huge downticks in 2009, 2013 and 2016. The reason might exist in the fact that compare to the industrial output, the R&D intensity can be more deeply influenced by government incentive policies which was volatile from year to year. Unlike the industrial output and the R&D intensity, the CR_{ii} remained almost unchanged with a slight fluctuation around 4.54 from 2006 to 2015, which reflected that the CR_{ii} was difficult to change because of the industrial inertia. Remarkably, the CR_{ii} soared to 7.14 in 2016. That might be explained by the fact that there were only data of 14 provinces presented by China Civil Aircraft Industry Yearbook in 2016, while there were data of 20–23 provinces given in other years. The fewer numbers of provinces involved in aircraft manufacturing, the higher CR_{ii} would be. The development score maintained stable and fluctuated around the zero, which indicated that the comprehensive development level improved little during the study period. That might be explained by the unbalanced development of the aircraft industry in different regions. Although the aircraft industry developed rapidly in some provinces, such as Tianjin and Guangdong, while in other provinces the aircraft industry was in decline. Thus, the average development level for the whole country seemed unchanged. It was worth noting that there had been a sharp increase of the industrial output since 2010. The reason might exist in the fact that in 2010, the central government implemented a series of incentive policies and introduced several important projects. For example, in 2010, Opinions on China's Low Altitude Airspace Management Reform was implemented. In the same year, the Airbus A320 assembly project was introduced in Tianjin, the helicopter headquarters of Aviation Industry Corporation of China also established in Tianjin, and Xizi Aircraft spare parts project with an annual industrial output of 1 billion RMB settled in Hangzhou Dajiangdong Industrial Cluster Area, Zhejiang.

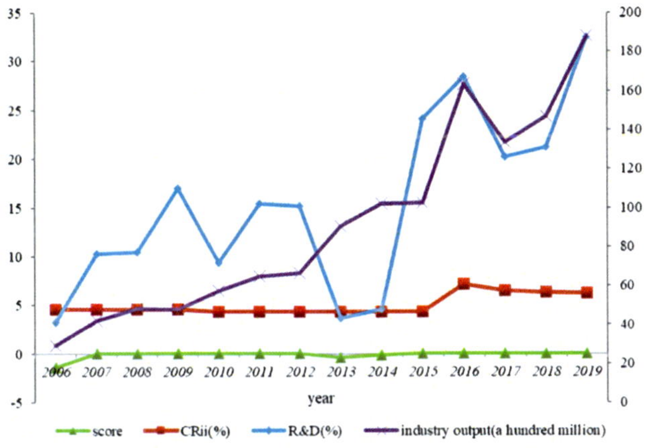

Fig. 1. The trend of the development of the aircraft industry during 2006–2019

3.2 The Spatial Temporal Pattern Evolution of the Aircraft Agglomeration

The civil aircraft industry had a trend of agglomeration because of its increasing return of scale nature. To investigate the agglomeration of the aircraft industry, the average CR_{ii} and CR_{ir} from 2006 to 2019 have been calculated in this study. There were totally three provinces' annual average CR_{ii} above 10% during the study period, which indicated the agglomeration degree of aircraft had been formed in those three provinces, i.e., Shaanxi, Liaoning and Sichuan. Jiangxi ranked the forth with a CR_{ii} of 6.92. Table 2 compared four aircraft industry production centers in China.

As shown in Table 2, Shaanxi, Liaoning and Sichuan were three traditional production centers, for in 2006 the industrial output of the three provinces were all above 10%. On one hand, the industrial output of the first three production centers accounted for 44.74% of the total industrial output, which revealed the tendency of preliminary agglomeration of the aircraft industry in China. On the other hand, the agglomeration degree of the aircraft industry was still at a relatively low level. Among all the production centers, Shaanxi, as the largest production center (measured by CR_{ii}), it only accounted for 22% of the total industrial output. Table 2 also indicated the eastward movement of the aircraft industry in China. The industrial output of Shaanxi was significantly higher than that in Liaoning and Sichuan in 2006. In 2006, the industrial output of Shaanxi was two times higher than that in Liaoning, 1.6 times higher than that in Sichuan, which indicated that Shaanxi was the most important production center in terms of industrial output in 2006. However, at the end of the study period, the agglomeration degree in western regions was decreasing. In 2019, the industrial output of Sichuan ranked first among four production centers, the industrial output in Sichuan only 1.29, 1.95 and 2.37 times than that of Shaanxi, Liaoning and Jiangxi. Respectively. During the study period, CR_{ii} of Shaanxi decreased remarkably, CR_{ii} of Liaoning maintained almost unchanged. In contrast, CR_{ii} of Jiangxi increased sharply from 4.06 to 8.68, with an increase rate of 214%.

Compared with CR_{ii}, CR_{ir} in the four production centers were even much lower. The annual average CR_{ir} was far below 10 for all those four productions. In 2006, CR_{ir} in Shaanxi and Liaoning was slightly above than 10.

Table 2 revealed two facts: (1) During the study period, the advantage of traditional aircraft productions, i.e. Shaanxi, Liaoning and Sichuan had been diminishing, while the emerging aircraft production, i.e., Jiangxi developed rapidly. (2) Although being the largest production centers, R&D input in Shaanxi, Liaoning, Sichuan and Jiangxi were still low.

3.3 The Spatial Temporal Pattern Evolution in the Comprehensive Development Level

Employed the Factor Analysis Method, we calculated the comprehensive development score (F) to measure the aircraft industry development level, then visualized it by GIS in Fig. 2. Generally speaking, during 2006–2019, only 10 from 23 provinces and municipalities' comprehensive scores were above 0, which indicates the development of the aircraft industry in those 7 provinces were better than the mean level. Those 10 provinces and municipalities were Shaanxi, Sichuan, Guizhou, Jiangsu, Liaoning, Fujian, Tianjin, Beijing, Henan and Guangdong. Among which, the score in Shaanxi, Sichuan, Jiangsu and Liaoning had a decreasing tendency and the score in Guangdong and Tianjin had a tendency of increase during the study period.

As shown by Fig. 2a, according to the comprehensive score, the 22 aircraft producing provinces and municipalities were divided into 5 levels. Specifically, Shaanxi ranked the first level; Liaoning, Gansu, Sichuan, Yunnan and Shaanxi belonged to the second level. The other provinces or municipalities were all below the second level. Among which, Shanghai and Zhejiang with comparatively abundant capital and technology belonged to the lowest development level. It reflected the fact that the aircraft industry developed much better in the previous "Third Line Construction" provinces than in other provinces in 2006. Until 2006, the spatial distribution of the aircraft industry had still been influenced by the "Third Line Construction Campaign".

Table 2. Comparison of the aircraft industry productions in China from 2006 to 2019

Region	Industrial output (million RMB)			R&D intensity			CRii (%)			CRir (%)		
	2006/	2019	Mean	2006/	2019	Mean	2006/	2019	Mean	2006	2019	Mean
Shaanxi	16819	54158	36926	7.07	19.76	10.48	27.34	16.01	21.33	12.16	3.38	3.85
Liaoning	8273	35831	22772	8.23	6.92	7.48	13.45	10.59	13.09	8.23	1.19	1.61
Sichuan	10151	69698	26470	3.00	10.86	3.00	16.50	20.60	13.95	4.44	1.86	2.81
Jiangxi	2501	29383	14381	0.66	57.88	10.09	4.06	8.68	6.92	0.98	9.90	4.15

As shown by Fig. 2d, in 2019, the spatial distribution of the aircraft industry had changed a lot compared to that of 2006. In 2019, Shaanxi, Guangdong and Sichuan ranked the first level. Yunnan, Fujian, Jiangxi, Henan, Hebei, Shaanxi, Beijing, Liaoning and Jilin belonged to the second level. Guangzhou, Hunan, Shanghai and Jiangsu belonged to the third level. Hubei, Shandong and Heilongjiang belonged to the fourth level. It revealed the fact that the previous key aircraft industrial bases constructed during the "Third Line Construction" were losing their advantage, as the comprehensive scores in Shaanxi, Liaoning, Sichuan and Guizhou had been decreasing. It also indicated that the aircraft industry in eastern regions developed rapidly.

In addition, the comprehensive development scores of provinces around the 4 production centers—Shaanxi, Sichuan and Guangdong, were relatively low which showed the already formed production centers, neither the traditional centers nor the emerging centers had played a leading role and promoted the development of aircraft industry in their surrounding area. This phenomenon was similar to the aircraft agglomeration in Mexico where a bunch of aviation manufacturing enterprises shared a common locality but without close interaction [3].

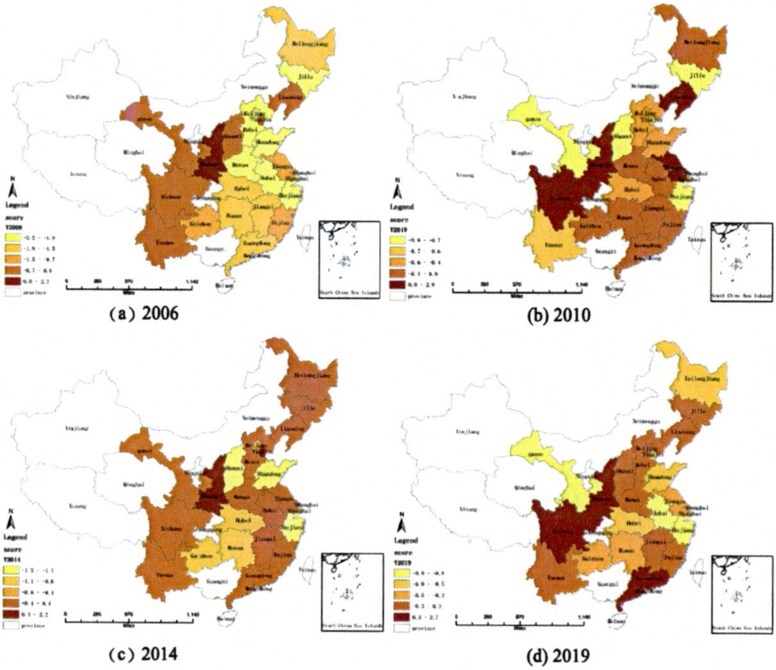

Fig. 2. Spatial temporal pattern of the aircraft industry in China

3.4 The Spatial Temporal Evolution in the Aircraft Industry Agglomeration and R&D Intensity

As shown by Fig. 3 and Table 3, the standard deviation ellipses of the agglomeration degree and the R&D intensity of the aircraft industry changed a lot from 2006 to 2019. In terms of the agglomeration degree, the standard deviation ellipses were mainly concentrated in the southwest, the central and the northeast China, which indicated most of the aircraft industry production centers located in the southwest and the central China. Its scope changed little from 2006 to 2009, but expanded from the southwest to the northeast after 2009, which reflected that the spatial distribution of the aircraft production centers remained unchanged before 2009 and began to expand from the southwest to the east after 2009. The reason lied behind this transformation was that around 2009 the central government took a series of incentive policies to foster the new aircraft center in eastern regions. For example, in 2008, the national Civil Aircraft Industrial Park in Tianjin Binhai New Area has been established. Airbus A320 Aircraft Assembly Line has been in operation in Tianjin since 2008. The rotation angle θ changed little, violating from 38.46 degrees in 2006 to 34.48 degrees in 2019, which reflected that during 2006–2019 the spatial distribution of aircraft industry production centers was distributed northeast-southwest. The main reason might be that as for the great industrial inertial, it was difficult for the aircraft industry to change its space distribution in a short time. As for the shape of the standard deviation ellipse, both the short and the long axis of the standard deviation ellipse of the industry agglomeration showed a cyclic trend. More specifically, the short axis decreased first (2006–2010) and then increased (2010–2014). Contrarily, the long axis first increased (2006–2010) and then decreased (2010–2014), which indicated that the spatial distribution of the production centers of aircraft industry presented a trend of contracting first and then expanding in the East-West direction, and a trend of expanding first and then contracting in the South-North direction.

Table 3. Standard deviation ellipse parameters of the aircraft industry distribution

		CR_{ii}			R&D	
Year	XStdDist	YStdDist	Rotation	XStdDist	YStdDist	Rotation
2006	633509.72	1107591.34	38.46	524326.76	947888.42	24.96
2010	577064.77	1230313.3	40.25	682944.80	1021299.97	31.09
2014	610250.61	1060846.04	29.2	530189.57	636992.42	13.09
2019	668554.45	1133444.79	34.48	539480.36	725924.75	38.80

Compared with the industry agglomeration degree, the rotation, the long axis and the short axis of the standard deviation ellipses of the R&D intensity were much more volatile during the study period. Specifically, the short axis of the R&D intensity was decreasing sharply from 2006 to 2016, which indicated a very rapid concentration of the R&D intensity of the aircraft industry in the East-West direction. The long axis of the R&D intensity increased slightly from 2006 to 2012 and then decreased sharply from 2012 to 2016, which reflected a trend of expanding first and then contracting in the South-North direction. In addition, the rotation of R&D intensity experienced even much larger fluctuation than that of the industrial output. It swelled from 7.02 in 2012 to 143.44 in 2016, which reflected unlike the industrial output, the spatial distribution of aircraft industry R&D centers can be changed in a short time. Furthermore, Fig. 3 also showed the production center was separated from the R&D center and the eastward trend of the R&D center was much faster than the industrial output center. Especially in 2016, Shanghai has become the undoubted R&D center. The reasons might exist in the following two facts: (1) Although the established industrial equipments, plants as well as the labor pool cannot be relocated in a short run, the R&D input was more likely to fluctuate with the incentive policy; (2) the Eastern China had more abundant capital and technology resources, which made it easier for the aircraft industry in the eastern region to access and acquire these resources. Therefore, it was an obvious separation of the production center from the R&D center.

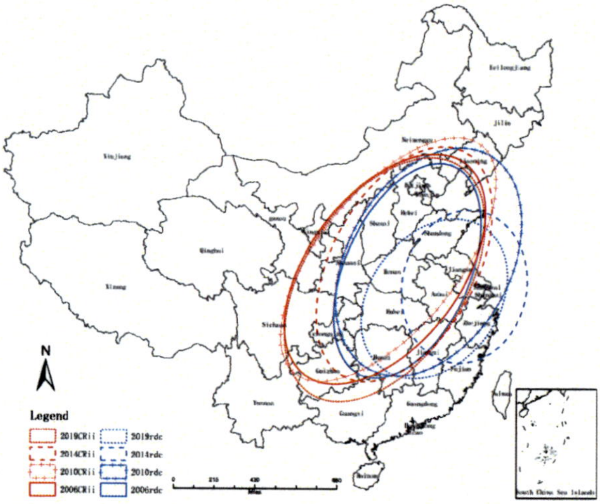

Figure. 3. Comparison between CR_{ii} and R&D input of the aircraft industry in China

After analysis the spatial temporal evolution of aircraft industry, we found the following spatial temporal evolution characteristics for the development level, agglomeration and R&D intensity of China's aircraft industry: (1) The overall agglomeration degree of the aircraft industry presented a low level during the study period, although that in some eastern regions were increasing. (2) The development level of the West was higher than the East in 2006, while this advantage of the West had been diminishing during the study period; (3) The production center and the R&D center had a trend of moving from the West to the East; (4) The production center centers and R&D centers were separated from each other. The production center centers were located in the western and the central China, while the R&D centers were more likely located in the East, represented by Shanghai.

4 Conclusions and Implications

The development of the aircraft industry has raised many concerns. Using the GIS technology and the panel data from 2006 to 2019, this paper visualized the spatial temporal characteristics of aircraft industry in China from the perspective of the following three aspects:

(1) Comparative with the US, the agglomeration degree of the aircraft industry in China was still in a low level and improved little during the study period. As the largest aircraft production center, Shaanxi's agglomeration degree of the aircraft industry was only 21.33% during the study period. In 2000, the agglomeration degree of the US' aircraft industry was 49.9% from a global perspective view [11]. Theoretically, the higher degree of agglomeration, the aircraft industry will be more likely to benefit from economies of scale, labor pool and knowledge spillovers. There was still a long way for China to go before it can break the duopoly of Boeing from the US and Air bus from the EU. In addition, from a spatial perspective, the agglomeration degree experienced an uneven development among different regions. Specifically, the agglomeration degree in western regions was decreasing during the study period and that in eastern regions were increasing.
(2) The R&D intensity went up and down from year to year and varied significantly among different regions. As a high technology industry, the R&D intensity of the aircraft industry was much higher than that of other industries. In 1994, the Gross Expenditure on R&D (% of GDP, GERD) in the UK was 2.19, while its R&D intensity in the aircraft industry in the UK reached 11.1. In 2016, the Gross Expenditure on R&D (% of GDP) in China was 2.11 which was almost equal to the level of UK's GERD in 1994, its R&D intensity in aircraft industry in that year reached 27.12 which was much higher than that of the UK. However, this figure was only 3.07 and 6.68 in 2014 and 2015, respectively. In addition, the R&D intensity was extremely unbalanced among regions. Generally speaking, the R&D intensity of eastern regions was much higher than western and central regions. During the study period, Shanghai (111.62) and Hunan (17.78) ranked first and second in terms of R&D intensity, while Yunnan (0.48) was among the least R&D intensity regions.
(3) The development level of the aircraft industry was somewhat like the agglomeration degree. During the study period, the average development level for the whole country remained unchanged. However, the development level among different regions presented quite a different trend. At the beginning of the study period, the development level of the West was higher than the East, while the gap between the west and the east had been diminishing at the end of the study period, which also indicated that the diffusion of the aircraft industry production centers of the aircraft industry in China were moving from the West to the East.

References

1. Pritchard, D.J.: The global decentralization of commercial aircraft production: implications for US based manufacturing activity (Doctoral dissertation, State University of New York at Buffalo) (2002)
2. Chu, B., Zhang, H., Jin, F.: Identification and comparison of aircraft industry clusters in China and United States[J]. Chin. Geogra. Sci. **20**(5), 471–480 (2010)
3. Romero, J.M.: Centripetal forces in aerospace clusters in Mexico[J]. Innov. Dev. **1**(2), 303–318 (2011)
4. Spreen, W.E.: Marketing in the International Aerospace Industry[M]. Routledge (2016)
5. Steenhuis, H.J., Kiefer, D.: Early stage cluster development: a manufacturers-led approach in the aircraft industry[J]. Compet. Rev. **26**(1), 41–65 (2016)
6. McGuire, S.: Sectoral innovation patterns and the rise of new competitors: the case of civil aerospace in Asia[J]. Ind. Innov. **6**(2), 153–170 (1999)
7. Clifton, N., David, R., Ehret, O., Pickernell, D.: An analysis of actual and potential clustering structures, stakeholder governance activities and cross-locality linkages in the Welsh aerospace industry[J]. Eur. Plan. Stud. **19**(2), 279–309 (2011)
8. McGuire, S., Islam, N.: Indigenous technological capabilities, emerging market firms and the aerospace industry[J]. Technol. Anal. Strat. Manag. **27**(7), 739–758 (2015)
9. Lawrence, P.K., Braddon, D.: Aerospace Strategic Trade: How the US Subsidizes the Large Commercial Aircraft Industry[M]. Routledge (2017)
10. Hagedoorn, J.: Inter-firm R&D partnerships: an overview of major trends and patterns since 1960[J]. Res. Policy **31**(4), 477–492 (2002)
11. Niosi, J., Zhegu, M.: Aerospace clusters: local or global knowledge spillovers? [J]. Ind. Innov. **12**(1), 5–29 (2005)
12. Vértesy, D.: Preconditions, windows of opportunity and innovation strategies: successive leadership changes in the regional jet industry[J]. Res. Policy **46**(2), 388–403 (2017)
13. Wagner, S.M., Walton, R.O.: Additive manufacturing's impact and future in the aviation industry[J]. Prod. Plan. Control. **27**(13), 1124–1130 (2016)
14. Garrette, B., Castañer, X., Dussauge, P.: Horizontal alliances as an alternative to autonomous production: product expansion mode choice in the worldwide aircraft industry 1945–2000[J]. Strateg. Manag. J. **30**(8), 885–894 (2009)
15. Jin, B., Li, G., Chen, Z.: An empirical analysis on international competitiveness of China manufacturing since WTO Accession[J]. China Ind. Econ. **10**, 5–14 (2006). (In Chinese)
16. Duan, J., Yang, H.F.: Research on the evaluation and upgrade strategy of the competitiveness of aviation industry in Shaanxi province[J]. Sci. Technol. Manag. Res. **24**, 82–103 (2010). (In Chinese)
17. Zhou, J., Yang, P.E.: Empirical analysis of evaluate the performance of the cluster of the aviation industry in Shaanxi province[J]. Sci. Technol. Prog. Policy **28**(23), 54–59 (2011). (In Chinese)

Flow Physics of a Top-Mounted Bump Inlet at Subsonic Cruise Condition

Longkai Guo, Yuan Yi, Peibo Yuan, Yaowu Zhu, and Xingya Da(✉)

Aerodynamics Research and Development Center, Institute of High Speed Aerodynamics, Mianyang, China
dxingya@163.com

Abstract. In order to explore the aerodynamic performance of the bump inlet with different angles of attack, a supersonic top-mounted bump inlet is designed. The objective of this study is to investigate the performance of the bump inlet under the subsonic cruise condition, especially including the flow characteristic with different angles of attack (AOA) in experiments and simulations. It is clear that the bump inlet could exclude the wall low energy boundary layer from both sides of the bump compression intake to ensure the airflow higher quality. Although the bump-induced flow filed is complex, it exhibits similar characteristics with smaller AOA. The total pressure recovery coefficient is decreasing with increase of AOA. In addition, the total pressure distortion at the aerodynamic interface plane is enhanced, but it has a tendency to decrease at small AOA. The turbulent intensity at the outlet enhances with increase of AOA. Moreover, under the same value of AOA, the internal airflow occurs separation at high back pressure, the bump inlet has higher total pressure recovery performance at low mass flow rate.

Keywords: Bump inlet · Angle of attack · Pressure recovery · Mass flow

1 Introduction

As an important part of modern fighter's power system, the supersonic inlet not only has to realize the deceleration and pressurization of the incoming airflow to match the engine intake conditions, but also to provide a homogeneous and high-quality airflow to the compressor face of the engine over a wide range of speeds, altitudes, and maneuvering conditions [1]. The complexity and importance of inlet become more and more obviously with the development of the aircraft design technique, the new generation aircraft has more strict requests on the aerodynamics of the inlet, invisibility, concise and low weight [2, 3]. The bump inlet is generally composed of a fixed bump and an expanded forward extension of the intake fairing. The basic idea of the bump inlet is based on the conical flow theory with shock wave ride principle, using the fuselage shape surface to intercept the conical flow field to construct a compression surface. Due to the function of the isentropic compression, there are normal and transverse pressure gradients on the compression surface of the bump inlet. The combined effect of the two pressure gradients is equivalent to the existence of a passive bleed device, which could

Chinese Society of Aeronautics and Astronautics: CASTYSF 2022, LNEE 972, pp. 675–689, 2023.
https://doi.org/10.1007/978-981-19-7652-0_59

diverter most of the fuselage boundary layer out of the inlet entrance [4]. Therefore, the bump inlet obtains a better performance without taking the surface layer spacer and blowing out suction measures.

To now, the investigation on the bump inlet is focused on several main aspects, such as the design of the bump inlet, the aerodynamics of the bump inlet, the mechanism of boundary layer, the RCS performance etc. However, the mechanism of bump inlet is still unrevealed. In this study, experimental wind tunnel tests and numerical simulations are conducted on a top-mounted bump inlet. All tests presented here, are performed in the presence of the forebody under the cruise condition Ma = 0.8 and with zero degrees sideslip angle. The experiments show that the inlet design can provide good performance through high total pressure recovery and low-pressure distortion index. Detailed description of the experiments and simulations are presented in the following sections.

2 Initial Configuration

Figure 1 shows 3D schematic views of bump inlet with forebody from a flying-wing aircraft. In order to remove the traditional boundary layer diverter and suppress the massive flow separation, a bump is merged with the forebody of the aircraft at the entrance of the top-mounted inlet. The preliminary diameter of the diffuser outlet is defined as D, which is matched with the engine inlet diameter. The length and width of the model are 14D and 4D, respectively.

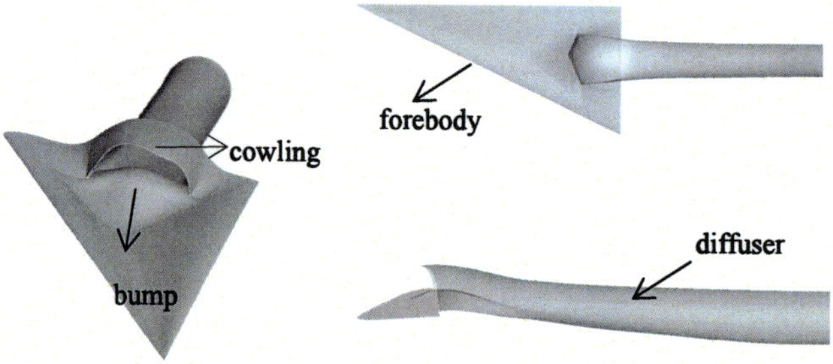

Fig. 1. Schematic of bump inlet with forebody from different views

3 Methodology

3.1 Experimental Approach

Bump inlet with forebody has been investigated by experiments in a transonic wind tunnel. The test section of the wind tunnel is $1.2 \times 1.2\ m^2$, and the estimated running time is typically 60 s. The AIP diameter of the inlet in experiments is defined as D, which has a ratio of 12.5 compared with real model size. It should be noted that the simulation cases use full size model to do the calculation, therefore, the AIP diameter in simulation is 12.5D. The blockage of the test section of the wind tunnel is 0.85%. The test model is installed on the C-type mechanism by with the model angles of attack and sideslip angles are varied. This mechanism has a C-shaped metallic arc with a strut that maintains the model in the wind-tunnel test section. The inlet mass flow rate is controlled via the movement of a conical plug located at the outlet of a sonic nozzle, which is connected to the AIP of the bump inlet. The conic flow plug is driven by a small electric motor, which could provide different throttling ratios through changing the exit area.

In order to obtain the air flow characteristics in the experiments at different conditions, plenty of pressure measurement points were placed on the test model at the aerodynamic interface plane (AIP), as seen in Fig. 2. In the experiments, two different types of pressure transducers were used. The first one is a low-frequency one with a maximum sample rate of about 1 kHz, and the second types were high-frequency miniature ones with 20 kHz installed inside the model as close as possible to the pressure ports and probes. 41 pressure transducers were used for both static and total pressure measurements at various locations of the model. In addition, 8 pressure transducers were located at the AIP position to measure the mean value of the static pressure at this station for various cases. Four dynamic pressure taps were also placed on the duct AIP to obtain the dynamic turbulent intensity of the AIP. The dynamic pressure transducers were standardized and give the coefficient of the calibration value before doing the experiments.

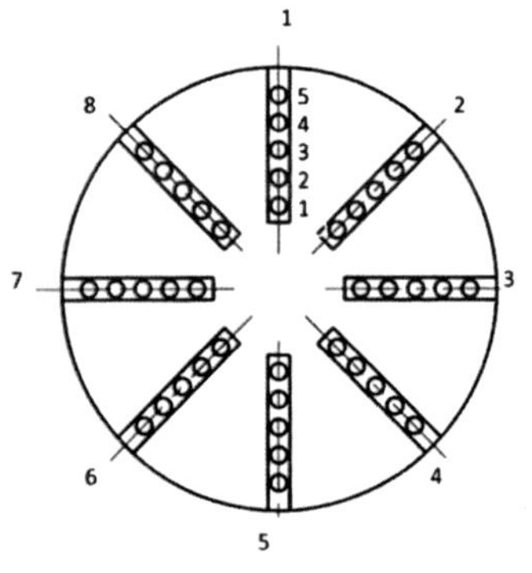

Fig. 2. Total pressure rake at the AIP

In this study, three most common parameters describing inlet performance are introduced, namely the total pressure recovery coefficient σ, the total pressure circumferential distortion index $\Delta\overline{\sigma}_0$ and the area-averaged turbulent intensity ε_{avg}. The total pressure recovery coefficient is defined as

$$\sigma = \frac{p^*_{avg}}{p^*_\infty}$$

where p^*_{avg} is the area-weighted average total pressure of the AIP; p^*_∞ is the freestream total pressure. The total pressure circumferential distortion index $\Delta\overline{\sigma}_0$ is calculated as

$$\Delta\overline{\sigma}_0 = 1 - \frac{\sigma_0}{\sigma_{avg}}$$

σ_0 is the average total pressure recovery coefficient in the low-pressure region, which is the ratio of the average total pressure in low region over the freestream total pressure as follows.

$$\sigma_0 = \frac{1}{\theta_2 - \theta_1}\int_{\theta_1}^{\theta_2} \sigma_{r,i}(\theta)d\theta$$

σ_{avg} is the area-averaged total pressure recovery coefficient at the AIP, which is indicated as

$$\sigma_{avg} = \frac{\int_0^{2\pi} \sigma_{r,i}(\theta)d\theta}{2\pi}$$

$\sigma_{r,i}(\theta)$ is the function of the radial averaged total pressure recovery coefficient and the circumferential angle θ. In this study, i denotes 8 total pressure taps in circumferential direction and j denotes 5 total pressure taps in radial direction (Fig. 3). The expression of $\sigma_{r,i}(\theta)$ can be expressed as

$$\sigma_{r,i}(\theta) = \frac{1}{5}\sum_{j=1}^{5} \sigma_{avg,(i,j)}$$

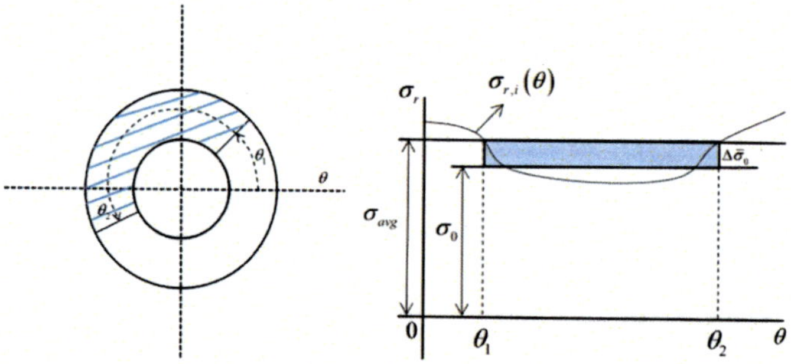

Fig. 3. D σ_r distribution along circumferential direction

The area-averaged turbulent intensity ε_{avg} represents the quantitative characteristics of the total pressure pulsation at the AIP. This is a type of flow field distortion, namely the total pressure spatial inhomogeneity changes rapidly with time, which is also called dynamic distortion. Since the total pressure inhomogeneity at the interface and the distribution of the low-pressure region during dynamic distortion are a function of time, the parameters of simple random statistics can be used to represent dynamic distortion of the AIP. In our experiments, four dynamic pressure taps are used to describe the pressure pulsation at the AIP, the averaged dynamic turbulent intensity ε_{avg} is expressed as follows.

$$\varepsilon_{avg} = \frac{\Delta \overline{p}_{rms}}{p_{avg}}$$

in which $\Delta \overline{p}_{rms} = \frac{\sum_{i=1}^{4}\sqrt{\frac{1}{T_u}\int_0^T (p_i(t)-p_{avg})^2 dt}}{4}$. $p(t)$ is the total pressure value evolution with time at the AIP. T_u is the sample time of pulsation airflow and p_{avg} is the averaged total pressure in Tu time.

3.2 Numerical Approach

3.2.1 Governing Equation

The numerical method supplements more information to investigate the internal flow structures inside the bump inlet comparing with experiments. A three-dimensional Reynolds-averaged Navier-Stokes (RANS) solver is adopted to perform the simulation, as follows.

$$\frac{\partial \boldsymbol{u}}{\partial t} + \nabla \cdot \boldsymbol{F}_I + \nabla \cdot \boldsymbol{F}_V = \boldsymbol{Q}$$

where $\boldsymbol{F}_I$ and $\boldsymbol{F}_V$ represent the inviscid and viscous matrix, $\boldsymbol{Q}$ is the source term vector.

The Spalart-Allmaras [14] model is used to close the RANS equation because of its good simulation results in solving boundary layer problems with wall-influenced flow and inverse pressure gradients. The governing equations were solved by a finite volume method. Moreover, a wall function is used to improve the accuracy of the wall boundary layer region [15]. The wall function method saves memory and computational time and is widely used in turbulence engineering calculations. The inclusion of the wall function can give a reasonably accurate explanation for high Re number, wall-limited flows.

Figure 4 shows the computational domain of the present simulations. In order to eliminate the effect of the boundary on the calculation result accuracy, a large computational domain, 30D in width, 53D in height and 88D in length, was used in the calculations. The diffuser outlet is extended linearly by 2D to uniform the air flow. In order to solve the conservation equations, boundary conditions are required for the computational domain. In this study, three types of boundary conditions are used. The total pressure and temperature of the freestream are estimated according to the experiments, which are treated as pressure Fairfield boundary condition. The diffuser exit static pressure appears to be common in the experiments. This result was also reported in the work of Menzies [16] and Lee [17]. The inlet mass flowrate could be controlled through changing the back pressure at the exit of the diffuser. The back pressure is set corresponding to a fixed position of the flow plug in the experiments. The other regions in the computation domain, such as fore-body, bump surface, duct surface etc., are non-slip adiabatic boundary conditions.

The bump inlet compression surface has a large curvature and complex typologies, which could bring a big challenge to construct the boundary layers near the walls. If the unstructured grid was used, there is a problem for the matching of the grid with the surface model. Therefore, the structured grid is used in the present study to improve the calculation accuracy and the veracity. The grid distribution of the bump inlet and combination of the forebody is displayed in Fig. 5. The grid has 92 blocks and the nodes number is about 8×10^6.

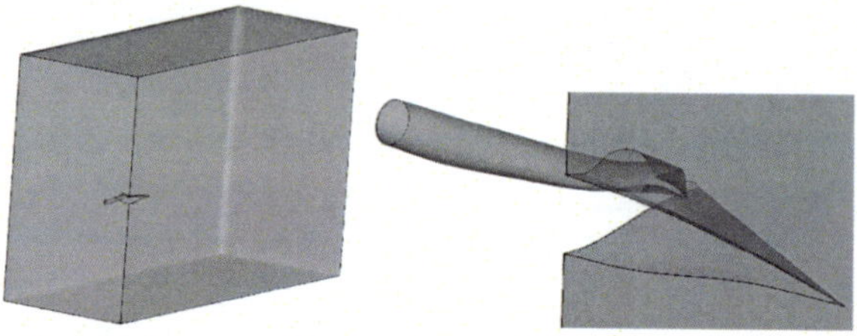

Fig. 4. Computational domain used in the calculation

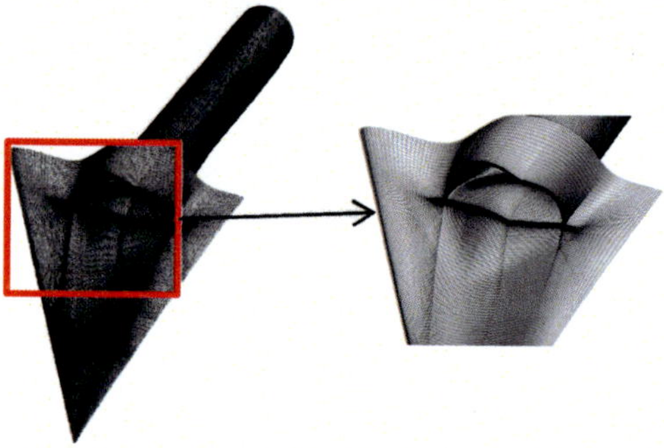

Fig. 5. Inlet surface mesh

3.2.2 Grid Independency Analysis

To validate the independency of the grid systems, three different sizes of grid are performed in this study. The total grid nodes are 6×10^6,8×10^6 and 1.6×10^7 for coarse grid, medium and fine grid, respectively. For the three grids, the grid is refined in the entire computational domain except for the wall boundary layer region. The fine grid is used as a reference simulation result to assess the convergence rate of calculation. The convergence of each calculation is decided by the residual history of the solution. In this study, convergence of the solutions is considered to be achieved when the L2 norm of the maximum residual reached 10^{-4}. The first point from the wall at $y+$ is less than unity for all grids. The total pressure distribution is important in assessing the performance of the bump inlet, because the loss of total pressure will degrade the engine performance and may severely reduce the lifetime of an aircraft. Table 1 shows the performance of the simulation and experiment results of the AIP at Ma = 0.8, AOA = 0°, $p_{AIP}/p_{\infty} = 1.2$. The total pressure recovery σ_{AIP} decreased with increasing number of grid cells. However, the error of σ_{AIP} compared with experiments for three types of grids is small.

Table 1. Inlet performance at $Ma = 0.8, AOA = 0°, p_{AIP}/p_{\infty} = 1.2$ with different grids

Data source	σ_{AIP}	Error of σ_{AIP}
Experiment	0.968	–
Coarse grid	0.965336651	2.663×10^{-3}
Medium grid	0.965405772	2.594×10^{-3}
Fine grid	0.965699324	2.301×10^{-3}

In experiments, one dimensionless number mass flow rate Φ is used, which is defined according to the flow plug positions. The maximum position of the plug in experiments far away from the AIP is defined overall mass flow rate, the minimum position of the plug is the nearest position to the AIP. It should be noted in this study, all different angles of attack have same plug position variation. Figure 6 shows the relationship between the dimensionless mass flow rate (Φ, which is defined by the (convertedmassflowrate)/(overallmassflowrate)) and the mean value of total pressure recovery coefficient for experiments and simulations. With decrease of the mass flow rate, namely the flow plug moves towards the AIP, σ value is increasing. Because of the advantage of the structure grids, all grids' results match the experimental results well. At $\frac{p_{AIP}}{p_{\infty}} = 1.2$ condition, the difference of the grids can hardly be noticed. At lower mass flow rate, the performance of coarse grid is not as good as the others. Furthermore, the AIP total pressure recovery contours of experiments and simulations are shown in Fig. 7 at $\frac{p_{AIP}}{p_{\infty}} = 1.2$ condition. The major part of the AIP was occupied by a mushroom high total pressure region. The distribution of the total pressure recovery coefficient in AIP illustrates the complexity of the flow structure inside the duct. However, no significant large separation region is found, which revealed that the bump-type inlet is able to produce good airflow field quality. From the experimental and numerical simulation results, the simulation and experiment results have similar contour distribution. Therefore, the medium grid is used in the following simulations.

$$\frac{p_{AIP}}{p_{\infty}} = 1.2$$

4 Results and Discussions

4.1 Performance Parameters and Surface Flow Patterns at Ma = 0.8, AOA = 0°

Figure 8 shows the Mach number contours distribution of slices at $p_{AIP}/p_{\infty} = 1.2$ with medium grid. The effect of bump duct is obvious that decelerate the speed of the air flow. The air speed near the wall is lower than the center line one, because of the work of

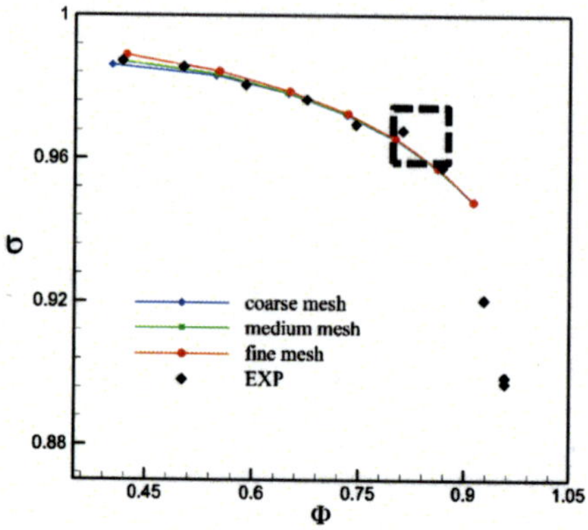

Fig. 6. Relations of AIP mean total pressure recovery with the dimensionless mass flow rate

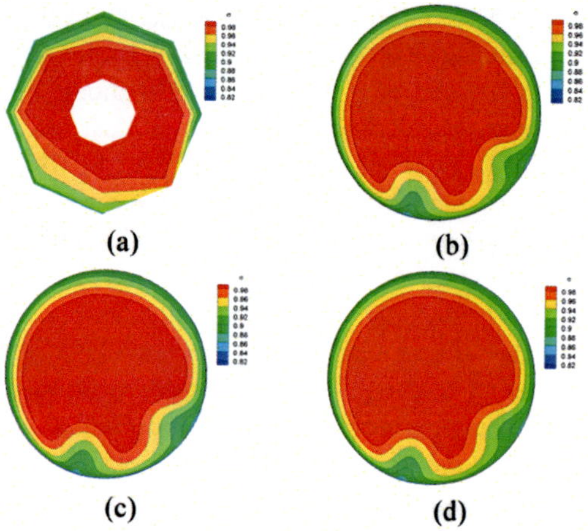

Fig. 7. Total pressure recovery contours at the AIP by: **a** experiment; **b** fine grid; **c** medium grid; **d** coarse grid

the viscous wall. There is no large airflow separation region due to the smooth changes in the intake duct. Figure 9 shows the velocity streamlines at middle surface along the freestream direction, the result shows the apparent deceleration of velocity in the duct. However, at the blunt lip position, an oblique shock wave is found. Meanwhile, there is a relative low Mach number region at the top walls. Figure 10 shows the skin friction streamlines on the surface of the forebody and the bump compression, which could describe the boundary layer trajectory. As the result shows, most part of boundary layer is removed from the entering of the inlet. It is generally believed that the principle of bump inlet to exclude the boundary layer is due to the bump compression has a higher pressure at the middle part and a lower pressure on both sides. Therefore, the airflow boundary layer is divertered from the inlet with the function of pressure gradient.

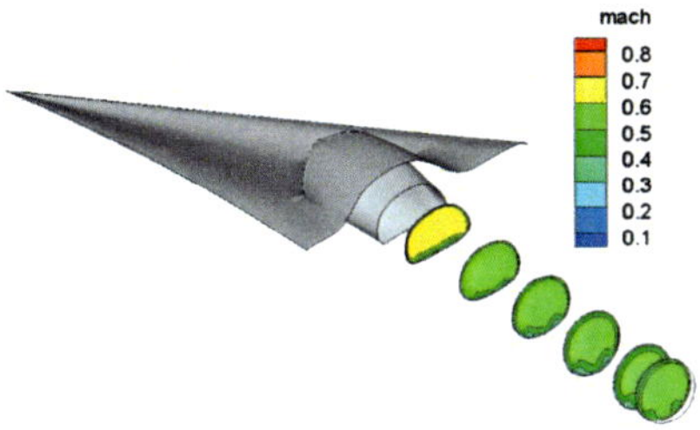

Fig. 8. Mach number contours of different slices in the bump diffuser

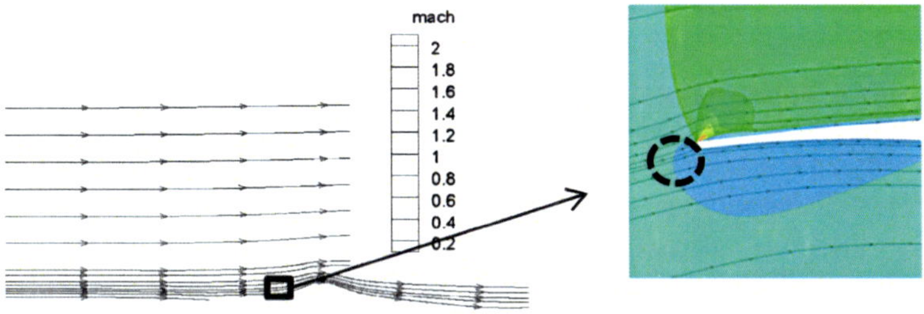

Fig. 9. Mach number contour and velocity streamlines along free stream with $p_{AIP}/p_{\infty} = 1.2$

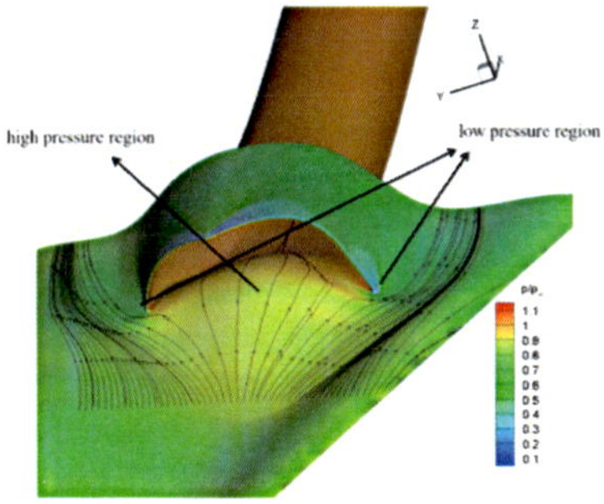

Fig. 10. The boundary layer trajectory on the bump compression surface with $p_{AIP}/p_{\infty} = 1.2$

In order to analyze the pressure distribution on the surface of bump compression in details, Fig. 11 illustrates the pressure contours distribution of a cross-section before the lip of the bump inlet at $p_{AIP}/p_{\infty} = 1.4$. As Fig. 11 shows, due to the function of the bump compression, the pressure isolines are distributed around the bump. A high-pressure point exists above the bump, the pressure distribution is centered on this point and distributed outward. However, the pressure isoline values are higher near the bump surface. Figure 11 fully illustrates the pressure distribution of the bump along the spreading direction with higher value in the middle part and lower ones on both sides.

Figure 12 indicates the streamlines of the bump inlet internal flow at Ma = 0.8 with zero angle of attack and zero angle of slide. The streamlines near the bump occur to turn, the core flow is more uniform. However, it is not possible to determine whether there

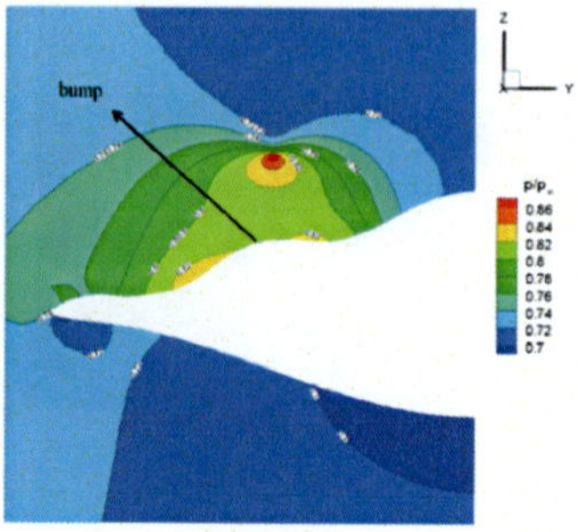

Fig. 11. Pressure contours on cross section of bump inlet with $p_{AIP}/p_{\infty} = 1.4$

is a separation region inside the diffuser from this figure. Moreover, the details of the airflow characteristics on the bump are shown in Fig. 12. There are two saddle points at both sides of the bump compression near the cowl. When the airflow crosses the bump, the saddle points make it together in the middle of the inlet channel. It is clear that the bump structure could change the direction of the airflow and has a significant effect on the movement of the airflow before entering the intake.

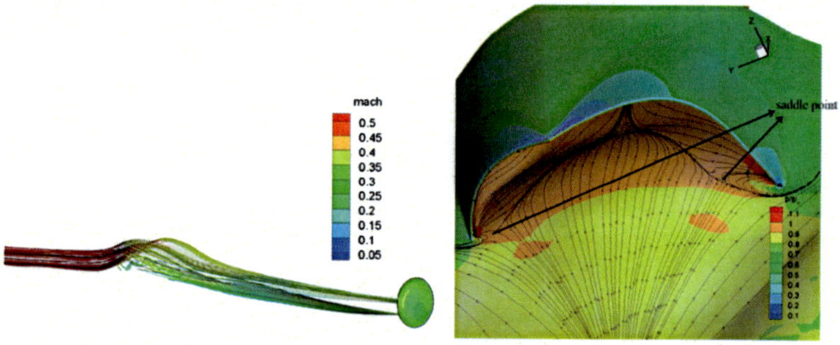

Fig. 12. Streamlines of internal flow and bump compression

4.2 Effect of AOA on the Performance of the Bump Inlet

When an aircraft is maneuvering at a high altitude, the change of AOA values has a great effect on the airflow intake performance. Therefore, it is necessary to analyze the flow characteristics of the bump inlet at different angles of attack. In the following parts, experiments and simulations of the bump inlet with different angles of attack are investigated. Several typical parameters, such as total pressure recovery coefficient (σ), $\Delta\overline{\sigma}_0$ and the flow patterns at AIP or the characteristics of the internal flow are analyzed to assess the bump inlet robust and reliability.

Figure 13 shows the experimental results of pressure recovery coefficient with different AOA values. It is clear that with increasing of the AOA values, σ is larger at the same mass flow rate condition. However, AOA in the range of [$-2°$,$5°$] has minor effect on t he total pressure recovery. In details, for smaller AOA, the pressure recovery of AIP seems identical at different mass flow rates. For larger AOA, with larger mass flow rates, the pressure recovery of AIP changes more obviously. The number index also indicates different outlet back pressure in experiments at AOA = 0° and the AIP σ contour map with different back pressure condition at AOA = 0° is given in Fig. 14. The total pressure distortion is decreasing with increase of the outlet pressure condition. At the beginning position of the flow plug, the channel has some flow separation at the left top of the wall, this phenomenon has been eliminated after moving the plug.

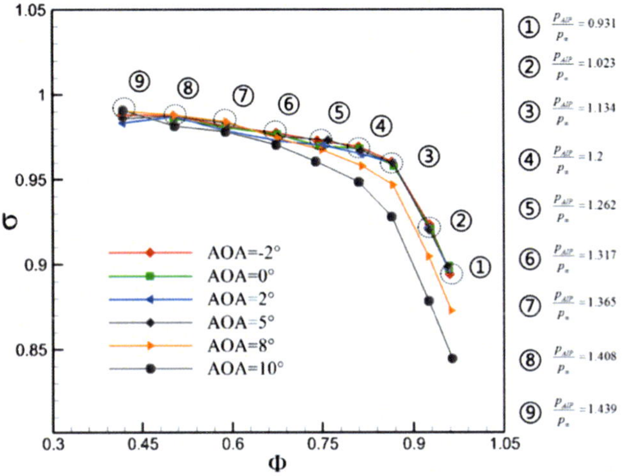

Fig. 13. Experiment results of total pressure recovery with different AOA values

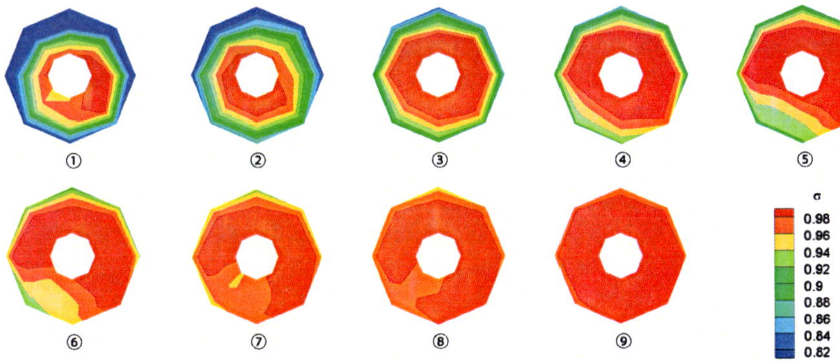

Fig. 14. Experiment results of total pressure recovery contours at the AIP with AOA = 0°

Figure 15a shows that as the mass flow rate increasing, the total pressure circumferential distortion index $\Delta\overline{\sigma}_0$ has a increasing tendency from the overall view, which indicates that the total pressure at the AIP is worse. Moreover, at Φ around 0.75, when AOA < 5°, the pressure distortion index has a sharp descent. However, this phenomenon is not apparent at AOA = 8° or AOA = 10°. For AOA = 10°, the pressure distortion index is enhancing. For describing the total pressure recovery performance at the AIP under Φ around 0.75, Fig. 14b is the contours of total pressure recovery coefficient at the AIP for AOA = 5° and AOA = 10°. From the view of the Fig. 15b, under AOA = 10°condition, the airflow at the AIP has a deflection and σ is decreased, hence $\Delta\overline{\sigma}_0$ increased. However, under AOA = 5°condition, the airflow is more uniform and the $\Delta\overline{\sigma}_0$ value is decreased. The presence of this phenomenon is that at large angle of attack, more low energy boundary layer fluid enters the intake, which leads to the airflow separation inside the bump inlet and has a significant effect on the quality of the AIP airflow.

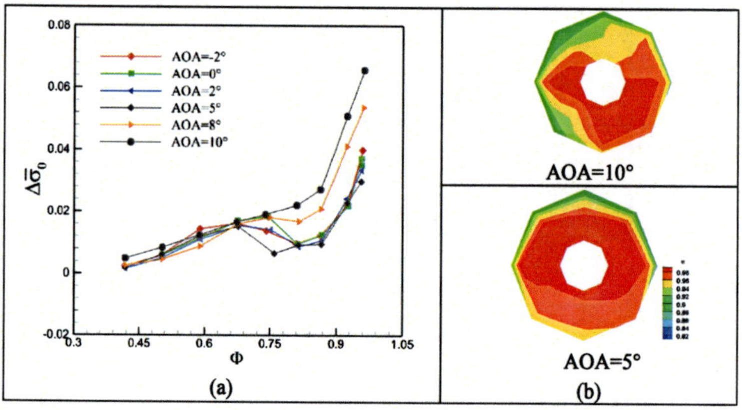

Fig. 15. Experiments of the total pressure circumferential distortion index $\Delta\overline{\sigma}_0$

Figure 16 shows the total pressure pulsation on the AIP. The change of airflow turbulent intensity at different angles of attack shown in the figure can be concluded that the trend is similar to the overall trend of the total pressure distortion index, but it does not show the same inflection point position as the total pressure distortion index. This indicates that the turbulent intensity is mainly affected by two aspects. The first one is the intake mass flow rate, the larger the intake flow rate, the larger the turbulent intensity, and the more uneven the airflow; the second one is the angle of attack, the small angle of attack has less influence on the turbulence, the turbulence changes strongly under the large angle of attack, and the aerodynamic performance of the bump inlet is worse.

In order to further analyze the influence of the inlet airflow under different angles of attack by the bump inlet, Fig. 17 gives the friction skin streamlines of the bump and forebody under different angles of attack. The comprehensive analysis leads to the following conclusions, (1) the airflow separation effect of the bump inlet on the internal airflow of the low-energy boundary layer is obvious, effectively avoiding the low-energy airflow into the bump inlet and mainstream mixing, forming the problem of air inlet distortion, affecting the quality of the exit airflow; (2) as the angle of attack increases, the airflow of the aircraft forebody converges and disperses to the outer sides of the bump at the position of the leading edge. The change of angle of attack makes the outer cover of the lip of the bump produce a low-pressure area, this situation has a

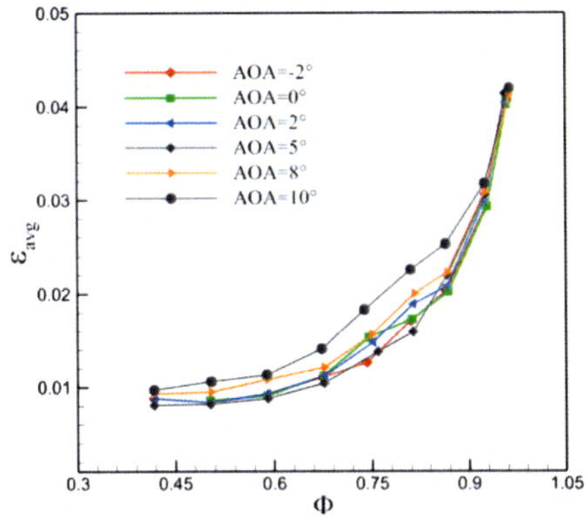

Fig. 16. Experiments of the area-averaged turbulent intensity ε_{avg}

certain impact on the airflow along the wall of the bump and the outer cover of the lip into the intake.

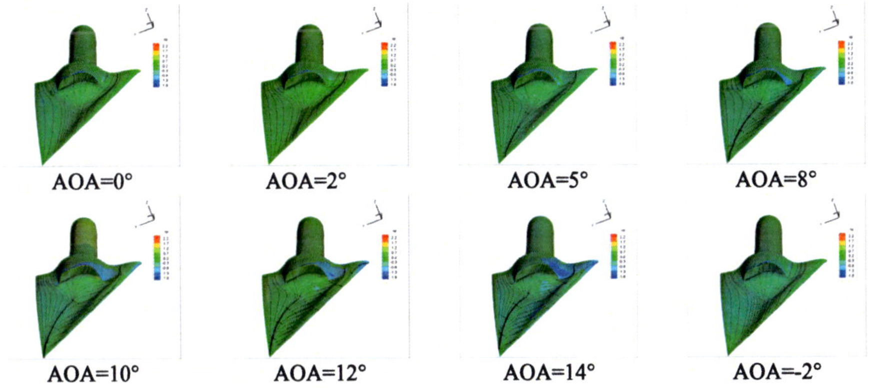

Fig. 17. Friction skin streamlines of bump inlet at different AOA values

5 Conclusion

In this paper, a top-mounted bump inlet for high-speed aircraft is designed and studied. The performance of the bump inlet, including the total pressure recovery coefficient, the total pressure distortion index and the turbulent intensity the AIP, is analyzed by experiments at different angles of attack. The airflow inside the bump inlet and the flow patterns of the bump compress surface are performed by numerical simulation. According to the analysis, the following conclusions can be obtained.

(1) The bump inlet can effectively diverter the low-energy airflow in the boundary layer, so that the quality of airflow into the diffuser is enhanced and the incoming strong distorted airflow is avoided to mix with the mainstream.
(2) There is a higher-pressure region on the surface of the bump inlet, and a central high-pressure area appears directly above it, the pressure drops around this center.
(3) Under the same angle of attack, increasing the outlet pressure, changing the inlet air flow rate, airflow is separated in the case of high back pressure and eventually blocking the internal bump inlet; under small mass flow rate, the bump inlet has a higher total pressure recovery.
(4) The total pressure recovery performance of the bump inlet changes less at small angles of attack. However, as the angle of attack increases, the overall total pressure recovery coefficient at the outlet decreases. In addition, the total pressure distortion at the outlet is enhanced, but there is a tendency to reduce the total pressure distortion at small angles of attack. The outlet turbulent intensity enhances with increasing angles of attack.

References

1. Kim, S.D.: Aerodynamic design of a supersonic inlet with a parametric bump. J. Aircr. **46**(1), 198–202 (2009)
2. Gridley, M.C., Walker, S.H.: Inlet and nozzle technology for 21st century fighter aircraft[R]. ASME Paper 96-GT-244,1996
3. Zhou, Y., Wang, Q., Li, T., et al.: A numerical simulation method for aircraft infrared imaging[J]. Infrared Phys. Technol. **83**, 68–77 (2017)
4. Mcfarlan, J.D.: Lockheed Martin's Joint Strike Fighter Diverterless Supersonic Inlet[M]. National Press Club (2000)
5. Seddon, J.M., Goldsmith, E.L.: Intake Aerodynamics[M]. Blackwell Science (1999)
6. Simon, P.C., Brown, D.W., Huff, R.G.: Performance of external-compression bump inlet at Mach numbers of 1.5 to 2.0[M]. National Advisory Committee for Aeronautics (1957)
7. Hamstra, J.W., Sylvester, T.G.: System and method for diverting boundary layer air. US, 5779189[P], 1998
8. Hamstra, J.W., Mccallum, B.N., Sylvester, T.G., et al.: Transition shoulder system and method for diverting boundary layer air. US, 5749542[P], 1998
9. Kim, S.D., Song, D.J.: A numerical analysis on three-dimensional flow field in a supersonic bump-type inlet[J]. J. Mech. Sci. Technol. **21**(2), 327–335 (2007)
10. Kim, S.D., Song, D.J.: Numerical study on performance of supersonic inlets with various three-dimensional bumps[J]. J. Mech. Sci. Technol. **22**(8), 1640–1647 (2008)
11. Kim, S.D.: A computational study on the influence of a bleed slot on the flowfield of a bump-type inlet[J]. J. Mech. Sci. Technol. **29**(12), 5245–5250 (2015)
12. Loth, E.: Meso-flap and bleed flow control for a Mach 2 Inlet. In: 42nd AIAA Aerospace Sciences Meeting and Exhibit, AIAA Paper 2004–855, 2004
13. Tillotson, B.J., Loth, E., Dutton, J.C., et al.: Experimental study of a Mach 3 bump-compression
14. flowfield[J]. J. Propuls. Power, **25**(3), 545–554 (2009)
15. Spalart, P.R., Allmaras, S.: a one-equation turbulence model for aerodynamic flows[C]. AIAA 92–0439 (1992)

16. Knopp, T., Alrutz, T., Schwamborn, D.: A grid and flow adaptive wall-function method for RANS turbulence modelling. J. Comput. Phys. **220**(1), 19–40 (2006)
17. Menzies, R.D.D.: Investigation of S-shaped intake aerodynamics using computational fluid dynamics. Ph.D. thesis, University of Glasgow, Glasgow, Scotland (2002)
18. Lee, B.J., Kim, C.: Automated design methodology of turbulent internal flow using discrete adjoint formulation. Aerosp. Sci. Technol. **11**(2–3), 163–173 (2007)

Research and Implementation of Temperature Control Method for Large Space Climatic Laboratory

Jingtao Wu, Binwen Wang, Xiangfu Wu(✉), and Yao Wang

AVIC Aircraft Strength Research Institute, Xi'an 710065, China
wuxiangfu1203@163.com

Abstract. The temperature control of aircraft climatic laboratory has the feature of big time delay, large temperature range and non-linear. A temperature control method of feedforward-cascade PID is proposed. Firstly, the model for temperature change of climatic laboratory is established through mechanism analysis, and the parameters of model are obtained based on step response curve identification method; Secondly, on the basis of analyzing the dynamic characteristics of the temperature control system, a two-stage refrigerant temperature control strategy based on constant flow and variable temperature is proposed, and a feedforward-cascade PID controller is designed; Finally, the controller parameters are obtained by the critical proportion method and trial-and-error method, then the controller is applied to the temperature control of the climatic laboratory. The results show that the control method can improve the dynamic response of the system and accurately track the desired temperature curve. This method has the advantages of fast response, no overshoot, high steady-state precision, and achieves the expected goal of temperature control.

Keywords: Climatic laboratory · Temperature control · Large time delay · Feed-forward control · Cascade PID

1 Introduction

In order to meet the experimental needs of aviation, aerospace and other industries for the products' climate adaptability, China has established a super large space climatic laboratory to realize the test for large-scale test pieces such as aircraft and vehicles. The climate laboratory can simulate various climatic environments which can help engineers obtain the function and performance data by putting the products in the laboratory. These data can expose the products' design or process defects, and provide the basis for improving the design of products' climate adaptability [1–4]. Temperature test is the most important test in products' climatic environment test. The national military standard proposes strict requirements for temperature regulation rate, stability and other indicators in the temperature control process. The temperature must be accurately controlled to ensure the authenticity of test data.

Chinese Society of Aeronautics and Astronautics: CASTYSF 2022, LNEE 972, pp. 690–699, 2023.
https://doi.org/10.1007/978-981-19-7652-0_60

The climatic laboratory has a large space and a wide range of temperature, so the temperature control has the characteristics of large time delay and nonlinearity. The test piece, auxiliary equipment, lights and floor will bring heat load interference to the temperature control process. It is difficult to achieve good dynamic and steady state characteristics at the same time by using the classical PID control method [5]. In order to solve the above problems, the paper analyzes the dynamic characteristics of temperature control process, and the transfer function is obtained by the step response identify method. A feedforward cascade PID temperature control method is proposed and applied to the climatic laboratory. This method makes the temperature rise-fall rate and steady-state fluctuation meet the requirements, and it achieves the goal of accurate temperature control.

2 Temperature Control System Model of Climatic Laboratory

2.1 The Composition of Temperature Control System

Super large space climatic laboratory is often used for temperature test of aircraft, automobile and other products. According to the temperature extreme values recorded in GJB 1172.2 “extreme climatic values of military equipment” and the actual measurement results of the temperature when the aircraft is parked at the airport, the low temperature extreme value experienced by the aircraft during operation and storage is −55°C and the high temperature extreme value is + 71 °C [6]. In order to cover the extreme temperature, the temperature regulation range of the climate laboratory is −55°C ~ + 71 °C. The temperature change rate during the heating or cooling process in the laboratory is less than 3°C/h, and the temperature control accuracy is ±3 °C. Temperature environment simulation equipment includes centrifugal fan, high-temperature heat exchanger, low-temperature heat exchanger, refrigeration unit and steam heat exchanger. The schematic diagram of temperature control process for climatic laboratory is shown in Fig. 1.

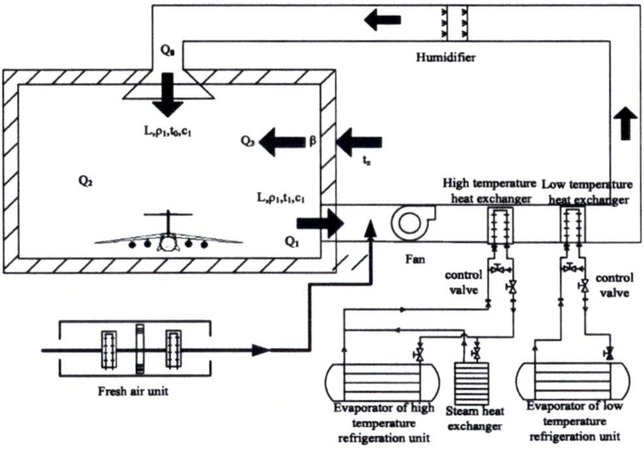

Fig. 1. The schematic diagram of temperature control process for climatic laboratory

The climatic laboratory adopts the air distribution form of top-air supply and side-air return. The air inside the laboratory enters the air-handling-unit from the return air outlet for temperature regulation, and then is sent to the laboratory from the top air outlet. After the air temperature in the laboratory achieves the expected value, the temperature test can start[7, 8]. In order to realize the wide temperature range (−55 °C ~ + 74 °C) regulation of the laboratory, two-stage heat exchangers with different refrigerant are arranged in the air-handling-unit to realize different temperature control. The first stage heat exchanger with high temperature refrigerant is used to realize the temperature regulation of −25 °C ~ + 71 °C. The cold source used in the temperature range of −25 °C ~ + 20 °C is the R507 high-temperature refrigeration unit, and the steam is used for heat source in the temperature range of + 20 °C ~+ 71 °C. The second stage heat exchanger with low temperature refrigerant is used to realize the temperature regulation of −55°C ~ +25°C and the cold source used in the temperature range is R23 low-temperature refrigeration unit. The two-stage heat exchangers are parallel arranged in the air-handling-unit [9]. The refrigerant pipeline of the two-stage heat exchanger is equipped with defrosting electric heating, which can also be used as the heat source in the heating process.

2.2 Modeling of Temperature Control System

The temperature changes of the climatic laboratory are closely related to the heat source, the enclosure structure, outdoor temperature, the indoor working equipment, etc. According to the principle of conservation of energy, the energy gradient of the laboratory is equal to the entering energy minus the outflow energy within unit time. The entering energy includes the heat emitted from working equipment in the laboratory(Q_2), the heat from air inlet(Q_0) and the heat transmitted from the outside of the laboratory(Q_3). The outflow energy is air flowing through the air outlet(Q_1). It can assume that the heat capacity of the laboratory is C_1 and the temperature is t_1, so the energy gradient can be expressed as follow:

$$C_1 \frac{dt_1}{d\tau} = Q_0 + Q_2 + Q_3 - Q_1 \tag{1}$$

Assuming that: the supply air density is ρ_1; the supply air volume is L; air specific heat is C_0; the supply air temperature is t_0; the laboratory wall's thermal resistance is R; outside temperature is t_z; the enclosure structure's heat transfer attenuation coefficient is β.

Equation (1) can be expressed as

$$C_1 \frac{dt_1}{d\tau} = L\rho_1 c_0 (t_0 - t_1) + Q_2 + \frac{1}{R}\beta(t_z - t_1) \tag{2}$$

Transposing the formula (2),

$$\frac{C_1}{L\rho_1 c_0 + \frac{1}{R}\beta} \frac{dt_1}{d\tau} + t_1 = \frac{L\rho_1 c_0}{L\rho_1 c_0 + \frac{1}{R}\beta} t_0 + \frac{Q_2 + \frac{1}{R}\beta t_z}{L\rho_1 c_0 + \frac{1}{R}\beta} \tag{3}$$

Assuming that:

$$T = \frac{C_1}{L\rho_1 c_0 + \frac{1}{R}\beta}, K = \frac{L\rho_1 c_0}{L\rho_1 c_0 + \frac{1}{R}\beta} \tag{4}$$

$$t_f = t_{f1} + t_{f1} = \frac{Q_2}{L\rho_1 c_0} + \frac{1}{L\rho_1 c_0} \cdot \frac{\beta t_z}{R} \tag{5}$$

The formula (3) can be expressed as

$$T\frac{dt_1}{d\tau} + t_1 = K\left(t_0 + t_f\right) \tag{6}$$

In Eq. (6), t_0 is the control variable while t_f is the interference variable, and t_1 is the controlled variable.

Assuming the input is a time function and the laboratory's delay-time is τ_1, Eq. (6) can be expressed as

$$T\frac{dt_1(\tau)}{d\tau} + t_1(\tau) = K[t_0(\tau - \tau_1) + t_f(\tau - \tau_1)] \tag{7}$$

Assuming the initial condition is zero, the laplace transformation of Eq. (7) is

$$G(s) = \frac{T_1(s)}{T_0(s) + T_f(s)} = \frac{K}{Ts + 1}e^{-\tau_1 s} \tag{8}$$

Therefore, the climatic laboratory's temperature control process can be expressed as a first-order inertial system with delay component.

2.3 Identification of the Model Parameters

The main purpose of establishing the system model is to design the controller and evaluate the dynamic and static characteristics and control strategy of the system, while the acquisition of the system model requires parameter identification. In order to improve the fitting degree of the model, the model obtained by system identification is usually a high-order model, while the commonly used Ziegler Nichols PID controller tuning method is based on the low-order model. As we all know, by adjusting the parameters of the low-order model to change the shape of the step response curve, we can constantly approach the step response curve of the high-order system. As long as the fitting degree of the step response curve between the identification model and the original system is well enough, the identification model can reflect the original system's characteristics [10]. Therefore, the system model can be obtained by step response identification method.

The aircraft climatic laboratory's temperature control process can be expressed as formula (8). The gain K affects the steady-state response value; the constant T affects the step response speed; the delay time of the system is τ_1. In order to obtain the temperature step response curve of the climatic laboratory, start the high-temperature refrigeration unit, adjust the refrigerant to—10 °C, fully open the refrigerant regulating valve of the heat exchanger, and then start the fan at the normal operating frequency of 30 Hz. After the laboratory temperature reaches the stable state, the temperature step response curve of the climate laboratory is obtained. On the step response curve, the corresponding time when the temperature reaches to 95% of the final value is the adjustment time T_s, and the time constant T is 1/3 of T_s; the steady-state value of the step response curve is K; the time from the start of fan operation to the start of temperature drop is the delay time.

From the step response curve, the temperature control process model of the climatic laboratory is

$$G(s) = \frac{0.72}{688s + 1}e^{-90s} \tag{9}$$

3 Designation of Temperature Control System

3.1 Control Strategy

According to the schematic diagram of temperature control for climatic laboratory, the temperature control strategies are as follow:

(1) Reduce the laboratory's temperature from room temperature (generally +20 °C) to [−25 °C, + 20 °C]: the cold source is the R507 high-temperature refrigeration unit, and the first stage heat exchanger works.
(2) Reduce the laboratory's temperature from −25 °C to [−55 °C, −25 °C): the cold source is the R23 low-temperature refrigeration unit,and the second stage heat exchanger works.
(3) Raise the laboratory's temperature from −55 °C to (−55 °C, −25 °C]: the heat source is the electric heater,and the second stage heat exchanger works.
(4) Raise the laboratory's temperature from -25°C to (−25 °C, + 20 °C]: the heat source is the electric heater, and the first stage heat exchanger works.
(5) Raise the laboratory's temperature from + 20°C to (+20 °C, +71 °C]: the heat source is steam, and the first stage heat exchanger works.
(6) Reduce the laboratory's temperature from +71 °C to [+20 °C, +71 °C): the cold source is R507 high-temperature refrigeration unit, and the first stage heat exchanger works.

According to the above control strategies, the temperature of the refrigerant needs to be adjusted first in the temperature control process. In the primary refrigerant circulation system, the refrigerant is cooled by refrigeration unit and is heated by steam. The process schematic diagram of the refrigerant temperature control in the secondary refrigerant circulation system is shown in Fig. 2: T_1 is the outlet pipeline temperature of primary circulation of refrigerant, and the difference between T_1 and target temperature is a constant value; T_2 is the temperature of the heat exchanger inlet; T_3 is the temperature of the heat exchanger outlet; T_4 is the inlet air temperature; T_5 is the outlet air temperature.

The secondary refrigerant circulation pipeline adopts the constant flow variable temperature control method. By controlling the opening of the return valve and bypass valve, the temperature of the heat exchanger inlet is adjusted, so as to realize the control of the air supply temperature T5 at the outlet of the heat exchanger.

3.2 The Analysis of Temperature Control Algorithm

For the climate laboratory shown in Fig. 1, if an ordinary single-loop PID controller is used for laboratory's temperature control, when the laboratory temperature deviates from

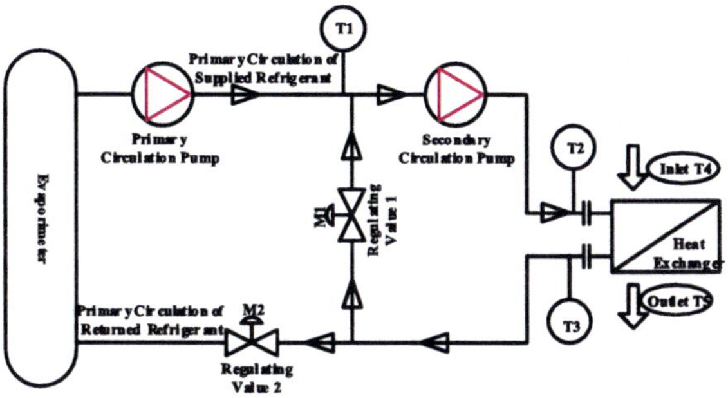

Fig.2. Refrigerant temperature control in the secondary refrigerant circulation system

the set value, the PID will adjust the temperature by changing the opening degree of the refrigerant regulating valve according to the magnitude and the positive and negative of the deviation value. However, this control accuracy of this control scheme is greatly low far from achieve the control of the climate laboratory. The reason is that the temperature control of the climate laboratory can be approximated as a first-order inertial system with a lag [12, 13]. If the temperature of the refrigerant fluctuates, the temperature of the supply air after the heat exchanger will change, and the circulating air will be recirculated. Enter the laboratory from the swirl tuyere of the air supply section. After a period of mixing, the temperature of the laboratory will deviate, and the control system will find the existence of interference, and then change the opening of the refrigerant regulating valve. Temperature is controlled. But this process has to go through the heat transfer of the heat exchanger, the process air entering the laboratory, and the air remixing, which is a link with a big lag. When the adjustment process takes effect, the temperature of the laboratory deviates far from the set value, resulting in large fluctuations in the laboratory temperature and reduced system stability [14].

Based on the above analysis, the feedforward-cascade PID control system as shown in Fig. 3 can be used, which has two closed loops. The closed loop of the inner loop selects the supply air temperature that is closer to the regulating valve and has a smaller lag time as the secondary controlled variable, and the main controlled variable is the laboratory temperature of the outer loop. In this way, the leading effect of the inner secondary loop can be used to timely reflect the temperature of the supply air before the temperature fluctuation of the refrigerant interferes with the laboratory temperature, and the secondary controller can timely overcome it. Due to the short loop, minor lag and timely control of the inner secondary loop, the control quality of the supply air temperature has been improved [15]. At the same time, for the thermal load interference caused by the test piece, test equipment, lighting, etc. in the laboratory, the feedforward control can be used to offset or reduce the effect of laboratory temperature through the control function of the feedforward controller [16].

As the feedforward-cascade PID control system shown in Fig. 3, the input of the outer loop PID is the expected temperature of laboratory, and the deviation between it and the laboratory temperature feedback value is calculated by the PID to calculate the supply air temperature, and used as the input of the inner loop PID, the inner loop PID calculates the opening of the refrigerant return valve and the bypass valve, and adjusts

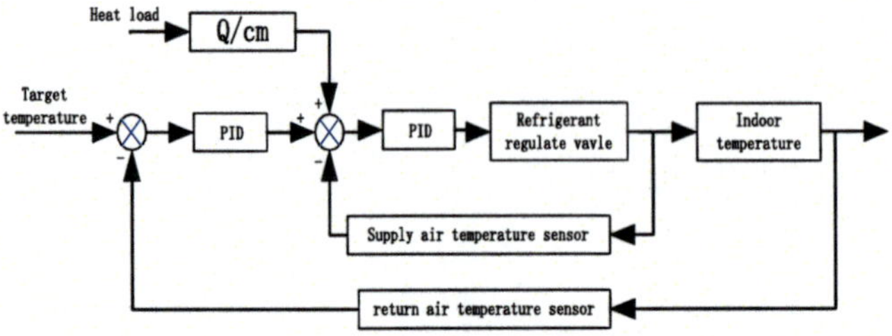

Fig. 3. Feedforward-cascade PID temperature controller

the temperature of the refrigerant at the inlet of the heat exchanger, thereby changing the supply air temperature and achieving the purpose of controlling the indoor temperature of laboratory.

A static feedforward controller [17] can be used in the feedforward link. The static feedforward controller calculates the temperature change value of the supply air caused by the heat load disturbance, and introduces it into the calculation of the supply air temperature deviation of the inner loop controller, so as to achieve the purpose of offsetting the heat load disturbance.

The supply air temperature change value is calculated by Eq. 10.

$$T_l = Q_l / c_a m_f \tag{10}$$

T_l is the change value of the supply air temperature. Q_l is the estimated heat brought by the indoor test piece and workin equipment. The c_a is the specific heat capacity at the current indoor temperature. The m_f is the air mass flowing through the circulating air heat exchanger per unit time.

In order to control the indoor temperature according to a certain heating and cooling rate, the laboratory temperature control belongs to the follow-up control system. The temperature target value (the input of the controller) is calculated by Eq. 11.

$$T_n = T_i + \frac{T_f - T_i}{t_c} \cdot t_k \tag{11}$$

T_n represents the current temperature target value. T_i represents the initial temperature. T_f represents the indoor final target temperature. The t_c represents the adjustment time from the initial temperature to the indoor final target temperature, and the t_k represents the adjusted time.

4 Application and Analysis

4.1 Parameters Tuning of PID Controller

The temperature control process of super large space climatic laboratory is a large time-delay object. As long as there is a temperature deviation, the function of the integral will continue. When the temperature deviation is large, integral saturation will exists and results overshoot of the temperature control curve. Therefore, integral separation PID is adopted for temperature control. When the deviation value is large, the integral effect is removed and PD control is adopted; when the deviation value is small, PID control is adopted [18]. The algorithm of incremental integral separation PID controller is

$$\Delta u(k) = K_p[e(k) - e(k-1)] + \beta(K_p/T_i)e(k)T + K_p * T_d[e(k) - 2e(k-1) + e(k-2)]/T \quad (12)$$

In formula (12), β is the integral separation coefficient, and ε is the integral separation threshold. If $e(k) > \varepsilon$, β is set to 0; If $e(k) < \varepsilon$, β is set to 1. In the temperature control of climatic laboratory, the integral separation threshold is set to 3.

According to the simulation method in reference [19], the MATLAB Simulink model is built, and the critical proportional method is used to turn the PID parameters of the system. The sampling period T must meet the Shannon sampling theorem. For the temperature control system with delay, the sampling period can be 1/10 of the pure delay time, so the sampling period is 9s. Set the integral coefficient and differential coefficient of PID controller to zero, so that the control system has only proportional component. Gradually increase the proportional coefficient $K_p^{'}$ until the system response appears equal amplitude oscillation. At this time, the s proportional coefficient $K_p^{'}$ is the critical scale δ_k, and the time taken for one oscillation is T_k.

According to the calculation method of critical proportional method, the empirical formula for the calculation of PID parameters is shown in Table 1.

Table 1. Empirical formula of critical proportion method

Parameters	*Kp*	*Ti*	*Td*
PID	$1.7\delta_k$	$0.5T_k$	$0.125T_k$

This method is used to turn the PID parameters of the sub controller and the main controller, and then a set of PID parameters are obtained (as shown in Table 2.) through fine-tuning trial based on the parameters.

Table 2. PID tuning parameters of main-loop and vice-loop

Parameters	*Kp*	*Ti*	*Td*
Sub controller	0.12	81	9.5
Main controller	0.35	410	110

4.2 Result Analysis

The feedforward cascade PID temperature controller is applied to the super large space climatic laboratory. The temperature process control includes cooling process

(+21°C→−20°C→-40°C→-45°C→−55°C) and heating process (−55°C→−40°C→-45°C→ + 21°C), and the temperature control curve is shown in Fig. 4. In each stage of heating and cooling process, the temperature regulation rate is consistent with the target rate, and the temperature fluctuation is less than ±2°C.

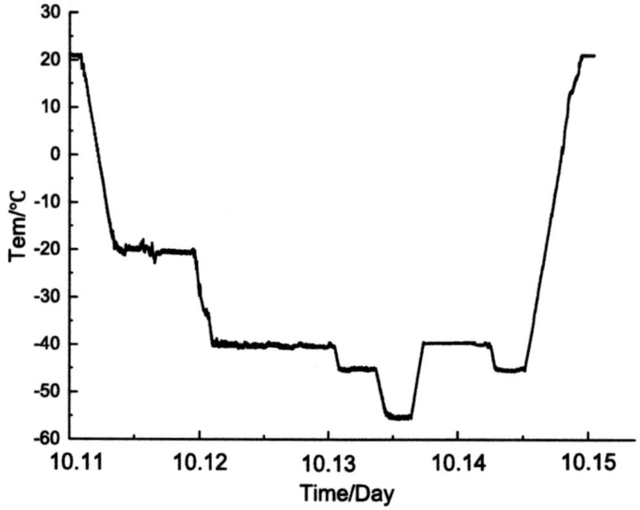

Fig.4. The temperature control curve

5 Conclusion

The large space climatic laboratory adopts feedforward-cascade PID for temperature control. The cascade PID controller overcomes the temperature fluctuation of refrigerant, and the feedforward controller overcomes the heat load interference caused by aircraft and working test equipment in the laboratory. The test results show that the established model can reflect the characteristics of the temperature control system of the climatic laboratory. The control method can effectively overcome the thermal disturbance, and it has fast dynamic response, high steady-state accuracy and no overshoot. The test results show that the temperature fluctuation is less than ±2°C.

References

1. Ran, Y.Q., Li, J., Yang, J.: Research and implementation of temperature control algorithm for environment simulation system based on PLC[J]. Comput. Meas. Control **23**(10), 3377–3384 (2015)
2. Wang, J., Huang, B.C., Wan, C.D., et al.: Environmental Simulation Technology [M]. National Defense Industry Press, Beijing (1996)
3. Deng, D.Q., Zhang, J.H., Gao, F.: Fuzzy-PID control theory for high/low temperature environmental simulation system[J]. Comput. Meas. Control **23**(1), 277–283 (2015)

4. Meng, Q.L., Wang, Y., Li, Y.P.: Temperature controller design for environment simulation laboratory[J]. Control Eng. China **22**(1), 170–175 (2015)
5. Li, K., Pang, L.P., Liu, W.K., Wang, J.: System model simulation and control method used in environmental simulation chambers[J]. J. Beijing Univ. Aeronaut. Astronaut. **33**(5), 535–538 (2007)
6. Wu, X.F.: Tailoring of laboratory temperature environment test for aviation equipment[J]. Equip. Environ. Eng. **15**(1), 74–79 (2018)
7. Liu, J.W., Wei, D., Dai, Z.W., et al.: Research and realization of fuzzy PID control-based for variable air volume control system [J]. Trans. Beijing Inst. Technol. **30**(8), 920–924 (2010)
8. Wu, X.F., Li, D.M.: Study on the control method of fresh air system for climatic environmental test facilities[J]. Control. Eng. China **24**(S0), 30–35 (2017)
9. Liu, H.Y., Ma, J.J., Zhang, H.: Discussion on design of the air handing system in large climatic environmental test laboratory[J]. Equip. Environ. Eng. **11**(5), 107–113 (2014)
10. Wu, X.F., Xu, J.: Study on identification method for linear system based on the feature of step response curve[J]. Meas. Control. Technol. **38**(5), 127–133 (2019)
11. Song, Z., Li, J., Qian, S.H.: Simulation of temperature control system for large climatic environmental test laboratory[J]. Comput. Simul. **35**(4), 200–203 (2018)
12. Gruber, M., Trüschel, A., Dalenbäck, J.-O.: Model-based controllers for indoor climate control in office buildings-complexity and performance evaluation[J]. Energy Build. **68**(Part A), 213–222 (2014)
13. Sanaye, S., Dehghandokht, M., Fartaj, A.: Temperature control of a cabin in an automobile using thermal modeling and fuzzy controller[J]. Appl. Energy **97**(9), 860–868 (2012)
14. Li, M., Li, Y.Z., Zhu, S.J.: Cold-side temperature control strategy in power measuring system of radioisotope heat source[J]. J. Aerosp. Power **32**(1), 203–212 (2017)
15. Zhang, J., Zhang, F., Ren, M., et al.: Cascade control of su perheated steam temperature with neuro-PID controller[J]. ISA Trans. **51**(6), 778–785 (2012)
16. Golbahar Haghighi, M.R., et al.: Temperature control of functionally graded plates using a feedforward–feedback controller based on the inverse solution and proportional-derivative controller[J]. Energy Convers. Manag. **51**(1), 140–146 (2010)
17. Ma, X.S.: Process Instrumentation and Automation[M]. China Machine Press, Beijing (2013)
18. Meng, Q.L., Wang, Y.: Modelling and control of CFD-based distributed parameter system[J]. J. Zhejiang Univ. (Eng. Sci.) **46**(8),1478–1484 (2012)
19. Wu, X.F., Li, D.M.: Modeling of climatic environment test facility and temperature control system simulation[J]. Meas. Control. Technol. **34**(10), 63–66 (2015)

Equivalent Thermal Conductivity Improvement of Stator Winding and Thermal Benefit for Air-Cooled Electrical Aviation Machine

Xuanyang Hu[1], Chaofan Dong[1], Yuping Qian[1], Yangjun Zhang[1](✉), Changjiang Wang[2], and Weilin Zhuge[1]

[1] School of Vehicle and Mobility, Tsinghua University, Beijing, China
yangjunzhang@tsinghua.edu.cn
[2] School of Electrical Engineering, Shandong University, Jinan, China

Abstract. Electrical machines with high power density are key enabling components of propulsion systems for flying cars. Air cooling system is widely utilized for electrical aviation machine for its compact and reliable composition. For air-cooled electrical machine, the equivalent thermal conductivity of stator winding is the main parameter in determining the thermal resistance inside the slot winding. To get a further understanding, the analytical evaluation method and improvement of the equivalent thermal conductivity of stator winding is studied. Furthermore, the influence of the improvement in equivalent thermal conductivity of stator winding on heat transfer path is investigated and thermal benefit and performance benefit are evaluated. The analysis shows that with the equivalent thermal conductivity of stator winding increases to 3 W/m·K, a 100 °C-temperature reduction can be achieved in the stator winding.

Keywords: Thermal conductivity · Electrical machine · Air cooling

1 Introduction

Flying cars as a new type of vehicle for urban air transportation and future travel, are getting more and more attention in the automotive and aeronautical fields, it has become an important development trend for the transborder integration of automotive and aeronautical technology and industry [1]. Electrical machines with high power density are key enabling components of propulsion systems for flying cars. The power density of electrical machines has gained great improvement with the rapid technological development of electric vehicles [2]. However, for flying cars applications, the power density of electrical machines needs to be further pushed to a higher level, like 10 kW/kg [3].

To increase power density of electrical machines, the technologies can be divided into three different approaches accordingly. First approach to improve power density is to adopt new electromagnetic designs, such as YASA topology [4], or the Halbach magnet array etc. [5]. The second approach is to utilize new type of materials to enable a higher magnetic loading with Fe-Co alloy and soft magnetic composite materials, or a higher electric loading with ultra-copper conductor composed of nanotube [6, 7].

Chinese Society of Aeronautics and Astronautics: CASTYSF 2022, LNEE 972, pp. 700–707, 2023.
https://doi.org/10.1007/978-981-19-7652-0_61

Lastly, through effective thermal management of air-cooled electrical machines, the current density of stator winding can be increased under temperature limits of winding materials [8–10]. Compared to liquid cooling system, air cooling system eliminates the pipes and pumps and consequently forms a neat and simple but robust cooling system which is suitable for flying cars applications [11].

In modern high power density electrical machine for aviation applications, copper loss is the dominant part in all losses including copper loss, core loss, eddy current losses etc. [12]. The heat dispassion path for copper loss in motor stator assembly can be divided into three parts. Firstly, the heat generated by winding at the bottom of slot is transferred to stator yoke via slot liner and then to the housing. Secondly, the heated generated by winding at the center of slot is transferred to stator teeth via slot liner and then to the stator yoke, finally to the housing. Thirdly, the heat generated in end-winding part is partly transmitted axially to winding inside slot and partly transferred to housing by heat convection or conduction process. Typically, the hotspot in stator winding lies in the end-winding and the winding in the center of the slot. The winding is composed of random-placed round copper conductor with insulation film and impregnation material. The radial and tangential equivalent thermal conductivity of winding is much lower than that of axial direction. Normally the equivalent thermal conductivity of winding (ETCSW) without impregnation material is 0.4 W/m·K in radial and tangential direction compared with 235 W/m·K in axially direction. The lower radial and tangential ETCSW causes large thermal resistance in heat path and consequently causes a high temperature rise inside slot winding. The calculation of ETCSW is proposed and researched by many studies [13–16]. The ETCSW can be increased by various ways, such as using impregnation materials with higher thermal conductivity or increasing the slot fill factor.

In this paper, the influence of the improvement in ETCSW on the thermal heat transfer in the investigated propulsion motor is analyzed based on the thermal modelling. And its benefits on thermal and electromagnetic performance are evaluated. This paper is structured as follows. In Sect. 2, the ETCSW is explained and its improvement method is given out. In Sect. 3, the analytical calculation of the ETCSW is carried out based on different material characteristics. In Sect. 4, the thermal and electromagnetic performance benefits of improvement for ETCSW are analyzed based on an investigated propulsion motor. Finally, the conclusions are drawn (Fig. 1).

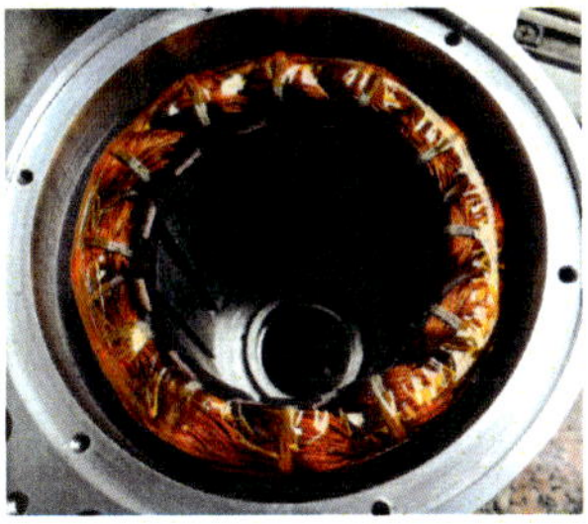

Fig. 1. Slot winding of an electrical aviation machine

2 Equivalent Thermal Conductivity of Stator Winding

The ETCSW is a critical parameter for evaluating the thermal performance of stator winding. The calculation model of the ETCSW is usually complex because of the random placement of winding composition inside slot including impregnation material, insulation film and conductor etc. The ETCSW is determined by the physical properties and the proportion of the materials used to form the winding. Accurate calculation of the ETCSW enables a more authentic thermal analysis result of the winding inside slot. The analytical prediction of the ETCSW has already been studied by many studies. In this paper, the analytical calculation of the ETCSW is based on the method proposed in [17]. The subscripts c and p represent the conductors and the impregnation materials. The υ_c is the volume ratio equal to the slot fill factor.

$$k_e = k_p \frac{(1 + \upsilon_c)k_c + (1 - \upsilon_c)k_p}{(1 - \upsilon_c)k_c + (1 + \upsilon_c)k_p} \tag{1}$$

As is predicted in analytical calculation Eq. (1), the slot liner and impregnation material play great influence on the equivalent thermal conductivity of slot winding. Therefore, this analysis is limited to the slot liner and impregnation materials which are selected to improve the ETCSW.

In selecting the material for improving the ETCSW, thermal conductivity of winding material is a key parameter. Other important parameters are insulation thickness, the dielectric strength, the viscosity and also the working temperature [18]. Traditionally, the slot liner insulation material used in electric machines is composed of polymeric coating, polymeric films and papers, such as polyamide and polyaramid. These materials provide perfect electric insulation but poor thermal conduction because of their low thermal conductivity of 0.14 W/m·K. A highly thermally conductive insulation for electric machine is reported in [19], with thermal conductivity up to 1.8 W/m·K. Table 1 lists the parameters of the slot liner materials.

Table 1. Slot liner insulation material characteristics

Material	Nomex 410	APTIV 1102 film	AlN coating
Thickness	0.25	–	0.127
Dielectric strength (kV/mm)	8.25	200	650
Thermal conductivity	0.139	0.43	1.75
Tensile strength	29.6 kN/m	100 MPa	–
Insulation class	220 °C	–	300 °C

Table 2 lists the electrical, thermal and mechanical characteristics of three selected impregnation materials. In order to improve ETCSW, the thermal conductivity of impregnation material should be as high as possible. At the same time, in order to withstand the high voltage that may occur between coils, the dielectric strength and volume resistivity are also remarkably important parameters.

Table 2. Impregnation material characteristics

Material	Varnish	EpoxyLite	SbTCM
Thermal conductivity (W/m·K)	0.25	0.85	3.2
Dielectric strength (kV/mm)	80	20	10
Volume resistivity at 25 °C (Ω·cm)	10^{15}	10^{14}	10^{14}
Viscosity (Pa·s)	–	3.5	25

3 Analytical Calculation of Equivalent Thermal Conductivity of Stator Winding

In this section, The ETCSW is calculated using analytical model proposed in Sect. 2. Both radial and tangential thermal conductivity are verified over large range to confirm the proposed analytical method prediction accuracy. The geometry, material physical properties and the boundary conditions are assigned in the analytical calculation according to the model. Figure 2 shows the analytical results comparison of ETCSW under different impregnation materials. The thermal conductivity of impregnation material and slot fill factor are selected as study parameters.

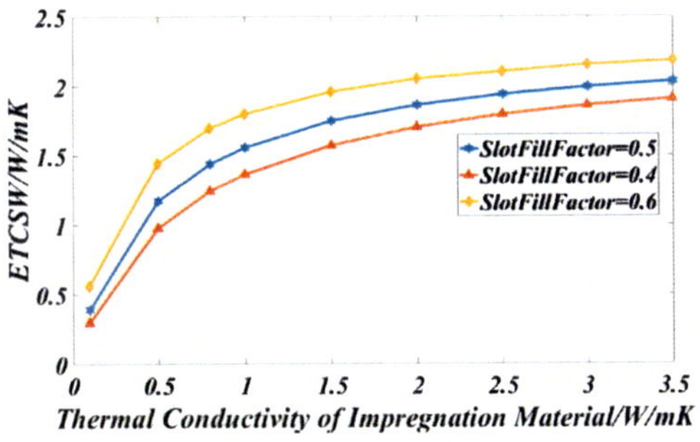

Fig. 2. Equivalent thermal conductivity of the slot winding

4 The Investigated Propulsion Motor and Thermal Modelling

4.1 The Investigated Propulsion Motor

This section analyzes the thermal benefits of the improvement of the ETCSW. A high speed PMSM for electric propulsion system is taken as a case study as shown in Fig. 3 and relevant parameters listed in Table 3. The machine employs a 10-pole 12-slot configuration with concentrated winding. The winding inside slot is composed of 0.15 mm slot liner and 45 strands of copper conductor with a 0.45 mm diameter. The slot fill factor is approximately 0.45. The machine is air-cooled by convection of 36 radial cooling fins placed at the outside of the housing. The fin thickness is 1.2 mm, height is 10 mm and with 35 mm length along the machine axial direction.

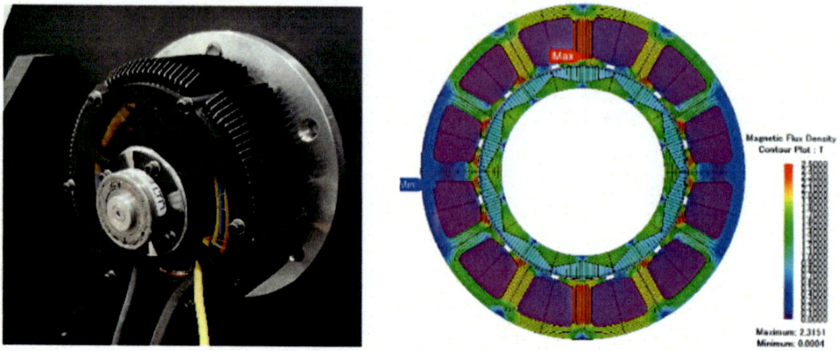

Fig. 3. Investigated propulsion motor.

Table 3. Table captions should be placed above the tables.

Machine data	Value	Machine data	Value
DC voltage	270 V	Rated speed	9000 rpm
Rated torque	4.5 N·m	Peak torque	9.35 N·m
Rated power	4.24 kW	Peak power	8.81 kW
Efficiency	95%	Working time at peak power	>120 s
Cooling	Air-cooled	Working temperature	−40–55 °C
Stator diameter	106 mm	Rotor diameter	56 mm
Active length	25 mm	Housing	Aluminum

4.2 Thermal Modelling of the Propulsion Motor

Lumped-parameter thermal network (LPTN) is widely used in thermal analysis of electric machines for its fast evaluation of electric motor with parameterized cooling and structure conditions. A LPTN thermal model in MotorCAD is developed to evaluate the thermal performance of the above-mentioned electric propulsion machine under varied ETCSW. A three-dimensional LPTN thermal model is established to account for the heat transfer in both radial and axial directions. In the axial direction, the motor active core is divided into three parts equally and one part for each end winding. In order to obtain a more accurate temperature distribution inside a stator slot, 20 thermal nodes are included with 5 layers in the radial direction. Figure 4 shows the thermal distribution of the motor under investigation in MotorCAD. The housing is cooled by forced air-convection; the corresponding heat transfer coefficient is assumed to be 200 W/m^2·K. A thermal isolation condition is applied to the two sides of the stator core.

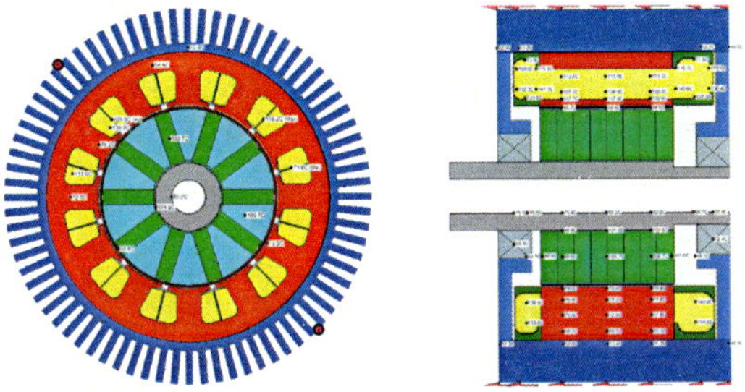

Fig. 4. Investigated motor thermal model in MotorCAD.

5 Thermal Benefits and Performance Benefits

The improvement of ETCSW enables the reduction of thermal resistance inside the slot winding, and a reduction of temperature rise in slot winding. The temperature reduction mechanism is that more heat flux inside the slot winding is dissipated through the slot liner and the stator teeth and yoke. In Fig. 5, the heat flux is divided into the heat flux through stator teeth (HF-Teeth-Winding), heat flux through stator yoke (HF-Yoke-Winding). It is clearly to see that with the increase of the ETCSW, the HF-Teeth-Winding and HF-Yoke-Winding shows the gradual growth. The TotalHF is the sum of HF-Teeth-Winding and HF-Yoke-Winding. The TotalHF grows from 188 to 201 W. However, the CopperLoss-TotalHF is the difference of copper loss and TotalHF. The CopperLoss-TotalHF decreased from 139 to 78 W with the ETCSW increased from 0.4 to 3 W/m·K. The heat flux reduction is the main reason behind the temperature rise reduction as shown in Fig. 5.

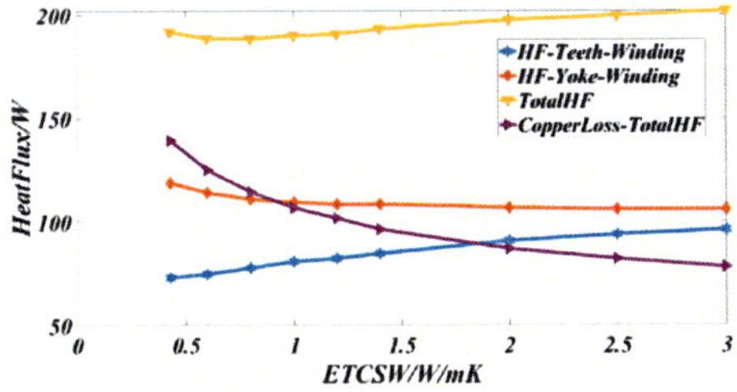

Fig. 5. Heat flux variation with ETCSW

The thermal benefit of the stator winding is shown in Fig. 6. With the ETCSW increases from 0.4 to 3 W/m·K, the maximum temperature decreases from 201 to 103 °C, which means the temperature rise in the slot winding is reduced. However, the minimum temperature remains nearly 85 °C. The temperature difference between maximum temperature and minimum temperature decreases from 120 to 18 °C, which means a more uniform temperature distribution inside the slot winding. For copper loss reduction, the copper resistance is positively correlated to the temperature. With the maximum temperature decreases, the winding resistance also decreases accordingly. Consequently, this results in a reduction in copper loss as shown in Fig. 7. The copper loss reduces from 330 to 280 W with the ETCSW improvement. The 50 W copper loss reduction implies a 1.1% efficiency improvement for the propulsion motor under investigation.

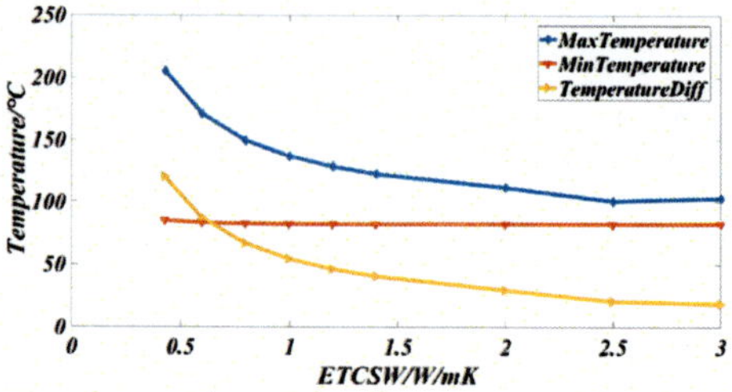

Fig. 6. Winding temperature variation with ETCSW

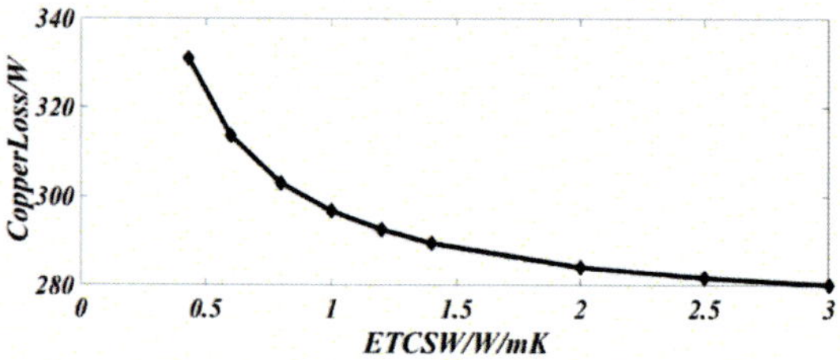

Fig. 7. Copper loss variation with ETCSW

6 Conclusion

The ETCSW has an important influence on determination of the temperature rise and distribution in the motor slot winding. The analytical calculation is widely studied by several researchers to provide fast and accurate evaluation of the ETCSW under various winding compositions. To get a further understanding of influence on the heat transfer path, this paper analyzed the improvement of the ETCSW. The heat flux for different heat transfer path was investigated under a case study motor. It shows that with the improvement of the ETCSW, the temperature rise and temperature difference is significantly reduced, for that nearly 100 °C temperature reduction is achieved when the ETCSW is increased to 3 W/m·K. The improvement of the ETCSW is proved to be an effective way to reduce the temperature rise inside the slot winding and consequently an effective way to increase the efficiency and power density of the electric motor.

References

1. Rajendran, S., Srinivas, S.: Air taxi service for urban mobility: a critical review of recent developments, future challenges, and opportunities. Transp. Res. Part E: Logist. Transp. Rev. **143**, 102090 (2020)
2. El-Refaie, A., Osama, M.: High specific power electrical machines: a system perspective. CES Trans. Electr. Mach. Syst. **3**(1), 88–93 (2019)
3. Chen, Y., et al.: Mechanical design considerations of an "ironless", high-specific-power electric machine. IEEE Trans. Transp. Electrif. **3**(4), 855–863 (2017)
4. Woolmer, T.J., McCulloch, M.D.: Axial flux permanent magnet machines: a new topology for high performance applications. Institution of Engineering and Technology, Stevenage (2006)
5. Schmidt, E., Lechner, C.: Design and assembling of a permanent magnet synchronous machine with a Halbach magnet array for the application in an electric driven race car. IEEE (2015)
6. Fan, T., Li, Q., Wen, X.: Development of a high power density motor made of amorphous alloy cores. IEEE Trans. Ind. Electron. **61**(9), 4510–4518 (2014)
7. Al-Timimy, A., et al.: Design and losses analysis of a high power density machine for flooded pump applications. IEEE Trans. Ind. Appl. **54**(4), 3260–3270 (2018)
8. Zhang, F., et al.: A thermal modeling approach and experimental validation for an oil spray-cooled hairpin winding machine. IEEE Trans. Transp. Electrif. **7**(4), 2914–2926 (2021)
9. Fawzal, A.S., et al.: Air inlet/outlet arrangement for rotor cooling application of axial flux PM machines. Appl. Therm. Eng. **130**, 1520–1529 (2018)
10. Zhang, F., et al.: Back-iron extension thermal benefits for electrical machines with concentrated windings. IEEE Trans. Ind. Electron. **67**(3), 1728–1738 (2020)
11. Popescu, M., et al.: Modern heat extraction systems for power traction machines—a review. IEEE Trans. Ind. Appl. **52**(3), 2167–2175 (2016)
12. Wrobel, R., et al.: Power loss analysis in thermal design of permanent magnet machines - a review. IEEE Trans. Ind. Appl. 1
13. Wrobel, R., et al.: Characterizing the in situ thermal behavior of selected electrical machine insulation and impregnation materials. IEEE Trans. Ind. Appl. **52**(6), 4678–4687 (2016)
14. Liu, X., et al.: Effective thermal conductivity calculation and measurement of litz wire based on the porous metal materials structure. IEEE Trans. Ind. Electron. **67**(4), 2667–2677 (2020)
15. Yi, X., et al.: Equivalent thermal conductivity prediction of form-wound windings with litz wire including transposition effects. IEEE Trans. Ind. Appl. **57**(2), 1440–1449 (2021)
16. Ayat, S., et al.: Estimation of equivalent thermal conductivity for electrical windings with high conductor fill factor. IEEE (2018)
17. Simpson, N., Wrobel, R., Mellor, P.H.: Estimation of equivalent thermal parameters of impregnated electrical windings. IEEE Trans. Ind. Appl. **49**(6), 2505–2515 (2013)
18. Nategh, S., et al.: Evaluation of impregnation materials for thermal management of liquid-cooled electric machines. IEEE Trans. Ind. Electron. **61**(11), 5956–5965 (2014)
19. Woodworth, A.W., et al.: Potential of materials to impact megawatt-scale electric machines. In: 2021 AIAA/IEEE Electric Aircraft Technologies Symposium (EATS) (2021)

The Research Status and Development of Military Aircraft Ground Support Equipment

Hongwei Liu(✉), Hanlu Zhong, Jiayuan Wu, Bo Cheng, Zhengyu Zhou, and Fanghua Cao

AVIC Chengdu Aircraft Industrial (Group) CO., LTD, Chengdu 610091, China
hitliuhongwei@163.com

Abstract. With the rapid development of aviation weapon equipment and the introduction of advanced technologies, new aviation support system concepts and support modes are emerging in recent years. Military aircraft support equipment is not only the material basis of an integrated support system for aviation weapon equipment, but also a significant factor affecting the combat effectiveness and the support cost. The improvement of support equipment performance and utilization rate contributes to the performance and the safety of weapon equipment. Ground support equipment carries a lot of basic support work and plays an increasingly important role in the ground support system. Combined with the research status of military aircraft ground support equipment, the configuration optimization of domestic military aircraft support resources and the status of ground support equipment standard system, this paper points out the main problems existing in military aircraft ground support equipment in China, and briefly describes the development trend of future aircraft support equipment, providing help and enlightenment for aviation support research.

Keywords: Ground support equipment · Research status · Configuration optimization · Standard system · Development trend

1 Introduction

According to standard GJB3872 [1], aircraft support equipment refers to the special support equipment that provides the environment and the corresponding conditions for use, detection and maintenance of aircraft systems and airborne equipment in the specified environment, including ground support equipment (GSE), air support equipment and shipboard support equipment. Among them, GSE carries a large number of basic support work [2, 3], including the implementation of gas, liquid, electricity, etc. GSE can be divided into power equipment (power supply vehicle, hydraulic pump vehicle), filling equipment (refueling vehicle, oxygen vehicle), thermal processing equipment (air conditioning vehicle, heating vehicle) and lifting installation equipment (tractor, crane). According to the structure form, GSE can also be divided into three types, vehicle type, fixed type and traction type [4].

Aircraft ground support equipment is not only the material basis of aviation weapon equipment ground support system (GSS), but also an important factor restricting the

Chinese Society of Aeronautics and Astronautics: CASTYSF 2022, LNEE 972, pp. 708–717, 2023.
https://doi.org/10.1007/978-981-19-7652-0_62

development of weapon equipment and affecting the life cycle cost. Different from the engine support mode of the air flight test mission, the ground static test of the whole aircraft is to continuously provide ground energy for the debugging and functional inspection of the aircraft hydraulic, flight control, avionics and other systems relying on GSE when the engine is not working, to verify whether the aircraft systems and airborne products can meet the flight standards. Therefore, the development of support equipment and the configuration optimization of resources have become the key to the effective development of aircraft support capability, which is of great significance to the product delivery and the rapid formation, continuous maintenance and boom of military combat effectiveness.

2 Development Process of Ground Support Equipment for Military Aircraft

Since the 1990s, aviation weapon equipment has entered a stage of rapid development. Countries are not only committed to the technical development of aviation equipment, but also pay attention to the construction of GSS. In terms of military GSS, foreign countries attached great importance to the control of GSE as early as 20 years ago. In order to achieve the operational ideas of rapid deployment, expeditionary warfare and the scale reduction of support equipment at that time, the US Air Force's headquarters established the "Air Force Support Equipment Working Group" and the "Support Equipment Vehicle Management Office", which were responsible for the development policy, development planning, strategic issues and the management of support equipment, respectively, to achieve the comprehensive management of air force logistics support equipment [5].

The early GSS is composed of a variety of single-function GSE. Different types of aircraft require different quality power supply, different flow and pressure of oxygen, nitrogen and hydraulic power energy to provide the support of equipment. Therefore, strict resource allocation and scheduling are required according to the actual functional requirements. To improve the support efficiency, the reliability of aircraft ground equipment and reduce the life cycle cost of equipment, in 1996, the United States Air Force proposed the concept of Modular Aircraft Support System (MASS) for a variety of single-function ground equipment and conducted a lot of research work. The resource allocation cost and the life cycle cost of equipment were reduced by 40% and 20%, respectively, and the reliability of the support system was improved by 15% [6]. The development and production of different models have changed the configuration requirements of GSS from single-function support equipment to multi-function support equipment, and the concept of support has also converted relying on increasing the scale of support to pursuing the modularization of support equipment. However, military non-standard equipment needs to be serialized and modularized to achieve the generalization of support equipment components. Such generalization components can be exchanged among similar support equipment, so as to reduce the scale of support equipment, improve the transfer capability and provide support for a variety of aircrafts. For example, the A/M27T hydraulic vehicle commonly used by the U.S. military can support different types of aircraft such as B2, C17 and F-22. The power supply vehicle can supply power to all transport aircraft and fighter aircraft, and the maintenance of troops and bases is applicable [7]. In the

1990s, BAE developed a new type of multi-functional ground support system (MAGSS). The system integrated lighting, environmental control, nitrogen supply and air compression functions on a portable device, and the fuel used was consistent with the fuel of the aircraft, which met the daily support requirements of multi-aircraft [1].

In 1997, Mashey first proposed the concept of "big data" [8], and the whole military aircraft system was gradually combined with big data. Since then, the development of GSS has begun to lag behind that of airborne system. The United States Department of Defense issued the "Product Assurance Guide for the 21st Century" guide in 2003, and proposed the concept of Performance Based Logistics (PBL), which took the support as a comprehensive and affordable performance package to purchase. Driven by the integrity and continuous mission capability of wartime equipment, it emphasized user-centered, constantly adapted to the requirements of new combat environment and combat style for equipment support [9]. In 2008, the F-22 project carried out supportability and self-support capacity design using PBL technology, focusing on reducing life-cycle support costs. The F-22 has a 50% reduction in the transport time required to deploy a squadron compared with the F15 [7].

With the continuous development of support equipment, the support mode and the support system are changing. The traditional ground support mode follows the passive reactive support process of "Failure occurrence - Detection isolation - Failure location - Resource scheduling - Equipment Maintenance", and the passive equipment support system cannot effectively support the rapid response ability required for future wars [10]. At the beginning of the development of the fourth-generation aircraft, the U.S. military put forward the concept of autonomic logistics support. This kind support technology is a logistics support system based on knowledge, which can identify and integrate a series of related information such as logistics support demand, supply chain management, reliability and safety of components to support and strengthen the implementation of tasks. The core of the autonomic logistics support system is the prognostic and health management (PHM) system and the joint distribution information system (JDIS). The PHM system applies advanced sensors and algorithm models to manage the health status of equipment and form decisions. The JDIS system informationizes equipment, personnel and other technical data to form real-time interoperability between systems [11–13]. The operation of autonomic logistics support system changes the traditional maintenance support strategy from post-maintenance and prevention to maintenance and a small amount of preventive maintenance based on real-time status, forming a two-level maintenance system, as shown in Fig. 1 [14].

In 2014, Lockheed Martin developed Autonomic Logistics Information System (ALIS) for the F-35 project, used for pre-maintenance and supply chain issues, as shown Fig. 2 [15, 16]. More than 400 F-35 aircrafts have been equipped with ALIS, which realizes the autonomy and decision support of logistics support procedures and reduces the cost of life cycle support [17].

In 2020, the U.S. Department of Defense launched a new support system - Operation Data Integration Network (ODIN) system for the problems of occasional database conflicts, outdated operating systems and cumbersome system hardware in ALIS [18]. ODIN system planned to host part of the original ALIS system to the cloud, saving hardware costs, enhancing data sharing between different departments, and reducing data

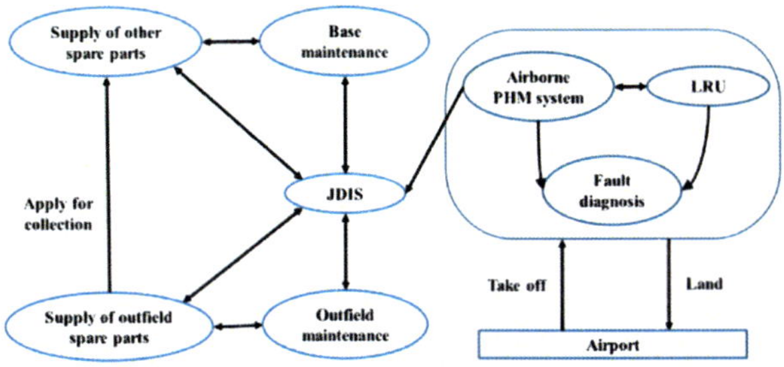

Fig. 1. Aircraft autonomic logistics support system [14]

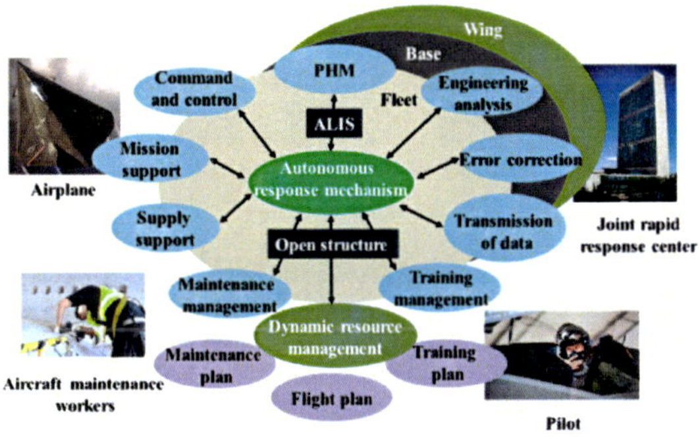

Fig. 2. The F-35 concept map of autonomic support [16]

communication costs. In essence, ODIN system adopts a new organizational model and development model, which is a major change for ALIS. Compared with ALIS, the hardware cost and the volume are reduced by 30% and 75%, respectively, and the enhanced computing ability shortens the processing time by 50%.

Advanced foreign military aircraft support mode promotes the development of GSS towards digitalization and automation. However, China starts late in the control of GSE. At first, some support technologies are introduced in accordance with the characteristics and needs of Chinese aviation industry. In 1981, Aviation Industry Group and Air Force introduced reliability centered maintenance (RCM) technology [19], which was applied in the ground support of J6 aircraft and extended the maintenance period under the premise of ensuring combat performance. Since 2003, RCM technology has been widely used in the field of support and maintenance of major weapon equipment [20, 21]. In the face of the large-scale and rapid development of domestic aviation weapon equipment, the traditional ground support mode causes more life cycle costs, which increases the possibility of false alarms and false alarms. The investment of a large amount of manpower and material resources may also not improve the aircraft's mission rate.

In the past, military aircraft ground equipment had the problems of single support function, backward support mode and low resource utilization rate. These factors seriously restricted the development of GSE and could not satisfy the needs of modern air force operations. In 2007, a certain station of the domestic air force completed the construction task of the first batch of multi-aircraft comprehensive support bases. The smooth construction of the base marked a historic leap in the flight support mode of the air force air force station, from supporting fewer aircrafts to multiple aircrafts in a single airport, greatly improving the support efficiency [22].

With the continuous development of Chinese aviation technology, the PHM concept was introduced during the "Tenth Five-Year Plan" period. The basic research on PHM of airborne system for aviation equipment was started, and the experimental research was carried out combined with model technology in the "Twelfth Five-Year Plan" period. At present, the fault diagnosis and fault isolation of line replaceable unit (LRU) for flight control, hydraulic energy, avionics, life insurance [23] and other systems are mainly focused on the model research technology, which makes the airborne system support farewell to the backward maintenance era. However, the development of GSS is slow and lagged behind the current support demand. There is still a certain gap with the construction goal [24] required by the comprehensive support of domestic weapon equipment, and it is not adapted to the development of the new generation of fighter support towards the direction of life-cycle and system management.

3 Configuration Optimization of Ground Support Equipment for Military Aircraft

Scientific optimization of resource allocation can effectively solve the problem of insufficient and excess resource allocation, greatly improving the efficiency of military aircraft support and reducing the cost. Existing optimal allocation methods mainly take the average satisfaction rate of equipment as the optimization objective, and take the equipment utilization rate, equipment operation cost and equipment scale as the constraint conditions [25]. The Petri net [26], the queuing theory [27] and other theories are used to model, and the intelligent algorithms such as genetic algorithm, particle swarm optimization algorithm [28] and related combinations are used to solve the problem, so as to obtain the optimal allocation scheme of support equipment.

Zhu established the resource optimization allocation model of maintenance support equipment based on variable precision rough set, and realized the resource optimization allocation of multi-resource points and multi-demand points under different precisions combined with the particle swarm optimization algorithm [29]. Han analyzed the constraints of aircraft support process and resources under integrated support mode, and established a multi-aircraft integrated aircraft support scheduling model [30]. Su studied the support resource constraints and the process constraints of carrier-based aircraft surface under different man-machine matching modes, and compared the advantages and disadvantages of various matching modes through simulation tests [31].

Bu established a Monte Carlo queuing model with the utilization rate and the cost of support equipment as the main constraint conditions and the aircraft departure rate as the objective function. Genetic optimization algorithm was used to optimize the configuration scheme of support equipment [32]. Zhang proposed the optimal allocation model of grassroots-level test equipment based on the types, methods and characteristics of grassroots-level equipment maintenance support tasks and the shortest average maintenance time of equipment [33]. Through genetic algorithm, Cui systematically analyzed the constraints such as operation process, process execution mode, resource transfer and allocation, and incorporated mobile support vehicles into the scope of support resources [34]. To optimize the configuration of integrated support vehicles, Cui analyzed the demand and parameter index system of mobile combat support, and constructed the evaluation model of mobile combat support capability by multi-level fuzzy comprehensive evaluation method [35]. In recent years, with the continuous development of information network and the wide application of big data and artificial intelligence, the role of information resource management in the configuration optimization of military aircraft support equipment has gradually become prominent [36].

4 Standard System of Ground Support Equipment for Military Aircraft

In the early years, the United Kingdom, the United States and Australia signed the "Defense Military Standardization Cooperation Agreement", and jointly carried out standardization construction. Subsequently, China established a relatively perfect standard system of safeguard equipment, including basic, management and technical standards [37]. The basic standards are the supporting materials for management standards and technical standards. Technical standards are in the form of hierarchical classes, including top-level standards, generic standards and dedicated standards. As shown in Fig. 3, the top-level standard is GJB5967 《Support equipment planning and development requirements》. The general class standard includes 16 items, such as GJB1132 《General specification for aircraft ground support equipment》 and the special class standard includes 50 items, such as GJB1125 《General specification for aircraft cloth cover》 [38]. The establishment of military aircraft ground support equipment standard system has greatly standardized the development and maintenance of support equipment.

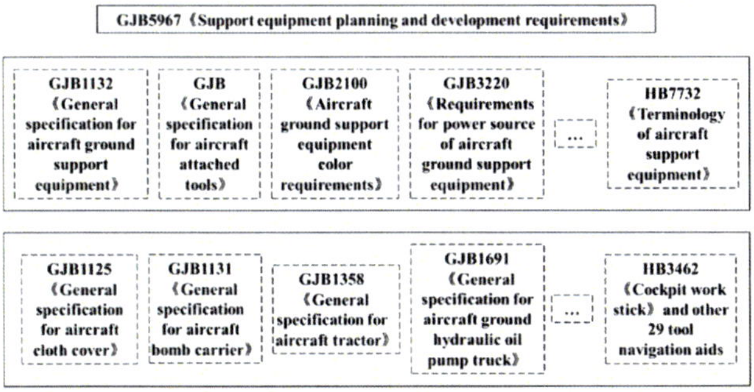

Fig. 3. Military aircraft support equipment standard system diagram [38]

5 Main Problems Existing in Ground Support Equipment

With the progress of technology and the improvement of user requirements, the design and development of military aircraft support equipment have made progress in functional modularization and integration, volume miniaturization and operation convenience. However, the comprehensive development and optimal configuration of support equipment are still lagging, and some problems need to be solved urgently.

5.1 Insufficient Generalization of Support Equipment

Although some support equipment integrates multiple functions, reducing the number of support equipment and facilitating the aircraft support work in the outfield, the support equipment after function integration has worse usability and lower efficiency than before. For example, a certain type of air tightness tester integrates many inspection functions such as cabin air tightness inspection, air tightness inspection of air tightness belt and air tightness inspection of radar pressurization system. However, when checking the air tightness of radar pressurization system, the pressurization process is too slow [38]. In addition, some function integration is only the integration of hardware, and the software has not been really realized, which is lack of universality.

5.2 Poor Environmental Adaptability

Since the newly developed support equipment has been gradually applied, some equipment has not carried out environmental identification tests or insufficient environmental identification tests, leading to more faults in the outfield, especially in some harsh environments. For example, a certain type of charging and discharging test system exits the faults of no output current and black screen of the monitor in the coastal high temperature and humidity environment [38].

5.3 Difficult Transportation

The transportability of support equipment is a comprehensive performance of the adaptability of railway, highway, waterway and air transport modes. The mobile combat mode puts forward higher requirements for the transfer of GSE. Due to the large volume and weight of common support equipment, the equipment transition is difficult. Some support equipment is designed without considering the needs of transfer transportation, without setting hanging points or special packing boxes, and without considering the impact of vibration, temperature and pressure changes on the bearing capacity of the equipment under various transportation conditions, all of which bring great inconvenience to transfer transportation [39].

5.4 Lack of Effective Information Management in Resource Allocation

Although a large number of domestic support equipment resource allocation optimization work has been carried out [40], the level of informatization and intelligence of support equipment needs to be improved, and the instant communication and information management of support resources with support needs have not been realized. The visual support system based on the Internet of Things, satellite positioning, radio frequency identification technology can shorten the response time. In the future, the configuration optimization system can be combined to output the required support resource allocation scheme, and provide decision makers with a choice.

6 Development Trend of Ground Support Equipment for Military Aircraft

With the continuous development of aviation weapon equipment, the support equipment has experienced a qualitative change process of "Completely relying on imports - Imitation research and development - Independent research and development". Although the research and development production capacity of domestic weapon equipment can meet the needs of the army, the leapfrog development of weapon equipment puts forward higher requirements for the support capacity of GSE. The ground support mode of the second and third generation aircraft belongs to the point-to-point energy output structure, and there is a decentralized control between the support equipment and the debugging resources such as the aircraft position. However, in the new international situation, the demand for GSS of the new aircrafts is no longer limited to the improvement of the quality and quantity of support equipment, but it is necessary to improve the cognition and attention of the demand for the economic, comprehensive and intelligent ability of the support system, and centralize the control and multi-degree allocation of various ground resources. It is committed to improving the interactive sharing, resource allocation rate and information management and control ability of the support equipment.

The development trend of military aircraft ground support equipment is not only the improvement of support equipment performance and system standards, but also the transformation of support mode and the optimization of support resource allocation. With the continuous improvement of the level of weapon equipment, the prediction of equipment data and resource demand will be more accurate, and the efficiency of resource allocation will be higher. The development direction of support resources is also moving closer to the demand for scientific management, the continuous maintenance of equipment combat effectiveness and the reduction of equipment life cycle cost.

References

1. GJB 3872-1999 General requirements for materiel integrated logistics support
2. Liu, B.X., Fan, B.: The direction of airplane ground support equipment. J. Shenyang Aerosp. Univ. (02), 84–86 (2000) (in Chinese)
3. Hu, Z.S., Huang, Z.J., Liu, S.Y.: Safety evaluation of aviation four stations support based on dynamic weight and both-branch fuzzy set. J. Mil. Transp. Univ. **19**(12), 80–84 (2017) (in Chinese)
4. Zhang, L.Y.: On the development of aircraft ground support equipment in China. China Sci. Technol. Overv. (20), 1 (2014) (in Chinese)
5. Ding, L.L.: Development status and trend of military aircraft ground support equipment. In: The 4th Yangtze River Delta Science and Technology Forum Aerospace and Yangtze River Delta Economic Development Sub Forum and the 3rd National Aviation Maintenance Technology Academic Annual Meeting, Ningbo, Zhejiang, China (2007) (in Chinese)

6. Xiao, A.H.: Development status and trend of military air-craft ground support equipment. China Sci. Technol. Overv. (5), 1 (2015) (in Chinese)
7. Li, S.H., Wu, S.J.: Development and enlightenment of American military aviation support equipment. Adv. Aeronaut. Sci. Eng. **9**(A01), 5 (2018) (in Chinese)
8. Mashey, J.R.: Big data and the next wave of infra-stress. In: Computer Science Division Seminar, University of California, Berkeley (1997)
9. Zhang, H.M., Bi, Y.Y., Guo, X.L.: Analysis on the research status of performance-based equipment support theory. Mil. Econ. Res. **31**(10), 3 (2010) (in Chinese)
10. You, M., Liu, Q., Sun, H.: New communication strategy for spectrum sharing enabled smart grid cyber-physical system. IET Cyber-Phys. Syst.: Theory Appl. **2**(3), 136–142 (2017)
11. Hess, A., Calvello, G., Dabney, T.: PHM a key enabler for the JSF autonomic logistics support concept. In: IEEE Aerospace Conference (2004)
12. Hess, A., Fila, L.: The joint strike fighter (JSF) PHM concept: potential impact on aging aircraft problems. In: Aerospace Conference. IEEE (2003)
13. Wu, N., Li, B., Wang, H., et al.: Parametric message passing-based relative navigation in joint tactical information distribution system. J. Syst. Eng. Electron. **27**(01), 81–89 (2016)
14. Shi, W., Chen, C., Wang, Z.L., et al.: Research on the autonomic logistics system. Logist. Sci-Tech (05), 93–96 (2008) (in Chinese)
15. Zhou, X.W., Li, X.D., Zhao, R.X.: Development of ground support equipment for military aircraft under autonomous support mode. In: Development of Aviation Support Equipment - The First Aviation Support Equipment Development Forum in 2017, Chengdu, Sichuan, China (2017) (in Chinese)
16. Wang, P.: The latest development of F-35 fighter's autonomous support information system. http://www.360doc.com/content/21/0119/09/38805566_957743974.shtml (2001). Accessed 6 Mar 2022
17. Sullivan, M.J.: F-35 joint strike fighter: observations on program progress (2015)
18. Wang, P.: F-35 ALIS redesigned. Int. Aviat. (7), 3 (2020) (in Chinese)
19. Brauer, D.C., Brauer, G.D.: Reliability-centered maintenance. IEEE Trans. Reliab. **R-36**(1), 17–24 (1987)
20. Chen, Y., Huang, L.J.: Application and development of reliability-centered maintenance (RCM) in world wide nuclear power field. In: Progress Report on China Nuclear Science & Technology, vol. 1 (2009) (in Chinese)
21. Cheng, Z.H., Jia, Y.X., Liu, L., et al.: RCM analysis decision-making system for weapon and radar equipment. J. Ordnance Eng. Coll. **16**(1), 5 (2004) (in Chinese)
22. China Youth News: Air force logistics department explores a new model of aviation logistics support base. http://zqb.cyol.com/content/2008-01/04/content_2019596.htm (2008). Accessed 6 Mar 2022
23. Bao, X.N., Zhao, P.L., Zhang, B.Z., et al.: Life support system in advanced fighter. Acta Aeronaut. Astronaut. Sin. **41**(06), 253–265 (2020) (in Chinese)
24. Chen, F., Zhang, Z.J., Hong, D.P.: Comprehensive support mode and development exploration of weapons and equipment. Aerosp. Ind. Manag. (08), 50–55 (2020) (in Chinese)
25. Wen, J.: Joint optimization model for support equipment's allocation under multi-constraint. Telecommuni. Eng. **53**(6), 6 (2013) (in Chinese)
26. Bu, B., Wang, Z.J., Pan, H.S.: Optimal location of multi-aircraft universal support equipment based on Petri nets in grass-roots level. Syst. Eng. Electron. **41**(2), 9 (2019) (in Chinese)
27. Guo, H.F., Liu, F.S.: Use queue theory mode to optimize the quantity of the support equipment. J. Armored Force Eng. Inst. **19**(1), 3 (2005) (in Chinese)
28. Hong, Z.F., Dong, B.L.: Study on optimization of equipment maintenance scheduling based on modified particle swarm optimization algorithm. J. Zhejiang Sci-Tech Univ. (Nat. Sci. Ed.) **37**(6), 7 (2017) (in Chinese)

29. Zhu, Y.H., Cao, J.P., Wang, Z.Y., et al.: Optimal deployment of equipment maintenance support resources based on variable precision rough set in wartime. Fire Control Command Control **39**(6), 5 (2014) (in Chinese)
30. Han, W., Su, X.C., Chen, J.F.: Integrated maintenance support scheduling method of multi-carrier aircrafts. Syst. Eng. Electron. **37**(4), 809–816 (2015) (in Chinese)
31. Su, X.C., Han, W., Zhang, Y., et al.: Scheduling algorithm for maintenance and service support of carrier-based aircraft on flight deck with different man-aircraft matching patterns. Acta Aeronaut. Astronaut. Sin. **39**(12), 19 (2018) (in Chinese)
32. Bu, B., Wang, Z.J., Pan, H.S.: Optimal allocation of support equipment based on aircraft sortie generation rate in the wartime. J. Air Force Eng. Univ. (Nat. Sci. Ed.) **19**(3), 6 (2018) (in Chinese)
33. Zhang, L., Cha, C.D., Chang, T.Q., et al.: Research on optimization of allocation model of armor equipment guarantee test equipment. Fire Control Command Control **43**(6), 4 (2018) (in Chinese)
34. Cui, R.W., Han, W., Su, X.C., et al.: Integrated optimization of carrier-based aircraft flight deck operations scheduling and resource configuration for pre-flight preparation stage. Syst. Eng. Electron. **43**(7), 10 (2021) (in Chinese)
35. Cui, L., Zhang, H.R.: Configuration optimization of integrated support vehicle based on mobile warfare support capability evaluation. Missiles Space Veh. (6), 8 (2021) (in Chinese)
36. Zhang, X.S., Yan, P.C., Sun, J.S., et al.: On storage and sharing of maintenance support information based on XML parsing technique. Comput. Appl. Softw. **29**(3), 3 (2012) (in Chinese)
37. Zhang, D.F., Liu, X.Y., Zhao, R.X.: Thinking about the standard construction of ground support equipment for military aircraft. In: Proceedings of the First Aviation Support Equipment Development Forum in 2017 (2017) (in Chinese)
38. Hu, Y., Li, Y.F., Feng, K.: Some issues and thoughts during use certification of military aircraft ground support equipment. Aviat. Maint. Eng. (2), 4 (2019) (in Chinese)
39. Xie, Z.F.: Research on transportability evaluation method of military aircraft support equipment. Technol. Innov. Appl. (23), 2 (2015) (in Chinese)
40. Gao, Y.J., Cai, J.P., Lei, N., et al.: Review of study on equipment maintenance support resource allocation optimization. J. Gun Launch Control **41**(3), 5 (2020) (in Chinese)

Design of an Electronically Beam-Scanned High-Power Capacity Reflectarray Antenna Based on Liquid Crystals for Airborne Radar Application

Gexing Kong[1], Wei Du[1], Peng Cui[1], Yingfeng He[1], and Rui Chen[2(✉)]

[1] Chengdu Aircraft Industrial (Group) Co., Ltd, Chengdu 610091, China
[2] Southwest Jiaotong University, Chengdu 610031, China
813034364@qq.com

Abstract. In this paper, an electronically beam-scanned high-power capacity airborne radar reflectarray antenna based on liquid crystals is presented. To achieving fast phase-tuning and high-power capacity performances of the airborne radar reflectarray antenna, the thoroughly simulated analysis of the structure of the patch and the position of the bias line is made. The optimized patch is rectangular patch with three rectangular slit, which shows good power capacity performance. And the bias line is placed at the center distance. Simulation shows that the introduction of bias line has little effect on the reflected amplitude and phase of the reflecting unite. A 11 × 11 airborne radar reflectarray antenna are designed and analyzed. The simulated results show that the power capacity of the reflectarray antenna is about 3.6 MW in air condition and 30° beam-scanning can be obtained. The results verify the electronically beam-scanned and high-power capacity performance of the airborne radar reflectarray antenna. The antenna shows good environmental adaptability and electronically beam-scanning performance.

Keywords: Airborne radar antenna · Electronically beam-scanned · High-power capacity · Reflectarray antenna

1 Introduction

The airborne radar antenna is an important part of a fighter [1–5]. In recent years, with the rapid development of aerospace technology and high-power microwave technology [6], the electromagnetic environment of the battlefield in which fighters located is extremely complex. In order to improve the battlefield survivability of the airborne radar antenna, the antennas should not only meet the basic radiation characteristics requirements, but also meet the high-power capacity performance. In addition, the antenna is also required to have fast beam scanning performance.

Reflectarray antenna has the potential for high-power capacity applications [7]. At the same time, based on the liquid crystals with the thickness of 100 μm, the response time of nematic liquid crystal molecules is within 30 ms [8], the dielectric constant of the liquid

Chinese Society of Aeronautics and Astronautics: CASTYSF 2022, LNEE 972, pp. 718–724, 2023.
https://doi.org/10.1007/978-981-19-7652-0_63

crystal is changed correspondingly. In addition, the available study shows that liquid crystal material has the performance of high-power capacity. Therefore, based on the liquid crystal material, an electronically beam-scanned high-power capacity reflectarray antenna is designed in this paper. The antenna shows good environmental adaptability and electronically beam-scanning performance.

2 Design of Reflecting Unite

The configuration of the proposed reflecting unite is shown in Fig. 1. The radiating antenna is three rectangular slit patch, which forms a sandwich structure with the upper Rogers 5880 dielectric plate layer and the lower liquid crystal material layer. The bias lines at both ends of the patch are connected to the bias line at the bottom of the PCB by the metallized through hole I, and the ground plane by the metallized through hole II. The metallized through hole I is not connected to the ground plane. The positive pole of the DC bias voltage source is connected to the metallized through hole I, and the negative pole is connected to the metallized through hole II. Thereby a bias loop combined with the unite structure is formed. The reflecting unites are arranged in a rectangular grid. In order to obtain relatively good reflecting amplitude and phase response results at normal incidence and 30° oblique incidence, the size parameters of the reflecting unite are optimized and designed, and the optimized key parameters of the patch are shown in Table 1. The unite spacing is 4.5 mm (about 0.53 times the wavelength of the center frequent). The patch thickness is 0.15 mm, the liquid crystal layer thickness is 100 μm, and the upper dielectric plate thickness is 2 mm.

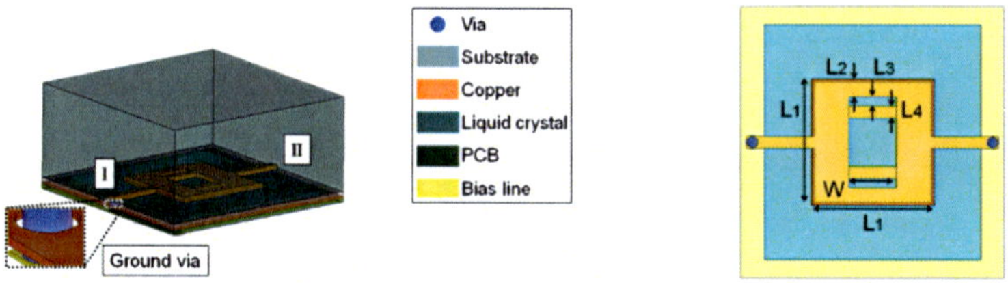

(a) 3D view

(b) Top perspective view

Fig. 1. The configuration of the proposed airborne radar reflecting unite.

Table 1. Optimized parameters.

Parameter	L_1	L_2	L_3	L_4	W
Value (mm)	2.1	0.3	0.15	0.2	0.8

In order to design a reflecting unite with high-power capacity performance, it is necessary to reduce the surface electric field of the radiating patch. The main means to reducing the electric field is to make the electric field distribute over a larger effective area of the patch. Based on this theory, the simulations of the reflect array unite based on the rectangular patch, the rectangular patch with a rectangular slit and the rectangular patch with three rectangular slit are carried out respectively. The input power is 0.5 W, based on the same liquid crystal materials with dielectric constant and loss tangent ($\varepsilon = 3.4$, $\tan\delta = 0.02$), the simulated electric field distributions of reflecting unites with different patch shapes are obtained are shown in Fig. 2. The electric field strength on the surface of the patch is mainly concentrated on both sides of the patch perpendicular to the polarization direction of the excitation electric field, the maximum electric field strength value of the rectangular patch with three rectangular slit unite is relatively lower, which mainly because the effective electric field area of the structure is relatively larger. The reflecting unite based on the patch structure has high power capacity performance.

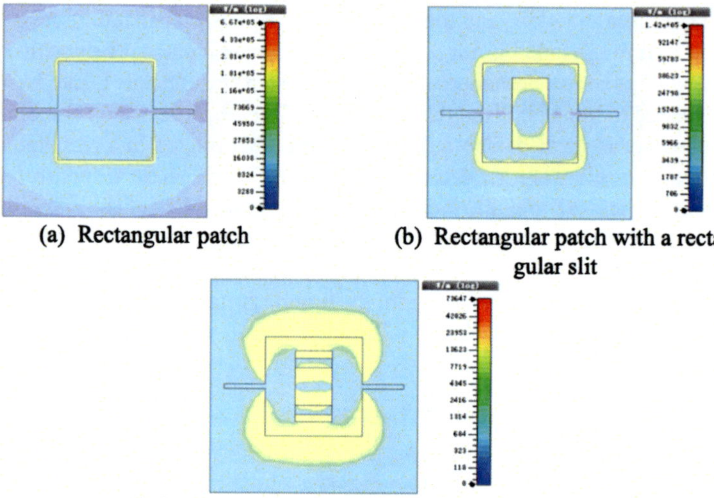

(a) Rectangular patch (b) Rectangular patch with a rectangular slit

(c) Rectangular patch with three rectangular slit

Fig. 2. Electric field distributions of different reflecting unites.

The reflecting unite with high-power capacity needs to load the bias line to connect with the voltage source. As a small structure, the bias line may cause the problem of electric field concentration, thus limiting the power capacity of the reflecting unite. With the goal that the introduction of the bias line has small impact on the power capacity performance of the unite, the position of the bias line is analyzed. For the convenience of analysis, based on the fixed dielectric constant and loss tangent ($\varepsilon = 3.4$, $\tan\delta = 0.02$), the unite with no-bias line, with horizontal bias line at the corner of the patch, the unite with horizontal bias line at the 1/4 distance, and the unite with horizontal bias line located at the center distance are simulated and analyzed. The input power is 0.5 W, the obtained electric field distribution on the surface of each patches are shown in Fig. 3. From the electric field distribution, it can be seen that the electric field distribution on the patch surface of the unite where the horizontal bias line is located at the center of the patch is similar to that of the unite without the bias line. As for the unite with the horizontal bias line at the corner and the horizontal bias line at the 1/4 distance, the introduction of the bias line makes the surface mode of the patch change significantly. The reflected amplitude and phase response results of different unites are shown in Fig. 4. The reflected amplitude and phase response results of the reflecting unite with horizontal bias line located at the center of the patch are consistent with the results of the unite without bias line. According to the above analysis results, placing the horizontal bias line at the center of the patch is the optimal position, which has little effect on the field strength distribution and reflection amplitude-phase response performance of the antenna.

By adjusting the dielectric constant, the change of the dielectric constant caused by the change of the director of the liquid crystal molecules under different applied voltages is simulated. And the reflected amplitude and phase response of the reflecting element under different conditions are simulated and analyzed based on unite cell boundary. The reflecting unite is excited by TM linearly polarized plane wave. The obtained reflected amplitude and phase response results corresponding to different dielectric constants under normal incidence and 30° oblique incidence are shown in Fig. 5. The reflected amplitude and phase response stability is good under under normal and 30° oblique incidence. When the dielectric constant changes from 2.42 to 3.4, the phase compensation of 305° can be achieved, and the reflection amplitude remains greater than −8.6 dB. Under different incident angles, the unite can achieve good reflected amplitude and phase response at the operating frequency of 35 GHz. The reflected magnitude is more than −3 dB among the 35% of the dielectric constant. So for a large antenna arrays, the element loss results are relatively good.

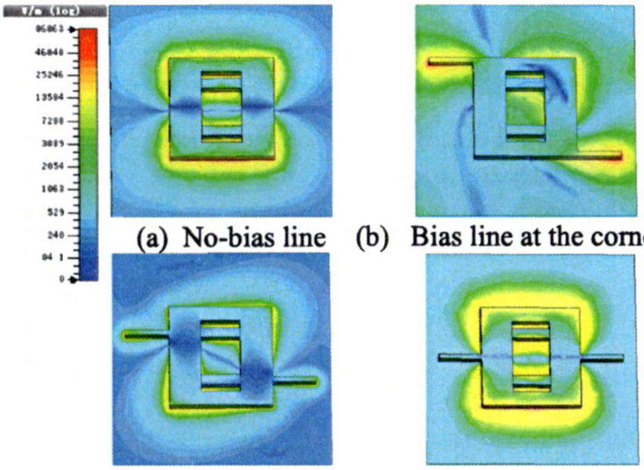

(a) No-bias line (b) Bias line at the corner of the patch

(c) Bias line at the 1/4 distance (d) Bias line at the center distance

Fig. 3. Electric field distribution on the surface of different patches.

3 Design of Reflectarray

The simulated results of the reflecting unite show that the reflecting unite based on liquid crystal material has stable reflected phase and amplitude response performance. And the unite has the potential for high-power microwave applications. In order to

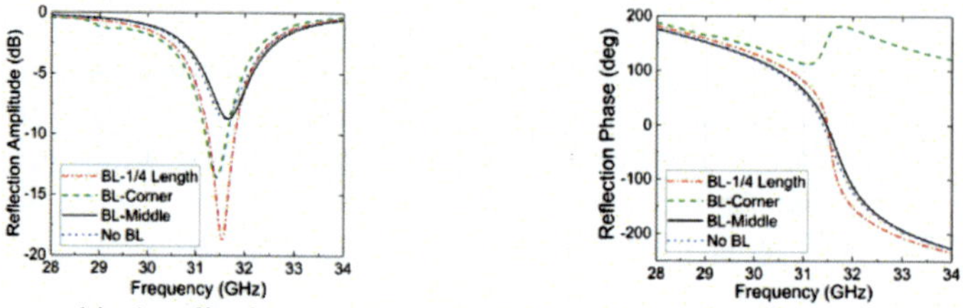

(a) Amplitude (b) Phase

Fig. 4. Reflected phase and amplitude responses of different unites.

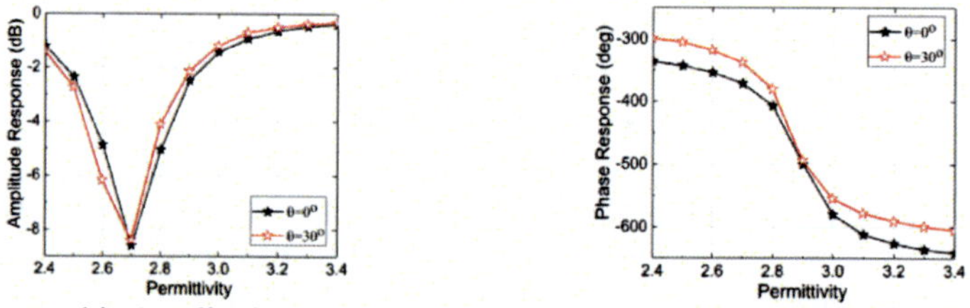

(a) Amplitude (b) Phase

Fig. 5. Reflected phase and amplitude responses under normal and 30° incidence.

verify the above performance, a 11 × 11 reflectarray antenna is designed with this unite. The reflectarray is fed by a linearly polarized corner cone antenna with 10° offset. The height of the feed was adjusted to meet the requirements of edge tapper of −8 dB. The calculated phase compensation required by each element of the array is shown in Fig. 6. The dielectric constant of each element is adjusted according to the phase response curve of Fig. 5(b). And the full-wave simulation analysis of the reflectarray antenna is performed based on the electromagnetic simulation software.

The simulated results show that the antenna gain with the main beam direction of ($\theta = 10°$, $\varphi = 0°$) is 20.7 dB at the operating frequency of 35 GHz. The corresponding aperture efficiency is 28%, and the sidelobe level is −18.1 dB, which achieves better directional beam radiation. Under different beam scanning angles, the 3D pattern and gain results are shown in Figs. 7 and 8. Within the scanning range of 0–40°, the pencil beam radiation pattern is achieved in the specified direction, and the gain variation is about 3 dB, When the main beam is at 20° direction, the antenna achieves the maximum gain of 21.7 dB, and the corresponding aperture efficiency is 35.3%. The relatively good efficiency is obtained.

Taking the main beam direction of ($\theta = 10°$, $\varphi = 0°$) as an example, the power capacity performance of the reflectarray antenna is analyzed. The input power is 0.5 W, and the electric field strength distribution of the key part of the reflectarray antenna is shown in Fig. 9. The maximum field strength of the internal patch is 11118 V/m. The internal breakdown threshold of the liquid crystal is 30 MV/m [8], the calculated power capacity is about 3.6 MW. And the maximum field strength on the antenna surface is 750 V/m. Calculated with a breakdown threshold of 3 MV/m in air, the surface power capacity is 8 MW. So the main limitation of the power capacity of the reflectarray antenna

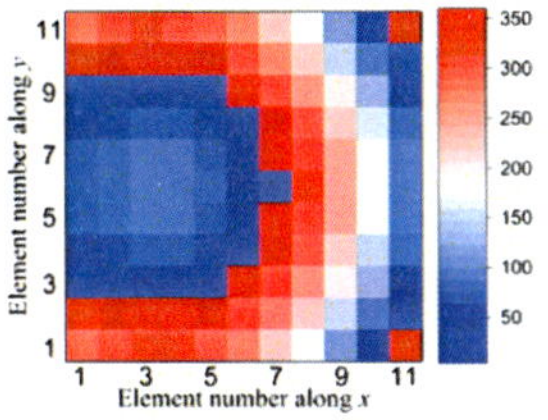

Fig. 6. The phase compensation required by each element

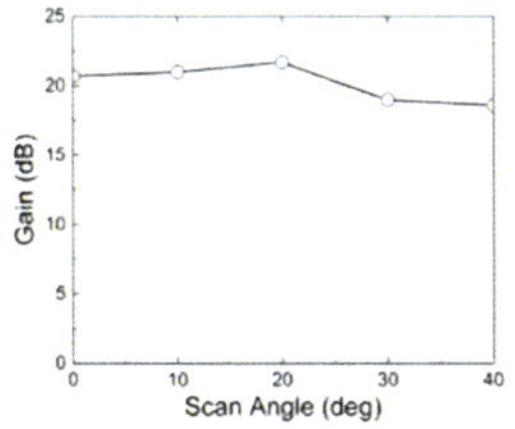

Fig. 7. Gains of different scanning angles.

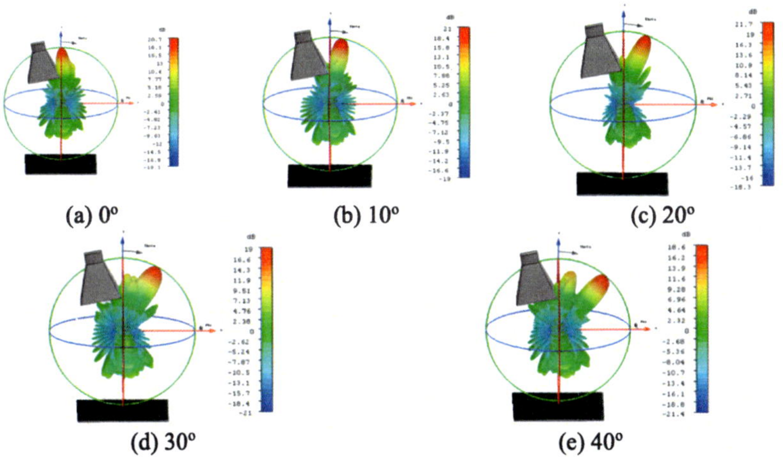

Fig. 8. 3D pattern of different scanning angles.

is on the internal patch, and its power capacity is 3.6 MW. The results show that the reflectarray antenna has high-power capacity performance.

4 Conclusion

In this paper, a electronically beam-scanned high-power capacity airborne radar reflectarray antenna based on nematic liquid crystal is proposed and designed. An innovative reflecting unite is designed with good reflecting amplitude and phase response, and

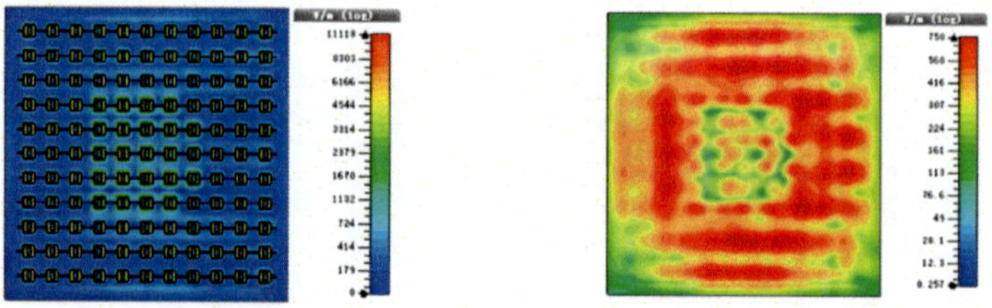

(a) Internal patch (b) Antenna surfacE

Fig. 9. Electric field strength distribution of the key part of the reflectarray antenna

electric field strength performances. The above performances are basically not affected by the introduction of bias lines. A Ka-band 11 × 11 reflectarray antenna is designed, and its electronically beam-scanned and high-power capacity performance are verified. The antenna shows good environmental adaptability and electronically beam-scanning performance.

References

1. Stimson, G.W.: Introduction to Airborne Radar (1998)
2. Seward, T.: Technology and character of phased array radar. J. Electron. Deference **6**, 54–58 (1998)
3. Cook, N.: Development tendency of future airborne radar military technology. (8), 76–82 (1999)
4. AVIC Leihua Electronic Technology Research Institute: Airborne Radar Handbook, 4th edn. National Defense Industry Press, Beijing (2013)
5. Jahn, A.M.: F-22 radar development. United States Air Force F-22 System Program Office, Aeronautical Systems Center, Wright Patterson AFB, Ohio (1997)
6. Barker, R.J.: High-power microwave sources and technologies. IEEE Xplore (2001)
7. Huang, J., Encinar, J.A.: Reflectarray Antennas. Wiley, Hoboken Piscataway, NJ, USA (2007)
8. George, H.H., Philadelphia, P.: Decreasing response time of liquid crystals. U.S. Patent 3575491, 20 Apr 1971

An Controller Workload Forecasting Method Based on Flight Plan

Heng Yuming[1,2](✉), Wu Minggong[1,2], Wen Xiangxi[1,2], and Lin Fugen[1,2]

[1] Air Traffic Control and Navigation College, AFEU, Xi'an 710051, China
13419886627@163.com
[2] National Key Laboratory of Air Traffic Collision Prevention, Xi'an 710051, China

Abstract. The significant growth of air traffic volume in the next few years requires a more accurate method of controller load assessment. Under the current framework of pre flight plan management, the most important thing is to quantify the control work of controllers according to the flight plan and find a reasonable method to evaluate their work. We present a controller load evaluation method based on flight plan. The first part aims to analyze the elements of the flight plan that affect the work of the controller, and the second part will be devoted to evaluating the controller load according to the flight plan. Finally, the practical application will be discussed.

Keywords: Controller workload · NNET · SVM · Random forest · Flight plan

1 Introduction

In recent years, with the rapid increase of air traffic flow, air traffic management (ATM) system is also constantly improving its capacity to cope with the increasing flight flow. The control seat has been overloaded for a long time. Accurate evaluation and prediction of controller load has become an urgent research task. In recent years, the frequency and frequency of air traffic unsafe incidents have increased significantly, most of which are caused by control reasons. Therefore, the prediction of controller load is of great significance to accurately grasp air traffic safety. We take the evaluation of controller load as the gateway premise, eliminates the hidden dangers in advance and increases the response ability.

The research on controller load is mainly different in the evaluation methods of workload, mainly focusing on the following two aspects: subjective evaluation and objective evaluation. The subjective evaluation method is to evaluate the work of the controller, the number of calls per unit time, duration and other behaviors. In this regard, the "DORATASK" method [1] proposed by the British operations research and Analysis Council and the "MBB" method [2] proposed by German scientist Messerschmidt are listed as publicly recommended methods by ICAO [2]. Reid and Nygren [3] proposed SWAT method in 1988, and Hart and Staveland [4] of NASA Ames Research Center proposed NASA-TLX (National Aeronautics and space administration task loading index) method. With the development and maturity of related technologies, the research on objective

Chinese Society of Aeronautics and Astronautics: CASTYSF 2022, LNEE 972, pp. 725–731, 2023.
https://doi.org/10.1007/978-981-19-7652-0_64

evaluation methods is also developing and improving. Zhao and others [5] analyzed and studied the variation law of ECG signal with working time by using principal component analysis, and came to the conclusion that rrmean and SDNN can evaluate the fatigue state. In 2020, Wang and Chen [6] proposed the method of evaluating load by EEG signal. Friedrichs and Yang [7] used eye movement recognition technology to identify and evaluate the controller load. However, the above subjective method has the problems of strong subjectivity and weak timeliness, and the objective method will interfere with the normal work of the controller because of wearing measuring equipment. However, from the perspective of cognition, as the core of the air traffic system, the controller needs to understand the airspace and its situation evolution including aircraft location and other information with the help of equipment, and make adjustments in case of emergencies, which is also the main cause of the workload of the controller. Therefore, Delahaye et al. [8, 9] proposed to establish a controller load evaluation system based on four indicators: traffic density, convergence, dispersion and sensitivity through the analysis of aircraft position and heading in airspace.

In recent years, with the development of machine learning and its application in the field of air traffic control, scientific research achievements continue to emerge. Zhang and others [10] took the lead in applying BP neural network to the field of air traffic control, and built a universal and feasible air traffic control operation quality evaluation system based on the universal collectability of data; Wen and Wang [11] used the ridge regression BP neural network model to predict and evaluate the controller load by analyzing and calculating the air traffic complexity evaluation index. Subsequently, Yue and others [12] proposed to use the machine learning model to cluster analyze and predict the health status of the control seat through the exercise data.

To sum up, domestic scholars' research on the workload of controllers mostly starts from objective data and does not consider the subjective factors of controllers in the process of control. We analyze the flight plan data and forecasts the controller load through the of machine learning model. It is helpful to fully grasp the operation of air traffic control system and provide basis for controller scheduling management and dynamic capacity management.

2 Flight Plan Index Selection

According to the requirements of relevant specifications [13], the flight plan usually needs to include dozens of items, including route direction, flight altitude and waypoints in and out of the flight area. The actual data summarized by the air traffic control department also includes the number of aircraft and the flow of sectors at peak hours. In actual control, there are many indicators that will not affect the controller load, such as the maximum takeoff and landing weight of aircraft. Therefore, through the questionnaire survey of front-line controllers, 10 factors in the flight plan that will affect the controller load are obtained as the analysis data of this paper, as shown in Table 1.

3 Controller Load Evaluation Method

The European aviation safety organization [14] puts forward the relationship model between controller workload and control work, as shown in Fig. 1.

Table 1. Flight plan index and meaning.

Index	Index name	Index meaning
C1	Number of aircraft in sector	Total number of aircraft operating in sector
C2	Peak value of aircraft number in time period	The maximum number of aircraft in the sector in all time nodes in the time period
C3	Number of aircraft whose control power is transferred	The number of aircraft entering and leaving sectors handed over within the time period
C4	Number of aircraft with altitude change	The number of altitude changes of all aircraft in the sector in the time period
C5	Number of aircraft in conflict	The number of times all aircraft in the sector have conflicts in the time period
C6	Number of aircraft with changed course	The number of heading changes of all aircraft in the sector within the time period
C7	Number of aircraft with speed change	The number of speed changes of all aircraft in the sector in the time period
C8	Number of inbound and outbound paths	The number of arrival and departure paths used by all aircraft in the sector within the time period
C9	Number of inbound and outbound points	The number of arrival and departure points used by all aircraft in the sector within the time period
C10	Model mixing factor	The number of aircraft types in operation in the sector within the time period

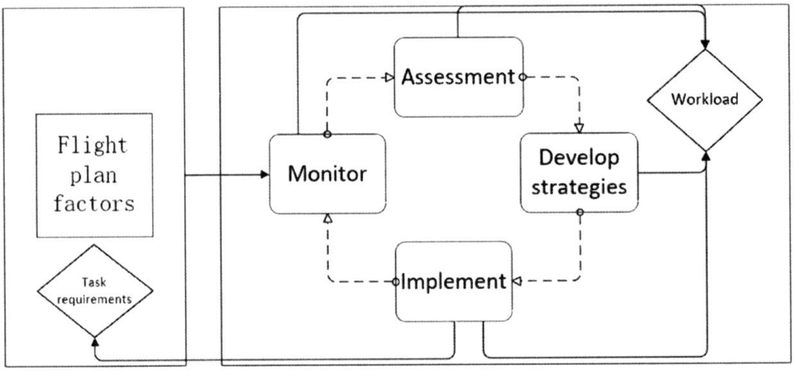

Fig. 1. Controller workload model

The model divides the generation of controller workload into two parts: task demand of external factors (i.e. the influencing factors of controller workload) and the processing process of internal factors. The approved flight plan triggers the mission requirements. The controller circularly monitors, evaluates, formulates control strategies and executes four processes according to the mission requirements obtained from observation. It can be seen that the controller's physiological and thinking load generated by executing four processes through observing task requirements is the controller's workload.

Based on the above analysis, "Doratask" method is selected as the evaluation method of controller load. In this method, the work of the controller is divided into three categories. One category is the invisible work, that is, the executive part in the figure above, including the routine control work of the controller, the communication of conflict resolution and filling in the progress sheet, which can be recorded and timed by the observer; The second category is invisible work, that is, the part of monitoring, evaluation and strategy formulation in the above figure, which mainly refers to the work of the controller monitoring the radar screen, comparing the progress sheet and thinking; These two types of work occupy most of the working time. In addition, the work of the controller also includes the third part, namely the recovery time. The workload evaluation for specific sectors and air traffic conditions is the sum of visible and invisible working hours, plus the recovery time, which is the workload of the controller. The visible part should account for about 50% of the whole workload, that is, the total workload is about twice the visible workload.

4 Controller Load Forecasting Based on Machine Learning

4.1 Data Acquisition

Xiamen Gaoqi International Airport is located in Xiamen, Fujian Province. As a 4E Civil International Airport, it plays an important role as a regional aviation hub in the southeast coastal area of China. In the whole year of 2020, Gaoqi airport has completed 16 710 197 passenger throughput, 278 336.4 tons of cargo and mail throughput, and 139 827 take-off and landing sorties, all of which are in the forefront of China. Its relevant data are highly representative.

We select the complete flight activity data in Xiamen air traffic control station system for statistics. Among them, the flight plan data includes the number of aircraft in the sector executing the flight plan, the peak number of aircraft in the time period, etc.; The radar track mainly records the longitude and latitude, altitude, heading, flight speed and other flight information during flight; The land air communication data is the audio file for command transmission and information transmission between the controller and the pilot. After analyzing such data, the controller workload can be obtained according to the "DORATASK" method. In this paper, the actual time span is two hours, and the data in the actual work of the controller in this stage are collected. 2 043 groups of effective data are obtained from the system, of which 80% and 1 634 items are used for the construction of the training set of the machine learning model, and the remaining 20% and 409 items are used for the verification and analysis of the model.

4.2 Model Building

Machine learning refers to automatically summarizing and sorting out some laws from massive data by computer, and obtaining a certain prediction model. The unknown data can be predicted by constructing a machine learning model. In the process of building machine learning model, we need to preprocess the collected data, and divide the data set into two parts: training set and test set. The training set is used to build the machine learning model, and the test set is used to verify the model. Because this paper has collected data and built the model through the data, this method belongs to supervised learning. Therefore, this paper selects three common supervised learning models: support vector machine [15], random forest [15] and neural network [15]. The model construction process can be summarized as follows:

(1) Data collection
The data collected in this article has been introduced in the previous section.
(2) Data processing
Because the collected data will inevitably have problems such as incompleteness and inconsistent format, it is necessary to process the data, including normalization, standardization, discretization and so on. In this paper, the normalization method is used to preprocess the data, and the formula is:

$$X_{norm} = \frac{X - X_{\min}}{X_{\max} - X_{\min}}$$

where $X_{\max}$ and $X_{\min}$ are the maximum and minimum values respectively; X is the original value; X_{norm} is the normalized value.
(3) The training and testing model
The training set and test set are constructed for the processed data. In this paper, 10 indicators collected from the flight plan are selected as the input and the controller load as the output to build the model. In the construction of the SVM, the Gaussian inner product kernel function is selected, and the parameter gamma in the kernel function is 0.14 and the penalty coefficient is 1; The parameters of random forest are the default values; The number of neurons in the hidden layer of NNET is 30, and the weight attenuation parameter is 0.001.
(4) Model evaluation
Through the analysis of the test results of the model, the prediction effect of the model is evaluated.

4.3 Comparative Analysis of Models

Through the analysis of the calculation results, the accuracy of the training set and test set of the three models can be obtained, as shown in Fig. 2.

The results of training set show that the prediction accuracy of support vector machine is lower than that of neural network and random forest. The results of the test set show that the prediction effect of neural network is the best among the three models.

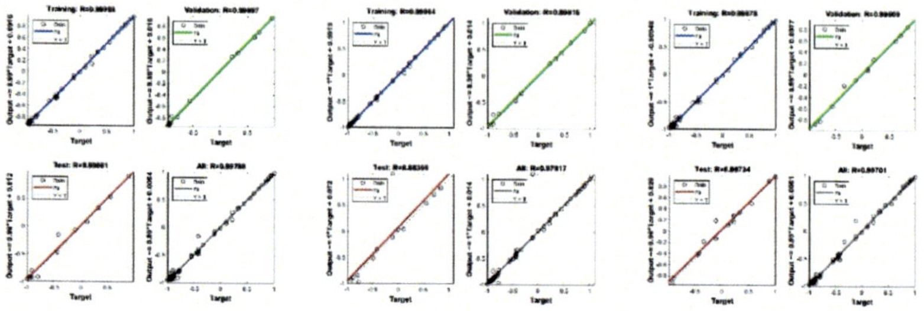

(1) Prediction results of NNET (2)Prediction results of SVM (3)Prediction results of random Forest

Fig. 2. Prediction results of machine learning models

5 Conclusion

This paper proposes to evaluate the controller load based on flight plan. Firstly, the factors affecting the controller load in the flight plan are screened in the form of questionnaire. Secondly, the "DORATASK" method is used to evaluate the controller workload. Finally, three machine learning models of SVM, random forest and NNET are introduced to predict the controller load. The effect of neural network is the best in the overall performance. In the process of model construction, we choose to build the model based on the data in the actual flight control process, and the results have a good guiding role for the actual control. The prediction results of the model prove the feasibility and rationality of the prediction method.

References

1. Stamp, R.G.: The DORATASK method of assessing ATC sector capacity an overview. DORA Communication 8924, Civil Aviation Authority, London (1992)
2. ICAO Doc. 9426: Air Traffic Services Planning Manual. ICAO, Montreal (1992)
3. Reid, G.B., Nygren, T.E.: The subjective workload assessment technique: a scaling procedure for measuring workload. North - Holland: Adv. Psychol. **52**, 185–218 (1988)
4. Hart, S.G., Staveland, L.E.: Development of NASA-TLX (task load index): results of empirical and theoretical research. Adv. Psychol. **52**(6), 139–183 (1988)
5. Zhao, X., Fang, R., Rong, J., Mao, K.: Research on travel time of basic road based on cellular automata. J. Beijing Univ. Technol. (10), 1511–1516
6. Wang, L., Chen, F.: Study on relationship between controllers' cognitive behavior and fatigue based on EEG. China Saf. Sci. J. **28**(07), 1–6 (2018)
7. Friedrichs, F., Yang, B.: Camera-based drowsiness reference for driver state classification under real driving conditions. In: Intelligent Vehicles Symposium. IEEE (2010)
8. Delahaye, D., Puechmorel, S.: Air traffic complexity: towards intrinsic metrics. In: Proceedings of the 3rd USA Europe Air Traffic Management Research and Development Seminar (2000)
9. Delahaye, D., Puechmorel, S., Hansman, R.J.: Airtraffic complexity map based on non linear dynamical systems. Air Traffic Control **12**(4), 367–388 (2004)
10. Zhang, J., Hu, M., Wu, Z., et al.: An improved integrated evaluation method on operation performance of air traffic control based on BP network. J. Southwest Jiaotong Univ. **48**(03), 553–558 (2013)

11. Wen, R., Wang, H.: A forecasting method of controller's workload based on ridge regression—BP neural network. J. Transp. Syst. Eng. Inf. Technol. **15**(01), 123–129 (2015)
12. Yue, R., Han, N., Zhao, Y.: Clustering analysis and forecasting of the health control position status-in-situ based on the practice data
13. CAAC: Measures for the administration of civil aviation advance flight plans. CAAC, Beijing (2006)
14. Hilbum, B.: Cognitive complexity in air traffic control: a literature review. Eurocontrol, Brussels (2004)
15. Li, H.: Data Mining with R: Learning with Case Studies, pp. 101–147. China Machine Press, Beijing (2018)

Research of Lateral Guidance Law for Small Aircraft

Wenqi Lyu(✉)

AVIC The First Aircraft Institution, Xi'an 710089, Shaanxi, China
lwq2018avic@outlook.com

Abstract. X-20 project designed by Cranfield University in England in 2021 is a supersonic business jet to meet the needs of commercial transports and private airlines. In this project, the flight management system is equipped as a critical part of the avionics system. Guidance laws in this system are always, decomposed into a lateral profile and a vertical profile. In the lateral profile, a guidance law is developed to track the straight path and the fix-radius transition which are components of a traditional flight path. The validation work of the guidance laws is carried out by simulation in MATLAB software. Then, the results show the efficiency and reasonability of the guidance algorithm.

Keywords: 3 degree of freedom model · Cross-track error · Lateral guidance law · RNP1 specification · Guidance law

1 Introduction

X-20 project was designed at Cranfield University in 2021 as a group design project. It was raised as a supersonic business jet to meet the needs of commercial transports and private airlines. Aiming at commercial and private requirements, one-pilot operation and supersonic design were employed in this project. As a result, one pilot would raise the workload in flight.

One way to reduce the workload in operation is to design the guidance law for in-flight management systems. A general guidance law is proposed for a lateral profile and a vertical profile of X-20. In this paper, a lateral guidance law was proposed to track lateral desire flight paths.

Proportional navigation (PN) derived from missile guidance applications is typically used in space vehicles as well as automobiles. As well as some new guidance laws are derived from this guidance law such as Ideal PN, Pure PN [1], Ture PN, PID PN [2] and switched bias PN [3]. These traditional navigation methods can be used in the guidance of non-maneuvering and maneuvering targets pursue.

A linear trajectory control was employed to guide a fixed-wing UAV in tracking the desired path [4]. The control command of this algorithm is the minimum distance from the current position to the desired path. Another liner analysis was proposed to analyse a motion on a circular path [5]. This method was derived from the PN principle, and a reference point was chosen from the desired path. A vector field was involved in the

Chinese Society of Aeronautics and Astronautics: CASTYSF 2022, LNEE 972, pp. 732–741, 2023.
https://doi.org/10.1007/978-981-19-7652-0_65

lateral guidance in [6]. In the vector field study, each vector in the vector field indicates a direction towards the desired path.

In this paper, a guidance law in lateral guidance law will be proposed. In Sect. 2, an aircraft model will be built aiming at validating the results of the algorithm. In Sect. 3, an algorithm for generating the cross-track error and heading error. In Sect. 4, the simulation will be employed to validate the algorithm. In the last section, there will be a brief conclusion to this paper.

2 Aircraft Model

To calculate the trajectory bias and prediction for the ATM and FMS, the point mass model can be acceptable accuracy at a low computational cost. Hence, in this algorithm design and validation, the 3DOF model is carried out.

It is noted that the earth is flat as well as the atmosphere is at rest relative to the earth. As well, the engine is fixed on the plane and flight is symmetric.

The kinematics of aircraft describes the motion of aircraft in a three-dimension coordinate. It derives the differential equations of the position information of the aircraft.

$$\begin{aligned} \dot{x} &= V \cos\gamma \cos\psi \\ \dot{y} &= V \cos\gamma \sin\psi \\ \dot{h} &= V \sin\gamma \end{aligned} \tag{1}$$

where ψ is called yaw angle or heading angle, γ is called pitch angle, and V denotes the velocity. $\dot{x}\dot{y}\ \dot{h}$ denote the differentiation of aircraft position aligned to x, y, and z axes respectively. Equation (2) is used to describe the dynamic model of aircraft.

$$\begin{aligned} \dot{\mathrm{V}} &= \left(\frac{g}{\mathrm{W}}\right)[T\cos\varepsilon - D - W\sin\gamma] \\ \dot{\psi} &= \left(\frac{g}{WV\cos\gamma}\right)[T\sin\varepsilon + L]\sin\mu \\ \dot{\gamma} &= \left(\frac{g}{WV}\right)[(T\sin\varepsilon + L)\cos\mu - W\cos\gamma] \end{aligned} \tag{2}$$

where μ is called roll angle or bank angle, g denotes the acceleration of gravity, W denotes the weight of aircraft, ε denotes the angle of attack, T denotes the specific fuel consumption, D denotes the drag, L denotes the lift.

The differentiation of aircraft weight is described as

$$\dot{W} = -CT \tag{3}$$

where C denotes specific fuel consumption, and T denotes the thrust of the aircraft. The specific fuel consumption is determined by engine characters and flight states.

In conclusion, the 3DOF model can be described by Eqs. (1)–(3).

3 Lateral Guidance Law

The lateral guidance law is aiming at tracking the path plan which is also named the desired path from the flight plan function. The basic is to calculate the cross-track error from the current position to the desired path followed by the algorithm in which the lateral guidance commands are determined aiming at reducing the cross-track error to as small as possible. Application of this guidance law means an automatic operation is processed onboard. To support it, the waypoint sequencing and leg switching are highlighted to reduce the workload of the crew.

The fly-by turn allows tangential interception of the next segment of a route, while the fly-over turn starts from the waypoint and joins the next segment of the route. As guidance laws for straight path and circle path obey different principles and calculate different errors in the flight, the straight path guidance is analysis firstly. Then, the guidance law for a circle path is designed. Lastly, the transition logic from the current tracking segment to the following segment is analyzed to realize the whole tracking process.

3.1 Straight Path Guidance Law Design

Let waypoint (n − 1) ($\phi(n-1), \lambda(n-1)$), waypoint (n) ($\phi(n), \lambda(n)$), waypoint (n + 1) ($\phi(n+1), \lambda(n+1)$) be three waypoints of a given path in NED frame. The path from waypoint (n − 1) to waypoint (n) and path from waypoint (n) to waypoint (n + 1) are path line. The transition from the path from waypoint (n − 1) to waypoint (n) to path from waypoint (n) to waypoint (n + 1) is a fly-by turn. Figure 2 shows the relationship of the path. The current position of the aircraft is denoted by M (Fig. 1).

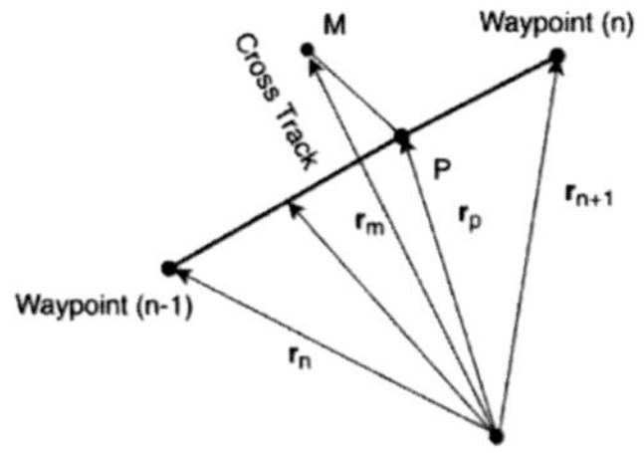

Fig. 1. Straight path schematic

Distance Between Waypoints

In ECEF, the vector from the center of the earth to the waypoint (n) is denoted by $\vec{r}_n$. And the vector from the center of the earth to the waypoint (n − 1) is denoted by $\vec{r}_{n-1}$. The distance between two waypoints is presented as

$$r = \frac{1}{2}(|\vec{r}_n| + |\vec{r}_{n-1}|)\cos^{-1}(\widehat{r}_n \cdot \widehat{r}_{n-1}) \tag{4}$$

where, $\widehat{r}_n$ and $\widehat{r}_{n-1}$ are unit vectors of $\vec{r}_n$ *and* $\vec{r}_{n-1}$

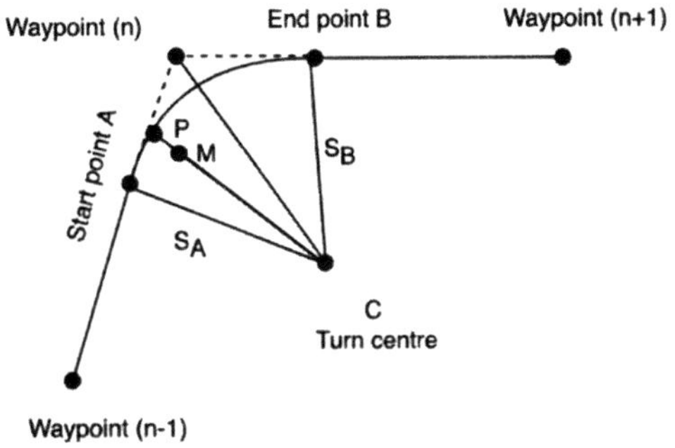

Fig. 2. Fly-by turn

Cross-Track Error

Cross-track error is the instantaneous normal displacement from the desired path of aircraft. Considering the relationship in a vector field, a new vector denoted by $\vec{r}_N$, is normal to the desired path.

$$\vec{r}_N = \widehat{r}_n \times \widehat{r}_{n-1} \tag{5}$$

The cross-track error is written as

$$crosstrac\ker ror = -\widehat{r}_M \cdot \widehat{r}_N \tag{6}$$

Heading Error

Heading error denoted by ψ_{error} is defined by the difference between the current heading angle, ψ_c and the desired heading angle, ψ_d.

$$\psi_{error} = \psi_c - \psi_d \tag{7}$$

The current heading angle is determined by

$$\psi_c = a\tan(\frac{V_e}{V_n}) \tag{8}$$

where V_e denotes the velocity component in the east direction and V_n denotes the velocity component in the north direction.

Let point P be the projection of the current position on the desired path. Then the desired heading angle of the aircraft is defined as the angle between the path plane and a meridian plane passing through point P.

The position vector of P denoted by $\vec{r}_p$ is given by

$$\vec{r}_p = \vec{r}_M - (\vec{r}_M \cdot \widehat{r}_N) \cdot \widehat{r}_N \tag{9}$$

The desired heading angle ψ_d was designed as

$$\psi_d = a\tan\left[\frac{(\vec{r}_p \times \widehat{k}) \times \hat{r}_N}{(\vec{r}_p \times \widehat{k}) \cdot \hat{r}_N}\right] \tag{10}$$

where $\widehat{k}$ denotes the unit vector of the spin axis.

3.2 Circle Path

In the analysis of the circle path, a new frame is defined to simplify the mathematical equations of the circle path. It is a frame with the original at waypoint (n), x-axis along East direction, y-axis aligned to North direction and z-axis aligned to upside. The position x, (φ, λ) is expressed in ENU as

$$\begin{aligned} x &= R_N(\lambda - \lambda_{(n)})\cos\phi_{(n)} \\ y &= R_M(\phi - \phi_{(n)}) \\ z &= 0 \end{aligned} \tag{11}$$

where R_N and R_M are the normal and meridian radii of curvature respectively.

$$\begin{aligned} R_N &= \frac{a}{\sqrt{1 - e^2 \sin\phi_{(n)}}} \\ R_M &= \frac{a(1 - e^2)}{\left(1 - e^2 \sin\phi_{(n)}\right)^{3/2}} \end{aligned} \tag{12}$$

where, e denotes the eccentricity of the ellipsoidal earth.

In the ENU frame, the position vector of the waypoint (n − 1) and the waypoint (n + 1) denoted by, can $\vec{q}_{n-1}$,$\vec{q}_{n+1}$ be written as

$$\begin{aligned} \vec{q}_{n-1} &= [R_N(\lambda_{(n-1)} - \lambda_{(n)})\cos\phi_{(n)}, R_M(\phi_{(n-1)} - \phi_{(n)}), 0] \\ \vec{q}_{n+1} &= [R_N(\lambda_{(n+1)} - \lambda_{(n)})\cos\phi_{(n)}, R_M(\phi_{(n+1)} - \phi_{(n)}), 0] \end{aligned} \tag{13}$$

The course change is given by

$$\alpha = a\cos(\vec{q}_{n-1} - \vec{q}_{n+1}) \tag{14}$$

Assuming that $\alpha < 120°$.

Turn Center

The distances from the start point and endpoint of turn to the original are given as well as the course changes. The distance from the turn center to the original is defined as

$$C = \frac{R}{\sin\left(\frac{\alpha}{2}\right)} \tag{15}$$

$$\vec{C} = \frac{R}{\sin\left(\frac{\alpha}{2}\right)}(\vec{q}_{n-1} - \vec{q}_{n+1}) \tag{16}$$

Turn Criterion

Aircraft position projected on earth is expressed as

$$\vec{M} = [R_N(\lambda_M - \lambda_{(n)})\cos\phi_{(n)}, R_M(\phi_M - \phi_{(n)}), 0] \tag{17}$$

The vector from center to M is expressed as

$$\vec{S}_M = \vec{M} - \vec{C} \tag{18}$$

Likewise, the start point position vector is defined as

$$\vec{S}_A = \frac{R}{\sin\left(\frac{\alpha}{2}\right)}\vec{q}_{n-1} - \vec{C} \tag{19}$$

The endpoint position vector from center is defined as

$$\vec{S}_B = \frac{R}{\sin\left(\frac{\alpha}{2}\right)}\vec{q}_{n+1} - \vec{C} \tag{20}$$

The start time and end time are critical in the turn logical implementation. If the sign of the $\vec{S}_M \times \vec{S}_A$ changes in two sequencing computation steps, the aircraft will change the guidance model from a straight line to a circular path. In Fig. 3, in the (t − 1) moment, $\vec{S}_M \times \vec{S}_A$ is positive while in the following (t + 1) moment, $\vec{S}_M \times \vec{S}_A$ is negative.

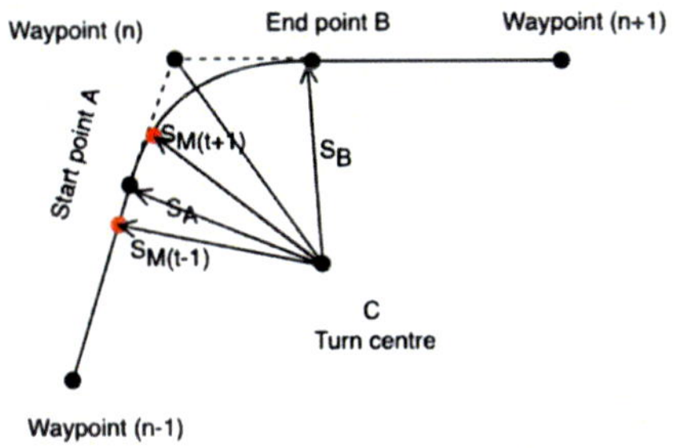

Fig. 3. Start turn criterion

Likewise, the sign changes of $\vec{S}_M \times \vec{S}_B$ is used to determine the end of the transition.

Cross-Track

The cross-track is derived by

$$crosstrack = R - \left|\vec{S}_M\right| \tag{21}$$

Heading Error

The heading error can be determined by a similar principle as (9)–(11). As it is the circle path, the specific desire calculation is designed as

$$\psi_d = a\tan\left[\frac{(\vec{r}_p \times \widehat{k}) \times \hat{r}_N}{(\vec{r}_p \times \widehat{k}) \cdot \hat{r}_N}\right] + a\cos(\hat{S}_A \cdot \widehat{S}_M) \tag{22}$$

3.3 Guidance Command

The previous section presents the error in the lateral plane which is a cross-track error and heading error. These errors are transmitted to the guidance law part to determine the commands of bank angle. The bank angle commands are executed to reduce the position error in the lateral profile resulting in good tracking in flight. In Fig. 5, the relationship between error and bank angle is shown. The guidance law employed in Fig. 4 is researched on $L1$ guidance law which is given in [8].

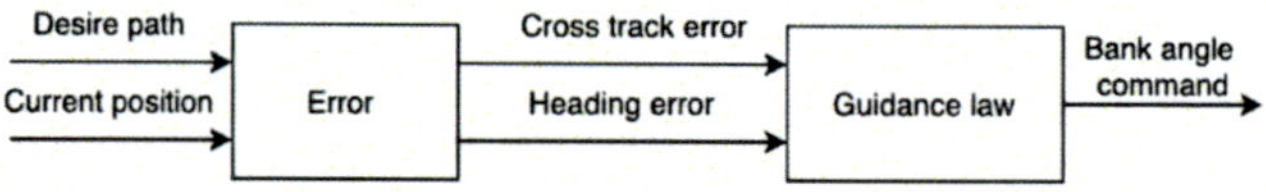

Fig. 4. Block diagram for guidance

4 Simulation and Verification

To verify guidance laws, the aircraft simulation model is built and employs the parameters of X-20. This aircraft is a one-pilot supersonic demonstrator that belonged to Cranfield University in 2021. Firstly, the aircraft model is defined and described as Eqs. (1)–(3). Furthermore, the guidance block is built to generate the control variables for the aircraft models which are not given in detail in this paper.

4.1 Model Construction

In the simulation, the 3DOF model is built in the Simulink environment. This is a high-fidelity and visual model plane. It includes the 3DOF, aerodynamics calculation and guidance blocks. The guidance models in vertical profile and horizontal profile execute separately. The simplified model block diagram is shown in Fig. 5.

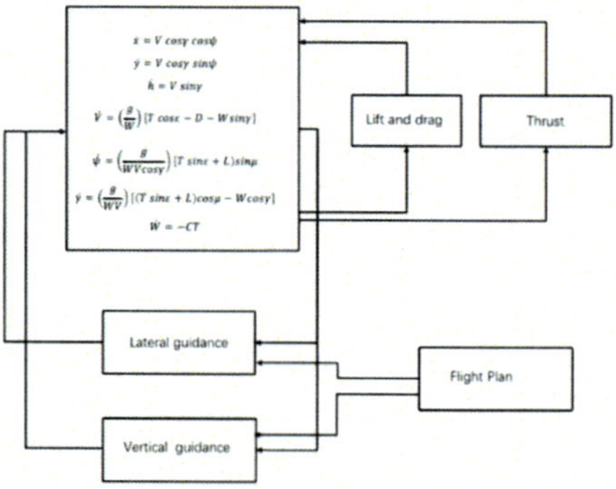

Fig. 5. Simulation model block diagram

4.2 Lateral Guidance Law Simulation

Lateral Segments

From the previous sections, the separate types of segments are verified. In this section, these straight paths with fly-by-turn transitions are validated with the desired path defined in Table 1. There are 4 waypoints to define a given trajectory for aircraft trajectory simulation. The course changes require the transitions between the nearby segments designated as fly-by turns with a fixed turn radius.

Table 1. Flight path definition by waypoints

Waypoints	Latitude (°)	Longitudinal (°)	Altitude (m)	Type
Waypoint 1	50	100	16,760	–
Waypoint 2	50	102.1	16,760	RF
Waypoint 3	52.5	102.1	16,760	RF
Waypoint 4	52.5	106	16,760	–

From the analysis in previous sections, L1 is designated as 1941 m (Fig. 6).

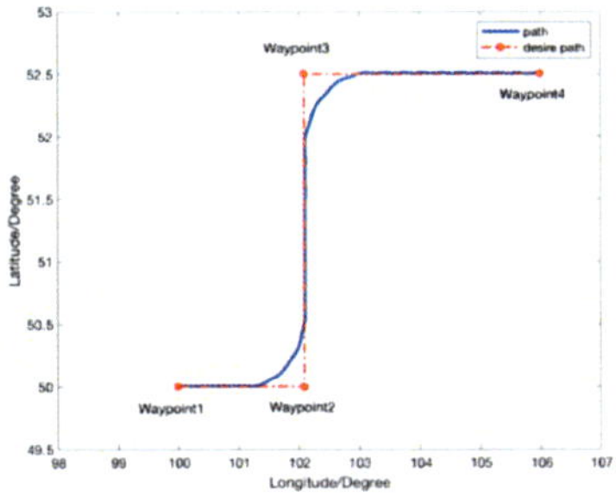

Fig. 6. Segments with transitions schematic

Three segments are tracked by the aircraft in the sequence of waypoint 1, waypoint 2, waypoint 3 and waypoint 4. Two transitions are carried around waypoint 2 and waypoint 3. This simulation time is 1500 s. It shows the cross-track error is in the range from − 870 to 1020 m. All of them meet the RNP1 specification (Fig. 7).

In Fig. 8, the process of switching the waypoint is shown. The horizontal axis is the time of the simulation. The vertical axis is the position in degree. The red line is the longitude. And the blue one is the altitude. The waypoint (n) denotes the middle waypoint in each group of three active waypoint sets. The dash vertical line denotes the transition time of active waypoints.

In the initial state, n is equal to 2, which means the first three waypoints are active. When the aircraft finish the first transition, n is equal to 3 resulting that waypoint 2,

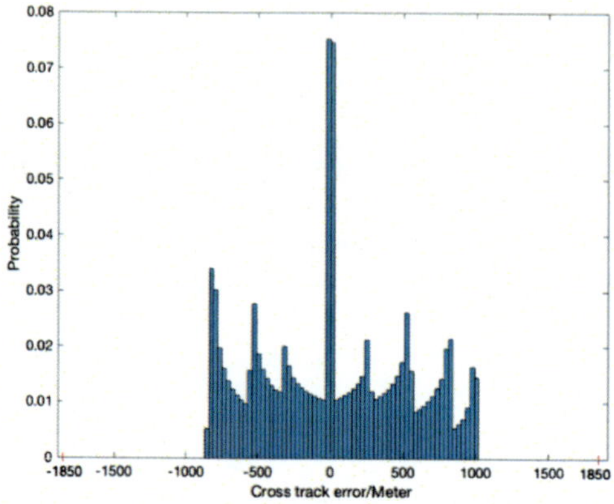

Fig. 7. Cross-track error histogram

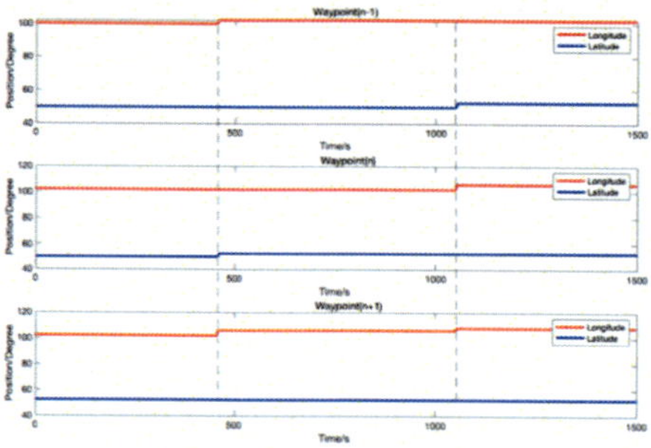

Fig. 8. Switching active waypoints

waypoint 3 and waypoint 4 being followed in the forthcoming process. From these simulations, the validation of the lateral algorithm is completed in MATLAB.

5 Conclusion

In this paper, the guidance algorithms for the lateral profiles are discussed based on the 3DOF motion model of aircraft. In the lateral guidance strategy, the transition is a fly-by turn which is the trendy method. The simulation was introduced to validate the algorithms. The results are used as the reference in the real-time software design work which generates the Ada code or C code into a flight management system device.

References

1. Murtaugh, S.A., Criel, H.E.: Fundamentals of proportional navigation. IEEE Spectr. **3**(12), 75–85 (1966). https://doi.org/10.1109/MSPEC.1966.5217080
2. Yaghi, M., Efe, M.O.: H2H∞-neural-based FOPID controller applied for radar-guided missile. IEEE Trans. Ind. Electron. **67**(6), 4806–4814 (2020). https://doi.org/10.1109/TIE.2019.2927196

3. Cho, N., Kim, Y.: Modified pure proportional navigation guidance law for impact time control. J. Guid., Control, Dyn. 852–872 (2016). https://doi.org/10.2514/1.G001618
4. Xing, X., Chen, X., Fan, D., Hou, X.: On the four-dimensional trajectory control of UAV. In: Chinese Control Conference, CCC, July 2018, pp. 4750–4755. Technical Committee on Control Theory, Chinese Association of Automation (2018). https://doi.org/10.23919/ChiCC.2018.8484025
5. Park, S.: Rendezvous guidance on circular path for fixed-wing UAV. Int. J. Aeronaut. Space Sci. **22**(1), 129–139 (2020). https://doi.org/10.1007/s42405-020-00281-8
6. Flores, A., Lugo, I., Gonzalez, I., Lozano, R.: Vector field guidance law for fixed wing UAV. In: 2017 21st International Conference on System Theory, Control and Computing, ICSTCC 2017, pp. 366–371 (2017). https://doi.org/10.1109/ICSTCC.2017.8107061
7. Hull, D.G.: Fundamentals of Airplane Flight Mechanics. Springer, Berlin (2007). https://scholar.google.co.uk/scholar?q=fundamental+of+airplane+flight+mechanics+david+hull&btnG=&hl=en&as_sdt=0,5#0
8. Park, S., Deyst, J., How, J.P.: A new nonlinear guidance logic for trajectory tracking. Collect. Tech. Pap. - AIAA Guid., Navig., Control Conf. **2**, 941–956 (2004). https://doi.org/10.2514/6.2004-4900
9. Samar, R., Shah, M.Z., Nzar, M.: Lateral control implementation for an unmanned aerial vehicle. ISRN Aerosp. Eng. **2013**, 1–14 (2013). https://doi.org/10.1155/2013/905865
10. ICAO: Doc 9613 Performance-Based Navigation (PBN) Manual. Montréal, Quebec, Canada (2013)

Development of Wind Tunnel Dynamic Test System Driven by Electro-hydraulic Servo Motor

Hao Tan[1,2](✉), Jianfeng Wang[1,2], Chen Bu[1,2], Yanjie Shen[1,2], and Hao Chen[1,2]

[1] AVIC Aerodynamics Research Institute, Harbin, China
tanhao_avic@163.com
[2] Low Speed High Reynolds Number Aeronautical Key Laboratory, Harbin, China

Abstract. The nonlinear and unsteady dynamic aerodynamic characteristics are often associated with aircraft maneuver flight at high angle of attack and post stall flight. In view of the urgent need of the nonlinear and unsteady aerodynamic characteristics research for the development of advanced layout aircraft, a new wind tunnel dynamic test system driven synchronously by dual electro-hydraulic servo motors is developed based on FL-51 wind tunnel of AVIC Aerodynamics Research Institute, which can conduct dynamic derivative test, large amplitude oscillation test and post stall maneuver simulation test at high angle of attack with test model at the scale of 1.5 m under the wind speed range of 20–50 m/s. Based on the load's rotational inertia and motion index of the test system, we design the hardware scheme of the control system and the position servo control algorithm, debug the performance index of the test system, and carry out the wind tunnel validation test with a low aspect ratio flying wing layout aircraft model. The debugging results show that the test system has good static/dynamic control precision, and the performance index and function achieve the design requirements. The Wind tunnel test results indicate that the test system can obtain the regular and rational test data of dynamic derivatives test and maneuver process simulation test, which can be used for the research of special aerodynamic characteristics of aircraft at high Angle of attack and post stall maneuver.

Keywords: Electro-hydraulic servo motor · Servo control · Wind tunnel dynamic test · Post stall maneuver · Wire-driven suspension

1 Introduction

In modern air warfare, the requirements of aircraft's maneuverability and agility are increasing, high maneuver-ability, high agility and post-stall maneuverability at high angles of attack have become important indexes to measure tactical performance and combat efficiency of modern fighter jets. When the fourth-generation fighter jet was developed, the emphasis of maneuverability, agility, and post-stall maneuverability was focused. The aerodynamic characteristics at high angle of attack is the key factor to determine the aircraft's features above and the flight performance. Therefore, accurately acquiring the nonlinear and unsteady aerodynamic data under high angle of attack

Chinese Society of Aeronautics and Astronautics: CASTYSF 2022, LNEE 972, pp. 742–756, 2023.
https://doi.org/10.1007/978-981-19-7652-0_66

maneuver is extremely important to the aerodynamic characteristics research and flight control system design of the modern fighter jet [1–3].

At present, the wind tunnel dynamic test is still the most important and reliable technical means to carry out the research work related to the aerodynamic characteristics of the aircraft at high angle of attack. In order to satisfy the requirements of nonlinear and unsteady test data at high angle of attack for the development of advanced layout aircraft, aviation developed countries in the world have developed corresponding wind tunnel dynamic test technology, just as oscillating cone motion test technology, wing rock test technology, multi-degree-of-freedom coupling test technology with large amplitude and high angle of attack, and maneuver motion process simulation test technology etc. The relevant research results provide a valuable basis for the design of high maneuverability and high agility aircraft and the analysis of flight dynamics at high angle of attack [2, 4–8].

The United States is the country whose wind tunnel dynamic test technology is the most advanced with the most developed dynamic test equipment. In addition to the large amplitude oscillation dynamic test technology, the United States has developed various comprehensive dynamic test equipment in recent years. For example, the Dynamic Plunge-Pitch-Roll test device developed by the Virginia State University can realize the coupling motion of Plunge, Pitch and Roll in any given function form used to simulate the time course of maneuver flight of the aircraft. These devices have played an important role in the development of the F-22 aircraft.

Since the 1980s, China has carried out a large amount of pre-research work accordingly, made great progress in the field of dynamic test technology, successfully developing the dynamic derivative test system, multi-degree-of-freedom coupling test technology with large amplitude and six degree-of-freedom test system [9–13]. However, the test systems above usually exist a lot of problems, such as the limited range of the angle of attack, frame interference of support device and the simplex test capability.

Aiming at the limitations of the conventional wind tunnel dynamic test equipment, in this paper, we developed a new dynamic test system synchronously driven by dual electro-hydraulic servo motors based on FL-51 wind tunnel (4 m scale low-speed wind tunnel) of AVIC Aerodynamics Research Institute, and the test models are supported with wire-driven suspension. Using this test system, dynamic test such dynamic derivative test, large amplitude oscillation test and post stall maneuver simulation test can be carried out, with the characteristics of large angle of attack range and no frame interference.

2 System Scheme Design

The synchronous drive system by dual electro-hydraulic servo motors has the characteristics of symmetrical load force and small size of hydraulic motor, widely used in high-precision and large-load servo control field. For example, the three-axis attitude hydraulic simulation turntable and large crawler crane often use the synchronous drive scheme with dual electro-hydraulic servo motors [14–16]. In this paper, we use two identical hydraulic motors as the drive unit who can move synchronously. On the drive shaft of each motor, we install a "X" bracket, then the test model's base is connected with two "X" brackets through drive wire. One end of the internal balance is connected with the model's base, and the other end is connected with the balance sleeve of the test model, so as to achieve the purpose of supporting the test model. The structure diagram of the test system is shown in Fig. 1.

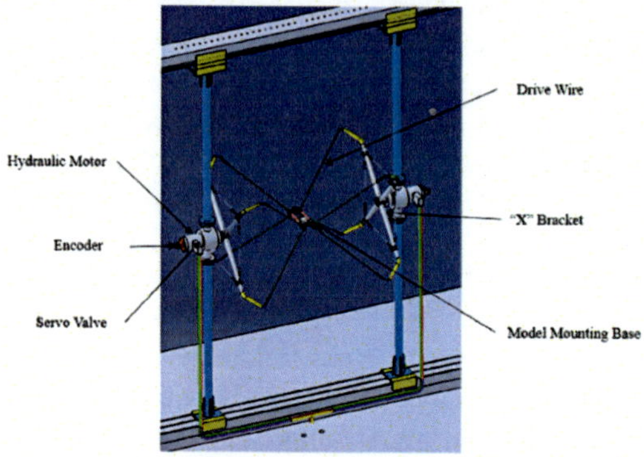

Fig. 1. Structure diagram of test system.

Figure 2 shows the hardware schematic diagram of the control system. The control system controls the synchronous motion of the electro-hydraulic servo motors and drives the test model to follow the motion in real time.

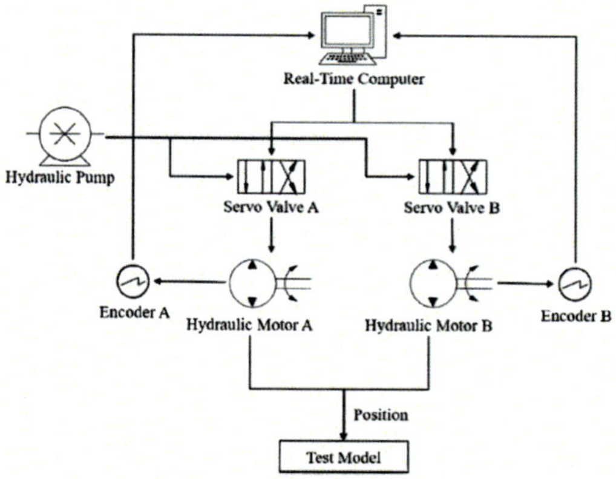

Fig. 2. Hardware schematic diagram of control system.

2.1 Hydraulic Motor Parameter Calculation

Taking the motor's torque as the design input condition, the theoretical displacement of motor can be expressed as

$$D_m = \frac{T_m}{2/3P_s} \tag{1}$$

where, D_m is the theoretical displacement of motor, L/rad; T_m is the design torque, N·m; P_s is the pressure of hydraulic pump station, MPa.

According to the principle of equal load distribution of dual motors, the design torque of the motors can be expressed as

$$T_m = \frac{1}{2}\left[\left(J_E\ddot{\theta}_{\max} + T_A\right)\cdot\xi\right] \tag{2}$$

where, J_E is the moment of inertia of the load, $\mathrm{kg}\cdot\mathrm{m}^2$; $\ddot{\theta}_{\max}$ is the maximum angular acceleration of the load, °/s^2; T_A is the aerodynamic load of the test model during the wind tunnel test, N·m; ξ is the safety factor designed for the motors, $\xi = 1.5$.

Via Eqs. (1) and (2), we can derive D_m

$$D_m = \frac{3\xi\left(J_E\ddot{\theta}_{\max} + T_A\right)}{4P_s} \tag{3}$$

According to the moment of inertia of the system, the maximum angular acceleration, the aerodynamic load and the working pressure of the hydraulic pumping station, the displacement of the motor can be calculated. The specific parameters are shown in Table 1.

Table 1. Basic parameters of hydraulic motor.

Design torque (N·m)	Radian displacement (L/rad)
3200	0.48

2.2 Electro-hydraulic Servo Valve Selection

The maximum load flow of the hydraulic system is

$$Q_M = D_M\dot{\theta}_{\max}K_l \tag{4}$$

where, Q_M is the maximum load flow of the hydraulic system, L/min; $\dot{\theta}_{\max}$ is the maximum angular velocity of the load, °/s; K_l is the leakage coefficient, $K_l = 1.2$.

Since the working pressure of the hydraulic pump station is 10 MPa, not 21 MPa, during the selection of the servo valve, we need to calculate the servo valve's rated flow (under 7 MPa pressure drop) according to the actual working pressure of the pump station and the actual flow demand of the hydraulic system.

$$Q_N = Q_M\sqrt{\frac{\Delta P_N}{\Delta P_L}} \tag{5}$$

where, Q_N is the rated flow of the servo valve, L/min; ΔP_N is the rated pressure drop of the servo valve, MPa; ΔP_L is the actual pressure drop of the servo valve, ΔP_L=2/3P_s, MPa.

It should be noted that we need to convert the units of the physical quantities before calculation. According to the calculation of the maximum design index, the actual flow demand of the load is shown in Table 2.

Table 2. Load flow parameter.

Maximum angular velocity (°/s)	Rated flow under 10 MPa (L/min)	Rated flow under 21 MPa (L/min)
300	181	262

Therefore, the rated flow of the selected servo valve under 21 MPa working pressure (7 MPa pressure drop) must meet: $Q_N \geq 262\,\text{L/min}$. Finally, we select the three-stage flapper-nozzle electro-hydraulic servo valve of Rexroth 4WSE series. The basic parameters of this servo valve are shown in Table 3.

Table 3. Basic parameters of servo valve.

Model	Maximum working pressure (MPa)	Rated flow under 21 MPa (L/min)
4WSE3E25V300E30 /VPT7/15K31C1	21	300

2.3 Position Feedback Element Selection

The position feedback element adopts the single-turn absolute encoder whose model is AFS60-S1AC262144 produced by SICK company. The encoder is installed and fixed at the end of the motor, and is connected to the motor's rotary shaft through an elastic coupling, ensure that it can feedback the position information of the motor in real time. The basic parameters of the encoder are shown in Table 4.

Table 4. Basic parameters of position feedback element.

Model	Communication protocol	Resolution	Repeatability precision
FS60A-S1AC 262144	SSI	18 bit	±0.002°

2.4 Control System Hardware

The hardware of the control system is developed based on industrial computer and embedded PCI card, and is equipped with AD data acquisition card, DA output card, digital I/O card and encoder signal acquisition card. The control system software is developed based on the Visual Studio programming environment, and the RTX real-time operating system (RTOS) is used to run the control program, and the cycle period can reach 0.5 ms.

In order to realize the high-precision and high-frequency response servo motion control, in addition to the advanced control algorithm, all the above-mentioned high-performance and high-precision hardware are also indispensable.

3 Control Algorithm Design

The electro-hydraulic servo system has strong nonlinearity, such as the pressure pulsation of the hydraulic pump, the change of the elastic modulus caused by the rise of oil temperature, the dead zone of the servo valve, the nonlinear control of the flow, the nonlinear friction and the internal leakage of the actuator. Therefore, it is difficult for the electro-hydraulic servo system to achieve excellent static/dynamic performance, excellent control accuracy and robustness only by conventional PID control. In view of the nonlinear characteristics of electro-hydraulic servo system, in recent years, lots of researchers all over the world have proposed a variety of effective control methods based on classical control theory and modern control theory, just as multi-state feedback and feedforward control strategy [17, 18], variable structure control [19], robust adaptive control [20], sliding mode control [21, 22], dynamic inverse control [23] and neural network adaptive control [24] etc. The control algorithm used in this system is as follows: based on the conventional PID control strategy, in order to improve the dynamic performance of the system and the synchronization performance of the dual motors, feedforward control is introduced, and fuzzy adaptive control is adopted to improve the adaptability and robustness of the electro-hydraulic servo system, which is characterized by nonlinear, time-varying and model uncertainty.

3.1 Feedforward Control

Since the introduction of feedforward controller does not change the characteristic equation of the system, that is, it does not affect the pole distribution of the system, so the feedforward control does not affect the stability of the system. The feedforward controller structure is shown in Fig. 3.

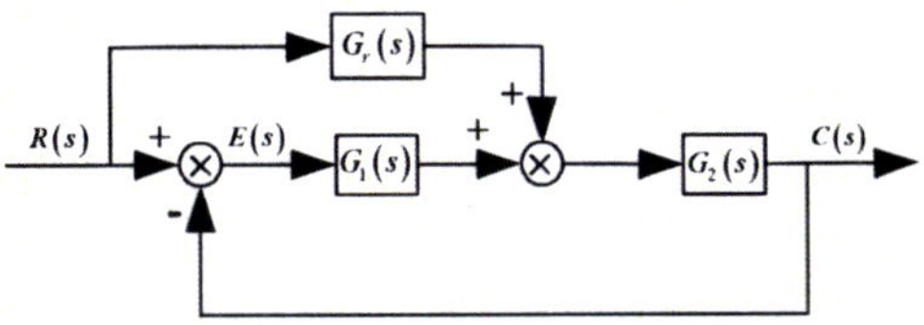

Fig. 3. Feedforward controller structure diagram.

From the controller structure diagram in Fig. 3, the error function of the system can be expressed as

$$E(s) = \frac{1 - G_2(s)G_r(s)}{1 + G_1(s)G_2(s)}R(s) \tag{6}$$

It can be seen from Eq. (6) that if we want to achieve full compensation of the system error, that is, you only need to satisfy

$$G_r(\mathrm{s}) = \frac{1}{G_2(s)} \tag{7}$$

However, in the electro-hydraulic servo control system, $G_2(s)$ has an extremely complex expression form. In the engineering field, it is difficult to achieve full compensation for systematic errors, and usually only partial compensation is performed. In this paper, the first-order velocity feedforward and second-order acceleration feedforward compensation are used, and the expressions are

$$G_{r1}(\mathrm{s}) = \frac{K_{\mathrm{r}1}s}{T_1 s + 1} \tag{8}$$

$$G_{\mathrm{r}2}(\mathrm{s}) = \frac{K_{\mathrm{r}1}s^2}{(T_2 s + 1)^2} \tag{9}$$

where, $K_{\mathrm{r}1}$ is the first-order differential coefficient, T_1 is the first-order differential feedforward time constant, $K_{\mathrm{r}2}$ is the second-order differential coefficient, T_2 is the second-order differential feedforward time constant.

Under the premise of not affecting the system stability, the feedforward controller can improve the response speed of the control system, meanwhile compensate the amplitude and phase of dynamic motion very well, so as to achieve the purpose of improving the control accuracy of the system [25–27]. In addition, since the dual electro-hydraulic motors use independent control mode, in order to ensure the synchronization of motors' movement, it is necessary to adjust the tracking error of the two motors to the command signal through the first-order speed feedforward control. After the signals can achieve the ideal tracking effect, the synchronization of the two motions can meet the design requirements of the system. Once each motor can achieve the ideal tracking precision to the command signal, the synchronization of the dual motors' motion can meet the requirements of the system design index.

3.2 Fuzzy Adaptive Control

Electro-hydraulic servo system is a typical nonlinear system, with the characteristics of time-varying and control model uncertainty. It is difficult for the conventional PID control algorithm to achieve the ideal control effect, but the fuzzy adaptive PID control can solves this problem effectively [28, 29].

The structure diagram of the fuzzy adaptive PID controller is shown in the Fig. 4. In the fuzzy control rules, we select control system error e and rate of error δ_e as input variables, and k_p as output variable. According to the value of e and δ_e, the fuzzy

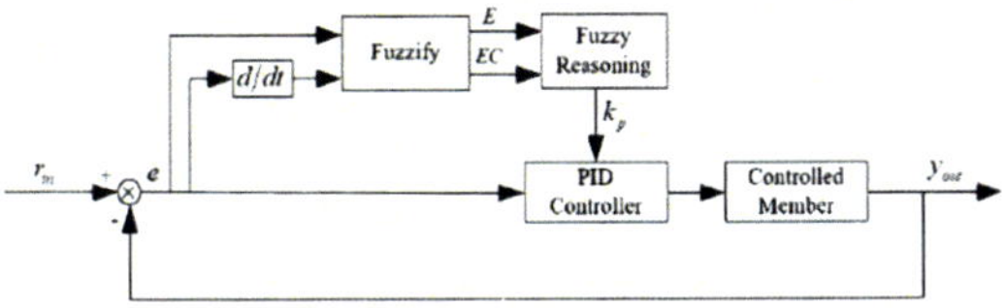

Fig. 4. Fuzzy adaptive PID controller structure diagram.

controller conducts fuzzy reasoning through the set fuzzy control rules, and queries the fuzzy matrix table, finally obtains the corresponding value of the output variables of k_p to realize the adaptive tuning of the PID controller parameters.

The domain of e and δ_e is taken as [−3, 3], and the domain of k_p is taken as [0.6, 1.4]. The initial proportional coefficient of the PID controller is K_0, then the proportional coefficient adjusted by the fuzzy controller can be expressed as

$$K_P = K_0 \cdot k_p \tag{10}$$

The fuzzy subsets of e, δ_e and k_p are taken as {NB, NM, NS, ZO, PS, PM, PB}, and the fuzzy control rules are shown in Table 5.

Table 5. Fuzzy control rules.

E	EC						
	NB	NM	NS	ZO	PS	PM	PB
NB	PB	PB	PM	PM	PS	ZO	ZO
NM	PB	PB	PM	PS	PS	ZO	NS
NS	PM	PM	PS	PS	ZO	NS	NS
ZO	PM	PM	PS	ZO	NS	NM	NM
PS	PS	PS	ZO	NS	NS	NM	NM
PM	PS	ZO	NS	NM	NM	NM	NB
PB	ZO	ZO	NS	NM	NM	NB	NB

In conclusion, the overall block diagram of the composite control algorithm based on multi-variable feedforward PID + fuzzy adaptive control of the dynamic test system is shown in Fig. 5.

4 System Verification

4.1 Control System Verification

Figure 6 shows the physical picture of the dynamic test system driven by dual electro-hydraulic servo motors with wire-driven suspension.

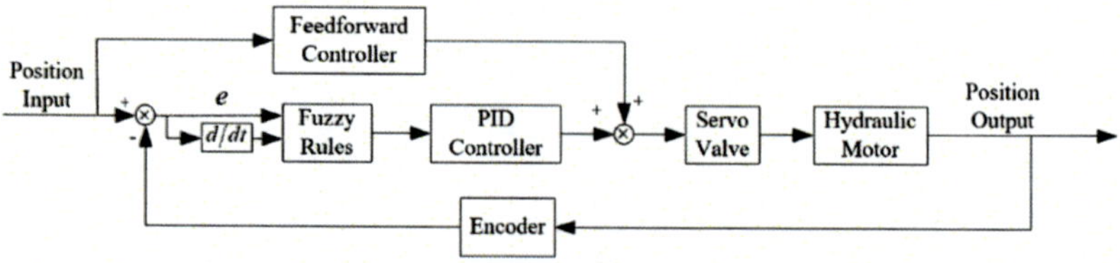

Fig. 5. General diagram of position servo control system.

Fig. 6. Wire-driven dynamic test system.

Table 6. Sinusoidal motion test index.

Amplitude (°)	Frequency (Hz)
5	2.5
10	2
15	1

According to the sinusoidal motion index commonly used in the dynamic derivative test and the large amplitude oscillation test listed in Table 6, we test the performance index of the test system, and the position following curves of the dual electro-hydraulic servo motors are shown in Figs. 7, 8, and 9.

The amplitude difference and phase difference corresponding to the sinusoidal motion index in Table 6 are shown in Table 7.

In addition, in terms of motion function, a function that can read the discretized time-position curve file and refit the curve in the control system is developed, then, the control system can control the motors driving the test model to simulate the curve motion. Through the test function developed, nonlinear and unsteady aerodynamic test data of the aircraft model during high angle of attack and post stall maneuver flight can be obtained in the wind tunnel, which can be used for aircraft aerodynamic modeling and model verification. Figures 10 and 11 are the simulated test curves of the Cobra maneuver and the high angle of attack holding maneuver based on the test system.

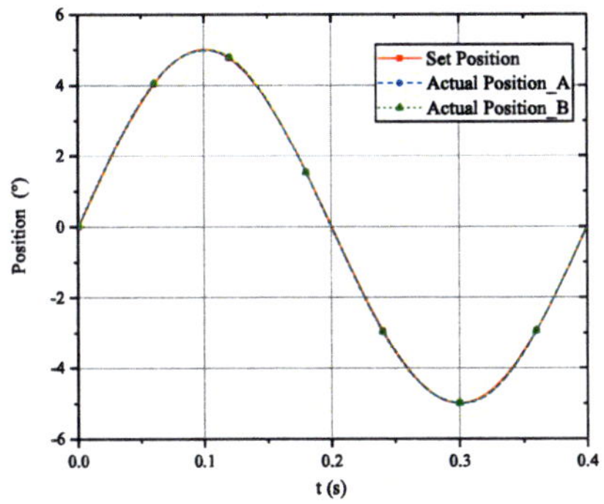

Fig. 7. Sinusoidal motion test curve of 5°/2.5 Hz.

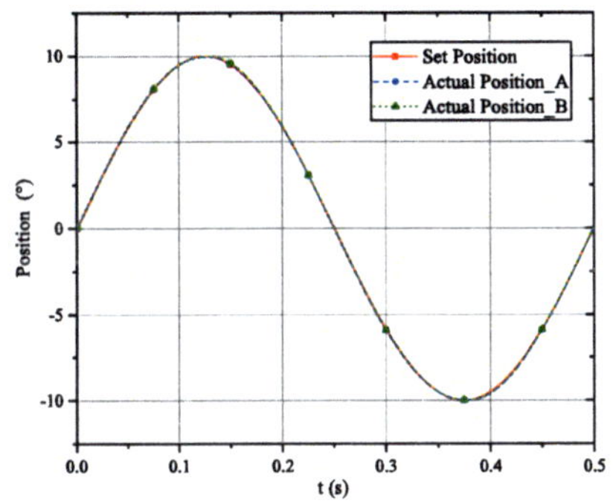

Fig. 8. Sinusoidal motion test curve of 10°/2 Hz.

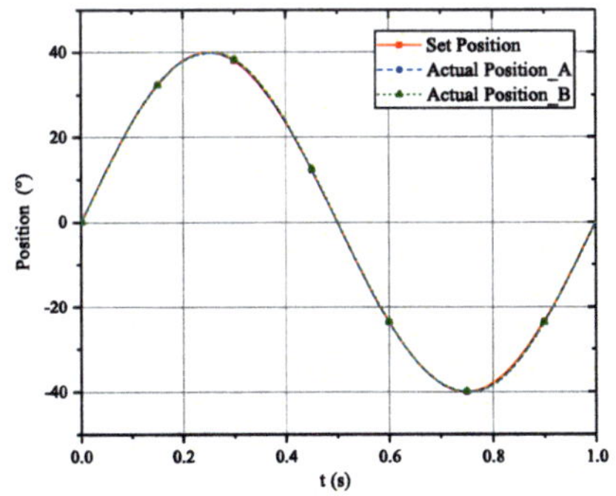

Fig. 9. Sinusoidal motion test curve of 40°/1 Hz.

According to the test curves of Figs. 7, 8, 9, and 10 and the results of the amplitude-frequency and phase-frequency characteristics shown in Table 7, it can be known that:

Table 7. Amplitude difference and phase difference.

	Test index	Motor A	Motor B
Amplitude difference	5°/2.5 Hz	0.56%	0.52%
	10°/2 Hz	0.32%	0.29%
	40°/1 Hz	0.15%	0.15%
Phase difference	5°/2.5 Hz	0.46°	0.42°
	10°/2 Hz	0.29°	0.27°
	40°/1 Hz	0.21°	0.15°

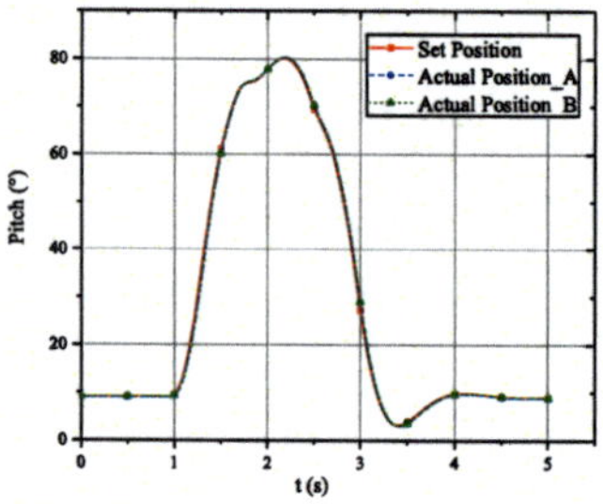

(a) Cobra Maneuver Simulation Test Curve

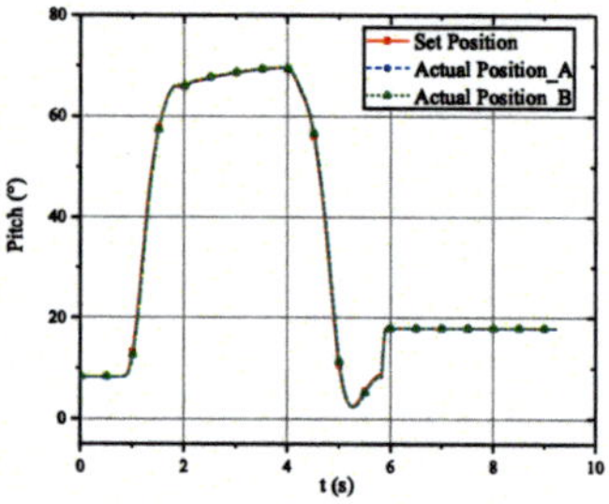

(b) High Angle of Attack Holding Maneuver Simulation Test Curve

Fig. 10. Typical post stall maneuver simulation test curve.

(1) The dual electro-hydraulic servo motors of the dynamic test system have excellent amplitude-frequency/phase-frequency characteristics when performing the sinusoidal motion index given in Table 6, and can satisfy the amplitude difference of less than 1% and the phase difference of less than 1°, to meet the system design requirement in term of dynamic tracking accuracy index;
(2) The synchronization of the two drive motors is very good, and the time difference corresponding to the phase difference is less than the cycle period of the control system (0.5 ms);
(3) The dynamic test system can realistically simulate typical post stall maneuver at high angle of attack, such as the Cobra maneuver and the high angle of attack holding maneuver.

4.2 Control System Verification

In order to verify the wind tunnel test data accuracy of the wire-driven dynamic test system, we carry out a wind tunnel validation test using a standard aircraft model with a low aspect ratio flying wing layout at the scale of 1.5 m, which has established a relatively complete aerodynamic database based on many low-speed wind tunnels in China. The test model adopts the structural form of metal skeleton covered with carbon fiber skin, and has control surfaces such as flaps, ailerons, leading edge flaps, full-moving wing tips, and upper and lower spoilers. Aerodynamic force and moment are measured by the balance with a 50 mm diameter inside the test model. The main geometric parameters of the model are shown in Table 8.

Table 8. Main geometric parameters of aircraft model.

Item	Value
Proportion	1:11
Average aerodynamic chord length	0.87 m
Reference area	0.7 m^2
Weight	12 kg
Composition	Metal skeleton covered with carbon fiber skin

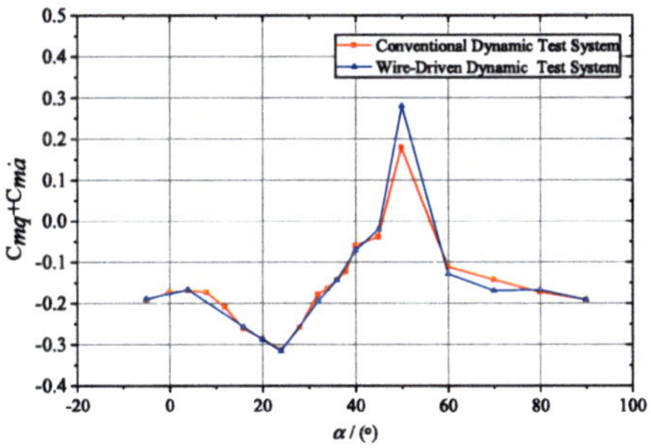

Fig. 11. Test data comparison curve of pitch dynamic derivative with different test system (under 5°/1.5 Hz).

According to the test results in Fig. 11, the overall curve magnitude and change trend of the dynamic derivative data obtained by the wire-driven dynamic test system and the conventional dynamic test system are consistent, and the overall deviation is very small, showing that the dynamic derivative test data obtained by this test system can satisfy the accuracy requirement.

According to the test results in Fig. 12, in the simulation test of Cobra maneuver process, the normal force and the pitching moment form an obvious hysteresis loop curve with the change of the angle of attack, especially in the high angle of attack range, the area of the hysteresis loop is large, and the hysteresis effect is strong. The overall pitching moment curve presents a double "8" characters, indicating that the damping characteristics have changed twice, which fully reflects the nonlinear dynamic characteristics of the model used in the test. Therefore, the typical maneuver process simulation wind tunnel test carried out by the wire-driven dynamic test system designed in this paper can effectively obtain the aerodynamic force of the aircraft during the maneuver flight process, and the test results are reasonable and credible. The aerodynamic data obtained through this type of wind tunnel test can be used for the analysis of the aircraft's

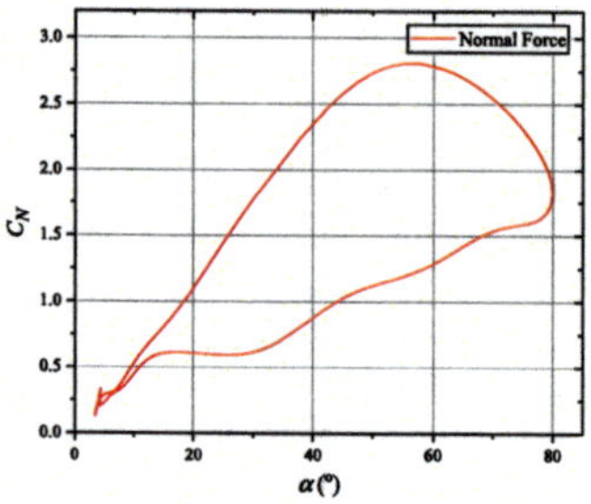

(a) Normal Force of the Cobra Maneuver Simulation Test

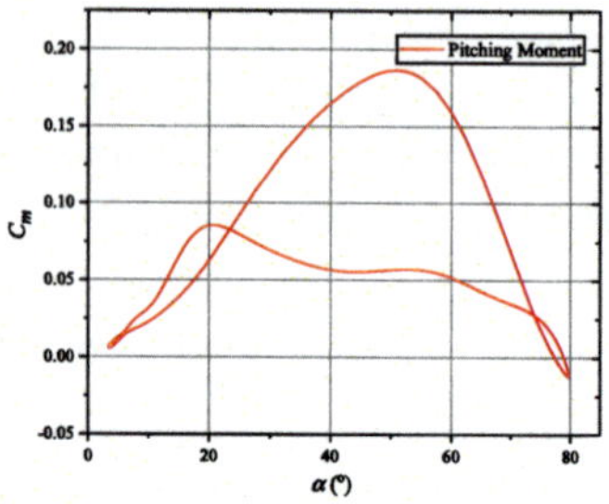

(b) Pitching Moment of the Cobra Maneuver Simulation Test

Fig. 12. Cobra maneuver simulation test data.

high angle of attack maneuver flight characteristics and the verification of the aircraft's aerodynamic modeling.

5 Conclusion

In this paper, based on FL-51 wind tunnel of AVIC Aerodynamics Research Institute, a dynamic test system driven by dual electro-hydraulic motors with wire-driven suspension is developed and verified through wind tunnel validation tests. The following conclusions are drawn:

(1) This test system has excellent static/dynamic performance, and the dual drive motors have excellent synchronization performance, which meets the system's performance index requirement of the dynamic test;
(2) Based on this test system, wind tunnel dynamic tests such as dynamic derivative test, large amplitude oscillation test and maneuver process simulation test with aircraft model at the scale of 1.5 m can be carried out;
(3) The dynamic test data obtained based on this system has reasonable regularity, and the data accuracy meets the index requirement;
(4) Based on this system, abundant and various wind tunnel dynamic test data can be obtained, which can be used to carry out research work such as the analysis of the aircraft's high angle of attack post stall maneuver flight characteristics and aerodynamic modeling verification.

References

1. Wang, H.F., Zhan, J.X., Chen, K., et al.: Development and application of aerodynamic research technologies for fighters at high angle of attack. Acta Aerodyn. Sin. **40**(1), 1–25 (2022)
2. Bridges, D.H.: The asymmetric vortex wake problem-asking the right question. AIAA-2006-3553, AIAA, Reston (2006)

3. Williams, B., Landman, D., Flory, I., et al.: The effect of systematic error in forced oscillation testing. In: 50th AIAA Aerospace Sciences Meeting including the New Horizons Forum and Aerospace Exposition, Nashville, Tennessee. AIAA, Reston, Virginia (2012)
4. Wang, B., Huang, C.D., Mang, B.F., et al.: The control method of precise reproduction of the wing rock motion. J. Exp. Fluid Mech. **23**(1), 79–84 (2009)
5. Yang, M., Huang, D.: The unsteady aerodynamic characteristics of coupled motion at high angle of attack. J. Exp. Fluid Mech. **25**(6), 19–22 (2011)
6. Owens, D.B., Brandon, J.M., Croom, M.A., et al.: Overview of dynamic test techniques for flight dynamics research at NASA LaRC (invited). AIAA 2006-3146, AIAA, San Francisco, California (2006)
7. Dan, D.V., Thomas, D.L., Andreas, S.: SACCON forced oscillation tests at DNW-NWB and NASA Langley 14x22-feet tunnel. In: 28th AIAA Applied Aerodynamics Conference, Chicago, Illinois, AIAA 2010-4394 (2010)
8. Lutze, F.H., Fan, Y.G., Stagg, G.: Multi-axis unsteady aerodynamic characteristics of an aircraft. AIAA Paper 99-4011 (1999)
9. Liu, C.M., Zhao, Z.J., Bu, C., et al.: Double degree-of-freedom large sample oscillation test technology in low-speed wind tunnel. Acta Aeronaut. Astronaut. Sin. **37**(8), 2417–2425 (2016)
10. Ma, S., Zhao, Z.L., Jiang, M.H., et al.: Dynamic test technique with large sample in 2.4m transonic wind tunnel. Acta Aerodyn. Sin. **37**(6), 998–1003 (2019)
11. Xie, Z.J., Sun, X.Y., Sun, H.S., et al.: Mechanism design and dynamics analysis of high-speed parallel robot for dynamic test in low-speed wind tunnel. Acta Aeronaut. Astronaut. Sin. **34**(3), 487–494 (2013)
12. Huang, D.: Unsteady aerodynamic characteristics for the aircraft oscillation in large amplitude. Nanjing University of Aeronautics and Astronautics, Nanjing (2007)
13. Zheng, Y.Q., Lin, Q., Liu, X.W.: Design methodology of wire-drive parallel support systems in the low-speed wind tunnels and attitude control scheme of the scale model. Acta Aeronaut. Astronaut. Sin. **26**(6), 774–778 (2005)
14. Guo, B.J., Peng, B.B., Li, Y.S., et al.: Research of position synchronization control for dual electro hydraulic servo motor. Mach. Tool Hydraul. **42**(24), 47–52 (2014)
15. Wang, C.L., Guo, Z.F., Dong, Y.L., et al.: Research of equal-status synchronizing controller for the dual hydraulic driven motors based on QFT. Chin. Hydraul. Pneum. **12**, 18–23 (2011)
16. Liu, X.F., Liu, X.H., Wang, L.S., et al.: Speed synchronization control of dual-motor in large crawler crane based on fuzzy PID control. J. Jilin Univ. (Eng. Technol. Ed.) **41**(3), 559–664 (2011)
17. Zhang, T.L.: Design and servo control study of single-axis high-frequence electro-hydraulic shaking table. Zhejiang University, Hangzhou (2018)
18. Rui, G.C., Hou, D.D., Shen, G.: A combined control based on feedback and feedforward compensation for 2-DOF electro-hydraulic shaking table. Chin. Hydraul. Pneum. **5**, 21–27 (2017)
19. Bonchis, A., Corke, P.I., Rye, D.C.: Variable structure methods in hydraulic servo systems control. Automatica **37**, 589–595 (2001)
20. Cheng, X.H., Yao, J.Y., Le, G.G.: Multiple model robust adaptive control of electro-hydraulic servo systems. J. Xi'an Jiaotong Univ. **52**(11), 156–162 (2018)
21. Zhu, Z.H., Li, W.J., Li, H.G.: Electro-hydraulic position servo system with high speed and large inertia using sliding mode control. Mach. Electron. **34**(9), 55–59 (2016)
22. Hao, X.X., Wang, X.P.: The low-speed performance improvements of electro-hydraulic position servo system based on sliding mode adaptive control. Chin. Hydraul. Pneum. **1**, 39–43 (2015)
23. Yao, J.Y., Jiao, Z.X., Huang, C.: Compound control for electro-hydraulic positioning servo system based on dynamic inverse model. J. Mech. Eng. **47**(10), 145–150 (2011)

24. Dang, X.J., Yuan, Y.Q., Jiang, H., et al.: Neural network based adaptive sliding mode control for electro-hydraulic position servo system. Mach. Tool Hydraul. **46**(1), 126–129 (2018)
25. Mohamed, K.M., Heath, W.P., Lanzon, A.: Enhanced tracking for nano positioning systems using feedforward/feedback multivariable control design. IEEE Trans. Control Syst. Technol. **23**(3), 1003–1013 (2015)
26. Zhao, J.S., Shen, G., Zhu, W.D., et al.: Robust force control with a feed-forward inverse model controller for electro-hydraulic control loading systems of flight simulators. Mechatronics **38**, 42–53 (2016)
27. Shen, G., Zhu, Z.C., Zhang, L., et al.: Adaptive feed-forward compensation for hybrid control with acceleration time waveform replication on electro-hydraulic shaking table. Control Eng. Pract. 1128–1142 (2013)
28. Pratumsuwan, P., Thongchai, S., Tansriwong, S.: A hybrid of fuzzy and proportional integral derivative controller for electro hydraulic position servo system. Int. J. Energy Res. **1**(2), 62–67 (2010)
29. Shao, J.P., Wang, Z.W., Li, J.Y., et al.: Rule self-tuning fuzzy PID controller of electro-hydraulic position servo system. J. Cent. South Univ. **41**(3), 960–965 (2010)

Study on Risk Control of the Horizontal Tail Coupling Flight Test

Dawei Zhang, Yi Su(✉), and Zhubin Fan

Flight Test Operation Department, COMAC Flight Test Center, Shanghai 200135, China
suyixiangfei@163.com

Abstract. Parameter identification flight test is mainly to establish an accurate aerodynamic parameter model. The flight test method and concern is different with that in the certification flight test, which is to show compliance with the regulations. The parameter identification flight test risk is likely to be high. This paper discusses the flight test hazards of horizontal tail elevator coupling subject, and analyse the flight test risk. The horizontal tail jammed and horizontal tail stall are main risk in this flight test. The reason of those risks are analyzed in detail. The risk control method and risk mitigation plan, as well as flight test limitation and pre-flight test due to the risk mitigation plan, are introduced. In this process, the method of flight risk control is presented in this paper.

Keywords: Parameter identification flight test · Risk control method · Horizontal tail jammed · Horizontal tail stall

1 Introduction

Different from the typical airworthiness certificate flight test subjects such as airspeed calibration, stall, flutter, etc., the horizontal tail elevator coupling flight test is a parameter identification type flight test subject. The main purpose of the certification flight test subjects is to verify the compliance of the aircraft design with the airworthiness provisions or design requirements by means of flight tests. For this type of flight test subjects, domestic and foreign practitioners have carried out a lot of research work, the research content includes flight test methods, data analysis methods, and risk control methods, etc. For example, Zhou et al. studied the flight quality flight test of traditional flight control system and fly-by-wire flight control system [1], Tang et al. introduced the data processing method of lift curve and polar curve [2], Liu et al. proposed a quantitative risk assessment method for flight test subjects based on the control characteristics of test pilots [3]. Therefore, the flight test method of this type of subject is relatively mature, and the risk control method also has generality to a certain extent. In order to facilitate relevant practitioners, NASA has specially developed a database of relevant risk assessments, and the FAA has issued Directive 4040.26B [4] to provide a recommended method for risk assessment of aircraft in the airworthiness certification stage.

Parameter identification flight test is mainly to establish an accurate aerodynamic parameter model. The core is to obtain accurate flight data, identify various factors that

Chinese Society of Aeronautics and Astronautics: CASTYSF 2022, LNEE 972, pp. 757–764, 2023.
https://doi.org/10.1007/978-981-19-7652-0_67

affect aerodynamic efficiency, and establish an accurate aerodynamic model. At present, articles in regard to parameter identification mainly focus on simulation, data processing, modeling and so on. For example, Yu's "Flight Data Processing and Aerodynamic Parameter Identification" [5], Wang's "Identification of Aircraft Aerodynamic Derivatives Using High Angle of Attack flight test Data" [6], and Cai's "Progress in Aircraft Aerodynamic Parameter Identification" [7]. Some subjects of parameter identification flight test require special flight test methods, which have special flight test risks, and some subjects even have high risks. However, very few articles focused on special flight test methods and risk control methods of parameter identification. This paper selects a typical subject in the parameter identification flight test - the horizontal tail elevator coupling flight test as the starting point, and elaborates on how to carry out the risk identification and risk control of this subject. This paper examines the parameter identification flight test from the perspective of flight test participants, which can be used as a reference for relevant practitioners.

2 Flight Test Method

Generally, the main principle of longitudinal full parameter identification of an aircraft with an all-movable horizontal tail design is shown in formula (1). In the formula, C_{L0}, C_{D0}, and C_{M0} are constants. $C_{L\alpha}$, $C_{D\alpha}$, $C_{M\alpha}$ are functions of the angle of attack α, and $C_{L\delta H}$, $C_{M\delta H}$ are functions of the horizontal tail angle. $C_{L\delta e}$, $C_{M\delta e}$ are functions of elevator angle. The rest are damping terms due to the pitch rate and angle of attack rate. The main purpose of the horizontal tail-elevator coupling flight test is to establish the corresponding relationship between the horizontal tail deflection and the elevator deflection. The main parameter to identify are $C_{L\delta H}$, $C_{M\delta H}$, $C_{L\delta e}$, $C_{M\delta e}$.

$$\begin{cases} C_L = C_{L_0} + C_{L_\alpha}\alpha + C_{L_{\delta_H}}\delta_H + C_{L_{\delta_e}}\delta_e + C_{L_q}\dfrac{qc}{2v} + C_{L_{\dot\alpha}}\dfrac{\dot\alpha c}{2v} \\ C_D = C_{D_0} + C_{D_\alpha}\alpha \\ C_m = C_{m_0} + C_{m_\alpha}\alpha + C_{m_{\delta_H}}\delta_H + C_{m_{\delta_e}}\delta_e + C_{m_q}\dfrac{qc}{2v} + C_{m_{\dot\alpha}}\dfrac{\dot\alpha c}{2v} \end{cases} \tag{1}$$

In the formula (1): α is the angle of attack, δ_H is the deflection of horizontal tail, δ_e is the deflection of elevator, q is the pitch rate, $\dot\alpha$ is the rate of angle of attack, v is the true airspeed.

The flight test method of the horizontal tail-elevator coupling flight test is to trim the aircraft in a stable level flight at a certain altitude and speed. The trim is then adjusted in a stepwise method while adjusting the elevators to maintain the speed/Mach number of the aircraft and establish different out-of-trim conditions. Thus, the elevator deflection required to maintain a stable flight under different horizontal tail deflections is obtained, and then the relationship between $C_M\delta_H$ and $C_M\delta_e$ is calculated.

3 Flight Test Risk Analysis

Compared with other subjects, the flight test method of this subject is not complicated. The aircraft is in a longitudinally out-of-trim condition, and the test pilot needs to exert a certain longitudinal control force for a long time to stabilize the aircraft. For aircraft with center yoke design, the longitudinal control force may even reach 20–30 kg when deviated from the trim condition, which will highly increase the workload of the test pilot. During the technical preparation of this subject, when using the simulator for training, it was found that when the horizontal tail deflection increased to a certain value, the aircraft appeared longitudinally uncontrollable. Further in-depth analysis found that in the case of large horizontal tail deflection, there may be two risks: horizontal tail jamming and horizontal tail stall.

On one hand, during the execution of the test point, the aircraft is in large longitudinal out-of-trim condition, and the horizontal tail deflection may exceed the mistrim requirements specified in CCAR 25.255 [8], so as to exceed the safety envelope of transport aircraft. The horizontal tail area of a common large civil aircraft is about twice that of the elevator, and the excessively large horizontal tail mistrim cannot be compensated and corrected by the elevator. The Boeing 737NG manual stipulates that if the horizontal stabilizer is out of control, the autopilot should be disconnected to manually control the aircraft. If the out-of-control persists, the stabilizer trim switch needs to be turned off to avoid an excessive out-of-trim condition. Once the horizontal tail is blocked, the aircraft may gradually lose pitch control or the maneuvering margin may not be sufficient to support a safe back to base and landing, resulting in loss of aircraft control and a catastrophic accident.

On the other hand, for large civil aircraft, the horizontal tail mainly provides negative lift. In order to reduce the trim resistance, the horizontal tail often adopts an asymmetric airfoil with reverse camber, that is, the horizontal tail provides negative lift when the angle of attack is 0 degree [9]. In the approach landing configuration, the large angle flaps will produce a strong downwash effect. With the downwash and large nose-down mistrim, the local angle of attack of the horizontal tail may be a large negative value, which will be far beyond the normal flight envelope. In extreme cases, a horizontal tail stall may occur, resulting in loss of control of the aircraft and a catastrophic accident.

In the following, the two risks of horizontal tail jamming and horizontal tail stall will be deeply analyzed, and the measures to control and reduce the above risks during the horizontal tail elevator coupling flight test and the emergency plan after the occurrence of the danger are given.

4 Risk Management of Horizontal Tail Jammed

For the aircraft with common layout, horizontal tail is mainly provides longitudinal stability of the aircraft. According to the speed stability of the aircraft from low speed to high speed in level flight, it is necessary to increase nose-down trim and the leading edge of horizontal tail needs to deflect upward. While it is opposite from high speed decelerate to low speed, the leading edge of horizontal tail needs to deflect downward. Normally, the horizontal tail doesn't need to be deflected to the travel limit. However, due to the needs of the subject, the horizontal tail may be adjusted to the deflection limit during the horizontal tail elevator coupling flight test. A large amount of elevator deflection to balance the longitudinal control will be required (the elevator deflection is about twice deflection of the horizontal tail mistrim). At this moment, the margin of elevator will be greatly reduced or even to zero in a certain direction. If the horizontal tail jammed, the remaining elevator margin cannot support the aircraft safety returning and landing.

In order to confirm our inference, we used a large civil aircraft simulator to carry out the horizontal tail jammed program recommended by manual. The simulator test showed that the aircraft can return and land safely when the horizontal tail was jammed at nose-down limitation. However, when the horizontal tail was jammed at nose-up limitation, the aircraft cannot return to the field and land safely. So, when planning the horizontal tail-elevator coupling flight test, if the remaining control margin can support the aircraft safe return and landing in case of the horizontal tail jam must be considered. With our own flight test experience, safe return and landing can be divided into the following capabilities:

a. Under the landing configuration and approach speed recommended by the horizontal tail jammed procedure, the aircraft has longitudinal maneuverability of 1.3 and 0.8 g to complete the necessary maneuvers such as turning, aligning with the runway and flare.
b. Under the cruise configuration and selected cruise speed, the aircraft has longitudinal maneuverability of 1.15 and 0.85 g to finish the necessary return maneuvers such as flight under turbulence, descent and heading correction.
c. Under the transition configuration and the selected transition speed between cruise and landing configuration, the aircraft is controllable during the flap/slat change.
d. The aircraft is controllable under other flight conditions and has the ability to adjust to the above three states.
e. Go around capability. On one hand, when the aircraft is jammed, the aircraft handling margin is very critical, so it is necessary to ensure that the aircraft has go around ability to ensure safety. On the other hand, the thrust is very large during go around, which deviates far from all the above situations, so it must be evaluated respectively.

The above-mentioned capabilities need to be determined and evaluated by special flight tests. The capacity of a and b mentioned above can be determined by push over and pull up maneuver or stable turning, with gradually increasing the horizontal tail trim to obtain the maximum horizontal tail up and down limited values respectively. The capabilities of c and e are completed on the basis of the results of a and b. The trim is chosen by the smaller of the limited value determined by the capabilities of a and b. The capability c is completed through the qualitative comments of the pilot during flap/slat retraction and retraction, and e is completed through the qualitative comments of the pilot during simulated go around. It should be noted that the flap and slat configuration cannot be changed during horizontal tail jammed go around, because the flap and slat configuration change affects the available longitudinal control margin. For capability d, due to the stability of the aircraft, over the above selected reference speed, the aircraft tends to return to the reference speed by changing the thrust. We need to focus on evaluation of the dynamic change of speed when back to the reference speed. It is suggested that the above-mentioned flight tests be completed on the existing simulator or desktop simulation model, and the flight test should be carried out on the basis of simulation.

Based on the completion of the above-mentioned tests, the horizontal tail jammed procedure for the horizontal tail elevator coupling flight test is as follows:

a. In case of horizontal tail jammed in the cruise configuration and jammed in the nose-down direction, the speed cannot be less than the selected cruise speed during returning to the field. When the aircraft is approaching, adjust the configuration to the landing configuration with the speed not less than the selected approach speed, and also the flap/slat cannot be retracted during go around. In case of horizontal tail jammed in the cruise configuration and jammed in the nose-up direction, the speed cannot be greater than the selected cruise speed required for return to the field. During approach, adjust the configuration to the landing configuration with the speed not greater than the selected approach speed, and the flap/slat should not be retracted during go around;
b. In case of horizontal tail jammed in other configurations, first we should adjust to the transition speed and retract the flaps step by step to the cruise configuration. Then return to the field and land according to the horizontal tail jammed in cruise configuration procedure.

Note: If horizontal tail jammed at nose-down direction in the cruise configuration, the aircraft deviates from the selected speed too much may be a risk that the control margin in the nose-up direction is insufficient and cannot be recovered. Therefore, if horizontal tail is jammed at nose-down direction, the return speed is specified as not less than the selected speed; vice versa. For horizontal tail jammed in other configurations, adjusting the flap, the speed strategy is similar to that of cruise configuration. It also means that the speed is larger than the selected speed if the horizontal tail jammed at nose-down direction, and the speed is smaller than the selected speed if the horizontal tail jammed at nose-up direction.

5 Risk Management of Horizontal Tail Stall

Horizontal tail stall is a more recent problem, seldom attention was paid in the early days. With the deepening understanding of the aircraft design, study was emerging from 1990s. The main consideration is the stall of the horizontal tail in the icing condition, such as Xue's "Influence of Horizontal Tail Ice Accumulation on Aircraft Longitudinal Aerodynamic Parameters" [10], Zhong's "Influence of Icing on Horizontal Tail Performance and Horizontal Tail Stall Analysis" [11], "NASA/FAA Tailplane Icing Program: Flight Test Report" [12] and etc. Certification flight test subjects - ice contamination horizontal stabilizer stall has now become a subject that must be completed for CCAR25 aircraft flight test certification. However, since the deflection of the horizontal tail generally does not reach a state close to the stall during normal flight (under non-icing conditions), no specific consideration is required for the certification flight test.

The horizontal tail of the aircraft is located behind the wing and will be affected by the downwash of the wing. Therefore, the main factors affecting the local angle of attack at the horizontal tail include the angle of attack of the aircraft fuselage, the deflection of the horizontal tail, the downwash of the wing, and other dynamic factors, as shown in the following formula (2). The downwash angle of the horizontal tail is mainly generated by the wing. Generally, the greater the flap deflection, the greater the corresponding horizontal tail downwash angle.

$$\alpha_H = \alpha_b + \delta_H - \varepsilon + (q + \dot{\alpha}\varepsilon_\alpha)\frac{\mathrm{l_H}}{V_T} \tag{2}$$

In the formula (2), α_H is the local angle of attack of horizontal tail, α_b is the body angle of attack of fuselage, ε is the downwash angle of the horizontal tail, q is the pitch rate, $\dot{\alpha}$ is the rate of fuselage angel of attack, $\mathrm{l_H}$ is the arm between horizontal tail and aircraft gravity center, V_T is the true airspeed.

The ice-contaminated horizontal stabilizer stall flight test mainly considers the dynamic effects of icing-induced horizontal tail stall advance and rapid stick push. For the horizontal tail-elevator coupling flight test, the large downwash angle and the large horizontal tail deflection are the main influencing factors. During the flight test of the horizontal tail-elevator coupling subject, when the horizontal tail is trimmed to the vicinity of the downward deflection limit position or when the calculated horizontal tail angle of attack (see more details in the following) approach a given value, it is also necessary to avoid rapid push stick maneuvers to prevent the horizontal tail from accidentally stall.

In stablized flight, the dynamic effect term can be ignored. According to formula (2), the estimated value of the horizontal tail angle of attack of a certain type of aircraft with approach and landing configuration is calculated and shown in the following Table 1. It can be seen from the table that the local angle of attack of the horizontal tail reaches a maximum of 18.5 degrees, which basically reaches the stall point of the conventional airfoil. In this state, the possibility of the aircraft accidentally entering the horizontal tail stall is very high.

Table 1. Estimation of the horizontal tail limit angle of attack in the horizontal tail elevator coupling subject of a certain type of aircraft in the approach landing configuration

No.	Speed (kts)	Horizontal tail		
		Deflection (°)	Local angle of attack (°)	Limitation of local angle of attack (°)
1	125	−6.2	−3.2	−10.5
2	150	−2.8	−4.8	−15.5
3	180	−1.0	−5.9	−18.5

Aircraft entering a horizontal tail stall is very dangerous. First of all, when the aircraft is in the horizontal tail stall, due to the separation of airflow at the trailing edge of the horizontal tail, the elevator will be in the air flow separation area, and the elevator may fail and enter the locked state, that is, the aircraft may not have obvious pitch motion when the elevator is manipulated. Secondly, after the horizontal tail enters the stall, the slope of the horizontal tail lift line is reversed, and the aircraft will be quickly nose down, and the additional dynamic effect of nose-down motion will in turn further aggravate the horizontal tail stall, thus entering an irresistible nose-down motion and dive state. Since the elevator is in the stall wake region, the stability moment it provides is limited, and it is difficult to recover through longitudinal control. Therefore, the danger of a horizontal tail stall far exceeds that of a wing stall. During the flight test, it is necessary to avoid entering a horizontal tail stall.

The most effect way to avoid entering the horizontal tail stall during the flight test is to measure and monitor the local angle of attack of the horizontal tail in real time. However, the civil aircraft design generally does not install the angle of attack sensor near the horizontal tail, and angle of attack sensor may be installed for this subject specificlly if required.

According to the principle of formula (2), the monitoring algorithm of the local angle of attack of the horizontal tail can be designed, which provides the real-time estimated local angle of attack of the horizontal tail for the flight test crew and the monitoring personnel. In the real-time estimation algorithm of the angle of attack, the fuselage angle of attack, horizontal tail deflection, pitch angle rate, and true airspeed are obtained through the bus data, the horizontal tail force arm is obtained through the real-time center of gravity conversion, and the rate of the angle of attack is obtained through the difference of the fuselage angle of attack. The coefficients related to the downwash angle are obtained using the experimental data of the previous wind tunnel test.

Considerable parameters are involved in the above algorithm, and the calculation of the rate of the angle of attack and the tail arm are also relatively complex. For subjects with small maneuvering such as horizontal tail-elevator coupling, the influence of dynamic effects is negligible, thus the variables involved in the above algorithm can be simplified to the fuselage angle of attack, horizontal tail deflection and flap position. Considering that the stall risk of the horizontal tail is pretty high, the simplified algorithm needs to leave a sufficient safety margin for the local angle of attack of the horizontal tail. Three major aspects are considered:

1. The limitation of the angle of attack of the horizontal tail during the test should be less than the stall angle of attack of the horizontal tail determined in wind tunnel test;
2. Certain error exists in the horizontal tail stall angle of attack and the downwash angle given by wind tunnel test. This part of the error cannot be corrected during the flight test process, and a sufficient margin needs to be remained;
3. The dynamic correction of the local angle of attack of the horizontal tail caused by the turbulent air flow and recovery maneuver may be large, and there is a risk of instantaneously exceeding the stalling angle of attack of the horizontal tail.

6 Conclusion

Parameter identification flight test is a very complex subject, in which different measures formulated due to different risks are the key to ensure the safe implementation of flight test. Not only for parameter identification flight test, but also for flight test of all subjects, risk identification and control of flight test are the most important things. The core of risk assessment is the identification and pre-planning before the test. It is not only based on a large number of quantitative data such as theoretical analysis and simulator tests, but also relies heavily on the flight experience of flight test engineers. Only through a series of systematic risk assessments, the formulation of good prerequisite subjects and emergency plans, can the smooth implementation and safety of the flight test be truly ensured.

References

1. Zhou, Z.Q., et al.: Flight test techniques on flight dynamics and flight control. Flight Dyn. **16**(3), 1–6 (1998)
2. Tang, H.R., et al.: A method to process flight test data. Civil Aircr. Des. Res. **2**, 21–25 (2011)
3. Liu, D.L., et al.: Quantitative risk evaluation of flight testing subject based on operational characteristics of test pilot. Flight Dyn. **28**(3), 82–88 (2010)
4. FAA 4040.26B.2001
5. Yu, C.J.: Flight data processing and aerodynamic parameter identification. Northwestern Polytechnical University, Xi'an (2007)
6. Wang, Q.: Aircraft aerodynamic parameter identification by using HAOA flight test data. Flight Dyn. **23**(1), 69–72 (2005)
7. Cai, J.S.: Advance in identification of aircraft aerodynamic parameters. Adv. Mech. **17**(4), 467–475 (1987)
8. CCAR part 25. http://www.caac.gov.cn. Last accessed 07 Mar 2016
9. Cheng, B.S.: Aircraft Design Manual, vols. 5 and 6, 1st edn. Aviation Industry Press, Beijin (2005)
10. Xue, Z.D.: Effects of stabilizer plane icing on aircraft longitudinal aerodynamics parameters. Acta Aeronaut. Astronaut. Sin. **34**(7), 1563–1570 (2013)
11. Zhong, C.S.: Influence of icing on horizontal stabilizer performance and horizontal stabilizer stall analysis. J. Civ. Flight Univ. China **15**(6), 6–12 (2004)
12. NASA/FAA stabilizer plane icing program: flight test report (2000)

Research on Precise Servo Control Technology for Integrated Aircraft Utility System

Mingliang Hu[1,2](✉), Wen Yan[1], and Jionghui Che[1]

[1] Xi'an Aeronautics Computing Technique Research Institute, AVIC, Xi'an 710068, China
minglhu@126.com

[2] School of Computer Science, Northwestern Polytechnical University, Xi'an 710072, China

Abstract. In this paper, stepper motor servo valve is selected as the execution unit to adjust environmental variables, such as temperature, pressure and flow of the aircraft environmental control system and fuel thermal management system, to meet new generation aircraft requirements of the reliability of the servo valve, the controllability of the opening and closing angle, as well as the controllability of the opening and closing angle requirements. Since there are a large number and great variety of stepping motor controllers in the airborne system, a unified architecture design is adopted, and it is integrated into the Remote Execution Unit (REU) of the supporting equipment of the aircraft utility system, reducing the aircraft weight and improving the integration level. To enhance the accurate positioning and rapid isolation of stepper motor faults, a fault diagnosis method for inverse-time protection power drive circuit and hierarchical interface are designed to improve system safety and testability. To tackle the temperature rise, noise, torque ripple, start out-of-step and stop overshoot, caused by the stepper motor drive current distortion, high-deterministic command transmission technology and constant current drive technology based on fixed duty cycle and segmented pneumatic technology are developed to perfect the control accuracy of stepper motor servo valve. The test results show that the stepper motor servo control system designed in this paper has high control accuracy and better engineering application value.

Keywords: Utility system · Servo control · Stepper motor · Inverse-time protection · Constant current drive

1 Introduction

There is a large number of click-driven valves in the aircraft environmental control system and the fuel thermal management system. These are able to adjust control parameters in the system, to adjust the temperature, to adjust the pressure and to control the flow and other functions within a time limit [1–3]. Stepper motor servo valve has become the first choice for electric servo valve because of its high control precision with flexible and convenient control method [4]. Stepper motor is a servo motor that can realize precise angle control. Through the cooperation of controllers, the valve angle can be adjusted with time limit. Its essence is an actuator that converts the electrical pulse signal into the corresponding angular displacement [5]. By controlling the frequency

Chinese Society of Aeronautics and Astronautics: CASTYSF 2022, LNEE 972, pp. 765–776, 2023.
https://doi.org/10.1007/978-981-19-7652-0_68

and quantity of the pulse signal, the adjustment of speed and position can be fulfilled, which can meet requirements of reliability, controllability of opening and closing angle and environmental resistance of new generation aircraft servo valve [6].

To deal with the large number and variety of stepping motor controllers in the airborne system, a unified architecture design is proposed and integrated into the Remote Execution Unit (REU) of the supporting equipment of the aircraft utility system to reduce the aircraft weight and improve the integration level. Aiming at accurate positioning and rapid isolation of stepper motor faults, a fault diagnosis method for inverse-time protection power drive circuit and hierarchical interface are designed to perfect system safety and testability. Targeting at the temperature increase, noise, torque ripple, start out-of-step and stop overshoot caused by the stepper motor drive current distortion, high-deterministic command transmission technology, constant current drive technology based on fixed duty cycle and segmented pneumatic technology are developed to improve the control accuracy of stepper motor servo valve. The test results show that the stepper motor servo control system designed in this paper has high control accuracy and better engineering application value, which therefore can be widely applied in aerospace and other industrial fields.

2 Overall Design

Figure 1 shows the block diagram of the stepper motor servo control system. Three isomorphic Control Computers (CC) constitute the computing center of the utility system, which is used to deal with the motor rotation direction, speed and position information. There are five isomorphic Remote Interface Unit (RIU), distributed in various locations of the aircraft to collect regional temperature, pressure, flow information, and etc., 10 isomorphic REUs are used to receive control commands, to generate a fixed pulse sequence, and to drive the motor to rotate to the predetermined position, to realize the real-time adjustment of the aircraft am temperature, pressure and thermal energy. Each REU contains three stepping motor interfaces. CC, RIU and REU utilize the IEEE-1394 bus for data and command interaction to achieve highly deterministic data transmission and action control of the control system. The internal stepping motor servo control system of REU mainly includes servo controller, double H power bridge driver with inverse time protection function, double H power MOSFET bridge, phase current, voltage acquisition circuit, output feedback and position feedback.

The system adopts the structure of DSP + FPGA, in which DSP is able to realize the algorithm control and bus communication, while FPGA can fulfill the state information acquisition and BIT calculation. The dual H power bridge driver with inverse time protection function can generate the driving signal of the dual H power MOSFET bridge, and realize the inverse time characteristic protection of the current position range, and the output feedback are realized to complete the BIT detection of the system based on the current.

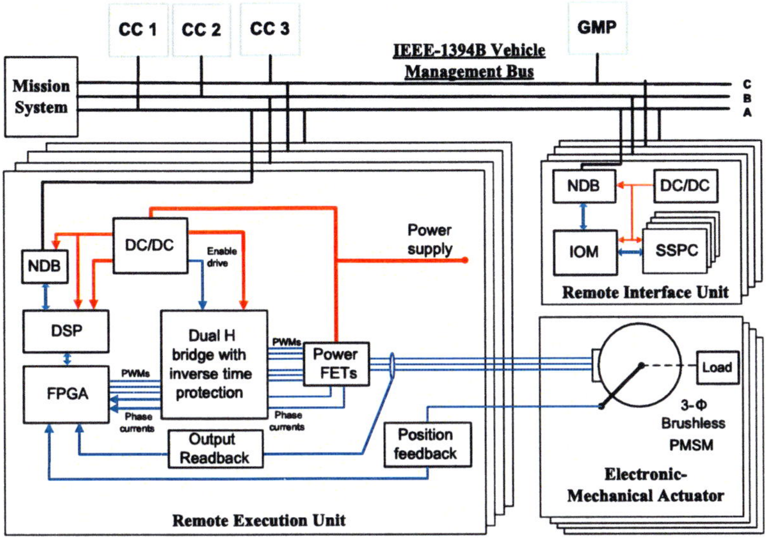

Fig. 1. Block diagram of stepper motor servo control system.

3 Drive System Design with High-Reliability

3.1 Hardware Design

Figure 2 illustrates the hardware functional block diagram of the stepper motor control system. The system mainly includes stepper motor DSP controller, FPGA controller, upper bridge power drive with inverse time protection, lower bridge power drive, power main loop, and current monitoring. The phase current acquisition circuits are fulfilled by current Hall, and the output acquisition circuits are implemented by an 8-channel configurable discrete input chip. The FPGA controller receives motor drive instructions sent by the DSP controller. The tube MOSFET is turned on and off by the driving circuit, and the driving current acts on the motor winding to drive the motor to rotate. The main power circuit can be configured as 4 groups of 28 V/open discrete output interface or 4 groups of ground/open discrete output interface individually to achieve the interface usage, reducing diversification within products, and reducing aircraft weight. The upper-bridge power drive circuit with inverse time protection function is able to monitor the current of the single-phase bridge arm in real time, to calculate the energy accumulation and dissipation of the current, and to realize Trip protection when the duration of the current amplitude reaches the protection threshold. The Trip state will be reported to CC through the bus in real time, while CC will judge the Trip state comprehensively and therefore generates a command to clear the Trip state or lock the interface.

3.2 Power Drive Design Based on Inverse Time Protection

Both stepper motor interface over current and short circuit protection are able to prevent abnormal short circuit damage to the motor or interface. However, traditional current

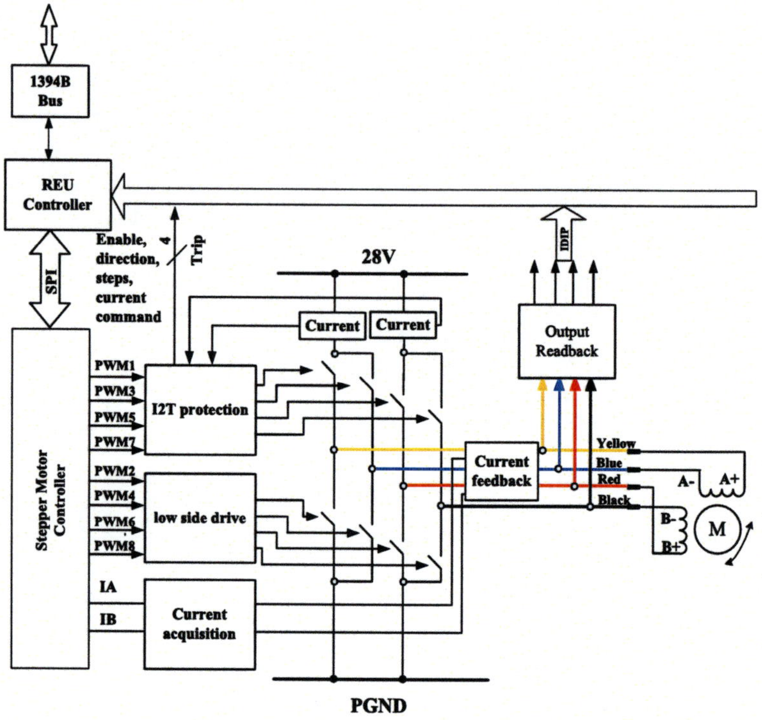

Fig. 2. Hardware functional block diagram of stepper motor control system.

protection hardware circuit is complex and the protection action is hysteretic, which reduces the reliability and security of the system [7]. Figure 3 shows the block diagram of the inverse time limit protection of the 28 V/on discrete output interface, which is mainly composed of sampling resistors, current amplification, current comparison arrays, inverse time limit controllers and power drive parts. The control commands are sent from the DSP controller to the inverse-time controller. The inverse-time controller accumulates energy according to the output state of the current comparison array. When the energy accumulation reaches the threshold value, the output is turned off, and a Trip signal is sent to set the hardware in a safe state until CC contact Trip. In order to improve the real-time performance and reduce resource overhead, the traditional A/D conversion circuit is not applied in this paper to collect the voltage corresponding to the power current, but it is compared with a given reference voltage matrix to generate a current indication signal array. The level inversion state is an overcurrent state. The system has 11 inversion thresholds. Different inversion thresholds represent different energy accumulation step values. When the energy accumulation is greater than the protection threshold, the Trip protection signal is output.

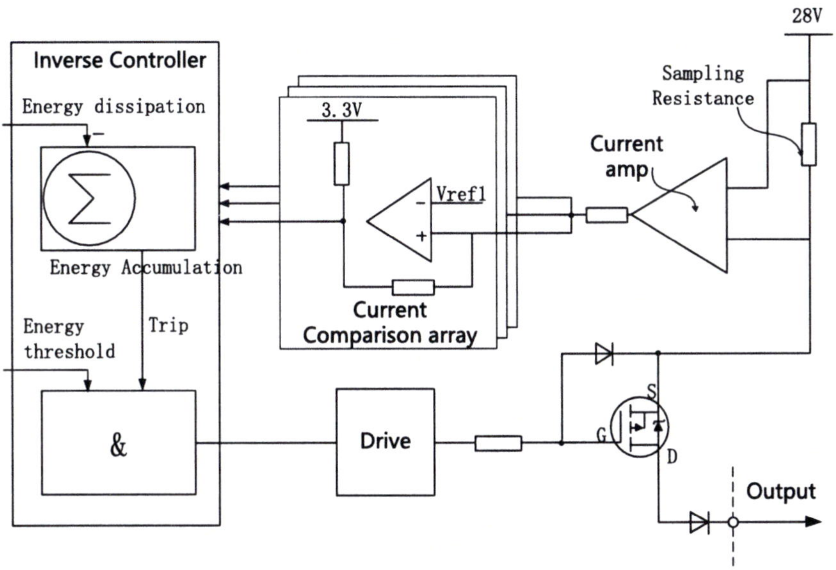

Fig. 3. 28 V/ON discrete output interface inverse time protection principle block diagram.

3.3 Fault Diagnosis Technology of Hierarchical Interface

Stepper motor fault diagnosis technology mainly includes hardware electric furnace detection, motor online status monitor and isolation as well as location of short-circuit fault [8]. When the stepper motor interface fails, the system tube adopts the fault diagnosis technology of fault detection and isolation of the graded interface to realize the rapid location and isolation of the fault. The basic principle of the hierarchical interface fault diagnosis technology is to divide the stepper motor faults into three types that is, the internal interface fault, the external load fault, and the three-chamber REU fault. The internal interface fault is used to diagnose the internal double H-bridge fault, the external load fault is used to detect the motor disconnection and short-circuit fault, while REU fault is used to Fig. out the control part fault. The system utilizes the internal interface of REU to realize fault detection through table configuration technology.

The timing of fault detection is when the interface is in normal operation. There is no need to apply special fault detection operation to the fault detection of the interface, and the normal operation of the old interface is carried out synchronously. The fault detection method is low power consumption, short path and high coverage. Table 1 shows the fault judgment logic of the stepper motor interface of the hierarchical interface.

4 High-Precision Stepper Motor Control Technology

4.1 Command Response Technology with High Deterministic

REU utilizes the IEEE-1394 bus to realize the highly reliable and deterministic transmission of control commands and responses [9]. The IEEE-1394 bus pre-assigns fixed

Table 1. Hierarchical interface stepper motor interface fault judgment logic.

Fault level	Fault type	Detection method
Internal interface failure	28 V/on	Output readback
	Ground/open	Output readback
	Readback circuit failure	Readback circuit BIT
	Current monitoring circuit failure	Voltage acquisition
	Current compare array failure	Array voltage self-test
External interface failure	Short	Abnormal current
	Disconnection	Single upper arm output, H-bridge readback status is inconsistent
Remote actuation unit failure	Electricity failure	Voltage acquisition judgment
	Processor failure	Processor BIT
	Bus communication failure	Bus BIT
	FPGA failure	Answer handshake

transmit offset, receive offset and data pump offset to each node, and plans the number and size of asynchronous flow packets, as well as provides the ability to change the time offset of remote nodes [10]. In the first bus cycle, REU can change its own predefined time offset according to the packet tail information of the asynchronous stream sent by the effective bus controller, and can arrange the system communication sequence reasonably with the help of the time offset and pre-allocated bandwidth, to ensure that the key data has a deterministic time delay that can be designed during the transmission process. Figure 4 shows the description of the de-qualification delay of the system. It is assumed that it takes $\Delta T1$ for RIU to collect flow pressure information to complete the asynchronous flow packet grouping, and $\Delta T2$ for CC to calculate the special and rotational speed position commands from receiving RIU data, the time spent by REU from receiving CC command to starting to drive the actuator is $\Delta T3$, and the transmission delay of asynchronous flow packets between bus nodes is $\Delta T4$. By rationally designing the time offset of RIU, CC and REU, it can be guaranteed that the response time delay of the motor control command in the utility system is able to be controlled at ($\Delta T1 + \Delta T2 + \Delta T3 + 2*\Delta T4$).

4.2 Constant Current Drive Technology Based on Fixed Duty Cycle

Figure 5 shows a schematic diagram of a two-phase four-beat stepper motor with a fixed duty cycle chopper drive. In each continuous shooting control, the first beat implements a 100% duty cycle drive signal, and the second beat implements a fixed duty cycle drive signal. The fixed duty cycle is 0.4–0.6, and the frequency is 6–20 kHz. The first beat

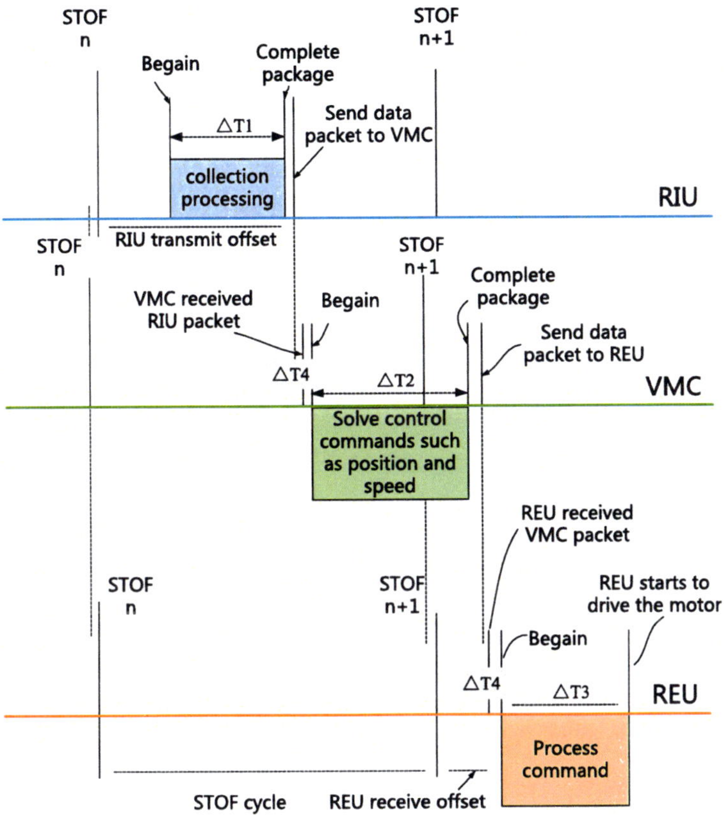

Fig. 4. Description of system deterministic delay.

control is used to increase the start-up capability. When the click turns the equilibrium position, a fixed duty cycle driving signal is applied, which can reduce the overshoot current as well as the torque ripple, and improve the control accuracy and reliability of the system effectively. The constant current drive of the stepper motor based on the fixed duty cycle is realized by an all-digital controller, omitting the complex PID control algorithm. The hardware is simple, the software is flexible with strong portability and expansibility.

4.3 Segmented Startup Technology

The traditional motor startup generally requires the design of complex acceleration and deceleration curves and rotor position positioning methods [11]. According to the motor load condition, this paper proposes a low-complexity segmentation determination method, which includes:

The first stage: Execute the last rotation command of the last rotation, the starting frequency is half of the rated operating frequency, and the number of starting step is 1,

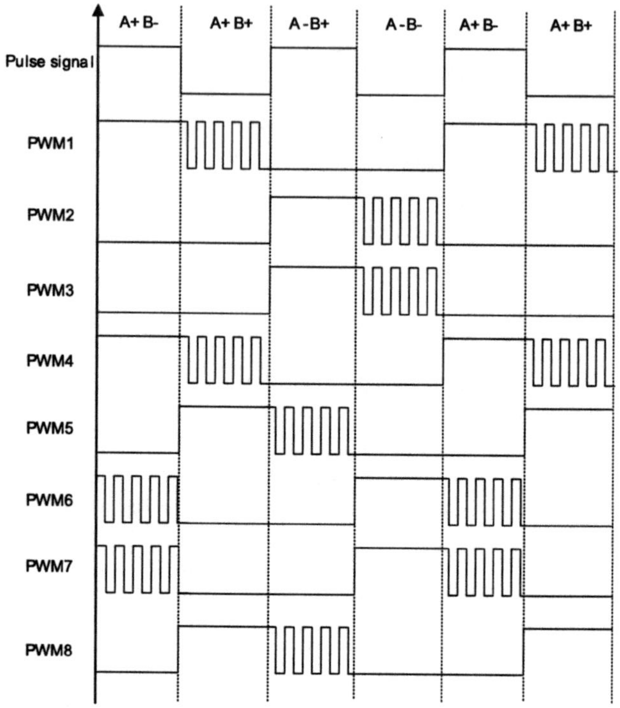

Fig. 5. Schematic diagram of a two-phase four-beat stepper motor with a fixed duty cycle chopper drive.

by which the motor can realize the starting positioning and obtain the low-frequency oscillation kinetic energy;
The second stage: the execution start frequency is 3/4 of the rated frequency, and the number of start step is 4;
The third stage: The motor runs at the rated frequency, and the number of rotation step is (total steps-4).

After the above simplification, the minimum starting step of the motor are only 5, and the rotation out-of-step probability of small steps can be controlled below 0.1%, which improves the control accuracy of the system.

5 Software Design

The stepping motor software sets three different working modes, which are normal working mode, safe working mode and testing mode. In the normal working mode, the software generates the driving sequence according to the pulse distributor. In the safe working mode, it is used for fault handling, and the interface is in a safe state. In the test mode, the 28 V/on discrete output test, ground/on discrete output test are completed. The system adopts stepper motor when appearing and mixing, which supports two-phase four-beat control. The forward rotation logic of the motor is AB-AB'-A'B'-A'B, while the reverse logic is AB-A'B-A'B'-AB'. Figure 6 shows the operation diagram of the motor control software in normal operating mode. After the initiation of system, the valid of the rotation enabling signal is checked. When the rotation enable is valid, the PWM peripheral module is turned off, and the step-by-step command is further judged. When the command is valid, different pulse distributors are called according to the rotation direction, and send out driving signals. During the rotation, when the output of two consecutive beats is valid, the first beat implements a 100% duty cycle drive signal, and the second beat fulfills a fixed duty cycle drive signal, in which the fixed duty cycle is 0.4–0.6, and the frequency is 6–20 kHz. If the above rotation process is disabled or the step-by-step command is invalid, all drive signals will be turn off to ensure that the motor work is in a safe state.

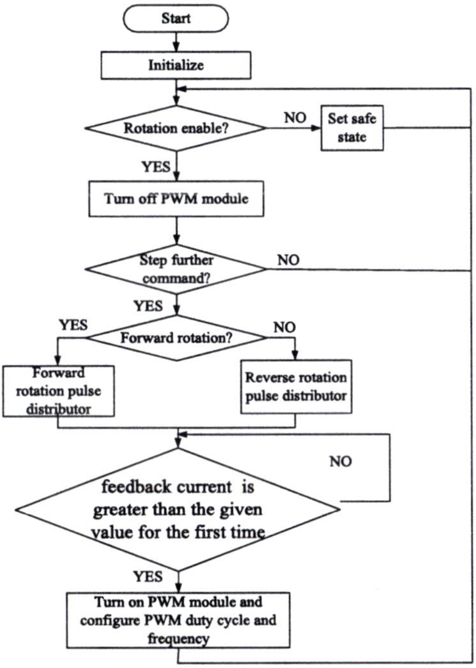

Fig. 6. Software running flow chart.

6 Experiments

A two experimental verifications are carried out in this paper. In order to verify the inverse-time protection strategy of the power output circuit, the experimental environment shown in Fig. 7 is constructed. The working voltage of REU is provided by an external DC power supply with an amplitude of 28 V. The power electronic load is used to simulate different Real load under current, and IEEE-1394 bus monitor is applied to monitor REU Trip status. The time-current inverse-time protection test waveform fitting curve is shown in Fig. 8. When the load current is more than 8 times the rated current, the turn-off protection time is 90 μs. The higher the current value, the shorter the protection time, ensuring the safe and reliable operation of the system.

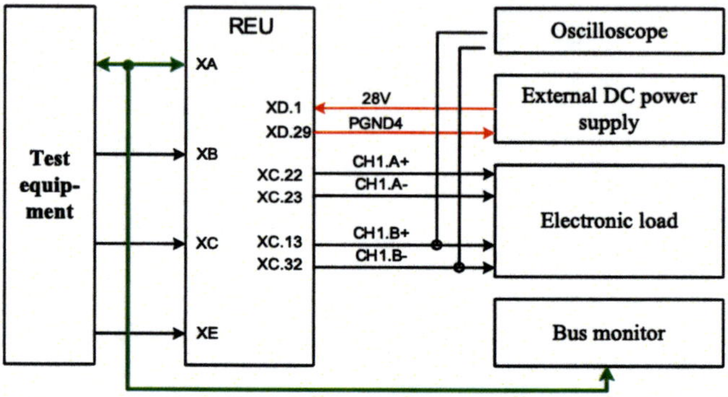

Fig. 7. Inverse time protection test configuration.

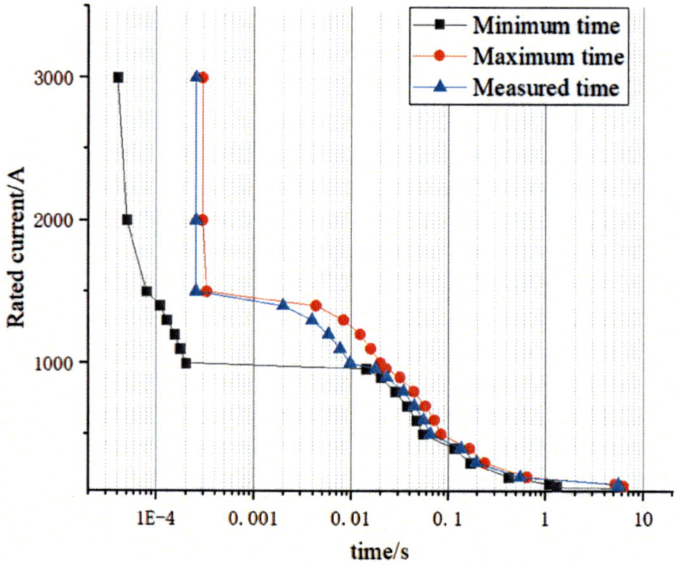

Fig. 8. Time-current protection fitting curve.

In order to verify the control accuracy of the stepper motor servo valve, the torque test environment shown in Fig. 9 is established. The parameters of the stepper motor used in the test are that the maximum speed is 750 steps/s, the rated speed is 600 steps/s, the rated current 3 A, the step angle is 5°, the reduction ratio is 600:1, A-phase winding resistance is 4.6 Ω (measured 4.88 Ω), and B-phase winding resistance is 4.6 Ω (measured 4.52 Ω). Figure 10 demonstrates the 50% fixed duty cycle phase current waveform. The fixed duty cycle frequency is 10 kHz and the duty cycle is 50%. Test results show that the stepper motor servo control strategy developed in this paper can limit the current fluctuation within 10%, and the current fluctuation and torque fluctuation can be decreased, with high control accuracy.

Fig. 9. Torque torque test bench.

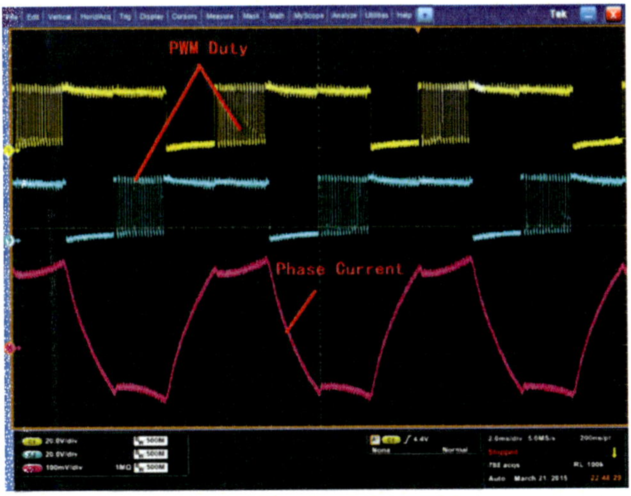

Fig. 10. 50% fixed duty cycle phase current waveform.

7 Conclusion

In this paper, stepper motor servo valve is selected as the execution unit to adjust environmental variables. A unified architecture design is adopted, and stepping motor controllers are integrated into REU of the supporting equipment of the aircraft utility system, reducing the aircraft weight and improving the integration level. To enhance the accurate positioning and rapid isolation of stepper motor faults, a fault diagnosis method for inverse-time protection power drive circuit and hierarchical interface are designed to improve system safety and testability. To tackle the temperature rise, noise, torque ripple, start out-of-step and stop overshoot, the high-deterministic command transmission technology and constant current drive technology based on fixed duty cycle and segmented pneumatic technology are developed. The test results show that the stepper motor servo control system designed in this paper has high control accuracy and can be widely applied in aerospace and other industrial fields.

References

1. Hu, M., Yu, E., Yan, W., et al.: Research on high cohesion and low coupling distributed remote interface technology of aircraft utility system. In: CSAA/IET International Conference on Aircraft Utility Systems 2020, AUS 2020, pp. 153–158. Beijing, China (2020)
2. Hu, M., Yu, E., Yan, W.: Design of highly deterministic and reliable aircraft electronic-mechanical actuator servo control system. In: 2021 IEEE International Conference on Electrical Engineering and Mechatronics Technology (ICEEMT), pp. 459–463. Qingdao, China (2021)
3. Koeln, P., Pangborn, C., Williams, A., et al.: Hierarchical control of aircraft electro-thermal systems. IEEE Trans. Control Syst. Technol. **28**(4), 1218–1232 (2019)
4. Yang, T., Foulkes, T., Kwon, B., et al.: An integrated liquid metal thermal switch for active thermal management of electronics. IEEE Trans. Compon., Packag. Manuf. Technol. **9**(12), 2341–2351 (2019)
5. Fatima, F., Noureddine, B.: Harmonic elimination by SPWM and THIPWM techniques applied in photovoltaic inverters. Int. J. Appl. Power Eng. **10**, 159–172 (2021)
6. Chaoji, C., Na, Z., Hongtao, J., et al.: Design and Implementation of the servo control system based on DSP. Comput. Inf. Sci. **3**(4), 131–134 (2010)
7. Izquierdo, D., Barrado, A., Fernández, C., et al.: SSPC active control strategy by optimal trajectory of the current for onboard system applications. IEEE Trans. Ind. Electron. **60**(11), 5195–5205 (2012)
8. Betin, F., Pinchon, D., Capolino, G.A.: Fuzzy logic applied to speed control of a stepping motor drive. IEEE Trans. Ind. Electron. **47**(3), 610–622 (2000)
9. Melin, P., Castillo, O.: Intelligent control of a stepping motor drive using an adaptive neuro–fuzzy inference system. Inf. Sci. **170**(2–4), 133–151 (2005)
10. Matsui, N., Nakamuram, M., Kosakam, T.: Instantaneous torque analysis of hybrid stepping motor. IEEE Trans. Ind. Appl. **32**(5), 1176–1182 (1996)
11. Zribi, M., Chiasson, J.: Position control of a PM stepper motor by exact linearization. IEEE Trans. Autom. Control **36**(5), 620–625 (1991)

Integral Sliding Mode Control for Antiskid Braking System of Unmanned Aerial Vehicles Based on Extended State Observer

Mingjun Xie[1], Yuhong Jia[2(✉)], and Song Ding[3]

[1] Large Aircraft Advanced Training Center, Beihang University, Beijing 100083, China
[2] School of Aeronautic Science and Engineering, Beihang University, Beijing 100083, China
jiayuhong@buaa.edu.cn
[3] Ningbo Institute of Technology, Beihang University, Ningbo 315800, China

Abstract. This article addresses an integral sliding mode control (ISMC) method for antiskid braking system (ABS) of unmanned aerial vehicles (UAV) based on extended state observer (ESO) to track the real-time optimal slip ratio. First, based on reasonable simplification, a comprehensive dynamic model of the UAV braking system is established. Then, an ESO is designed to guarantee the precise estimation of uncertainty disturbance during aircraft ground taxiing, an ABS controller is generated by combining ISMC and ESO. Finally, simulation is carried out under different working conditions, including representative road, step reduction of braking torque, road changed suddenly, and parameter high-low bias. The simulation results verify the effectiveness of the proposed slip ratio control strategy.

Keywords: Antiskid braking system · Slip ratio · Extended state observer · Integral sliding mode control

1 Introduction

Ground taxiing is a critical stage in the safe take-off and landing of UAV, and efficient ABS control strategy plays a vital role in ground taxiing [1]. Compared with the traditional wheel deceleration control strategy, the sliding ratio control strategy has been regarded as a better ABS control strategy [2]. The essence of slip rate control is that the actual slip ratio tracks its optimal value, so as to obtain the maximum adhesive coefficient.

Various control algorithms have been proposed to track the optimal slip ratio, such as PID control [3], backstepping control [4], fuzzy control [5], neural network control [6], and model predictive control [7] etc. Although the above methods can improve the ground taxiing performance of UAV in a certain extent, there are still problems such as high accuracy requirements for the system model and weak adaptability of the perturbation parameters. Besides, it is generally assumed that the optimal slip ratio is a constant value, but in fact it is susceptible to runway status and UAV speed. The optimal slip rate of UAV is constantly changing when it is taxiing on the ground.

Chinese Society of Aeronautics and Astronautics: CASTYSF 2022, LNEE 972, pp. 777–786, 2023.
https://doi.org/10.1007/978-981-19-7652-0_69

ESO proposed by Han [8] can not only observe the state of the system, but also observe the uncertainty factors of the system. It extends the total disturbance of the model uncertainty part and the external disturbance into the state variables of the system, and can estimate the total disturbance. ISMC eliminates process of sliding mode control from the initial state to the sliding surface, and retains the advantages of sliding mode control converging to the target value in a finite time [9]. With the purpose of tracking the real-time optimal slip ratio, this article proposes an ISMC method based on ESO, which combines the advantages of ESO and ISMC.

2 Mathematical Model of ABS for UAV

2.1 UAV Mathematical Model

The following reasonable assumptions are made when establishing UAV mathematical model:

Assumption 1: UAV is not affected by side wind during ground taxiing;
Assumption 2: The friction coefficient of UAV front wheel is a constant;
Assumption 3: Both sides of the main wheel braking effect is consistent.

Based on the above assumptions, force analysis of fuselage during ground taxiing is shown in Fig. 1.

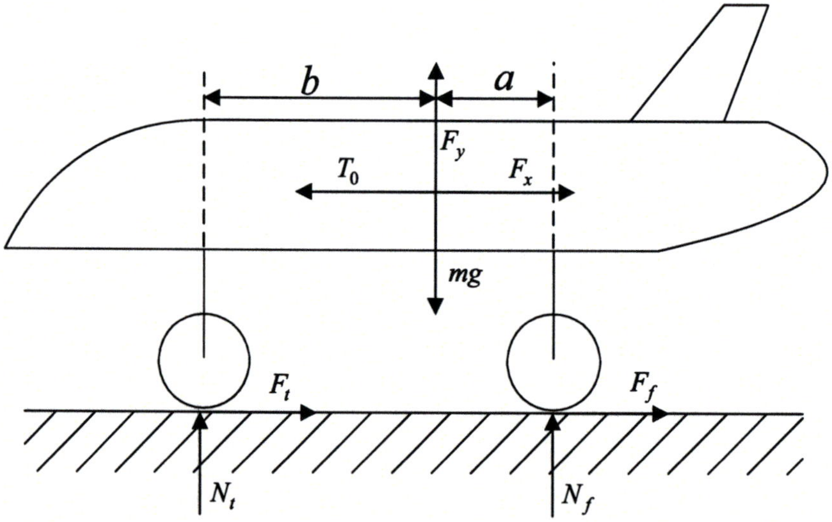

Fig. 1. Analysis of the force on the UAV during ground taxiing.

The model has simple structure but retains the essential characteristics of the braking process, which is suitable for the design of antiskid braking controller of UAV. According to Newton's second law of motion, the UAV's longitudinal motion equation, vertical direction balance equation and centroid moment balance equation can be obtained by Eq. (1):

$$\begin{cases} m\dot{V} + F_x + n_1F_t + n_2F_f - T = 0 \\ mg - F_y - n_1N_t - n_2N_f = 0 \\ n_1N_tb - n_2N_fa - n_2F_fh - n_1F_th = 0 \end{cases} \tag{1}$$

The aerodynamic drag F_x, the lift force F_y, the adhesion between the wheel and the runway F_t and F_f can be generated as

$$F_x = \frac{1}{2}JC_xSV^2,\ F_y = \frac{1}{2}JC_ySV^2,\ F_t = \mu_fN_t,\ F_f = \mu N_f \tag{2}$$

where m is the quality of UAV, V is the longitudinal sliding speed of the aircraft, n_1 and n_2 are the number of front wheel and main wheel, μ_f is the adhesive coefficient between the front wheel and the runway, μ is the adhesive coefficient between the main wheel and the runway, J is air density, C_x is air resistance coefficient, C_y is lift coefficient, S is wing area of UAV, T is engine thrust, g is gravitational acceleration, a is the horizontal distance between the center of gravity of the aircraft and the center of the main wheel, b is the horizontal distance between the center of gravity of the aircraft and the center of the front wheel, and h is the distance between the center of gravity of the aircraft and the runway.

The force analysis of the main wheel is shown in Fig. 2.

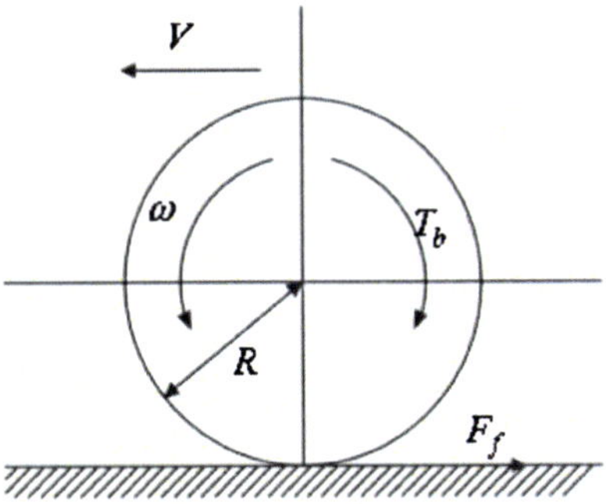

Fig. 2. Analysis of the force on the main wheel

The motion equation of the main wheel during braking can be described as follows:

$$I\dot{\omega} = F_fR - T_b \tag{3}$$

where I is the rotational inertia, ω is angular velocity, R is wheel radius, T_b is brake torque.

2.2 Slip Ratio Model

The main wheel is applied with braking torque, and the wheel slip ratio reflects the motion characteristics of aircraft during ground taxiing. The slip rate λ can be defined as:

$$\lambda = (V_x - \omega R)/V_x \tag{4}$$

According to UAV mathematical model, the derivative of slip ratio λ is:

$$\dot{\lambda} = -\frac{\mu N_f R^2}{IV} + \frac{1-\lambda}{mV}(T - F_x - n_2\mu N_f - n_1\mu_f N_t) + \frac{R}{VI}T_b \tag{5}$$

To facilitate the later design of the ABS controller, Eq. (5) is expressed as:

$$\dot{\lambda} = f(\lambda) + \frac{R}{IV}T_b \tag{6}$$

where $f(\lambda) = -\frac{\mu N_f R^2}{IV} + \frac{1-\lambda}{mV}(T - F_x - n_2\mu N_f - n_1\mu_f N_t)$.

With the aircraft taxiing status and ground conditions changing, the adhesive coefficient is nonlinear variation. In this paper, the Burckhardt model [10, 11] is described as:

$$\mu(\lambda) = \left[c_1\left(1 - e^{-c_2\lambda}\right) - c_3\lambda\right]`e^{-c_4\lambda V} \tag{7}$$

In the formula, the model under different runway states can be established by changing the value c_1, C_3, c_3 and c_4. Table 1 lists the values under three different runways.

Table 1. Burckhardt model parameters.

Runway states	c_1	c_2	c_3	c_4
Dry	1.2801	23.99	0.52	0.04
Wet	0.857	33.822	0.347	0.04
Snow	0.1946	94.129	0.0646	0.04

3 Extended State Observer Design

By Considering that there are uncertain disturbances such as parameter perturbation in the anti-skid braking process of UAV and defining $x_1 = \lambda$, $u = p$, $b_0 = \frac{R}{IV}$, system (6) can be designed as follows:

$$\dot{x}_1 = f(x_1) + f_d + b_0 u \tag{8}$$

where f_d represent the uncertain disturbance.

Considering $f(x_1)$ and f_d as an extended state and defining $d = f(x_1) + f_d$, then system (8) can be described by the following extended state system:

$$\begin{cases} e_z = x_1 - \hat{x}_1 \\ \dot{\hat{x}}_1 = \hat{x}_2 + b_0 u + \beta_1 e_z \\ \dot{\hat{x}}_2 = \beta_2 e_z \end{cases} \tag{9}$$

where $\hat{x}_1$ and $\hat{x}_2$ are the estimates of x_1 and d, e_z denotes an estimation error, β_1 and β_2 are constant gain parameter.

Configure the poles of ESO characteristic equation at the same location, which requires the observer gain to satisfy the following equation:

$$s^2 + \beta_1 s + \beta_2 = (s + w_0)^2 \tag{10}$$

where w_0 represent the observer bandwidth. Choosing appropriate w_0 can achieve the precise estimation performance of the designed ESO.

4 Controller Design

The control objective of the anti-skid braking system is to ensure that the actual slip rate can track the optimal slip rate, and the slip rate error is defined as $e = x_1 - \lambda_d$, so the integral sliding mode surface can be designed as follows:

$$s_p = c\int_0^t edt + \mathrm{e} \tag{11}$$

where $c > 0$ and c is a constant.

According to sliding function (10) and system (8), the derivative of s_p is

$$\dot{s}_p = c\mathrm{e} + f(x_1) + f_d + b_0 u - \dot{\lambda}_d \tag{12}$$

the following reaching law is used to obtain sliding mode control law:

$$u_n = -k_1 s_p - \eta sgn(s_p) \tag{13}$$

where $k_1 > 0$, $\eta > 0$ are reaching law designable parameters.

According to (12) and (13), the control input is constructed as

$$u = -\frac{1}{b_0}\left(ce + f(x_1) + f_d - \dot{\lambda}_d\right) - k_1 s_p - \eta sgn(s_p) \tag{14}$$

Now, considering the Lyapunov function

$$V_1 = \frac{1}{2}s_p^2 \tag{15}$$

The derivative of V_1 is

$$\dot{V}_1 = -b_0\left(k_1 s_p^2 + \eta\left|s_p\right|\right) \tag{16}$$

From Eqs. (15) and (16), the complete system is asymptotic.

Due to the unknown disturbance variable f_d in formula Eq. (14), the output of the controller cannot be obtained directly. Combined with the extended state observer, the control law is designed as follows:

$$u = -\frac{1}{b_0}\left(ce + \hat{x}_2 - \dot{\lambda}_d\right) - k_1 s_p - \eta sgn\left(s_p\right) \tag{17}$$

5 Simulation Results

To verify the reliability of the designed ABS control law, MATLAB was used for numerical simulation under different working conditions. The initial UAV speed and wheel speed are both 50 m/s, the controller parameters are selected as follows: $w_0 = 500$, $c = 0.1$, $k_1 = 1500$, $\eta = 0.0001$. Then, other main simulation parameters are listed in Table 2.

Table 2. Main simulation parameters of the UAV model

Name	Value	Name	Value	Name	Value
m	860 kg	n_1	1	n_2	2
a	1.34 m	b	2.09 m	h	0.84 m
R	0.25 m	I	0.3 kg·m^2	g	9.8 m/s^2
J	1.2 kg/m^3	C_x	0.3 N·s^2/m^2	C_y	0.6 N·s^2/m^2

5.1 Ideal Condition

Firstly, the numerical simulation is carried out on a dry runway and a snow runway with low friction coefficient, the simulation results are shown in Fig. 3.

It can be seen from the Fig. 3 that the optimal slip rate of real-time change is well tracked, the adhesive coefficient remains the maximum value of Eq. (7), and the aircraft and the wheel speed decelerate smoothly.

5.2 Step Reduction of Braking Torque

The braking torque step of the braking system is reduced by 200 N·m at 6 s to simulate the working condition of the braking efficiency mutation. On a dry runway, the simulation results of ideal and mutation conditions are compared as shown in the Fig. 4.

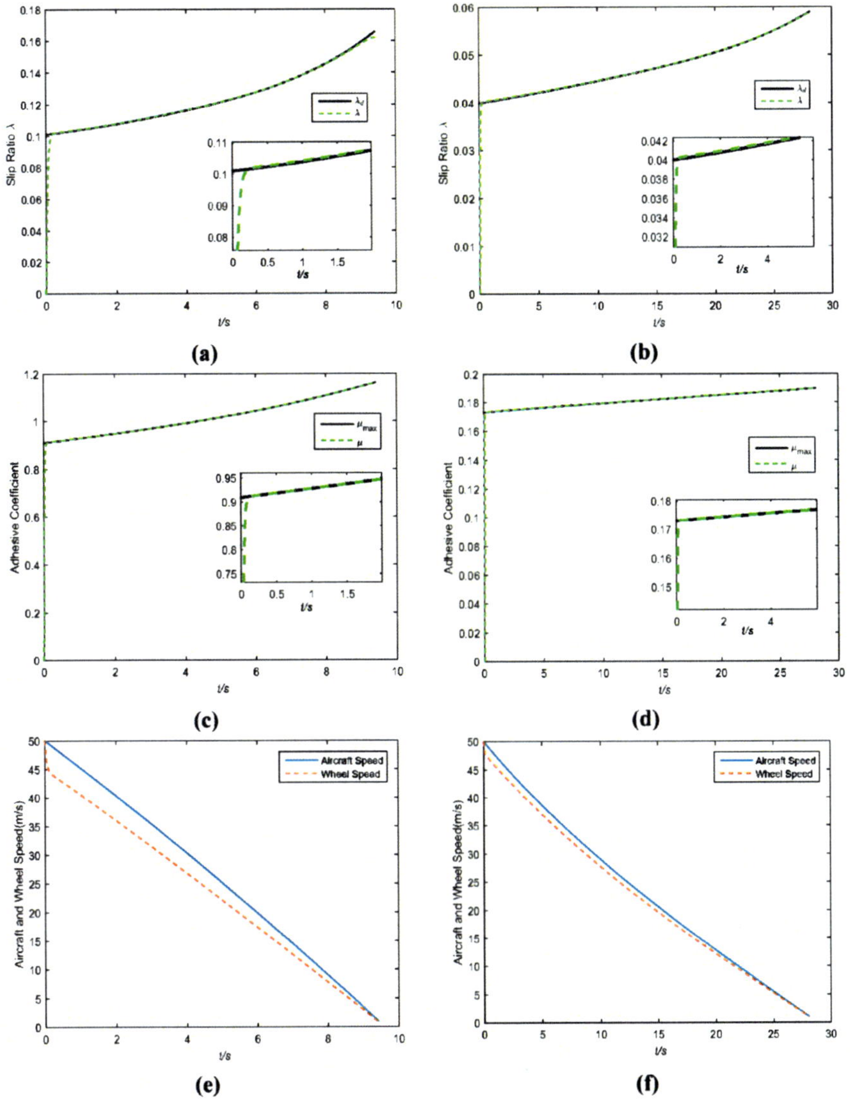

Fig. 3. Simulation results of the ABS controller under different runway condition. **a** slip rate under dry runway; **b** slip rate under snow runway; **c** adhesive coefficient under dry runway; **d** adhesive coefficient under snow runway; **e** aircraft and the wheel speed under dry runway; **f** aircraft and the wheel speed under snow runway; **g** brake torque under dry runway; **h** brake torque under snow runway

It can be seen from the Fig. 4 that the braking torque is suddenly reduced by 200 N·m at 6 s, then, the braking torque is quickly compensated, which is basically consistent with the ideal working condition. Besides, the slip rate mutates at 6 s and the real-time

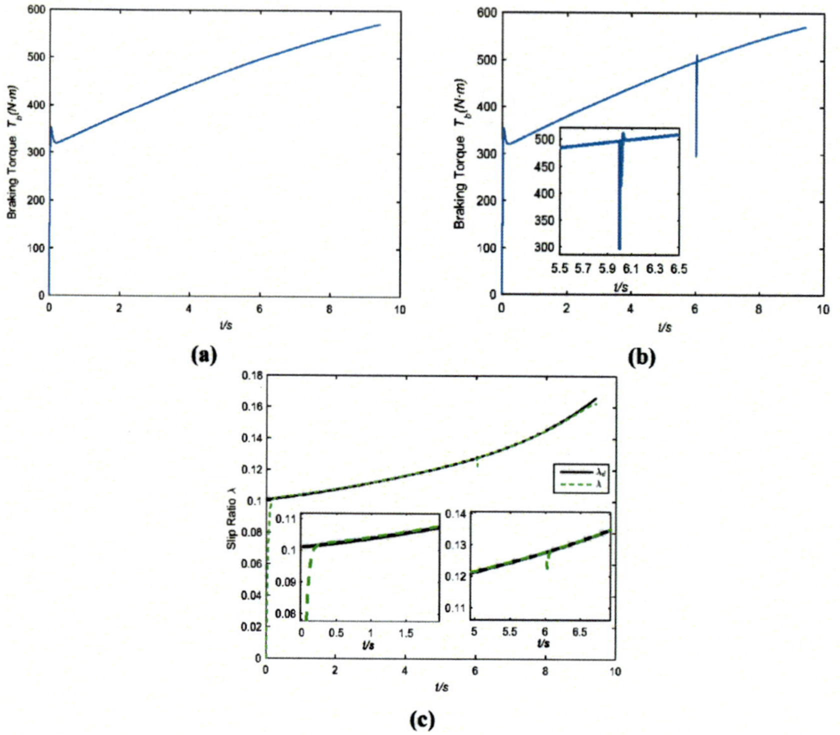

Fig. 4. Simulation results of the ABS controller under dry runway condition. **a** braking torque under ideal condition; **b** braking torque under mutation condition; **c** slip rate under mutation condition.

optimal slip rate is well tracked at other times. It is proved that the designed control law can adaptively compensate the disturbance uncertainty caused by input.

5.3 Road Changed Suddenly

The numerical simulation is carried out from a dry runway to a wet runway, and this happens at 5s, the simulation results are shown in Fig. 5.

From the Fig. 5, although the road is suddenly changed at 5 s, the real-time optimal slip rate is tracked quickly. This shows that the controller design is adaptable.

5.4 Parameter High-Low Bias

To verify the adaptability of the control law to parameter perturbation, considering the deviation range of UAV mass and wheel inertia are both ±5%, and the maximum negative values are taken in the simulation. On a dry runway, Fig. 6 shows the comparison of simulation results between ISMC based on ESO and ISMC.

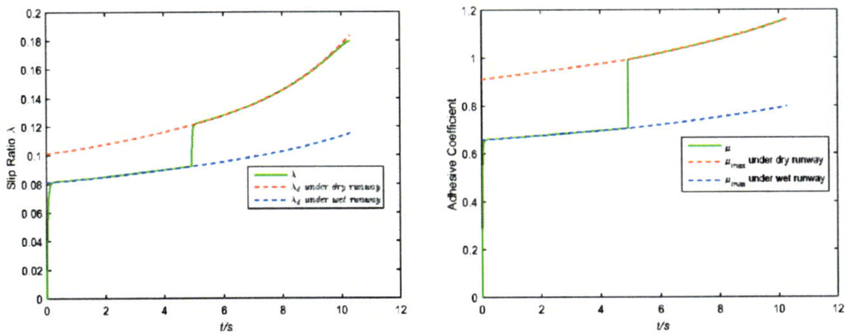

(a) **(b)**

Fig. 5. Simulation results of the ABS controller with road changed suddenly. **a** Slip rate; **b** adhesive coefficient

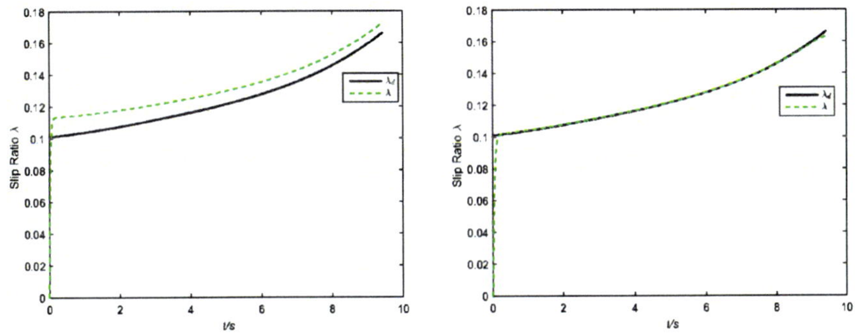

(a) **(b)**

Fig. 6. Simulation results of the ABS controller with two control methods under dry runway condition. **a** Slip rate with ISMC; **b** slip rate with ISMC based on ESO.

It can be seen from the Fig. 6 that the optimal slip rate cannot be track only under the action of ISMC in the case of parameter high-low bias, and there is always a steady-state error. The controller with ESO compensation can achieve the control target. It is proved the controller designed has a significant effect on improving the performance of the control system.

6 Conclusions

To ensure the safety of take-off and landing of UAV, a new control strategy based on slip ratio is designed. The simulation results show that:

1. This control method can adapt to different runway. When ground taxiing on a dry runway state or a snow runway with low friction coefficient, the real-time optimal slip ratio can be well tracked and the braking efficiency can be improved.

2. The control method has good robustness and high fault tolerance, and can maintain the optimal slip ratio state under different working conditions, including representative road, step reduction of braking torque, road changed suddenly, and parameter high-low bias.

References

1. Yang, W., Cai, W., Song, Y.: A barrier Lyapunov function (BLF) based approach for antiskid traction/braking control of high speed trains. In: The 26th Chinese Control and Decision Conference (2014 CCDC), pp. 5023–5028. IEEE (2014)
2. Zhang, X., Lin, H.: Backstepping fuzzy sliding mode control for the antiskid braking system of unmanned aerial vehicles. Electronics **9**(10), 1731 (2020)
3. Bo, L., Li, Y.: Research on simulation of aircraft electric braking system. In: Recent Advances in Computer Science and Information Engineering, pp. 301–309. Springer, Berlin, Heidelberg (2012)
4. Qiu, Y., Liang, X., Dai, Z.: Backstepping dynamic surface control for an anti-skid braking system. Control Eng. Pract. **42**, 140–152 (2015)
5. Ćirović, V., Aleksendrić, D.: Adaptive neuro-fuzzy wheel slip control. Expert Syst. Appl. **40**(13), 5197–5209 (2013)
6. Poursamad, A.: Adaptive feedback linearization control of antilock braking systems using neural networks. Mechatronics **19**(5), 767–773 (2009)
7. Li, S., Guo, L., Zhang, B., et al.: MPC-based slip control system for in-wheel-motor drive EV. IFAC-PapersOnLine **51**(31), 578–582 (2018)
8. Han, J.Q.: From PID technique to active disturbances rejection control technique. Basic Autom. (2002)
9. Guermouche, M., Ahmed Ali, S., et al.: An adaptive integral sliding mode control design for internal combustion engine air path. IFAC Proc. Vol. **46**(25), 87–94 (2013)
10. Burckhardt, M.: Fahrwerktechnik: Radschlupfregelsysteme. Vogel-Verlag, Würzburg (1993).(in German)
11. Baffet, G., Charara, A., et al.: Sideslip angle, lateral tire force and road friction estimation in simulations and experiments. In: Proceedings of the 2006 IEEE International Conference on Control Applications, Munich, Germany, pp. 903–908 (2006)

Study on Mechanical Environment Analysis Method of Aircraft Cabin

Xiaoguang Zhang and Bin Li(✉)

School of Aeronautics, Northwestern Polytechnical University, Xi'an 710072, China
leebin@nwpu.edu.cn

Abstract. In order to ensure the high reliability of aircraft, it is necessary to study the mechanical environment of aircraft. In this paper, the vibration and noise sources of the aircraft in cruise state are sorted out, and the external dynamic load of the typical mission profile of the aircraft is calculated by computational fluid dynamics simulation analysis combined with empirical formula method to obtain the aerodynamic noise load of the aircraft surface. Then, the low frequency response of the aircraft cabin is analyzed based on the finite element method, and the high frequency response of the aircraft cabin is analyzed based on the statistical energy method. The full frequency vibration and noise environment of the aircraft cabin is predicted, and the internal vibration and noise spectrum of the aircraft cabin is obtained. The detailed design method of cabin mechanical environment is established, which lays the foundation for the establishment of complete mechanical environment of the whole aircraft, and also provides a reference for the mechanical environment analysis of aircraft in high mobility and large overload environment.

Keywords: Aircraft cabin · Mechanical environment · Aerodynamic noise · Acoustic vibration coupling · Statistical energy method

1 Introduction

Acoustic vibration analysis is a necessary step in the virtual design of aerospace structures. In order to reduce the design cost and maximize the acoustic performance of aerospace structures, it is necessary to make a robust and mature prediction of the internal noise level. Internal noise is a problem that must be considered in the design and operation of aircraft. Noise is caused by the combination of different sound sources, such as the powerful propulsion system, the high-speed aerodynamic flow on the surface of the aircraft and the operation of the internal system of the aircraft (air conditioning and pressurization system). High noise level may cause communication interference, unit fatigue or sensitive electronic equipment failure.

At present, many studies have been carried out on acoustic vibration coupling [1–6]. For example, taking the hypersonic vehicle X-43A as the research object, Zhang et al. [7] conducted the aircraft structure noise induced vibration test in the high sound intensity reverberation room, and obtained the acceleration power spectral density of

Chinese Society of Aeronautics and Astronautics: CASTYSF 2022, LNEE 972, pp. 787–796, 2023.
https://doi.org/10.1007/978-981-19-7652-0_70

the aircraft structure measuring point and the noise sound pressure level of the cabin sound field. Through the comparison between the acoustic vibration coupling numerical simulation results and the test values, the finite element analysis of the structural vibration response and the cabin noise response is more consistent with the test results. Sun et al. [8] predicted the high-frequency dynamic environment of the instrument cabin of a certain type of missile by using the statistical energy method, and compared the simulation results with the test values. It was found that the simulation results were in good agreement above 300 Hz, and the errors were controlled within 3 dB, meeting the engineering requirements. Luo et al. [9] established acoustic and vibration models of spacecraft under different working conditions and different conditions based on hybrid FE-SEA method, and studied the influence of support vehicle on the acoustic and vibration response of spacecraft structure, which can provide guidance for the simplification of acoustic and vibration model of complex spacecraft. Petrone et al. [10] studied the sound pressure level inside the aircraft fuselage under cruise state by using the statistical energy method.

When calculating the acoustic and vibration response of the model, the description of the sound pressure load of the sound source is not clear, especially the surface sound pressure load of the aircraft is not described in detail. From the current research, most of them use empirical formula, but the empirical formula does not accurately describe the surface sound source of aircraft. In this paper, the method of CFD calculation combined with empirical formula is used to accurately describe the aircraft surface sound source, so as to completely calculate the aircraft surface sound load. Then, based on FEM-SEA methods, the vibration and noise of the whole cabin in the whole frequency band are predicted, and the internal vibration and noise spectrum of each cabin of the whole aircraft are obtained.

2 Calculation of Aerodynamic Noise Load on Cabin Surface

2.1 Nonlinear Acoustic Equation and Its Numerical Solution

In the three-dimensional Cartesian coordinate system, it is assumed that a disturbance is added to the N-S equation, and each original variable is decomposed into statistical average variable and random disturbance variable, i.e., $\varphi = \varphi + \varphi'$, The nonlinear acoustic equation can be obtained by substituting it into the N-S equation and reorganizing the N-S equation:

$$\frac{\partial Q'}{\partial t} + \frac{\partial F_i'}{\partial x_i} - \frac{\partial (F_i^v)'}{\partial x_i} = -\frac{\partial \overline{Q}}{\partial t} - \frac{\partial \overline{F}_i}{\partial x_i} + \frac{\partial \overline{F}_i^v}{\partial x_i} \tag{1}$$

where, Q' is the disturbance; Q is the transient average; F_i' is linear inviscid momentum; F_i is the average amount of inviscid; $(F_i^v)'$ is the viscous disturbance; F_i^v is the average viscosity.

Omit the density pulsation and time average it:

$$\overline{LHS} = \overline{RHS} = \frac{\partial R_i}{\partial x_i} \tag{2}$$

$$R_i = \begin{bmatrix} 0 \\ \overline{\rho u_i' u_j'} \\ c_p \overline{\rho T' u_i'} + \overline{\rho u_i' u_k'} \overline{u_k} + \frac{1}{2} \overline{\rho u_k' u_k' u_i'} + \overline{u_k' \tau_{ki}} \end{bmatrix} \tag{3}$$

To solve the nonlinear acoustic equation, we need to obtain the values of these unknowns first. It can usually be obtained by solving the RANS equation, such as the N-S equation which is closed by cubic k-ε equation, and the small quantities that cannot be solved can be obtained by the artificial reconstruction method of turbulence. After calculating the statistical average variable, the nonlinear acoustic equation can be time advanced.

2.2 Calculation Process

Taking the cabin shape of an aircraft as the object, this paper applies the nonlinear acoustic calculation method and empirical formula introduced above to calculate the aerodynamic noise on the cabin surface. The geometric model of the cabin is shown in Fig. 1.

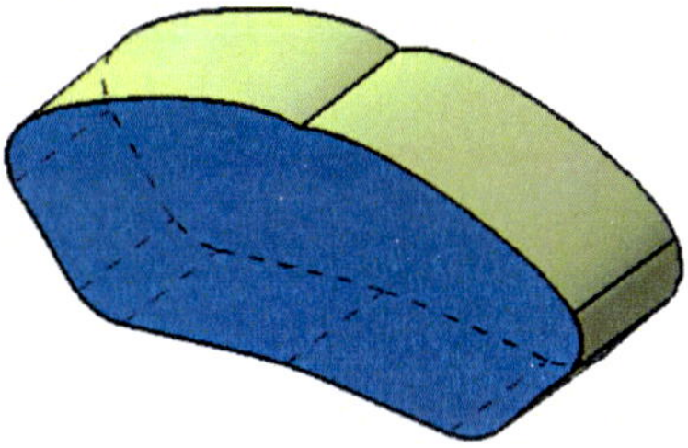

Fig. 1. Cabin geometry model.

In this paper, the surface pressure of aircraft at Mach number Ma = 2.5 is calculated. The aircraft wall adopts the non-slip wall condition. The noise calculation needs to start the unsteady calculation, and the time step is 1×10^{-5} s, the flow solution time is 1.0 s, the maximum number of iterations in each time step is 10, and the new boundary is set as the absorption layer boundary. The calculated cabin monitoring points are shown in Fig. 2 to measure the pressure change at the monitoring points. The calculation results are superimposed with the radiation pressure of engine noise to obtain the pressure time history curve of some monitoring points, as shown in Fig. 3 (the curve in the figure is dimensionless).

3 Calculation of Cabin Low Frequency Response

3.1 Structural Vibration Calculation

The structural vibration is carried out in the three-dimensional solid mechanics transient analysis. The main calculation process is as follows:

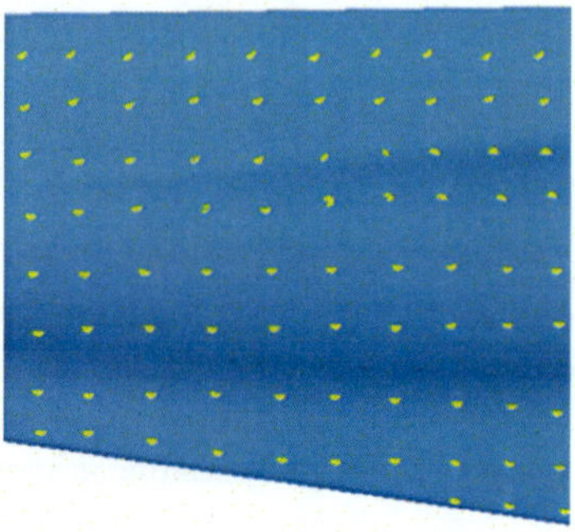

Fig. 2. Monitoring points on surface.

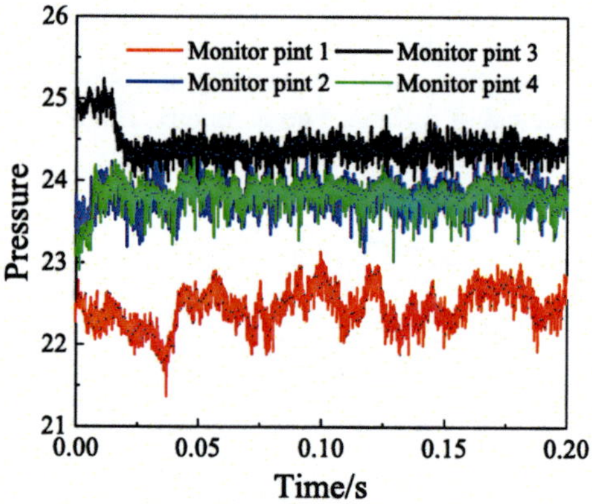

Fig. 3. Pressure time curve.

1. Input structural model. Remove unnecessary edges from the model, merge the whole surface and define load points.
2. Define the interpolation function. The external load data is input into the structure through the function and loaded on the structure through the load variable.
3. Select the inner surface of the structure as the "display boundary", and define the selection as "generalized tension", so as to map the vibration acceleration data.
4. The given structural material is aluminum, the density is 2700 kg/m^3, the young's modulus is 70 GPa and the Poisson's ratio is 0.33. The end faces on both sides of the structure are given as fixed end constraints.
5. Divide the grid, and the structural grid shall be processed according to refinement.
6. The transient model is used for research, and the solution time is 0.2 s.

Under the external load, the shell of the cabin structure has vibration response, and the maximum vibration response position of the structure changes gradually from 0 to 0.2 s.

Under load excitation, the maximum vibration of the structure in the initial stage will occur on the upper and lower surfaces of the cabin structure at the same time. With the extension of time, the vibration mainly affects the lower surface of the structure. The vibration response on both sides of the cabin section is small compared with the vibration generated by the upper and lower middle surfaces, and the main vibration contribution comes from the middle part of the upper and lower wall panels.

3.2 Acoustic Response Calculation

The steps of acoustic physical field analysis are as follows:

1. Import the acoustic solid domain model, simplify the edge of the model and form a composite surface.
2. Add 'pressure acoustics' physical field and 'steady state' research, and define the whole domain as air.
3. Select the external surface of the acoustic domain as the display boundary, and define the normal acceleration boundary on the interface to load the particle acceleration obtained by the generalized stretching of the structural domain.
4. The front and rear sides are defined as the air characteristic impedance boundary, and the impedance value is the product of air density and velocity 411.6 Pa·s/m, indicating the non reflective flow of sound waves on both sides.

The vibration data in time domain is transferred to frequency domain through Fourier transform, and the vibration data using generalized stretching mapping is mapped to sound field domain. Under the excitation disturbance, the sound field in the cabin constitutes its independent modal vibration mode, and the vibration is represented by the change of internal sound pressure. Figure 4 (a), (b) show the sound pressure level distribution in the internal sound field of the cabin at 500 Hz and 800 Hz respectively.

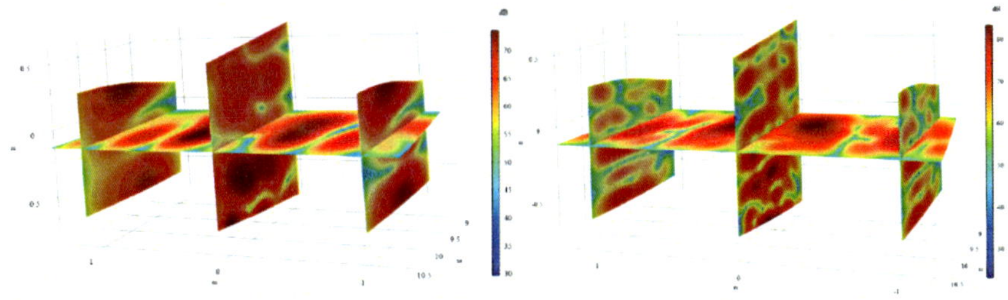

(a) In 500Hz (b) In 800Hz

Fig. 4. Sound pressure level distribution.

In order to describe the sound field more accurately, different monitoring points are set to explore the sound pressure level of the internal sound field. Select three points at the center of the upper surface of the sound field, as shown in Fig. 5. The sound pressure levels of the three points are shown in Fig. 6.

As can be seen from Fig. 7, the sound pressure level of point 3 is large in most frequency bands, such as 400 and 800 Hz. The sound pressure level of point 1 is small as

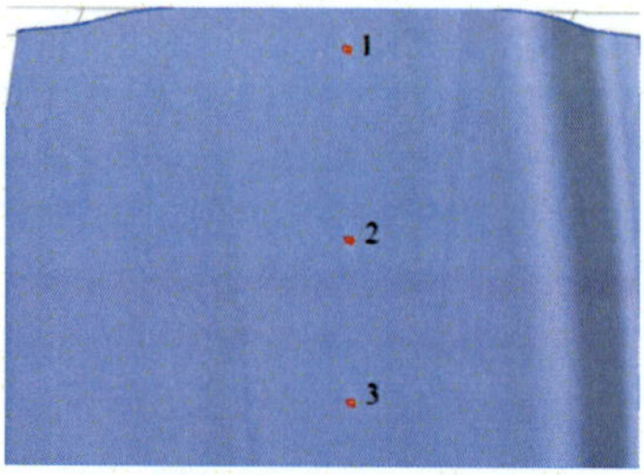

Fig. 5. Monitoring points on the upper surface of sound field.

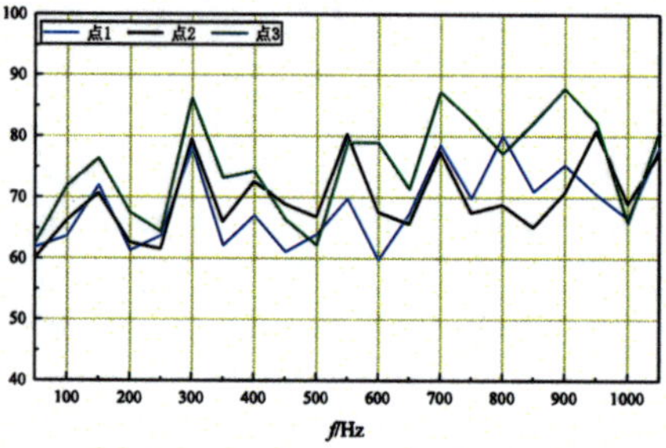

Fig.6. SPL of upper surface monitoring points in sound field.

a whole. Comparing the maximum sound pressure level, point 1 is 87.83 dB at 900 Hz, point 2 is 80.36 dB at 550 Hz, and point 3 is 80.15 dB at 800 Hz. There is little difference between the maximum sound pressure levels of point 2 and point 3. The average sound pressure level in the calculated frequency band is 70.26 dB for point 1, 69.81 dB for point 2 and 68.69 dB for point 3. Therefore, the sound pressure level difference of the three monitoring points is small.

4 Calculation of Cabin High Frequency Response

4.1 Basic Theory of Statistical Energy Method

When establishing statistical energy analysis model or applying statistical energy method to analyze energy flow between subsystems, the following basic assumptions need to be made:

1. The subsystems are linearly coupled and follow the principle of energy conservation;
2. The energy flow exists between subsystems with resonant modal frequency in the frequency band;
3. The external excitations are broadband random excitations, which have modal incoherence and accord with the principle of linear superposition of energy;
4. In a certain frequency band, energy is evenly distributed among subsystems with resonant modes;

5. Each subsystem meets the reciprocity principle;
6. The energy flow between any two subsystems is directly proportional to the average coupling mode energy difference.

4.2 Calculation Results

The cabin model consists of four parts: cabin surface, cabin end face, stiffener and internal sound field. The cabin surface is divided into 10 curved plate subsystems, as shown in Fig. 7.

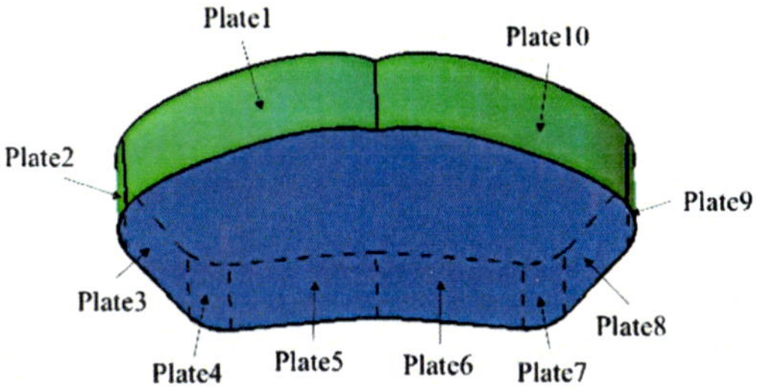

Fig. 7. Cabin surface curved plate subsystem.

The modal number and vibration response energy in each frequency band of plate1–plate6 on the cabin surface are shown in Figs. 8 and 9 respectively. It can be seen from the figure that the modal number of each curved plate subsystem on the surface of the cabin is greater than 5 in each 1/3 octave frequency range, which meets the calculation requirements of the statistical energy method.

The sound pressure level of the cavity subsystem in the frequency band range of 1000–20,000 Hz is calculated, as shown in Fig. 10. As can be seen from the figure, the sound pressure level in the cabin generally shows a downward trend with the increase of frequency.

5 Conclusions

1. The vibration of the cabin structure is mainly in the middle part of the structure, and the bottom vibration is the largest;
2. For the sound field, the sound pressure level of the lower surface monitoring point is generally greater than that of the upper surface monitoring point;
3. The sound fields on the left and right sides of the structure are symmetrically distributed, and the variation amplitude of sound pressure level at the symmetrical monitoring points on both sides is the same at each frequency point;
4. In the calculated frequency band, the total average sound pressure level in the internal sound field is 95.86 dB;
5. The sound pressure level in the cabin generally decreases with the increase of frequency.

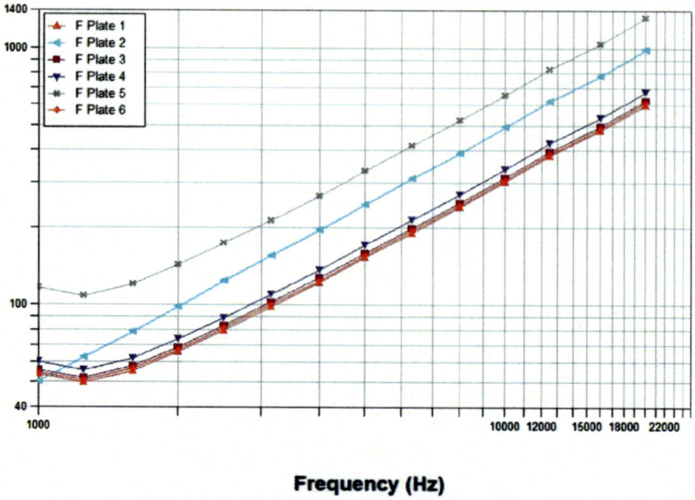

Fig. 8. Mode number of plate1–plate6 in each 1/3 octave.

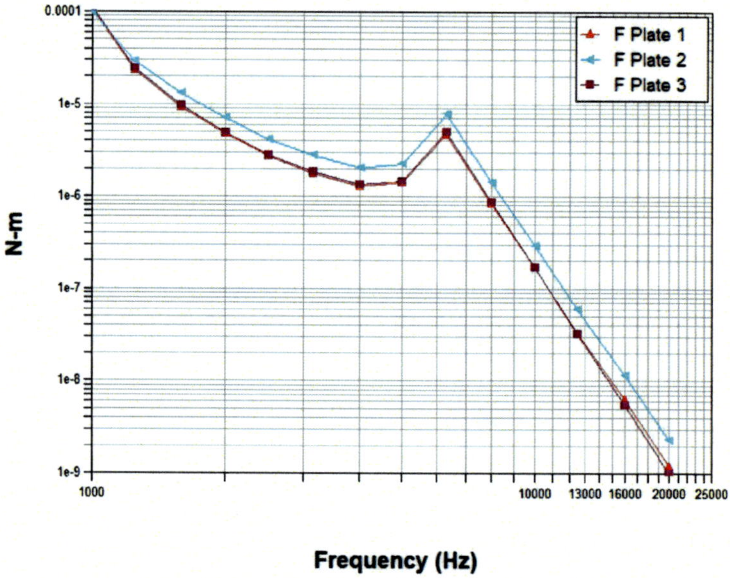

Fig. 9. Vibration response energy of plate1–plate3.

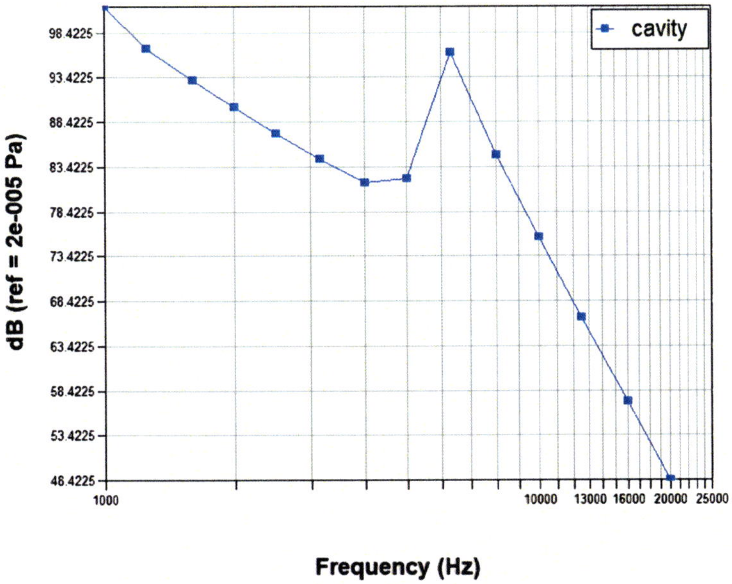

Fig. 10. Cavity noise.

Acknowledgements. This work was sponsored by Innovation Foundation for Doctor Dissertation of Northwestern Polytechnical University (number CX2022002) and the National Natural Science Foundation of China (number 11872312). The authors are grateful to the anonymous reviewers for their critical and constructive review of the manuscript.

References

1. Wang, K.: The research of modeling of aircraft structures by SEA and vibro-acoustic analysis. Master Thesis, Harbin Institute of Technology (2010)
2. Hu, X.L., Chen, H.L., Zhao, J.P.: Rationalization and experiments of current vibroacoustic analytical models. J. Xi'an Jiaotong Univ. **31**(7), 1–6 (1997)
3. Yin, Z.J., Yue, H.C., Chen, B., et al.: Vibration and noise analysis of an exhaust piping system based on statistical energy analysis. J. Vib. Shock **29**(2), 159–162 (2010)
4. Desmet, W., Sas, P., Vandepitte, D.: Numerical Acoustics Theoretical Manual. LMS International, Leuven (2000)
5. Cremre, L., Heckl, M., Ungar, E.E.: Structure-Borne Sound. Springer-Verlag, Berlin (1988)
6. Yin, L.Z., Xu, X.C., Tan, Z.Y.: Response analysis and experimental verification for reentry vehicle shell subjected to aerodynamic noise. Struct. Environ. Eng. **29**(2), 5–9 (2002)
7. Zhang, G.J., Yan, Y.J., Li, P.B.: Applications of statistical energy analysis in influencing factors analysis of aircraft vibro-acoustic response characteristics. J. Air Force Eng. Univ. (Nat. Sci. Ed.) **14**(2), 23–27 (2013)
8. Sun, M., Wang, X.J., Pan, Z.W., et al.: Applications of statistical energy analysis in prediction of flight dynamics environment. Missiles Space Veh. **301**(3), 11–14 (2009)

9. Luo, Y.Z., Wang, C., Zhang, H.: Acoustic response predication for a spacecraft unit with bracket based on FE-SEA method. Spacecr. Environ. Eng. **31**(3), 262–266 (2014)
10. Petrone, G., Melillo, G., Laudiero, A., et al.: A statistical energy analysis (SEA) model of a fuselage section for the prediction of the internal sound pressure level (SPL) at cruise flight conditions. Aerosp. Sci. Technol. **88**, 340–349 (2019)

Author Index

Chinese Society of Aeronautics and Astronautics: CASTYSF 2022, LNEE 972, pp. 797–800, 2023.
https://doi.org/10.1007/978-981-19-7652-0